MATERIALS SCIENCE AND TECHNOLOGIES SERIES

SHAPE MEMORY ALLOYS: MANUFACTURE, PROPERTIES AND APPLICATIONS

MATERIALS SCIENCE AND TECHNOLOGIES SERIES

Magnetic Properties of Solids
Kenneth B. Tamayo (Editor)
2009. ISBN: 978-1-60741-550-3

Mesoporous Materials: Properties, Preparation and Applications
Lynn T. Burness (Editor)
2009. ISBN: 978-1-60741-051-5

Physical Aging of Glasses: The VFT Approach
Jacques Rault
2009. ISBN: 978-1-60741-316-5

Physical Aging of Glasses: The VFT Approach
Jacques Rault
2009. ISBN: 978-1-61668-002-2 (Online Book)

Graphene and Graphite Materials
H. E. Chan (Editor)
2009. ISBN: 978-1-60692-666-6

Dielectric Materials: Introduction, Research and Applications
Ram Naresh Prasad Choudhary and Sunanda Kumari Patri
2009. ISBN: 978-1-60741-039-3

Handbook of Zeolites: Structure, Properties and Applications
T. W. Wong
2009. ISBN: 978-1-60741-046-1

Strength of Materials
Gustavo Mendes and Bruno Lago (Editors)
2009. ISBN: 978-1-60741-500-8

Photoionization of Polyvalent Ions
Doris Möncke and Doris Ehrt
2009. ISBN: 978-1-60741-071-3

Building Materials: Properties, Performance and Applications
Donald N. Cornejo and Jason L. Haro (Editors)
2009. ISBN: 978-1-60741-082-9

Concrete Materials: Properties, Performance and Applications
Jeffrey Thomas Sentowski (Editor)
2009. ISBN: 978-1-60741-250-2

Corrosion Protection: Processes, Management and Technologies
Teodors Kalniņš and Vilhems Gulbis (Editors)
2009. ISBN: 978-1-60741-837-5

Corrosion Protection: Processes, Management and Technologies
Teodors Kalniņš and Vilhems Gulbis (Editors)
2009. ISBN: 978-1-61668-226-2 (Online Book)

Handbook on Borates: Chemistry, Production and Applications
M.P. Chung (Editor)
2010. ISBN: 978-1-60741-822-1

Smart Polymer Materials for Biomedical Applications
Songjun Li , Ashutosh Tiwari, Mani Prabaharan and Santosh Aryal (Editors)
2010. ISBN: 978-1-60876-192-0

Definition of Constants for Piezoceramic Materials
Vladimir A. Akopyan, Arkady Soloviev, Ivan A. Parinov and Sergey N. Shevtsov
2010. ISBN: 978-1-60876-350-4

Definition of Constants for Piezoceramic Materials
Vladimir A. Akopyan, Arkady Soloviev, Ivan A. Parinov and Sergey N. Shevtsov
2010. ISBN: 978-1-60876-350-4

Handbook of Photocatalysts: Preparation, Structure and Applications
Geri K. Castello (Editor)
2010. ISBN: 978-1-60876-210-1

Organometallic Compounds: Preparation, Structure and Properties
H.F. Chin (Editor)
2010. ISBN: 978-1-60741-917-4

Surface Modified Biochemical Titanium Alloys
Aravind Vadiraj, M . Kamaraj (Authors)
2010. ISBN: 978-1-60876-581-2

Composite Laminates: Properties, Performance and Applications
Anders Doughett and Peder Asnarcz (Editors)
2010. ISBN: 978-1-60741-620-3

Shape Memory Alloys: Manufacture, Properties and Applications
H. R. Chen (Editor)
2010. ISBN: 978-1-60741-789-7

MATERIALS SCIENCE AND TECHNOLOGIES SERIES

SHAPE MEMORY ALLOYS: MANUFACTURE, PROPERTIES AND APPLICATIONS

H. R. CHEN
EDITOR

Nova Science Publishers, Inc.
New York

For permission to use material from this book please contact us:
Telephone 631-231-7269; Fax 631-231-8175
Web Site: http://www.novapublishers.com

Library of Congress Cataloging-in-Publication Data

Shape memory alloys : manufacture, properties, and applications / H.R. Chen.
p. cm.
Includes index.
ISBN 978-1-60741-789-7 (hardcover)
1. Shape memory alloys. 2. Shape memory effect. I. Chen, H. R.
TA487.S495 2009
620.1'65--dc22

2009018837

Published by Nova Science Publishers, Inc. ✦ New York

Contents

Preface		**ix**
Chapter 1	Constitutive Models for Shape Memory Alloys: A Two-Phase Mixture Model and a Microstructure-Based Model *Xianghe Peng, Wenli Pi and Xuesong Long*	**1**
Chapter 2	Control of SMA Systems: Review of the State of the Art *Mohammad H. Elahinia and Shuo Wang*	**49**
Chapter 3	Fabrication and Surface Modification of Porous Nickel-Titanium Shape Memory Alloy for Bone Grafts *Shuilin Wu, Xiangmei Liu, Paul K. Chu, Tao Hu, Kelvin W. K. Yeung and Jonathan C. Y. Chung*	**69**
Chapter 4	Structure and Properties of a Monocrystalline Cu-Al-Ni Alloy Submitted to Thermal Cycling Under Load *L. A. Matlakhova, E. C. Pereira, A. N. Matlakhov and S. N. Monteiro*	**113**
Chapter 5	Interaction between Martensitic Transformations and Defects During Thermal and Pseudoelastic Cycling In Cualni Single Crystals *R. Gastien, M. Sade and F. C. Lovey*	**145**
Chapter 6	Importance of the Study of Thermal Properties in the Research of Ni-Mn-Ga Shape Memory Alloys *Alexandra Rudajevova*	**181**
Chapter 7	Chracteristic Deformation and Structural Changes of Ni-Ti Alloys by Molecular Dynamics: Phase Transformation and Amorphization *Tomohiro Sato, Ken-Ichi Saitoh, Keisuke Kubota and Noboru Shinke*	**231**
Chapter 8	Evolution of the Stress-Induced Martensitic Transformation in a Cualbe Shape Memory Alloy Under Uniaxial Tension *F. M. Sánchez-Arévalo, G. Pulos and R. Pérez-Cabeza de Vaca*	**247**
Chapter 9	Formation of Transitory Bainite as a Precursor of α-Phase During Tempering of Martensitic Cu-Zn-Al SMAs *Leandru-Gheorghe Bujoreanu*	**267**
Chapter 10	The Shape Memory Effect in Antiferromagnetic Alloys *J. H. Zhang*	**287**

Chapter 11	Phase Transformation in NiTi Shape Memory Alloy Under Thermomechanical Conditions *Manjunatha Pattabi and K. Ramakrishna*	**317**
Chapter 12	Dependence of Transformation Temperatures of Shape Memory Alloys on the Number and Concentration of Valence Electrons *Mehrdad Zarinejad and Yong Liu*	**339**
Chapter 13	Some Factors Affecting the Transformation Hysteresis in Shape Memory Alloys *Yong Liu*	**361**
Chapter 14	Micromechanical Modeling of Shape Memory Alloy Composites *Jacob Aboudi and Yuval Freed*	**371**
Chapter 15	The Numerical Analysis of Two - Shape Memory Model *N. Chemetov*	**425**
Chapter 16	Shape Memory Alloys in Micropositioning Applications *E. Asua, J. Feuchtwanger, V. Etxebarria and A. García-Arribas*	**441**
Chapter 17	Analysis of a Thermomechanical Model of Shape Memory Alloys *Aida Timofte and Vlad Timofte*	**487**
Index		**537**

PREFACE

A shape memory alloy (SMA, also known as a smart metal, memory alloy, or muscle wire) is an alloy that "remembers" its shape, and can be returned to that shape after being deformed, by applying heat to the alloy. When the shape memory effect is correctly harnessed, this material becomes a lightweight, solid-state alternative to conventional actuators such as hydraulic, pneumatic, and motor-based systems. Shape memory alloys have numerous applications in the medical and aerospace industries.This new book presents the latest research in the field from around the globe.

Chapter 1 – Two models for the constitutive behavior of polycrystalline shape memory alloys (SMAs) are presented. One is a two-phase mixture model, and the other is a microstructure-based two-phase model, accounting for the typical constitutive behavior of Ni-Ti SMAs under different stress states. In both models, it is assumed that in the interested ranges of stress and temperature, the austenite phase is linearly elastic while the martensite phase is elastoplastic.

Chapter 2 – Shape memory alloy (SMA) actuators can be made very compact and simple. An effective control algorithm is an integral part of every SMA actuator. This article presents a review of the state of the art research on control of SMA systems. It is worth noting that almost exclusively all control systems applied to SMA actuators are not model-based control algorithms. In other words, the control system is designed based on the intuitive knowledge of the system, through a combination of simulations and experiments. Therefore, the parameters of the controller are not derived based on the system's parameters. As a result, stability analysis, a major step in designing control systems, has rarely been done for SMA-actuated systems. It is also important to recognize the co-dependent relationship between stress, martensite fraction, and transformation temperatures of the SMA materials. This complex nonlinear relationship causes many difficulties with regards to control system design for SMA actuators. It is also worth noting that SMAs are one way actuators and they are sensitive to changes in the ambient temperature.

Chapter 3 – Equiatomic porous nickel-titanium shape memory alloys (NiTi SMAs) are becoming one of the most promising biomaterials for bone grafts because of their unique advantages over currently used biomaterials. For example, they have good mechanical properties and lower Young's modulus relative to dense NiTi, Ti, and Ti-based alloys. Porous NiTi SMAs are relatively easy to machine compared to porous ceramics such as hydroxyapatite and calcium phosphate that tend to exhibit brittle failure. The porous structure with interconnecting open pores can also allow tissue in-growth and favors bone osseointegration. In addition, porous NiTi alloys possess good shape memory effect (SME) and superelasticity (SE) similar to dense NiTi alloys. In order to optimize porous NiTi SMAs used in bone grafts, current research focuses on the fabrication methods and surface modification techniques to obtain adjustable bone-like structures with good mechanical properties and excellent superelasticity as well as a bioactive passivation layer on the entire exposed surface to block nickel leaching and enhance the surface biological activity. This chapter describes the recent progress in the fabrication process and surface modification methods of porous NiTi SMAs and discusses subsequent effects of these processes and techniques on the materials properties from the perspective of bone grafts applications.

Chapter 4 – Monocrystalline Cu-Al-Ni alloys are promising alternatives to be used in shape memory effect (SME) actuators, in which the material suffers under temperature and applied load a combined action of many factors including thermal cycling as well as elastic and plastic deformations that modifies the initial structure and characteristic properties. In this chapter, a monocrystalline Cu-13.5wt.%Al-4wt.%Ni (Cu-13.5Al-4Ni) alloy displaying SME was submitted to thermal cycling treatments, load-free and under load, within the critical range of temperatures associated with reversible martensitic transformations (RMT). The investigation was conducted in terms of number of cycles, magnitude of applied load, resulting structure changes, characteristics of the RMT and mechanical behavior. The alloy was produced by the "*Memory Crystal Group*" in Russia and its characterization was carried out using optical and electron microscopy, X-rays diffraction, differential scanning calorimetry and mechanical tests. A special device was developed to perform the hundreds of thermal cycling treatments under load. It was found that the thermal cycling treatment promotes changes in the amount of phases - a martensite γ'_1, a high temperature stable β_1 and an intermediate R phase - associated with the RMT. These changes depend on the way the treatment ends, either by heating or by cooling. Load-free thermal cycling contributes to enlarge the temperature interval of RMT occurrence, but does not change the interval of hysteresis. After 300 cycles, if the end occurs by cooling below M_f, an intermediate R phase, related to the $\gamma'_1 \leftrightarrow \beta_1$, is obtained. The alloy submitted to thermal cycling under load reveals a tendency towards a decrease in the critical intervals for both direct and reverse martensite transformation. This tendency is accentuated for 300 cycles, when the critical intervals smaller than 10 °C, are recorded. The applied load promotes the reverse transformation at lower temperatures as well as the direct martensitic transformation at higher temperatures, reorienting the structure and stabilizing the R phase, coherent with γ'_1 and β_1. The alloy's mechanical behavior is characterized by a complex process involving both its RMT and phase reorientation. The results of compressive tests of the initial state revealed a "pseudo-yielding" plateau followed by elastic deformation up to 1350 MPa and subsequent fracture, disclosing its brittleness. Thermal cycling under load reduces the pseudo-yield stress by 20-25 MPa and decreases the ultimate stress to 950 MPa. The alloy shows good resistance to irreversible changes during all thermal cycling tests. These results have a particular relevance for the practical application of the monocrystalline Cu-13.5Al-4Ni alloy as SME actuator.

Chapter 5 – CuAlNi shape memory materials are promising for high temperature technological applications over other Cu based alloys, as they show low diffusion at temperatures close to room temperature or higher [1, 2]. For the development of actuator devices using CuAlNi alloys, it is necessary to study the thermal and mechanical response of the material due to cycling through the martensitic transformation. Additionally, the use of a single crystal material increases the life to fracture, and finally enhances the performance of future devices based on these materials.

In some CuAlNi alloys, a metastable β-bcc ordered structure obtained at room temperature can undergo martensitic transformations by cooling or applying stress (σ). Three types of martensitic transformations can be induced for chemical compositions between 13-15Al and 0-6Ni (wt.%): β↔γ', β↔β' and β↔β'+γ'.

In this chapter, the authors review the more recent research made on thermal and pseudoelastic tensile cycling in CuAlNi single crystals, and the interaction between martensitic transformations and defects which causes a novel effect in these alloys.

Chapter 6 – The thermal properties of Ni-Mn-Ga based shape memory alloys were studied in the temperature range from room temperature up to 650 K. Both studied polycrystalline alloys, $Ni_{53.6}Mn_{27.1}Ga_{19.3}$ and $Ni_{54.2}Mn_{29.4}Ga_{16.4}$, are high-temperature shape memory alloys.

Martensitic phase transformation in Ni-Mn-Ga-based shape memory alloys is an athermal phase transformation that starts practically immediately after reaching a certain transient temperature. This temperature is given in each point of the sample by two processes: the thermal conductivity and the phase transformation. Both processes take place in cooperation and their dynamics determines properties of the shape memory material in the transient temperature range. The complex process of the propagation of the phase transformation in the sample is analyzed. The influence of the heat transfer in the sample as well as in the measuring system on the transformation temperatures is also discussed.

Chapter 7 – In this study, molecular dynamics simulation (MDS) is conducted by using embedded atom method (EAM), which is an interatomic potential function suitable for Ni-Ti system. The EAM potential reproduces the experimental results of the energetic for stable Ni-Ti phases (B2, B19') and the Ti precipitates which are consisted of bct lattices. In the simulation model, the authors put disc-shaped Ti precipitates inside Ni-Ti parent alloy. Firstly, in the model with the disc-shaped precipitates, martensite is maintained around the interfaces because inevitable local residual strain can be balanced, when the whole simulation model has no strain (completely relaxed). Under tensile loading, in the case with disc precipitates, martensite region propagates from the interface plane. In the propagation, the unit lattices show structurally two pathways to complete transformation. The split into these pathways is strongly depending on the angle between tensile direction and the orientation of the martensitic unitcell. The results suggest that the interface is a preferred site for nucleation of martensite and that there is a certain crystallographic condition for the next variants to form smoothly. Secondly, stress-induced amorphization is calculated by MDS with a simple potential function derived from modified EAM (MEAM) framework. Relation between deformation (elastic and/or plastic) and structural changes into amorphous phase is characterized by the method of common neighbor analysis (CNA).It is found that the amorphization caused by shear deformation is well enhanced especially when slip motion is restricted due to the choice of shear direction. The obtained amorphous is found energetically meta-stable. Besides, it is found that martensite is observed around the amorphous region.

Chapter 8 – The evolution of the stress-induced martensitic transformation was determined for a Cu-Al 11.2 wt%-Be 0.6 wt% shape memory alloy by digital image analysis. The CuAlBe tensile specimens were subjected to uniaxial tension; during the tension test, digital images of the polished and chemically etched surface's specimens were taken. From the tension test, the stress-strain curve representing the macromechanical behavior of the material was obtained. From the acquired images, the stress-induced martensitic transformation was observed and the number of grains, the area of the grains and the martensitic plates in the grains were estimated. Additionally, the orientation of the martensitic plates was determined for each image associated to the stress-strain curve.

Image analysis showed that, during the tension test, the martensitic variants appeared and disappeared due to the interaction between a grain and its neighbors, which is caused by the stress state of the grain, the original variant or variants, and the state of transformation of the neighboring grains. With the results from the micromechanical behavior, the strain measured with the strain gage could be resolved into its elastic and transformational components.

Chapter 9 – The paper assumes that, in the case of a martensitic Cu-Zn-Al shape memory alloy subjected to heating above 770 K, the endothermic peaks revealed by differential scanning calorimetry (DSC) and differential thermal analysis (DTA) and located around 630 K would correspond to the transitory formation of α_1 bainite. This endothermic phenomenon has been regarded as transitory since heating proceeding causes exothermic precipitation of equilibrium α-phase which is copper-rich and therefore its formation augments both Zn and Al amounts in β_2 austenitic matrix.

Chapter 10 – The effects of thermal cycling and mechanical operations like cutting and polishing on the phase transformation behavior of heat-treated NiTi shape memory alloy (SMA) have been investigated by means of Powder X-Ray Diffraction (XRD) and Differential Scanning Calorimetry (DSC). The results of studies carried out on the effect of thermal cycling on SMA heat-treated at 660 °C and effect of mechanical cutting and polishing on SMA heat-treated at 560 and 660 °C, are presented. In NiTi alloy heat-treated at 660 °C, the integrated peak area (integrated intensity) corresponding to $(110)_A$ plane of the austenitic phase increases and the area corresponding to martensitic phase decreases with increasing temperature. Austenitic start and finish temperatures change during the first few thermal cycles. The former decreases by ≈15 °C and the latter decreases by ≈35 °C during the first cycle, whereas martensitic start and finish temperatures do not get affected by thermal cycling. The transformation characteristics determined through XRD are compared with those obtained from DSC. The thermal width observed from XRD studies is larger than that from DSC measurements due to the better sensitivity of the former in detecting smaller volume fractions of the phases in the martensitic transformations as compared to the thermal methods like DSC. Mechanically cut and polished samples show broad thermal width and lesser heat flow associated with the M↔A transformation compared to chemically etched samples due to the internal stresses generated during mechanical operation. Among the samples heat-treated at 560 °C, the sample subjected to mechanical cutting and polishing (HT560MC) shows a single step M→A on reverse phase transformation and two step A→R→M on forward transformation whereas chemically etched sample (HT560CE) shows the intermediate R-phase both during forward and reverse transformations. In the case of HT560MC, the dislocations created by the mechanical cutting and polishing may be responsible for the non-appearance of R-phase in the reverse transformation even up to 150 cycles, possibly due to the entanglement of dislocations at high density of dislocations. Heat treatment at a higher temperature of 660 °C does not facilitate the formation of R-phase in the reverse transformation due to the reduced dislocation density. However, dislocations created due to mechanical operations (in HT660MC) help the formation of R-phase to a very small extent during forward transformation. Both the heat-treated samples show the effect of mechanical cutting and polishing even at 150th thermal cycle.

Chapter 11 – The effects of thermal cycling and mechanical operations like cutting and polishing on the phase transformation behavior of heat-treated NiTi shape memory alloy (SMA) have been investigated by means of Powder X-Ray Diffraction (XRD) and Differential Scanning Calorimetry (DSC). The results of studies carried out on the effect of thermal cycling on SMA heat-treated at 660 °C and effect of mechanical cutting and polishing on SMA heat-treated at 560 and 660 °C, are presented. In NiTi alloy heat-treated at 660 °C, the integrated peak area (integrated intensity) corresponding to $(110)_A$ plane of the austenitic phase increases and the area corresponding to martensitic phase decreases with increasing temperature. Austenitic start and finish temperatures change during the first few thermal cycles. The former decreases by ≈15 °C and the latter decreases by ≈35 °C during the first cycle, whereas martensitic start and finish temperatures do not get affected by thermal cycling. The transformation characteristics determined through XRD are compared with those obtained from DSC. The thermal width observed from XRD studies is larger than that from DSC measurements due to the better sensitivity of the former in detecting smaller volume fractions of the phases in the martensitic transformations as compared to the thermal methods like DSC. Mechanically cut and polished samples show broad thermal width and lesser heat flow associated with the M↔A transformation compared to chemically etched samples due to the internal stresses generated during mechanical operation. Among the samples heat-treated at 560 °C, the sample subjected to mechanical cutting and polishing (HT560MC) shows a single step M→A on reverse phase transformation and two step A→R→M on forward transformation whereas chemically etched sample (HT560CE) shows the intermediate R-phase both during forward and reverse transformations. In the case of HT560MC, the dislocations created by the mechanical cutting and polishing may be responsible for the non-appearance of R-phase in the reverse transformation even up to 150 cycles, possibly due to the entanglement of dislocations at high density of dislocations. Heat treatment at a higher temperature of 660 °C does not facilitate the formation of R-phase in the reverse transformation due to the reduced dislocation density. However, dislocations created due to mechanical operations (in HT660MC) help the formation of R-phase to a very small extent during forward transformation. Both the heat-treated samples show the effect of mechanical cutting and polishing even at 150th thermal cycle.

Chapter 12 – Dependence of transformation temperatures of shape memory alloys on the number (e_v/a) and concentration (c_v) of valence electrons is investigated. The alloys are categorized into low (e_v/a<5), medium (5≤e_v/a≤7.50), and high (e_v/a>7.50) valence electron groups. For majority of alloys in medium and high valence electron group, clear correlations between transformation temperatures and c_v are found. M_s and A_s both decrease from 1100 and 1150°C to as low as -206°C and -153°C respectively, with increasing c_v from 0.10 to around 0.30. When e_v/a>7.50, the transformation temperatures show much less dependence on c_v and can be lower than the medium group temperatures even at medium or low c_v values. High electron NiMnX (X=Ga, In, Al) Heusler alloys containing Mn (>30 at%) show high transformation temperatures at high c_v values due, most likely, to anti-ferromagnetism of excess Mn which reduces the elastic moduli. The low valence electron group alloys show significantly lower transformation temperatures for their c_v values compared to the main group. The influence of c_v is discussed on the basis of the variations of elastic constants due to composition change.

Chapter 13 – Transformation hysteresis is an important property of shape memory alloys (SMAs) which requires careful consideration during the materials selection for targeted technical applications. When being used for repeated active actuation purposes (for example in MEMS, morphing wing of aerial vehicles, robotics, etc.) a small hysteresis is required since it strongly influences the operational frequency achievable and the power consumption needed for actuation. On the other hand, some applications may require very large transformation hysteresis in order to maintain a stable austenite phase within a large temperature range for retaining the predefined shape (for example in deployable space structures, pipe joining, etc). In order to obtain SMAs with either very low or very high hysteresis, a fundamental understanding on its influencing factors is ultimately important. This short communication highlights several recent results on the factors affecting the transformation hysteresis in shape memory alloys. This aricle is not aimed to provide a complete spectrum on the transformation hysteresis of shape memory alloys. It is hoped to encourage further investigation to understand the fundamental factors controlling the transformation hysteresis, thus enabling an effective alloy design.

Chapter 14 – This chapter reviews several approaches for the mathematical modeling of shape memory alloys to be embedded as constituents (e.g. fibers) thus forming shape memory alloy composites. To this end, shape memory alloy fibers that exhibit pseudoelasticity, one and two-way shape memory effects, rate dependence and transformation induced plasticity are considered. The behavior of the shape memory composites is determined by employing a micromechanical analysis which is capable of predicting the response of such composites that consist of polymer as well as metallic matrices. Since the shape memory alloy fibers behavior is dominated by the temperature, this micromechanical model accounts for the full thermomechanical coupling between the fibers and matrix in which the mechanical and thermal effects mutually affect each other. The micromechanical modeling is also applied to predict the behavior of shape memory alloy honeycombs, which form a promising new type of smart materials. The possibility of increasing the tensile strength of concrete by embedding activated shape memory wires is investigated. This is carried out by a micromechanical modeling in which both evolving damage effects as well as coupled damage-plasticity effects are incorporated in the prestressed concrete.

Chapter 15 – In this chapter, the authors discuss the implicit time discretization for Fr'emond's model of two-shape memory phenomenon and the error estimates between the approximate solution and the exact solution of Fr'emond's model. The authors show optimal order error estimates on the whole time interval. Also, the authors formulate the continuous dependence result of the solution from the data of the problem.

Chapter 16 – Shape memory alloys (SMA) are a special kind of smart materials whose dimensions can be modified as the result of a temperature-dependent structural phase transition. This property can be used to generate motion or force in electromechanical devices and micromachines. The use of shape memory alloys in actuators offers the opportunity to develop robust, simple and lightweight elements for application in a multitude of different industries. Despite all these advantages, the accuracy of SMA actuators is severely limited by their highly nonlinear stimulus-response characteristics. Traditionally, they have been used as ``on-off'' electromechanical actuators due to the non-linear and hysteretic nature of the martensite-austenite transformation. The ideal solution to this problem is to model the hysteresis mathematically to compensate for it in the control loop. However, useful models of phase transitions are difficult to obtain. In this chapter, a Nickel-Titanium alloy (Nitinol) wire is considered as the active element in micropositioning actuators. An electric current, applied through the wire, heats it to induce the phase transition and the consequent contraction. The purpose of this investigation is to finely control the wire contraction. The wire needs to be able to contract and relax to an intermediate position within its range of movement and the gradual contraction and relaxation needs to be controllable. An experimental micropositioning device, based on a Ni-Ti wire, is controlled using several control strategies. To account for the hysteresis effects, an inverse hysteresis model is obtained using two different compensating methods. When used in a controller, the inverse models nicely compensate the non-linear and hysteretic behavior of the wire, and a proportional-integral with antiwindup control works adequately, obtaining position accuracies around 3 micrometers. Once the real possibility of using these materials as micrometric actuators is analyzed and tested, a SMA based actuator prototype grip for a flexible robot is developed, so that it could be used in light applications. For this purpose, the bulky LVDT (linear variable differential transformer) used in the previous experiments to provide the position feedback to the controller must be replaced by lighter alternatives. The first solution consists of deducing the strain of the SMA wire from the readings of a set of strain gauges that are placed in the fingers of the grip. Although the precision is limited by the strain gauge accuracy, position errors about

Chapter 17 – This chapter is devoted to the mathematical study of a thermomechanical model describing the macroscopic behavior of shape memory alloys (SMA). Various experiments (slow uniaxial traction-compression tests on cylindrical SMA samples) emphasized the temperature changes during hysteresis loops and relaxation and creep phenomena. They also established the absence of permanent deformations at the end of a loading-unloading cycle, the so-called pseudoelastic behavior of SMA. The macroscopic phenomena observed during the experiments can be predicted by using a classical (Gibbs) thermomechanical model founded on a free energy, which is convex with respect to the strain and to the martensitic volume fraction, and concave with respect to the temperature. The model takes into account the non-isothermal character of the phase transition, as well as the existence of the intrinsic dissipation. The first law of thermodynamics, the balance of momentum in its quasistatic form, together with the evolution equation for the internal variables, form a partial differential equation system, whose solutions fulfill the second principle of thermodynamics (the entropy inequality). In the circular cylindrical case with non-negligible radius, this system reduces to an ordinary differential system. In this case, the authors prove the uniqueness of solutions in a large class of spaces (abstract differential structures). Existence and regularity of solutions are then established in various functions spaces endowed with natural derivatives. At the end of the chapter, for a particular traction-compression test the authors present the exact solution, together with the associated hysteresis loop.

In: Shape Memory Alloys
Editor: H. R. Chen, pp. 1-47

ISBN 978-1-60741-789-7

Chapter 1

CONSTITUTIVE MODELS FOR SHAPE MEMORY ALLOYS: A TWO-PHASE MIXTURE MODEL AND A MICROSTRUCTURE-BASED MODEL

Xianghe Peng, Wenli Pi and Xuesong Long
Department of Engineering Mechanics, Chongqing University,
Chongqing, 400044, China

ABSTRACT

Two models for the constitutive behavior of polycrystalline shape memory alloys (SMAs) are presented. One is a two-phase mixture model, and the other is a microstructure-based two-phase model, accounting for the typical constitutive behavior of Ni-Ti SMAs under different stress states. In both models, it is assumed that in the interested ranges of stress and temperature, the austenite phase is linearly elastic while the martensite phase is elastoplastic.

In the first model, an SMA is considered to be composed of austenite and martensite phases, and its constitutive behavior is the combination of the individual behavior of each of the two phases. Making use of the simple Tanaka's phase-transformation rule, the main features of SMAs, such as shape memory effect and pseudoelasticity, can be successfully described. The constitutive behavior of SMA Au–47.5 at.% Cd subjected to uniaxial tension/compression and the pseudoelasticity of a polycrystalline Cu–Al–Zn–Mn SMA subjected to proportional and nonproportional complex stress/strain histories were simulated and compared with experimental results.

Experiments showed that the pseudoelastic behavior of NiTi SMAs under pure tension and that under pure torsion were distinctly different, due to the different deformation-induced-transformation microstructures. SEM observation showed parallel texture on the surface of the NiTi SMA microtubes subjected to pure torsion, indicating alternatively arranged parallel fine lamellas of austenite and martensite phases. During pure tension, it was observed that martensite initiated from the parent phase and grew to macroscopic bands, and, correspondingly, a typical stress drop or macroscopic instability was observed in the tensile stress–strain curve. FE analyses based on these microstructures replicated the experimental results under pure tension and under pure torsion, respectively, indicating the important role of the microstructure in the macroscopic constitutive response.

A microstructure-based two-phase model was developed for the pseudoelastic behavior of NiTi SMAs, which includes three stages: elastoplastic model for the martensite phase and the linearly elastic model for the austenite phase; the model for a representative volume element (RVE) of a cell composed of alternatively arranged parallel austenite and martensite lamellas; and the model for an NiTi SMA by considering the material as an aggregate of numerous cells and making use of the Hill's self-consistent scheme. The model could satisfactorily describe the main characteristics of the pseudoelastic behavior of the NiTi SMA under pure tension and pure torsion. The pseudoelastic behavior of the NiTi SMA under biaxial tension-torsion were also analyzed and compared with experimental results.

1. INTRODUCTION

Shape memory alloys (SMAs) are composed of austenite and martensite phases, the forward and inverse thermoelastic martensitic transformation determines the amazing properties of SMAs. When an SMA is cooled down to below some critical temperature, it will mainly be composed of martensite phase and can be manipulated easily in a very large strain range. When it is heated to above some critical temperature, it will change back to the austenite phase, the previous inelastic deformation will be recovered, and it will resume the shape it normally has at the temperature (shape memory effect, SME). At higher temperature, applied stress may cause martensitic transformation and an SMA may deform inelastically, but the inelastic deformation may be recovered as the applied stress is removed, which is referred to as pseudoelasticity (PE). These particular properties attracted extensive attention in the past decades.

Great progress has been made in the constitutive modeling of SMAs. In the early constitutive models, attention was mainly paid to the phenomenological description for the behavior under simple tension and compression with thermodynamic approaches or statistical micromechanics, as those by Falk (1980) and Achenbach *et al* (1986). Graesser and Cozzarelli (1991) proposed a phenomenological and practical model by introducing an error function into the back stress, which was further extended to the case of general state of stress (Graesser and Cozzarelli, 1994). Tanaka (1986) obtained an expression of stress rate by making use of thermodynamic laws, which was then simplified and integrated by Liang and Rogers (1990) so that the stress can be expressed as a function of strain, temperature and martensite volume fraction. Boyd (1994) proposed a model by decomposing strain into three parts corresponding respectively to elasticity, temperature variation and phase transformation, which could describe the pseudoelasticity and shape memory effect of SMAs. A constitutive model based on an internal variable formalism and the framework of generalized plasticity was proposed by Auricchio *et al* (1997), with which the basic features of SMAs could be reproduced. Levitas (1995) discussed the phase transitions in elastoplastic materials and the corresponding conditions. It was suggested that the phase transformation in a microregion should be related to a critical value of the Gibbs free energy, and a description for the phase transformation in polycrystalline SMAs was obtained with a statistical method (Fischer, 1997). Goo and Lexcellent (1997) proposed a kind of micromechanically based Helmholtz free energy, based on which a kinetic relationship, the nucleation criterion of martensitic phase and the reorientation criterion of martensitic variants were obtained. Based on a generic form of Gibbs energy over a representative volume element (RVE) as well as a set of internal variables, a thermomechanical model for polycrystalline SMAs was proposed by Bo and Lagoudas (1999) and applied to the analysis of the response of SMAs under cyclic loading (Bo and Lagoudas, 1999; Lagoudas and Bo, 1999).

Being cognizant that the complex constitutive behavior of SMAs could hardly be well described with phenomenological models, Tanaka *et al* (1986) and Patoor *et al* (1988) investigated the constitutive behavior of SMAs through a microscopic analysis and the corresponding averaging procedure. However, it was limited to stress induced forward martensitic transformation and could not work well in the cases involving inverse martensitic transformation and non-proportional loading. A micro-macro thermomechanical modelization of the grain behavior was proposed by Siredey *et al* (1999), in which the average martensite fraction in each domain was used as an internal variable, and the description for the evolution of the material was suggested. The overall behavior of polycrystalline SMAs was then obtained with a self-consistent scheme. Sun and Hwang (1993) proposed a model by making use of micromechanics and thermodynamics, which can describe both forward and inverse martensitic transformation and the main features of SMAs under complex deformation and temperature histories. Shaw and Kyriakides (1998) presented finite element (FE) simulations to demonstrate a continuum approach for the nucleation and propagation phenomena and the severe thermo-mechanical interaction in pseudoelastic NiTi SMAs. This approach was then extended to include the thermal interactions between a mechanically loaded NiTi specimen and its environment to capture the nonisothermal response at higher loading rates and in a less convective medium (Shaw, 2000), and the trends of localized nucleation and propagation phenomena for a wider range of loading rates and ambient thermal conditions (Iadicola and Shaw, 2004). A rate-independent and finite-deformation-based crystal mechanics model was proposed for martensitic reorientation and detwinning in SMAs (Thamburaja, 2005), which was further shown to be able to quantitatively predict the experimental response of an initially textured and martensitic polycrystalline NiTi rod under a variety of uniaxial and multiaxial stress states (Pan, *et al*, 2007). Based on the concept that an SMA is transformably composed of the austenite and martensite phases, and the constitutive behavior of an SMA is substantially the combination of that of each of the two phases, a kind of two-phase mixture models for SMAs was proposed by Peng *et al* (2005, 2000), incorporating conventional and non-conventional theory of plasticity, respectively. In these models, the complicated constitutive behavior of an SMA is separated into the simple responses of its two phases, with which the forward and inverse martensitic transformation, shape memory effect (SME) and pseudoelasticity (PE) of SMAs under complex thermomechanical loadings can be described. With the concept of two-phase mixture, Müller and Bruhns (2006) proposed a thermodynamic finite-strain model for the pseudoelastic response of SMAs based on a self-consistent Eulerian theory of finite deformations.

Two models for the constitutive behavior of SMAs are introduced in this chapter. In section 2, a two-phase mixture model (Peng *et al*, 2005) is introduced. It is based on the mixture theory incorporating the conventional theory of plasticity. In this model, an SMA is considered to be composed of austenite and martensite phases, and its constitutive behavior is the combination of the individual behavior of each of the two phases. In the interested ranges of stress and temperature, the behavior of the austenite is linearly elastic while that of martensite is elastoplastic. Making use of the simple Tanaka's phase-transformation rule, the main features of SMAs, such as shape memory effect and pseudoelasticity, can be successfully replicated. The constitutive behavior of SMA Au–47.5 at.% Cd subjected to uniaxial tension/compression and the pseudoelasticity of Cu–Al–Zn–Mn SMA polycrystal subjected to proportional and nonproportional complex stress or strain histories are simulated and verified with the experimental results, demonstrating the validity of the model. This model is simple, of distinct physical sense, and can easily be applied in practice.

In section 3, a microstructure-based constitutive model is introduced for the unique constitutive behavior of a kind of NiTi SMAs. Experiments showed that the pseudoelastic behavior of NiTi shape memory alloys under pure tension and pure torsion is distinctly different, which can be attributed to the different deformation-induced-transformation microstructures. SEM observation showed uniformly distributed parallel texture on the surface of the NiTi SMA microtubes subjected to pure torsion, indicating alternatively arranged parallel lamellas of austenite and martensite phases. During pure tension, martensite initiates from the parent phase and grows to macroscopic bands, accounting for the typical stress drop or macroscopic instability in the tensile stress–strain curve. Finite element analyses based on the two kinds of microstructures and variation well replicated respectively the responses of the SMA microtubes under pure tension and pure torsion, indicating the important effect of the transformation microstructure on the macroscopic constitutive response. Assuming purely elastic austenite phase and elastoplastic martensite phase, the model for a representative volume element (RVE) of the cell composed of alternatively arranged parallel lamellas of austenite and martensite can be obtained, and then the description for the NiTi SMAs can be derived by considering the material as an aggregate of numerous cells with different orientations and making use of the self-consistent scheme. The corresponding numerical algorithm was developed and the responses of an NiTi SMA under pure tension, pure torsion and biaxial tension-torsion stressing were analyzed, respectively. The comparison with the experimental results showed that the model can satisfactorily describe the main characteristics of the pseudoelastic behavior of the NiTi SMAs, demonstrating the validity of the microstructure-based two-phase constitutive model.

2. A Two-Phase Mixture Model for Shape Memory Alloys

A comprehensive description for the constitutive behavior of shape memory alloys (Peng *et al*, 2005) is introduced in this section, which is based on a two-phase mixture model incorporating the conventional theory of plasticity. The main features of SMAs, such as shape memory effect and pseudoelasticity, can be successfully replicated with the constitutive model. The constitutive behavior of SMA Au-47.5at.%Cd subjected to uniaxial forward and inverse loading and the pseudoelasticity of Cu-Al-Zn-Mn SMA polycrystal subjected to proportional and nonproportional complex stress or strain histories are analyzed and verified with the experimental results.

2.1. A Two-Phase Mixture Model

An SMA is composed of austenite and martensite phases and the volume fraction of each phase may vary during a thermal-mechanical loading process. At a lower temperature, it is mainly composed of martensite phase and may behave inelastically; if the temperature is sufficiently high so that it is mainly composed of austenite phase and behaves almost elastically. The typical macroscopic behavior of an SMA, such as pseudoelasticity and shape memory effect, can be regarded as the combination of the individual behavior of each of the two phases.

In this section, the discussion is restricted to the case of small deformation, the material is assumed inelastically incompressible and the transformation lattice volume change can be neglected (Sun and Hwang, 1993). The strain deviation **e**, in general, can be assumed to consist of elastic strain $\mathbf{e}^e$, inelastic strain e^p and phase transformation strain $\mathbf{e}^T$, i.e.,

$$\mathbf{e} = \mathbf{e}^e + \mathbf{e}^p + \mathbf{e}^T , \tag{1}$$

where e^T is related to the temperature induced recovery of inelastic distortion. Assuming that the strains in both the martensite and austenite phases are identical and equal to the overall strain in the SMA, without considering the detailed microstructure, the macroscopic stress deviation s and volumetric stress σ_{kk} of the SMA can, therefore, be expressed as

$$\mathbf{s} = \bar{\mathbf{s}}_A + \bar{\mathbf{s}}_M, \qquad \sigma_{kk} = (\bar{\sigma}_{kk})_A + (\bar{\sigma}_{kk})_M, \tag{2}$$

where $\bar{\mathbf{s}}_A$, $(\bar{\sigma}_{kk})_A$ and $\bar{\mathbf{s}}_M$, $(\bar{\sigma}_{kk})_M$ denote respectively the contributions of the austenite phase and that of the martensite phase to the corresponding overall stress components.

It should be noted that, although some kind of substructure should exist in an SMA under thermal-mechanical loading, for example, lamellar microstructures were found in NiTi SMA specimens when they were stretched (Shaw and Kyriakides, 1998; Sun and Li, 2002), for a phenomenological model, in order to avoid complicity, we would like to temporarily ignore the detailed substructure and assume the strain in both the martensite and austenite phases to be identical and equal to the overall strain in the SMA if the interlaminar spacing is sufficiently small (Peng, *et al*, 2002). With this assumption, Eq. (2) can easily be derived from the equivalency of the energy determined by the overall stress and strain in a representative volume element (RVE) of an SMA and that determined by the stress and strain in both the martensite and austenite phases. On the other hand, for a given overall strain the overall stress in an SMA usually lies between those in its two phases, which implies that the overall stress in the SMA should be some combination of those in its two phases. It would be shown that such a disposal could make the model simple but satisfactorily accurate when it was applied to practical engineering problems.

From the analysis of the constitutive behavior of SMAs, it is acceptable to assume that the austenite phase is elastic while the martensite phase is elastoplastic within the interested ranges of stress and temperature. It is in accordance with the constitutive characteristics of an SMA loaded at lower temperature when it is mainly composed of martensite phase and at higher temperature when it is mainly composed of austenite phase, respectively. Letting ξ_A and ξ denote respectively the volume fractions of the austenite and martensite phases, s_M and $(\sigma_{kk})_M$ denote respectively the stress deviation and the volumetric stress in the martensite phase, s_A and $(\sigma_{kk})_A$ denote those in the austenite phase, and keeping in mind the assumption that the strain and the phase transformation strain in the martensite and austenite phases are identical to the overall **e** and $\mathbf{e}^T$, respectively, the conventional mixture theory gives

$$\bar{\mathbf{s}}_M = \xi \cdot \mathbf{s}_M, \qquad (\bar{\sigma}_{kk})_M = \xi \cdot (\sigma_{kk})_M, \\ \bar{\mathbf{s}}_A = \xi_A \mathbf{s}_A, \qquad (\bar{\sigma}_{kk})_A = \xi_A \cdot (\sigma_{kk})_A \tag{3}$$

with $\xi + \xi_A = 1$. (4)

For materials involving mixed isotropic and kinematic hardening, the yield function as well as the yield surface can be expressed as

$$f = \frac{3}{2}(\mathbf{s}_M - \boldsymbol{\varphi}):(\mathbf{s}_M - \boldsymbol{\varphi}) - [\psi(\zeta)]^2 = 0, \tag{5}$$

where $\boldsymbol{\varphi}$ is the back stress tensor defining the location of the yield surface, ψ is the size of the yield surface. $\boldsymbol{\varphi}$ and ψ take the following evolution laws (Armstrong and Frederick, 1966)

$$d\boldsymbol{\varphi} = c\,d\boldsymbol{\varepsilon}^p - \gamma\boldsymbol{\varphi}d\zeta, \text{ with } \boldsymbol{\varphi}(0) = 0 \tag{6}$$

$$d\psi = \alpha[p\sigma_{Ms} - \psi]d\zeta. \text{ with } \psi(0) = \sigma_{Ms} \tag{7}$$

$$d\zeta = \sqrt{\frac{2}{3}d\mathbf{e}^p : d\mathbf{e}^p} = d\bar{\varepsilon}^p, \tag{8}$$

where σ_{Ms} is the initial yield stress of the martensite phase, $d\zeta$ denotes the increment of equivalent plastic strain. It can be seen that recovering terms are introduced in both Eqs. (7) and (8), indicating the existence of the saturated states of both $\boldsymbol{\varphi}$ and ψ.

The following normal flow rule is adopted

$$d\boldsymbol{\varepsilon}^p = d\lambda\frac{\partial f}{\partial \boldsymbol{\sigma}_M}. \tag{9}$$

Making use of the condition of consistency, $df = 0$, one obtains

$$3(\mathbf{s}_M - \boldsymbol{\varphi}):(d\mathbf{s}_M - d\boldsymbol{\varphi}) - 2\psi\psi' d\zeta = 0. \tag{10}$$

It can be solved from Eqs. (9) and (10) that

$$d\lambda = \frac{(\mathbf{s}_M - \boldsymbol{\varphi}):d\mathbf{s}_M - [\frac{2}{3}\alpha\psi(p\sigma_{Ms} - \psi) - \gamma(\mathbf{s}_M - \boldsymbol{\varphi}):\boldsymbol{\varphi}]d\zeta}{2c\psi^2}. \tag{11}$$

The plastic flow can be determined as follows by combining Eq. (9) with Eq. (11)

$$d\boldsymbol{\varepsilon}^p = \frac{3(\mathbf{s}_M - \boldsymbol{\varphi})\otimes(\mathbf{s}_M - \boldsymbol{\varphi}):d\mathbf{s}_M - b(\mathbf{s}_M - \boldsymbol{\varphi})d\zeta}{2c\psi^2}, \tag{12}$$

where $b = 2\alpha\psi(p\sigma_{Ms} - \psi) - 3\gamma(\bar{\mathbf{s}}_M - \boldsymbol{\varphi}) : \boldsymbol{\varphi}$. (13)

Eq. (8) can also be expressed as

$$d\zeta = \frac{2}{3}\frac{d\boldsymbol{\varepsilon}^p}{d\zeta} : d\boldsymbol{\varepsilon}^p. \tag{14}$$

Substituting Eq. (14) into Eq.(12) yields

$$d\boldsymbol{\varepsilon}^p = \frac{3b'}{2c\psi^2}(\mathbf{s}_M - \boldsymbol{\phi}) \otimes (\mathbf{s}_M - \boldsymbol{\phi}) : d\mathbf{s}_M, \tag{15}$$

where $b' = \dfrac{1}{1+\omega}$, $\quad \omega = \dfrac{b}{3c\psi^2}(\mathbf{s}_M - \boldsymbol{\varphi}) : \dfrac{d\boldsymbol{\varepsilon}^p}{d\zeta}$. (16)

The elastic behavior of the martensite phase can be expressed as

$$\mathbf{s}_M = 2G_M(T)(\mathbf{e} - \boldsymbol{\varepsilon}^p - \mathbf{e}^T), \tag{17}$$

where G_M is the elastic shear modulus of the martensite phase.
The differential form of Eq. (17)

$$d\mathbf{s}_M = 2G_M(d\mathbf{e} - d\boldsymbol{\varepsilon}^p - d\mathbf{e}^T) + \frac{\mathbf{s}_M}{G_M}\frac{\partial G_M}{\partial T}dT. \tag{18}$$

By substituting Eq. (18) into Eq. (14), one has

$$d\mathbf{s}_M = 2G_M\left[\mathbf{I}_4 - \frac{\mu(\mathbf{s}_M - \boldsymbol{\varphi})\otimes(\mathbf{s}_M - \boldsymbol{\varphi})}{1+\mu(\mathbf{s}_M - \boldsymbol{\varphi}):(\mathbf{s}_M - \boldsymbol{\varphi})}\right] : (d\mathbf{e} - d\mathbf{e}^T - \frac{\mathbf{s}_M}{G_M}\frac{\partial G_M}{\partial T}dT) \tag{19}$$

where $\mu = a\dfrac{3G_M b'}{c\psi^2}$, (20)

and $a = \begin{cases} 1 & f = 0 \quad \text{and} \quad (\mathbf{s}_M - \boldsymbol{\varphi}) \cdot d\mathbf{e} > 0 \\ 0 & \text{otherwise} \end{cases}$ (21)

is the parameter distinguishing between loading and unloading.

It is known that a inverse martensitic transformation causes the recovery of inelastic strain, which is related to both the current state of inelastic strain and the temperature induced change of martensite volume fraction during a inverse martensitic transformation. Letting $\boldsymbol{\eta} = \mathbf{e}^p + \mathbf{e}^T$, $d\mathbf{e}^T$ was assumed to take the following simple relation

$$d\mathbf{e}^T = \frac{1}{2} k_T \boldsymbol{\eta} \left(\frac{\partial \xi}{\partial T} dT - \left| \frac{\partial \xi}{\partial T} dT \right| \right) \tag{22}$$

where k_T is a material constant. It can be seen that e^T changes only when both $\eta \neq 0$ and $\frac{\partial \xi}{\partial T} dT < 0$, i.e., when inverse martensitic transformation proceeds.

Keeping in mind that the austenite phase is linearly elastic, its constitutive relationship can be expressed as

$$\mathbf{s}_A = 2G_A(T)(\mathbf{e} - \mathbf{e}^T), \tag{23}$$

where G_A is the elastic shear modulus of the austenite phase.

For simplicity, the transformation lattice volume change was neglected in this model, since for most SMAs it is negligible compared with the lattice shear deformation (Sun and Hwang, 1993). The volumetric stress in the martensite and the austenite can, therefore, be determined respectively with

$$\begin{aligned} (\sigma_{kk})_M &= 3K_M(T)[\varepsilon_{kk} - 3\alpha_M(T)(T - T_0)] \\ (\sigma_{kk})_A &= 3K_A(T)[\varepsilon_{kk} - 3\alpha_A(T)(T - T_0)] \end{aligned} \tag{24}$$

where T_0 is the reference temperature, ε_{kk} is the overall bulk strain, K_M and α_M are the bulk modulus and the coefficient of thermal expansion of the martensite phase, and K_A and α_A are the bulk modulus and the coefficient of thermal expansion of the austenite phase, respectively. The differential of Eq. (24) is

$$\begin{aligned} (d\sigma_{kk})_M &= 3K_M(T)d\varepsilon_{kk} + \left[\frac{1}{K_M} \frac{dK_M}{dT} (\sigma_{kk})_M - 9 \frac{d\alpha_M}{dT} K_M(T - T_0) - 9\alpha_M(T)K_M \right] dT \\ (d\sigma_{kk})_A &= 3K_A(T)d\varepsilon_{kk} + \left[\frac{1}{K_A} \frac{dK_A}{dT} (\sigma_{kk})_A - 9 \frac{d\alpha_A}{dT} K_A(T - T_0) - 9\alpha_A(T)K_A \right] dT. \end{aligned} \tag{25}$$

In a stress-free state the volume fraction of the martensite phase is determined by temperature, and applied stress may increase the transformation temperature. Tanaka (1986) suggested the following relationship for the determination of the volume fraction ξ,

$$\xi = \begin{cases} 1 - \exp[-A_m < M_s + B_m \sigma_e - T >] & A \to M \\ \exp[-A_a < T - (A_s + B_a \sigma_e) >] & M \to A \end{cases} \tag{26}$$

in which

$$<\Phi> = \begin{cases} \Phi & \text{if} \quad \Phi > 0 \\ 0 & \text{if} \quad \Phi \le 0, \end{cases} \tag{27}$$

M_s and A_s denote respectively the martensitic transformation start temperature and the austenitic transformation start temperature, A_m, B_m, A_a and B_a are material constants related to the phase transformation, and σ_e is equivalent stress defined by

$$\sigma_e = \sqrt{\frac{3}{2}\mathbf{s}:\mathbf{s}} \cdot \tag{28}$$

The incremental form of Eq. (26)

$$d\xi = \begin{cases} A_m(1-\xi)(B_m d\sigma_e - dT) & A \rightarrow M \\ A_a\xi(dT - B_a d\sigma_e) & M \rightarrow A. \end{cases} \tag{29}$$

2.2. Analytical Approach in the Case of Biaxial Loading

2.2.1. Identification of the Material Parameters

The material parameters involved in the two-phase mixture model can be separated into three groups: parameters of the martensite phase, those of the austenite phase, and those related to forward and inverse martensitic transformation. These material constants can be identified by the following procedure:

- Keeping temperature sufficiently low ($T < M_f$) so that the SMA to be tested is composed of pure martensite phase, the following parameters can be identified with a simple elastoplastic deformation test:

1) Young's modulus E_M (or shear modulus G_M) and Poisson's ratio ν_M
2) Plastic parameters, including initial yield stress σ_{Ms}, isotropic hardening parameters p and α, and the kinematic hardening parameters c and γ.
3) The temperature-dependence of the above material parameters, by adjusting temperature within the range $T < A_s$.

- Keeping temperature sufficiently high so that no martensitic transformation can occur when applying the stress to the SMA, noticing that the SMA is composed of pure austenite phase during testing, the following parameters can be identified with a simple test:

1) Young's modulus E_A (or shear modulus G_A) and Poisson's ratio ν_A
2) The dependence of E_A and ν_A on temperature, by adjusting temperature within the range in which no martensitic transformation can occur.

- M_f, M_s, A_s, A_f are usually provided by material provider, and the rest parameters related to the forward and inverse martensitic transformation can be identified by using pseudoelasticity and shape memory effect:

1) A_m, B_m, in a loading process involving stress induced martensitic transformation
2) A_a, B_a,, in the subsequent unloading process involving stress induced inverse martensitic transformation

k_T with a shape memory effect (temperature induced inverse martensitic transformation that induces the recovery of previous residual strain).

2.2.2. Numerical Algorithm for the Case of Biaxial Loading

The model can describe the constitutive response of SMAs under three-dimensional thermal-mechanical loading, involving forward and inverse thermoelastic martensitic transformation. In this section, it is applied to the case of biaxial loading. For this purpose, Eq. (19) can be rewritten as

$$d\mathbf{s}_M = 2G_M\left[d\mathbf{e} - d\mathbf{e}^T - \frac{\mathbf{s}_M}{G_M}\frac{\partial G_M}{\partial T}dT - dg(\mathbf{s}_M - \boldsymbol{\varphi})\right], \tag{30}$$

where $$dg = \frac{\mu(\mathbf{s}_M - \boldsymbol{\varphi}):(d\mathbf{e} - d\mathbf{e}^T - \frac{\mathbf{s}_M}{G_M}\frac{\partial G_M}{\partial T}dT)}{1 + \mu(\mathbf{s}_M - \boldsymbol{\varphi}):(\mathbf{s}_M - \boldsymbol{\varphi})}. \tag{31}$$

In the case of biaxial tensile-shear loading, the components of stress increment can be derived by combining Eqs.(30) with (25) as

$$ds_{11M} = \frac{6K_M G_M}{3K_M + G_M}[d\varepsilon_{11} + h_M dT - de_{11}^T - \frac{s_{11M}}{G_M}\frac{dG_M}{dT}dT - dg(s_{11M} - \varphi_{11})], \tag{32}$$

$$ds_{12M} = 2G_M[de_{12} - de_{12}^T - \frac{s_{12M}}{G_M}\frac{dG_M}{dT}dT - dg(s_{12M} - \varphi_{12})],$$

where $$h_M = \frac{1}{K_M}\left[\frac{s_{11M}}{6K_M}\frac{dK_M}{dT} - \frac{d\alpha_M}{dT}K_M(T - T_0) - \alpha_M K_M\right]. \tag{33}$$

Strain-controlled biaxial loading

Given an increment of strain and temperature, $d\boldsymbol{\varepsilon}$ and dT, $d\xi$ and $d\mathbf{e}^T$ can be determined respectively with Eqs. (29) and (22), and then the corresponding overall stress $\mathbf{s}$ can be computed by the following procedures:

1) Determining loading/unloading parameter a with Eq. (21);
2) Computing $d\mathbf{s}_M$ with Eq. (32);
3) Superimposing the increment $d\mathbf{s}_M$ on the $\mathbf{s}_{M,n\text{-}1}$ obtained in the (n-1)th incremental loading and obtaining $\mathbf{s}_{M,n}$;

4) Computing the stress $\mathbf{s}_{A,n}$ with Eq. (23);
5) Computing the overall stress with Eqs. (2) through (4).

The isotropic and kinematic hardening parameters, ψ and $\boldsymbol{\varphi}$, can be determined with Eqs. (6) and (7), respectively, or the following more accurate expressions in an incremental loading process provided the related material parameters σ_{Ms}, p, α, c and γ are constant,

$$\psi_n = \sigma_{Ms}\left[p-(p+1)\exp(-\alpha\zeta_n)\right] \tag{34}$$

$$\boldsymbol{\varphi}_n = \boldsymbol{\varphi}_{n-1} + \Delta\boldsymbol{\varphi}_n,\ \Delta\boldsymbol{\varphi}_n = k(c\Delta e^p - \gamma\,\boldsymbol{\varphi}_{n-1}\Delta\zeta_n) \tag{35}$$

where $k = \dfrac{1-\exp(-\gamma\,\Delta\zeta_n)}{\gamma\,\Delta\zeta_n}$. (36)

Stress-Controlled Biaxial Loading

Given an increment of stress and temperature, $d\mathbf{s}$ and $\mathrm{d}T$, $\mathrm{d}\xi$ and $d\mathbf{e}^T$ can be determined respectively with Eqs. (29) and (22). Substituting Eqs. (3), (4) and (19) into Eq. (2), one obtains

$$\begin{aligned} d\mathbf{s} &= d\xi(\mathbf{s}_M - \mathbf{s}_A) + \xi d\mathbf{s}_M + (1-\xi)d\mathbf{s}_A \\ &= d\mathbf{R} + \left[2G_M\xi(\mathbf{I}_4 - \frac{\lambda(\mathbf{s}_M-\boldsymbol{\varphi})\otimes(\mathbf{s}_M-\boldsymbol{\varphi})}{1+\lambda(\mathbf{s}_M-\boldsymbol{\varphi}):(\mathbf{s}_M-\boldsymbol{\varphi})}) + 2G_A(1-\xi)\mathbf{I}_4\right]: d\mathbf{e} \end{aligned} \tag{37}$$

where

$$\begin{aligned} d\mathbf{R} &= d\xi(\mathbf{s}_M - \mathbf{s}_A) - 2G_M\xi(\mathbf{I}_4 - \frac{\lambda(\mathbf{s}_M-\boldsymbol{\varphi})\otimes(\mathbf{s}_M-\boldsymbol{\varphi})}{1+\lambda(\mathbf{s}_M-\boldsymbol{\varphi}):(\mathbf{s}_M-\boldsymbol{\varphi})}):(d\mathbf{e}^T + \frac{\mathbf{s}_M}{G_M}\frac{dG_M}{dT}dT) \\ &\quad - 2G_A(1-\xi)\left[d\mathbf{e}^T + \frac{\mathbf{s}_A}{G_A}\frac{dG_A}{dT}dT\right] \end{aligned} \tag{38}$$

$$d\mathbf{s} - d\mathbf{R} = 2G\left[\mathbf{I}_4 - q(\mathbf{s}_M-\boldsymbol{\varphi})\otimes(\mathbf{s}_M-\boldsymbol{\varphi})\right]: d\mathbf{e}\,. \tag{39}$$

It can be solved from Eq. (37) that

$$d\mathbf{e} = \frac{1}{2G}\left[\mathbf{I}_4 + \frac{q(\mathbf{s}_M-\boldsymbol{\phi})\otimes(\mathbf{s}_M-\boldsymbol{\phi})}{1-q(\mathbf{s}_M-\boldsymbol{\phi}):(\mathbf{s}_M-\boldsymbol{\phi})}\right]:\left[d\mathbf{s}-d\mathbf{R}\right], \tag{40}$$

where $2G = 2G_M\xi + 2G_A(1-\xi)$, $q = \dfrac{G_M\xi}{G}\cdot\dfrac{\lambda}{1+\lambda(\mathbf{s}_M-\boldsymbol{\varphi}):(\mathbf{s}_M-\boldsymbol{\varphi})}$. (41)

After $d\mathbf{e}$ is solved, the components of the increment of the stress in the austenite phase can be computed with the incremental form of Eq. (23), i.e.,

$$d\mathbf{s}_A = 2G_A(T)(d\mathbf{e} - d\mathbf{e}^T) + \frac{\mathbf{s}_A}{G_A}\frac{dG_A}{dT}dT \tag{42}$$

and then the stress in the austenite phase can be computed. The stress in the martensite phase can be obtained with Eqs. (2) and (3).

Application and Verification

2.3.1. Comprehensive Description of the Behavior of SMA Au-47.5 At.% Cd

Nakanishi *et al* (1973) made a detailed investigation to the uniaxial behavior of SMA Au-47.5 at.%Cd. The main features of its constitutive behavior were described with the two-phase mixture model.

It could be seen from the experimental results (Nakanishi *et al*, 1973) that, in the interested range of temperature, the effect of temperature on the material constants is less significant, and the involved material constants were, therefore, assumed to be independent of temperature to simplify the analysis. The finish and start temperatures of martensitic transformation, M_f and M_s, as well as the start and finish temperatures of austenitic transformation, A_s and A_f , were given as 308, 331K, 345K and 358K, respectively (Nakanishi *et al*, 1973). The other material constants were identified and listed in Table 1.

Table 1. Material constants of Au-47.5 at.% Cd

G_A (GPa)	ν_A	G_M (MPa)	ν_M	σ_{Ms} (MPa)	p	α
2.48	0.28	1.50	0.28	10	1.1	24
c (MPa)	γ	A_a	B_a (MPa)$^{-1}$	A_m	B_m (MPa)$^{-1}$	
550	60	0.59	0.216	0.25	0.31	

Figure 1 shows the description of the shape memory effect of Au-47.5at%Cd. The specimen was stretched at 287K up to σ= 20.7MPa and then unloaded (Figure 1(a)). It can be seen that inelastic deformation occurred to the material and large residual strain existed after unloading. Under the given temperature and loading condition, the material kept martensitic during this process. Then the material was heated and the variation of strain against temperature is plotted in Figure 1(b). It can be seen that the residual strain almost kept unchanged before the temperature reached A_s (345K) but it was recovered quickly once the temperature exceeded this criterion when inverse martensitic transformation took place. The residual strain gradually decreased and tended to vanish when the temperature was sufficiently high and the material was mainly composed of the austenite phase. Then if the temperature fell below M_s, martensitic transformation would occur and the material would return to its original state.

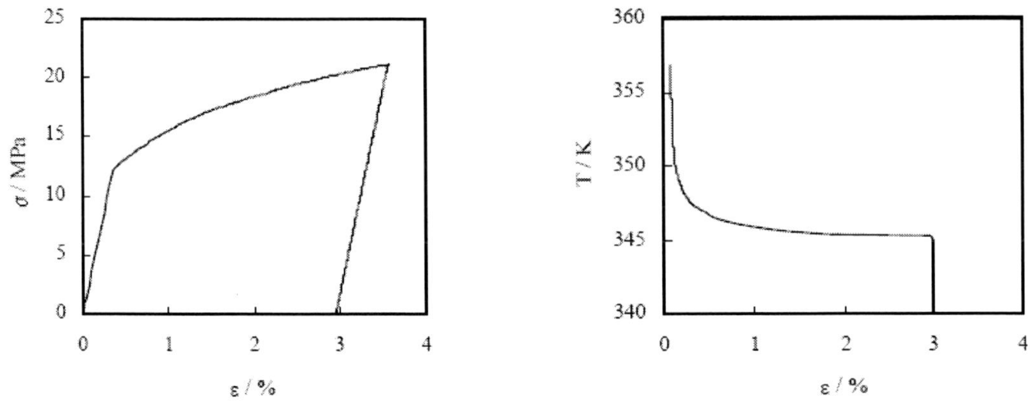

(a) Loading and unloading (b) Heating after unloading

Figure 1. Analysis for SME of SMA Au-47.5at%Cd.

Figs. 2(a) and 2(b) show respectively the tensile/compressive curve and the contributions of both the austenite and martensite phases of Au-47.5at%Cd at 357K. It can be seen from Figure 2(b) that, although the two phases were assumed respectively linearly elastic and elastoplastic, the contributions of the two phases $\overline{\sigma}_A$ and $\overline{\sigma}_M$, in which the change of martensite volume fraction has been taken into account, to the overall stress look like "pheudoelasticity". As the sum of $\overline{\sigma}_A$ and $\overline{\sigma}_M$, the overall stress is, therefore, typically pseudoelastic. It is known that the material is almost composed of the austenite phase at the given temperature ($T \approx A_f = 358$K), the applied stress would increase $M_s+B_m\sigma_e$, and martensitic transformation would take place as $M_s+B_m\sigma_e > T$. During unloading, $A_s+B_a\sigma_e$ decreased and inverse martensitic transformation would occurre as $A_s+B_a\sigma_e < T$. Compared with experimental results (Nakanishi *et al*, 1973), it can be seen that the pseudoelastic behavior could be well described with the model.

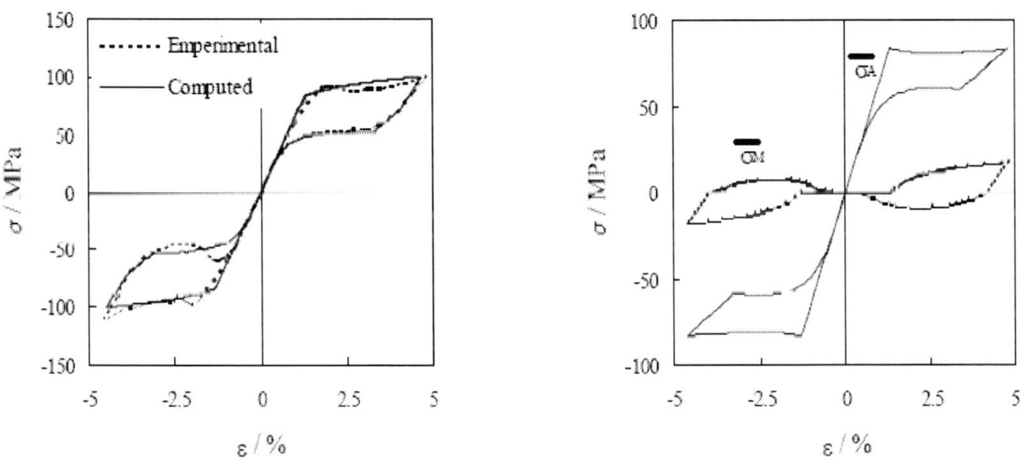

(a) Macroscopic seudoelasticitic response (b) Contributions from each phase

Figure 2. Pseudoelasticity of Au-47.5at%Cd at 357K under reversed strain.

The tensile/compressive curves of Au-47.5at%Cd at different temperatures are shown in Figure 3(a). At 330K the constitutive behavior seemed elastoplastic. It is known that at this temperature ($T<A_s<A_s+B_a\sigma_e$), the material kept mainly martensitic during the loading and unloading process, and the behavior of the SMA was, in fact, the inelastic behavior of the martensite phase. At 366K the material is mainly composed of the austenite phase ($T>A_f=$ 358K), but martensitic transformation may occur when it is subjected to applied stress, and the characteristics of the constitutive behavior should be similar to that shown in Figure 2(a), but the critical stress should increase due to the increase of temperature. At a higher temperature T = 385K, the critical stress became larger due to that higher stress is needed for martensitic transformation, and the hysteresis loop became much smaller because the contribution of the martensite phase became less significant. It can be conjectured that if temperature is sufficiently high so that the no stress is able to induce martensitic transformation, the material would be composed of the austenite phase and the behavior of the material would be determined by the austenite phase.

Figure 3(b) shows the dependence of the critical stress on temperature. It can be seen that at lower temperature the critical stress almost keeps constant with the increase of temperature, while at higher temperature the increase of critical stress is almost linearly proportional with the increase of temperature. It is known from the previous discussion that at lower temperature the behavior of the SMA is mainly determined by the elastoplastic property of the martensite phase. With the increase of temperature the volume fraction of the austenite phase increases, the contribution of the austenite phase to the overall stress increases and, therefore, the critical stress of the material increases. This tendency agrees well with the experimental result (Nakanishi et al, 1973).

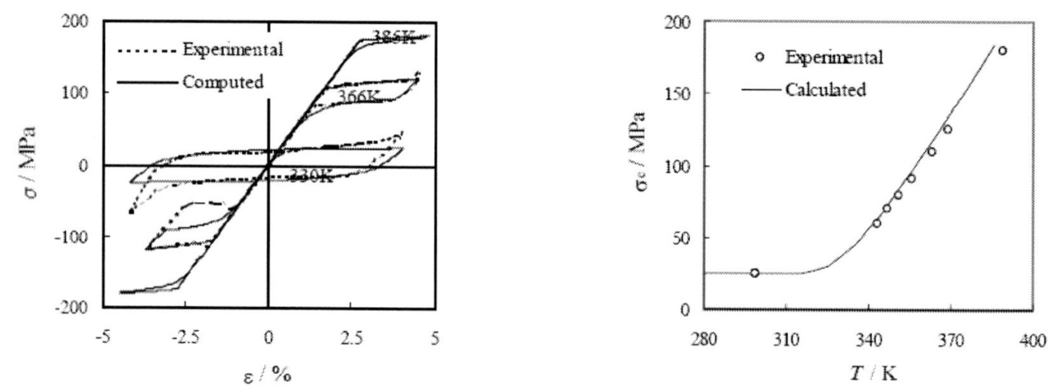

(a) σ–ε curves at typical temperatures (b) Variation of critical stresses vs. temperature

Figure 3. Response of Au-47.5at%Cd at different temperatures.

2.3.2. Analysis of the Pseudoelastic Behavior of Cu-Al-Zn-Mn SMA Polycrystal

Sittner *et al* (1995) and Sittner and Takuda (1995) systematically investigated the pseudoelastic behavior of a Cu-Al-Zn-Mn polycrystalline SMA at temperature $T = A_f$+25 K. Thin-walled specimens with d_{ext}= 8 mm and d_{int}= 5 mm were used for the combined tension-torsion tests. The specimens were finally heat-treated at 873K for two hours and quenched in ice water. The measured grain size of the material was less than 120μm in diameter. The transformation temperatures was determined by electric resistivity measurement as M_f= 223 K, M_s = 239 K, A_s = 248 K and A_f= 260 K. The experiment was performed in a stress- or strain-controlled mode. The shear strain and stress were calculated with (Sittner and Takuda, 1995)

$$\gamma = \frac{d_{ext}\theta}{2l}, \qquad \tau = \frac{16Md_{ext}}{\pi(d_{ext}^4 - d_{int}^4)}, \tag{43}$$

where θ is the twist angle over the length l and M is the measured torque. Analysis shows that if inelastic deformation takes place, Eq. (43) might overestimate the shear strain and shear stress.

Since the experiments were performed at a constant temperature (Sittner *et al*, 1995; Sittner and Takuda, 1995), the effect of temperature on the material properties and the temperature induced phase-transformation strain $\mathbf{e}^T$ (see Eq. (22)) need not be considered in the corresponding analysis. The Young's modulus and elastic shear modulus of austenite were given as E_A=53 GPa and G_A=19.5 GPa (Sittner *et al*, 1995), and p=1 due to the insignificant isotropic hardening, the other material constants were identified and shown in Table 2.

Table 2. Material constants of Cu-Al-Zn-Mn polycrystalline SMA

E_M (GPa)	G_M (GPa)	σ_{Ms} (MPa)	c (MPa)	γ	A_a	B_a (MPa)$^{-1}$	A_m	B_m (MPa)$^{-1}$
16.2	6	160	1560	60	0.045	0.11	0.035	0.24

Figure 4 shows the pseudoelastic behavior of the Cu-Al-Zn-Mn polycrystalline SMA subjected to uniaxial forward and reverse straining. Experimental observation (Sittner *et al*, 1995; Sittner and Takuda, 1995) showed that there existed small residual strain after unloading, which is attributed to: (1) plastic deformation might also occur to the austenite phase, or (2) the inverse martensitic transformation might not be complete. The latter can be described with the model without difficulties, and the former can also be described by simply giving the austenite phase elastoplastic property. Since the residual strain is very small, we tend to ignore it or equivalently describe it with the model in order to reduce complexity.

The computed response to symmetrically tensile/compressive straining is shown in Figure 4(a), which is in satisfactorily agreement with the experimental result. The comparison between the computed and the experimental responses to the symmetrically reversed torsion does not show satisfactory agreement, as shown in Figure4 (b), where the computed shear stress is smaller than the experimental measurement. It can be attributed to that: (1) Eq. (43) might overestimate the shear strain and shear stress if inelastic deformation takes place, and (2) for this material, the equivalent stress and strain does not follow Mises equivalent rule (Sittner *et al*, 1995), as used in the model (Eqs. (5) and (8)). In the case of combined tensile-torsional loading, the following equivalent rules were suggested by Sittner *et al* (1995) for the material,

$$\sigma_{eq} = \sqrt{\sigma^2 + (C^S\tau)^2}, \qquad \varepsilon_{eq} = \sqrt{\varepsilon^2 + (\gamma / C^E)^2}, \tag{44}$$

in which C^S and C^E are the ratios between the yield strengths and between the yield strains of the material under pure tension and under pure torsion, respectively. Such kind of anisotropic phenomenon could also be described by the two-phase mixture model incorporating the non-conventional model of plasticity and making use of an anisotropic rank-four tensor **P** (Peng, *et al*, 2000). Substantially, such kind of anisotropy should be related to the difference of the martensitic transformation substructure formed during a tensile/compressive process and during a pure torsional process, as was reported by Sun and Li (2002). A more detailed description should, therefore, take into account the transformation substructure, and use a micro-macro approach, which will be discussed in the following section.

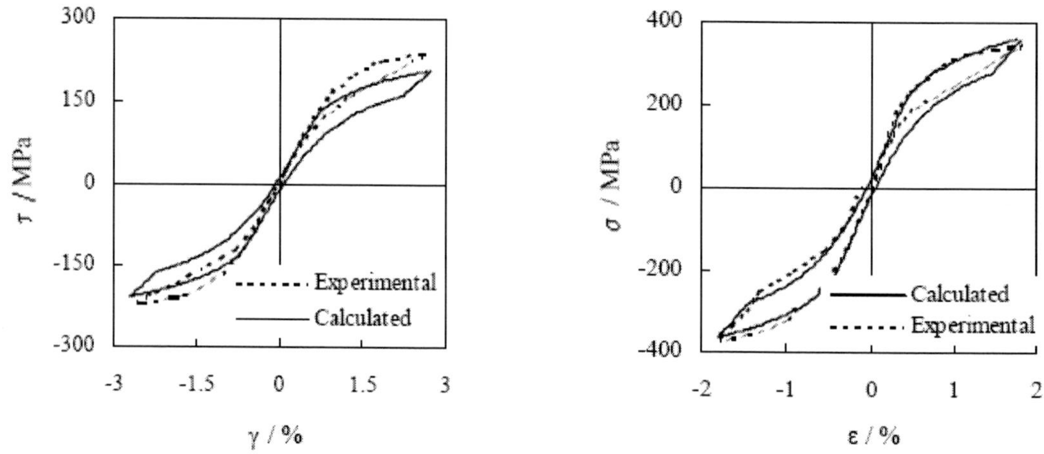

(a) Symmetrical tension/compression (b) Symmetrical torsion

Figure 4. Uniaxial stress-strain curves of Cu-Al-Zn-Mn polycrystalline SMA, strain-control, 285K.

The response of the Cu-Al-Zn-Mn polycrystalline SMA subjected to nonproportional stress along a rectangular path in the σ–τ plane is shown in Figure 5. The rectangular stress path is shown in Figure 5(a), the four vertexes of which are (0, 0), (250MPa, 0), (250MPa, 200MPa) and (0, 200MPa) in sequence. It was observed in experiment that, on one hand, the pseudoelastic strain in a nonproportional stress loop is also reversible, and on the other hand, the equivalent stress-strain loop corresponds relatively well to the uniaxial pseudoelastic loop (Sittner *et al*, 1995; Sittner and Takuda, 1995). Figure 5(b) and (c) show respectively the tensile and torsional response, and it can be seen that the experimental result is reasonably replicated. During the stress cycle, the material suffered from a severe inelastic deformation, but it was almost completely recovered after unloading (in Figure 5(b) the small circle represents the last point of the computed result).

The response of the Cu-Al-Zn-Mn polycrystalline SMA subjected to nonproportional strain along a rectangular path in the $\varepsilon-\gamma$ plane is shown in Figure 6. The four vertexes of the rectangle are (0, 0), (0, 0.0264), (0.0178, 0.0264) and (0.0178, 0) in sequence. Figure 6(a) shows the relation between the tensile stress and strain, and Figure 6(b) shows that between the shear stress and strain. It can be seen that the computed stress response of the material is in reasonable agreement with the experimental result (Sittner *et al*, 1995; Sittner and Takuda, 1995).

The difference may partly be attributed to the difference between the control variables. In the computation an ideal rectangular strain path in the $\varepsilon-\gamma$ plane was used, but in the experiment (Sittner *et al*, 1995; Sittner and Takuda, 1995) an invariant value of shear strain was defined and used to control the experiment.

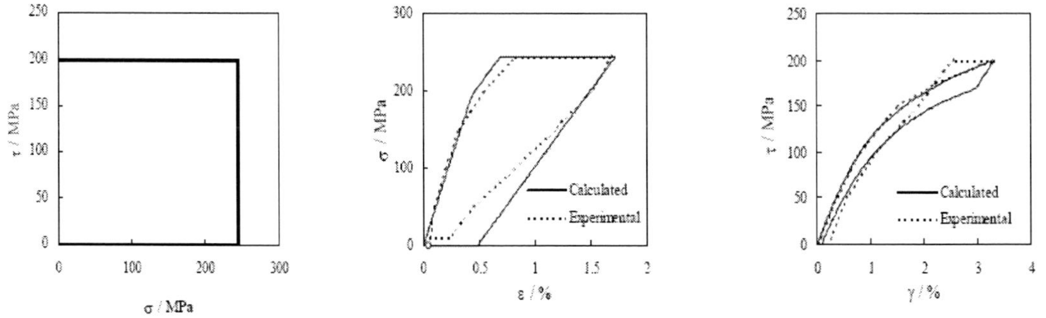

(a) Biaxial square stress path (b) Tensile response (c) Torsional response

Figure 5 Response of Cu-Al-Zn-Mn SMA polycrystal subjected to combined biaxial stress along a rectangular path in σ–τ plane, T=285 K.

There is a distinct difference between the path adopted in the computation and that used in the experiment. It should also be noted that there are marked differences in the coefficients $C^s=\sigma^F/\tau^F$ and $C^E=\gamma^F/\varepsilon^F$ obtained in different experiment (Sittner *et al*, 1995; Sittner and Takuda, 1995), but in the computation, the following Mises equivalent rule was adopted throughout the computation

$$\varepsilon_{eq}=\sqrt{\varepsilon^2+\gamma^2/3}, \qquad \sigma_{eq}=\sqrt{\sigma^2+3\tau^2}\,. \tag{45}$$

Figure 6(c) shows the comparison between the σ_{eq}-ε_{eq} relation and σ–ε curve corresponding to pure tension. The similarity shows that the pseudoealstic behavior of the material under nonproportional loading can be characterized by the equivalent stress-strain relation obtained from uniaxial or simple experiment.

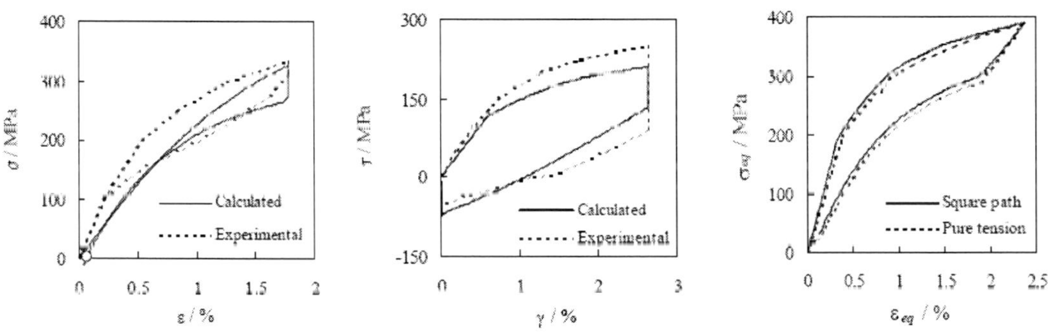

(a) Tensile response (b) Torsional response (c) σ_{eq}-ε_{eq} curves

Figure 6. Response of Cu-Al-Zn-Mn SMA polycrystal subjected to combined biaxial strain along a rectangular path in ε–γ plane, T=285 K.

2.4. Brief Summary and Discussion

A two-phase mixture model for SMAs was introduced, which is based on the concept that an SMA is composed of austenite and martensite phases, and the overall constitutive behavior of the SMA is the combination of the individual behavior of each of the two phases. Within the interested ranges of stress and temperature, the behavior of the austenite phase was assumed linearly elastic while that of the martensite phase elastoplastic.

The pseudoelastic behavior and shape memory effect of SMA Au-47.5at.%Cd subjected to tension / compression at constant or varying temperature were successfully replicated. The pseudoelastic behavior of a polycrystalline Cu-Al-Zn-Mn SMA subjected to proportional and nonproportional stress and strain histories was also analyzed. The comparison between the computed and experimental results showed satisfactory agreement.

In the two-phase mixture model, the strains in the two phases of an SMA were assumed identical to the overall strain in the SMA, while the overall stress in the SMA was the sum of the contribution from the two phases. The complicated constitutive behavior of a SMA can, therefore, be separated into the simple constitutive behavior of each of its two phases, which provides a simple but comprehensive description for the thermomechanical behavior of SMAs.

It should be noted that plastic flow may take place in the austenite phase when the applied stress is sufficiently large, which can be taken into account in the constitutive framework by simply assuming that the austenite phase is also elastoplastic. At lower temperature, the critical stress decreases with the increase of temperature, which can be described by simply considering the temperature-dependence of the material parameters of the martensite phase (Eqs. (24) and (25)), thus, the "V" shape of the relation between critical stress and temperature (Sun and Hwang, 1993) can be replicated.

It has been found that there might be distinctly some kinds of substructures in polycrystalline SMAs during stress-induced martensitic transformation (Shaw and Kyriakides, 1998; Sun and Li, 2002; Fung and Sun, 2006), which strongly affects the macroscopic response of the SMAs. It implies the necessity to develop microstructure-based constitutive models to describe more accurately the constitutive behavior of SMAs.

3. A Microstructure-Based Constitutive Model for NiTi SMAs

In the family of SMAs, NiTi polycrystalline SMAs have been receiving extensive attention due to their excellent thermal-mechanical properties. Besides the common properties of SMAs, an interesting phenomenon was reported by Shaw and Kyriakides (1998), and Sun and Li (2002). In the former, it was observed that, during a displacement controlled uniaxial tension of a NiTi SMA strip that was initially austenitic, martensite nucleates on the surface of the strip in sharp bands inclined at about 55° to the axis of loading. With the increase of deformation, these bands spread either by steady-state propagation of inclined transformation fronts or via a criss-cross pattern of finger-like features until covering the whole working section of the specimens. During unloading, austenite nucleates in sharp bands with the same orientation, grows with unloading and finally covers the whole surface of the working section of the strips. The strips returned to their original austenitic state after complete unloading, as is shown in Figure 7 (Shaw and Kyriakides, 1998).

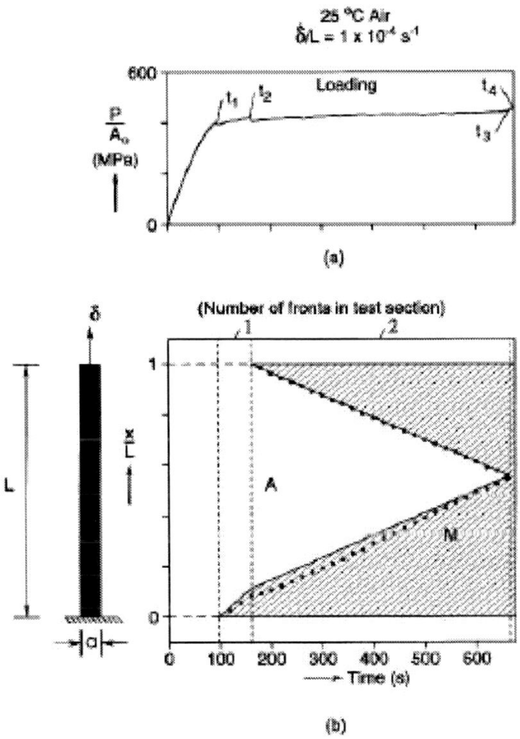

Figure 7. σ-ε curve and martensite transformation in a NiTi SMA strip under tension (Shaw and Kyriakides, 1998).(a) σ-ε curve, (b) variation of transformation section during tension.

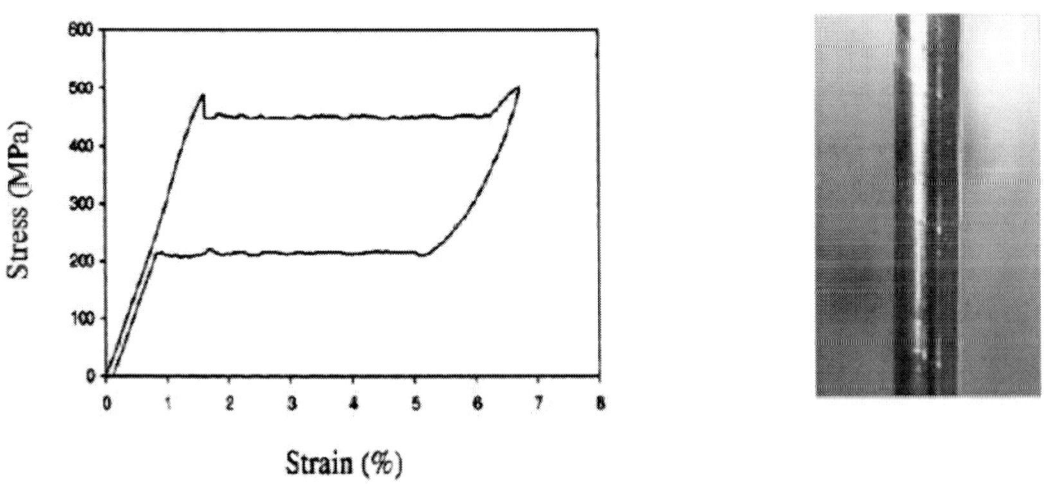

a b

Figure 8 Response of a nano-grained polycrystalline NiTi SMA microtube under pure tension (Sun and Li, 2002)(a) nominal σ-ε curve, (b) photo of the surface morphology.

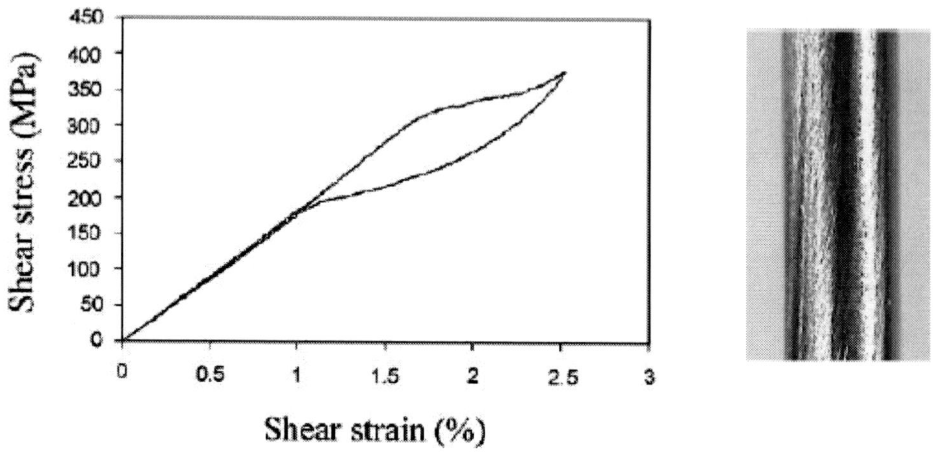

a) b)

Figure 9. Response of a nano-grained polycrystalline NiTi SMA microtube under pure torsion (Sun and Li, 2002) (a) nominal τ-γ curve, (b) photo of the surface morphology.

3.1. Observation of the Surface Morphology of the Deformed Niti SMA Microtube Specimens

In the work by Sun and Li (2002), the initiation and the growth of the spiral martensite band on the surface of the stretched NiTi SMA microtube were observed, which accounted for the stress drop or the macroscopic instability in the tensile stress responses of the material (Zhong, *et al*, 2000; Sun and Zhong, 2000). However, few evidences related to the stress-induced martensitic transformation microstructure on the torsional specimens were reported.

In order to investigate the morphology related to the stress-induced martensitic transformation microstructure of the deformed NiTi SMA microtubes, a testing device was designed for the SEM observation, as shown in Figure 10. This device was designed to fit the size of the chamber of scanning electron microscope JSM-6460 LV, including tension and torsion loading units. In order to meet the testing requirement, the maximum axial load and torque were designed as ±500N and ±500N-mm, the maximum elongation and the twist angle were 3mm and ±30° (or 0~60°) (corresponding to ε_{max}=0.18, γ_{max}=0.053), respectively. During a test, the specimen is loaded to prescribed load (or displacement), "freeze" the load on specimen, and then put the specimen together with the device into the chamber of the SEM for observation.

The specimens were commercially available nano-grained polycrystalline NiTi microtubes, with 1.7 mm in outer diameter and 1.4 mm in inner diameter. The composition of the material is shown in Table 3. The transformation temperatures of the material were measured with a differential scanning calorimeter DSC 200-PC as A_f= -0.5°C, A_s= -23°C, M_s= -49°C and M_f=-76°C, indicating that in a stress-free state the material should be austenitic at room temperature. Deformation may induce martensitic transformation, and during unloading, inverse martensitic transformation will take place and the material will transform back to austenite (patent phase) provided $T > A_f$. In other words, pseudoelastic behavior of the material can be guaranteed at room temperature.

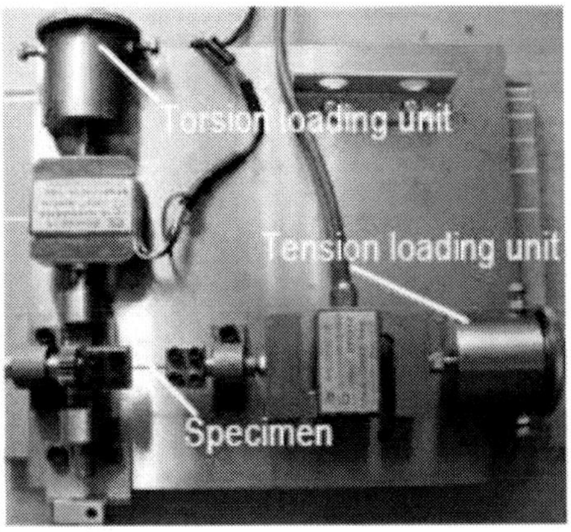

Figure 10. Tension-torsion loading device for SEM observation.

Before experiment each of the specimens was polished with 0.5μm polishing powder, and the SEM observation showed that the surface of the polished specimen is sufficiently smooth.

Table 3. Composition of the testing material

Ni	Ti	C	O	A.S.M.*	T.T.E.**
55.5%	rest	357ppm	220ppm	<0.1%	<0.4%

* A.S.M.: Any Single Element; ** T.T.E.: Total Trace Elements.

Under pure tension the test results and the evolution of the surface morphology were similar to that reported by Sun and Li (2002). With the increase of elongation, macroscopic spiral martensite bands initiated, grew and connected with each other and finally covered the whole surface of the working section. The surface morphology with distinct martensite bands is shown in Figure 11, where the spiral angle is about 60° to the axis of loading. It should be noted that no distinct texture could be observed on the surface of either the martensite band or the austenite section, indicating that no detectable local austenite exists in the martensite bands, and vice versa.

When the microtube specimen was twisted, the surface appeared overall uniform and no distinct macroscopic martensite band could be observed, as was reported by Sun and Li (2002). However, with a SEM, distributed stripes could be observed on the surface of the microtube after shear strain $\gamma \approx 2.5\%$ (Figure 12). Keeping in mind that, for the NiTi SMA microtube, the maximum transformation shear strain is about 0.7% (Sun and Li, 2002), and for the applied shear strain no permanent residual strain exists after unloading, the observed distributed stripes should, therefore, corresponds to the change of local volume or local orientation due to martensitic transformation. It can be seen that the stripes are almost parallel, and the angle between the stripes and the axial direction is about 26.5°, which is similar to that of the texture on the torsional microtube surface shown by Sun and Li (2002). Keeping in mind that the applied strain was strictly constrained within the range of pseudoelasticity of the NiTi SMA, the deformation in the austenite phase should be elastic. Thus the boundaries between a stripe and the neighboring flute can be regarded as the boundary between the lamellas of the austenite and martensite phases. For this kind of microstructure the interphase constraint can be very strong (Peng, *et al*, 2002), which accounts for the larger equivalent stress without distinct stress drop under pure torsion.

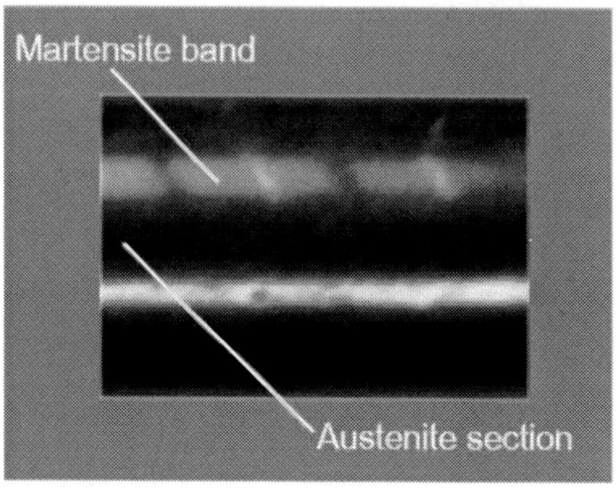

Figure 11. Surface morphology under tension

Figure 12 Surface morphology under torsion.

The comparison between the tension and torsion testing results showed the following typical features: (a) The plateau tensile stress is about 450MPa, while the corresponding shear stress is about 350 MPa, they could not be simply related to each other with any conventional equivalent rule, e.g., Mises or Tresca rule; (b) There is a drop in the tensile stress response, corresponding to some critical width of macroscopic martensite bands, but such macroscopic instability does not appear in pure torsion; (c) The microstructure of the NiTi SMA under either pure tension or pure torsion can be regarded as alternatively arranged parallel lamellas of the austenite and martensite phases. In the case of pure torsion, the lamellas distribute uniformly, with very small interlamellar spacing and strong interphase constraint. While in the case of pure tension, the nucleation and the growth of martensite phase is highly localized, the thickness of which can be sufficiently large so that the interphase constraint is greatly reduced. The difference in the microstructure of the materials can adequately account for the differences in the macroscopic constitutive behavior and should be taken into account in the constitutive description.

3.2. Numerical Investigation to the Effects of Transformation Microstructures

It was supposed from experiment that there should be a kind of lamellar microstructure in the NiTi SMA under either pure tension or pure torsion when martensite transformation takes place, but the interlamellar spacing or the thickness of martensite lamellas could be quite different, which might account for the remarkable difference between the macroscopic stress-strain responses under these two loading conditions.

In order to verify the above conjecture, the effect of the observed microstructures on the macroscopic responses was investigated numerically with commercially available finite element code ABAQUS. For simplicity, in the numerical investigation, the austenite phase is assumed elastic, while the martensite phase is assumed elastic-ideally plastic. The Young's modulus of the austenite phase, and the Young's modulus and the yield strength of the martensite phase were identified with the stress-strain curve obtained experimentally and given in Table 4. The geometry of the specimens used in experiment and the stress-induced transformation microstructures observed in experiment (Sun and Li, 2002) were adopted in computation.

In the case of pure tension, the angle of the spiral martensite band to the axis of loading (61°) and the variation of the its width against applied strain were the same as that measured in experiment (Sun and Li, 2002). For each set of applied strain and the corresponding width (related to martensite volume fraction) of martensite band, the finite element mesh could be made (as shown in Figure 13(a) for example), and the corresponding stress can be computed. Fourteen sets of applied strain and the corresponding width of martensite bend were used, and the computed σ-ε curve is shown in Figure 13(b), which agrees well with the experimental result. It can be seen that at the strain corresponding to very small width of the martensite bend (or very small thickness of the martensite lamella), the stress increases with the applied stress, which can be attributed to the strong constraint from the austenite (stronger phase) lamella. With the increase of the thickness of the martensite (weaker phase) lamella, such kind of constraint is reduced sharply. When it is increased to a critical size, such constraint would vanish and the response of the microtube will be mainly determined by the martensite phase, accounting for the drop of stress on the σ-ε curve (or macroscopic instability) and the following response on the σ-ε curve.

Table 4. Material constants identified and used in numerical simulation

Austenite phase		*Martensite phase*		
E (GPa)	v	*E* (GPa)	v	σ_y (MPa)
30	0.25	30	0.25	446

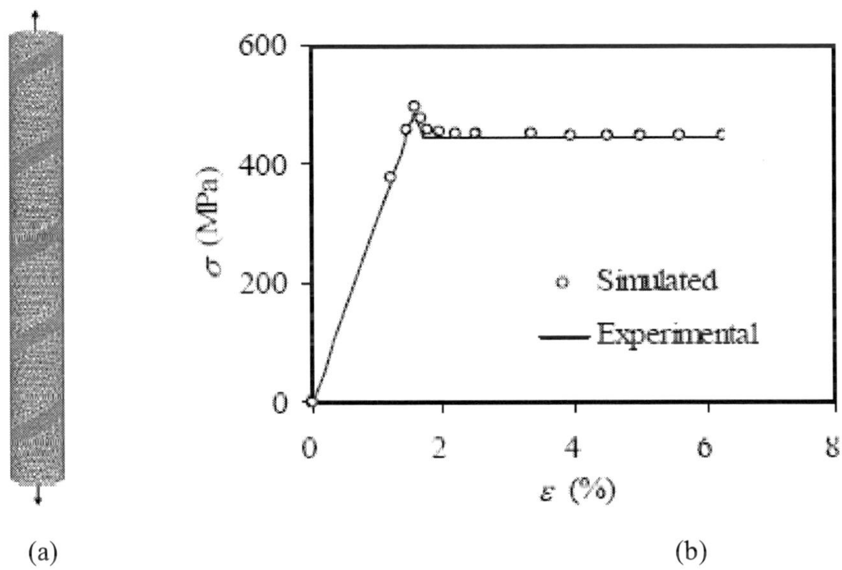

Figure 13. FE simulation of the response of a NiTi SMA microtube under pure tension (a) FE mesh, (b) σ-ε curves.

In the case of pure torsion, the angle of the spiral martensite bands to the axial direction 26.5° was adopted, as was measured in experiment. The interlamellar spacing was fixed at 10 μm (Figure 12), and the thickness of the martensite lamellas were determined by its volume fraction related to the applied shear strain. For each set of applied shear strain and the corresponding martensite volume fraction, the finite element mesh could be made (as shown in Figure 14(a) for example), and the corresponding shear stress could be computed. The computed τ-γ curve is shown in Figure 14(b), which agrees reasonably with the experimental result, the qualitative difference between the experimental and the computed results as γ>2.2% could be attributed to re-orientation hardening that was not considered in computation. It can be seen that the stress increases monotonically with the increase of strain, without any stress drop of macroscopic instability observed in the computation. It can be attributed to the strong constraint from the austenite phase on the martensite phase due to the small thickness of the martensite lamellas. However, such constraint would also be gradually reduced with the increase of the thickness of the martensite lamellas and the decrease of the thickness of the austenite lamellas, and tend to vanishes as the austenite phase tends to vanish.

(a)

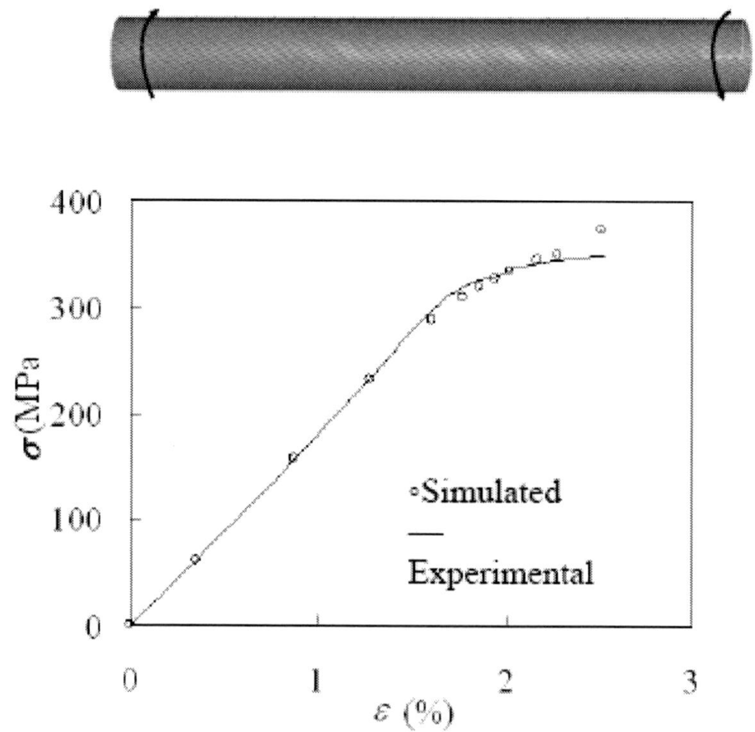

(b)

Figure 14. FE simulation of the pure tension of a NiTi SMA microtube under pure tension (a) FE mesh, (b)τ-γ curves.

It can be concluded from the numerical investigation that:

1) FE approach can satisfactorily replicate the responses of the SMA under pure tension and pure torsion, respectively.
2) It is reasonable to assume that the SMA is composed of alternatively arranged laminas of martensite and austenite. In the ranges of interested stress and temperature, the austenite is linearly elastic and the martensite is elastoplastic.
3) The thickness of the martensite (weaker phase) lamina plays a key role in the difference between the constitutive behavior of the NiTi SMA under pure tension and that under pure torsion. Specifically, the constraint from the austenite phase (stronger phase) due to the inconsistency of deformation decreases with the increase of the thickness of the martensite lamellas.

3.3. A Microstructure-Based Constitutive Model

3.3.1. The Constitutive Models for the Martensite and Austenite Phases

A NiTi SMA is composed of martensite and austenite phases with transformable martensite volume fraction. In order to validate its shape memory effect or pseudoelasticity, austenite phase, as the parent phase, should be elastic in the interested ranges of stress and temperature, while the martensite phase can be elastoplastic. In the following, the elastoplastic model for the martensite phase and the model for the austenite phase are introduced.

1. An Introduction to the Thermomechanically Consistent Elastoplastic Model

For simplicity, the discussion is restricted to the case of pseudoelasticity of the NiTi SMAs. The martensite phase is assumed initially isotropic and plastically incompressible, neglecting transformation lattice volume change (Sun and Hwang, 1993) and temperature induced phase transformation strain, in the case of isothermal and small deformation, the simple mechanical model shown in Figure 15 was adopted to obtain the constitutive description for inelastic materials (Fan and Peng, 1991; Peng and Ponter, 1994; Peng *et al*, 1997). It consists of n Maxwell-type models arranged in parallel together with one spring element in series. The elastic response of a material is represented with spring G (with macroscopic elastic shear modulus G) and the inelastic response with the various parallel elements. The rth dissipative mechanism is described with spring C_r (with stiffness C_r) and the dashpot-like block a_r (with plastic damping coefficient a_r). The energy stored in C_r during an inelastic deformation process corresponds to that stored in the microstress fields determined by the substantial heterogeneity of the material and the respective pattern of microdefects, etc. It is easily obtained from Figure 15 that

$$\mathbf{s} = \sum_{r=1}^{n} \mathbf{s}^{(r)} \tag{46}$$

$$\mathbf{s}^{(r)} = C_r (\mathbf{e}^i - \mathbf{p}^{(r)}) \tag{47}$$

with $\mathbf{e}^i = \mathbf{e} - \mathbf{e}^e, \quad \mathbf{e}^e = \dfrac{\mathbf{s}}{2G}$ (48)

where $\mathbf{e}^i$, $\mathbf{e}^e$ and $\mathbf{e}$ represent inelastic, elastic and total deviatoric strain tensors, respectively; $\mathbf{s}$ is the deviatoric stress tensor, $\mathbf{p}^{(r)}$ and $\mathbf{s}^{(r)}$ are respectively the rth deviatoric internal variable and the corresponding generalized force, which satisfy the dissipation inequality,

$$\mathbf{s}^{(r)} : d\mathbf{p}^{(r)} \geq 0, \quad (r = 0,1,\ldots,n). \tag{49}$$

$\mathbf{s}^{(r)}$ was assumed to relate the generalized rate of $\mathbf{p}^{(r)}$ through b_r by the following simple relationship,

$$\mathbf{s}^{(r)} = b_r \frac{d\mathbf{p}^{(r)}}{d\zeta}, \tag{50}$$

where $d\zeta$ can be defined in terms of the distance between two adjacent states of inelastic strain as

$$d\zeta^2 = d\mathbf{e}^i : d\mathbf{e}^i \tag{51}$$

and ζ can be regarded as a generalized time, which possesses the characteristics of time and can serve as some kind of "time" to measure an inelastic deformation process.

It can be seen in the above relationships that C_r and b_r ($r = 0,1,\ldots,n$) should be positive due to the non-negative property of both the stored and dissipated energy. Considering inelastic strain induced hardening, it was reasonable to assume

$$b_r = f b_r^0, \tag{52}$$

where f is a hardening parameter that is closely related to the change in the internal structure of the material, inelastic deformation history induced residual microstress fields, etc. Keeping in mind Eqs. (50) and (52), the differential of Eq. (47) gives

$$d\mathbf{s}^{(r)} = C_r d\mathbf{e}^i - \alpha_r \mathbf{s}^{(r)} dz \tag{53}$$

where $\alpha_r = \dfrac{C_r}{b_r^0}$ and $dz = \dfrac{d\zeta}{f}$. (54)

On the other hand, the elastic deviatoric response can be expressed as (Figure 15 and Eq. (48)),

$$\mathbf{s} = 2G(\mathbf{e} - \mathbf{e}^i), \; d\mathbf{s} = 2G(d\mathbf{e} - d\mathbf{e}^i). \tag{55}$$

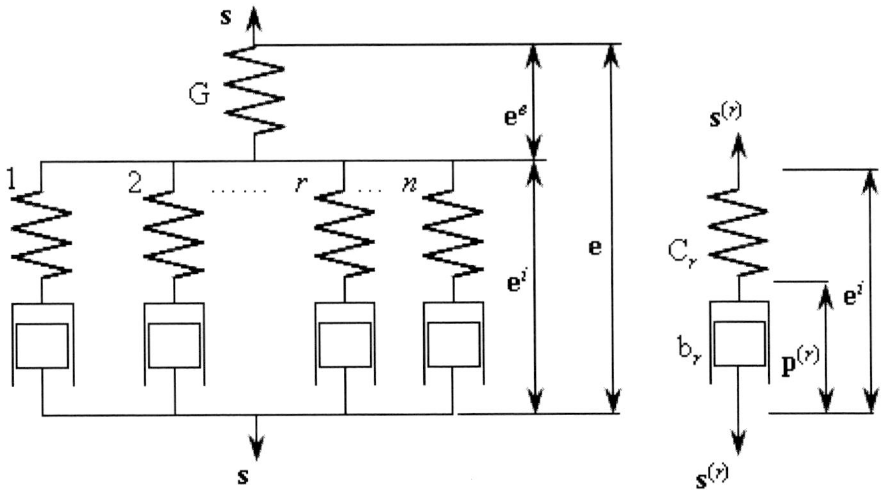

Figure 15. A simple mechanical model for the constitutive description.

The bulk response is elastic and can be described with

$$\sigma_{kk} = 3K\varepsilon_{kk},\; d\sigma_{kk} = 3Kd\varepsilon_{kk} \tag{56}$$

where σ_{kk} and ε_{kk} are the volumetric stress and strain, K is the bulk modulus.

The incremental form of the constitutive equation, Eq. (53), can be expressed as (Fan and Peng, 1991; Peng and Fan, 1993)

$$\Delta\mathbf{s}(z) = A\Delta\mathbf{e}^i + \mathbf{B}(z_n)\Delta z \tag{57}$$

where $\mathbf{s}$ and $\mathbf{e}^i$ are deviatoric stress and plastic strain, respectively,

$$A = \sum_{r=1}^{n} k_r C_r \quad \mathbf{B}(z_n) = -\sum_{r=1}^{n} k_r \alpha_r \mathbf{s}^{(r)}(z_n), \tag{58}$$

$$\Delta\mathbf{s}^{(r)}(z) = k_r(C_r\Delta\mathbf{e}^i - \alpha_r\mathbf{s}^{(r)}(z_n))\Delta z,\; k_r = \frac{1-e^{-\alpha_r\Delta z}}{\alpha_r\Delta z}, \quad z = z_n + \Delta z. \tag{59}$$

Combining Eq. (51) with Eq. $(54)_2$ gives

$$\Delta z = \frac{\Delta\mathbf{e}^p}{f^2\Delta z} : \Delta\mathbf{e}^p. \tag{60}$$

Making use of the following relationships

$$\Delta\mathbf{s} = \Delta\boldsymbol{\sigma} - \frac{1}{3}tr(\Delta\boldsymbol{\sigma})\mathbf{I}_2 \quad \Delta\mathbf{e} = \Delta\boldsymbol{\varepsilon} - \frac{1}{3}tr(\Delta\boldsymbol{\varepsilon})\mathbf{I}_2, \tag{61}$$

where $\mathbf{I}_2$ is the identity tensor of rank two, the incremental constitutive equation can be derived from Eqs. (57), (60) and (61) as (Appendix A)

$$\Delta\boldsymbol{\sigma} = \mathbf{L} : \Delta\boldsymbol{\varepsilon}, \tag{62}$$

with $$\mathbf{L} = 2G_p\mathbf{I}_4 + \left\{K - \frac{2}{3}G_p\right\}\mathbf{I}_2 \otimes \mathbf{I}_2 + \frac{(2G)^2}{(2G+A)^2 f^2(z)}\frac{1}{a}\mathbf{B} \otimes \frac{\Delta\mathbf{e}^i}{\Delta z}, \tag{63}$$

where $$G_p = \frac{GA}{2G+A}, \quad a = 1 + \frac{\mathbf{B} : \Delta\mathbf{e}^p}{(A+2G)f^2\Delta z}, \tag{64}$$

and $\mathbf{I}_4$ is the identity tensor of rank four.

For the austenite phase, linearly elastic behavior is assumed and the constitutive relationship can easily be obtained as follows by ignoring the terms related to inelastic deformation and noticing that for an elastic material, A is sufficient large and G_p tends to G, (Eqs. (63) and (64))

$$\mathbf{L} = 2G\mathbf{I}_4 + \left\{K - \frac{2}{3}G\right\}\mathbf{I}_2 \otimes \mathbf{I}_2 . \tag{65}$$

In a rectangular coordinate frame, Eq. (63) can be expressed in the following matrix form for the martensite phase

$$\Delta\sigma^M = \left[L^M\right] \Delta\varepsilon^M . \tag{66}$$

The expression for the austenite phase can be obtained by replacing the superscript "*M*" with "*A*".

3.3.2. Constitutive Model for a Representative Volume Element of the NiTi SMAs

The representative volume element (RVE) of a cell in an NiTi SMA is shown in Figure 16, in which the martensite and austenite lamellas are parallel with each other, alternatively arranged and bonded perfectly, h_M and h_A are the thickness of the martensite lamellas and that of the austenite lamellas, respectively, and the interlamellar spacing $h=h_A+h_M$. The ratios h_M/h and h_A/h change synchronously with the change of the volume fraction of martensite ξ, and can be represented with ξ and ξ_A=1-ξ, respectively. A coordinate frame is fixed on the RVE, with x_1x_2 plane parallel with the lamellas and x_3 perpendicular to the lamellas. For a compact expression in the adopted coordinate frame, we let

$$\sigma^R = \begin{Bmatrix} \bar{\sigma}^R \\ \hat{\sigma}^R \end{Bmatrix}, \quad \varepsilon^R = \begin{Bmatrix} \hat{\varepsilon}^R \\ \bar{\varepsilon}^R \end{Bmatrix}, \tag{66}$$

where

$$\bar{\sigma}^R = \{\sigma_{11}^R, \ \sigma_{22}^R, \ \tau_{12}^R\}^T, \qquad \hat{\sigma}^R = \{\sigma_{33}^R, \ \tau_{23}^R, \ \tau_{13}^R\}^T$$
$$\hat{\varepsilon}^R = \{\varepsilon_{11}^R, \ \varepsilon_{22}^R, \ \gamma_{12}^R\}^T, \qquad \bar{\varepsilon}^R = \{\varepsilon_{33}^R, \ \gamma_{23}^R, \ \gamma_{13}^R\}^T, \tag{67}$$

σ and **ε** represent stress and strain, the superscript "*R*" can be replaced with "*A* ", "*M*" and "*R* " to obtain the expressions for the austenite phase, the martensite phase and the RVE, respectively. If the thickness of each lamella is sufficiently small so that the variations of the stress and the strain in a lamella can be neglected, the in-plane components of strain, $\hat{\varepsilon}^A$ and $\hat{\varepsilon}^M$, should be equal to the corresponding components of strain in the RVE, $\hat{\varepsilon}^R$, to meet the condition of compatibility; and the out-of-plane components of stress, $\hat{\sigma}^A$ and $\hat{\sigma}^A$, should be equal to the corresponding components of stress in the RVE $\hat{\sigma}^R$ to meet the condition of equilibrium, i.e.,

$$\{\hat{\varepsilon}^M\} = \{\hat{\varepsilon}^A\} = \{\hat{\varepsilon}^R\}, \qquad \{\hat{\sigma}^M\} = \{\hat{\sigma}^A\} = \{\hat{\sigma}^R\}. \tag{68}$$

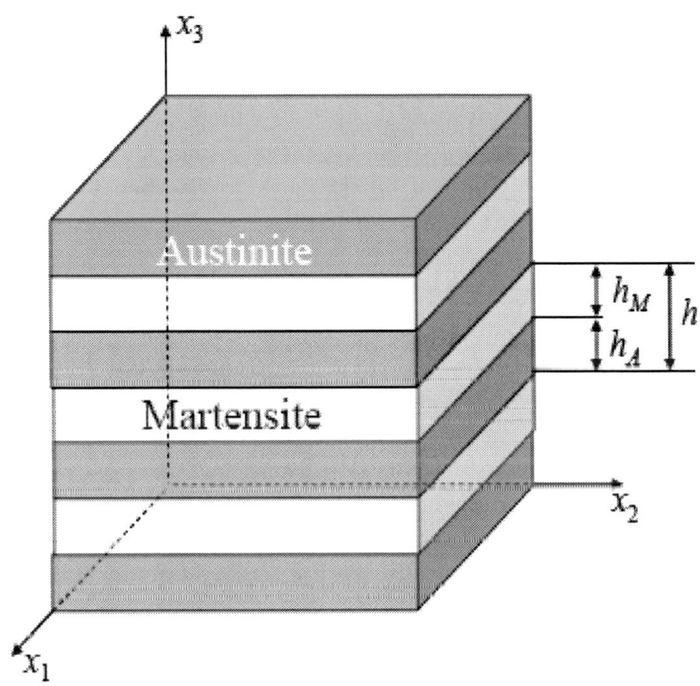

Figure 16. A representative volume element.

The in-plane components of stress, $\{\bar{\sigma}^R\}$, and out-of-plane components of strain, $\{\bar{\varepsilon}^R\}$, in a RVE can be determined by volume average as

$$\{\bar{\sigma}^R\}=\xi\{\bar{\sigma}^M\}+(1-\xi)\{\bar{\sigma}^A\}, \qquad \{\bar{\varepsilon}^R\}=\xi\{\bar{\varepsilon}^M\}+(1-\xi)\{\bar{\varepsilon}^A\}. \tag{69}$$

The increment of Eq. (69) can be expressed as

$$\begin{Bmatrix}\Delta\bar{\sigma}^R\\ \Delta\bar{\varepsilon}^R\end{Bmatrix}=\xi\begin{Bmatrix}\Delta\bar{\sigma}^M\\ \Delta\bar{\varepsilon}^M\end{Bmatrix}+(1-\xi)\begin{Bmatrix}\Delta\bar{\sigma}^A\\ \Delta\bar{\varepsilon}^A\end{Bmatrix}+\Delta\xi\begin{Bmatrix}\bar{\sigma}^M-\bar{\sigma}^A\\ \bar{\varepsilon}^M-\bar{\varepsilon}^A\end{Bmatrix}. \tag{70}$$

It should be noted that, for the two-phase lamellar material, the uniformity of stress and strain in each phase and the relationships shown in Eqs. (68) and (69) are based on the assumption that the interlamellar spacing (especially the thickness of the weaker phase lamellas) is sufficiently small. It is known that in this case the constraint from the stronger phase to the weaker phase can be very strong due to the forced compatibility (Eq. (68)) of the deformation in the two phases through the interphase interaction. But such kind of constraint can be greatly reduced with the increase of the thickness of the martensite lamellas as well as the interlamellar spacing. Peng *et al* (2002) investigated the effect of the thickness of weaker phase lamellas, h_M, on the load bearing capability of the adopted RVE with an FE approach. It was found that the load bearing capability decreases monotonically with the increase of h_M and tends to that of the pure weaker phase material as h_M is sufficiently large. It was also concluded that the interphase constraint depends strongly on the variation of the volumetric strain (or volumetric stress) in each phase, which decreases sharply with the increase of h_M and tends to the asymptotic value corresponding to that without considering interphase interaction as the thickness of the weaker phase lamellas is sufficiently large (Peng *et al*, 2002).

The above discussion is significant for the pseudoelastic behavior of the NiTi SMAs, in which the martensite band may grow to a macroscopic one during a pure tension process so that the constraint from the austenite lamella will be greatly reduced. It accounts for the macroscopic instability and the subsequent reduction of the load bearing capability. In order to extend the result obtained from the ideal RVE (Figure 16) to the case of finite thickness of the martensite lamellas, an additional volumetric strain was introduced to relax the over-constraint in the adopted ideal RVE model (Figure 16) as the thickness of martensite lamellas is no longer infinitesimal.

Suppose the RVE is subjected to external loading, its initial and deformed configurations are sketched in Figure 17. The initial configuration is plotted with dot lines, the ideal deformed configuration is plotted with dashed lines, and the actual deformed configuration is plotted with solid lines. The comparison between the deformed configuration of the ideal RVE and that of an actual RVE shows that both phases in the ideal RVE model may involve unrealistic deformation. FE analysis shows that the unrealistic volumetric strain plays a dominant role in the deviation of the response of the actual RVE from that of the ideal RVE. In order to reduce the error induced by the ideal configuration of the adopted RVE, an additional volumetric strain κ^I was introduced so that the volumetric response of each phase can, therefore, be expressed as follows without increasing much complexity,

$$tr(\boldsymbol{\sigma}^I) = 3K^I\left[tr(\boldsymbol{\varepsilon}^I) - \kappa^I\right] \text{ or } tr(\Delta\boldsymbol{\sigma}^I) = 3K^I\left[tr(\Delta\boldsymbol{\varepsilon}^I) - \Delta\kappa^I\right], \tag{71}$$

where K^I is bulk modulus of phase I, κ^I is related to the volumetric strain and can be expressed as follows for simplicity

$$\kappa^I = g(\xi)tr(\boldsymbol{\varepsilon}^I), \tag{72}$$

$$\Delta\kappa^I = \Delta g(\xi)tr(\boldsymbol{\varepsilon}^I) + g(\xi)tr(\Delta\boldsymbol{\varepsilon}^I), \tag{73}$$

where the superscript "*I*" can be replaced with "*A*" and "*M*" for the expressions of the austenite and martensite phases, respectively, and $g(\xi)$ decreases with the decrease of ξ and vanishes if ξ is sufficiently small.

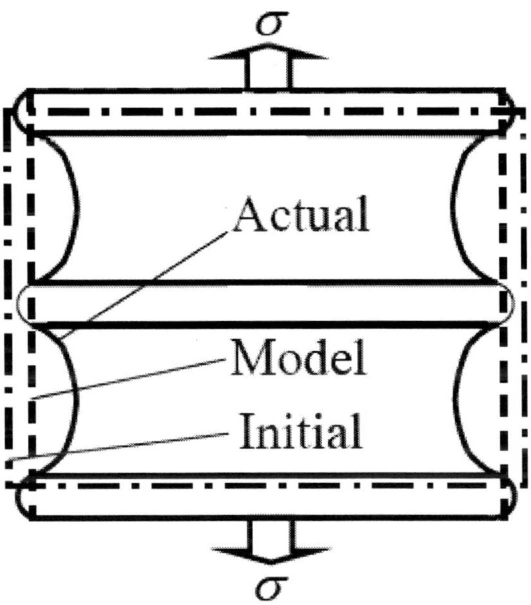

Figure 17. Initial and deformed configurations of the RVE.

It should be emphasized that the introduced additional volumetric strain κ^I is not at all related to the volume change during martensite phase transformations in the NiTi SMAs, which is normally very small and has already been neglected in the model. Replacing Eq. (56) with Eq. (71), Eq. (62) can be rewritten as (Appendix A)

$$\Delta\boldsymbol{\sigma}^M = \mathbf{L}^M : \Delta\boldsymbol{\varepsilon}^M + \mathbf{H}^M(\boldsymbol{\varepsilon}^M)\Delta\xi \tag{74}$$

for the response of the martensite phase, where

$$\mathbf{L}^M = 2G_p^M \mathbf{I}_4 + \left\{ K^M \left[1 - g^M(\xi) \right] - \frac{2}{3} G_p^M \right\} \mathbf{I}_2 \otimes \mathbf{I}_2 + \frac{(2G^M)^2}{(2G^M + A)^2 f^2(z)} \frac{1}{a} \mathbf{B} \otimes \frac{\Delta \mathbf{e}^p}{\Delta z}$$
$$\mathbf{H}^M(\boldsymbol{\varepsilon}^M) = -K^M \varepsilon_{kk}^M \frac{dg^M(\xi)}{d\xi} \mathbf{I}_2 \tag{75}$$

For the austenite phase,

$$\Delta \boldsymbol{\sigma}^A = \mathbf{L}^A : \Delta \boldsymbol{\varepsilon}^A + \mathbf{H}^A(\boldsymbol{\varepsilon}^A) \Delta \xi, \tag{76}$$

with

$$\mathbf{L}^A = 2G_p^A \mathbf{I}_4 + \left\{ K^A \left[1 - g^A(\xi) \right] - \frac{2}{3} G_p^A \right\} \mathbf{I}_2 \otimes \mathbf{I}_2, \qquad \mathbf{H}^A(\boldsymbol{\varepsilon}^A) = -K^A \varepsilon_{kk}^A \frac{dg^A(\xi)}{d\xi} \mathbf{I}_2 \tag{77}$$

Eq. (74) can also be specified in the following matrix form

$$\begin{Bmatrix} \Delta\sigma_{11}^M \\ \Delta\sigma_{22}^M \\ \Delta\tau_{12}^M \\ \hdashline \Delta\sigma_{33}^M \\ \Delta\tau_{23}^M \\ \Delta\tau_{31}^M \end{Bmatrix} = \left[\begin{array}{ccc:ccc} D_{1111}^M & D_{1122}^M & D_{1112}^M & D_{1133}^M & D_{1123}^M & D_{1131}^M \\ D_{2211}^M & D_{2222}^M & D_{2212}^M & D_{2233}^M & D_{2223}^M & D_{2231}^M \\ D_{1211}^M & D_{1222}^M & D_{1212}^M & D_{1233}^M & D_{1223}^M & D_{1231}^M \\ \hdashline D_{3311}^M & D_{3322}^M & D_{3312}^M & D_{3333}^M & D_{3323}^M & D_{3331}^M \\ D_{2311}^M & D_{2322}^M & D_{2312}^M & D_{2333}^M & D_{2323}^M & D_{2331}^M \\ D_{3111}^M & D_{3122}^M & D_{3112}^M & D_{3133}^M & D_{3123}^M & D_{3131}^M \end{array} \right] \begin{Bmatrix} \Delta\varepsilon_{11}^M \\ \Delta\varepsilon_{22}^M \\ \Delta\gamma_{12}^M \\ \hdashline \Delta\varepsilon_{33}^M \\ \Delta\gamma_{23}^M \\ \Delta\gamma_{31}^M \end{Bmatrix} + \begin{Bmatrix} h_{11}^M \\ h_{22}^M \\ h_{12}^M \\ \hdashline h_{33}^M \\ h_{23}^M \\ h_{31}^M \end{Bmatrix} \Delta\xi \tag{78}$$

Keeping in mind Eqs. (66) and (67), Eq. (78) can further be expressed as

$$\begin{Bmatrix} \Delta\bar{\sigma}^M \\ \Delta\hat{\sigma}^M \end{Bmatrix} = \begin{bmatrix} A_1^M & A_2^M \\ A_3^M & A_4^M \end{bmatrix} \begin{Bmatrix} \Delta\hat{\varepsilon}^M \\ \Delta\bar{\varepsilon}^M \end{Bmatrix} + \begin{Bmatrix} h_1^M \\ h_2^M \end{Bmatrix} \Delta\xi, \tag{79}$$

where

$$A_1^M = \begin{bmatrix} D_{1111}^M & D_{1122}^M & D_{1112}^M \\ D_{2211}^M & D_{2222}^M & D_{2212}^M \\ D_{1211}^M & D_{1222}^M & D_{1212}^M \end{bmatrix}, \quad A_2^M = \begin{bmatrix} D_{1133}^M & D_{1123}^M & D_{1131}^M \\ D_{2233}^M & D_{2223}^M & D_{2231}^M \\ D_{1233}^M & D_{1223}^M & D_{1231}^M \end{bmatrix},$$
$$A_3^M = \begin{bmatrix} D_{3311}^M & D_{3322}^M & D_{3312}^M \\ D_{2311}^M & D_{2322}^M & D_{2312}^M \\ D_{3111}^M & D_{3122}^M & D_{3112}^M \end{bmatrix}, \quad A_4^M = \begin{bmatrix} D_{3333}^M & D_{3323}^M & D_{3331}^M \\ D_{2333}^M & D_{2323}^M & D_{2331}^M \\ D_{3133}^M & D_{3323}^M & D_{3131}^M \end{bmatrix} \tag{80}$$

$$h_1^M = h_{11}^M, h_{22}^M, h_{12}^M, {}^T, \qquad h_2^M = h_{33}^M, h_{23}^M, h_{31}^M, {}^T. \tag{81}$$

Replacing the superscript "*M*" in Eqs. (78) through (81) with "*A*" gives the expressions for the austenite phase.

The combination of Eq. (79) with Eqs. (68) and (70) yields the following incremental constitutive equation for the RVE (Appendix B)

$$\Delta\boldsymbol{\sigma}^R = \mathbf{L}^R : \Delta\boldsymbol{\varepsilon}^R + \mathbf{H}^R \Delta\xi \text{ or } \Delta\sigma^R = \left[L^R\right] \Delta\varepsilon^R + H^R \Delta\xi. \tag{82}$$

Keeping in mind Eq. (66), Eq. (82) can be specified as

$$\begin{Bmatrix} \Delta\bar{\sigma}^R \\ \Delta\hat{\sigma}^R \end{Bmatrix} = \begin{bmatrix} (g_2 - g_1 g_3^{-1} g_4) & g_1 g_3^{-1} \\ -g_3^{-1} g_4 & g_3^{-1} \end{bmatrix} \begin{Bmatrix} \Delta\hat{\varepsilon}^R \\ \Delta\bar{\varepsilon}^R \end{Bmatrix} + \begin{Bmatrix} g_5 - g_1 g_3^{-1} g_6 \\ -g_3^{-1} g_6 \end{Bmatrix} \Delta\xi, \tag{83}$$

where

$$\begin{aligned} g_1 &= \xi A_2^M (A_4^M)^{-1} + (1-\xi) A_2^A (A_4^A)^{-1} \\ g_2 &= \xi[A_1^M - A_2^M (A_4^M)^{-1} A_3^M] + (1-\xi)[A_1^A - A_2^A (A_4^A)^{-1} A_3^A] \\ g_3 &= \xi (A_4^M)^{-1} + (1-\xi)(A_4^A)^{-1} \\ g_4 &= -\xi (A_4^M)^{-1} A_3^M - (1-\xi)(A_4^A)^{-1} A_3^A \\ g_5 &= \xi[h_1^M - A_2^M (A_4^M)^{-1} h_2^M] + (1-\xi)[h_1^A - A_2^A (A_4^A)^{-1} h_2^A] + (\bar{\sigma}^M - \bar{\sigma}^A) \\ g_6 &= -\xi (A_4^M)^{-1} h_2^M - (1-\xi)(A_4^A)^{-1} h_2^A + (\bar{\varepsilon}^M - \bar{\varepsilon}^A) \end{aligned} \tag{84}$$

Eq. (82) is able to describe the inelastic behavior of the RVE, taking into account the effects of the microstructure and the volume fraction of the martensite phase as well as its variation.

With the constitutive description for the RVE, the constitutive description for the NiTi SMAs can further be obtained by making use of some kind of mean-field scheme.

3.3.3. The Hill's Self-Consistent Scheme

The Hill's self consistent scheme was adopted for the evaluation of the effective properties of the NiTi SMAs. Suppose an SMA material is an aggregate of numerous cells, keeping in mind the assumption of inelastic incompressibility, under the condition of isothermal and small deformation, the Hill's self-consistent scheme (Hill, 1965) gives

$$\Delta\boldsymbol{\sigma}^R - \Delta\boldsymbol{\sigma} = -\mathbf{L}^* : (\Delta\boldsymbol{\varepsilon}^R - \Delta\boldsymbol{\varepsilon}) + \mathbf{H}(\boldsymbol{\varepsilon}^R - \boldsymbol{\varepsilon} - \boldsymbol{\varepsilon}^*)\Delta\xi \tag{85}$$

where $\Delta\boldsymbol{\sigma}^R$ and $\Delta\boldsymbol{\varepsilon}^R$ denote respectively the increments of the stress and strain tensors in a RVE, $\Delta\boldsymbol{\sigma}$ and $\Delta\boldsymbol{\varepsilon}$ are the increments of the stress and strain in the aggregate, and $\mathbf{L}^*$ is a tangential "constraint" tensor determined by

$$\mathbf{L}^* = \mathbf{L}:(\mathbf{S}^{-1} - \mathbf{I}_4), \tag{86}$$

where $\mathbf{L}$ is the tangential elastoplastic tensor relating $\Delta\boldsymbol{\sigma}$ and $\Delta\boldsymbol{\varepsilon}$ by

$$\Delta\boldsymbol{\sigma} = \mathbf{L}:\Delta\boldsymbol{\varepsilon} + \mathbf{H}(\boldsymbol{\varepsilon})\Delta\xi\,, \tag{87}$$

in which $\mathbf{S}$ is the Eshelby tensor. Assuming isotropic matrix and spherical inclusions, the Eshelby tensor can approximately be expressed as

$$\mathbf{S} = (1-b)\,\mathbf{I}_2 \otimes \mathbf{I}_2 + b\,(\mathbf{I}_4 - \mathbf{I}_2 \otimes \mathbf{I}_2) \tag{88}$$

with $b = \dfrac{2(4-5\nu)}{15(1-\nu)}$. (89)

For general anisotropic matrix and rotationally ellipsoidal inclusions, the Eshelby tensor can also be obtained (Mikata, 2000). $\Delta\boldsymbol{\varepsilon}$ is related to $\Delta\boldsymbol{\varepsilon}^R$ by the following relationship

$$\Delta\boldsymbol{\varepsilon}^R = \mathbf{A}^R : \Delta\boldsymbol{\varepsilon} - [\mathbf{L}^* + \mathbf{L}^R]^{-1} : \bar{\mathbf{H}}^R \Delta\xi \tag{90}$$

where $\mathbf{A}^R = [\mathbf{L}^* + \mathbf{L}^R]^{-1} : [\mathbf{L}^* + \mathbf{L}]$, $\bar{\mathbf{H}}^R = \mathbf{H}^R(\boldsymbol{\varepsilon}^R) - \mathbf{H}(\boldsymbol{\varepsilon}^R - \boldsymbol{\varepsilon} - \boldsymbol{\varepsilon}^*) - \mathbf{H}(\boldsymbol{\varepsilon})$, (91)

$$\dot{\boldsymbol{\varepsilon}}^* = \mathbf{S}^{-1} : (\dot{\boldsymbol{\varepsilon}}^r - \dot{\boldsymbol{\varepsilon}}), \tag{92}$$

$\mathbf{L}^R$ is the tangential elastoplastic tensor relating $\Delta\boldsymbol{\sigma}^R$ and $\Delta\boldsymbol{\varepsilon}^R$ of a single RVE by Eq. (83), and

$$\Delta\boldsymbol{\sigma}^R = \mathbf{L}^R : \mathbf{A}^R : \Delta\boldsymbol{\varepsilon} + \mathbf{L}^R : [\mathbf{L}^* + \mathbf{L}^R]^{-1} : [\mathbf{H}(\boldsymbol{\varepsilon}) + \mathbf{H}(\boldsymbol{\varepsilon}^R - \boldsymbol{\varepsilon} - \boldsymbol{\varepsilon}^*) - \mathbf{H}^R(\boldsymbol{\varepsilon}^R)] + \mathbf{H}^R(\boldsymbol{\varepsilon}^R)\ \Delta\xi\,. \tag{93}$$

Assuming that the increment of the overall stress, $\Delta\boldsymbol{\sigma}$, is determined by the $\Delta\boldsymbol{\sigma}^R$ of all single cells through a certain averaging procedure, the relationship between $\Delta\boldsymbol{\sigma}$ and $\Delta\boldsymbol{\varepsilon}$ can be obtained as

$$\Delta\boldsymbol{\sigma} = \langle\Delta\boldsymbol{\sigma}^R\rangle = \langle\mathbf{L}^R : \mathbf{A}^R\rangle : \Delta\boldsymbol{\varepsilon} + \langle\mathbf{L}^R : [\mathbf{L}^* + \mathbf{L}^R]^{-1} : [\mathbf{H}(\boldsymbol{\varepsilon}) + \mathbf{H}(\boldsymbol{\varepsilon}^R - \boldsymbol{\varepsilon} - \boldsymbol{\varepsilon}^*) - \mathbf{H}^R(\boldsymbol{\varepsilon}^R)] + \mathbf{H}^R(\boldsymbol{\varepsilon}^R)\rangle\Delta\xi, \tag{94}$$

in which $\langle\theta^R\rangle$ represents the averaging of θ^R in all single cells. Comparing Eq. (94) with Eq. (87) gives

$$\mathbf{L}=\langle\mathbf{L}^R:\mathbf{A}^R\rangle,\ \mathbf{H}(\boldsymbol{\varepsilon})=\langle\mathbf{L}^R:[\mathbf{L}^*+\mathbf{L}^R]^{-1}:[\mathbf{H}(\boldsymbol{\varepsilon})+\mathbf{H}(\boldsymbol{\varepsilon}^R-\boldsymbol{\varepsilon}-\boldsymbol{\varepsilon}^*)-\mathbf{H}^R(\boldsymbol{\varepsilon}^R)]+\mathbf{H}^R(\boldsymbol{\varepsilon}^R)\rangle. \quad (95)$$

Eq. (95) is highly implicit and should be solved with iterative algorithm. The combination of Eqs. (83), (85) and (87) gives the following invariant for each RVE

$$[\mathbf{L}^*+\mathbf{L}^R]:\Delta\boldsymbol{\varepsilon}^R+\left[\mathbf{H}^R(\boldsymbol{\varepsilon}^R)-\mathbf{H}(\boldsymbol{\varepsilon}^R-\boldsymbol{\varepsilon}-\boldsymbol{\varepsilon}^*)\right]\Delta\xi=[\mathbf{L}^*+\mathbf{L}]:\Delta\boldsymbol{\varepsilon}+\mathbf{H}(\boldsymbol{\varepsilon})\Delta\xi=\Delta\mathbf{q}, \quad (96)$$

which is important in the iterative computation.

3.4. Application and Verification

3.4.1. Determination of ξ and $g(\xi)$

In a stress-free state the volume fraction of martensite is determined by temperature. Applied stress may increase the transformation temperature. Tanaka (1986) suggested a kinetic rule for stress-induced forward and inverse martensitic transformation, which can work well in stress-controlled processes. But it cannot work well in the case involving macroscopic instability. Noticing that: (a) the response of the NiTi SMAs under pure tension involves macroscopic instability, its constitutive behavior during loading should better be investigated in strain space; and (b) the NiTi SMA under pure tension and pure torsion processes exhibits quite different transformation characteristics and inelastic behavior that cannot be well described with the conventional constitutive models, it was suggested that the pseudoelastic behavior of the material should be related to both the deviatoric and volumetric components of stress and strain. Tanaka's rule (Tanaka, 1986) was, therefore, modified as (Peng, et al, 2008)

$$\xi=\begin{cases}1-\exp[-A^M\langle M_s+B^M\theta_\varepsilon-T\rangle], & \theta_\varepsilon=\varepsilon_e-a^M\varepsilon_{kk}; \quad A\rightarrow M\\ \\ \exp[-A^A\langle T-(A_s+B^A\theta_\sigma)\rangle], & \theta_\sigma=(\sigma_e+a^A\sigma_{kk})/G; \quad M\rightarrow A,\end{cases} \quad (97)$$

in which M_s and A_s are respectively the start temperature of the martensitic transformation and that of the austenitic transformation, a^M and a^A are material constants related to the effect of bulk stress on transformation, A^M, B^M, A^A and B^A are material constants related respectively to forward and inverse martensitic transformation processes, G is the elastic shear modulus of the material,

$$<\Phi>=\begin{cases}\Phi & if \quad \Phi>0\\ 0 & if \quad \Phi\le 0\end{cases}, \quad (98)$$

$$\text{and } \varepsilon_e=\sqrt{\frac{2}{3}\mathbf{e}:\mathbf{e}}, \qquad \sigma_e=\sqrt{\frac{3}{2}\mathbf{s}:\mathbf{s}}, \quad (99)$$

with **e** and **s** the deviatoric components of the overall strain and stress.

In Eq. (97), θ_ε and θ_σ were introduced to consider the effect of different states of stress and strain on the deformation induced forward and inverse martensitic transformation, respectively. In computation the following equivalent incremental form was adopted,

$$\Delta\xi = \begin{cases} A^M B^M (1-\xi)\Delta\theta_\varepsilon, & \text{if} \quad M_s + B^M\theta_\varepsilon - T > 0; \quad A \to M \\ A^A B^A \xi\Delta\theta_\sigma, & \text{if} \quad M_s + B^A\theta_\sigma - T > 0; \quad M \to A, \end{cases} \tag{100}$$

If h_M (Figure 16) is small, $g(\xi)$ in Eq. (72) will be small, and the response of the SMA will be strongly affected by the properties the austenite phase) due to strong interphase constraint. With the increase of h_M, the contribution of the martensite phase tends to be dominant, and the response of the material will be determined by the properties of the martensite phase if h_M is sufficiently large. In general, $g(\xi)$ is an increasing function of ξ with a saturated value and was assumed to take the following expression

$$g(\xi) = \beta \ 1 - \exp(-\alpha\xi) \tag{101}$$

with the evolution

$$g'(\xi) = \beta\alpha \exp(-\alpha\xi) = \alpha\beta[1-(1-\exp(-\alpha\xi))] = \alpha[\beta - g(\xi)]\,. \tag{102}$$

3.4.2. Identification of the Material Parameters

The material parameters related to the pseudoelastic behavior of a SMA can be separated into four groups: the parameters of the austenite phase, those of the martensite phase, the parameters related to forward and inverse martensitic transformation, and those related to stress drop or macroscopic instability. These material parameters can be identified at room temperature ($T > A_f$) by the following procedures:

1) Before stress-induced martensitic transformation occurs, the SMA to be tested is composed of pure austenite, the following parameters can be identified with a simple elastic tension (or torsion) test: Young's modulus E^A (or shear modulus G^A) and Poisson's ratio ν^A;
2) With the development of deformation when the martensite volume fraction is sufficiently large, the response of the SMA is mainly determined by that of the martensite phase. The parameters related to the inelastic behavior, i.e., C_r and α_r and $f(z)$ can be determined with the test result in this stage; the elastic property of the martensite phase, E^M and ν^M can be determined at the onset of unloading;
3) M_f, M_s, A_s, A_f are usually provided by material provider, and the parameters related to forward and inverse martensitic transformation can be identified by using pseudoelastic effect: (a) A^M and B^M in a loading process involving deformation induced martensitic transformation, (b) A^A *and* B^A in the subsequent unloading process

involving deformation induced inverse martensitic transformation, and (c) a^M and a^A describe the effect of volumetric stress (or strain) and can be determined with the difference between the responses along different loading paths, e.g., shearing deformation and tensile deformation;

4) The constants β and α in Eq. (101), a^M and a^A in Eq. (97) can be determined with the difference between the inelastic responses of pure tension and pure torsion.

3.4.4. Numerical algorithm

A numerical procedure for the analysis of the response of the SMAs subjected to given strain histories was suggested as follows:

With the results obtained in the kth iteration of the mth increment of loading $\Delta\boldsymbol{\varepsilon}_{(m)}$, such as $\mathbf{L}^{(k)}_{(m)}$, $\mathbf{H}^{(k)}_{(m)}$ of the material, $\Delta\boldsymbol{\varepsilon}^{R(k)}_{(m)}$ and $\Delta\boldsymbol{\sigma}^{R(k)}_{(m)}$ of each single cell, $\Delta\boldsymbol{\varepsilon}^{A(k)}_{(m)}$ and $\Delta\boldsymbol{\sigma}^{A(k)}_{(m)}$ of the austenite phase and $\Delta\boldsymbol{\varepsilon}^{M(k)}_{(m)}$ and $\Delta\boldsymbol{\sigma}^{M(k)}_{(m)}$ of the martensite phase in each cell, and $\mathbf{L}^{*(k)}_{(m)}$ of the material can be obtained with Eq. (86), $\mathbf{L}^{R(k)}_{(m)}$, $\mathbf{H}^{R(k)}_{(m)}$ and $\boldsymbol{\Lambda}^{R(k)}_{(m)}$ of each single cell with Eqs. (83) and (91), respectively. For the given increment of strain, $\Delta\mathbf{q}^{(k+1)}_{(m)}$ can be computed, $\Delta\boldsymbol{\varepsilon}^{R(k+1)}_{(m)}$ can be solved from Eq. (90), $\Delta\boldsymbol{\sigma}^{R(k+1)}_{(m)}$ in each single cell can be computed with Eq. (83), and the unknown parts of the strain and stress in each phase in the single cell, i.e., $\Delta\bar{\boldsymbol{\varepsilon}}^{A(k+1)}_{(m)}$ and $\Delta\bar{\boldsymbol{\sigma}}^{A(k+1)}_{(m)}$ in the austenite phase and the $\Delta\bar{\boldsymbol{\varepsilon}}^{M(k+1)}_{(m)}$ and $\Delta\bar{\boldsymbol{\sigma}}^{M(k+1)}_{(m)}$ in the martensite phase, can be computed respectively with

$$\begin{aligned}
\begin{Bmatrix} \Delta\bar{\sigma}^M \\ \Delta\bar{\varepsilon}^M \end{Bmatrix} &= \begin{bmatrix} A_2^M\left(A_4^M\right)^{-1} & A_1^M - A_2^M\left(A_4^M\right)^{-1}A_3^M \\ \left(A_4^M\right)^{-1} & -\left(A_4^M\right)^{-1}A_3^M \end{bmatrix} \begin{Bmatrix} \Delta\hat{\sigma}^M \\ \Delta\hat{\varepsilon}^M \end{Bmatrix} + \begin{Bmatrix} h_1^M - A_2^M\left(A_4^M\right)^{-1}h_2^M \\ -\left(A_4^M\right)^{-1}h_2^M \end{Bmatrix}\Delta\xi \\
\begin{Bmatrix} \Delta\bar{\sigma}^A \\ \Delta\bar{\varepsilon}^A \end{Bmatrix} &= \begin{bmatrix} A_2^A\left(A_4^A\right)^{-1} & A_1^A - A_2^A\left(A_4^A\right)^{-1}A_3^A \\ \left(A_4^A\right)^{-1} & -\left(A_4^A\right)^{-1}A_3^A \end{bmatrix} \begin{Bmatrix} \Delta\hat{\sigma}^A \\ \Delta\hat{\varepsilon}^A \end{Bmatrix} + \begin{Bmatrix} h_1^A - A_2^A\left(A_4^A\right)^{-1}h_2^A \\ \left(A_4^A\right)^{-1}h_2^A \end{Bmatrix}\Delta\xi
\end{aligned}, \quad (103)$$

which is derived from Eq. (79). Then, $\Delta\mathbf{e}^{p(k+1)}_{(m)}$ (Eq. (55)), $\Delta\zeta^{(k+1)}_{(m)}$ and $\Delta z^{(k+1)}_{(m)}$ (Eqs. (51) and (54)), $f^{(k+1)}_{(m)}$, $\Delta\mathbf{s}^{(r)(k+1)}_{(m)}$ and $\mathbf{s}^{(r)(k+1)}_{(m)}$ (Eq. (59)), $A^{(k+1)}_{(m)}$ and $\mathbf{B}^{(k+1)}_{(m)}$ (Eq. (58)) can be computed sequentially, $\mathbf{L}^{(k+1)}_{(m)}$ and $\Delta\boldsymbol{\sigma}^{(k+1)}_{(m)}$ can be obtained from Eqs. (95) and (94), respectively. The iteration continues until the following defined maximum relative error δ satisfies

$$\delta = \max_{j=1}^{N}\left|\frac{\Delta\mathbf{q}^{(k+1)}_{(m)} - \Delta\mathbf{q}^{(k)}_{(m)}}{\Delta\mathbf{q}^{(k+1)}_{(m)}}\right| \le \delta_T,, \quad (104)$$

where N is the number of cells involved in computation and δ_T is the tolerance, δ_T=0.001 being prescribed in computation. Then, after superimposing respectively the obtained increments on the corresponding results up to the (m-1)th increment of loading, one can start the computation for the next increment of loading.

3.4.5. Analyses of the Responses of the NiTi SMAs Under Tensile/Torsional Loadings

In the following analyses, n=2 (Eq. (58)) was adopted to meet the requirement of both the accuracy and efficiency (Valanis and Fan, 1983), and the hardening function $f(z)$=1 was used for the material that does not exhibit marked isotropic hardening. M_s and A_s were provided as 230K and 247K, respectively (Sun and Li, 2002). The Young's modulus E^A =30 GPa and the elastic shear modulus G^A=18 GPa obtained experimentally (Sun and Li, 2002) were directly used respectively for the pure tension and the pure torsion analyses. It could be seen in the experimental σ-ε curve of pure tension and the τ-γ curve of pure torsion by Sun and Li (2002) that, at the onset of unloading where the material is almost martensitic, the slope E^M and G^M are almost identical with E^A and G^A, respectively. The other material constants involved were identified and are shown in Table 5. During deformation, the ambient temperature is fixed at 293K (20°C).

Table 5. Material Constants

$C_{1,2}$ (GPa)	$\alpha_{1,2}$	A^M, A^A (K^{-1})	B^M, B^A (K)	α, β, a^M, a^A
2010, 2.50	5000, 60	0.025, 0.30	5330, 1940	3.0, 0.42, 0.67, 0.40

The main purpose of this work is to describe the responses of the NiTi SMA under pure tension and pure torsion with a unified constitutive model and investigate the corresponding mechanisms related to the marked difference in the constitutive responses under the two loading conditions.

During pure tension or pure torsion, the macroscopic nominal stress and strain in the working section of the specimens were averagely represented with that of a material element. If the in-plane properties of a lamella in a RVE (Figure 16) can be approximately assumed homogeneous and the axis x_3 of the RVE (Figure 16) coincides with the axis 3 of the Euclidian frame (Figure 18), then ϕ and θ can determine the orientation of the RVE.

For pure tension with applied ε_{III}, the computed tensile σ–ε curves are shown in Figure 19. The solid curve marked with computed_1 corresponds to ϕ=29° and θ=ω= 0° (i.e., the angle between the spiral orientation and the axis of loading is 61°, or the angle between the outwards normal of the martensite lamellas and the axis of loading is 29°, as was observed experimentally).

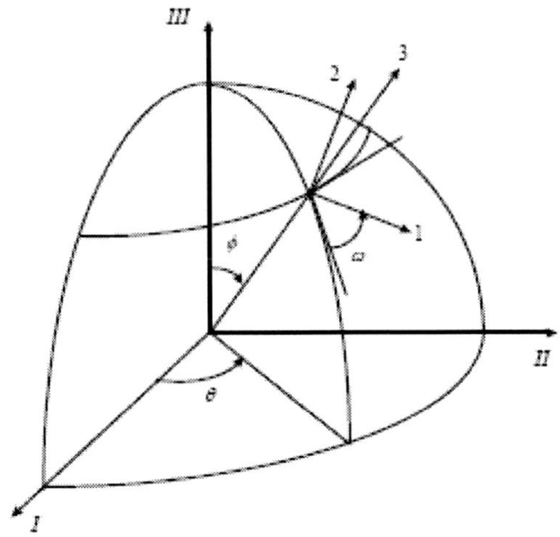

Figure 18. Global and local coordinate frames.

It can be seen that at first the SMA deforms elastically, corresponding to the response of the austenite phase (parent phase). With the increase of deformation, martensite phase initiates from the parent phase and grows. At the stage when the thickness of the martensite lamella is very small, the constraint from the parent phase is strong and the macroscopic behavior is less affected by the appearance of the martensite phase. With the development of martensitic transformation, the increase of the thickness of the martensite lamella reduces the constraint from the parent phase and results in a sharp drop in the σ–ε curve. Thereafter, the load-bearing-capability of the material will be mainly determined by that of the martensite phase.

At the onset of unloading, the response is almost linear. With the progress of inverse martensitic transformation, it tends to be non-linear, the contribution of the martensite phase to stress decays and tends to vanish when the martensite phase transforms completely back to the parent phase, the mechanical property of the material will henceforth be determined by that of the austenite phase, which is elastic so that the stress tends to vanish as the strain vanishes. Compared with the experimental result (also shown in Figure 19), it can be seen that the model can satisfactorily describe the main characteristics of the pseudoelastic behavior of the material subjected to a tensile strain cycle.

The pseudoelastic behavior of the NiTi SMA subjected to a cycle of pure shear strain $\gamma_{\text{III-II}}$ at room temperature is shown in Figure 20. The solid curve marked with computed 1 corresponds to ϕ =63.5° and $\theta = \omega$ = 0° (*i.e.*, the angle between the outwards normal of the lamellas and the longitudinal axis is 63.5°, as was observed experimentally). It can be seen that, although distinct nonlinearity occurs during the strain cycle, the dissipation is not as remarkable as that in tension. Especially the instability or stress drop that appears in pure tension does not appear in pure torsion, which could be attributed to the fine martensite lamellas that keep strong constraint from the austenite phase. This characteristic can be described automatically with the microstructure-based constitutive model. Compared with the experimental results, it can be seen that the model can reasonably replicate the pseudoelastic behavior of the material subjected to a torsional strain cycle.

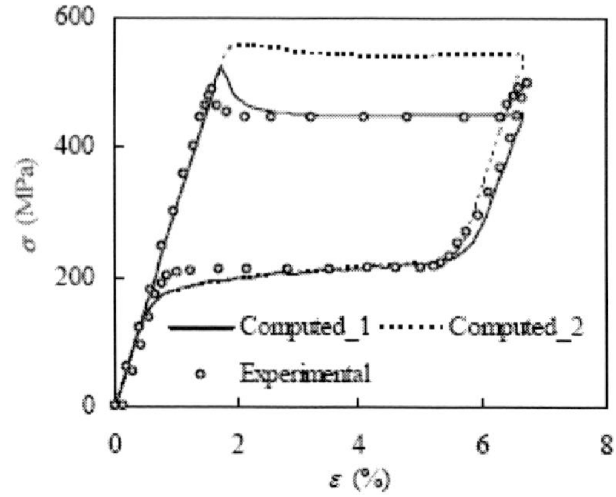

Figure 19. Tensile stress and strain curves

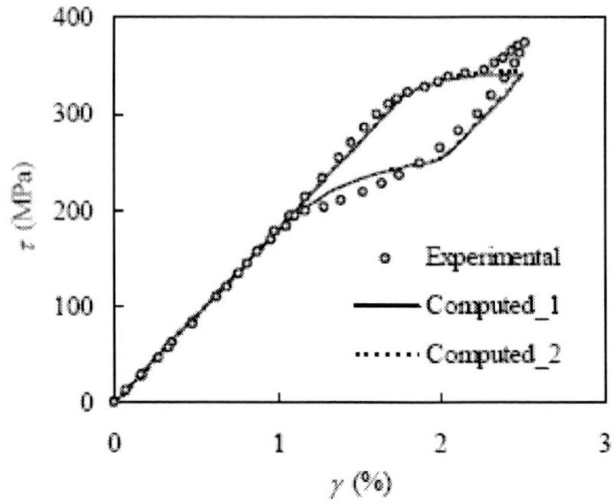

Figure 20 Shear stress and strain curves.

The simulations of the pseudoelastic behavior of the NiTi SMA subjected respectively to a pure tensile strain cycle and a pure torsional strain cycle, with the material constants shown in Table 5 but $\beta = 0$, i.e., without taking into account the possible localization of martensitic transformation and the macroscopic growth of the martensite band, are also shown with the dashed curves marked with "computed_2" in Figure 19 and Figure 20, respectively. In the case of pure tension, it can be seen that, the simulation with $\beta=0$ (or $g(\xi) = \kappa = 0$, Eqs. (101), (71) and (72)) not only fails to describe the macroscopic instability or stress drop in the tensile stress-strain curve, but also considerably overestimates the plateau stress in tensile deformation. But in the case of pure torsion (pure shear), the results with and without introducing the effect of the additional volumetric strain do not show distinct difference (Figure 20).

It should be noted that, in Figs. 19 and 20, the experimental curves have a marked jump in stress following the flat region, which can be attributed to the re-orientation hardening related to the activation of some new martensite variant deformation systems and the corresponding complicated mechanisms, which could be described by introducing the description based on the deformation of different martensite variants and will be discussed elsewhere.

The stress–strain curves during the tension processes corresponding to different preloaded shear stress (different values of applied torque) are shown in Figure 21(a). It can be seen that, at lower levels of preloaded shear stress, the subsequent tensile responses are unstable, demonstrated by the stress drops and plateaus in the subsequent tensile σ–ε curves. With the increase of the preloaded shear stress, such instability decreases gradually and vanishes as the preloaded shear stress is sufficiently large. Compared with the experimental results shown in Figure 21(b), it can be seen that the microstructure-based constitutive model can describe satisfactorily the main characteristics of the material under the given loading condition.

3.5. Brief Summary and Discussion

The surface morphologies of nano-grained NiTi SMA microtubes under pure tension and pure torsion were observed with a scanning electron microscope. It shows uniformly distributed parallel texture at the surface of the NiTi SMA microtubes subjected to pure torsion, indicating alternative austenite and martensite lamellar microstructure related to deformation-induced martensitic transformation. For the microtubes subjected to pure tension, martensite lamellas initiate and grow to macroscopic bands that can be observed even with naked eyes, accounting for the macroscopic instability or the stress drop during pure tension.

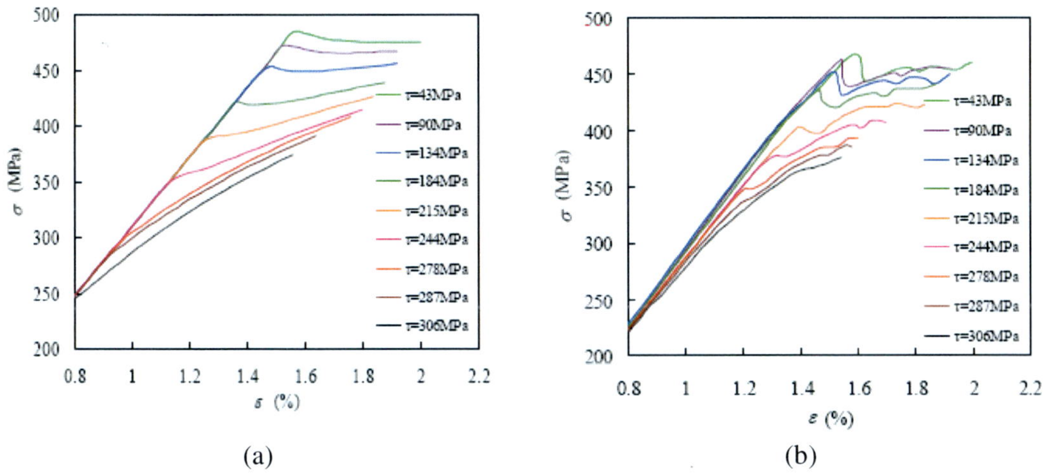

Figure 21. Tensile σ-ε curves at different pre-shear stresses, (a) Computational (b) Experimental (Sun and Li, 2002).

It was suggested that there are alternatively arranged lamellar microstructures in the SMA microtubes under pure tension and under pure torsion, respectively, but the localization of the transformation microstructure and the macroscopic growth of the martensite band during a tensile process may greatly reduces the constraint between the martensite and austenite lamellas, resulting in macroscopic instability or stress drop on the overall stress-strain curve, which was confirmed by the numerical simulation.

A microstructure-based constitutive model for the pseudoelastic behavior of the materials was introduced. An elastoplastic relationship was adopted for the martensite phase and a linearly elastic one for the austenite phase. The model for the representative volume element (RVE) of a cell of the material was then derived, which is composed of alternatively arranged parallel fine lamellas of martensite and austenite. An additional volumetric strain was introduced in each phase to exclude the unrealistic volumetric strain caused by the deviation of the actual deformed configuration of the RVE from the ideal one, due to the localization and growth of the martensite lamellas. The description for the pseudoelastic behavior of the NiTi SMAs was finally derived by making use of the Hill's self-consistent scheme.

The corresponding numerical algorithm was developed, and the pseudoelastic behavior of an NiTi SMA under pure tension and pure shear was analyzed. The computed results were in satisfactory agreement with the experimental results. The main characteristics of the pseudoelastic behavior, especially the macroscopic instability during pure tension and the distinct differences between the pseudoelastic behavior under pure tension and that under pure torsion, could be well described. In order to verify the validity of the model in a more complicated loading history, the tensile responses of the NiTi SMA, which was preloaded with different shear stress, were analyzed, and the main characteristics of the material behavior were also replicated satisfactorily.

In the above numerical examples, the responses of the NiTi SMA were analyzed separately with the orientations observed in experiment. It remains difficult for this approach to be applied in general triaxial loading cases where the orientations of the material elements may vary with the state of stress and the history of deformation. In order to overcome this difficulty, a simple approximation is to assume that the material subjected to deformation-induced martensitic transformation is an aggregate of numerous cells with randomly distributed orientations. However, the error caused by this assumption needs to be estimated. With this assumption, the pseudoelastic behavior of the material subjected to pure tension and pure torsion were analyzed using the material constants identical with those given in subsection 3.4.5. The corresponding results are shown with the dashed curves marked with "Computed_3" in Figs. 22 and 23, respectively. It can be seen that the results are in reasonable agreement with the results using the orientations observed experimentally (marked with "Computed_1", or those shown in Figs. 19 and 20) and the experimental results. It should be noted that the results shown in Figure 21(a) were also computed with this assumption. These results demonstrate the capability of the model in the analysis of the pseudoelastic behavior of the materials subjected to complicated triaxial deformation.

It should also be noted that the re-orientation hardening was not considered in the model, it involves the deformation of new martensitic variants that are activated at high level of stress. The corresponding mechanisms and modeling are complicated, which will be discussed elsewhere.

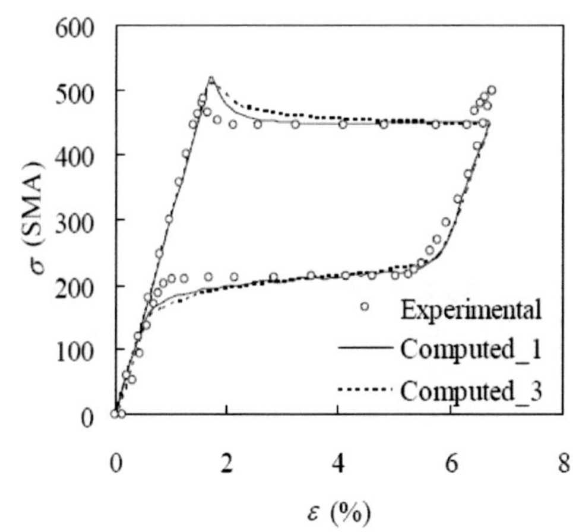

Figure 22. Tensile stress and strain curves

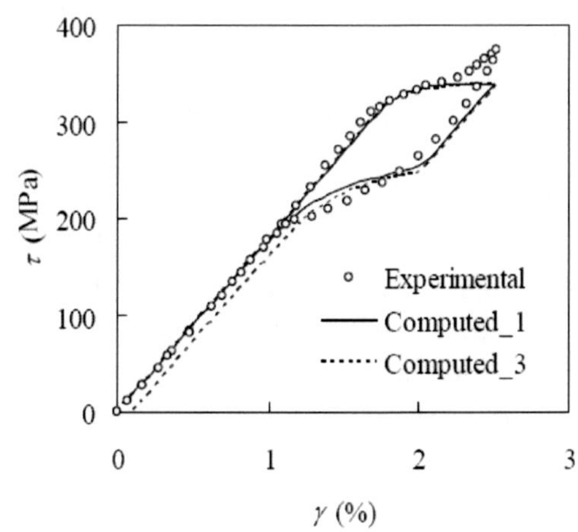

Figure 23 Shear stress and strain curves.

4. CONCLUSION

Two models for the constitutive behavior of SMAs were introduced in this chapter. Both the models are based on the concept that the constitutive behavior of an SMA is the combination of the individual behavior of each of its two phases, and in the interested ranges of stress and temperature, the austenite phase is linearly elastic while the martensite phase is elastoplastic.

The first model is based on a two-phase mixture theory, in which a conventional theory of plasticity was used for the constitutive behavior of the martensite phase. The strains in the two phases were assumed identical with the overall strain in the SMA, while the overall stress in the SMA was the sum of the contributions of the two phases. The complicated constitutive behavior of a SMA can, therefore, be separated into the simple constitutive behavior of each of its two phases, which yields a simple but comprehensive description for the constitutive behavior of SMAs.

The constitutive model is concise, physically distinct and easily applicable. It can describe the main constitutive behavior of SMAs, such as pseudoelastic behavior and shape memory effect under various thermal-mechanical loading conditions. The pseudoelastic behavior and shape memory effect of SMA Au-47.5at.%Cd subjected to tension/compression at constant or varying temperature, and the pseudoelastic behavior of a polycrystalline Cu-Al-Zn-Mn SMA subjected to proportional and nonproportional stress and strain histories were analyzed. The comparison between the computed and experimental results showed satisfactory agreement.

The second model was developed for the typical constitutive pseudoelastic behavior of some NiTi SMAs under different loading condition, which could not be well described with any of the current phenomenological constitutive models. Based on the analysis of the martensitic transformation microstructures under typical loading conditions, a microstructure-based constitutive model was developed for the pseudoelastic behavior of the materials by three steps: (a) the elastoplastic model for the martensite phase and a linearly elastic one for the austenite phase; (b) the model for the RVE of a cell composed of alternatively arranged parallel fine lamellas of martensite and austenite, by introducing additional volumetric strain to exclude the unrealistic volumetric strain caused by the deviation of the actual deformed configuration of the RVE from the ideal one; and (c) the model for the pseudoelastic behavior of the NiTi SMAs by making use of the Hill's self-consistent scheme.

The pseudoelastic behavior of a NiTi SMA under pure tension and pure shearing and the tensile responses of the NiTi SMA preloaded with different prescribed shear stress, were analyzed, the main characteristics of the pseudoelastic behavior were well described, and the main characteristics of the material behavior were also replicated satisfactorily.

The model could be approximately extended to the analysis of the pseudoelastic behavior of the materials subjected to complicated triaxial deformation with the assumption that the material is an aggregate of numerous cells with randomly distributed orientations, and the validity of this extension was also demonstrated.

It should be noted that the reorientation hardening was not considered in the two models. It involves the activation of some new martensite variant deformation systems and the corresponding complicated mechanisms, which will be discussed elsewhere.

APPENDIX A

Combining Eqs. (55) with (57) yields

$$A\Delta\mathbf{e}^p + \mathbf{B}(z_n)\Delta z = 2G(\Delta\mathbf{e} - \Delta\mathbf{e}^p)\,. \tag{A1}$$

Substituting Eq. (60) into Eq. (A1) gives

$$(A+2G)\left[\mathbf{I}_4 + \mathbf{B}(z_n)\otimes\frac{\Delta\mathbf{e}^p}{(A+2G)f^2\Delta z}\right]:\Delta\mathbf{e}^p = 2G\Delta\mathbf{e}\cdot \tag{A2}$$

It can be solved from Eq. (A2) that

$$\begin{aligned}\Delta\mathbf{e}^p &= \frac{1}{A+2G}\left[\mathbf{I}_4 + \mathbf{B}(z_n)\otimes\frac{\Delta\mathbf{e}^p}{(A+2G)f^2\Delta z}\right]^{-1}:2G(T)\Delta\mathbf{e}\\ &= \frac{1}{A+2G}\left[\mathbf{I}_4 - \frac{1}{a}\mathbf{B}(z_n)\otimes\frac{\Delta\mathbf{e}^p}{(A+2G)f^2\Delta z}\right]:2G(T)\Delta\mathbf{e},\end{aligned} \tag{A3}$$

where $a = 1 + \dfrac{\mathbf{B}:\Delta\mathbf{e}^p}{(A+2G)f^2\Delta z}\cdot$ (A4)

By substituting Eq. (A4) into Eq. (5), one obtains

$$\begin{aligned}\Delta\mathbf{s} &= 2G(\Delta\mathbf{e} - \Delta\mathbf{e}^p) = 2G\Delta\mathbf{e} - \frac{2G}{A+2G}\left[\mathbf{I}_4 - \frac{1}{a}\mathbf{B}(z_n)\otimes\frac{\Delta\mathbf{e}^p}{(A+2G)f^2\Delta z}\right]:2G\Delta\mathbf{e}\\ &= \left[\frac{2GA}{A+2G}\mathbf{I}_4 + \frac{1}{a}\frac{2G^{\ 2}}{(A+2G)^2 f^2}\mathbf{B}(z_n)\otimes\frac{\Delta\mathbf{e}^p}{\Delta z}\right]:\Delta\mathbf{e}.\end{aligned} \tag{A5}$$

Noticing that $\Delta\mathbf{e}^p:\Delta\mathbf{e} = \Delta\mathbf{e}^p:(\Delta\boldsymbol{\varepsilon} - \frac{1}{3}\Delta\varepsilon_{kk}\mathbf{I}_2) = \Delta\mathbf{e}^p:\Delta\boldsymbol{\varepsilon}$ (A6)

and $\Delta\boldsymbol{\sigma} - \frac{1}{3}\Delta\sigma_{kk}\mathbf{I}_2 = \Delta\mathbf{s} = \dfrac{2GA}{2G+A}\mathbf{I}_4:(\Delta\boldsymbol{\varepsilon} - \frac{1}{3}\Delta\varepsilon_{kk}\mathbf{I}_2) + \dfrac{(2G)^2}{(2G+A)^2 f^2(z)}\dfrac{1}{a}\mathbf{B}\otimes\dfrac{\Delta\mathbf{e}^p}{\Delta z}:\Delta\boldsymbol{\varepsilon}$, (A7)

one has $\Delta\boldsymbol{\sigma} = \dfrac{A}{1+A/(2G)}(\Delta\boldsymbol{\varepsilon} - \frac{1}{3}\Delta\varepsilon_{kk}\mathbf{I}_2) + \dfrac{(2G)^2}{(2G+A)^2 f^2(z)}\dfrac{1}{a}\mathbf{B}\otimes\dfrac{\Delta\mathbf{e}^p}{\Delta z}:\Delta\boldsymbol{\varepsilon} + \frac{1}{3}\Delta\sigma_{kk}\mathbf{I}_2$ (A8)

By substituting Eq. (56) into Eq. (A8) and noticing that $\Delta\varepsilon_{kk} = \mathbf{I}_2:\Delta\boldsymbol{\varepsilon}$, one derives Eqs. (62) and (63).

On the other hand, substituting Eq. (73) into $(71)_2$ gives

$$\Delta\sigma_{kk} = 3K \quad 1 - g(\xi) \quad \Delta\varepsilon_{kk} - g'(\xi)\varepsilon_{kk}\Delta\xi \ . \tag{A9}$$

Then substituting Eq. (A9) into Eq. (A8), one derives Eqs. (74) and (75).

APPENDIX B

Making use of the following incremental form of Eq. (68)

$$\{\Delta\hat{\varepsilon}^M\} = \{\Delta\hat{\varepsilon}^A\} = \{\Delta\hat{\varepsilon}^R\}, \qquad \{\Delta\hat{\sigma}^M\} = \{\Delta\hat{\sigma}^A\} = \{\Delta\hat{\sigma}^R\}, \tag{B1}$$

the unknown components of the stress and strain increments in the martensite phase $\{\Delta\bar{\sigma}^M\}$ and $\{\Delta\bar{\varepsilon}^M\}$ can be solved from Eq. (79) as

$$\begin{Bmatrix} \Delta\bar{\sigma}^M \\ \Delta\bar{\varepsilon}^M \end{Bmatrix} = \begin{bmatrix} A_2^M \ A_4^M \ ^{-1} & A_1^M - A_2^M \ A_4^M \ ^{-1} \ A_3^M \\ A_4^M \ ^{-1} & - \ A_4^M \ ^{-1} \ A_3^M \end{bmatrix} \begin{Bmatrix} \Delta\hat{\sigma}^M \\ \Delta\hat{\varepsilon}^M \end{Bmatrix} + \begin{Bmatrix} h_1^M - A_2^M \ A_4^M \ ^{-1} \ h_2^M \\ - \ A_4^M \ ^{-1} \ h_2^M \end{Bmatrix} \Delta\xi \ . \tag{B2}$$

The unknown components of the stress and strain increments in the austenite phase, $\{\Delta\bar{\sigma}^A\}$ and $\{\Delta\bar{\varepsilon}^A\}$, can be solved as

$$\begin{Bmatrix} \Delta\bar{\sigma}^A \\ \Delta\bar{\varepsilon}^A \end{Bmatrix} = \begin{bmatrix} A_2^A \ A_4^A \ ^{-1} & A_1^A - A_2^A \ A_4^A \ ^{-1} \ A_3^A \\ A_4^A \ ^{-1} & - \ A_4^A \ ^{-1} \ A_3^A \end{bmatrix} \begin{Bmatrix} \Delta\hat{\sigma}^A \\ \Delta\hat{\varepsilon}^A \end{Bmatrix} + \begin{Bmatrix} h_1^A - A_2^A \ A_4^A \ ^{-1} \ h_2^A \\ - \ A_4^A \ ^{-1} \ h_2^A \end{Bmatrix} \Delta\xi \ . \tag{B3}$$

Substituting Eqs. (B2) and (B3) into Eq. (70) yields

$$\Delta\bar{\sigma}^R = g_1\Delta\hat{\sigma}^R + g_2\Delta\hat{\varepsilon}^R + g_5\Delta\xi \tag{B4}$$

and $\Delta\bar{\varepsilon}^R = g_3\Delta\hat{\sigma}^R + g_4\Delta\hat{\varepsilon}^R + g_6\Delta\xi$. (B5)

It can be solved from Eq. (B4) that

$$\Delta\hat{\sigma}^R = g_3^{-1}\Delta\bar{\varepsilon}^R - g_3^{-1}g_4\Delta\hat{\varepsilon}^R - g_3^{-1}g_6\Delta\xi \ , \tag{B6}$$

and substituting Eq. (B6) into Eq. (B5) gives

$$\Delta\bar{\sigma}^R = g_1 g_3^{-1}\Delta\bar{\varepsilon}^R + (g_2 - g_1 g_3^{-1} g_4)\Delta\hat{\varepsilon}^R + (g_5 - g_1 g_3^{-1} g_6)\Delta\xi \ . \tag{B7}$$

Eqs. (B6) and (B7) can be rewritten in matrix form as Eq. (82), in which g_1 through g_6 are given in Eq. (84).

ACKNOWLEDGMENT

The authors gratefully acknowledge the financial support from the Natural Science Foundation of China and the Japanese Society for the Promotion of Sciences.

REFERENCES

Achenbach, M.; Atanackovic, T.; Muller, I. *Int. J. Solids Struct.* 1986, *22*, 171-193.
Armstrong, R. J.; Frederick, C.O. *CEGB Report/RD/B/N 731, General Electricity Generating Board,* 1966.
Auricchio, F.; Taylor, R. L.; Lubliner, J. *Comput. Methods Appl. Mech. Eng.* 1997, *146*, 281-312.
Bo, Z.; Lagoudas, D. C. *Int. J. Eng. Sci.* 1999, *37*, 1089-1249.
Fung, P.; Sun, Q. P. *J. Mech. Phys. Solids*. 2006, *54*, 1568-1603.
Fan, J.; Peng, X. *J. Eng. Mater. Tech.* 1991, *113*, 254-262.
Boyd, J. G. *J. Intell. Mater. Syst. and Struct.* 1994, *5*, 333-346.
Falk, F. *Acta metall.* 1980, *23*, 1773-1780.
Fischer, F. D.; Oberaigner, E.R.; Tanaka, K. *Comput. Mater. Sci.* 1997, *9*, 56-63.
Goo, B. C.; Lexcellent, C. *Acta Mater.* 1997, *45*, 727-737.
Graesser, E. J.; Cozzarelli, F. A. *J. Eng. Mech.* 1991, *117*, 2590-2608.
Graesser, E. J.; Cozzarelli, F.A. *J. Intell. Mater. Syst. Struct.*. 1994, 120, *5*, 78-89.
Hill, R. *J. Mech. Phys. Solids*. 1965, *13*, 89-101.
Iadicola, M. A.; Shaw J. A. *Int. J. Plast.* 2004, *20*, 577–605.
Lagoudas D. C.; Bo, Z. *Int. J. Eng. Sci.*1999, *37*, 1041-1173.
Liang, C.; Rogers, C. A. *J. Intell. Mater. Syst. and Struct.* 1990, *1*, 207-234.
Levitas, V. I. *Mech. Res. Commun.* 1995, *22*, 87-94.
Mikata, Y., *Int. J. Eng. Sci.* 2000, *38*, 605~641.
Müller, Ch.; Bruhns, O. T. *Int. J. Plast.* 2006, *22*, 1658–1682.
Nakanishi, N.; Mori, T.; Miura, S.; Murakami, Y.; Kachi, S. *Philos Mag.* 1973, *28*, 277-292.
Pan, H.; Thamburaja P.; Chau F.S. *Int. J. Plast.* 2007, *23*, 711–732.
Patoor, E.; Eberhardt, A.; Berveiller, M. *Arch. Mech.* 1988, *40*, 775-794.
Peng, X.; Fan, J. *Comput. Struct.* 1993, *47*, 313-320.
Peng, X.; Fan, J.; Zeng, X. *Int. J. Solids Struct.* 1998, *35*, 2733-2747.
Peng, X,; Fan, J.; Zeng, J. *Int. J. Solids Struct.* 2002, *39*, 435-448.
Peng, X.; Li, H.; Chen, W. *Smart Mater. Struct.* 2005, *14*, 425-433.
Peng, X.; Pi, W.; Fan, J. *Int. J. Plast.* 2008, *24*, 966-990.
Peng, X.; Ponter, A. R. S. *Int. J. Solids Struct.*. 1994, *31*, 1099-1111.
Peng, X., Yang, Y., Huang, S. *Int. J. Solids Struct.* 2000, *38*, 6925-6940.
Shaw, J. A., *Int. J. Plast.* 2000, *16*, 541-562.
Shaw, J. A.; Kyriakides, S. *Int. J. Plast.* 1998, *13*, 837-871.

Sittner, P.; Hara, Y.; Takuda, M. *Metall Mater Trans A*. 1995, *26A*, 2923-2935.
Sittner, P.; Takuda, M. *J. De Physique IV*. 1995, *5*, 1003-1008.
Siredey, N.; Patoor, E.; Berveiller, M.; Eberhardt, A. *Int. J. Solids Struct.* 1999, 36, 4289-4315.
Sun, Q. P.; Hwang, K. C. *J. Mech. Phys. Solids*. 1993, *41*, 1-33.
Sun, Q.P.; Li, Z.Q. *Int. J. Solids Struct.* 2002, *39*, 3797–3809.
Sun, Q. P.; Zhong. Z. *Int. J. Plast.* 2000, *16*, 1169-1187.
Tanaka, K.; Kobayashi, S.; Sato, Y. *Int. J. Plast.* 1986, *2*, 59-72.
Tanaka, K. *Res Mechanica*. 1986, *18*, 251-263.
Thamburaja, P., *J. Mech. Phys. Solids*. 2005, *53*, 825–856.
Valanis, K. C.; Fan, J. *J. Appl. Mech*. 1983, *50*, 789-793.
Zhong Z.; Sun, Q. P.; Tong P. *Int. J. Solids Struct.* 2000, *37*, 5943-5955.

In: Shape Memory Alloys
Editor: H. R. Chen, pp. 49-68
ISBN 978-1-60741-789-7

Chapter 2

CONTROL OF SMA SYSTEMS: REVIEW OF THE STATE OF THE ART

Mohammad H. Elahinia*[*1], *Ehsan Tarkesh Esfahani*[2] and *Shuo Wang*[1]

[1]Dynamic and Smart Systems Laboratory
Mechanical Industrial and Manufacturing Engineering Department
University of Toledo, OH, USA

[2]Inteligent Design and Manufacturing Lab
Department of Mechanical Engineering
University of California Riverside, CA, USA

ABSTRACT

Shape memory alloy (SMA) actuators can be made very compact and simple. An effective control algorithm is an integral part of every SMA actuator. This article presents a review of the state of the art research on control of SMA systems. It is worth noting that almost exclusively all control systems applied to SMA actuators are not model-based control algorithms. In other words, the control system is designed based on the intuitive knowledge of the system, through a combination of simulations and experiments. Therefore, the controller is not appropriately designed based on the system's parameters. As a result, stability analysis, a major step in designing control systems, has rarely been done for SMA-actuated systems. It is also important to recognize the co-dependent relationship between stress, Martensite fraction, and transformation temperatures of the SMA materials. This complex nonlinear relationship causes many difficulties with regards to control system design for SMA actuators. It should be also noted that SMAs are one way actuators and they are sensitive to changes in the ambient temperature.

* Author to whom all corresponding should be addressed, mohammad.elahinia@utoledo.edu

1. Introduction

Shape memory alloy (SMA) actuators are attractive due to their many potential applications in mechanical, electrical, civil, medical and aerospace systems. SMA actuation has many advantages over conventional actuation methods. Its significantly reduced weight, size, complexity and noiseless operation make them suitable as an actuator for wide range of application form micro robotics to medical application and aerospace systems [1-19]. The benefits over other smart materials such as piezoelectric materials, electrostrictive materials and magnetostrictive materials are the extremely high force to weight ratio, and large displacement. Among all the shape memory alloys such as Ag-Cd, Au-Cd, Cu-Sn, In-Ti, Ni-Al, Ni-Ti, Mn-Cu, Nickel-Titanium (Ni-Ti) has proven to be the most promising one for many applications [20-23]. This is due to its greater ductility, more recoverable motion, excellent corrosion resistance, stable transformation temperatures, high biocompatibility and convenience to be electrically heated. In these alloys, shape memory effect (SME) takes place due to transformation between the Martensite phase (less stiff and stable at lower temperatures) and Austenite phase (stiff and stable at higher temperatures).

When an SMA is in the 100% Martensite phase and it is heated from 'A_s' (Austenite start temperature) to 'A_f' (Austenite finish temperature), the crystalline structure will transform to fully Austenite. When the SMA in the Austenite phase is cooled down from 'M_s' (Martensite start temperature) to 'M_f' (Martensite finish temperature), the crystalline structure will transform to Martensite. In another mode of phase transformation, exertion of an external stress causes the "twinned" Martensite to change to "detwinned" Martensite below 'A_s'; a large strain will remain with the removal of the stress. When the "detwinned" Martensite is heated above 'A_f' the remaining strain is completely recovered which realizes the "memory" effect.

Hysteresis behavior is one of the main disadvantages of SMA actuation [24, 25]. Other nonlinear phenomena such as nonlinear heat transfer, nonlinear changes in temperature and stress also are undesired characteristics. Furthermore, SMA actuators have to operate at a low bandwidth, mainly due to the slowness of its cooling mechanism. All these factors make the control of SMA actuators a difficult topic and smart designs of control strategies are still challenging for these actuators.

The existing control issues for SMA actuators are presented in next section of this chapter. In section 3, various control designs are reviewed. Some of these controllers are applied to a SMA system and the results are compared in section 4. The last section summarizes the existing work for SMA and provides some insight for future work.

2. Various Control Issues for SMA

Shape memory alloy actuators can replace the conventional actuators due to its large mechanical strains upon heating and cooling. Large mechanical stress is consequently produced when SMA wires are heated or cooled beyond some critical temperatures. SMA actuators can apply up to 600 MPa. A 0.1mm diameter Nitinol wire can therefore apply a force of 4.7 N. This means that the wire can lift up an object about 100,000 times its own weight. Another reason that SMA actuators can be substituted for the conventional actuators is that they can be made compact and simple.

There are many advantages to employ SMA wires as actuators in various applications. Of course, shape memory alloy actuators have their own drawbacks to overcome in order to be fully put to use. First of all, one main disadvantage in controlling SMA actuators is the nonlinear relationship of the temperature dependent stress–strain curve. The hysteresis properties existing in SMA include the relationships between displacement and temperature, Young's modulus and temperature, displacement and voltage, strain and temperature as well as Martensite fraction and temperature. These nonlinearities present many difficulties in the design of control system for SMA actuators.

Another main disadvantage in controlling SMA is coming through the actuation mechanism (adding and removing heat). The SMA actuation is mostly initiated by inducing the current to the element and therefore is subject to Joule heating and natural convection. Therefore SMA system experience a large time constant depended on the heat transfer characteristics of the SMA cooling mechanism. Heat removal techniques are usually in the forms of using heat sink, water immersion, and forced convection. The temperature and type of environment, the convection of the environment and the surface to volume ratio of the SMA wires could all influence the heating and cooling time. These time delays for SMA actuators to respond to control signals make another nonlinear problem and present a challenge in controller design. To tackle this problem different heating and cooling method for actuation have been used.

Direct electrical current (Joule heating) and inductive heat transfer through microwaves or infrared light are common techniques of actuation [26]. Heat removal techniques are usually in the terms of using heat sink, water immersion, and forced convection. Shahin et al [27] used semiconductor "heat pump" modules for heating/cooling cycles of an SMA system to reduce the response time for complete thermal cycling. Tadesse et al studied the effect of active cooling and prestress on response time of SMA wires [28]. Sreekumar et al. suggested the use of 'Cool chips' for miniaturized application like micro robots [29].

Flemming [30]and Mascaro [31]used SMA wire as wet actuators by putting NiTi wires in latex rubber tube. Heating/cooling to the system is provided by passing hot or cold water through the tubes. Implementing all these different techniques does not solve the slow actuation and nonlinearity problems of the system. Therefore it requires a careful implementation of a control technique which can handle the nonlinearity and slow actuation.

The other challenge in controlling SMA actuators is a number of state variables which are not easy to measure. This lack of information makes the implementing of model-based controllers more difficult. For example, in a regular mechanical application, to have a model based controller, it is essential to have information about state variables such as the stress and temperature of the SMA, Martensite fraction and kinematic data of the system. Among these variables only the kinematics data can be easily measured.

To address this issue, Ikuta proposed an algebraic model which could ultimately estimate the phase transformation based on the electric resistance (neglecting the hysteresis) [32]. This technique is later shown not to be suitable for closed loop control system [33].

Elahinia et al. also proposed an Extended Kalman filter to estimate the actual stress based on the kinematics and dynamics of a system [34].

The final issue which should be noted in designing a controller is to avoid overheating and overloading. If an SMA wire is overheated it will take a longer time for the actuator to be cooled down also in medical device overheating may cause severe injuries [35].

3. Control Strategies for SMA Systems

Displacement control has been investigated extensively by different research groups, where accurate position has been the goal of control either in robotics or medical systems [10, 11, 35-49]. SMA systems are tied with numerous uncertainties. Therefore various controller designs are conducted in order for this system to achieve the desired state fast enough, stable enough, without overshooting and oscillating and to be robust to withstand any external disturbance [39, 40, 47-55].Various models such as Preisach model [24,25], constitutive model [56,57], phenomenological model [58,59] and heat transfer model had been integrated with conventional controllers for enhancing the SMA actuation. Generally different controller of SMA systems can be categorized into one of the following groups: 1-Linear controller, 2-Nonlinear controller, 3-Hybrid Controller.

3.1 Linear Control

The conventional linear proportional integral derivative (PID) controllers are naturally the first to be adopted. Majima et al. [44] combined a PID feedback controller with a feed forward loop. In the feed forward loop a static SMA model based on Preisach hysteresis determine the input to the system to reach the desired state. Singh et al. [60] implemented a PID controller with gains selected by Ziegler-Nichols for tracking the tab in a rotor blade system. Gorbet et al. [53] designed a PD controller based on the passivity of the Preisach hysteresis model. Moallem et al. [61] presented a PD controller in flexible beam to control the force to actuate the SMA and to further exert the position control on the endpoint. PI controller was designed by Jayender et al. [62] to control the strain in the SMA actuator and a proportional controller was designed by Madill et al. [33] to track the position of SMA actuator. Modified PID controller has been also investigated by some researchers. Shameli et al. [64] proposed a PID-P^3 controller to perform a precise position tracking. PID-P^3 controller has a cubic term of positioning error. This term has considerable effects in reducing the settling time and overshoot of the positioning system.

Ma et al. [42] presented the proportional plus derivative (PD) controller with the pulse width modulation (PWM) to track the position for a SMA actuator. The pulse width (PW) modulated PD and continuous PD control were tested and compared. The PWM controller saved energy (30% less energy than the continuous PD controllers) and was robust to the external disturbances and produced no oscillations though it had a long settling time. Song and Ma [65] also used a Pulse-Width Pulse-Frequency (PWPF) modulator to modulate a PD controller. Their experimental results showed that continuous to PD controller, PWPF is effectively save actuation while it has almost the same performance.

3.2 Nonlinear Control

Because of the hysteresis relations in SMA such as displacement and temperature, Young's modulus and temperature, displacement and voltage/current, strain and temperature and Martensite fraction and temperature, a nonlinear controller design can potentially improve the performance of the system by eliminating or reducing the nonlinear effect of the hysteresis. Many researchers have studied and designed nonlinear controllers to deal with the nonlinearity in SMA systems. Following are the main general approached used in designing nonlinear controllers for SMA systems.

3.2.1 Sliding Mode and Variable Structure Controller

Sliding mode control (SMC) is usually used to deal with uncertainties in nonlinear dynamic systems [36, 38, 40 and 63]. Choi et al. [38] adopted SMC to provide fast responses and overshoot free in the position control of a single-link mechanism activated by SMA springs. Lim et al. [63] also employed SMC to present a non-contact start/stop motion control for a hard disk drive (HDD) suspension. Ashrafiuon et al. [36] used the extended Kalman filter (EKF) to make it possible, by only measuring the position, to asymptotically stabilize a second-order SMC to compensate for the hysteresis. Elahinia et al. [40] used sliding mode stress based controller for a rotary SMA actuated manipulator. They used the dynamics of the system in the SMC controller to calculate a desired stress based on a desired position. Finally a PI controller regulates the voltage to achieve the desired stress. The block diagram of this controller is shown in **Figure 1**.

SMC shows high accuracy in tracking both stationary and variable input signals. For parameter uncertainties and nonlinearities in SMA, SMC demonstrate encouraging ability. Nevertheless, SMC designs are most of the time based on specific applications without rigorous stability analysis.

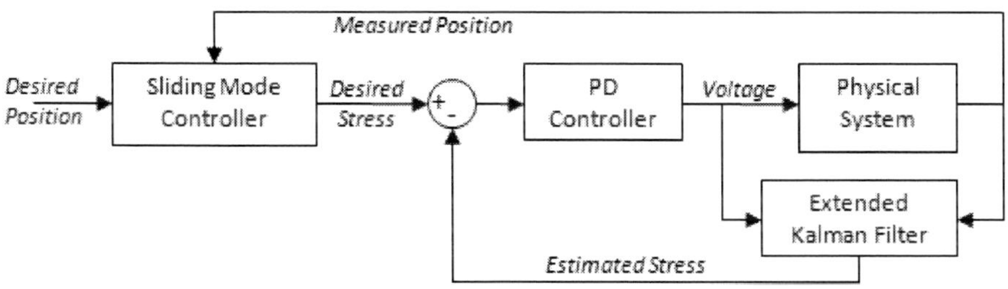

Figure 1. SMC in the experimental setup for non-CSS drive system

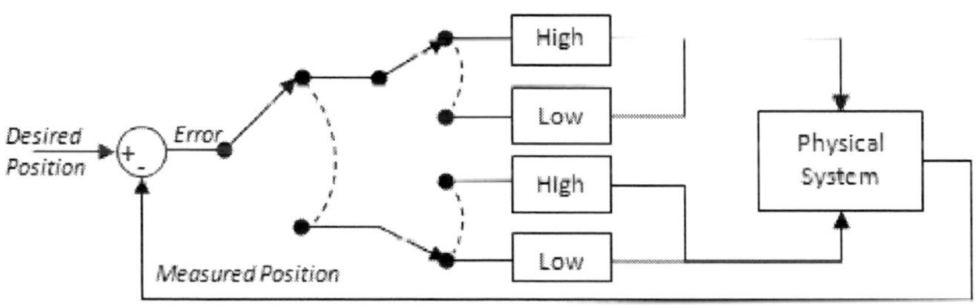

Figure 2. Block diagram for VSC

In Figure 2, Grant et al. [66] used two SMC to control a pair of antagonistic actuators. A switching control input created a sliding mode within a boundary layer in which a linear control was used and a multistage switching control was used to simplify amplifier construction. Elahinia et al. [39] presented a nonlinear, robust control algorithm for the displacement control. Modified switch with boundary layer designed in the VSC reduced the chattering which often appeared in many other VSC designs. The controller was shown to effectively track the position and to be robust to the modeling uncertainties and to changes of payload and temperature drop. Generally, VSC is robust to the parameter uncertainties and nonlinearities in SMA model and external disturbance. However, its discontinuous nature of switching may cause ringing and introduce high frequency components. Another problem with VSC is its inability for tracking when the error is large.

3.2.2 Adaptive Controllers

Neural network controller (NNC) has also been designed for the control of SMAs [10, 43, 49 and 51]. As shown in Figure 3, Song et al. [49] used neural network model and neural network inverse model to capture the hysteresis relationship between applied voltage and displacement. In the experiments an inverse model was put before the plant model and the results showed that the desired position signal and the plant model output are satisfyingly close. The NNC using the inverse model was found effective to compensate for the nonlinearity in SMA wire actuation and was able to eliminate the position sensor to reduce cost. Damle et al. [51] discussed both neural network method and observer formulation method to obtain Markov parameters for structure identification. The linear quadratic Gaussian with loop transfer recovery (LQG/LTR) controller was designed after the system identification and a modified structure for the controller was proposed. The neural network based robust controller was also adopted to imitate the modified LQG/LTR controller.

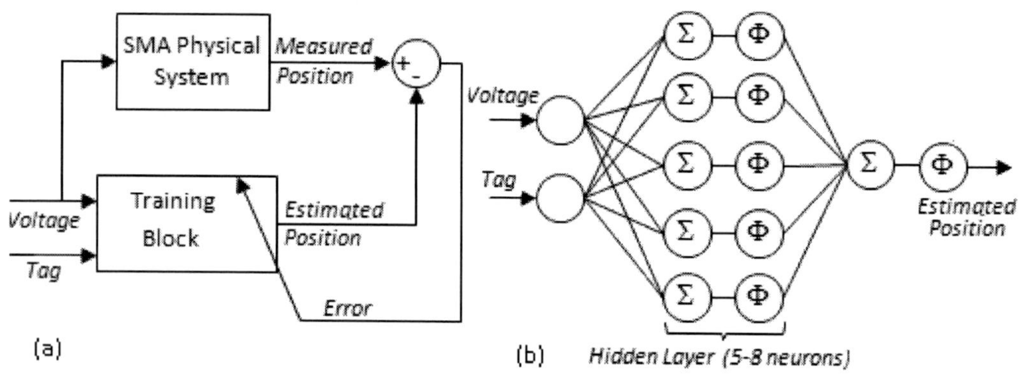

Figure 3. Block diagram for Neural Network inverse model as NNC

The experiment results showed that both controllers could achieve satisfactory performance for the nominal tip mass and both of them demonstrated robustness to tip mass changes. Stevens et al. [10] presented that the radial basis function network, one type of artificial network works better than the conventional PID controller in miniature robotic surgical systems with the manipulator actuated by SMA. Ma et al. [43] used neural networks to picture the nonlinear hysteresis between internal electrical resistance and the displacement. NNC showed great possibility in mapping and adaptation for capturing the hysteresis behavior of the SMA. However, neural networks need training to operate and therefore the training data and training methods are critical. This may require a large amount of data to secure the statistical accuracy.

Fuzzy logic controller (FLC) design is one of the many control algorithms used for nonlinear systems. Many researchers have used this method for the control of shape memory alloy hysteretic behavior [35, 54 and 67]. Rustighi, et al. [54] presented an adaptive tuned vibration absorber (ATVA) using SMA elements. An error signal was defined as the phase angle between the velocities of the host structure and that of the absorber mass which were measured by an accelerometer. As shown in Figure 4, Ahn et al. [35] designed a self tuning fuzzy PID controller to track the position of SMA actuators. The nonlinearity of the SMA and other uncertainties in the system and the surrounding environment were the major challenges and the fuzzy inference was introduced to tune the PID parameters to compensate for the hysteresis and therefore improve the control performance. The experimental results showed that the self tuning fuzzy PID controller demonstrated better tracking response than the conventional PID controllers and the nonlinear phenomena disappeared in the closed loop control. FLC is a comparatively simple method to construct a nonlinear controller and demonstrates a great ability to deal with imprecise or uncertain information in SMA models. Nonetheless, the fuzzy rules are designed based on intuition or practical experience.

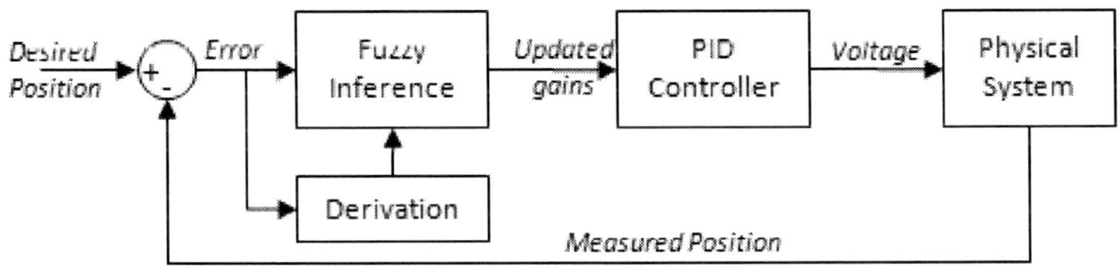

Figure 4. Block diagram for FLC

Esfahani et al. [68] implemented the adaptive PID controller introduced by Chang [69] for SMA systems. This adaptive PID controller is motivated from sliding mode control. In this approach, the PID control gains, are adjustable parameters and will be updated online with an adaptation mechanism to minimize a sliding condition. By introducing a supervisory controller, the stability of the closed-loop PID control system under the effect of plant uncertainty and external disturbance is guaranteed too.

The adaptation law that is based on the use of gradient method and minimizes a designed sliding condition for updating the PID control gains. An extra supervisory controller is also combined with the adaptive controller. It will be fired only when the states of the system exceed some bound in order to guarantee stability of the system [68-70]. The schematic of this controller is shown in Figure 5.

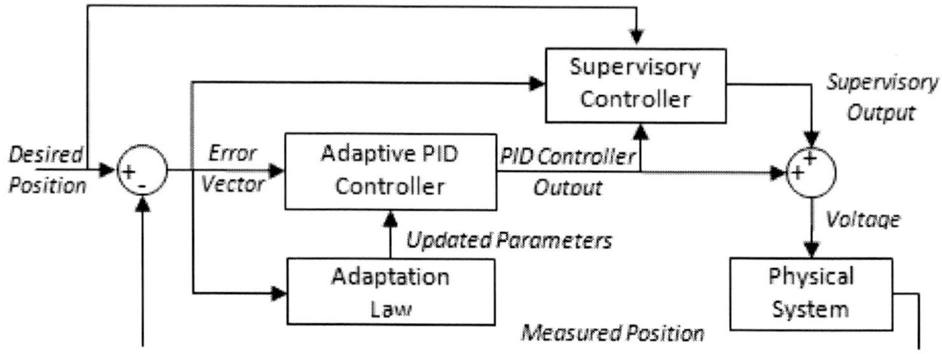

Figure 5. Block diagram of Adaptive PID controller [64]

3.3 Other Control Methods

Several other control strategies have also been investigated by researchers. Jayender et al. [62] proved that a H_∞ loop-shaping robust controller shows better performance when the disturbance is caused by blowing air over the SMA wire and when the experiment is carried out in milk which provides a good indication of the feasibility of SMA actuator in blood. Silva et al. [71] presented the beam shape control for the deflection of a flexible beam the heating or cooling process. The design is simple and step functions, sinusoidal functions as well as triangular functions as inputs were tested. Selden et al. [72] presented hysteresis loop control (HLC) for SMA based on the segmented binary control (SBC). The hysteresis nonlinearity captured by Schmidt-Trigger type bi-stable switching was exploited instead of being avoided. Less latency time (which means the increase of bandwidth) and lower energy consumption were achieved by adopting the intermediate 'hold' temperature. This allowed for thicker SMA wires to be used.

3.4 Hybrid Control Design

Some researchers have combined control goals and/or controller methodologies in the form of hybrid controllers for SMA system.

Song et al. [48] developed a neural network feed forward controller and a sliding-mode based robust feedback controller to compensate for the hysteresis in SMA actuators. As shown in Figure 6, the control system consists of three components: a neural network feed forward part which is used to offset for the environmental heat losses and provides a voltage to heat the SMA wires. The second part is a proportional derivative (PD) controller which is used to increase the damping and to stabilize the system. The third part is a sliding mode based robust compensator which is used for compensating for the hysteresis, external disturbance and modeling uncertainties.

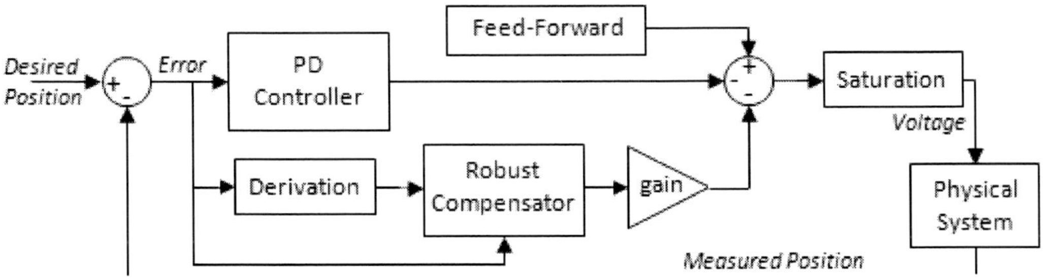

Figure 6. Block diagram for SMBRC

Iterative and feedback control were adopted by Leang et al. [41] and jointly to improve the position tracking performance in dealing with the hysteresis property in an SMA actuator. Preisach hysteresis model was employed to describe the hysteresis property. An iterative controller was designed to compensate for the hysteresis and a feedback controller was constructed to reduce the sensitivity of output to a small change in input. The implemented controller did not need an accurate model and the tracking error of an iterative controller diminishes as the number of the operating cycles increase.

Tarkesh and Elahinia [73] improved a stress-based SMC controller by combining it with a PID position controller. With the stress-based SMC controller, it takes more time to reach the desired position, but as it does, it can follow the trajectory very well. The reason for the delay in time is that while the temperature of the wire is less than Martensite starting transformation temperature, no actuation takes place. Therefore, to shorten this delay time it is important to apply a high voltage first and when the transformation begins, the controller takes its role. When the system is far from the sliding surface, a PID will control and as the system approached the sliding surface, the control system will switch to the sliding mode controller. This combined controller is shown to be both robust and precise in tracking. The proposed switching controller is shown schematically in Figure 7.

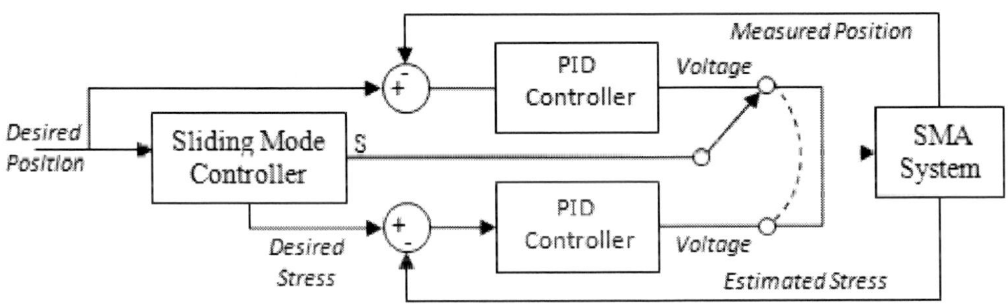

Figure 7. Block diagram for SMC-PID switching control

Moallem et al. [15] achieved tracking control of an antagonistic SMA actuator by using combination of torque and position control. In Figure 8, both the position feedback control and the torque feedback control are designed to collaboratively deal with the nonlinear property and thermal characteristics of the SMA wires. An inversion-based controller was used to serve as the inner-loop torque control and to take the outer-loop position tracking as an input. The experimental results showed good tracking on position and convergent tracking on torque. The proposed controller does not depend on the characteristics of SMA wire and only an approximate model was sufficient. In general, an antagonistic configuration provides better controllability than the bias-type actuators. Smooth torque signals were obtained even though there were rapid changes in the input.

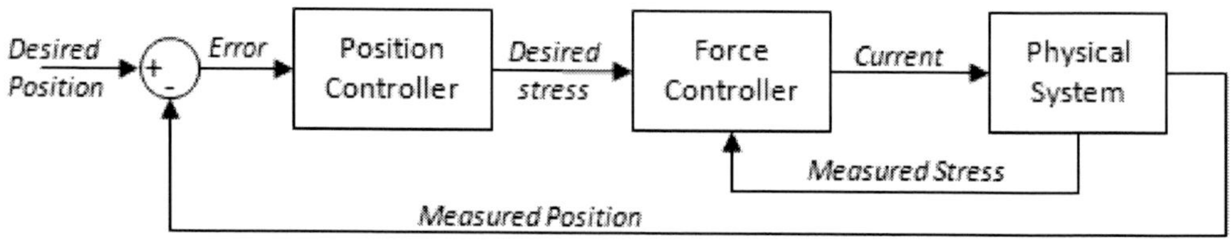

Figure 8. Block diagram of position and force/torque controllers

4. Comparison of the Control Strategies

Various controller designs have been investigated and simulations and experiments have been conducted in the past. It can be observed that the nonlinear controllers which take in to the account the nonlinear properties of the SMA provide better performance than the linear ones. Here in this section we have implemented some of the reviewed controllers on an SMA rotary manipulator which is shown in Figure 9-a.

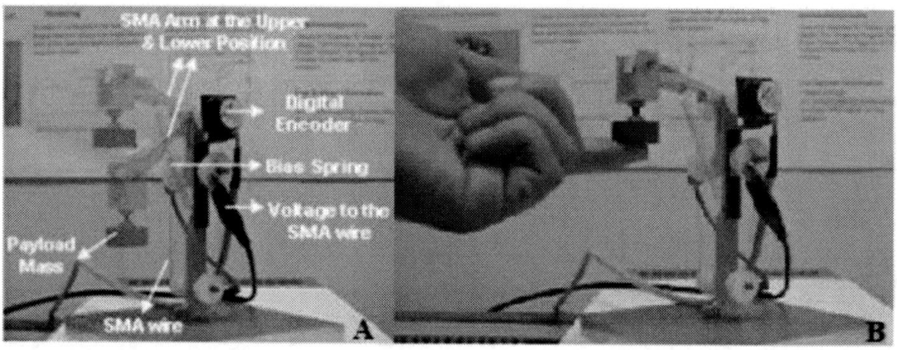

Figure 9. The experimental setup used for comparing the control systems.

This system shown in Figure 9-a can provide up to 135^0 angular displacement. By adjusting the electrical current to the SMA wire it is possible to rotate the mechanical linkage to track desired trajectories. The SMA wire is Flexinol with the diameter of 150 μm. Since single SMA wire provides one way action, a linear sprig is used to provide bias force.

To compare the selected controllers, two criteria are chosen; the first one is the performance of the controller in tracking the signal (measured by mean absolute value of the error over time) and the second one is the robustness of the controller in tracking a desired trajectory while subjected to external disturbances. To disturb the experimental system, as shown in Figure 9-b, we physically prevent the manipulator to move downward as SMA wire cools down for a short period of time.

Simulation result for an adaptive PID controller in tracking a sine wave trajectory with varying frequency (0.1 ~ 0.3 Hz) and different amplitude (0.25 ~0.5 rad) is shown in Figure 10.

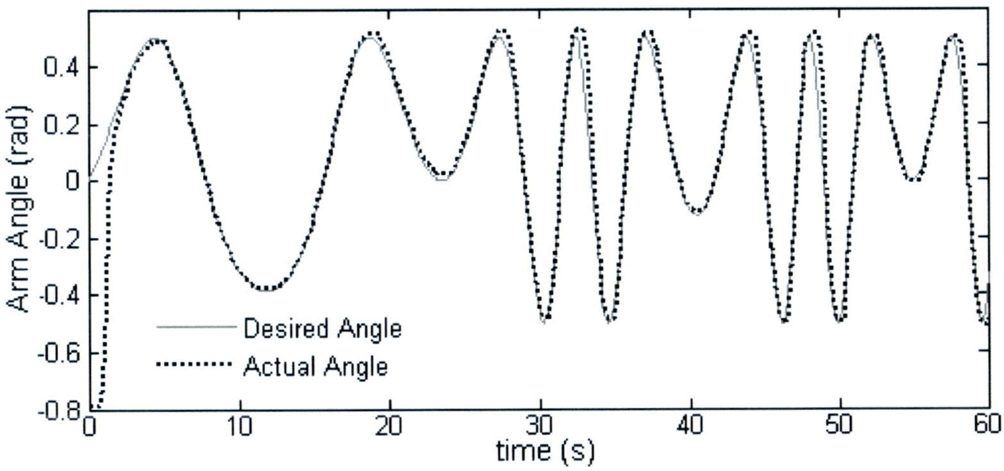

Figure 10. Adaptive PID tracking a sine wave with variable amplitude and frequency

For comparison purpose we have used the root square mean of the error (RMS Error) in tracking the same desired trajectory as shown in Figure 10. The selected controllers are 'adaptive PID' [68], 'Feadforward with Neural Network' [48], 'Hybrid Control' [45], 'PID-P3' [64] and 'SMC-PID' [73]. The comparison is shown in Figure 11.

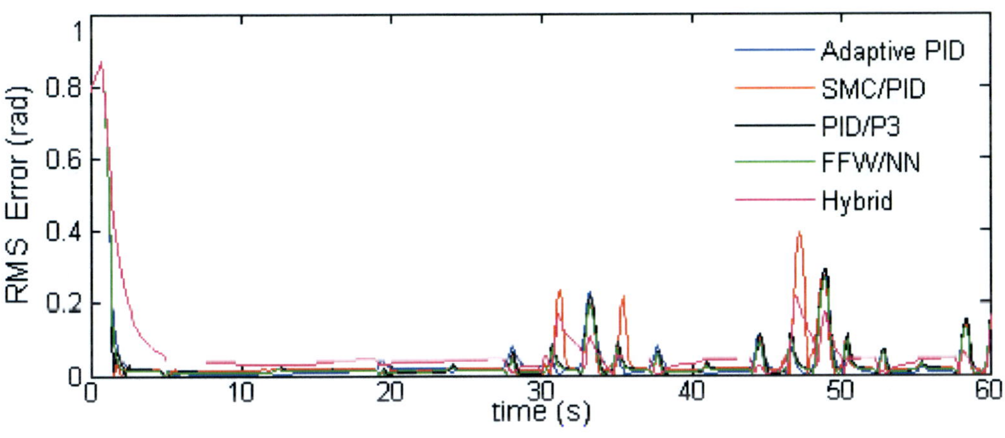

Figure 11. Comparison the Root Mean Square of Tracking Error

Figure 11 can be studied in three different regions: 1-the first 5 seconds where the system has an initial error (Figure 12). 2-at time range of 10~25 seconds where desired trajectory has a low frequency of 0.1 Hz (Figure 13) 3-the last 20 second where frequency of the desired trajectory increase to 0.3 Hz (Figure 14).

Performance of number of controllers subjected to large initial error is compared in Figure 12. During the first 0.8 seconds, temperature of the wire is less than starting transformation temperature, therefore error of the system increase due to no phase transformation in material structure. After this point, by increasing the desired angle, the performance of the controller will be highly depended on the speed of heating. As shown in Figure 12 controllers such as PID-P3, SMC-PID and FFW-NN which are equipped with some kind of preheating mechanism, apply very large input when they are far from desired states. Therefore the error in these controllers converges to zero faster than the other controllers.

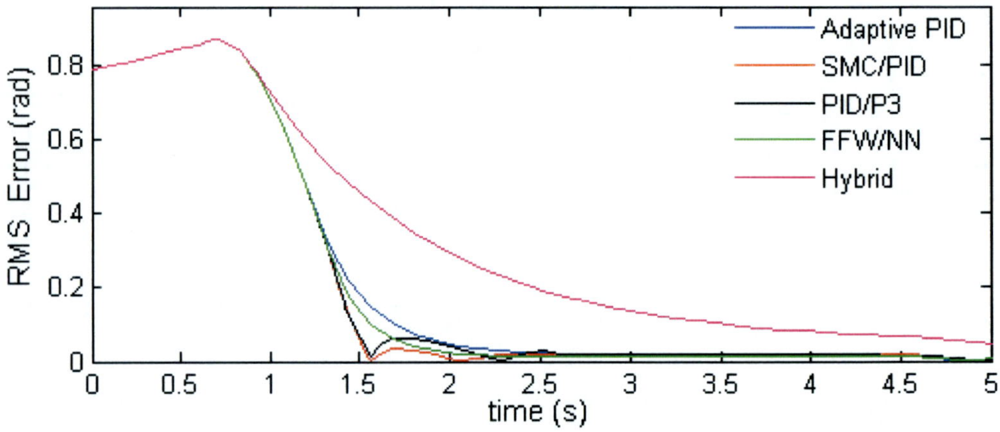

Figure 12. RMS Error at the starting time with initial error

Figure 13 and Figure 14 compare the error of controllers at two different time periods. In period of 10~25 seconds the frequency of the desired input (Figure 11) is equals to 0.1 Hz and as it appears in Figure 13 the maximum RMS error is around 0.03 rad. In the second period of time (40-60 seconds) by increasing the frequency of desired trajectory to 0.3 Hz the maximum error jumps to 0.3 rad (Figure 14). This is one of the main disadvantages of shape memory alloy actuators to which is referred as low bandwidth or slow actuation. Slow actuation is mostly due to passive cooling of shape memory alloys and one way actuation of SMA wires. However by using two SMA wires actuating in different direction we can successfully track desired trajectory with a frequency up to 2 Hz.

Adding another SMA wire will improve the bandwidth but at the same time we will need more electric energy for actuation. Since the actuation of shape memory alloys is very low energy efficiency mechanism (~2%) adding another set of wires will reduce the efficiency more and more.

Second criterion for comparing selected controllers is their robustness against external disturbance. External disturbance is applied as an external force that stop the rotation of SMA arm for a short period of time. As shown in Figure 9-b this disturbance is applied during the downward rotation (when SMA wire is cooling down) to avoid overheating of the SMA actuator. This disturbance causes an increase in angular error during the cooling process. Because of passivity of the system during the cooling process, the response of different controllers to increasing angle error is the same: applying no electric voltage. However after removing the external force (disturbance) position controllers and force controllers (or hybrid controllers) will responds differently with respect to each other.

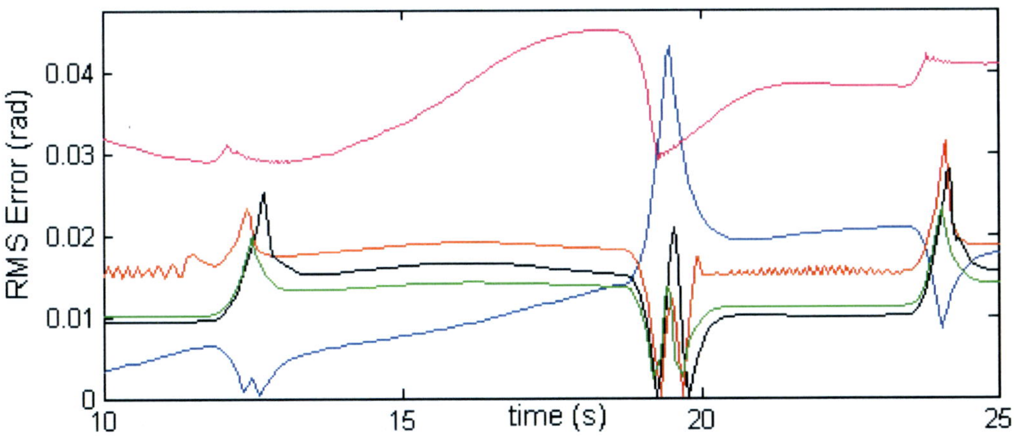

Figure 13. RMS Error during lower frequency (0.1 Hz)

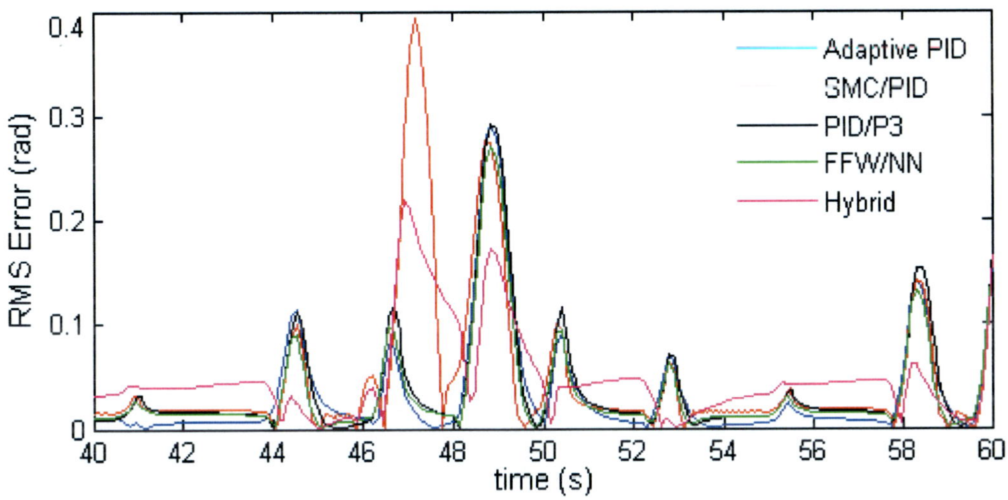

Figure 14. RMS Error during higher frequency (0.3 Hz)

By removing the external force in rising part of the desired trajectory, the spring forces the arm to go down quickly. During this sudden motion, the error of the angle decrease and therefore the output of position controllers stay zero (since there is no negative current to apply or any active cooling process). Output of position controllers remains zero until the actual angle gets smaller than the desired angle. From this point on the controller can adjust the electric current to track the desired angle as before (Figure 15). Figure 15 shows the performance of the nonlinear position controllers (adaptive PID, feed forward neural network, PID-P3) when the external disturbance is applied.

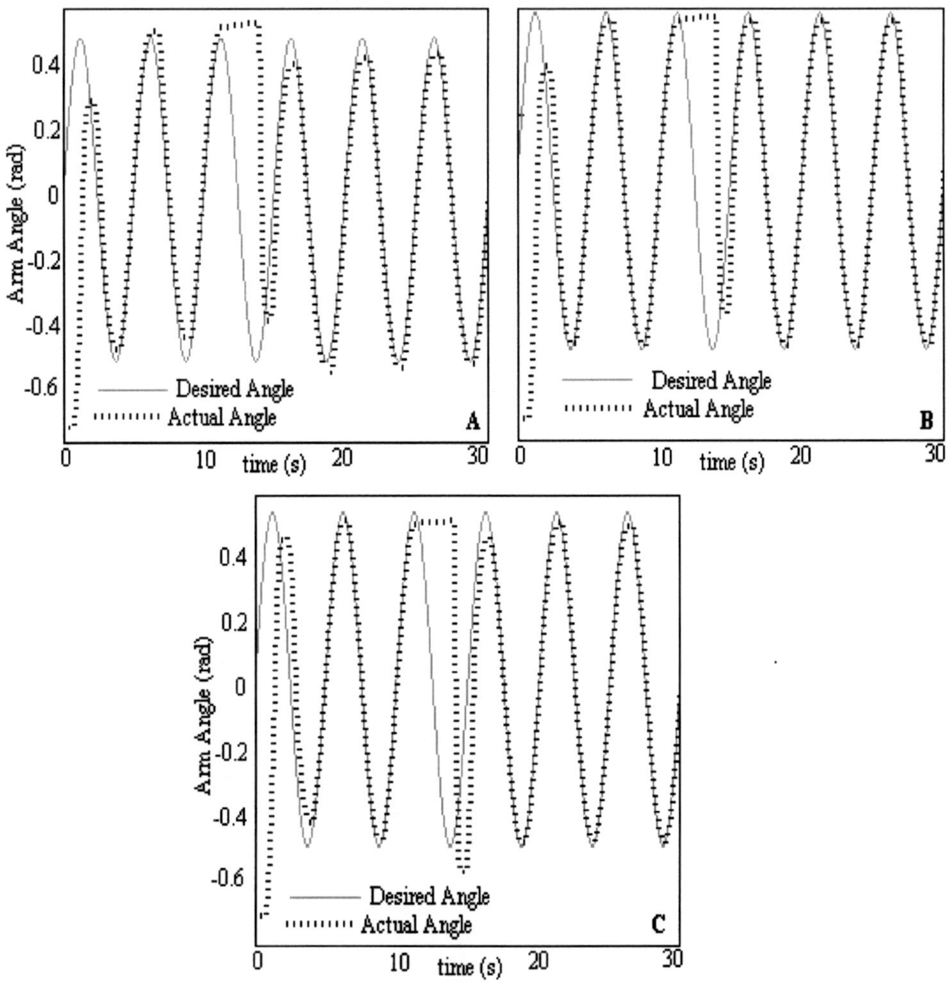

Figure 15. Performance of some nonlinear displacement controllers subjected to external disturbance. A) Adaptive PID controller, B) Feed forward neural network controller, C) PID-P3 Controller

While in position controllers the error is only depended on the position, in hybrid controllers, error is a function of both angle and force. Therefore with hybrid controllers by removing the external force, the arm goes down faster due to spring force. However during the sudden fall down, the force's error increases. This increase in force's error will balance the decrease in angle's error and set the total error to zero even if the desired angle still smaller than actual angle.

The performance of some of hybrid controllers subjected to external disturbance is shown in Figure 16.

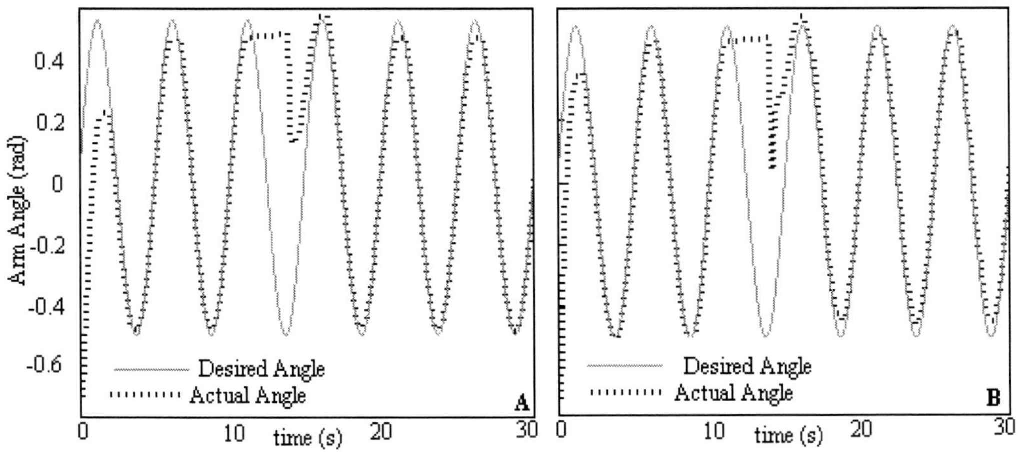

Figure 16. Performance of some nonlinear force-displacement controllers subjected to external disturbance. A) Hybrid controller B) Sliding mode-PID switching controller

5. Conclusion

Many valuable contributions have been made to address the control issues of SMA actuators but it is still a challenging topic. There are still fundamental difficulties open to the researchers: 1) modeling and controlling the nonlinear behavior of SMA actuators; 2) responding speed and 3) external disturbance even with complicated models and complex control strategies are presented.

This chapter investigated the various control problems for shape memory alloy actuators in many applications. A review of different controller designs is displayed. Some of the nonlinear controllers are compared in tracking a sine wave at different frequencies and amplitudes. It was shown that because of passive cooling structure controlling of shape memory alloys is limited to low frequencies. Considering all the novel works which are done in controlling shape memory alloys, there are some criteria which improve the performance of the controller. These criteria are as follow:

1. The actuation in shape memory alloys is due to the phase change in crystal structure. To start the phase transformation the system needs to reach to starting transformation temperature. Therefore a controller should include a rapid heating mechanism that reduces the respond time of the system. This rapid heating mechanism should be design in such a way to avoid overheating the SMA elements.

2. Conventional controllers such as PID may require online tuning involving a change of PID gains at different operating points. However using an adaptation law will help the system to adjust its parameter to maintain a stable behavior. Having an adapting mechanism in SMA controller is one of the most importance requirements that enable the SMA system to deal with uncertainties and unmodeled parameters of the system.

3. Because SMA materials mostly provide one way actuation, a biasing force is necessary to return them to the neutral position. During the cooling process where the bias force is acting the controller output is zero. This will cause the system to be sensitive to disturbances during cooling process. To reduce this sensitivity the control strategies should be more comprehensive by combining position control, force control, and temperature control. The design should take into account a more general picture of the whole system.

REFERENCES

[1] Epps, J.J. and Chopra, I., "In-flight tracking of helicopter rotor blades using shape memory alloy actuators," *Smart Materials and Structures*, vol. 10, 2001, pp. 104-111.

[2] Esfahani, E.T. and Elahinia, M.H., "Stable Walking Pattern for an SMA-Actuated Biped," *IEEE/ASME Transactions on Mechatronics*, vol. 12, 2007, pp. 534-541.

[3] Maclean, B., Patterson, G. and Misra, M., "Modeling of a Shape Memory Integrated Actuator for Vibration Control of Large Space Structures," *Journal of Intelligent Material Systems and Structures*, vol. 2, Jan. 1991, pp. 72-94.

[4] Maclean, B., Draper, J. and Misra, M., "Development of a Shape Memory Material Actuator for Adaptive Truss Applications," *Journal of Intelligent Material Systems and Structures*, vol. 2, Jul. 1991, pp. 261-280.

[5] Maeda, S., Abe, K., Yamamoto, K., Tohyama, O. and Ito, H., "Active endoscope with SMA (Shape Memory Alloy) coil springs," *Proceedings of the 9th IEEE MEMS workshop*, 1996, pp. 290-295.

[6] Roglin, R.L. and Hanagud, S.V., "A helicopter with adaptive rotor blades for collective control," *Smart Materials and Structures*, vol. 5, 1996, pp. 76-88.

[7] Romano, M., Agrawal, B. and Zazzera, F., "Experiments on Command Shaping Control of a Manipulator with Flexible Links," *Journal of Guidance, Control, and Dynamics*, vol. 25, 2002, pp. 232-239.

[8] Rustighi, E., Brennan, S.V. and Mace, B.R., "A shape memory alloy adaptive tuned vibration absorber: design and implementation," *Smart Materials and Structures*, vol. 14, 2005, pp. 19-28.

[9] Song, G., Lam, P., Srivatsan, T., Kelly, B. and Agrawa, B., "Application of shape memory alloy wire actuator for precision position control of a composite beam," *Journal of Materials Engineering and Performance*, vol. 9, Jun. 2000, pp. 330-333.

[10] Stevens, J.M. and Buckner, G.D., "Actuation and Control Strategies for Miniature Robotic Surgical Systems," *Journal of Dynamic Systems, Measurement, and Control*, vol. 127, Dec. 2005, pp. 537-549.

[11] Wang, G. and Shahinpoor, M., "Design of a knee and leg muscle exerciser for paraplegics using a shape memory alloy rotary joint actuator," *Proceeding of Smart Structures and Materials SPIE 1998*, San Diego, CA, USA, pp. 193-201.

[12] Wilde, K. Gardoni, P. and Fujino, Y., "Base isolation system with shape memory alloy device for elevated highway bridges," *Engineering Structures*, vol. 22, Mar. 2000, pp. 222-229.

[13] Williams, E. and Elahinia, M.H., "An Automotive SMA Mirror Actuator: Modeling, Design, and Experimental Evaluation," *Journal of Intelligent Material Systems and Structures*, vol. 19, No. 12, Sep. 2008, pp. 1425-1434.

[14] Williams, K., Chiu, G., and Bernhard, R., "Adaptive-Passive Absorbers Using Shape-Memory Alloys," *Journal of Sound and Vibration*, vol. 249, Jan. 2002, pp. 835-848.

[15] Slatkin, A.B., Burdick, J. and Grundfest, W., "The development of a robotic endoscope," *Proceedings of the 1995 IEEE/RSJ International Conference on Intelligent Robotic and Systems*, 1995 pp.162-171, Pittsburgh, PA

[16] Hashimoto, M., Tabata, T. and Yuki, T., "Development of electrically heated SMA active forceps for laparoscopic surgery," *Proceedings of IEEE International Conference on Robotics and Automation, 1999*, pp. 2372-2377.

[17] Peirs, J., Reynaerts, D. and Brussel, H.V., "Design of a shape memory actuated endoscopic tip," *Sensors and Actuators*, vol. 70, Oct. 1998, pp. 135-140.

[18] Pfeiffer, C., DeLaurentis, K. and Mavroidis, C., "Shape memory alloy actuated robot prostheses: initial experiments," *Proceedings of IEEE International Conference on Robotics and Automation, 1999*, pp. 2385-2391.

[19] Peng, F., Jiang, X., Hu, Y. and Ng, A., "Actuation precision control of SMA actuators used for shape control of inflatable SAR antenna," *Acta Astronautica*, vol. 63, Sep. 2008, pp. 578-585.

[20] Mabe, J.H., Ruggeri, R.T., Rosenzweig, E. and Yu, C.M., "NiTinol performance characterization and rotary actuator design," *Proceedings of Smart Structures and Materials SPIE 2004: Industrial and Commercial Applications of Smart Structures Technologies*, San Diego, CA, USA: 2004, pp. 95-109.

[21] Duerig, T., Pelton, A. and Stockel, D., "An overview of nitinol medical applications," *Materials Science and Engineering*, vol. A273-275, 1999, pp. 149-160.

[22] Es-Souni, M., Es-Souni, M. and Fischer-Brandies, H., "Assessing the biocompatibility of NiTi shape memory alloys used for medical applications," *Analytical and Bioanalytical Chemistry*, vol. 381, February. 2005, pp. 557-567.

[23] Saadat, S., Salichs, Y., Noori, M., Hou, Z., Davoodi, H., Bar-on, I., Suzuki, Y. and Masuda, A., "An overview of vibration and seismic applications of NiTi shape memory alloy," *Smart Materials and Structures*, vol. 11, 2002, pp. 218-229.

[24] Bernard, Y., Mendes, E. and Bouillault, F., "Dynamic hysteresis modeling based on Preisach model," *IEEE Transactions on Magnetics*, vol. 38, 2002, pp. 885-888.

[25] Hughes D. and Wen, J.T., "Preisach modeling of piezoceramic and shape memory alloy hysteresis," *Smart Materials and Structures*, vol. 6, 1997, pp. 287-300.

[26] Qiu, J., Tani, J., Osanai, V., Urushiyama, Y., Lewinnek, D., "High speed response of SMA actuators." *International Journal of Applied Electromagnetics and Mechanics*, vol. 12, 2000, pp 87-100.

[27] Shahin, A.R., Meckl, P.H., Jones, J.D., and Thrasher, M.A., "Enhanced Cooling of Shape Memory Alloy Wires Using Semiconductor "Heat Pump" Modules," *Journal of Intelligent Material Systems and Structures,* vol. 5, Jan. 1994, pp. 95-104.

[28] Tadesse, Y., Thayer, N. and Priya, S., "Tailoring the Response Time of Shape Memory Alloy Wires through Active Cooling and Pre-stress," *Journal of Intelligent Material Systems and Structures*, Nov. 2009.

[29] Sreekumar, M., Nagarajan, T., Singaperumal, M., Zoppi, M. and Molfino, R., "Critical review of current trends in shape memory alloy actuators for intelligent robots," *International Journal of Industrial Robot*, vol. 34, 2007, pp. 285 - 294.

[30] Flemming, L., and Mascaro, S., "Control of Scalable Wet SMA Actuator Arrays," *Proceedings of the 2005 IEEE International Conference on Robotics and Automation,* 2005, pp. 1338-1343.

[31] Mascar, S., and Asada, H., "Wet shape memory alloy actuators for active vasculated robotic flesh," *Proceedings of the 2003 IEEE International Conference on Robotics and Automation*, 2003, pp. 282-287.

[32] Ikuta, K., Tsukamoto, M., and Hirose, S., "Mathematical model and experimental verification of shape memory alloy for designing microactuator," *in Proc. IEEE Microelectromechanical Syst. (MEMS-91),* 1991, pp. 103-108.

[33] Madill, D., and Wang, D., "Modeling and L2-stability of a shape memory alloy position control system," *IEEE Transactions on Control Systems Technology*, vol. 6, 1998, pp. 473-481.

[34] Elahinia, M.H., Ahmadian, M., and Ashrafiuon, H., "Design of a Kalman filter for rotary shape memory alloy actuators," *Smart Materials and Structures*, vol. 13, 2004, pp. 691-697.

[35] Ahn, K.K., and Nguyen, B.K., "Position Control of Shape Memory Alloy Actuators Using Self Tuning Fuzzy PID Controller," *International Journal of Control, Automation, and Systems*, vol. 4, Dec. 2006, pp. 756-762.

[36] Ashrafiuon, H. and Jala, V.R., "Sliding Mode Control of Mechanical Systems Actuated by Shape Memory Alloy," *Journal of Dynamic Systems, Measurement, and Control*, vol. 131, Jan. 2009, pp. 011010-6.

[37] Ashrafiuon, H., Eshraghi, M. and Elahinia, M.H., "Position Control of a Three-link Shape Memory Alloy Actuated Robot," *Journal of Intelligent Material Systems and Structures*, vol. 17, May. 2006, pp. 381-392.

[38] Choi, S., "Position control of a single-link mechanism activated by shape memory alloy springs: experimental results," *Smart Materials and Structures*, vol. 15, 2006, pp. 51-58.

[39] Elahinia, M.H., and Ashrafiuon, H., "Nonlinear Control of a Shape Memory Alloy Actuated Manipulator," *Journal of Vibration and Acoustics*, vol. 124, Oct. 2002, pp. 566-575.

[40] Elahinia, M.H., Seigler, T.M., Leo, D.J. and Ahmadian, M., "Nonlinear Stress-Based Control of a Rotary SMA-Actuated Manipulator," *Journal of Intelligent Material Systems and Structures*, vol. 15, Jun. 2004, pp. 495-508.

[41] Leang, K.K., Ashley, S. and Tchoupo, G., "Iterative and Feedback Control for Hysteresis Compensation in SMA," *Journal of Dynamic Systems, Measurement, and Control*, vol. 131, Jan. 2009, pp. 014502-6.

[42] Ma, N. and Song, G., "Control of shape memory alloy actuator using pulse width modulation," *Smart Materials and Structures*, vol. 12, 2003, pp. 712-719.

[43] Ma, N., Song, G. and Lee, H., "Position control of shape memory alloy actuators with internal electrical resistance feedback using neural networks," *Smart Materials and Structures*, vol. 13, 2004, pp. 777-783.

[44] Majima, S., Kodama, K. and Hasegawa, T., "Modeling of shape memory alloy actuator and tracking control system with the model," *IEEE Transactions on Control Systems Technology*, vol. 9, 2001, pp. 54-59.

[45] Moallem, M. and Tabrizi, V., "Tracking Control of an Antagonistic Shape Memory Alloy Actuator Pair," *IEEE Transactions on Control Systems Technology*, vol. 17, 2009, pp. 184-190.

[46] Song, G., "Design and control of a Nitinol wire actuated rotary servo," *Smart Materials and Structures,* vol. 16, 2007, pp. 1796-1801.

[47] Song, G. and Ma, N., "Robust control of a shape memory alloy wire actuated flap," *Smart Materials and Structures*, vol. 16, 2007, pp. N51-N57.

[48] Song, G., Chaudhry, V. and Batur, C., "Precision tracking control of shape memory alloy actuators using neural networks and a sliding-mode based robust controller," *Smart Materials and Structures*, vol. 12, 2003, pp. 223-231.

[49] Song, G., Chaudhry, V. and Batur, C., "A Neural Network Inverse Model for a Shape Memory Alloy Wire Actuator," *Journal of Intelligent Material Systems and Structures*, vol. 14, Jun. 2003, pp. 371-377.

[50] Arai, K., Aramaki, S., and Yanagisawa, K., "Feedback linearization for SMA (shape memory alloy)," *Proceedings of the 34th SICE Annual Conference*. 1995, pp. 1383-1386.

[51] Damle, R., Lashlee, R., Rao, V. and Kern, F., "Identification and robust control of smart structures using artificial neural networks," *Smart Materials and Structures*, vol. 3, 1994, pp. 35-46.

[52] Elahinia, M.H., Koo, J. and Tan, H., "Improving robustness of tuned vibration absorbers using shape memory alloys," *Shock and Vibration*, vol. 12, 2005, pp. 349-361.

[53] Gorbet, R., Morris, K. and Wang, D., "Passivity-based stability and control of hysteresis in smart actuators," *IEEE Transactions on Control Systems Technology*, vol. 9, 2001, pp. 5-16.

[54] Rustighi, E., Brennan, M.J. and Mace, B.R., "Real-time control of a shape memory alloy adaptive tuned vibration absorber," *Smart Materials and Structures*, vol. 14, 2005, pp. 1184-1195.

[55] Song, G., Kelly, B. and Agrawal, B.N., "Active position control of a shape memory alloy wire actuated composite beam," *Smart Materials and Structures*, vol. 9, 2000, pp. 711-716.

[56] Liang, C. and Rogers, C. A. "One-dimensional thermo-mechanical constitutive relations for shape memory materials," *J. Intell. Mater. Syst. Struct.*, vol1, no. 2, pp 207-234, 1990.

[57] Brinson, L. C. and Huang, M. S. "Simplification and comparisons of shape memory alloy constitutive models." *Journal of intelligent material systems and structures*, 7: 108-114, January 1996

[58] Elahinia, M. H. and Ahmadian, M., "An Enhanced SMA Phenomenological Model. Part I. The Shortcomings of the Existing Models," *Journal of Smart Materials and Structures,* December 2005, 14 (6): 1297-1308.

[59] Elahinia, M. H. and Ahmadian, M., "An Enhanced SMA Phenomenological Model. Part II. The Experimental Study," *Journal of Smart Materials and Structures,* December 2005, 14(6): 1309-1319.

[60] Singh, K., Sirohi, J. and Chopra, I. "An improved shape memory alloy actuator for rotor blade tracking." *Journal of intelligent material systems and structures* 14, 2003, pp. 767-786.

[61] Moallem, M., and Lu, J. "Application of shape memory alloy actuators for flexure control: Theory and Experiments." *IEEE/ASME transactions on mechatronics* 10, 2005, pp. 495-501.

[62] Jayender, J., Patel, R.V., Nikumb, S., and Ostojic, M. "Modeling and control of shape memory alloy actuators." *IEEE transactions on control systems technology,* vol. 16, 2008, pp. 279-287.

[63] Lim, S.C-, Park, J. S., Choi, S. B., and Park., Y. P., "Non-contact start/stop motion control of HDD suspension using shape memory alloy actuators." *Smart materials and structures* 10, 2001, 1069-077.

[64] Shameli, E., Alasty, A. and Salaarieh, H. "Stability analysis and nonlinear control of a miniature shape memory alloy for precise application" *Mechatronics* 15 (2005) 471-486

[65] Song, G., Ma, N., "Control of Shape Memory Alloy Actuators Using Pulse-Width Pulse Frequency (PWPF) Modulation" *Journal of Inteligent Material System and Structures*, vol. 14, No. 1, 2003, pp. 15-22

[66] Grant, D. and Hayward, V. "Variable structure control of shape memory alloy actuator" *IEEE Control System Magazine*, vol.17, No. 3, 1997, 80-88.

[67] Nakazato, Y., Kato, T., and Masuda, T., "Position control of SMA actuator by fuzzy reasoning," Trans. Jpn. Soc. Mech. Eng., Ser. C, Vol. 59(565), 1993, 2733-2738

[68] Esfahani, E.T, Elahinia, M.H., "Developing an adaptive controller for a shape memory alloy walking assistive device" Journal of vibration and control, in Press.

[69] Chang, W. and Yan, J., 2005 "Adaptive robust PID controller design based on a sliding mode for uncertain chaotic systems" *Journal of Chaos, Solutions and Fractals,* 26:167-175.

[70] Kuo, T.C., Huang, Y.J. Chen, C.Y. and Chang, C.H., 2006 "Adaptive Sliding Mode Control with PID Tuning for Uncertain Systems" *Engineering Letters*, 16:3, EL_16_3_05

[71] Silva, E.P. D. "Beam shape feedback control by means of a shape memory actuator" *Materials & Design* 28, 2007, 1592-1596.

[72] Selden, B., Kyujin C., and Asada, H. H. "Segmented shape memory alloy actuators using hysteresis loop control" *Smart Material and Structures* 15 (2006): 642-652.

[73] Tarkesh, E., Elahinia, M. H. "Nonlinear control techniquies for a SMA active ankle foot orthosis" *In Proceding of ASME IMECE2007*, Seattle, WA, Nov 11-15, 2007.

In: Shape Memory Alloys
Editor: H. R. Chen, pp. 69-111
ISBN 978-1-60741-789-7

Chapter 3

FABRICATION AND SURFACE MODIFICATION OF POROUS NICKEL-TITANIUM SHAPE MEMORY ALLOY FOR BONE GRAFTS

Shuilin Wu[1, 2], Xiangmei Liu[1, 2], Paul K. Chu[1,1], Tao Hu[1], Kelvin W. K. Yeung[1] and Jonathan C. Y. Chung[1]

[1] Department of Physics and Material Science, City University of Hong Kong, Tat Chee Avenue, Kowloon, Hong Kong, China

[2] Ministry-of-Education Key Laboratory for the Green Preparation and Application of Functional Materials, School of Materials Science and Engineering, Hubei University, Wuhan, 430062, China

ABSTRACT

Equiatomic porous nickel-titanium shape memory alloys (NiTi SMAs) are becoming one of the most promising biomaterials for bone grafts because of their unique advantages over currently used biomaterials. For example, they have good mechanical properties and lower Young's modulus relative to dense NiTi, Ti, and Ti-based alloys. Porous NiTi SMAs are relatively easy to machine compared to porous ceramics such as hydroxyapatite and calcium phosphate that tend to exhibit brittle failure. The porous structure with interconnecting open pores can also allow tissue in-growth and favors bone osseointegration. In addition, porous NiTi alloys possess good shape memory effect (SME) and superelasticity (SE) similar to dense NiTi alloys. In order to optimize porous NiTi SMAs used in bone grafts, current research focuses on the fabrication methods and surface modification techniques to obtain adjustable bone-like structures with good mechanical properties and excellent superelasticity as well as a bioactive passivation layer on the entire exposed surface to block nickel leaching and enhance the surface biological activity. This chapter describes the recent progress in the fabrication process and surface modification methods of porous NiTi SMAs and discusses subsequent effects of these processes and techniques on the materials properties from the perspective of bone grafts applications.

1 Correspondence/Reprint request: Prof. Paul K. Chu, Department of Physics and Material Science, City University of Hong Kong, Tat Chee Avenue, Kowloon, Hong Kong E-mail: paul.chu@cityu.edu.hk.

ABBREVIATIONS AND SYMBOLS

SMAs	Shape memory alloys
SME	Shape memory effect
SE	Superelasticity
PM	Powder metallurgy
EPS	Elemental powder sintering
CS	Combustion synthesis
SHS	Self-propagating high temperature synthesis
SPS	Spark plasma sintering
CF-	HIP Capsule-free hot isostatic pressing
MIM	Metal injection molding
PIII	Plasma immersion ion implantation
EIS	Electrochemical impedance spectroscopy
SBF	Simulated body fluids
PBS	Phosphate buffered saline
CVD/PVD	Chemical/physical vapor deposition
XRD	X-ray diffraction
SEM	Scanning electron microscopy
XPS	X-ray photoelectron spectroscopy
OM	Optical microscopy
ICPMS	Inductively-coupled plasma mass
HA	Hydroxyapatite
EDS	Energy-dispersive spectroscopy
TF-XRD	Thin Film X-ray diffraction
MP	Mechanically polishing
SCE	Saturated calomel electrode
HBSS	Hanks' balanced salt solution
MPS	Mean pore size

1. INTRODUCTION

1.1. Necessity of Artificial Bone Implants

Bones in the skeletal system provide the supporting structure for the body. Bone is a structural composite composed of collagen fibers with hydroxyapatite nano-crystals precipitated along the collagen fibrils [1]. Bone also contains other constituents such as mucopolysaccharides, blood vessels, and bone cells. The low elastic modulus collagen fibers are aligned in bones along the main stress directions. The high elastic modulus hydroxyapatite minerals constitute approximately 70% of the dry bone mass and contribute significantly to the bone stiffness. Bone can remodel and adapt itself to the applied mechanical environment, which is generally known as Wolff's law [2]. However, bone will fracture under excessive loading or impact (schematically shown in Figure 1).

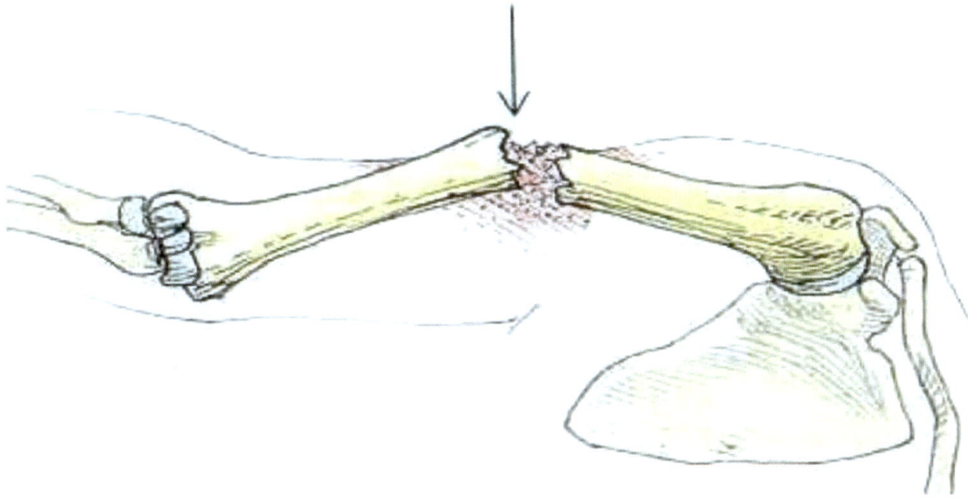

Figure 1. Schematic of bone fracture.

There are many types of bone fractures depending on the crack size, orientation, morphology, and location [3]. Ettinger, et al. have revealed that easy vertebral bone fractures in many postmenopausal women are due to osteoporosis [4]. Therefore, there is a necessity to replace bone substance that has been lost. The lost bone can be replaced by endogenous or exogenous bone tissues, but they are still not satisfactory. The use of an endogenous bone substance involves additional surgery [5]. In addition, the endogenous bone is available only in limited quantities [6]. The major disadvantage of exogenous bone implants is that they may be rejected by the human body and diseases may be transmitted together with these implants [6]. Furthermore, the clinical performance of exogenous bone is considerably inferior to fresh endogenous graft materials [5]. Therefore, there is a big potential market for artificial bone implants. According to the report of MedMarket Diligence [7], the worldwide market for orthopedic biomaterials such as allografts, synthetic bone substitutes, bone growth factors, polymers, ceramics, etc. is growing steadily due to a concurrent evolution of surgical procedures, the emergence of innovative products and a dynamic and growing patient population. The current orthopedic biomaterials market worldwide has an annual revenue of $5 billion with annual percentage increases in double digits and will reach $10 billion by 2011 (Figure 2). Although the orthopedic biomaterials sector is expanding, it is expected that the volume of materials required will never exceed tens of tons, compared to thousands of tons in other developing engineering markets [8].

1.2. Types, Structures and Properties of Bone

It is very important to know the types, structures, as well as physical, chemical, and mechanical properties of the bone tissues because they provide quantitative parameters necessary for the fabrication and design of artificial bone replacement implants. There are different types of bone in the human body, for example, long or short bone, regular or irregular bones, and compact or cancellous bones according to different standards [9].

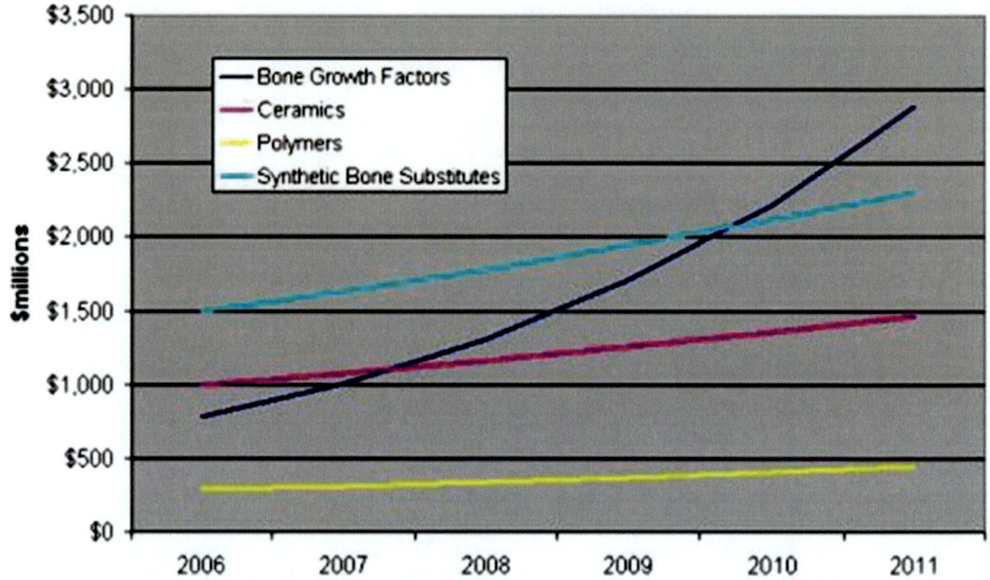

Figure 2. Growth of orthopedic biomaterials market segments (2006-2011) [7].

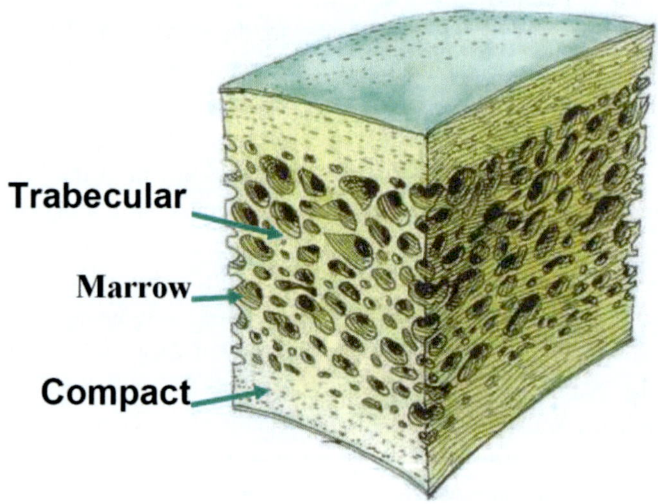

Figure 3. Schematic of osseous (bone) tissue [9].

Figure 3 depicts the structure of a typical bone structure. It is very difficult to generalize the bone structure because most bones have an elaborate construction. A "typical" bone is made up of an outer shell of dense compact bone enclosing a core of porous cellular, cancellous, or trabecular bone [10]. Human bone is mainly composed of collagen (20 wt. %), hydroxyapatite mineral (69 wt. %), and water (9 wt. %) [11]. In addition, other organic materials such as proteins, polysaccharides, and lipids are also present in small quantities [12]. Collagen, which can be considered as the matrix, is in the form of small microfibers. It is difficult to observe distinct collagen fibers because of its net-like appearance [12]. The diameter of the collagen microfibers varies from 100 to 2000 nm. Hydroxyapatite (HA) mineral provides stiffness to the bone. The HA crystals present in the form of plates or needles are about 40–60 nm long, 20 nm wide, and 1.5–5 nm thick [11-13], and the mineral phase present in the bone is not a discrete aggregation of the HA crystals, but rather is made of a continuous phase which is evidenced by a very good strength of the bone after complete removal of the organic phase [11].

The hierarchical levels of the structural organization in a human compact bone are illustrated in Figure 4. The mineral-containing fibers are arranged into lamellar sheets (3–7 mm thick). 4 to 20 lamella arranged in concentric rings around the Harversian canal form an osteon [11]. The cross sections of the compact bone showing cylindrical osteons with blood vessels running along Harversian canals (in the center of each osteon) are shown in Figure 5(A). The metabolic substances can be transported by the intercommunicating systems of canaliculi, lacunae, and Volkman's canals, which are connected with the marrow cavity [11]. The various interconnecting systems are filled with body fluids and their volume can be as high as 19% [11].

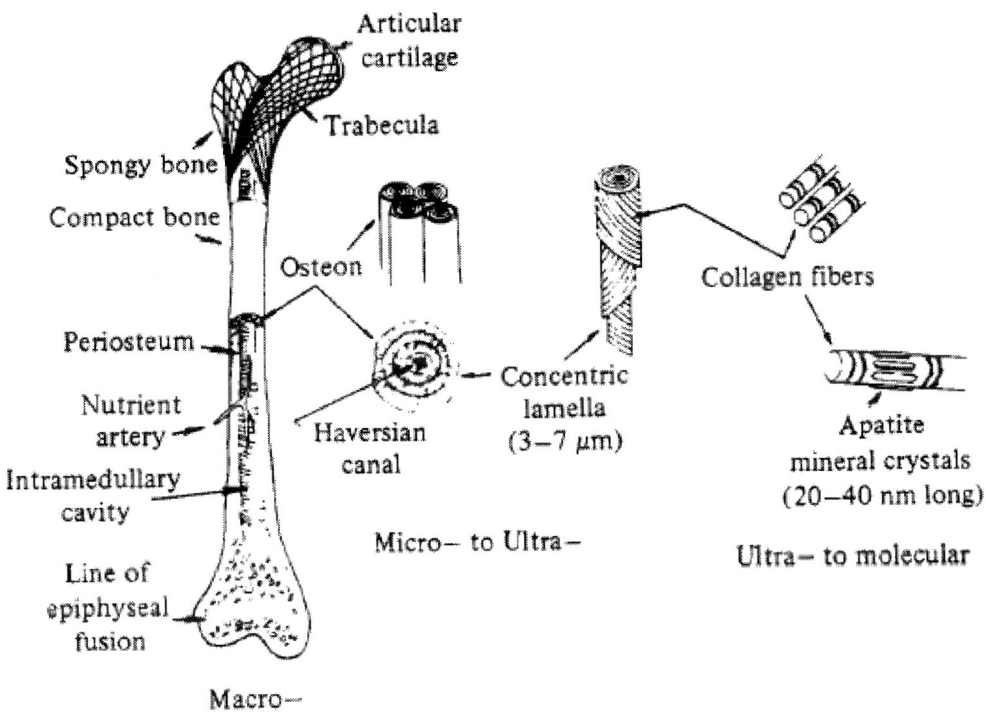

Figure 4. Hierarchical structures in a human long bone [8, 11].

The cellular structure of cancellous bone is shown in Figures 5(B), (C) and (D) [14]. It is known that cancellous bone is made up of an interconnected network of rods or plates. A network of rods produces low density, open cells whereas one of the plates gives high density, virtually closed cells. The relative density of cancellous bone varies from 0.05 to 0.7 (technically, any bone with a relative density less than 0.7 is classified as "cancellous") [14].

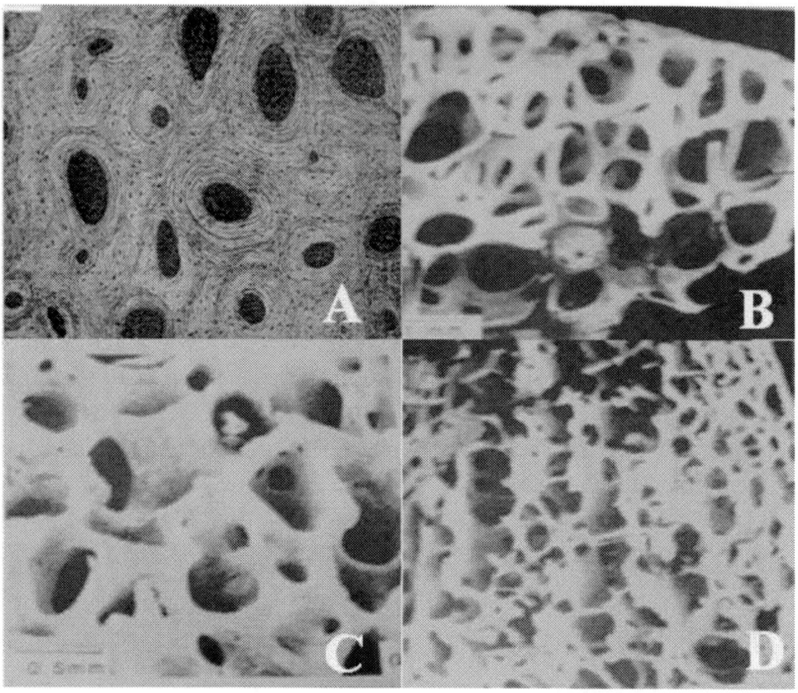

Figure 5. Images showing the structure of bones. (A) Optical micrograph of transverse cross section showing the microstructure of compact lamellar bone–human femora; (B) SEM image of a specimen taken from the femoral head, showing a low density, open-cell, rod-like structure of cancellous bone; (C) SEM image of a specimen from the femoral head, showing a higher-density, roughly prismatic cell structure of cancellous bone; (D) SEM image of a specimen from the femoral condoyle, of intermediate density, showing a stress-oriented, parallel plate structure of cancellous bone with rods normal to the plates [13, 14].

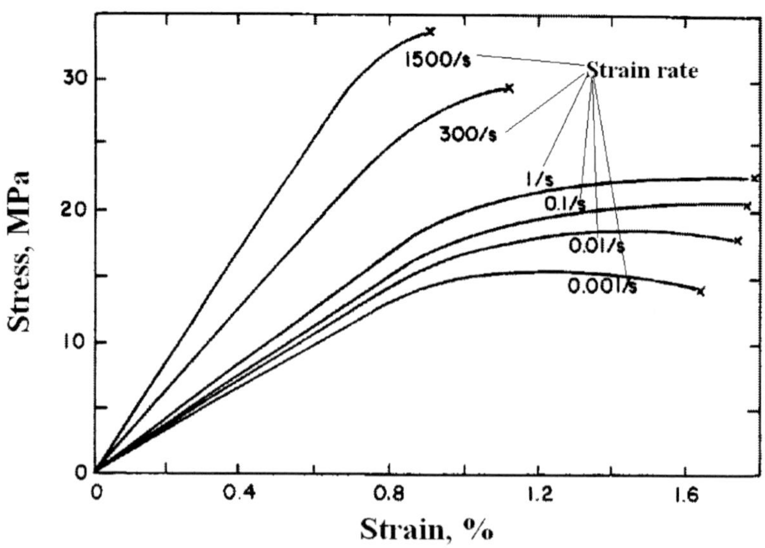

Figure 6. Stress-strain curve for human compact bone [11].

As shown in Figure 6, the stress-strain curves of compact bones exhibit a linear elastic region, followed by a flat plastic region at about 0.8% strain at lower strain rates. It can be found that bone exhibits a tough behavior at low strain rates but fractures more like a brittle material at high strain rates [11, 15]. The slope of the stress-strain curve, i.e., the elastic modulus of the bone, increases with increasing mineral content [12, 13]. At low strain rates, bone exhibits excellent toughness mostly due to its hierarchical structure (shown in Figure 4), which stops cracks after little propagation [12]. The properties of compact human bones are listed in Table 1.

Table 1. Mechanical properties of compact human bones [8]

	Test direction related to bone axis	
	Parallel	Normal
Tensile strength (MPa)	124–174	49
Compressive strength (MPa)	170–193	133
Bending strength (MPa)	160[a]	
Shear strength (MPa)	54	
Young's modulus (GPa)	17.0–18.9	11.5
	20–27 (random)	
Work of fracture (J/m^2)	6000 (low strain rate)	
	98 (high strain rate)	
K_{Ic} ($MPa \cdot m^{1/2}$)	2–12[a]	
Ultimate tensile strain	0.014–0.031	0.007
Ultimate compressive strain	0.0185–0.026	0.028
Yield tensile strain	0.007	0.004
Yield compressive strain	0.010	0.011

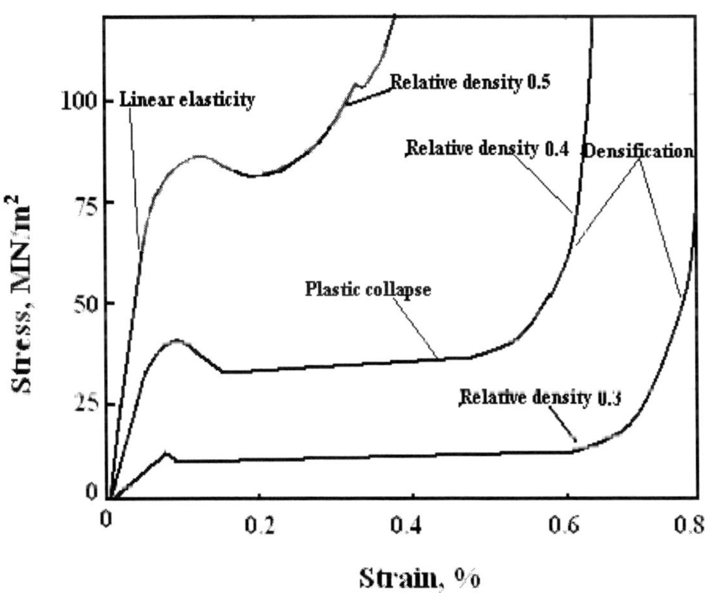

Figure 7. Compressive stress-strain curves for several relative densities of wet cancellous bone [19]

The compressive stress-strain curve of cancellous bone is shown in Figure 7. The curves exhibit three regimes of behavior, namely the linear-elasticity, plastic collapse-plateau, and densification areas. The small strain and linear-elastic response of low-density of the near-isotropic cancellous bone originate from the elastic bending of the cell walls [16-18]. The linear-elastic regime ends when the cells begin to collapse. Progressive compressive collapse gives the long, horizontal plateau of the stress-strain curve which continues until opposing cell walls meet and touch, causing the stress to rise steeply [19]. The main mechanical properties of a cancellous human bone are shown in Table 2.

Table 2. Solid cell-wall properties for a cancellous human bone [18]

Property	Value
Young's modulus	12GPa
Compressive strength	136MPa
Tensile strength	105MPa

1.3. Niti Shape Memory Alloys (Smas) in Bone/Hard Tissue Replacement

In recent years, rapid development of biomaterials such as biopolymers [2], bioceramics [20], composites materials [2], and hybrid materials [8] may falsely give us an impression that there is no longer any future for metals and alloys as biomaterials due to their intrinsic shortcomings such as corrosion, wear resistance, tissue response and systemic biocompatibility. However, in the past three decades, research and innovations in materials engineering have led to considerable improvement in metal and alloy properties as well as new types of products for a wide area of medical applications. Some medical applications of biometals are shown in Table 3 [21, 22].

Table 3. Examples of biometals and their applications [21, 22]

Material	Principal applications
Metals and alloys	
316L stainless steel	Fracture fixation, stents, surgical instruments
CP-Ti, Ti-Al-V, Ti-Al-Nb, Ti-13Nb-13Zr, Ti-Mo-Zr-Fe	Bone and joint replacement, fracture fixation, dental implants, pacemaker encapsulation
Co-Cr-Mo, Cr-Ni-Cr-Mo	Bone and joint replacement, dental implants, dental restorations, heart valves
Ni-Ti	Bone plates, stents, orthodontic wires
Gold alloys	Dental restorations
Silver products	Antibacterial agents
Platinum and Pt-Ir	Electrodes
Hg-Ag-Sn amalgam	Dental restorations

Among these biometals, near-equiatomic NiTi shape memory alloys have recently led to rather spectacular medical applications and achieved the largest commercial success [23-26]. The combination of good biocompatibility, good strength, and ductility with specific functional properties such as the shape memory effect (SME) and superelasticity (SE) has made it a unique biomaterial. In NiTi SMAs, SME is based on the temperature-induced transformation whereas SE is based on the stress-induced transformation [26]. In particular, the SE of NiTi SMAs results in a unique combination of high strength, high stiffness, and high pliability, and very few other materials or technology can offer such unique combinations. The important mechanical properties of NiTi SMAs are listed in Table 4. At present, NiTi SMAs are used in orthodontic arch wires [27], dental root implants [28], stents [29], orthopedic implants [30], and some medical instruments/tools [31]. However, their high Young's modulus causes mismatch with human bones, thereby hampering wider applications in bone implants. Therefore, it is necessary to decrease the Young's modulus of this alloy but not impair its good mechanical properties, unique SME and SE.

Table 4. Mechanical properties of NiTi SMAs [22]

Property	Young's Modulus	Fatigue strength N=106	Ultimate tensile strength	Transformation temperatures	Maximum recovery stress
Value	80GPa	350MPa	700-1100(annealed) 1300-2000(not annealed)	-100-100oC	600-900MPa

A low stiffness close to that of the natural bone can provide better load transfer whereby stimulation of new bone formation is released and high stiffness results in bone resorption. Therefore, in recent years, porous NiTi shape memory alloys are becoming one of the most promising biomaterials. Besides a lower Young's modulus and good mechanical properties, the porous structure with interconnected open pores is suitable for tissue ingrowth and transport of nutrition [32, 33]. In addition, this structure favors the fixation of bone implants [33-35]. Simske, Rhalmi and Assad et al. [34-38] have disclosed that this porous NiTi SMA has no apparent toxicity to cells and tissues. Therefore, various fabrication processes have been developed to produce porous NiTi SMAs products as discussed in Section 2.

2. Fabrication Methods

It can be found from the Ni-Ti phase diagram in Figure 8 [39] that equiatomic NiTi SMAs have a high melting point of about 1310 ^{o}C. Only powder metallurgical methods are thus suitable for the fabrication of porous NiTi SMAs.

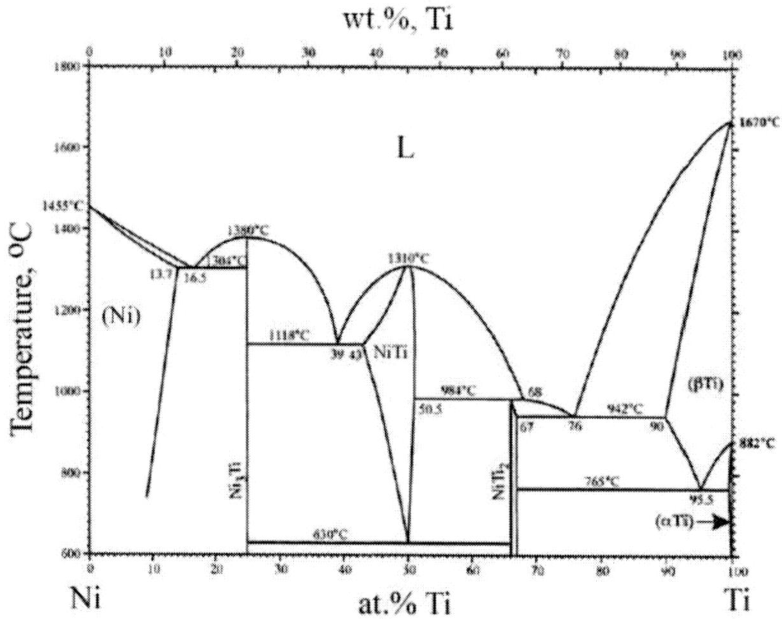

Figure 8. The Ni-Ti phase diagram [39].

All the PM processes are based on the following reactions [40].

$$Ni + Ti \rightarrow NiTi + 67\ kJ/mol\ (1),\ \ Ni + Ti \rightarrow Ti_2Ni + 83\ kJ/mol \qquad (2)$$

$$Ni + Ti \rightarrow Ni_3Ti + 140\ kJ/mol \qquad (3)$$

The reported PM processes can be described as followings.

2.1. Elemental Powder Sintering (EPS)

Traditional elemental powder sintering (EPS) in an argon atmosphere is one of the simplest methods. The process involves the use near-equiatomic elemental nickel and titanium as raw materials. The fully mixed powders are compressed at different cold pressures and the compacts are finally sintered under vacuum or argon atmosphere at higher temperatures for various time durations. The sintering temperature and sintering time influence the phase composition, pore size, and mechnical properties. The reported maximum porosity achieved by this method is below 45 vol%. However, the porous NiTi produced by EPS does not possess large recovery strain [40, 41]. In addition, the pore size cannot be controlled precisely, and the sintered products have more secondary phases which impair the SE and mechanical properties of these porous SMAs. The typical morphologies are displayed in Figure 9.

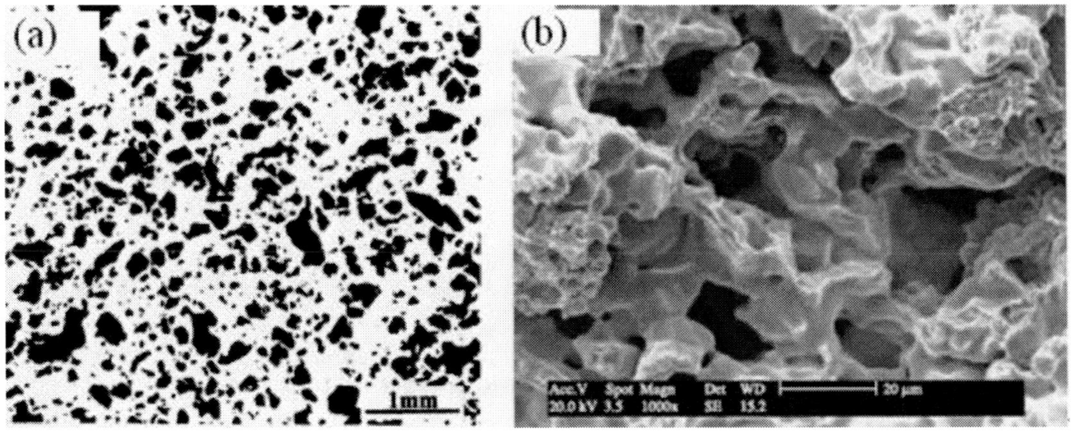

Figure 9. Typical images of porous NiTi produced by EPS. (a) Optical image [40]; (b) SEM image [41].

2.2. Combustion Synthesis (CS)

Combustion synthesis (CS) is often used to fabricate porous NiTi SMAs in the early stage due to its simplicity. In the CS method, the fully blended Ni and Ti powders with near equiatomic ratio are pressed into cylindrical pellets under different pressures and then put into a tube furnace under a protective gas preheated to various temperatures, and finally ignited. In comparison with EPS, the porosity of porous NiTi SMAs by CS increases in the range of 50-70 vol% and the size of most pores ranges from 200–600 μm. The typical images of porous NiTi SMAs produced by CS are shown in Figure 10.

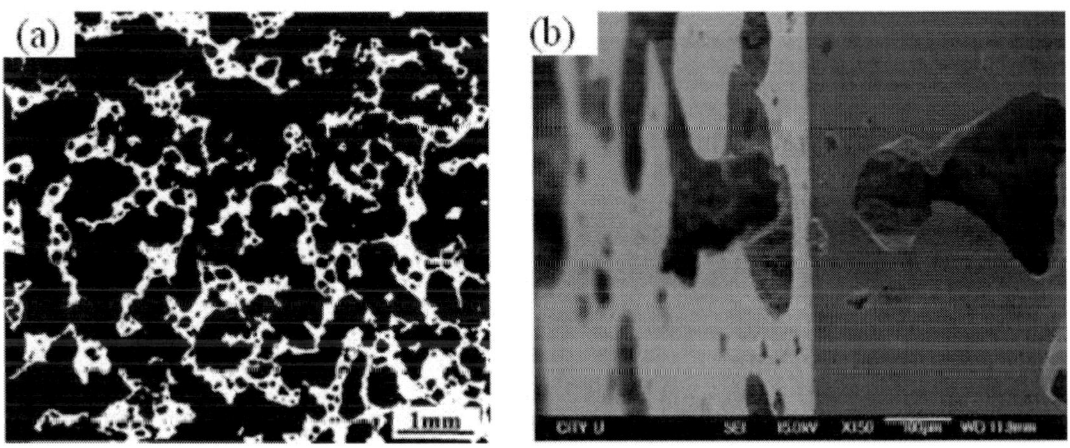

Figure 10. Typical images of porous NiTi produced by CS. (a) Optical image [42]; (b) SEM image [43].

The shortcoming of this process is that the compressive strength of the porous NiTi alloy produced by CS is low due to the irregular pore shape and distribution [43, 44]. The compressive behavior is described by the stress-strain curve in Figure 11.

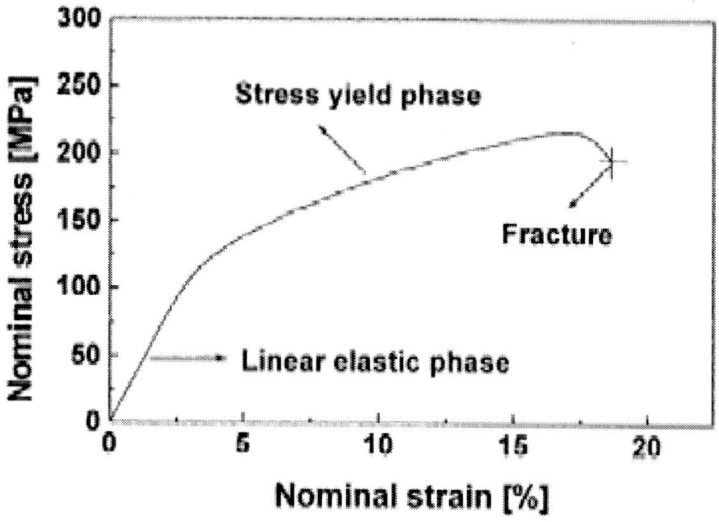

Figure 11. Compressive stress-strain curve of porous NiTi SMAs produced by CS [44].

2.3. Self-Propagating High Temperature Synthesis (SHS)

Self-propagating high-temperature synthesis (SHS) is often used to synthesize ceramics and some intermetallics [45-49] including NiTi intermetallic compounds. In comparison with conventional processes such as EPS and CS, the SHS method provides advantages with respect to time, energy savings, and easier processing, and the porosity and pore shape can be controlled easily. This method is similar to CS. The difference is that the exothermic reactions (1), (2) and (3) carried out along the green bar. Using this method, anisotropic pore structures can be easily achieved and it can be attributed to the convective flow of liquid and argon during combustion. Li et.al. used this method to synthesize porous NiTi SMAs with larger porosity and linear channels [50]. The typical micrographs of porous NiTi SMAs are depicted in Figure 12.

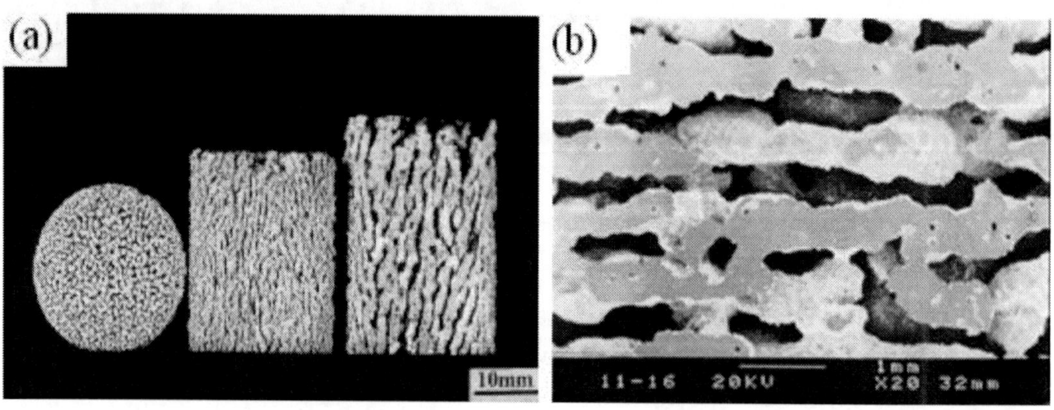

Figure 12. Typical micrographs of porous NiTi produced by SHS [50]. (a) Optical image; (b) SEM image.

Kim's results show that porous NiTi alloys made by SHS are brittle and the tensile strength is as low as 15-25 MPa [51]. This is possibly ascribed to the high porosity and formation of unidirectional porous channels. A recent report [52] reveals that a proper post-sintering heat treatment can induce the formation of a single NiTi phase which possesses better SE. The compressive behavior of porous NiTi SMAs produced by this method is illustrated by the stress-strain curve in Figure 13.

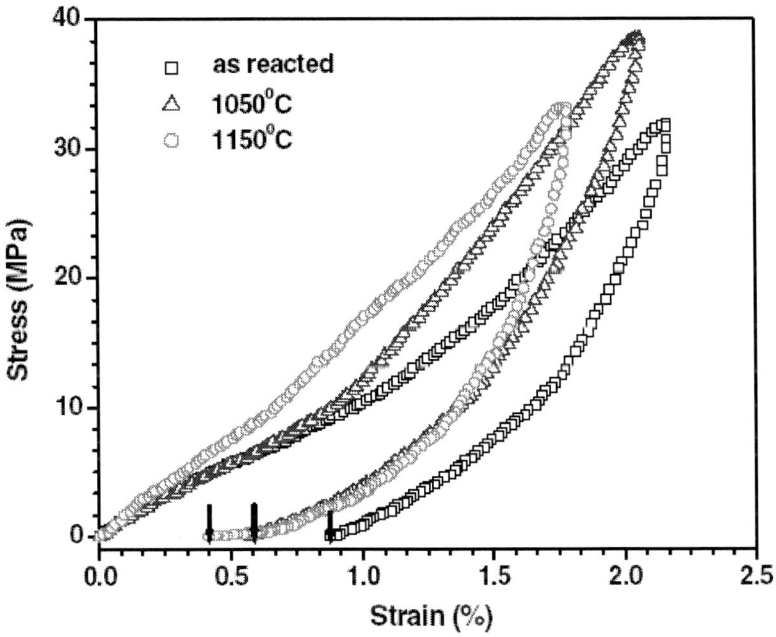

Figure 13. Compressive stress-strain curve of porous NiTi SMAs produced by SHS [52].

2.4. Spark Plasma Sintering (SPS)

Spark plasma sintering (SPS) is one of the spark plasma systems (SPS) [53]. Spark plasma system includes spark plasma sintering (SPS), spark plasma consolidation (SPC), spark plasma joining (SPJ), spark plasma growth (SPG) and spark plasma reaction (SPR) [54]. The spark plasma system is widely used for sintering of metals and ceramics, consolidation of polymers, joining of metals, crystal growth, and chemical reactions. SPS which is based on the idea of using the plasma on electric discharge machine to sinter metals and ceramics is proposed in the early 1960s by Inoue et al. [55, 56]. SPS is mainly characterized by the spark plasma created by a pulsed direct current during heat treatment of the powders in a graphite die. The high-energy plasma is generated between the gap of two electrodes in the electric discharge machine [54]. The reported porosity of porous NiTi SMAs prepared by SPS is lower than 30 vol% and their typical microstructure is exhibited in Figure 14.

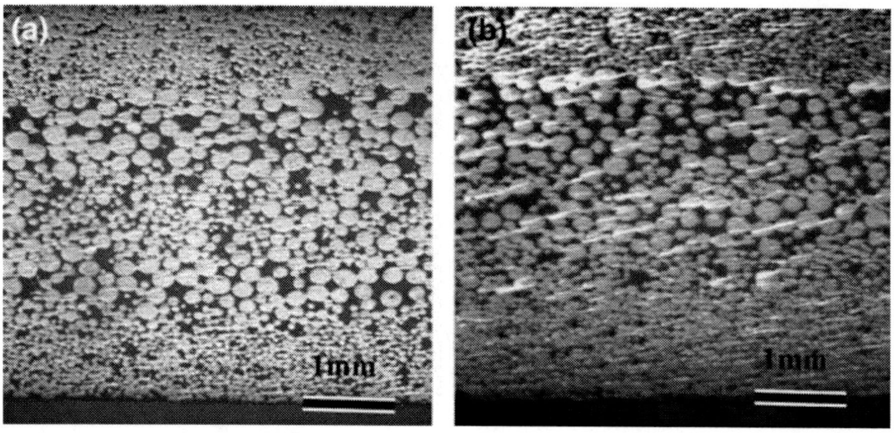

Figure 14. Microstructure of porous NiTi specimens: (a) 25% porosity; (b) 13% porosity [57].

Zhao et al. sinter porous NiTi alloys of two different porosities by spark plasma sintering (SPS) [57]. Their results show that the porous NiTi with 13% porosity has higher compressive strength and excellent SE similar to dense NiTi alloys. The 25% porosity sample exhibits lower compressive strength. The compressive stress-strain curve is shown in Figure 15.

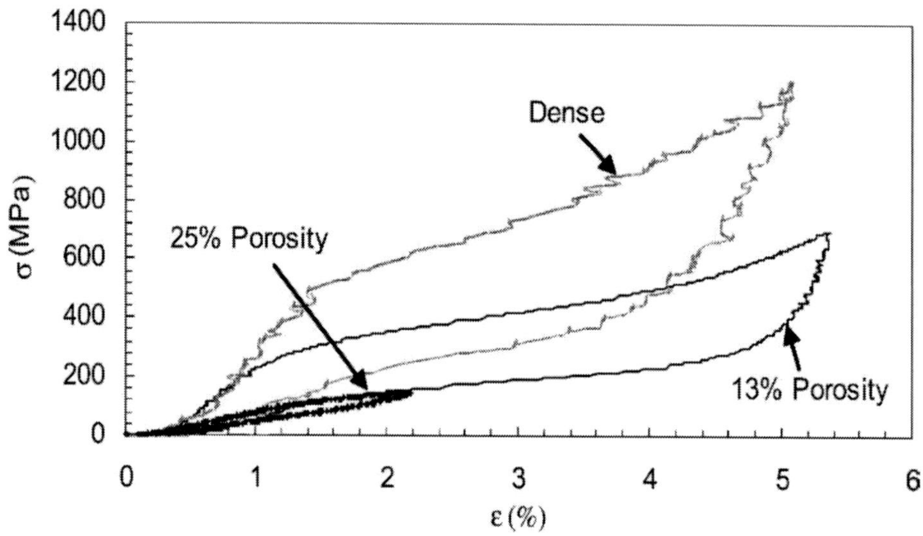

Figure 15. Compressive stress–strain curve of porous NiTi SMAs produced by SPS tested at room temperature [57].

2.5. Hot isostatic Pressing (HIP)

The traditional HIP process has unique characteristics such as consolidation of powders such as near-net-shape capability, ability to partially melt the compact without destroying the shape, and economical post-HIP heat treatment. The HIPing process can be illustrated by Figure 16.

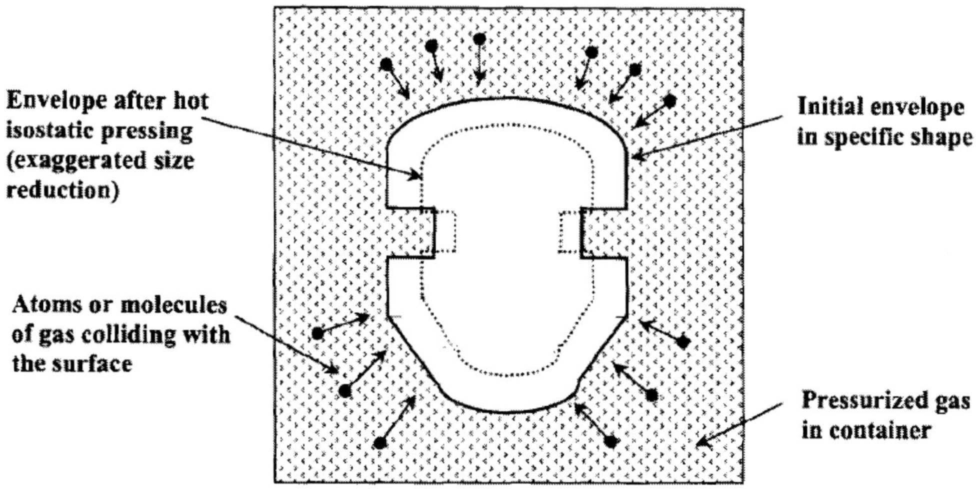

Figure 16. Schematic diagram of traditional HIP process [58].

Lagoudas et al. have prepared porous NiTi from mixed Ni and Ti powders filled with argon gas by hot isostatic pressing (HIP) [59]. However, the compressive strength is still low and it is possibly due to the irregular pore shape and pore distribution in the porous NiTi. Greiner et al. have fabricated porous NiTi alloys by mixing pre-alloyed martensitic NiTi powders and small amount of elemental Ni powders. The mixture is HIPed under an argon atmosphere [60] and they obtain porous NiTi by long time creep expansion at high vacuum and high temperature. As shown in Figure 17, these porous products show excellent SE and higher compressive strength.

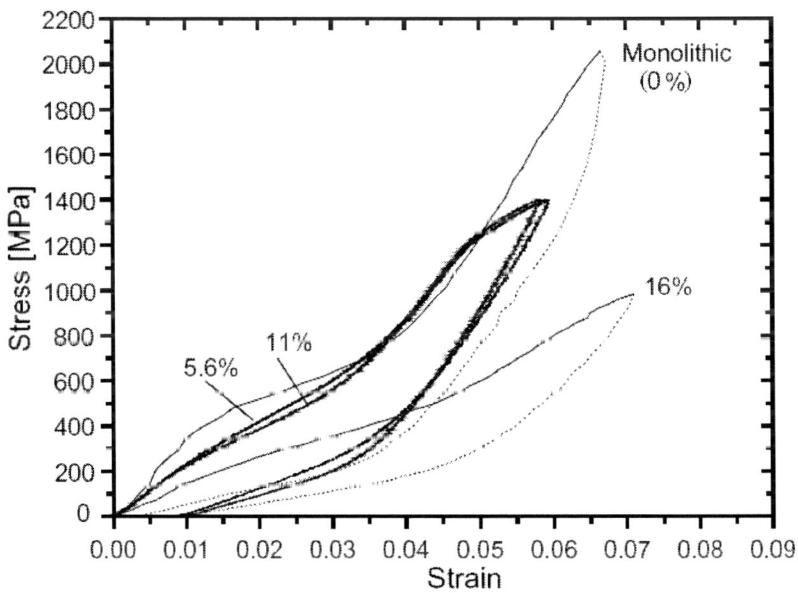

Figure 17. Compressive stress–strain curves for aged porous NiTi prepared by HIP with 0%, 5.6%, 11% and 16% porosity [60].

Unfortunately, the porosity of the porous NiTi prepared by this method is lower than 20% although the sintered porous NiTi has been creep expanded for 100 hrs. Besides, the long expansion time and high temperature heat treatment are not economically attractive. In addition, the size of most pores obtained by this process is only 10-50 μm (shown in Figure 18), which is still much smaller than the 50-500 μm pore size suitable for bone ingrowth and tissue integration [61].

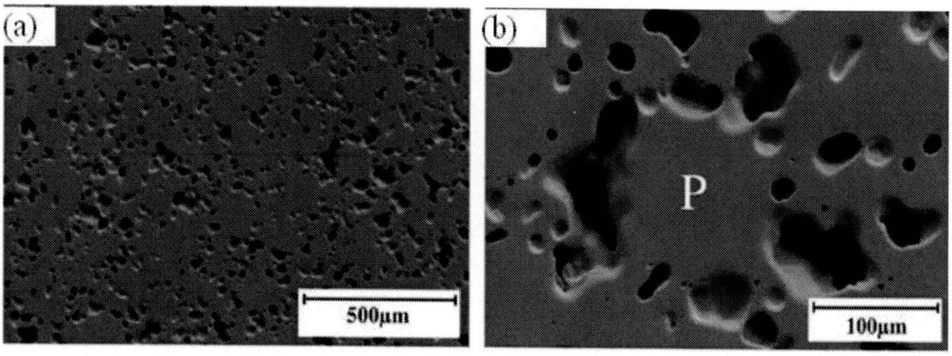

Figure 18. (a) SEM micrographs of 18.8% porous NiTi prepared by HIP; (b) high magnification detail of (a) [60].

2.6. Capsule-Free Hot Isostatic Pressing (CF-HIP).

Capsule-free hot isostatic pressing (CF-HIP) is quite different from traditional HIP. It generally induces the porous structure of sintered materials. CF-HIP is often used to improve the mechanical properties of ceramic materials while retaining adequate porosity [62, 63]. The typical schematic diagram of a CF-HIP apparatus is displayed in Figure 19. Using this method, Yuan et al. have fabricated porous NiTi SMAs with a porosity of 40% with pore size ranging from 50 to 200 μm [64]. Although the porous materials have good mechanical properties, it is relatively difficult to adjust the pore size, distribution, and porosity by CF-HIP. Our research group has fabricated porous NiTi SMAs with adjustable porous structure and good mechanical properties using CF-HIP with NH_4HCO_3 as the space holder, as shown in Figure 20 [65].

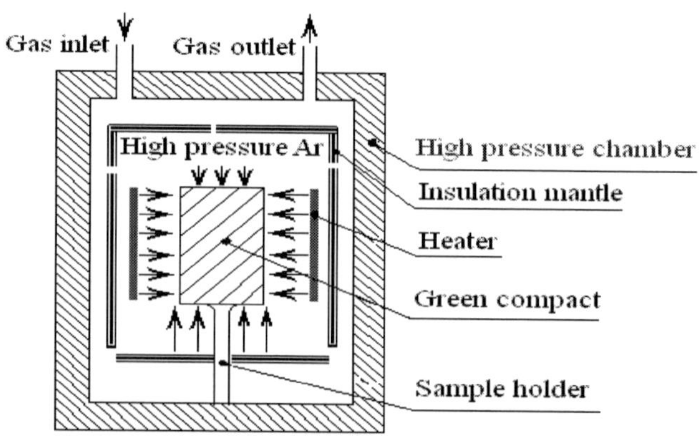

Figure 19. The schematic diagram of CF-HIP unit.

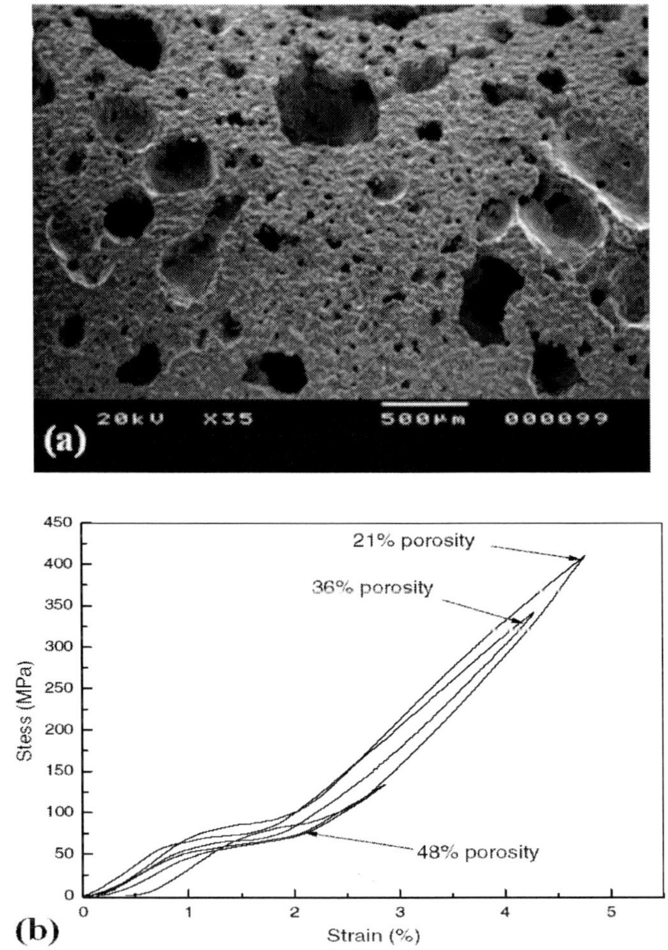

Figure 20. Porous NiTi fabricated by CF-HIP. (a) Longitudinal SEM micrographs (without grinding); (b) Compressive stress-train curves of porous products with various porosities at room temperature [65].

2.7. PM Processes with Space Holders

One of the most promising methods to produce adjustable porous NiTi SMAs is by sintering of compacted or extruded powder mixtures during PM processing that contains removable space-holder materials such as ammonium acid carbonate (NH_4HCO_3) with CF-HIP [65], NaF with HIP [66], NaCl with HIP [67] or with metal injection molding (MIM) [68], and TiH with EPS [42]. Some reports indicate that 25-80% porosity can be achieved in titanium by this method [69, 70].

3. Surface Modification

Nickel is a main component of NiTi. In order to realize SE, near-equiatomic NiTi SMAs are often used. More widespread applications of NiTi SMAs are limited by the toxicity and hypersensitivity caused by nickel ions released from alloys with a nickel concentration of about 50 at% [71-73]. The corrosion resistance of NiTi and toxicity of the by-products released from corroded surfaces have been reported [74, 75]. A surface barrier layer such as titanium oxide, titanium nitride, or titanium carbide has been shown to improve the wear resistance, corrosion resistance and biocompatibility of dense NiTi shape memory alloys (SMAs) [76-80]. Compared to conventional dense NiTi, the complex surface morphology and large surface areas of porous NiTi pose a more serious issue with regard to Ni release. Furthermore, the complex surface morphology and large surface area of the porous structure make the surface very difficult to be modified by line-of-sight techniques such as laser gas nitriding [81], ion beam modification [82], and arc discharge ion plating [83]. The demand for the proper surface modification techniques is thus more stringent. Considering the complex surface morphologies, the following modification techniques have been developed to treat the exposed surface of porous NiTi SMAs.

3.1. Oxidation

Heat treatment (Atmospheric oxidation) is one of the common methods to produce a layer of protective titanium oxide on the surface of NiTi to avoid the allergic and toxic effects stemming from the released nickel ions [84-87]. The samples are often put into an unsealed furnace and treated at various temperatures for different duration. Firstov et al. have calculated the equilibrium phase diagram of $Ni_{50}Ti_{50}$–O under the assumption that oxygen is insoluble in the compounds NiTi and Ni3Ti and that nickel is similarly insoluble in the oxides of titanium (shown in Figure 21) [87].

Chu and Wu have proposed that surface oxidation of equiatomic NiTi at higher temperature of 550 to 1000°C proceeds in four stages as illustrated in Figure 22 [85]. In the early stage of oxidation, Ti is oxidized first to form oxides while Ni remains unchanged due to the higher affinity of Ti with oxygen than that between Ni and oxygen. In the second stage, the rutile crystal grows resulting in the formation of a local Ni-rich zone beneath this layer. As oxidation proceeds, titanium atoms diffuse outward while oxygen atoms diffuse inward. The diffusivity of Ti in rutile is faster than that of oxygen [88], whereas Ni appears to follow a Kirkendall effect for its inter-diffusion with Ti. This results in an outward growth of ruffle and an inward growth of Ni-rich phases, such as Ni (Ti), inducing the formation of a mixed layer of TiO_2 and Ni (Ti) beneath the outer layer. The formation of large voids beneath the surface layer is possibly ascribed to the different growth rates of rutile crystals in the vertical and lateral directions. As the oxidation time increases, outward movement of Ti atoms gives rise to the formation of a Ni-rich phase $TiNi_3$ and Ti-depleted layer.

Based on these oxidation theories of dense NiTi and the non-line-of-sight nature of atmospheric oxidation, we have reported the characteristics of porous NiTi after oxidation in air [89]. The phase components and surface layer structure after high temperature oxidation are in good agreement with the results acquired from dense materials [85], as shown in Figure 23. However, lower temperature oxidation induces some differences. It can be observed from the XPS results (Figure 24) that the intermediate transition layer is composed of TiO_2, TiO, Ti_2O_3 and NiTi. In view of nickel release and cyto-compatibility, the oxidation temperature should not be higher than 450 °C because the nickel concentration significantly increases when the oxidation temperature exceeds 450°C (shown in Figure 25).

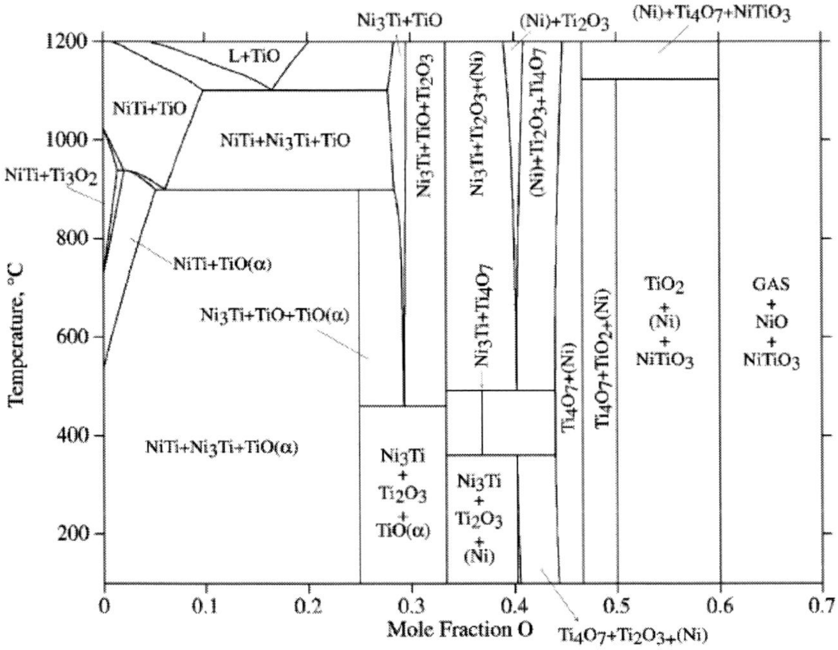

Figure 21. Calculated equilibrium phase diagram of NiTi–O [87].

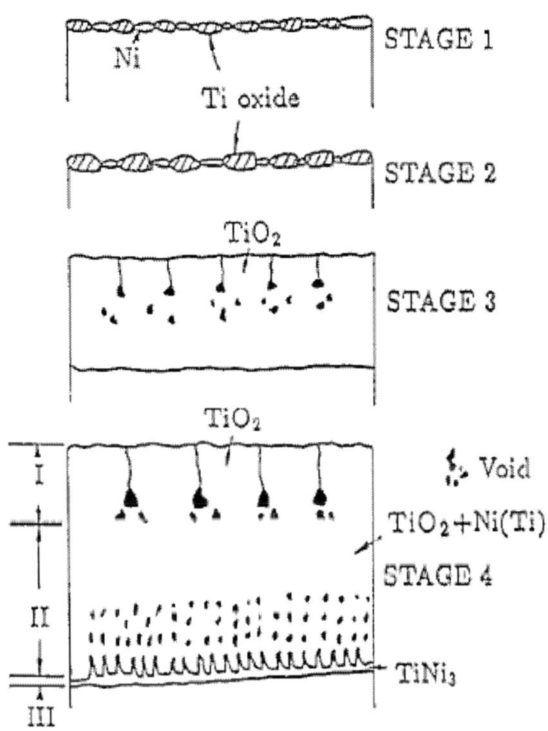

Figure 22. Schematic diagram of oxidation scale formation of equiatomic NiTi alloy [85].

The unexpected result is possibly due to the lamellar structure of the surface layer with a thicker porous intermediate mixture including nickel-rich Ni_3Ti, NiTi and rutile as well as larger voids when the temperature is higher than 450°C [85]. Therefore, nickel can diffuse more easily from the porous NiTi when the solution penetrates into the porous intermediate and nickel-rich layer in the immersion tests. In addition, the oxidized surface at a temperature higher than 450°C does not favor osteoblast attachment and proliferation. This topic will be discussed in more details in Section 4.

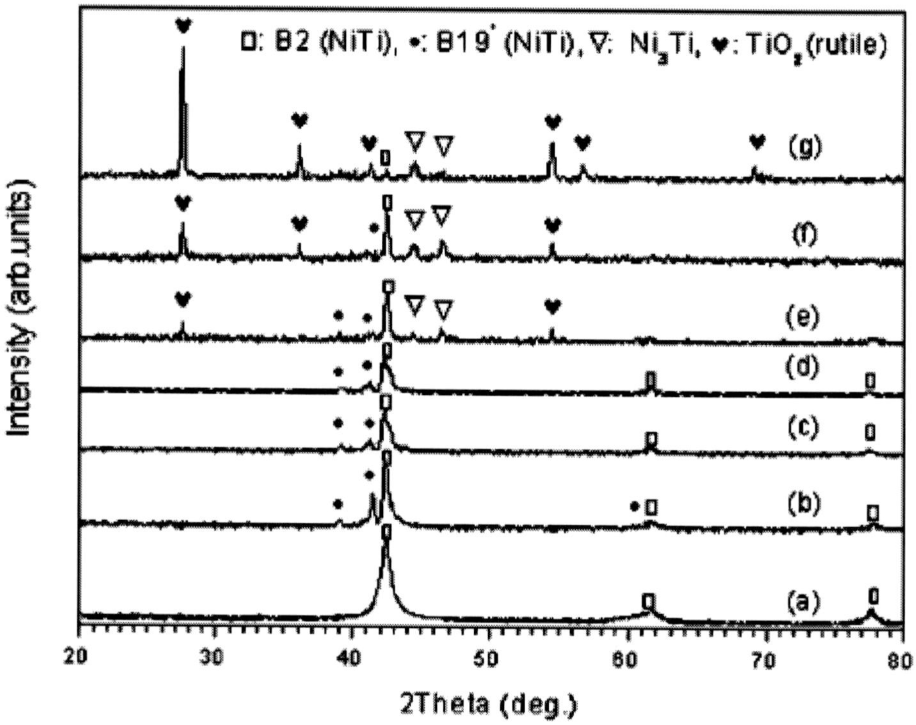

Figure 23. XRD patterns acquired from porous NiTi SMA after oxidation at different temperatures: (a) Unoxidized, (b) 300oC, (c) 400oC, (d) 450oC, (e) 550oC, (f) 600oC, and (g) 800oC [89].

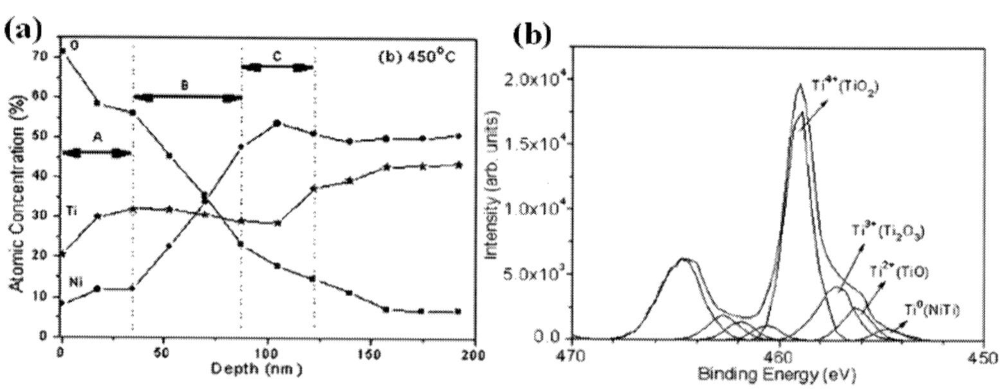

Figure 24. XPS analysis acquired from the wall of an internal pore of porous NiTi oxidized at 450oC. (a) small area XPS depth profile, (b) the deconvoluted Ti2p spectrum [89].

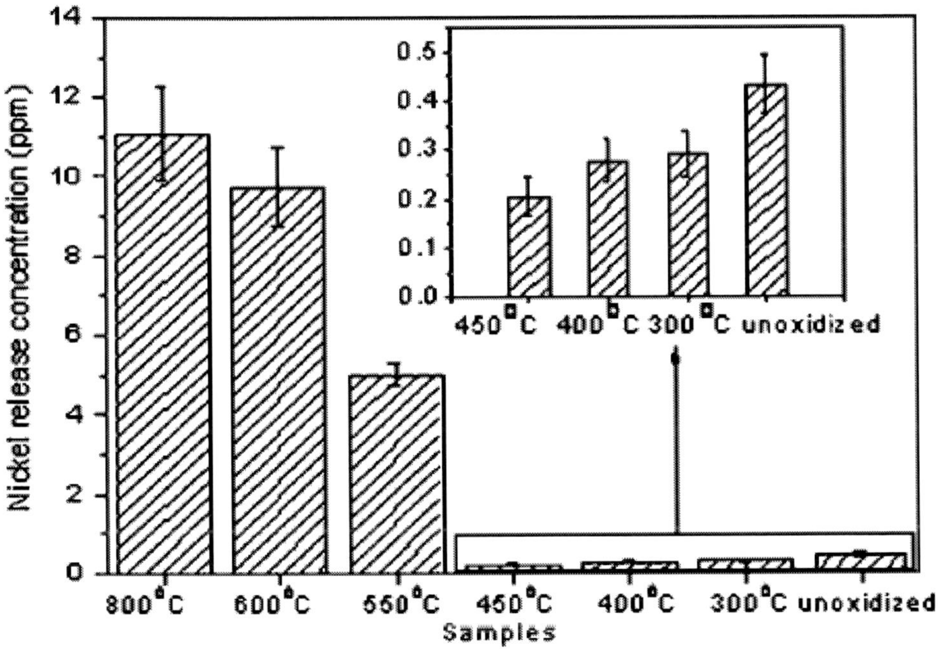

Figure 25. Amounts of nickel released from the oxidized porous NiTi samples immersed in SBF at 37±0.5oC for 6 days determined by ICPMS [89].

3.2. Oxygen Plasma Immersion Ion Implantation

A plasma which can be regarded as the fourth state of matter is composed of highly excited atomic, molecular, ionic, and radical species. It is typically obtained when gases are excited into energetic states by radio frequency (RF), microwave, or electrons from a hot filament discharge [90]. Plasma immersion ion implantation (PIII) which excels as a non-line-of-sight technique has been demonstrated on components with irregular geometries [80, 91-93]. In addition, a good theoretical understanding of PIII into complex-shaped samples such as mechanical gears and ball bearings has been developed and simulation software is available to determine ion trajectories and ion fluence distributions on different types of surfaces [94-97]. Furthermore, PIII has been successfully utilized to improve the surface mechanical and biological properties of dense NiTi SMAs [98-100]. The schematic drawing of the PIII instrument used in our work is shown in Figure 26. We have successfully treated porous NiTi SMAs using oxygen PIII using parameters listed in Table 5 [101, 102]. As shown in Figure 27, the small area XPS depth profiles reveal that Ti exists only as TiO_2 on the surface of the O-PIII sample whereas three chemical states of Ti can be observed on the surface of the untreated sample, namely, Ti^0 (NiTi), Ti^{2+} (TiO) and Ti^{4+} (TiO_2). It also implies that O-PIII is an effective technique to treat most of the exposed surface of porous NiTi SMAs with complex topographies. Moreover, O-PIII results in the formation of a gradient surface layer composed of outermost TiO_2, intermediate transition layer with components of TiO_2, Ti_2O_3, TiO, and a trace amount of NiTi, and NiTi substrate.

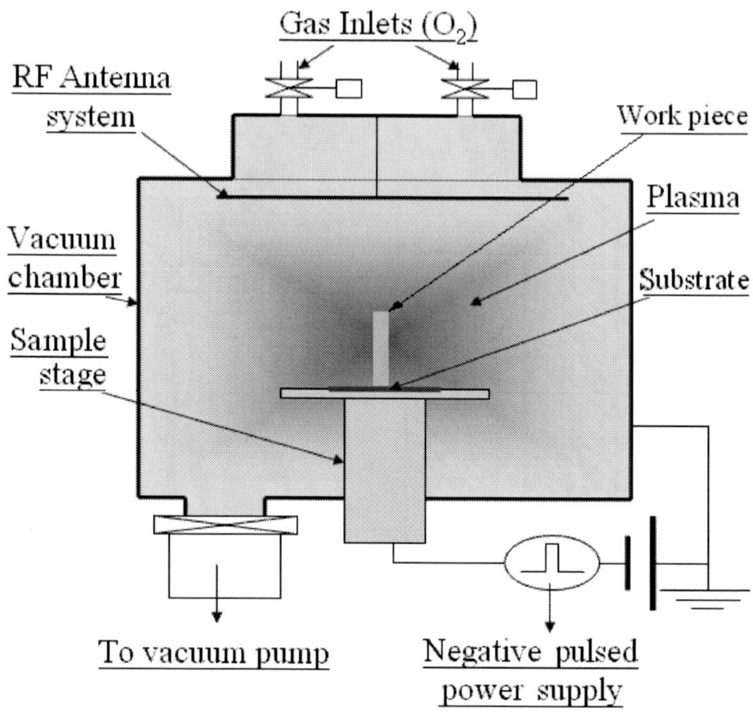

Figure 26. Schematic drawing of the PIII facility at City University of Hong Kong.

Table 5. Oxygen plasma immersion ion implantation (O-PIII) parameters [102]

RF Power	1000 W
Sample Voltage	40 kV
Pulse Width	50 μs
Frequency	200 Hz
Pulse Duration	120 min
Base Pressure	1×10^{-5} Torr
O_2 Working Pressure	6.0×10^{-4} Torr

The inductively-coupled mass spectrometry (ICPMS) results unequivocally indicate that O-PIII can suppress the leaching of Ni ions from the porous samples [101,102]. This is superior to other techniques which may have difficulty dealing with non-planar surfaces. In our work [102], the untreated (mechanically polished) sample shows higher nickel release, probably due to the thinner oxide layer on the surface. Another important factor is that the pores cannot be oxidized during mechanical polishing, whereas these pores are more effectively implanted during PIII. For pores with sizes from 50μm to 400μm, they are sufficiently big that the plasma can penetrate inside and so the sidewalls can be adequately implanted, as PIII has been shown to implant the sidewalls of sub-micrometer trench structures in integrated circuits having aspect ratios exceeding 30 [103]. However, we are not implying that all the sidewalls of the pores are always adequately implanted during PIII as small and deep pores are still difficult to be implanted. Nonetheless, our immersion tests reveal the unambiguous improvement offered by O-PIII.

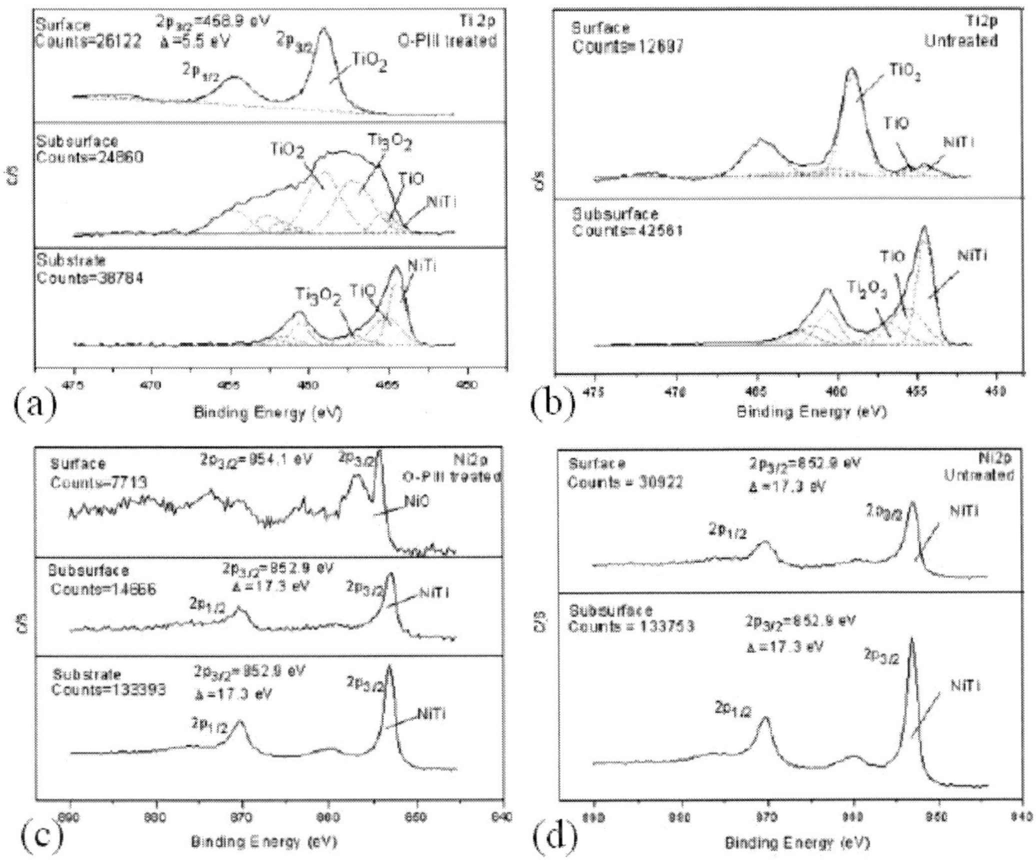

Figure 27. XPS spectra of Ti2p and Ni2p acquired from the untreated and O-PIII porous NiTi samples. The solid line (black color) indicates the experimental spectra and the dash dot line (green color) indicates the fitted spectra: (a) Ti2p from O-PIII sample; (b) Ti2p from untreated sample; (c) Ni2p from O-PIII sample; (d) Ni2p from untreated sample. [102]

3.3. Chemical Treatment

Chemical treatments have been reported to be effective in modifying the surface of Ti and Ti-based alloys to favor the formation of bioactive apatite on the surface in simulated body fluids (SBF) and prevent nickel ions from leaching from the dense NiTi SMAs [104-107]. It is believed that this method is suitable for the modification of the complex surface of porous NiTi due to its non-line-of-sight nature which is better than that in plasma immersion ion implantation (PIII) and air oxidation [89, 102, 108]. In the chemical treatment, different chemical solutions such as hydrogen peroxide solution (H_2O_2), H_2SO_4, and NHO_3 are often used to pre-treat titanium metal to form a layer of titania film followed by alkaline treatment [109-111]. However, these reports focus on the enhancement of surface bioactivity of dense materials, and few studies have been carried out on porous NiTi SMAs. A recent investigation by Jiang and Rong [112] reveals that HNO_3 solution and subsequent NaOH treatment can enhance the formation of hydroxyapatite (HA) on the surface of porous NiTi immersed in SBF. It also depresses nickel leaching and they ascribe the superior behavior to the formation of TiO_2 and $NaTiO_3$. We have achieved good results using a hydrogen peroxide solution in conjunction with subsequent sodium hydroxide treatment of porous NiTi SMAs [113-115].

After treatment in 10M NaOH solution with H_2O_2 preoxidation, a layer of spiculate structure is formed on the surface of porous NiTi SMA, as shown in Figure 28(a). After 5 days of immersion in SBF at 37°C, some precipitates are found on the surface with discontinuous distribution, Figure 28(b). The quantity and size of these precipitates increase as the immersion time increases. After immersion in SBF at 37°C for 15 days, the surface is almost completely covered by a layer of precipitates, as shown in Figure 28(c). The higher magnification image reveals that there are small pores and cracks on the precipitates. Figure 28(d) discloses that the apatite particles may nucleate and grow gradually at different regions. These particles merge with each other while growing as the immersion time increases. The pores and cracks are also formed simultaneously.

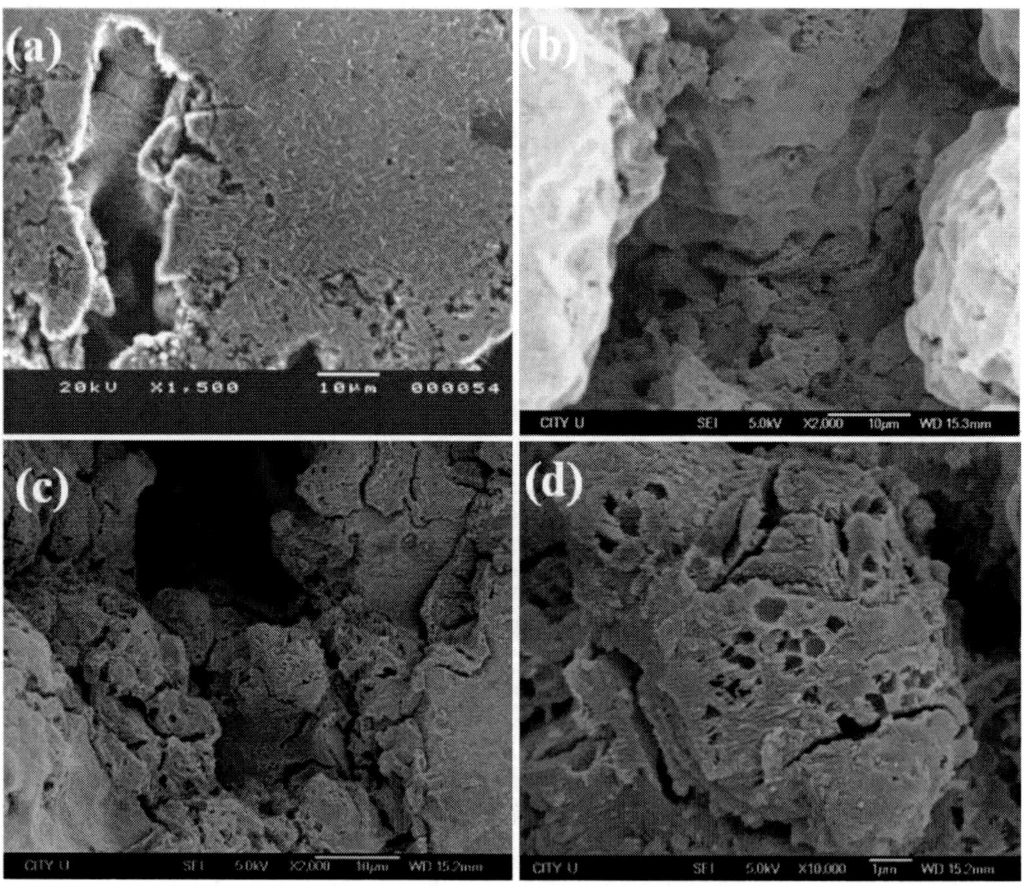

Figure 28. Morphologies of the porous NiTi treated by 30% H2O2 aqueous at 80oC for 4hrs and subsequent 10M NaOH aqueous treatment at 60oC for 24 hours: (a) without immersion; (b) after immersion in SBF for 5 days; (c) after immersion in SBF for 15 days; (d) higher magnification of (c). [113]

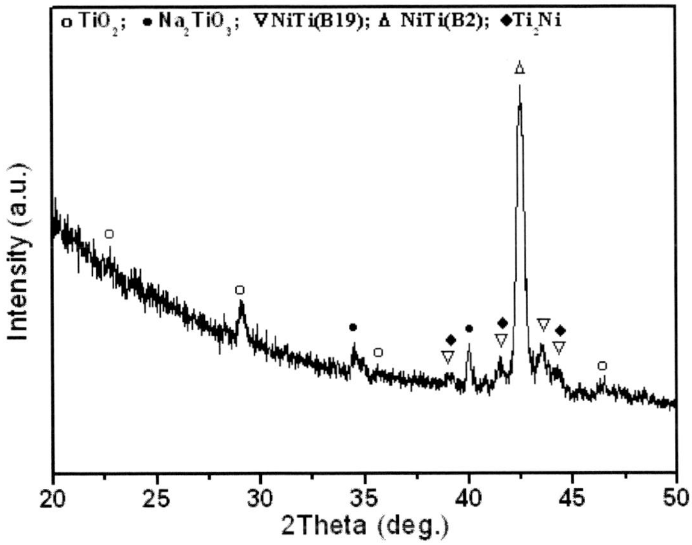

Figure 29. TF-XRD pattern of H2O2 and subsequent NaOH solution treated sample [113].

Figure 29 shows the TF-XRD pattern of porous NiTi alloys modified by H_2O_2 solution pre-oxidation and subsequent NaOH solution treatment. The XRD pattern reveals that the surface layer on the porous NiTi alloys is composed of predominantly Na_2TiO_3 and TiO_2 after H_2O_2 and NaOH modification. The NiTi phase and minor Ti_2Ni phase signal should come from the substrate. After immersion in SBF at 37°C for 15 days, the EDS results in Figure 30 suggest the formation of precipitates composed of calcium and phosphorus on the modified porous sample, as shown in Figure 28 (c). Na and O peaks are also detected and they probably originate from Na_2TiO_3 and TiO_2 below the precipitates. The Ti and Ni signals come from the substrate.

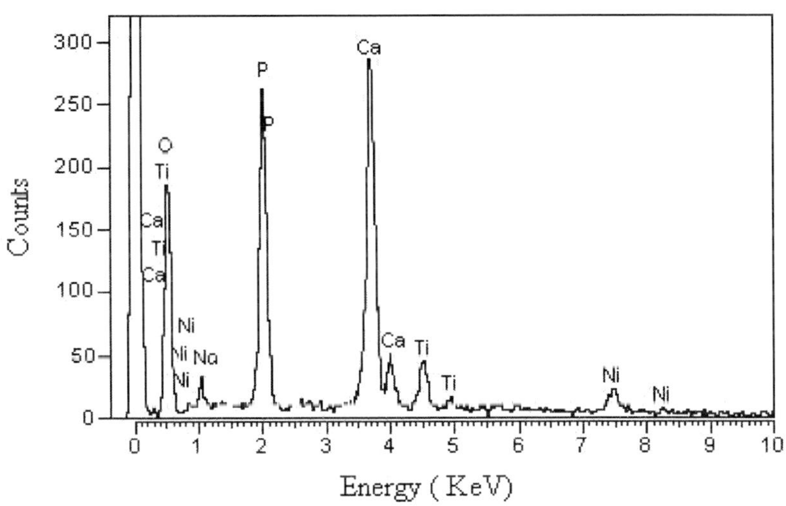

Figure 30. EDS profile of the porous NiTi alloy modified by H2O2 solution and subsequent NaOH solution after immersed in SBF for 15 days [113]

Figure 31 shows the TF-XRD pattern of chemically treated porous NiTi soaked in SBF for 15 days at 37°C. It can be observed that the precipitate layer is dominantly composed of apatite. Minor amounts of TiO_2 and Na_2TiO_3 exist in the surface layer, stemming from the chemical reaction during chemical modification [112] as confirmed by the TF-XRD pattern in Figure 29.

It can be found from Figure 32(a) that after treatment in H_2O_2 and subsequently NaOH, the surface layer on the porous NiTi alloy is about 800 nm thick and exhibits a graded structure. Figure 32(b) shows the high resolution XPS spectra of Ti2p of this porous sample. It is obvious that from the surface to the substrate, Ti2p3/2 exhibits three different binding energies of 458.3eV, 459.1eV and 454.5eV, corresponding to the surface layer, intermediate transition layer and substrate, respectively. According to the reported binding energies of Ti2p3/2 in different compounds [116], Ti2p in the surface layer should exist in TiO_3^{2-}. In the middle layer, Ti2p exists as TiO_2 Ti_2O_3, TiO and TiO_3^{2-}. It can be inferred that the outermost layer is composed of Na_2TiO_3 and the next transition layer is made up of Na_2TiO_3, TiO_2 Ti_2O_3 and TiO. As the depth increases, the quantities of TiO_2 and TiO_3^{2-} diminish while the amount of metal Ti (NiTi) increases. When analyzing the depth profile curves shown in Figure 32(a), it is evident that the surface layer on the chemically modified porous NiTi alloy is composed of mainly Na_2TiO_3 and the middle layer comprises TiO_2 and Na_2TiO_3 with trace amounts of NiTi, Ti_2O_3 and TiO. The quantities of TiO_2 and Na_2TiO_3 gradually decrease near the substrate. NiTi is predominant while TiO_2 and Na_2TiO_3 become less in concentration when the depth is over 800 nm.

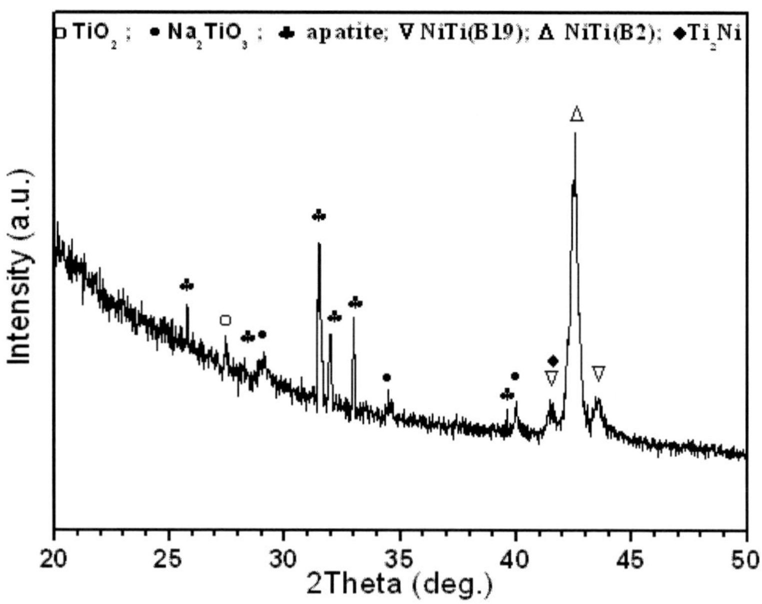

Figure 31. TF-XRD pattern of the porous NiTi alloy modified by H2O2 solution and subsequent NaOH solution after immersed in SBF for 15 days. [113]

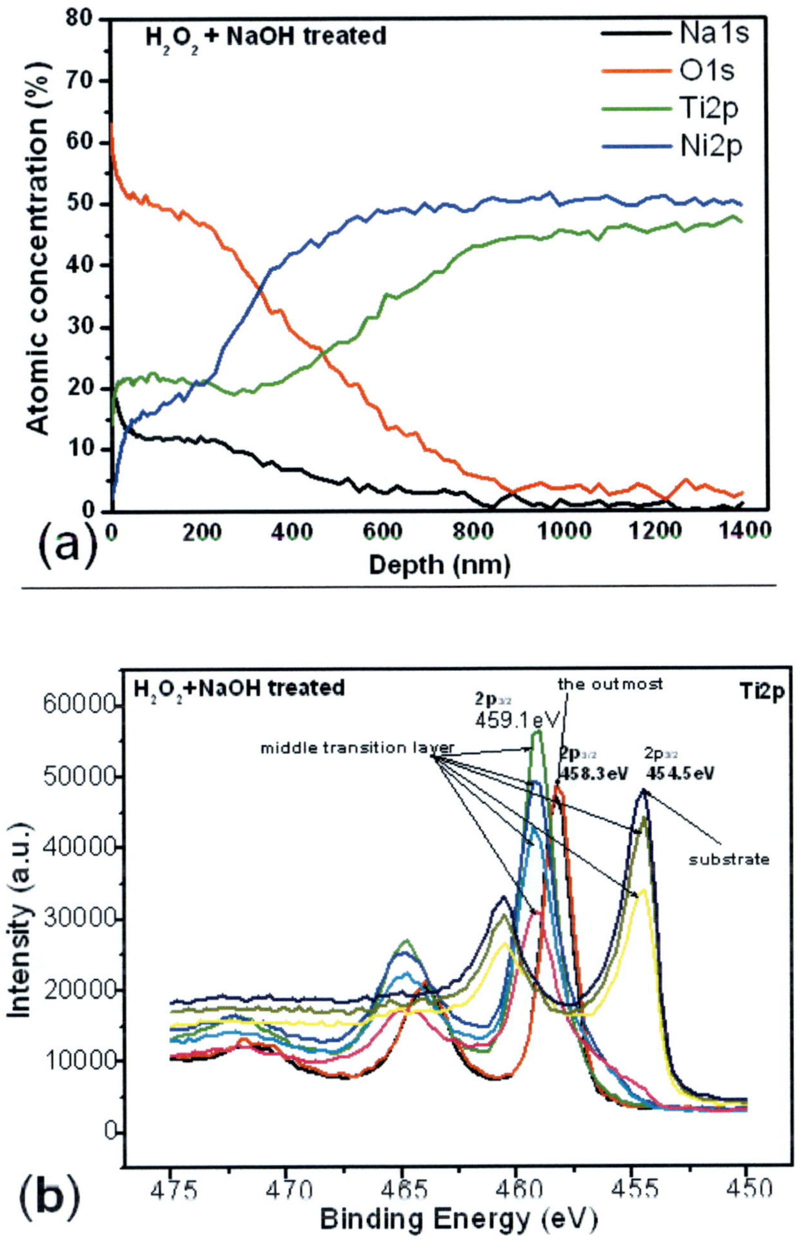

Figure 32. XPS analysis the porous NiTi samples modified by H2O2 solution and subsequent NaOH solution. (a) depth profile, (b) high resolution XPS narrow spectra of Ti2p. [113].

Our recent research reveals that long time alkali solution treatment can induce the formation of hierarchically nanostructured titanates on three-dimensional microporous Ti-based alloys [117].

3.4. Physical vapor Deposition (PVD)

PVD is fundamentally a vaporization coating technique, involving transfer of materials on an atomic level. It is an alternative process to electroplating. The raw materials/precursors used for deposition start out in a solid form. PVD processes are performed under a vacuum environment and generally include four steps, namely evaporation, transportation, reaction, and deposition. The schematic diagram of this process is shown in Figure 33 [118]. During the first stage, a target consisting of the material to be deposited is bombarded by a high energy source such as a beam of electrons or ions. The atoms are vaporized from the surface of the target. The second process is the movement of vaporized atoms from the target to the substrate. In some cases, we want to get coatings composed of metal oxides, nitrides, carbides and other such materials. In these cases, the target will consist of the metal. The atoms of metal will react with the selected gas during the transportation stage. The last stage is the formation of coatings on the substrate surface. In some cases, the reactions between the target materials and reactive gases may also take place at the substrate surface simultaneously during deposition.

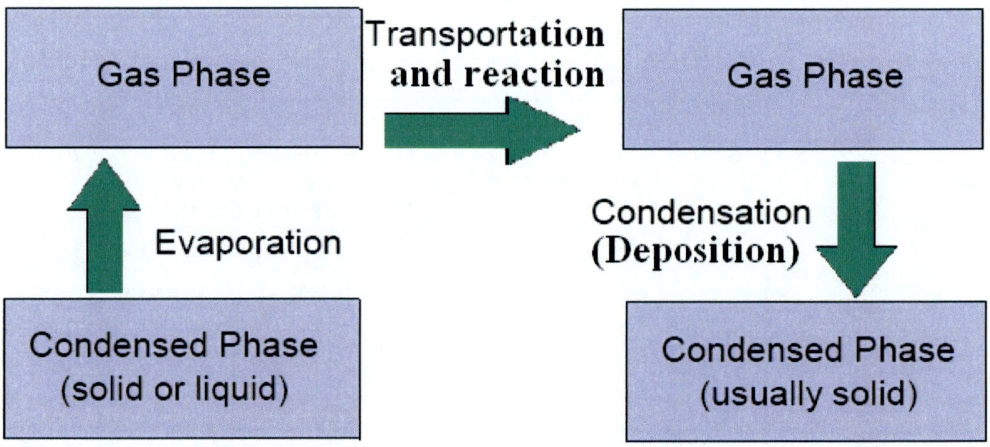

Figure 33. Schematic diagram of the PVD process [118].

This technique is different from the aforementioned three techniques due to its line-of-sight nature. Not many research groups have treated porous NiTi using this technique. Lemaire et al. have reported that TiN and TiO_2 coatings can be deposited on porous NiTi SMAs by PVD [119]. They claim that the coatings cover the entire exposed surface area, even in the center of 8 ×17 mm^2 samples with 210 ± 110 μm pores. Nickel ion release decreases relative to an untreated sample by factors of 10 and 14 for TiN and TiO_2 coatings, respectively. On a whole, the porous NiTi SMAs treated by PVD show higher nickel release compared to other surface modification techniques. The natural disadvantage of PVD for the treatment of porous NiTi with complex surface morphologies is its line-of-sight nature meaning that it is extremely difficult to coat workpieces with complex 3D topographies.

There are many other line-of-sight techniques such as chemical vapor deposition (CVD) and laser treatment. However, these processes are not suitable for the surface modification of porous NiTi with complex three-dimensional topographies.

Table 6. Nickel release concentrations of various porous NiTi products immersed in SBF

PM processes		Porosity (vol%)	Pore size (μm)	Surface modification methods	Ni release concentrations (ppm)
Our group's results [89, 101, 102, 113, 115]	CF-HIP	40	50-200	Oxygen-PIII	0.06–0.18 (untreated) 0.01–0.05 (treated) 7–28 days
	CF-HIP with Ar expansion	42	50-400	Oxygen-PIII	0.2–0.3 (untreated) 0.05–0.08 (treated) 70 days
	CF-HIP with space holder + homogenization	48	50-500	Atmospheric oxidation at 450 °C	0.45 (untreated) 0.2 (treated) 6 days
	CF-HIP with space holder	56	50-800	Chemical treatment	0.54 (untreated) 0.39 (30% II_2O_2, 80°C, 4 hrs) 0.35 (10M NaOH, 60°C, 24 hrs) 0.30 (30% H_2O_2, 80°C, 4 hrs plus 10M NaOH 60°C 24 hrs) 15days
Other group's results [123]	SHS + homogenization	65	100-320	TiN and TiO_2-PVD coating; SBF soaking	0.66–0.85 ppm (not homogenized) 0.17–0.30 ppm (untreated) 0.04–0.09 ppm (TiN) 0.06–0.07 ppm (TiO_2) 0.05–0.07 ppm (SBF pre-soak) 0.03–0.04 ppm (TiN + SBF pre-soak) 0.06–0.07 ppm (TiO_2 + SBF pre-soak) 1–16 days
	SHS	61	200-600	HNO_3/NaOH with HA coating	6.7 ppm (untreated) 0.48 ppm (treated) 50 days

4. In *Vitro* and in *Vivo* Studies

Since synthetic materials often integrate poorly with host tissues and result in infection or other adverse body response [120-122], porous NiTi SMAs shoule dbe fabricated by different processes and modified by different techniques and a large number of in *vitro* and in *vivo* tests must be conducted before clinical applications.

4.1. *In vitro* tests

4.1.1. In Vitro Nickel Release Test

According to the Bansiddhi *et al*'s survey [123] and our group's immersion test results, the nickel concentrations released from porous NiTi SMAs are summarized in Table 6. Earlier research indicates that the cell growth rate of cultured fibroblasts is depressed and the cell morphology also changes when the cobalt or nickel concentrations are higher than 15 ppm [124, 125]. Shih et al.'s result further confirm that for rat aortic smooth muscle cell, its growth inhibition becomes significant when the nickel concentration exceeds 9 ppm and the corrosion products also alter the cell morphology, induce cell necrosis, and decrease cell numbers [126]. As shown in Table 6, the short-term immersion tests indicate that released nickel concentrations from all the porous NiTi SMAs are much lower than 9 ppm, even for the untreated porous products with most pose size in the range of 200-600 μm and higher porosity of above 60 vol%. It is obvious that the surface layer formed by different modification techniques can significantly suppress the nickel leaching in short term tests. These surface layers generally exhibit a graded structure and are composed of titanium dioxide, titanium nitride, sodium titanate, or composites of these materials. It is noted that the released nickel concentration is different from porous NiTi prepared by different methods and various modification techniques possibly due to the different porous structure and various surface layer structure and compositions. In addition, deposition of HA on the surface is also favorable when nickel release is depressed.

4.1.2. In Vitro Corrosion Test

The corrosion performance of NiTi is closely related to nickel ion release. Some reports show the failure of corroded NiTi stents and grafts retrieved from patients after 5 to 43 months [127]. In comparison with dense NiTi SMAs, the complex surface morphology and larger surface area of porous NiTi make porous NiTi more susceptible to corrosion in the biological environment in which corrosion accelerates nickel ion leaching from the nickel-rich substance. In simulated body fluids (SBF), the corrosion behavior of porous NiTi alloys is directly related to the porous structure such as porosity, pore size, pore shape, and pore distribution. In addition, the surface roughness, crevices, complex geometry, internal angles, restriction of the flow of solution species, and residual stress may also influence the corrosion rate [127-131]. Other materials characteristics such as thickness, phase composition, and structure that are modified by different surface treatment techniques also substantially affect the corrosion behavior of the materials [76, 80, 132].

Electrochemical techniques such as potentiodynamic polarization measurements and electrochemical impedance spectroscopy (EIS) are commonly employed to monitor the corrosion and electrochemical behaviors of NiTi SMA. In general, a standard three-electrode apparatus is usually placed in a glass cell immersed in a 37°C water bath filled with different electrolyte solution such as SBF, Hank's solution and physiological solution. The working electrodes are the porous NiTi alloys. The counter electrodes usually consist of two graphite rods mounted diametrically opposite to each other, and all the potentials are expressed with reference to a standard calomel electrode (SCE) connected to the working electrode through a Luggin probe. Li et al.'s research reveals that porous NiTi has a much lower value of potential breakdown than dense NiTi SMAs either in a 0.9% NaCl solution or in Hank's solution, as shown in Figure 34 [133]. This is possibly caused by the large exposed surface area and complex morphology. These results are confirmed by other groups. Table 7 summarizes the corrosion results using the potentiodynamic polarization method [134]. The corrosion resistance of porous NiTi SMAs depends on the conditions. It can be found from Table 7 that the porous NiTi has higher corrosion resistance in Hank's solution and the porous NiTi from traditional powder sintering has lower corrosion resistance. In addition, the porous NiTi with pores filled has higher potential breakdown value than that without filling.

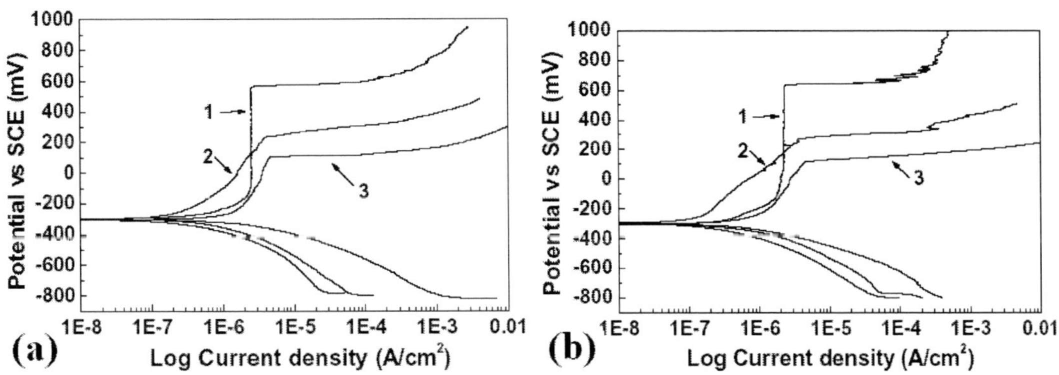

Figure 34. Polarization curves of porous and solid NiTi SMAs in solution of (a) 0.9% NaCl, and (b) Hank's; curve 1: Solid NiTi SMA; curve 2: porous NiTi SMA with pores filled by resin; curve 3: porous NiTi SMA without filling [133]

Table 7. In vitro corrosion results of porous/dense NiTi

NiTi material	Test medium	E_{corr} (mV)	I_{corr} (10^{-6} A cm^{-2})	E_{bd} (mV)	Conditions	Ref
Porous	0.9% NaCl	−300		60	Powder sintering	[134]
Solid		−400		1200	Polished disc	
Porous	0.9% NaCl	−300		101 to 131	SHS, pore size >100 μm pore unfilled	[133]
Porous	0.9% NaCl	−300		228 to 262	SHS, pore size >100 μm pore filled	
Solid	0.9% NaCl	−300		562 to 592		
Porous	Hanks'	−300		122 to 152	SHS, pore size >100 μm pore unfilled	
Porous	Hanks'	−300		162 to 198	SHS, pore size >100 μm pore filled	
Solid	Hanks'	−300		634 to 662		
Porous	Hanks'	−277/−281		173/16	SHS, pore size 100 μm Before/after surface treatment CR = 0.02 to 0.76 mm/year	[134]
Porous	Hanks'	−286/−153		136/15		
Porous	HBSS	638 ± 99	0.5 ± 0.7	270 ± 70	SHS,230± 130 μm and a porosity of 65 ± 10%	[134]
Solid	Hanks'	−457	0.009	888	Electropolished disc CR = 7.85×10^{-3} mm/year	[134]

Our recent research shows that EIS is also an effective technique to study the electrochemical behavior and stability of the surface layers on the exposed areas of porous NiTi SMAs fabricated by different surface modification techniques in SBF [135]. The results show that non-line-of-sight techniques such as PIII and chemical treatment are useful surface modification techniques to produce protective films on the surface of porous NiTi alloys and that the surface films produced by different modification techniques significantly influence the electrochemical behavior of porous NiTi electrodes.

4.1.3. In Vitro Cell Culture Test

In vitro cell culture tests have been carried out on various NiTi foams before and after surface modifications. Porous NiTi SMAs fabricated by both SHS and CF-HIP exhibit good cytocompatibility, even without any surface modification. As shown in Figure 35(a), for example, peripheral blood leukocytes can immediately attach to the surface of porous NiTi produced by SHS and exhibit rapid viability within 24 h [38]. Our short-term *in vitro* test (8 days) shows that osteoblasts can attach and proliferate well on almost the entire exposed surface, even the inside of the exposed pores of porous NiTi produced by CF-HIP (shown in Figure 35(b)). Porous NiTi samples made by other processes (e.g. MIM with space holder) also show strong proliferation and attachment of cultured cells [68]. Our cell culture tests show that mice osteoblast cells can attach and grow successfully on the modified surface of CF-HIP fabricated porous NiTi using either 450°C low temperature oxidation, O-PIII, or chemical process (shown in Figures 35(c), 35(d) and 35(e), respectively). However, the released nickel concentrations significantly influence how well osteoblast cells adhere and proliferate on the porous NiTi surfaces. As shown in Figure 34(f), the 600 °C oxidized surface does not favor attachment of osteoblasts because immersion tests show that higher temperature oxidation induces higher nickel concentration. Figure 25 confirms that a nickel concentration of 10 ppm released from the 600 °C oxidized porous NiTi. Our results are in agreement with the findings of Gu et al. [136]. In their study, no cell attachment can be observed on the porous NiTi surfaces containing unreacted, elemental Ni left from an incomplete SHS process, while fast proliferation and differentiation are found after surface treatment.

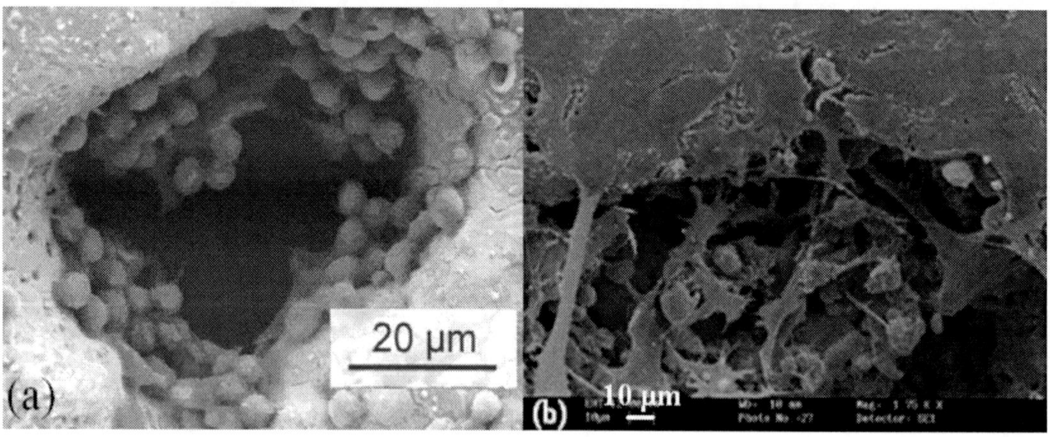

Figure 35. (Continues)

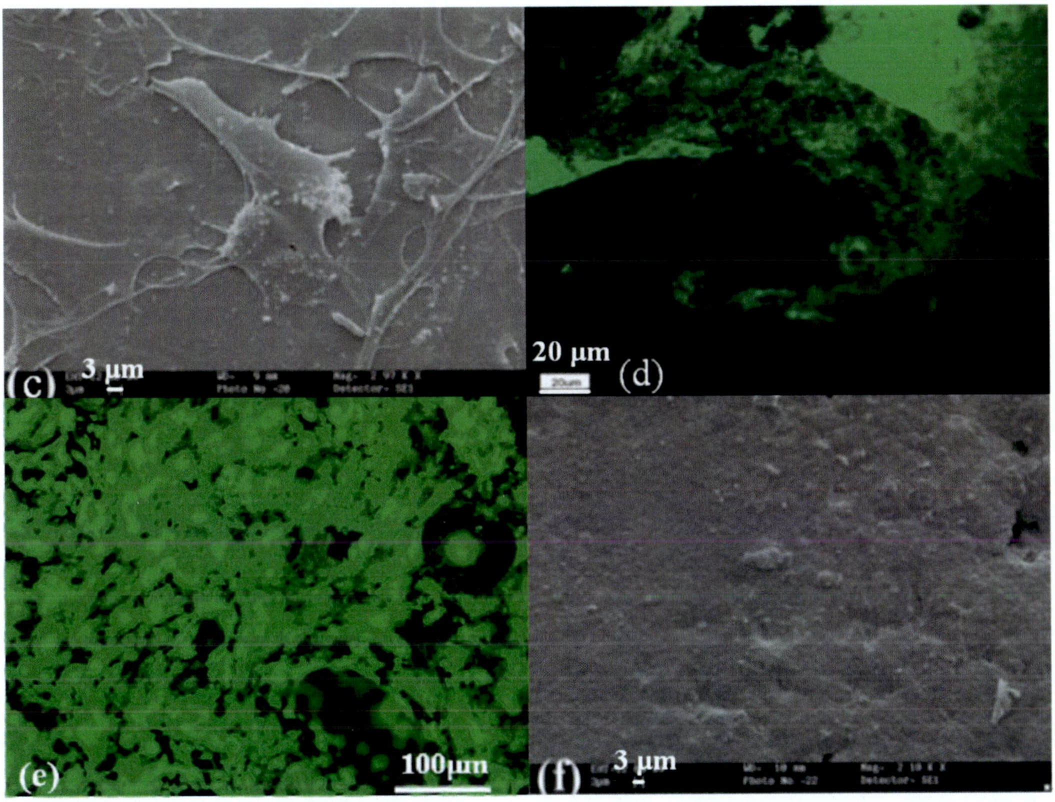

Figure 35. SEM/OM microscopic view of porous NiTi alloy after cell culturing. (a) SEM image of adherent peripheral blood leukocytes (PMN) at the surface of a porous NiTi sample. Isolated PMN were incubated with the NiTi sample [38], (b) SEM image of mice osteoblasts on untreated CF-HIP sample [65], (c) SEM image of mice osteoblasts on 450 oC oxidized CF-HIP sample [89], (d) OM image of mice osteoblasts on PIII treated CF-HIP sample [102], (e) OM image of mice osteoblasts on chemically modified CF-HIP sample [114], (f) SEM image of mice osteoblasts on 600 oC oxidized CF-HIP sample [89]

4.1.4. In Vivo Tests

In vivo study of porous NiTi implanted in animals is a critical step before the biometal can be clinically implanted into patients. Kim et al. show no apparent adverse reactions on or around the implant area in proximal tibia of rabbit after 6 weeks, and the in-grown bone has similar properties to the surrounding bone [51]. Other groups have also indicated that porous NiTi shows excellent bone implant contact and a high level of bone ingrowth in rabbits and rats without signs of loosening [137 138]. Our recent *in vivo* tests in rabbit's tibia for 105 days show that bone tissues can contact and grow well on the entire surface and bone tissues grow smoothly into the inside of the interconnected internal pores [117]. Our work also shows that a high porosity (48%) does not exhibit obvious superiority for bone contact and ingrowth compared with low porosity (42%). Kujala et al. have implanted 3 types of porous NiTi into rats for 30 weeks, and their results reveal that a porosity of 66.1% (MPS 259±30 μm) shows the best bone–implant contact (51%). However, a porosity of 46.6% (MPS 505±136μm) with bone–implant contact of 39% is not significantly inferior in this respect and shows significantly lower incidents of fibrosis within the porous implant. A Porosity of 59.2% (MPS

272±17μm) leads to lower bone–implant contact values [137]. It indicates that the porosity of implants is not the only predominant factor influencing bone ingrowth. The condition of bone ingrowth is possibly determined by the combination of porosity, pore size, shape and some other factors. More work should be carried out to investigate the action mechanism between bone tissues and porous NiTi scaffolds in the future.

5. Implant Applications

In 1997, Biorthex commenced the development of a porous nickel-titanium alloy, Actipore™, to address the challenge of host bone to implant fixation [139]. The Canadian company has developed three porous NiTi products derived from its Actipore™ technology. Two of them are advanced interbody fusion device indicated for lumbar fusion: the Actipore™ PLF system and the Actipore™ PLFx system, and the third Actipore™ ACF system is indicated for cervical fusion. The photos of these products are shown in Figures 36(a), (b) and (c), respectively. The new porous nitinol product via anterior approach is also been developed (Figure 36(d)). Actually, many research groups including our group in Hong Kong have been developing various porous NiTi implants and applying these products clinically to repair bone tissues in different parts of different animals such as cranial bone of rabbits [34], distal femoral metaphysis of the rats [137], two noncontiguous intervertebral lumbar sites in sheep [37], distal part of femur/ tibia of rabbits under weight-bearing [117], as shown in Figures 36(e), (f), (g) and (h), respectively. These clinical implantations in animals have yielded satisfactory results. That is, porous NiTi SMAs are suitable for bone tissue ingrowth and repair of various injured bones. However, few long-term clinical results have been reported, and the nickel-release behavior in animals after long-term implantation as well as subsequent effects on tissues has not been systematically investigated. The effects of the pore shape, size, and surface morphologies on bone tissue growth are also not known in details. The mechanism between surface layers produced by different surface modification techniques and bone tissues in clinical implantation is also not well understood. Therefore, more work must be done before these new porous NiTi SMAs scaffolds have more widespread clinical uses.

6. Conclusion

Since Itin et al. reported the excellent properties and promising applications in orthopedic fields in 1994 [33], porous NiTi SMAs have attracted the attention of a large number of biomaterials researchers. Many groups have achieved great progress in fabrication, surface modification, and biocompatibility research of porous NiTi SMAs. Besides discussing the research work of our group in this field, the present chapter systematically reviews the progress of this biometal, mainly focusing on the fabrication techniques and modern surface modification processes used in this area as well as the subsequent effects of these products and modification process on the *in vitro* and *in vivo* biological behavior of porous NiTi SMAs.

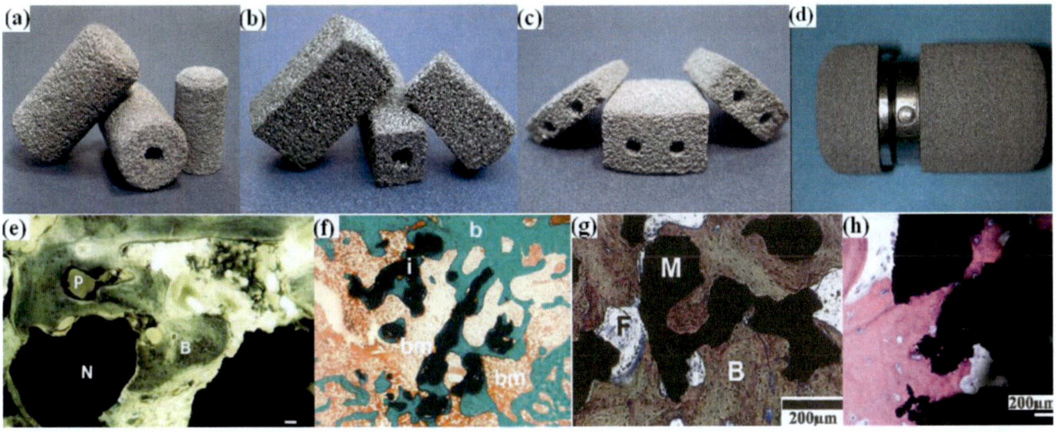

Figure 36. Porous NiTi implant products. (a) PLF and (b) PLFx for lumbar interbody devices fusion devices, (c) ACF implant for anterior cervical interbody fusion devices, and (d) Actipore™ Vertebral Body Replacement System [139], (e) porous nitinol implants to repair cranial bone of New Zealand White rabbits for 12weeks, Ingrown bone (B) is readily distinguished from porosities (P) and the nitinol (N) material. [34], (f) porous nitinol implanted in the distal femoral metaphysis of the rats for 30 weeks (b=bone, i=implant, f=fibrosis, bm=bone marrow) [137], (g) implanted at two noncontiguous intervertebral lumbar sites in sheep for 12 months. Metal (M), cartilaginous tissue (C), osteoid (O), bone tissue (B), and fibrous tissue (F). Stevenel's blue and Van Gieson's picro-fuchsin staining [37], (h) bone tissues can also grow into the exposed internal pores for 105 days (black indicating porous NiTi scaffold, and pink indicating bone tissues) [117].

The PM method is the main fabrication technique. EPS, CS, SHS, HIP, MIM, CF-HIP and PM processes with space holders have been developed for fabrication of the porous NiTi SMAs, and these methods are based on the exothermic reactions between nickel and titanium powders. Several surface modification techniques have been employed to produce bioactive protective layers on the surface of the porous alloy, and these techniques are proven to be effective in suppressing nickel release and improving the biocompatibility of this alloy. Among these techniques, PIII, chemical treatment, and low temperature atmospheric oxidation are the most effective processes due to their non-line-of-sight nature. They are suitable for the surface modification of products with complex morphologies. Both short-term *in vitro* cell culture and *in vivo* animal implantation tests confirm that the porous NiTi SMAs are biocompatible and have no obvious toxicity. Although some products have been developed and implanted in various animals, the mechanisms between this porous implant and bone tissues, such as effects of the porous structure, surface layer, and interface on bone tissue growth, are not well understood. More work must be done in this area as the market demand for bone implants increases in the near future.

References

[1] Profio, AE. Biomedical Engineering. New York: John Wiley and Sons; 1993.

[2] Ramakrishna, S; Mayer, J; Wintermantel, E; Leong, KW. Biomedical applications of polymer-composite materials: a review. *Composites Science and Technology,* 2001, 61, 1189-1224.

[3] http://www.webmd.com/osteoporosis/understanding-fractures-basic-information.

[4] Ettinger, B; Black, DM; Mitlak, BH; Knickerbocker, RK; Nickelsen, T; et al. Reduction of vertebral fracture risk in postmenopausal women with osteoporosis treated with raloxifene - Results from a 3-year randomized clinical trial. Multiple Outcomes of Raloxifene Evaluation (MORE) Investigators. *JAMA : the Journal of the American Medical Association,* 1999, 282, 637-645.

[5] Parsons, JR; Ricci, JL; Alexander, H; Bajpai, PK. Osteoconductive composite grouts for orthopedic use. *Annals of the New York Academy of Sciences*, 1988, 523,190-207.

[6] Willmann, G. Material Properties of Hydroxylapatite Ceramics, *Interceram*, 1993, 42, 206–208.

[7] http://medmarkets.blogspot.com/2006/11/orthopedic-biomaterials-market-segment.html.

[8] Suchanek, W; Yoshimura, M. Processing and properties of hydroxyapatite-based biomaterials for use as hard tissue replacement implants. *Journal of Materials Research*, 1998, 13, 94-117.

[9] William Lu. Bone Cement and Applications. Course AP4173 Lecture, City University of Hong Kong, 2006.

[10] Fung,YC. In Biomechanics. Mechanical Properties of Living Tissues. New York: Springer-Verlag Inc.; 1993.

[11] Park, JB. Biomaterials Science and Engineering. New York: Plenum Press; 1987.

[12] Gunderson, SL; Schiavone, RC. *In International Encyclopedia of Composites*. Lee, SM; New York: VCH Publishers Inc.; 1991; 324.

[13] Katz, JL. In Symposia of the Society for Experimental Biology, Number XXXIV: *The Mechanical Properties of Biological Materials*. England: Cambridge University Press; 1980; 99;

[14] Gibson, LJ; Ashby, MF. Cellular solids. Cambridge: Cambridge Press; 1997.

[15] Piekarsk, K. Fracture of bone. *Journal of Applied Physics, 1970*, 41, 215-223.

[16] Pugh, JW; Rose, RM; Radin, EL. Elastic and viscoelastic properties of trabecular bone-dependence on structure. *Journal of Biomechanics*, 1973, 6, 475-478.

[17] Stone, JL; Beaupre, GS; Hayes, WC. Multiaxial strength characteristics of trabecular bone. *Journal of Biomechanics*, 1983, 16, 743-747.

[18] Gibson, LJ. The mechanical behavior of cancellous bone. *Journal of Biomechanics*, 1985, 18, 317-328

[19] Hayes, WC; Carter, DR. Post-yield behavior of subchondral trabecular bone. *Journal of Biomedical Materials Research*, 1976, 10, 537-544.

[20] Hench, LL. Bioceramics: from concept to clinic. *Journal of the American Ceramic Society*, 1991, 74, 1487-1510.

[21] www.asminternational.org

[22] Helsen, JA; Jurgen Breme, H. Metals as Biomaterials. New York: John Wiley and Sons; 1998.

[23] Besselink, PA; Sachdeva, RCL. Applications of shape memory effects. *Journal De Physique IV*, 1995, C8, 111-116.

[24] Melton, KN. Engineering design and application of shape memory materials, *in Shape Memory Materials*, C.Yougi and T. Hailing, editor, Beijing: International Academic Publishers; 1994; 23-529.

[25] Firkins, P; Sanders, M; Gowland, C; Veldhuizen, A. A SMA-based scoliosis correction system first designs and clinic experience. *SMST-2003: The International Conference on Shape Memory and Superelastic Technologies*; Pacific Grove, CA; USA; 2004; 529-538.

[26] Stockel, D. Status and trends in shape memory technology, *in* Borgmann H. and Lenz, K, editor. *Proceedings of Actuator 92*, Berlin: VDI/VDE -Technologiezentrum Informationstechnik; 1992; 79-84.

[27] Duerig, TW; Melton, KN; Stockel, D; Wayman, CM. Engineering Aspects of Shape Memory Alloys. London: Butterworth-Heinemann; 1990.

[28] Asaoka, K; Yokoyama, K; Nagumo, M. Hydrogen embrittlement of nickel-titanium alloy in biological environment. *Metallurgical and Materials Transactions A-Physical Metallurgy and Materials Science*, 2002, 33, 495-501.

[29] Sigwart U. ed. Endoluminal Stenting. London: W.B. Saunders; 1996.

[30] Duerig, T; Pelton, A; Stöckel, D. An overview of nitinol medical applications. *Materials Science and Engineering A,* 1999, 273-275, 149-160.

[31] Pelton, AR; Hodgson, D; Duerig, T. Proceedings of SMST-94. Shape Memory and Superelastic Technologies. MIAS, Monterey, 1995, 527.

[32] Greiner, C; Oppenheimer, SM; Dunand, DC. High strength, low stiffness, porous NiTi with superelastic properties. *Acta Biomaterialia*, 2005, 1, 705-716.

[33] Itin, VI; Gyunter, VE; Shabalovskaya, SA; Sachdeva, RLC. Mechanical properties and shape memory of porous nitinol. *Materials Characterization,* 1994, 32, 179-187.

[34] Simske, SJ; Sachdeva, R. Cranial bone apposition and ingrowth in a porous nickel-titanium implant. *Journal of Biomedical Materials Research,* 1995, 29, 527-533.

[35] Simske, SJ; Ayers, RA; Bateman, TA. Porous materials for bone engineering. *Porous Materials For Tissue Engineering, Book Series: Materials Science Forum* 1997, 250, 151-182.

[36] Rhalmi, S; Odin, M; Assad, M; Tabrizian, M; Rivard, CH; Yahia, LH. Hard, soft tissue and *in vitro* cell response to porous nickel-titanium: a biocompatibility evaluation. *Bio-Medical Materials and Engineering,* 1999, 9, 151–162.

[37] Michel, A; Peter, J; Michel, AL; Christine, C; Alexandr, VC; Sylvie, C; Charles-H, R. Porous Titanium-Nickel for Intervertebral Fusion in a Sheep Model: Part 1. Histomorphometric and Radiological Analysis. *Journal of Biomedical Materials Research-Part B: Appiedl Biomaterials*, 2003, 64, 107–120.

[38] Prymak, O; Bogdanski, D; Koller, M; Esenwein, SA; Muhr, G; Beckmann, F; Donath, T; Assad, M; Epple, M. Morphological characterization and in vitro biocompatibility of a porous nickel-titanium alloy. *Biomaterials,* 2005, 26, 5801–5807.

[39] Massalski, TB; Okamoto, H; Subramanian, PR; Kacprzak, L. Binary Alloy Phase Diagrams. Materials Park, Ohio: ASM International; 1990; 2874.

[40] Li, BY; Rong, LJ; Li, YY. Stress-strain behavior of porous Ni-Ti shape memory Intermetallics synthesized from powder sintering. *Intermetallics*, 2000, 8, 643-646.

[41] Zhu, SL; Yang, XJ; Hu, F; Deng, SH; Cui, ZD. Processing of porous TiNi shape memory alloy from elemental powders by Ar-sintering. *Materials Letters,* 2004, 58, 2369-2373.

[42] Li, BY; Rong, LJ; Li, YY; Gjunter, VE. An investigation of the synthesis of Ti-50At. Pct Ni alloys through combustion synthesis and conventional powder sintering. *Metallurgical and Materials Transaction A- Physical Metallurgy and Materials Science*, 2000, 31, 1867-1871.

[43] Chu, CL; Chung, CY; Lin, PH; Wang, SD. Fabrication of porous NiTi shape memory alloy for hard tissue implants by combustion synthesis. *Materials Science and Engineering A,* 2004, 366, 114-119.

[44] Li, YH; Rong, LJ; Li, YY. Compressive property of porous NiTi alloy synthesized by combustion synthesis. *Journal of Alloys and Compounds*, 2002, 345, 271-274.

[45] Munir, ZA; Anselmi-Tamburini,U. Self-propagating exothermic reactions: The synthesis of high-temperature materials by combustion. *Materials Science Reports*, 1989, 3, 279-365.

[46] Moore, JJ; Feng, HJ. Combustion synthesis of advanced materials .1. Reaction parameters. *Progress in Materials Science*, 1995, 39, 243-273.

[47] Yi, HC; Moore, JJ. A novel technique for producing NiTi shape memory alloy using the thermal-explosion model of combustion synthesis. *Scripta Metallurgica,* 1988, 22, 1889-1892.

[48] Munir, ZA. Synthesis of high-temperature mateirals by self-propagating combustion methods. American Ceramic Society Bulletin, 1988, 67, 342-349.

[49] Yi, HC; Moore, JJ; Petric, A. Microstructural study of Ti50Pd50 and Ti50Ni50-XPDX alloy produced by combustion synthesis. *Metallurgical Transactions A-Physical Metallurgy and Materials Science*, 1992, 23, 59-64.

[50] Li, BY; Rong, LJ; Li, YY; Gjunter, VE. Synthesis of porous Ni-Ti shape memory alloys by self-propagating high-temperature synthesis: reaction mechanism and anisotropy pore structure. *Acta Materialia,* 2000, 48, 3895-3904.

[51] Kim, JS; Kang, JH; Kang, SB; Yoon, KS; Kwon, YS. Porous TiNi biomaterial by self-propagating high-temperature synthesis. *Advanced Engineering Materials*, 2004, 6, 403-406.

[52] Biswas, A. Porous NiTi by thermal explosion mode of SHS: processing, mechanism and generation of single phase microstructure. *Acta Materialia,* 2005, 53, 1415-1425.

[53] Tokita, M. Trends in Advanced SPS Spark Plasma Sintering Systems and Technology, *Journal Of The Society Of Powder Technology Japan,* 1993, 30, 790–804.

[54] Mamoru, O. intering, consolidation, reaction and crystal growth by the spark plasma system (SPS). *Materials Science and Engineering A*, 2000, 287, 183-188.

[55] Inoue, K. US Patent, 1966; No. 3 241 956.

[56] Inoue, K. US Patent, 1966; No. 3 250 892.

[57] Zhao, Y; Taya, M; Kang, YS; Kawasaki, A. Compression behavior of porous NiTi shape memory alloy. *Acta Materialia*, 2005, 53, 337-343.

[58] Atkinson, HV; Davies, S. Fundamental aspects of hot isostatic pressing: An overview. *Metallurgical and Materials Transactions A- Physical Metallurgy and Materials Science*, 2000, 31, 2981-3000.

[59] Lagoudas, DC; Vandygriff, EL. Processing and characterization of NiTi porous SMA by elevated pressure sintering. Journal of Intelligent Material Systems and Structures, 2002, 13, 837-850.

[60] Greiner, C; Oppenheimer, SM; Dunand, DC. High strength, low stiffness, porous NiTi with superelastic properties. *Acta Biomaterialia*, 2005, 1, 705-716.

[61] Tadic, D; Beckmann, F; Donath, T; Epple, M. Comparison of different methods for the preparation of porous bone substitution materials and structural investigations by synchrotron mu-computer tomography. *Materialwiss Werkst,* 2004, 35, 240-244.

[62] Hayashi, T; Sugihara, S; Okazaki, K. Preparation of porous PZT with high connectivity using capsule-free O_2-HIP and their dielectric-properties. *Ferroelectrics,* 1992, 131, 75-82.

[63] Tamai, M; Miki, S; Pezzotti, G. Fabrication and evaluation of porous hydroxyapatite by pressureless sintering and capsule-free HIP sintering. *Journal of the Ceramic Society of Japan,* 2000, 108, 915-920.
[64] Yuan, B; Chung, CY; Zhu, M. Microstructure and martensitic transformation behavior of porous NiTi shape memory alloy prepared by hot isostatic pressing processing. *Materials Science and Engineering A,* 2004, 382, 181-187.
[65] Wu, SL; Chung, CY; Liu, XM; Chu, PK; Ho, JPY; Chu, CL; Chan, Y L; Yeung, K WK; Lu, WW; Cheung, KMC; Luk, KDK. Pore Formation Mechanism and Characterization of Porous NiTi Shape Memory Alloys Synthesized by Capsule-Free Hot Isostatic Pressing. *Acta Materialia*, 2007, 55, 3437 – 3451.
[66] Bansiddhi, A; Dunand, DC; Shape-memory NiTi foams produced by solid-state replication with NaF. *Intermetallics*, 2007, 15, 1612-1622.
[67] Bansiddhi, A; Dunand, DC; Shape-memory NiTi foams produced by replication of NaCl space-holders. *Acta Biomaterialia*, 2008, 4, 1996–2007.
[68] Kohl, M; Bram, M; Buchkremer, P; Stover, D; Habijan, T; Koller, M. Production of highly porous near-net-shape NiTi components for biomedical applications. In: Metfoam Conference; 2007.
[69] Dunand, DC. Processing of Titanium Foams. *Advanced Engineering Materials,* 2004, 6, 369-376.
[70] Degischer, HP; Kriszt, B. Handbook of Cellular Metals: Production, Processing, Applications. Weinheim: Wiley-VCH; 2002.
[71] Blancodalmau, L; Carrasquilloalberty, DH; Silvaparra, J. A study of nickel allergy. *Journal of Prosthetic Dentistry*, 1984, 52, 116-119.
[72] Jia, WY; Beatty, MW; Reinhardt, RA; Petro, TM; Cohen, DM; Maze, CR; Strom, EA; Hoffman, M. Nickel release from orthodontic arch wires and cellular immune response to various nickel concentrations. *Journal of Biomedical Materials Research,* 1999; 48: 488-495.
[73] Savarino, L; Granchi, D; Ciapetti, G; Stea, S; Donati, ME; Zinghi, G; Fontanesi, G; Rotini, R; Montanaro, L. Effects of metal ions on white blood cells of patients with failed total joint arthroplasties. *Journal of Biomedical Materials Research,* 1999; 47: 543-550.
[74] Villermaux, F; Tabrizian, M; Yahia, L; Czeremuszkin, G; Piron, DL. Corrosion resistance improvement of NiTi osteosynthesis staples by plasma polymerized tetrafluoroethylene coating. *Bio-medical Materials and Engineering*, 1996, 6, 241-254.
[75] Heintz, C; Riepe, G; Birken, L; Kaiser, E; Chakfe, N; Morlock, M; Delling, G; Imig, H. Corroded nitinol wires in explanted aortic endografts: An important mechanism of failure?. Jounal of Endovascular Therapy 2001, 8, 248-253.
[76] Wang, DY; Chiu, MC. Characterization of TiN coatings post-treated by metal-plasma ion implantation process. *Surface and Coatings Technology*, 2002, 156, 201-207.
[77] Trepanier, C ; Tabrizian, M ; Yahia, L ; Bilodeau, L ; Piron, DL. Effect of modification of oxide layer on NiTi stent corrosion resistance. *Journal of Biomedical Materials Research,*1998, 43, 433-440.
[78] Kao, CT; Ding, SJ; Chen, YC; Huang, TH. The anticorrosion ability of titanium nitride (TiN) plating on an orthodontic metal bracket and its biocompatibility. *Journal of Biomedical Materials Research,* 2002, 63, 786-792.

[79] Cisse, O; Savadogo, O; Wu, M; Yahia, L. Effect of surface treatment of NiTi alloy on its corrosion behavior in Hanks' solution. *Journal of Biomedical Materials Research,* 2002, 61, 339-345.

[80] Liu, XY; Chu, PK; Ding, CX. Surface modification of titanium, titanium alloys, and related materials for biomedical applications. *Materials Science and Engineering: Reports*, 2004, 47, 49–121.

[81] Cui, ZD; Man, HC; Yang, XJ. Characterization of the laser gas nitrided surface of NiTi shape memory alloy. *Applied Surface Science,* 2003, 208, 388-393.

[82] Lotkov, AI; Meisner, LL; Grishkov, VN. Titanium nickelide-based alloys: Surface modification with ion beams, plasma flows, and chemical treatment. *Physics of Metals and Metallography,* 2005, 99, 508-519.

[83] Ohgoe, Y; Kobayashi, S; Ozeki, K; Aoki, H; Nakamori, H; Hirakuri, KK; Miyashita, O. Reduction effect of nickel ion release on a diamond-like carbon film coated onto an orthodontic archwire. T*hin Solid Films*, 2006, 497, 218-222.

[84] Gu, YW; Tay, BY; Lim, CS; Yong, MS. Characterization of bioactive surface oxidation. *Applied Surface Science,* 2005, 252, 2038–2049.

[85] Chu, CL; Wu, SK; Yen, YC. Oxidation behavior of equiatomic TiNi alloy in high temperature air environment. *Materials Science and Engineering A*, 1996, 216, 193-200.

[86] Chan, CM; Trigwell, *S*; Duerig, T. Oxidation of a NiTi Alloy. *Surface and Interface Analysis*, 1990,15, 349-354.

[87] Firstov, GS; Vitchev, RG; Kumar, H; Blanpain, B; Van Humbeeck, J. Surface oxidation of NiTi shape memory alloy. *Biomaterials*, 2002, 23, 4863–4871.

[88] Kekare, SA; Sheiton, DK; Aswath, PB. Oxidation of High-Temperature Intermetallics. Warrendale, PA: The Minerals, Metals and Materials Society; 1993; 325.

[89] Wu, SL; Liu, XM; Chan, Y L; Ho, JPY; Chung, CY; Chu, PK; Chu, CL; Yeung, K WK; Lu, WW; Cheung, KMC; Luk, KDK. Nickel Release Behavior, Cyto-Compatibility, and Superelasticity of Oxidized Porous Single-Phase NiTi, *Journal of Biomedical Materials Research A*, 2007, 81, 948–955.

[90] Chu, PK; Chen, JY; Wang, LP; Huang, N. Plasma Surface Modification of Biomaterials. *Materials Science and Engineering: Reports*, 2002, 36, 143 – 206.

[91] Chu, PK; Qin, S; Chan, C; Cheung, NW; Larson, LA. Plasma immersion ion implantation - a fledgling technique for semiconductor processing. *Materials Science and Engineering: Reports*, 1996, 17, 207-280.

[92] Zeng, XC; Kwok, TK; Liu, AG; Chu, PK; Tang, BY; Sheridan TE. Effects of the auxiliary electrode radius during plasma immersion ion implantation of a small cylindrical bore. *Applied Physics Letters,* 1997, 71, 1035-1037.

[93] Zeng, ZM; Chu, PK; Tian, XB; Tang, BY; Kwok, TK. Ion implantation into race surfaces of aerospace ball bearings in a plasma immersion configuration. *IEEE Trans Plasma Science,* 2000, 28, 394–402.

[94] Sheridan, TE; Kwok, TK; Chu, PK. Kinetic model for plasma-based ion implantation of a short, cylindrical tube with auxiliary electrode. *Applied Physics Letters,* 1998, 72, 1826-1828.

[95] Zeng, ZM; Kwok, TK; Tian, XB; Tang, BY; Chu, PK. Investigation of dose uniformity on the inner races of bearings treated by plasma immersion ion implantation. *Journal of Applied Physics,* 1999, 86, 120–123.

[96] Wood, BP; Rej, DJ; Anders, A; Brown, IG; Faehl, RJ; Malik, SM; Munson, CP. Fundamentals of plasma immersion ion implantation and deposition, in handbook of plasma immersion ion implantation, A. Anders (Ed.). New York: John Wiley and Sons; 2000; 243.

[97] Tian, XB; Kwok, DTK; Chu, PK. Modeling of incident particle energy distribution in plasma immersion ion implantation. *Journal of Applied Physics,* 2000, 88, 4961–4966.

[98] Tan, L; Dodd, RA; Crone, WC. Corrosion and wear-corrosion behavior of NiTi modified by plasma source ion implantation. *Biomaterials,* 2003, 24, 3931-3939.

[99] Green, SM; Grant, DM; Wood, JV. XPS characterization of surface modified NiTi shape memory alloy. *Material Science Engineering A,* 1997, 224, 21-26.

[100] Shevchenko, N; Pham, MT; Maitz, MF. Studies of surface modified NiTi alloy. *Applied Surface Science*, 2004, 235, 126–131.

[101] Ho, JPY; Wu, SL; Poon, RWY; Chung, CY; Tjong, SC; Chu, PKK; Yeung, KWK; Lu, WW; Cheung, KMC; Luk, KDK. Oxygen Plasma Treatment to Restrain Nickel Out Diffusion from Porous Nickel Titanium Orthopedic Materials. *Surface and Coatings Technology,* 2007, 201, 4893 – 4896.

[102] Wu, SL; Chu, PK; Liu, XM; Chung, CY; Ho, JPY; Chu, CL; Tjong, SC; Yeung, K WK; Lu, WW; Cheung, KMC; Luk, KDK. Surface Characteristics, Mechanical Properties, and Cytocompatibility of Oxygen Plasma-Implanted Porous Nickel Titanium Shape Memory Alloy, *Journal of Biomedical Materials Research A*, 2006, 79, 139–146.

[103] Chu, PK; Chan, C. Applications of plasma immersion ion implantation in microelectronics-a brief review. *Surface and Coatings Technology,* 2001, 136, 151–156.

[104] Nishio, K; Neo, M; Akiyama, H; Nishiguchi, S; Kim, HM; Kokubo, T; Nakamura, T. The effect of alkali- and heat-treated titanium and apatite-formed titanium on osteoblastic differentiation of bone marrow cells, *Journal of Biomedical Materials Research,* 2000, 52, 652-661.

[105] Han,Y; Fu, T; Lu, J; Xu, K. Characterization and stability of hydroxyapatite coatings prepared by an electrodeposition and alkaline-treatment process. *Journal of Biomedical Materials Research,* 2001, 54, 96-101.

[106] Wang, XX; Hayakawa, S; Tsuru, K; Osaka, A. A comparative study of *in vitro* apatite deposition on heat-, H_2O_2-, and NaOH-treated titanium surfaces. *Journal of Biomedical Materials Research,* 2001, 54, 172-178.

[107] Kokubo, T; Miyaji, F; Kim, HM; Nakamura, T. Spontaneous formation of bonelike apatite layer on chemically treated titanium metals. Journal of the American Ceramic Society, 1996, 79,1127-1129.

[108] Gu, YW; Tay, BY; Lim, CS; Yong, MS. Biomimetic deposition of apatite coating on surface-modified NiTi alloy. *Biomaterials,* 2005, 26, 6916-6923.

[109] Wen, HB; de Wijin, JR; Cui, FZ; de Groot, K. Preparation of calcium phosphate coatings on titanium implant materials by simple chemistry. *Journal of Biomedical Materials Research,* 1998, 41, 227-236.

[110] Xiao, F; Tsuru, K; Hayakawa, S; Osaka, A. In vitro apatite deposition on titania film derived from chemical treatment of Ti substrates with an oxysulfate solution containing hydrogen peroxide at low temperature. *Thin Solid Films,* 2003, 441, 271-276.

[111] Jonasova, L; Muller, FA; Helebrant, A; Strnad, J; Greil, P. Biomimetic apatite formation on chemically treated titanium. *Biomaterials*, 2004, 25, 1187-1194.

[112] Jiang, HC; Rong, LJ; Effect of hydroxyapatite coating on nickel release of the porous NiTi shape memory alloy fabricated by SHS method, *Surface and Coatings Technology,* 2006, 201, 1017-1021.

[113] Wu, SL; Liu, XM; Chan, YL; Chung, CY; Chu, PK; Chu, CL; Yeung, KWK. Biomimetic Deposition of Apatite on Surface Chemically Modified Porous NiTi Shape Memory Alloy. *Surface Review and Letters*, 2008, 15, 97–104.

[114] Wu, SL; Liu, XM; Chan, Y L; Chung, CY; Chu, PK; Chu, CL; Lam,KO; Yeung, K WK; Lu, WW; Cheung, KMC; Luk, KDK. *In vitro* Bioactivity and Osteoblast Response on Chemically Modified Biomedical Porous NiTi Synthesized by Capsule-Free Hot Isostatic Pressing. *Surface and Coatings Technology*, 2008, 202, 2458–2462.

[115] Wu, SL; Liu, XM; Chan, Y L; Chu, PK; Chung, CY; Chu, CL; Yeung, K WK; Lu, WW; Cheung, KMC; Luk, KDK. Nickel Release Behavior and Surface Characteristics of Porous NiTi Shape Memory Alloy Modified by Different Chemical Processes. *Journal of Biomedical Materials Research Part A*, *in press*, 2009.

[116] Moulder, JF; Stickle, WF; Sobol, PE; Bomben, KD; Chastain, J. Handbook of X-Ray Photoelectron Spectroscopy: A Reference Book of Standard Spectra for Identification and Interpretation of XPS Data. Minnesota: Perkin-Elmer, Physical Electronics Division; 1992.

[117] Wu, SL; Liu, XM; Hu, T; Chu, PK; Ho, JPY; Chan, YL; Yeung, K WK; Chu, CL; Hung, TF; Huo, KF; Chung, CY; Lu, WW; Cheung, KMC; Luk, KDK. A Biomimetic Hierarchical Scaffold: Natural Growth of Nano Titanates on Three-Dimensional Microporous Ti-based Metals. *Nano Letters*, 2008, 8, 3803–3808.

[118] http://www.ee.washington.edu/research/microtech/cam/PROCESSES/PDF%20FILES/PhysicalVaporDeposition.pdf

[119] Lemaire, V; Sicotte, B; Allard, S. Surface modification treatments to reduce Ni leaching from porous nitinol. In: Metfoam Conference; 2007.

[120] Hollister, SJ. Porous scaffold design for tissue engineering, *Nature Materials*, 2005, 4, 518-524.

[121] Sclafani, AP; Thomas, JR; Cox, AJ; Cooper, MH. Clinical and histologic response of subcutaneous expanded polytetrafluoroethylene (Gore-Tex) and porous high-density polyethylene (Medpor) implants to acute and early infection Archives of Otolaryngology Head and Neck Surgery, 1997, 123, 328-336.

[122] Chen, G; Ushida, T; Tateishi, T. Scaffold design for tissue engineering. *Macromolecular Bioscience*, 2002, 2, 67-77.

[123] Bansiddhi, A; Sargeant, TD; Stupp, SI; Dunand, DC. Porous NiTi for bone implants: A review. *Acta Biomaterialia*, 2008, 4,773-782.

[124] Bearden, LJ; Cooke, FW. Growth inhibition of cultured fibroblasts by cobalt and nickel. *Journal of Biomedical Material Research,* 1980, 14, 289–309.

[125] Wataha, JC; Hanks, CT; Craig, RG. The *in vitro* effects of metal cations on eukaryotic cell metabolism. *Journal of Biomedical Material Research,* 1991, 25, 1133–1149.

[126] Shih, CC; Lin, SJ; Chen, YL; Su, YY; Lai, ST; Wu, GJ; Kwok, CF; Chung, KH. The cytotoxicity of corrosion products of nitinol stent wire on cultured smooth muscle cells. *Journal of Biomedical Material Research,* 2000; 52: 395-403.

[127] Shabalovskaya, SA. Surface, corrosion and biocompatibility aspects of Nitinol as an implant material. *Bio-Medical Materials and Engineering,* 2002, 12, 69–109.

[128] Bundy, K J; Luedemann, R. Factors which influence the accuracy of corrosion rate determination of implant materials. *Annals Biomedical Engineering,* 1989, 17, 159-175.

[129] Seah, KHW; Thampuran, R; Teoh, SH. The influence of pore morphology on corrosion. *Corrosion Scienc*e, 1998, 40, 547-556.

[130] Azumi, K; Yasui, N; Seo, M. Changes in the properties of anodic oxide films formed on titanium during long-term immersion in deaerated neutral solutions. *Corrosion Scienc*e, 2000, 42, 885-896.

[131] Aziz-Kerrzo, M; Conroy, KG.; Fenclon, AM; Farrell, ST; Breslin, CB. Electrochemical studies on the stability and corrosion resistance of titanium-based implant materials. *Biomaterials,* 2001, 22, 1531-1539.

[132] Jacobs, JJ; Gilbert, JL; Urban, RM. Corrosion of metal orthopaedic implants. *Journal of Bone and Joint Surgery-American Volume,* 1998, 80, 268-282.

[133] Li, YH; Rao, GB; Rong, LJ; Li,YY; Ke, W. Effect of pores on corrosion characteristics of porous NiTi alloy in simulated body fluid, *Materials Science and Engineering A*, 2003, 363, 356–359.

[134] Schrooten, J; Assad, M; Humbeeck, JV; Leroux, MA. In *vitro* corrosion resistance of porous NiTi intervertebral fusion devices, *Smart Materials Structures,* 2007, 16, S145–S154.

[135] Wu, SL; Liu, XM; Jiang, J; Hu, T; Chu, PK; Yeung, KWK; Chu, CL; Chung, CY; Lu, WW; Cheung, KMC; Luk, KDK. Electrochemical stability of orthopedic porous NiTi shape memory alloys treated by different surface modification Techniques. *Journal of Electrochemical Society,* submission for publication.

[136] Gu, YW; Li, H; Tay, BY; Lim, CS; Yong, MS; Khor, KA. In vitro bioactivity and osteoblast response of porous NiTi synthesized by SHS using nanocrystalline Ni–Ti reaction agent. *Journal of Biomedical Materials Research Part A,* 2006, 78, 316-323.

[137] Kujala, S; Ryhanen, J; Danilov, A; Tuukkanen, J. Effect of porosity on the osteointegration and bone ingrowth of a weight-bearing nickel–titanium bone graft substitute. *Biomaterials,* 2003, 24, 4691-4697.

[138] Ayers, RA; Simske, SJ; Bateman, TA; Petkus, A; Sachdeva, RLC; Gyunter, VE. Effect of nitinol implant porosity on cranial bone ingrowth and apposition after 6 weeks. *Journal of Biomedical Materials Research*, 1999, 45, 42-47.

[139] http://www.biorthex.com/.

In: Shape Memory Alloys
Editor: H. R. Chen, pp. 113-143
ISBN 978-1-60741-789-7

Chapter 4

STRUCTURE AND PROPERTIES OF A MONOCRYSTALLINE CU-AL-NI ALLOY SUBMITTED TO THERMAL CYCLING UNDER LOAD

L. A. Matlakhova, E. C. Pereira, A. N. Matlakhov and S. N. Monteiro

State University of the Northern Rio de Janeiro, Advanced Materials Laboratory

ABSTRACT

Monocrystalline Cu-Al-Ni alloys are promising alternatives to be used in shape memory effect (SME) actuators, in which the material suffers under temperature and applied load a combined action of many factors including thermal cycling as well as elastic and plastic deformations that modifies the initial structure and characteristic properties. In this chapter, a monocrystalline Cu-13.5wt.%Al-4wt.%Ni (Cu-13.5Al-4Ni) alloy displaying SME was submitted to thermal cycling treatments, load-free and under load, within the critical range of temperatures associated with reversible martensitic transformations (RMT). The investigation was conducted in terms of number of cycles, magnitude of applied load, resulting structure changes, characteristics of the RMT and mechanical behavior. The alloy was produced by the "*Memory Crystal Group*" in Russia and its characterization was carried out using optical and electron microscopy, X-rays diffraction, differential scanning calorimetry and mechanical tests. A special device was developed to perform the hundreds of thermal cycling treatments under load. It was found that the thermal cycling treatment promotes changes in the amount of phases - a martensite γ'_1, a high temperature stable β_1 and an intermediate R phase - associated with the RMT. These changes depend on the way the treatment ends, either by heating or by cooling. Load-free thermal cycling contributes to enlarge the temperature interval of RMT occurrence, but does not change the interval of hysteresis. After 300 cycles, if the end occurs by cooling below M_f, an intermediate R phase, related to the $\gamma'_1 \leftrightarrow \beta_1$, is obtained. The alloy submitted to thermal cycling under load reveals a tendency towards a decrease in the critical intervals for both direct and reverse martensite transformation. This tendency is accentuated for 300 cycles, when the critical intervals smaller than 10 °C, are recorded. The applied load promotes the reverse transformation at lower temperatures as well as the direct martensitic transformation at higher temperatures, reorienting the structure and stabilizing the R phase, coherent with γ'_1 and β_1. The alloy's mechanical behavior is characterized by a complex process involving both its RMT and phase reorientation. The results of compressive tests of the initial state revealed a "pseudo-yielding" plateau followed by elastic deformation up to 1350 MPa and subsequent fracture, disclosing its brittleness. Thermal cycling under load reduces the pseudo-yield stress by 20-25 MPa and decreases the ultimate stress to 950 MPa. The alloy shows good resistance to irreversible changes during all thermal cycling tests. These results have a particular relevance for the practical application of the monocrystalline Cu-13.5Al-4Ni alloy as SME actuator.

Keywords: Cu-13.5Al-4Ni monocrystalline alloy; Thermal cycling treatment; Structural changes; Reversible martensite transformations; Effect of applied load

1. INTRODUCTION

Alloys that present shape memory effect (SME) and reversible martensitic transformations (RMT) [1-6] are still attracting considerable attention to research works in possible applications associated with the technology of intensive actuator devices [7-10]. In this respect, one of the challenges for the practical use of these devices is the extreme sensitivity of the SME parameters to processing conditions, which may affect the desirable properties by means of changes caused to the microstructure and mechanical behavior [10-17].

The relation between SME and microstructural modifications has been well documented [1-6,11-18]. In fact, SME was found to be exhibited by alloys that undergo RMT with relevant changes in the existing phases [2,3]. These are diffusionless phase transformations, in which atoms move cooperatively and often by a shear-like mechanism. Usually the parent phase, stable at high temperature (A) is cubic, mostly a β-ordered phase [19]. By contrast, the martensite (M), which is a low temperature metastable phase, has a lower symmetry [1-6]. It is possible to observe different types of martensitic phases depending on the alloy crystallographic system and chemical composition as well as the investigation procedure [2-6,17,18]. The RMT is crystallographically reversible and takes place with small variation of the system Gibbs energy (< 100 J/mol) [2-6] within a narrow interval of temperature, M_s-M_f on cooling, and A_s-A_f on heating [5,6,15-18]. Moreover, the A↔M cycle is associated with a relatively small temperature interval of thermal hysteresis during which the A/M interface remains coherent or semi-coherent with respect to both crystalline lattices [5,6,18]. As the RMT proceeds, with variation in temperature, the new phase tends to accommodate into the initial phase without any change in chemical composition. For RMT to occur, it, is important not to alter the metastable nature of the participant phases that suffer neither decomposition nor evolution to stable structure [5,6,20,21].

In many practical situations, SME devices perform under repetitive operational regimes [7-10]. This is associated with multiple heating and cooling cycles, within an interval that includes the critical RMT temperatures. In additional to thermal cycles, the SME may require an applied load, which is the typical case of actuator-type of devices, such as thermal motors. The applied load causes localized stresses that interfere with the RMT. A stress assisted SME may induce other phase transformations that result in complex changes in the structure and properties of the material [5,6].

Although different alloys were found to display SME associated with RMT, special attention has been given to these aspects in copper-based alloys [22-33]. Among the alloys with SME, the TiNi based ones are considered the most promising for technological applications owing to their superior SME parameters, associated with good mechanical properties and excellent corrosion resistance [1,5,6,34,35]. TiNi based alloys are, however, relatively expensive and hence other SME alloys have also been proposed. Less expensive Cu-based SME alloys present additional advantages in terms of superior electrical and thermal conductivities [36-38]. Moreover, their deformability, which is an important property for wire fabrication, needed in thermal motor coils, is superior than that of TiNi SME alloys [38]. In the case of polycrystalline Cu-Al-Ni, the alloy's low deformability is, however, an exception. These alloys are susceptible to intergranular cracking, which limits the scope of many potential applications. It was found [36] that the reason for Cu-Al-Ni intergranular cracking is the formation of stress-induced martensite along grain boundaries upon quenching.

Polycrystalline SME alloys, in general, are not as effective as single crystals in terms of SME parameters. The recoverable strains in the former are considerably reduced since an applied stress generates strains that differ from grain to grain and, on average for the bulk material, are close to zero. For this reason, and the above-mentioned occurrence of intergranular cracking, polycrystalline Cu-Al-Ni SME alloys present inherent limitations that impair their mechanical response for practical SME purposes [38,39].

A particular interest is currently being focused in monocrystalline Cu-Al-Ni alloys owing to their desirable SME parameters, in association with a tendency not to accumulate non-reversible plastic deformation [23,30-32, 39,40]. In comparison with similar polycrystalline alloys, monocrystalline Cu-Al-Ni presents higher pseudo elastic deformation, up to 12%, and SME back stress above 600 MPa, as well as a low level of residual deformation [6,8,10,26,31-33].

One of the limitations for the practical application of an SME alloy is the alteration in behavior, which is commonly developed after multiple thermal cycles. Cycling within the temperature interval of martensite transformation is known to cause changes in the substructure and phase composition as well as a consequent modification in the physical and mechanical properties of the alloys [23,30-33]. In fact, each thermal cycle may produce additional dislocations that, in turn, may interfere with the martensitic transformation. These dislocations might be responsible for shifts in the values of the characteristic temperatures associated with the RMT, as reported in the work of Nakata et al. [22]. In that work, however, the effect of a superimposed stress during a thermal cycling treatment had not been evaluated.

Among the copper-based alloys, the monocrystalline ones displaying RMT were found to exhibit superior SME parameters in association with a reasonable ductility [6,8,9,17,38]. In particular, the monocrystalline Cu-Al-Ni alloys, based on the intermetallic Cu_3Al, presents a stable high temperature β_1 type DO_3 ordered body centered cubic structure that, upon quenching, may transforms into martensite β'_1, with 18R orthorhombic structure, and martensite γ'_1, type Cu_3Ti ordered (2H) structure [2,3,6,17,22-30,41-45].

The practical advantages of the monocrystalline Cu-Al-Ni are their good conformability and elevate resistance to embrittlement, normally caused by an applied temperature variation [6,9,17,22]. Another positive aspect is that these alloys accumulate less quantity of crystalline defects during thermal cycling and present only small changes in their SME parameters [24-31]. However the influence of the stress due to an applied load has yet to be evaluated [21,24,31-33].

Therefore, the objective of this work was to investigate the modifications in the RMT parameters as well as the changes in the structure and mechanical behavior, that occurred in a monocrystalline copper based alloy with 13.5 wt.%Al and 4.0 wt.%Ni as a result of load-free thermal cycling and thermal cycling under two different levels of loading, within an interval covering the critical RMT temperatures. The thermal cycling treatments under applied load were conducted in tests using a compressive elastic stresses. After thermal cycling, the mechanical behavior was evaluated by compression tests performed up to fracture.

2. Material and Experimental Procedure

A monocrystalline copper based alloy with 13.5 wt.%Al and 4.0 wt.%Ni (Cu-13.5Al-4Ni) was obtained in the form of a 5 mm in diameter cylindrical bar. The alloy was fabricated by the "*Memory Crystal Group*" of the Technical University of Saint Petersburg, Russia, according to a procedure described elsewhere [8]. Disc-shaped samples were initially sectioned from the cylindrical monocrystalline alloy using a mechanical MINITON cutter. These samples were then prepared for metallographic examination by final polishing with 0.1 μm alumina paste. No chemical attack was used and the examination was carried in a NEOPHOT-32 optical microscope.

The phase composition of the alloy was determined by X-ray diffraction (XRD) in both a DRON-3M and a URD-65 diffractometers using Cu-K_α and Co-K_α radiations, respectively, for 2θ angles from 25 to 75° at scanning steps of 0.03° per 3s and 0.05° per 3s. Both equipments were operating at ambient temperature.

The determination of the RMT critical temperatures and associated thermal hysteresis as well as corresponding thermal effects were obtained by differential scanning calorimery (DSC) in a model DSC-2010, TA-Instrument equipment. The DSC curves were determinated under inert atmosphere between −100°C and +150 °C at a heating rate of 10 °C/min. Cooling was carried out using liquid nitrogen. From the DSC curves, the critical RMT temperatures were evaluated according to a methodology proposed by distinct research groups [6,22,29-31,46]. Initially, in this methodology, tangents were drawn at the inflexion points in the curves corresponding to the endothermic and exothermic peaks. It was then possible to evaluate the temperatures A_o, M_o of the "*on-set*" and A_{of}, M_{of} of the "*off-set*" associated with the interval were the martensitic transformation occurs in an intensive way. The peak temperatures, A_e and M_e, that correspond to the maximum heat flow during transformation were visually established. The conventional A_s, A_f, M_s and M_f temperatures were obtained by considering the point of deviation of the peak curves from the base heat flow line. These conventional temperatures A_s, A_f, M_s and M_f depend on human interpretation while the critical temperatures A_o, A_{of}, A_e, M_o, M_{of} and M_e can be automatically determined by computer calculation [46]. In the present work, these latter critical temperatures will be used to interpret the interval where the major A↔M martensitic transformation occurs.

Since the room temperature (RT), 25 ±5 °C, is inside the interval of critical temperatures, one should expect to observe the changes in the alloy structure as RT is reached after the cycling treatment. In order to do this, a sample of the initial alloy structure was immersed in boiling water, +100 °C and subsequently cooled in water to RT (half heating cycle), before being submitted to XRD. Then, the same sample was cooled down to −15 °C and heated in water to RT (half cooling cycle) before being submitted, again, to XRD.

The specimens were then submitted to 100, 200 and 300, load-free and under load thermal cycles. The load-free thermal cycling (TC) was performed for 1, 100, 200 and 300 cycles of heating to 100 °C (above A_f) in boiling water and cooling to −15 °C (below M_f) in a cryogenic brine mixture composed of crushed ice and NaCl.

Three distinct thermal cycling under load (TCL) involving 100, 200 and 300 cycles of heating up to 100 °C, above A_f, and cooling down to 0 °C, below M_f, were performed under a static load of 0.5 kg, corresponding to a compressive stress of 0.275 MPa. This stress is a fraction of the alloy's "pseudo-yield" stress, determined as 60–65 MPa [26,31-33]. In other tests, TCL were performed under different loads of 0.2 and 0.5 kg involving 300 cycles.

In order to observe the morphological changes that occur in the alloy's surface during a cooling and a heating cycle associated with the RMT interval of critical temperatures, a special apparatus was developed. This consists of a thermal insulated vessel adapted to the optical microscope [24,25]. A mixture of water/gel (heating) or water/ice (cooling) was used inside the vessel. The metallographically prepared alloy sample was held onto the vessel for direct observation under the NEOPHOT-32 microscope at a desired temperature.

The structure characterizations were performed at room temperature, after TC and TCL. The treatment ended either by cooling, with a half cooling cycle, or by heating, with a half heating cycle, as schematically shown in Figure 1.

The DSC analyses were made in samples of the alloy under the following conditions: in its initial state, after one cycle of heating and cooling without loading, and after 100, 200 and 300 cycles of both TC and TCL.

Compression tests after the 300 TC and 300 TCL were carried out at RT until fracture, in a model 5582 Instron machine operating with a crosshead speed of 0.1 mm/s. The fracture surfaces of these specimens were analyzed by scanning electron microscopy, SEM, in a model DSM-962 Zeiss microscope. SEM images were obtained with secondary electrons accelerated at 15 kV.

3. Results and Discussion

3.1. Initial State

According to the metallographic analysis of a transversal section, the monocrystalline Cu-13.5Al-4Ni alloy in its initial state, Figure 2(a,b), presents the typical martensitic structure formed by needles and laths, associated with the RMT. This structure is composed of a larger central square-shaped block surrounded by four other small blocks in the periphery of Figure 3(a). In each of these blocks, the typical martensitic relief can be seen with its own orientation. The martensitic laths are entirely twinned cross each other in a V shaped pattern, which is typical of the γ_1' martensitic morphology [2,3,6,24-31].

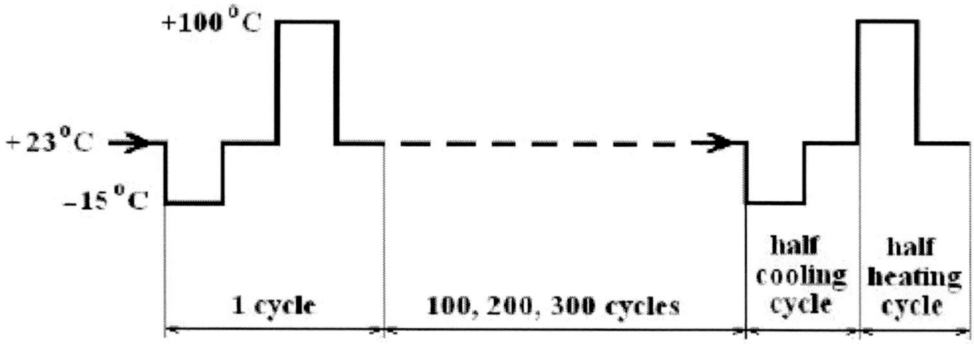

Figure 1. Schematic of cycling treatment.

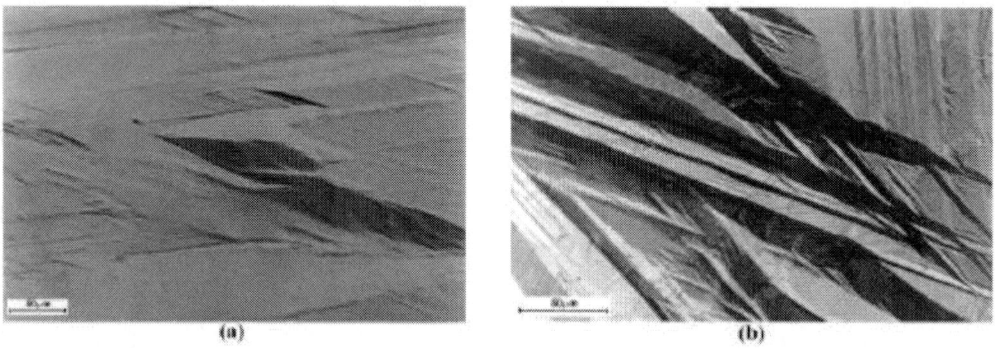

Figure 2 – Surface morphology of the monocrystalline Cu-13.5Al-4Ni alloy: (a) 200x; (b) 400x.

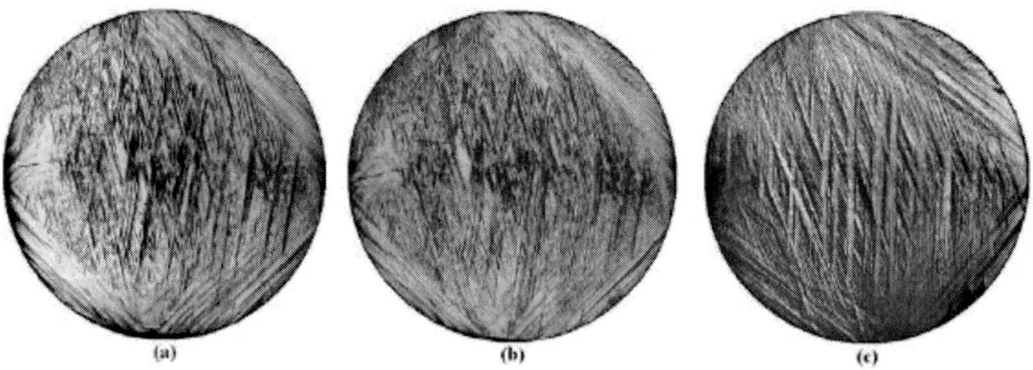

Figure 3. Micrographs of the whole transversal section of the monocrystalline Cu-13.5Al-4Ni alloy (total diameter of 5 mm): in its initial state (a) and in a first thermal cycle (b) a half cooling part of the cycle and (c) a half heating part of the cycle (see Figure 1).

The structure of the alloy in its initial state, identified by the XRD pattern in Figure 4(a), presents three phases [24-31]. The first one is a high temperature type DO_3 ordered β_1 phase [41] the second is a type Cu_3Ti ordered martensitic γ'_1 [42], and the third is a trigonal Al_7Cu_4Ni ordered rhombohedric phase, with a 166-R-3m space group [43-45].

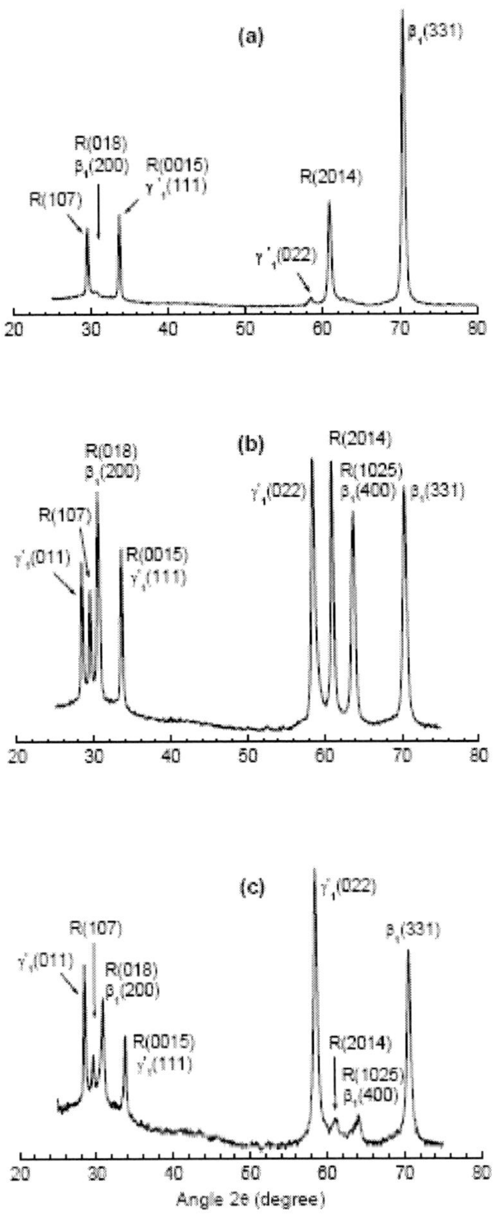

Figure 4. X-ray diffraction patterns of the monocrystalline Cu-13.5Al-4Ni alloy: in its initial state (a) and in the first thermal cycle: half cooling part of the cycle (b) and half heating part of the cycle (c).

In this work, this latter will be denominated as the R phase. The XRD in Figure 4(a) shows that the R phase has some peaks coinciding with those of the β_1 and γ_1' phases. Moreover, the β_1 presents an intense $(331)_{\beta 1}$ peak, while the martensitic phase γ'_1 two less intense $(022)_{\gamma'1}$ and $(111)_{\gamma'1}$ peaks. In addition to these peaks, as shown in Figure 4, at the angular positions (2θ) of approximately 29,6° and 61°, two peaks $(107)_R$ and $(2014)_R$ can be attributed to the rhombohedric R phase. Other peaks found for this phase such as $(200)_{\beta 1}|(018)_R$ and $(111)_{\gamma'1}|(0015)_R$ are coinciding with peaks belonging to β_1 and γ'_1 phases [41-45]. However, since the martensitic structure of the alloy shows no microscopic evidence for the presence of this rhombohedric phase, it is suggested that the R structure can be considered as a transient state, resulting from the ordering of Ni atoms incorporated into the alloy's crystalline lattice of Cu_3Ti. No indication of dispersed particles of R phase was found in the alloy microstructure. This, together with the coinciding XRD peaks, β_1 and γ'_1 in Figure 4, indicates that these phases are coherent with R, sharing the same needle-like morphology. In other words, the R phase is an intermediate of the β_1 and γ'_1 phases.

3.2. Load-Free Thermal Cycling Tests

As previously determined, [24-33], the monocrystalline Cu-13.5Al-4Ni alloy suffers a type A↔M RMT from its initial state. Upon heating, a reverse $\gamma'_1 \rightarrow \beta_1$ martensitic transformation takes place in the interval from +9 °C (A_s) to +52 °C (A_f). This is an endothermic process associated with a phase transformation enthalpy of 7.36 J/g. Moreover, this process develops from an onset at 25 °C (A_o) and presents an endothermic peak at +33 °C (A_e), which corresponds to a maximum heat flow. Upon cooling, the direct $\beta_1 \rightarrow \gamma'_1$ martensitic transformation takes place from +25 °C (M_s) to −9 °C (M_f) in association with an onset at +18 °C (M_o) and a peak at +14 °C (M_e). This exothermic process has an enthalpy of 6.04 J/g. During both, heating and cooling processes, the RMT critical temperatures (A_s, A_f, M_f and M_s) as well as the onset (A_o and M_o) and peak temperatures (A_e and M_e) do not superimpose. This behavior is characteristic of the phenomenon known as the RTM thermal hysteresis [2,3,6]. As a conventional measure of the temperature interval of thermal hysteresis, the present work adopted the difference between the peaks of heating and cooling, corresponding to the critical temperatures ($\Delta T = A_e - M_e$). For the alloy investigated, ΔT was found to be around 20 °C. The temperatures *off set*, on heating and cooling, were determined as 41 °C (A_{of}) and 0.5 °C (M_{of}).

After the first thermal cycle, which starts at RT proceeding by cooling to −15 °C and ending by heating from −15 °C to RT, denominated as a half cooling cycle, the alloy morphology, Figure 3(b) looks like the initial RT state, shown in Figure 3(a). On the other hand, by initially heating to 100 °C and ending by cooling from 100 °C to RT, a half heating cycle, it can be noted, Figure 3(c), a V shaped pattern and the large needle-like morphology of γ'_1 martensite variants in the optical microscopy contrast of a parent β_1 phase. Fine shear lines extending across the parent phase and the γ'_1 martensite variants could probably be associated with the intermediate R phase, Figure 3(b,c).

The influence of thermal cycling on the structure of this alloy is shown in Figure 4 by XRD analysis [30]. As above-mentioned, the initial state XRD pattern, Figure 4(a), reveals a greater participation of the high temperature stable β_1 phase in comparison to γ'_1 and R. In a first thermal cycling, the alloy presents different structural characteristics than those at RT, depending on the ending of the cycling treatment, whether by cooling, with half cooling cycle, Figure 4(b) or by heating, with half heating cycle, Figure 4(c). The basic structure is composed of γ'_1, β_1 and R with ordering of Ni. However, the planes that identify the same phases suffer a definite change. The additional XRD peaks, resulting from ordering of Ni atoms participating in the formation of the Al_7Cu_4Ni, the R phase, are more pronounced after a half cooling cycle Figure 4(c).

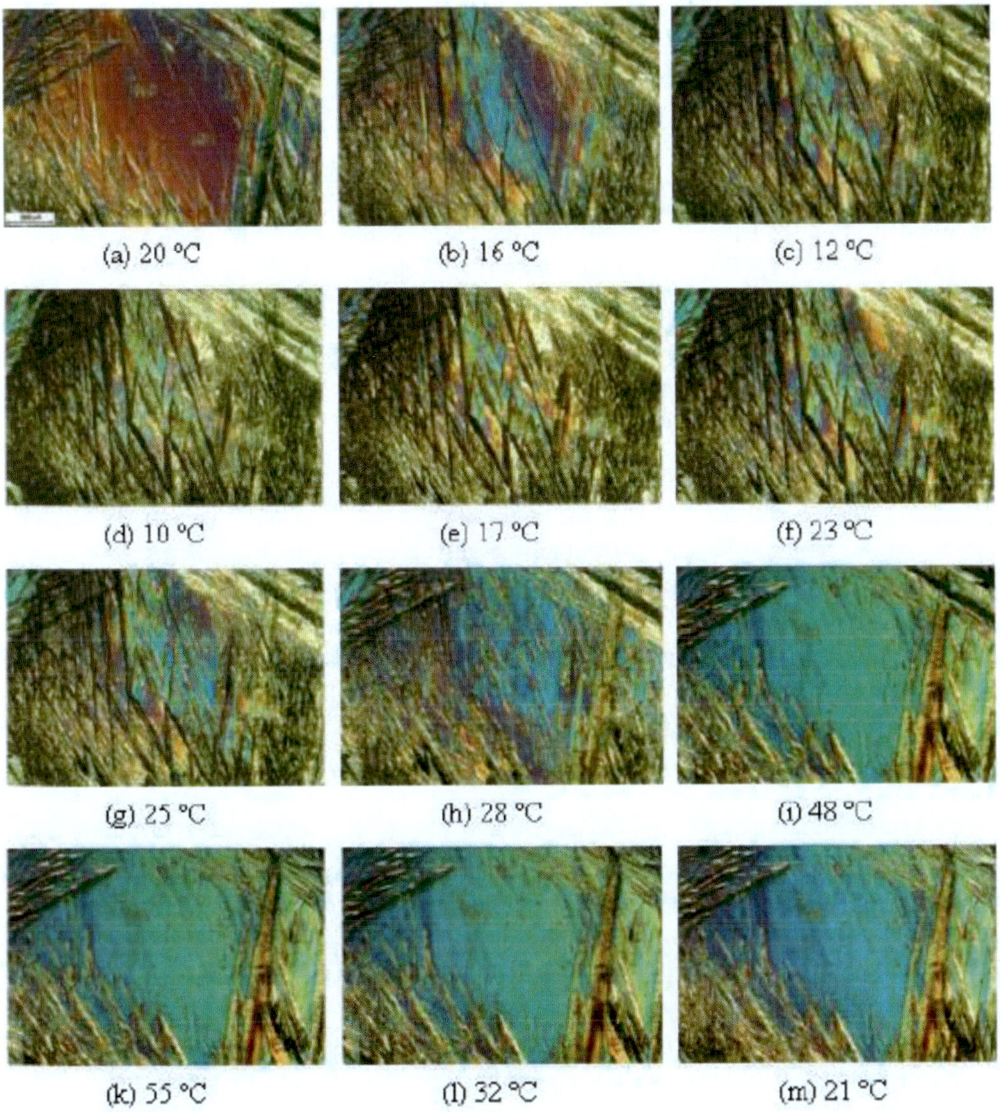

Figure 5. Morphology of the monocrystalline Cu-13.5Al-4Ni alloy during cooling (a,b,c,d), heating (e,f,g,h,i,k) and cooling (l,m) first cycle (Differential Interference Contrast; 25x).

Figure 5 presents the typical morphological changes in the surface of the monocrystalline Cu-13.5Al-4Ni alloy during a cooling cycle from RT to 10 °C followed by heating to 55 °C and then back to RT. The initial RT surface structure, Figure 5(a) is composed of a square-shaped block at the center of the 5 mm in diameter section. This central block is surrounded by four peripherical blocks with more pronounced participation of crossing needle-like martensite marks. The central block in Figure 5(a) has most of its area free of marks. By cooling below RT, Figure 5(b) to 5(d), significant changes occurred in the surface morphology. An intense martensite relief is developed on the central block, mainly on areas that were free of marks at RT. This lower temperature relief is formed by strong V-shaped γ'_1 martensite marks. The temperature of 10 °C was found to have the highest density of martensite. Therefore, by optical observation, this temperature can be considered the closest to the critical, M_f, for finishing the direct $\beta_1 \rightarrow \gamma'_1$ transformation.

By heating the previously cooled alloy, the surface morphology changed in a reverse manner, Figure 5(e) to 5(k). At 55 °C, Figure 5(k), most of the surface of the central block becomes free of marks, just like the initial state, Figure 5(a), at RT. This can be considered close to the critical, A_f, for finishing the reverse $\gamma'_1 \rightarrow \beta_1$ transformation. An important point to be mentioned is the fact that, different than the clear central morphology at RT, Figure 5(a), observed on cooling, the RT surface morphology on heating, Figure 5(g), is still covered with marks. This is due to the thermal hysteresis that causes distinct amounts of martensite observed at the same temperature, if the cycle is in its cooling, Figure 5(a), or in its heating, Figure 5(g), part.

In the last part of the cycle, by cooling from 55 °C to RT, Figure 5(k) to 5(m), the surface morphology is practically not changed. This means that the high temperature β_1 stable predominates during the cooling part of the hysteresis loop, from 55 °C to RT. In fact, the start, Figure 5(a), and the finish, Figure 5(m), of the thermal cycle display the same morphology with most of the central block surface being free of martensite marks.

The observations in Figure 5 revealed a significant influence of the temperature on the surface morphology. It confirms that the RMT occurring in the monocrystalline Cu-13.5Al-4Ni is crystallographically reversible within an interval of approximately 20 °C. As a consequence, the RMT thermal hysteresis causes changes in both the phasic composition and the alloy structure. As in any typical RMT, these changes depends on the part of the thermal cycle, either cooling or heating, in which the alloy is within the critical interval, M_s and A_f, of temperatures.

In practice, these changes are relevant for actuator devices made of a SME alloy, such as the monocrystalline Cu-13.5Al-4Ni. Attention should be paid to the thermal hysteresis involved during the device's operation, which is associated with modifications in the structure and composition of the material. Depending on the external factors, temperature and applied load, a distinct reaction can occur in the material's performance. The effect of the applied load will be investigated in the next section of this work.

The DSC analysis of the thermal cycling treated alloy without an applied load [24-31] revealed that changes in the critical temperatures are associated with the number of cycles. Figure 6 shows the variation of these critical temperatures during: (a) cooling in which γ'_1 (M) is directly transformed and (b) heating in which β_1 (A) is reversely transformed. As the number of cycles increases up to 300 TC, the total interval of RMT (M_s–M_f and A_s–A_f) is extended and the peaks (M_e and A_e) are displaced to lower temperatures. It was observed that the critical temperatures, A_e and M_e, (curve 2) decrease to values that are 5 to 7 °C lower at the end of the 300 TC treatment. However, the magnitude of the RMT thermal hysteresis stays the same, around 20 to 21 °C, even after 300 TC. Furthermore, it was noted that the M_o-M_e interval, curves 2 and 3 in Figure 6(a), which is 5 °C for the initial state, increases to 11 °C at the end of the 300 TC. By contrast, the A_o-A_e interval, curves 2 and 3 in Figure 6(b), which is 7 °C for the initial state, remains almost constant for any number of cycles up to 300 TC. It should also be noticed that M_s is displaced to higher values and so increase the M_s-M_o interval with the number of cycles [29].

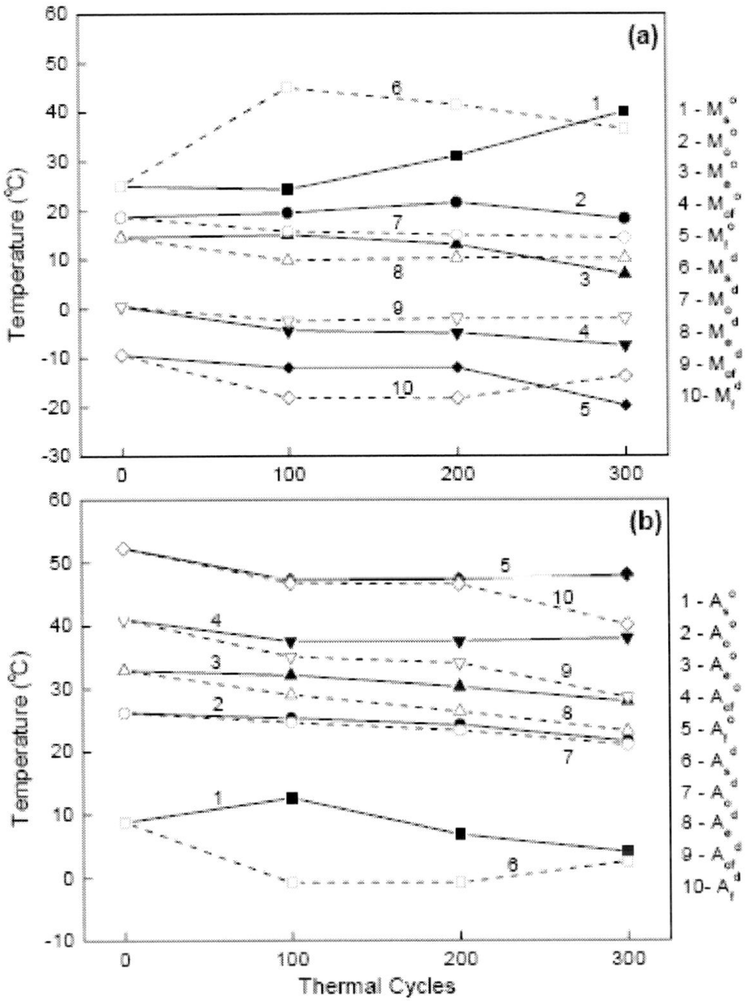

Figure 6. RMT critical temperatures obtained by DSC during (a) cooling, M, and (b) heating, A, of the monocrystalline Cu-13.5Al-4Ni alloy as a function of the number of cycles, load-free TC (1-5) and TCL under 0.5 kg load (6-10).

Figure 7 presents the results obtained by XRD after: (a-b) 200 TC and (c-d) 300 TC. Depending on the end of the cycling treatment, Figure 7(a,c) corresponds to a half cooling cycle, while Figure 7(b,d) to a half heating cycle. The results in Figure 7 were compared with those obtained, Figure 4(a) for the initial state and, Figure 4(b,c) after the first thermal cycle. This comparison was based on the intensity of the $(331)_{\beta 1}$ and $(111)_{\gamma' 1}$ peaks that exist in both the initial state and after the thermal cycling treatments. It can be seen that after the first cycle, Figure 4(b,c), upon cooling or heating, the intensity of the martensitic phase γ'_1 increases, while β_1 decreases. Regarding the R phase, apparently a re-orientation occurs. This can be inferred by the pattern corresponding to the first cooling cycle, Figure 4(b), in which β_1 presents additional $(400)_{\beta 1}$ and $(200)_{\beta 1}$ peaks very close to $(1025)_R$ and $(018)_R$, respectively, corresponding to the R phase.

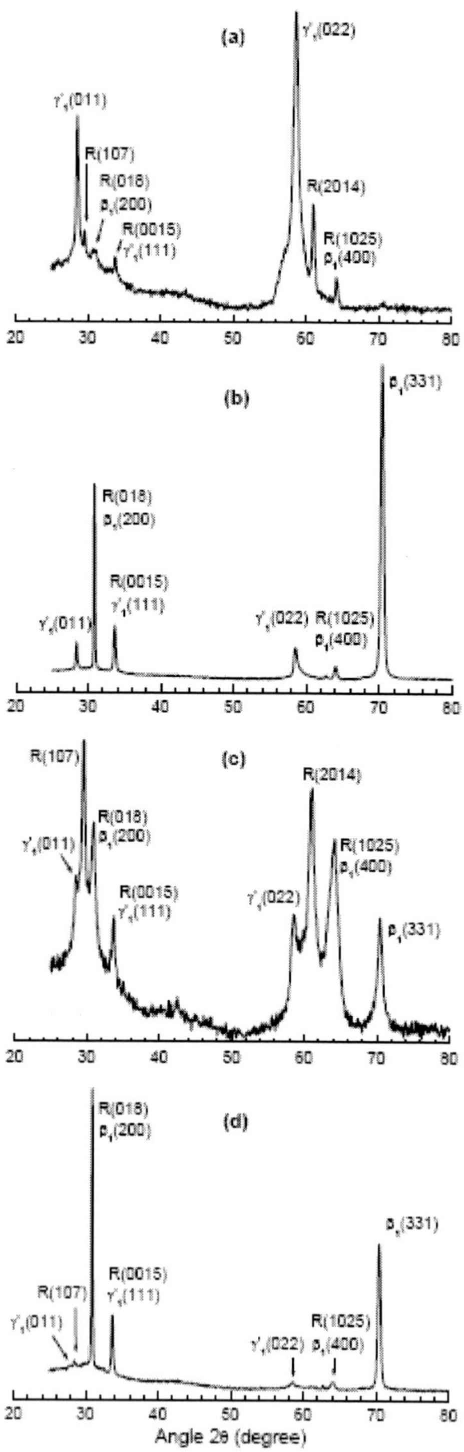

Figure 7. X-ray diffraction patterns for the monocrystalline Cu-13.5Al-4Ni alloy after load-free 200 TC (a,b) and 300 TC (c,d) ending with a half cooling cycle (a,c) and a half heating cycle (b,d).

It is worth mentioning that in the first cycle cooling cycle γ'_1 presents two peaks $(022)_{\gamma'1}$ and $(011)_{\gamma'1}$ that were not detected in the as-received initial state, Figure 4(a).

The first heating cycle, Figure 4(c), displays similar features with $(400)_{\beta1}|(1025)_R$, $(200)_{\beta1}|(018)_R$, $(022)_{\gamma'1}$ and $(011)_{\gamma'1}$ peaks. The same $(107)_R$ e $(2014)_R$ peaks for the R phase are also observed, Figure 4(c), with significant lower intensity in comparison not only to those observed, Figure 4(b), after the first cooling cycle but also to those, Figure 4(a), for the initial state.

After 200 TC ending with a half cooling cycle, Figure 7(a), the XRD pattern shows the presence of all three phases with a greater participation of the martensitic γ'_1. In comparison with the first half cooling cycle, Figure 4(b), the presence of both β_1 and R is less intense. By contrast, after 200 TC ending with a half heating cycle, Figure 7(b), the intensity of the β_1 peaks, $(200)_{\beta1}$ and $(331)_{\beta1}$ is sensibly higher than the corresponding, Figure 4(c), first half heating cycle. Furthermore, the martensite peaks $(022)_{\gamma'1}$ and $(011)_{\gamma'1}$ display a considerably lower intensity while the R peaks $(107)_R$ e $(2014)_R$ disappeared in the 200 TC.

The accumulation of 300 TC ending with a half cooling cycle, Figure 7(c), still shows the presence of all the three phases: γ'_1, β_1, and R. The main difference is that their peaks are considerably broader than the corresponding for the first half cooling cycle in Figure 4(c). Moreover, the β_1 peaks are displaced from the original position; $(200)_{\beta1}$ to the right, and $(331)_{\beta1}$ to the left. It is suggested that the broad peaks in Figure 7(c) could be a consequence of structural defects such as dislocations accumulated after 300 TC. These defects tend to retard the RMT process and stabilize transitions states like the R phase. On the contrary, after 300 TC ending with a half heating cycle, Figure 7(d), The XRD pattern depicts relatively sharp β_1 peaks with intensities comparable to the corresponding, Figure 7(b), for 200 TC. Following the tendency of the first, Figure 4(c), and 200 TC, Figure 7(b), half heating cycles, the γ'_1 and R peaks for the 300 TC were being diminished, Figure 7(d), in intensity.

Figure 8 presents micrographs of the alloy after 300 TC ending with a half cooling cycle, Figure 8(a), and a half heating cycle, Figure 8(b). In comparison with the results of the first cycle, Figure 3(b,c), the micrographs in Figure 8 illustrate the typical martensitic structure, needles and laths, associated with the RMT. These micrographs revealed that, after hundreds of thermal cycling within an interval that includes the critical temperatures, significant changes might occur [24-31]. Actually, the multiple reversible martensitic transformations, which take place during the thermal cycling treatment of the alloy, contribute to accumulate structural defects that retard the RMT process. As a consequence, the critical temperatures are displaced to other values while their range of occurrence is enlarged. Moreover, the high temperature phase, β_1 and R coherent with β_1 in the present case, become dominant with increasing thermal cycling.

3.3. Thermal Cycling Under Load

The variation of the critical RMT temperatures with the number of thermal cycles under a load (TCL) of 0.5 kg is presented through the curves 6 to10 in Figure 6 for: (a) cooling and (b) heating conditions. One should notice in Figure 6(a) a slight decrease in M_o (curve 7), M_e (curve 8) and M_{of} (curve 9) corresponding to the critical temperature values of the direct martensitic transformation, $\beta_1 \rightarrow \gamma'_1$, after 100 TCL.

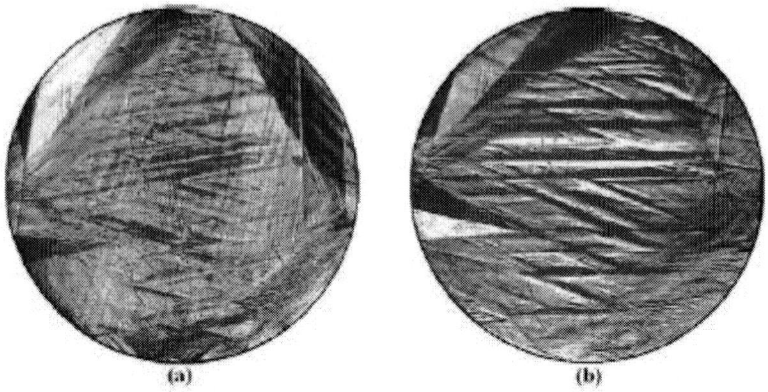

Figure 8. Structural aspects of the monocrystalline Cu-13.5Al-4Ni alloy after 300 cycles, ending with a half cooling cycle (a) and a half heating cycle (b).

However, after more accumulated cycles, up to 300 TCL, these temperatures do not change. By contrast, the critical temperatures, A_o (curve 7), A_e (curve 8) and A_{of} (curve 9) in Figure 6(b), associated with the reverse transformation $\gamma'_1 \rightarrow \beta_1$, continuously decrease with the amount of thermal cycles, while the intervals A_o-A_e and A_o–A_{of} increase up to 2.2 ^oC and 7.5 ^oC, respectively.

The results obtained in the present work are in agreement with those reported by Morin and Trivero [22] in a polycrystalline Cu-13.5wt.%Al-4wt.%Ni SME alloy submitted to thermal cycling under an applied static load, which results in a stress of 100 MPa. This treatment was found to promote structural modifications that tend to shift the critical temperatures (A_e and, principally, M_e) to lower values.

Regarding the curves 6 to 10 in Figure 6, it is worth noticing that the critical temperatures do not necessary follow the same tendency of the corresponding automatically calculated RMT critical temperatures. For instance, M_s increases after 100 thermal cycles, which is contrary to the behavior of all other critical temperatures. These results will be discussed further on this paper.

It was noted, that the temperature interval of the direct martensitic transformation (M_o–M_{of}), after more than 100 TCL, was stabilized, see curves 7 and 9 in Figure 6(a), while the interval of the reverse tranformation (A_o–A_{of}), was slowly decreased, curves 7 and 9 in Figure 6(b). The whole interval of RMT (M_s–A_f), after 100 TCL, became wider, as compared to the initial state of the alloy. Nevertheless, after this treatment, the whole interval of RMT became narrower.

After 100 TCL, the value of the thermal hysteresis, (ΔT=A_e-M_e) slightly increases to 19.2 ^oC. Meanwhile, after more than 100 TCL, its value decreased almost linearly, reaching ΔT=13 ^oC, after 300 TCL. This behavior is different than that of the load-free thermal cycling, in which the magnitude of the RMT thermal hysteresis stays the same, around 20 to 21 ^oC, even after 300 TC [29-31].

It was determined that the phase transformation enthalpies of the endothermic (direct) and the exothermic (reverse) transformation processes, for the initial state, and the first load-free thermal cycle, are about 6.8-7.4 J/g. After 100 TCL, these enthalpies slowly increase and, after more cycles, there is a tendency towards a decreasing in these values. In fact, upon cooling, the enthalpy of 7.3 J/g, after 100 TCL, is increased to 7.5 J/g, after 200 TCL, and reduced to 6.2

J/g, after 300 TCL. Upon heating, the enthalpy of 7.6 J/g, after 100 TCL, is slightly reduced to 7.4 J/g after 200 TCL and reduced to 5.9 J/g, after 300 TCL. This indicates that the thermal cycling treatment under load favors the RMT by decreasing both the direct and reverse transformation enthalpies [47].

The influence of thermal hysteresis alone on the structure of this alloy was observed by XRD analysis [28-31,47]. In the first cycle, even without an applied load, the alloy presents different structural behavior at RT, depending on the way the treatment ends, either by cooling, with half cooling cycle, or by heating, with half heating cycle, Figure 4(b,c).

The structural modifications observed in the monocrystalline Cu-13.5Al-4Ni alloy as a consequence of the thermal cycling treatments under a load of 0.5 kg are presented in Figure 9. As above mentioned, the structural phases developed in association with a treatment ending with a last half cooling cycle, differ significantly from those ending with a last half heating cycle. Figure 9(a) shows XRD pattern of the alloy after 200 TCL, ending with a half cooling cycle. In this pattern, there is a considerable participation of the β_1 phase with its preferential orientation revealed by an intense $(331)_{\beta1}$ peak. Peaks associated with the coherent γ'_1/R phase as well as the coherent β_1/R can also be noted. These dual crystallographic peaks are actually related to the coherent plane between the phases. One could identify a stabilizing tendency of the β_1 phase, Figure 9(a), after the 200 TCL ending with a half cooling cycle. After 300 TCL, ending with a half cooling cycle, Figure 9(b), the main peaks are now those corresponding to the R phase alone and R coherent with γ'_1. With less intensity, the β_1 $(331)_{\beta1}$ and $(411)_{\beta1}$ peaks are also detected in Figure 9(b). In summary, the structure after TCL, ending with a last half cooling cycle, tends to develop mainly the R phase as well as a relatively small amount of γ'_1, generally coherent with R.

Figure 9(c) shows the XRD pattern of the alloy after 200 TCL, ending with a half heating cycle. It is worth noting that, the main peaks are those associated with the γ'_1 as well as peaks corresponding to the coherent planes of R with γ'_1. A relatively small participation of β_1 and coherent planes of R with β_1 is also observed. After 300 TCL, ending with a half heating cycle, Figure 9(d), in addition to a small participation of γ'_1, the most intense peaks are those associated with the coherent planes of R with β_1. Different than the previous case of a last half cooling cycle, Figure 9(b), the existing phases after 300 TCL ending with a last half heating cycle, Figure 9(d), involve both γ'_1 and coherent planes of R with γ'_1 and R with β_1.

According to a microstructural analysis of the whole transversal section, the alloy, after 200 TCL, Figure 10(a,b), differs from its initial states (Figure 3) and is already exhibiting pits that result from oxidation related to the thermal treatment. The structure corresponding to a last half cooling cycle, Figure 10(a), displays martensite needles that are, however, not as clearly observed. The contrast of distinct blocks, one central and four peripherals can be easily seen owing to the shape deformation caused by martensite variants. In particular, two characteristic arrow-shaped martensite blocks make a visible contrast top right, against the surface background. The structure corresponding to a last half heating cycle, Figure 10(b), also shows fine needles, not only at the central region of the sample but, in greater number, at the periphery. The shape deformation of martensite blocks is also associated with different tones of gray as well as arrow-shaped γ'_1 martensite variants. According to the XRD, Figure 9(a,c), the structure corresponding to a last half heating cycle presents comparatively greater amounts of γ'_1 in addition to the coherent planes of R phase, in this case, with γ'_1.

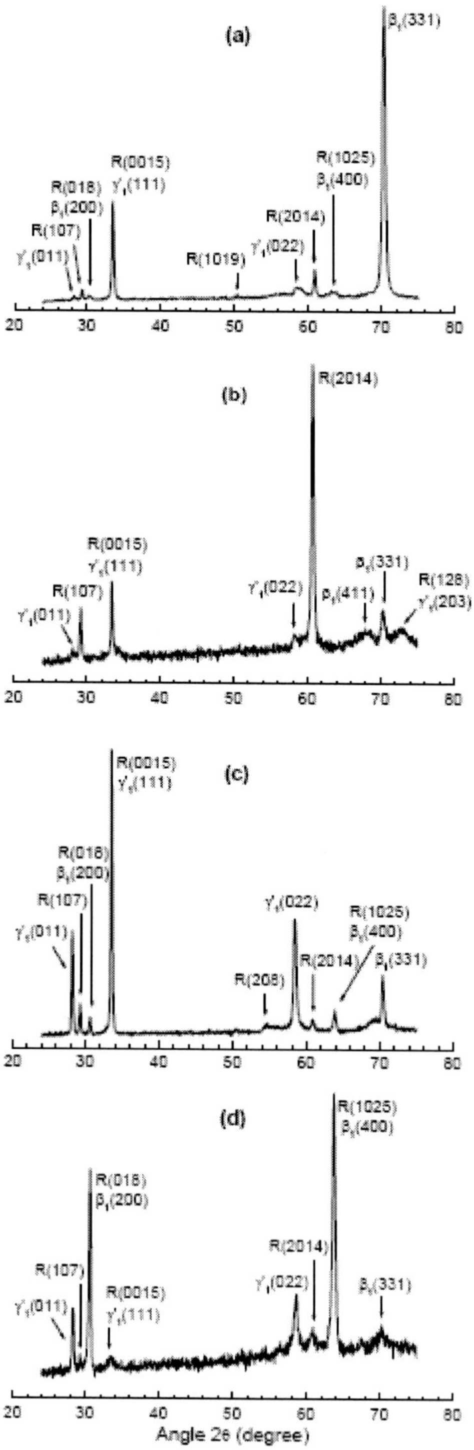

Figure 9. X-ray diffraction patterns of the monocrystalline Cu-13.5Al-4Ni alloy, after 200 TCL (a,c) and 300 (b,d) TCL under load a of 0.5 kg, ending with a half cooling cycle (a,b) and a half heating cycle (c,d).

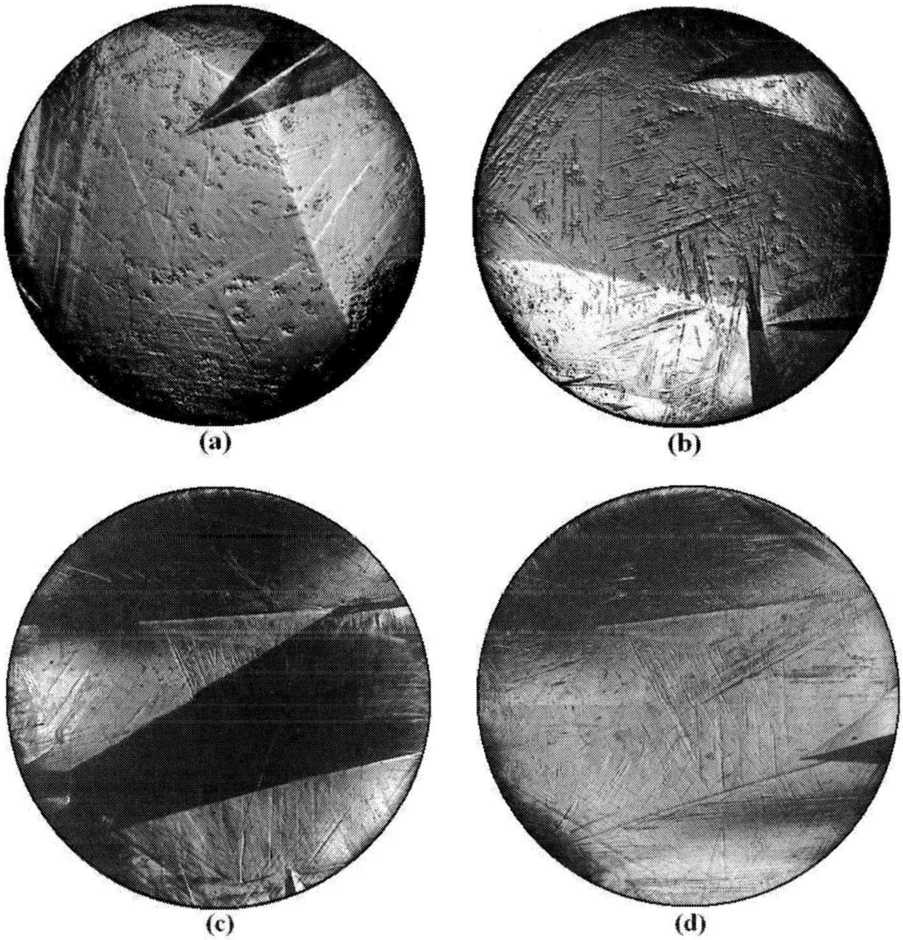

Figure 10. Micrographs of the whole transversal section of the monocrystalline Cu-13.5Al-4Ni alloy (total diameter of 5.0 mm), after 200 TCL (a,b) and 300 TCL (c,d), under 0.5 kg, ending with a half cooling cycle (a,c) and a half heating cycle (b,d).

After 300 TCL, the metallographic surface of the alloys reveals significant differences, whether the structure corresponds to a treatment ending with a last half cooling cycle, Figure 10(c), or half heating cycle, Figure 10(d). In the former, a strong contrast between martensite blocks, Figure 10(c), suggests a more intense shape deformations caused by the martensite variants. The XRD, Figure 9(b,d) detected the presence of peaks corresponding to coherent planes of R with γ'_1 and, in small intensity, peaks of β_1 [29,30,32-34]. In the latter condition, no clear martensite blocks can be seen, Figure 10(d), except for small arrow-shaped variants in the right middle region of the sample's periphery. The structural analysis revealed, mainly, γ'_1 martensite in addition to coherent planes of R with β_1.

3.4. Thermal Treatment of 300 TCL under Two Different Loads

In order to determine the influence of the applied load, during the thermal cycling treatment (TCL), on its structure and properties, the monocrustalline monocrystalline Cu-

13.5Al-4Ni alloy was submitted to 300 TCL, performed under two different applied loads, of 0.2 and 0.5 kg, corresponding to stresses of $11x10^{-2}$ and $28x10^{-2}$ MPa, respectively [31,48]. Both loads were applied in a device specially developed for this purpose. These loads and corresponding stresses were selected from values below the pseudo-yield of the alloy [26]. The 300 TCL consisted of heating the alloy above A_f, up to 100 °C, and cooling down to 0 °C, near the M_f. After the 300 cycles the process was ended with a last half cooling cycle to 0 °C and heating to RT [33].

Figure 11 presents the XRD patterns for the alloy in its state after the first cooling cycle (a), after a load-free 300 TC (b), and after 300 TCL under the two different applied loads of 0.2 kg (c) and 0.5 kg (d). After the load-free 300 TC, the structure, shown in Figure 11(b), resembles the structure of the alloy after a half cooling cycle, Figure 11(a). This 300 TC pattern exhibits peaks corresponding to β_1, γ'_1 and R phases, Figure 11(b). It is worth emphasizing that these peaks are relatively broader. Moreover, the peaks of the β_1 phase coherent with the peaks of the R phase are slightly displaced from their original position, namely, $(200)_{\beta 1}$ $|(018)_R$ to the right and $(400)_{\beta 1}$ $|(1025)_R$ to the left. This is apparently related to some transition associated with an intermediate state of the structure during the RMT $\gamma'_1 \leftrightarrow \beta_1$ transformation. The R phase, identified with the peaks $(107)_R$ and $(2014)_R$, corresponds to the suggested intermediate state. The relatively high intensity of the R peaks, coherent with β_1 and γ'_1, can be regarded as an indication of a greater stability of the RMT intermediate state, after the 300 TC, Figure 11(b), treatment. The changes observed in the XRD patterns, special the broadening of the peaks, are probably due to defects such as dislocations accumulated in the alloy during the 300 TC. These defects could promote stabilization of the intermediate state at RT, Figure 11(b), which causes the higher intensity peaks for R.

After the 300 TCL under an applied load of 0.2 kg, the pattern shown in Figure 11(c) exhibits peaks corresponding to β_1 and γ'_1. Moreover, a peak corresponding to the coherent planes $(111)_{\gamma 1}|(0015)_R$, also detected in its state after the first cooling cycle, Figure 11(a), as well as another peak for the coherent planes, $(222)_{\beta 1}|(027)_R$ are observed. The relatively high intensity peaks for the coherent planes can be regarded as an indication of a greater stability of the intermediate state, R phase, after the 300 TCL under 0.2 kg.

The alloy submitted to 300 TCL under an applied load of 0.5 kg, Figure 11(d), displays an intense $(2014)_R$ and a relatively moderate intensity $(107)_R$ peaks corresponding to the R phase. The participation of the martensite γ'_1 can now be detected by the moderate intensity, coherent planes $(111)_{\gamma 1}|(0015)_R$ peak. Here it should be noted that the $(331)_{\beta 1}$ peak of the β_1 phase has much lower intensity, as compared to corresponding peaks of the initial state, Figure 4(a).

The increase in the applied load for a 300 TCL promotes changes in the RMT critical temperatures. It was found that the greater the applied load, the higher the tendency to decrease the critical transformation intervals. Table 1 exemplifies this tendency for the most relevant intervals of critical temperatures, obtained from Figure 12, corresponding to the load-free 300 TC and the 300 TCL under an applied load of 0.5 kg.

This decrease in the critical temperature intervals can be associated with the effects of applied load on the development of the RMT both in direct and reverse conditions [31].

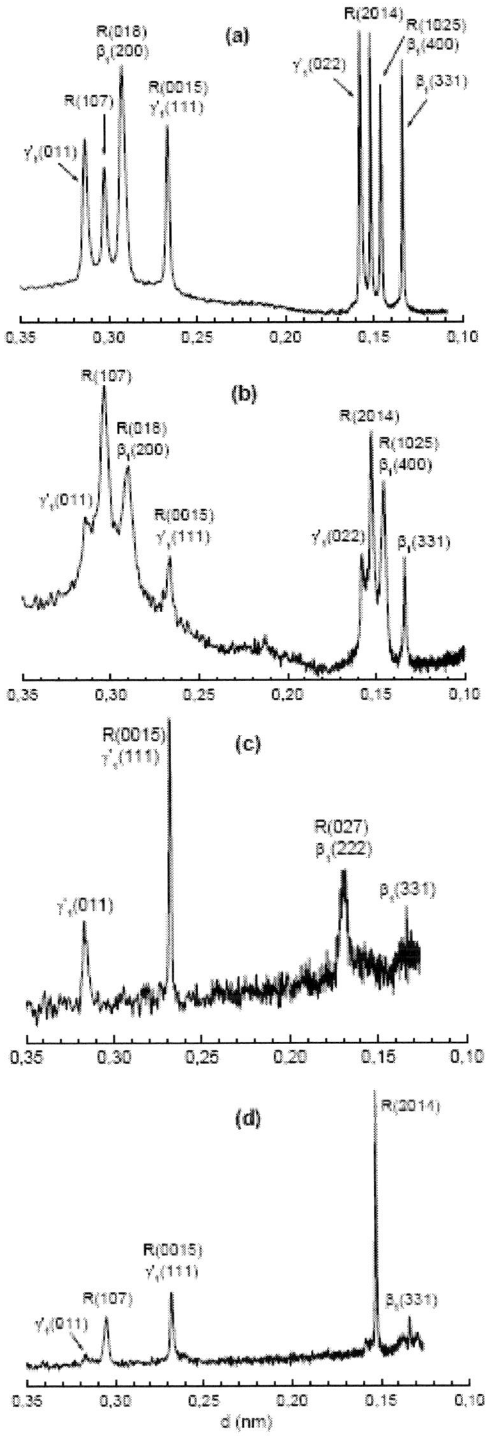

Figure 11. XRD patterns for the monocrystalline Cu-13.5Al-4Ni alloy in its state after the first half cooling cycle (a), after a load-free 300 TC (b) and after the 300 TCL under applied loads of 0.2 kg (c) and 0.5 kg (d).

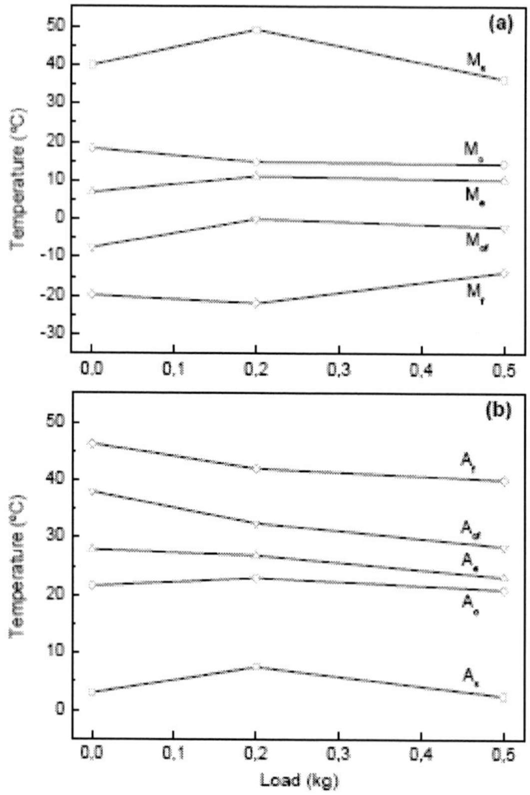

Figure 12. RMT critical temperature obtained by DSC during cooling (a), and heating (b) of the monocrystalline Cu-13.5Al-4Ni alloy as a function of the applied load on a 300 TCL.

Table 1. RMT critical temperature intervals for the monocrystalline Cu-13.5Al-4Ni alloy from the results in Figure 12

Critical Temperature Interval	Load-Free 300 TC (°C)	0.5 kg 300 TCL (°C)
$M_s - M_f$	60	50
$A_s - A_f$	43	38
$M_o - M_e$	11.4	4
$M_o - M_{of}$	26	16
$A_o - A_e$	6.3	2
$A_o - A_{of}$	16.4	7.5

3.5. Mechanical Behavior After Thermal Cycling Treatments

The stress-strain curves in Figure 13 show the mechanical behavior of the monocrystalline Cu-13.5Al-4Ni alloy in its initial state (curve 1) as well as after the 300 TCL under applied loads of 0.2 kg (curve 2) and 0.5 kg (curve 3) [26,31,33]. It should be noted in this figure that all curves display, approximately, the same general behavior. A very small initial elastic part of each curve is followed by a relatively extensive “pseudo-yield” plateau, which is typical of alloys presenting non-elastic effects [5,6,12-15,26,31-33]. Subsequently, another steeper elastic section continues with a higher slope up to fracture. The specific parameters related to each of these curves, however, show perceptible differences.

In its initial state, curve 1, the alloy displays around 8% strain for the “pseudo-yield” deformation at a 60 to 65 MPa level of stress. In this interval, an accumulation of reversible deformation occurs in relation to the SME and is accompanied by RMT reorientation of the initial structure.

For strains beyond the “pseudo-yield”, i.e. 8%, a steeper linear elastic behavior occurs up to rupture, which takes place at an ultimate stress of 1350 MPa for a total strain of 17.2%. The high strength attained by the alloy can be attributed to both the monocrystalline nature of the material and some number of dislocations introduced by the reverse martensite transformation [22]. Similar results were obtained in previous works [32] where Cu-Al-Ni monocrystalline alloys, with compositions comparable to that of the alloy in the present work, were investigated.

In Figure 13 it can also be seen that the alloy submitted to the 300 TCL under an applied load of 0.2 kg, curve 2, exhibits in compression a “pseudo-yield” strain, of 8%, a rupture stress of 1370 MPa, and a total strain of 15.8%, which are very close to the values of curve 1, corresponding to the initial state. Here it is worth noticing that the XRD patterns, Figure 11(a,b) also display an analogous situation with a greater incidence of the intermediate R phase coherent to the γ'_1 phase.

The alloy submitted to the 300 TCL under an applied load of 0.5 kg, curve 3 in Figure 13, shows significant differences associated with lower values of the “pseudo-yield” strain of 6.5%, rupture stress of 956 MPa, and total strain of 13.5%, as compared to curves 1 and 2. This corresponds to reductions of 41% and 30% in the maximum stress and total strain, respectively, as compared with the initial state. Apparently, under the applied load of 0.5 kg, the number of accumulated dislocations [22] is sufficient to interfere with the RMT causing a decrease in the SME parameters.

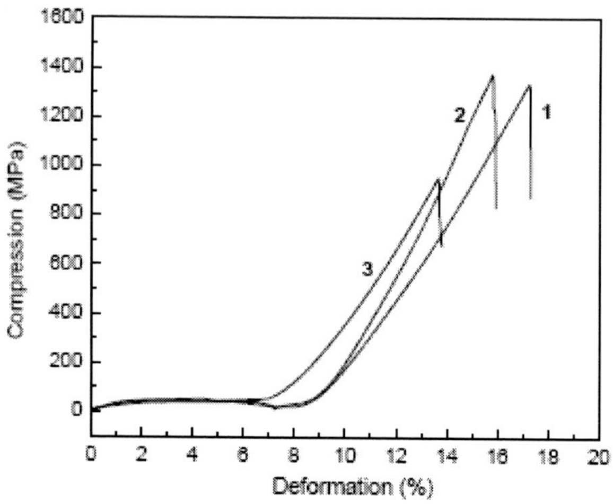

Figure 13. Compression stress-strain curves up to fracture for the monocrystalline Cu-13.5Al-4Ni alloy: (1) initial state and after 300 TCL under (2) 0.2 kg, (3) 0.5 kg of applied load.

Here it is important to explain the reason for the choice of both levels of load in the present work. The applied load of 0.2 kg was found to be the maximum for which no apparent change was caused to the mechanical behavior of the alloy by thermal cycling. On the other hand, the applied load of 0.5 kg was the minimum for which clear decreases were obtained in the mechanical parameters.

Figure 14 presents the XRD patterns of the alloy after compression until fracture for the alloy; (a) in its initial state, (b) 300 TCL under 0.2 kg and, (c) 300 TCL under 0.5 kg load. The XRD analysis of the alloy in its initial state after fracture, Figure 14(a), shows an intense peak corresponding $(111)_{\gamma'1}|(0015)_R$ for the martensitic γ'_1 and R phases. Another $(331)_{\beta 1}$ peak with lower intensity is related to the β_1 phase.

The specimen subjected to 300 TCL under 0.2 kg followed by compression until fracture, Figure 14(b), exhibits on the XRD patterns the same peaks as in the initial state, Figure 14(a) in addition to a very low intensity $(2014)_R$ peak associated with the R phase. The specimen subjected to 300 TCL under 0.5 kg after compression until fracture, Figure 14(c), presents an XRD pattern, which indicates a greater participation of the β_1 phase, as represented by its $(331)_{\beta 1}$ peak. Moreover, both the R phase, detected by its own $(2014)_R$ and $(107)_R$ peaks, as well as the martensitic γ'_1 phase, by its moderate intensity peak corresponding to the coherent planes $(111)_{\gamma'1}|(0015)_R$, are also revealed in Figure 14(c).

The XRD patterns in Figures 11 and 14 indicate that an increase in applied load during the 300 TCL favors a coherent interface between the R phase and the γ'_1 phase. At the end of the 300 TCL, during heating from 0 °C up to room temperature, the applied load promotes the reverse martensitic transformation $\gamma'_1 \rightarrow R \rightarrow \beta_1$ as noted in TiNi alloys [12-15]. However, by unloading the specimen to perform the XRD analysis, the structure may suffer a relaxation process followed by a transformation in the opposite direction, $\beta_1 \rightarrow R \rightarrow \gamma'_1$. This can be explained by the fact that the active stress is associated with an unstable thermodynamic state. It is then proposed in this work that structural imperfections accumulated during the 300 TCL under load, would retard all RMT reactions, not only retaining but also enhancing both the R and the martensitic γ'_1 metastable phases.

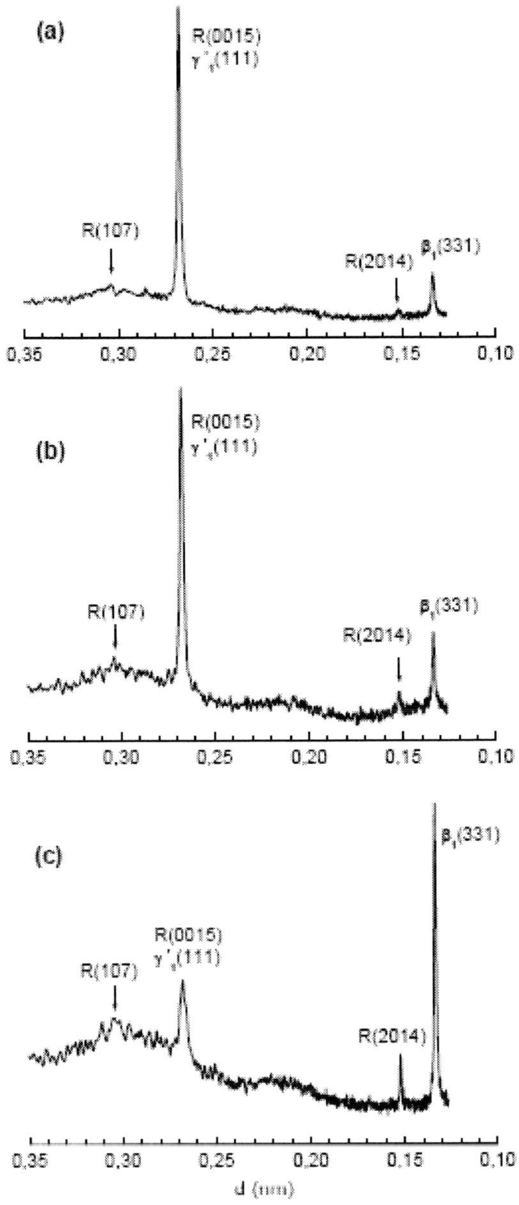

Figure 14. XRD patterns of the monocrystalline Cu-13.5Al-4Ni alloy after compression until fracture for the: (a) initial state, (b) 300 TCT under 0.2 kg and (c) 300 TCT under 0.5 kg.

Specimens that were subjected to the 300 TCL under the smaller load of 0.2 kg followed by compression until fracture, Figure 14 (b), showed a greater incidence of the transient R phase coherent to the γ'_1 phase. By contrast, the accumulation of more plastic deformation in the fracture of specimens submitted to 300 TCL under the larger applied load of 0.5 kg, resulted in a pronounced incidence of the stress induced β_1 phase, as depicted in Figure 14(c), due to the RMT $\gamma'_1 \rightarrow R \rightarrow \beta_1$ reaction.

The fracture characteristics of the monocrystalline Cu-13.5Al-4Ni alloy, corresponding to the three investigated conditions, are presented in Figure 15.

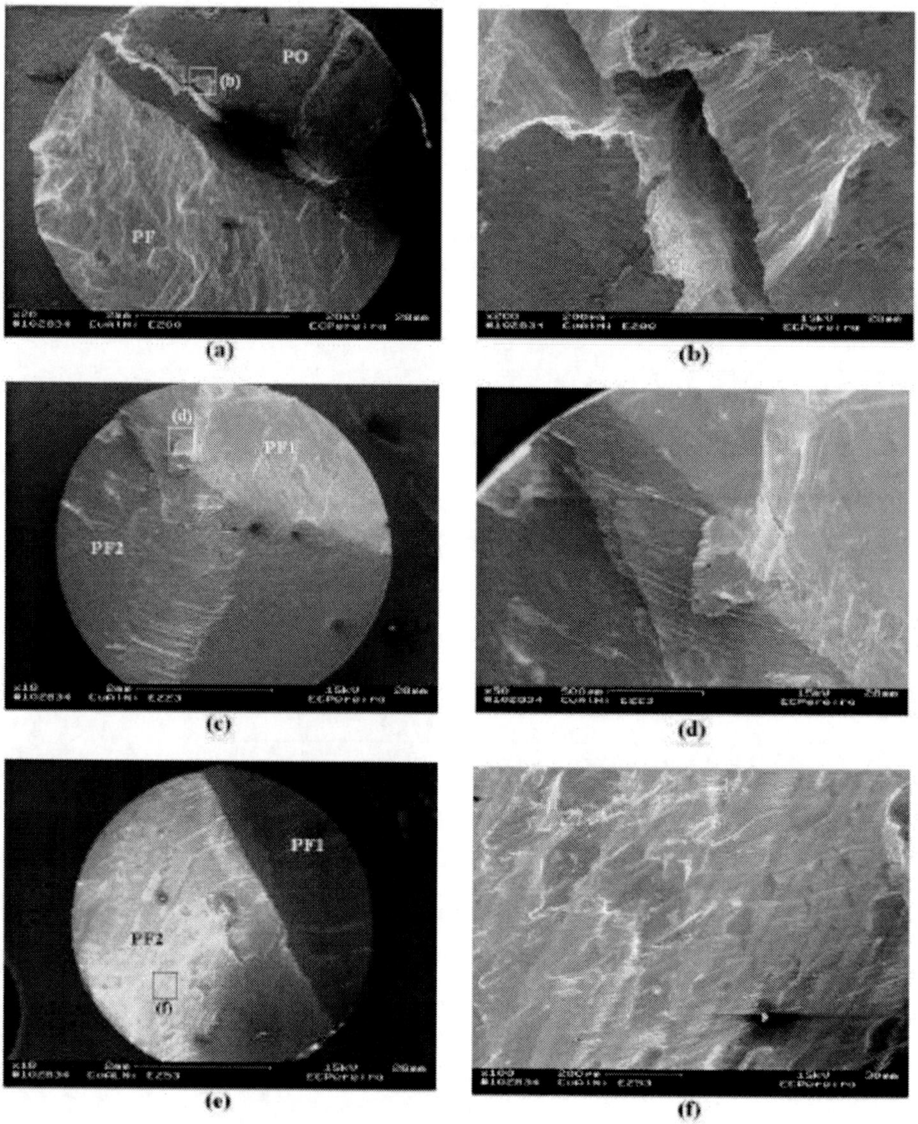

Figure 15. Fracture surface of the Cu-13.5Al-4Ni alloy in its initial state (a,b), after 300 TCT under applied load of 0.2 kg (c,d) and 0.5 kg (e,f).

A common aspect observed in all conditions is the brittle nature of the fracture, which was also reported in a previous work [30] investigating comparable Cu-Al-Ni alloys. As a general observation, it is important to bear in mind that a crack always initiates at a point on the specimen's flat surfaces that are in contact with the compressing plates of the Instron machine. Rupture then propagates following a single or different planes, inclined at 45° with respect to the direction of uniaxial compression.

During fracture in compression of the specimen corresponding to the initial state, Figure 15(a), a crack propagated in a single plane almost through the middle of the cross section as shown with low magnification in Figure 15(a). In this figure, the unbroken part of the original specimen's flat surface, which was in contact with the plate, is labeled as PO, while the plane of fracture as PF. In addition to the main crack that initiated the fracture, another relatively large secondary crack can be seen in the original flat surface. This secondary crack was already propagating, Figure 15(b), when the main crack broke the specimen. It was detached polyhedral-shaped block at the extremity of the secondary crack, apparently associated with a block of martensite variant, which was previously observed by optical microscopy at the periphery of similar alloys [30]. With higher magnification, the plane of fracture displays "river patterns" typical of the cleavage-like rupture associated with the propagation of the main crack [33].

The fracture aspect of the alloy subjected to 300 TCL under 0.2 kg, and then submitted to compression until rupture, is shown in Figure 15(c,d) [33]. With low magnification, Figure 15(c), one can see that the fracture occurred in two planes, PF1 and PF2. Similar to Figure 15(a,b), for the initial state, part of the original specimen's flat surface (PO) was left intact. By contrast, the fracture surface presents differences with respect to those of the initial state. The rupture process in this case is associated more with slip lines, Figure 15(c,d), rather than cleavage-like marks. This is apparently a consequence of the number of dislocations introduced by the thermal cycles [31]. A small polyhedral block was also observed, Figure 15(d), at the border between two fracture surfaces. The same interpretation of a martensite variant is suggested for this block [28]. In addition, the location of this block indicates that it may have served as an initiation point for crack nucleation. If this is the case, the change in martensite orientation within the structure of a monocrystalline alloy can play a relevant role in the fracture behavior.

The fracture in compression of the alloy after the 300 TCL under 0.5 kg is presented in Figure 15(e,f) [33]. At low magnification, Figure 15(e), two planes of fracture (PF1 and PF2) can be observed in a similar manner as the specimen subjected to 300 TCL under 0.2 kg previously shown in Figure 15(c,d). However, contrary to the initial state, Figure 15(a,b), and the specimen subjected to 300 TCL under 0.2 kg, Figure 15(c), no part of the original specimen's flat surface remains. With higher magnification, Figure 15(f), the fracture surface could be associated with slip lines and cross-slip from different deformation systems. This indicates a slippage process along preferential planes caused by dislocation movement. Moreover, other areas of the fracture surface, Figure 15(e), present a rough aspect typical of intense plastic deformation. This is in agreement with the influence of the 0.5 kg level of applied load, on the mechanical behavior of the alloy, as mentioned before.

To explain the reported structure alterations, Figures 6 and 9-12 for the thermal cycling with applied load, the following discussion takes into consideration the mechanical and thermal constrains that the investigated monocrystalline Cu-13.5Al-4Ni alloy suffer during the TCL. These constrain result in stress-induced martensite and anomalous behavior of the elastic modulus, with the variation in temperature, as previously reported [5,6,12-15,20,21,49-51] in other alloys with RMT.

Initially, it is worth mentioning that materials that undergo allotropic and polymorphic transformations, including alloys with RMT, exhibit a specific change in the elastic modulus (E), as well as the shear modulus (G), in the interval of critical temperatures [12-15, 20,21,49-51]. In this interval, E and G pass though a minimum value, which indicates a loss of stiffness. Actually, a significant drop has been noted during both heating and cooling [12]. After the minimum, as the high temperature phase is formed during heating, the elastic stability of the lattice is increased. Consequently, it becomes more rigid and the drop in E associated with RMT ceased. Above A_f^o, as determined by thermal analysis in load-free conditions, the elastic modulus maintains an increasing tendency, although at a smaller rate. At temperatures well above A_f^o a slight decrease in the modulus begin to occur [12-15,19,20], which is a typical behavior of stable materials when heated [19,20,51].

It is well known that an external mechanical constrain on the parent (A) phase may promote stress induced martensitic transformation at M_s^d temperature (under load/deformation), above A_f^o [5,6,11,51]. The value of the temperature M_s^d increases with the applied load up to a critical M_d temperature above which martensite no longer can be formed [5,6,11]. One may then infer that not only the temperature but also the stress, through the load, can alter the thermodynamic equilibrium, A↔M, and eventually favor the martensitic transformation above a critical temperature, A_f^o, corresponding to the beginning of the stable high temperature phase [2,3,6,11,20].

Since the elastic modulus behavior in RMT alloys is similar (symmetrical) during heating and cooling, one may suppose that at temperatures near M_f^o, the metastable martensite could also become unstable under a mechanical constraint. It is then possible to expect a stress induced reverse transformation ($M \rightarrow A_d$). This reverse transformation and the development of an intermediate phase, as a consequence of its incomplete return after unloading, was predicted by Wasilewski [5,11]. Indeed, the formation of the high temperature phase and/or an intermediate R phase, coherent with the other existing phases, were observed in previous publications [12-15,48], in TiNi and Cu-Al-Ni alloys.

One can then predict that an applied load during the TCL facilitates the formation of induced martensite. On cooling, the transformation will occur at a temperature higher than M_s^o, while, on heating, the high temperature stable phase will be formed at a temperature lower than A_s^o. In favor of this rational, the results in Figure 6 show up to 100 TCL a displacement of both M_s to higher values with increasing temperature and A_s to lower values with decreasing temperature. With greater number of TCL cycles, M_s and A_s tend to return to their initial values. It is suggested that the internal elastic stress field accumulated during the thermal cycles, promotes the reaction ($\beta_1 \rightarrow R \rightarrow \gamma'_1$) at higher temperatures on cooling. The occurrence of the R phase enlarges the M_s-M_f interval and consequently M_s^d is displaced towards higher temperatures. Furthermore, the plastic deformation introduced during each cycle by structural imperfections, mainly dislocations, obstructs the end of the transformation and, therefore, displaces M_f^d to lower temperatures. These processes favor the stability of the intermediate R phase coherent with the high temperature stable β_1, and the martensite, γ'_1.

The results can be explained in terms of reversible martensitic transformations induced by stress/strain [11-13]. The decrease in the critical temperatures (M_o, M_e and M_{of}) for direct martensite transformation during a last half cooling cycle can be associated with the reaction about this:

$$\beta_1 \rightarrow R \rightarrow \gamma'_1 \qquad (1)$$

Indeed, as more cycles are accumulated, Figure 7 and 9, the R phase and R coherent with γ'_1 martensite become predominant. On the other hand, the decrease in the critical temperatures (A_o, A_e and A_{of}) for reverse martensite transformation during a last half heating cycle, can be associated with the reaction

$$\gamma'_1 \rightarrow R \rightarrow \beta_1 \qquad (2)$$

Similar to equation (1), the intermediate formation of the R phase plays an important role for the stability of martensite at lower temperatures after the accumulation of thermal cycles under load.

One may then infer that an applied load during the thermal cycling treatment promotes the stabilization of the RMT at lower temperatures and reduces the energy expended during an accumulated cycle. The formation of the intermediate R phase in association with both, the direct and the reverse martensite transformation, is apparently responsible for this low temperature stabilization. This fact that the TCL treatment tends to promote the stabilization of the combined R and γ'_1 structure at room temperature was not observed after thermal cycles without applied load, TC [47,48].

In practice, this tendency for the load assisted RMT to occur at temperatures close to room temperature is an advantage for technological applications of Cu-Al-Ni based alloys on the SME actuators [8-10].

CONCLUSION

The monocrystalline Cu-13.5wt.%Al-4.0wt.%Ni alloy suffers reversible martensitic transformation, RTM, $\gamma'_1 \leftrightarrow \beta_1$ associated with a thermal hysteresis of about 20 °C when subjected to thermal cycling treatments. The phase transformation enthalpies of endothermic (direct) and exothermic (reverse) processes, in the initial state, and in a first free thermal cycle, are about 6.8–7.4 J/g. The thermal hysteresis of the RMT, promotes changes in the amount of phases associated with the RMT. These changes depend on the way the treatment ends, either by heating or by cooling.

After 300 TC, without load, a displacement of the critical temperatures M_o, M_f, A_o, A_s, as well as the peak temperatures A_e and M_e, is observed towards values that would be up to 10 °C lower while M_s increases more than 10 °C. This contributes to enlarge the temperature interval of RMT occurrence. By contrast, no change occurred in the thermal hysteresis. After a load-free 300 TC, if the end occurs by heating above A_f, a larger amount of β_1 is transformed. If the end occurs by cooling below M_f, an intermediate structure, the R phase, related to the $\gamma'_1 \leftrightarrow \beta_1$ is found. This is apparently a consequence of accumulated structural defects that retard the RMT and tend to stabilize the intermediate stage.

The alloy submitted to TCLs under a load of 0.5 kg revealed a tendency towards a decrease in the critical intervals for both direct and reverse martensite transformation with increasing number of cycles. This tendency is accentuated for treatments up to 300 cycles, in which critical temperature intervals of RMT smaller than 10 °C were recorded. Both the hysteresis and the enthalpy of the RMT also decrease. This can be attributed to a higher participation of the R phase, particularly above 200 cycles, either alone or coherent with γ'_1 (direct on cooling) and β_1 (reverse on heating). It is also suggested that the applied load contribute to the shape deformation observed in blocks of martensite variants.

The alloy's mechanical behavior in its initial state is characterized by a complex process involving RMT and phase reorientation under compression. This behavior is associated with a "pseudo-yielding" plateau at a 60-65 MPa level of stress, followed by a steeper elastic deformation up to 1350 MPa with total strain of 17.2% and subsequent fracture, disclosing its brittle aspect. The TCL under a load of 0.5 kg reduces the stress of the pseudo-yielding by 20-

25 MPa, decreases the ultimate stress to 950 MPa and the total strain to 13.5%. Moreover, under this TCL condition, the fracture surface is associated with slip lines indicating the participation of dislocations in the process of rupture. In all cases, the presence of martensite variants in the form of polyhedral blocks serves as preferential site for crack nucleation.

The accumulation of more plastic deformation in the fracture of specimens submitted to 300 TCL under the larger applied load of 0.5 kg, resulted in a pronounced incidence of the stress induced R/β_1 phases, due to the RMT $\gamma'_1 \rightarrow R \rightarrow \beta_1$ reaction. The applied load favors the existence of the R phase coherent to the γ'_1 and β_1 phases.

Finally, it is important to emphasize the fact that the alloy shows good resistance to irreversible changes during both TC and TCL tests. These results have a particular relevance for the practical application of the monocrystalline Cu-Al-Ni alloys..

ACKNOWLEDGMENTS

The authors thank the Brazilian agencies CNPq, CAPES and FAPERJ for supporting this research. It is also acknowledged the collaboration of Prof. Rubén Jesús Sánchez Rodríguez and Teresa Elígio Castilló (UENF) on the DSC tests; Prof. Eduardo Atem de Carvalho (UENF) on the deformation tests; Profa. Rosane Toledo (UENF) and Guenrikh Lubman (IMET, Moscow) on the X-ray diffraction analyses; and Prof. Ivan Viahhi (Technical University, Saint Petersburg) on the alloys fabrication.

REFERENCES

[1] Buehler, WJ; Gilfrich, JV; Wiley, RC. Effect of Low-Temperature Phase changes on themechanical properties of alloys near composition TiNi. *Journal of Applied Physics*, 1963, 34, 1475-1477.

[2] Kurdiumov, GV. Diffusionless (martensitic) transformations in alloys. *Journal of Technical Physics*, 1948, 18, 999-1025. (in Russian).

[3] Kurdiumov, GV; Handros, LG. About the thermoelastic equilibrium in martensite transformations. *Doklady Akademii Nauk SSSR*, 1949, 2, 214-221. (in Russian).

[4] Chang, LC; Read, TA. Plastic deformation and diffusionless phase changes in metals. The gold-cadmium beta phase. *Trans. AIME*, 1951, 189-1, 47-52.

[5] Perkins, J. *Shape Memory Effects in Alloys*. In: Proceeding International Conference on Shape Memory Effects and Applications. Toronto-Ontario, Canada, 470 p., 1975.

[6] Otusuka, K; Wayman, CM. *Shape Memory Materials*. Cambridge University Press. Cambridge, 285 p., 2000.

[7] Kohl, M. *Shape Memory Microactuators. Series: Microtechnology and MEMS*. Springer-Verlag Berlin Heidelberg. 247 p., 2004.

[8] Priadko, A; Pulnev, S; Viahhi, I; Vetrov, V; Yudin, V. Actuators and Drives Based on Cu-Al-Ni Shape Memory Single Crystals. In: *Proceedings of the Institute of Robotics and Technical Cybernetics*. St-Petersburg, Russia, 2000, 21 p.

[9] Dynnikov, VI; Santos, CML; Braz Fernandes, FM. Cu-Al-Ni shape memory actuators for anthropomorphic upper limb prostheses. In: *Meeting of Biomechanical*, Martinchel Abrantes, Portugal. University Abrantes, 2005, 295-299.

[10] Dynnikov, V; Santos, C; Matlakhova, LA; Pereira, EC. Cu-Al-Ni Shape Memory Actuator for Prostheses and other Applications: Response Time and Structural Stabilization. In: *Annals of DAAAM and Proceedings of the 18th International DAAAM Symposium*. Editor Katalinic, B. Vienna, Austria: Published by DAAAM International, 2007, 267-268.

[11] Wasilewski, RJ. Stress-assisted martensite formation in TiNi. *Scripta Metallurgica,* 1971, 2-5, 127-130.

[12] Kovneristyi, YK; Fedotov, SG; Matlakhova, LA. Effect of deformation on phase transformations and modules of elongation in titanium nickelide-based alloy. In: *Proceedings of the 5-th International Conference of Titanium. Titanium Science and Technology*. Munich, 1984, 3, 1675-1681.

[13] Kovneristyi, YK; Fedotov, SG; Matlakhova, LA. The Influence of plastic deformation on the structure, shape memory effect and other properties of TiNi alloy. In: *Proceedings of the International Symposium on Shape Memory Alloys*. Guilin, China, 1986, v. 1, p. 175.

[14] Kovneristyi, YK; Fedotov, SG; Matlakhova, LA. Dependence of shape memory and shape reversibility effects of the TiNi alloy on deformation. In: *J. Fizika Metallov i Metalovedenie. Sverdlovsk*. URSS, 1986, 2-62, 344-348 (in Russian).

[15] Fedotov, SG; Matlakhova, LA; Kovneristyi, YK; Zhebynieva NF. Structure changes in TiNi SME alloy under deformation. In: *J. Fizika Metallov i Metallovedenie*. Sverdlovsk, URSS, 1988, 3-65, 564-569 (in Russian).

[16] Brinson, LC; Moran, B. *Mechanics of phase transformations and shape memory alloys*. The American Society of Mechanical Engineers. 215 p., 1994.

[17] Fremond, M.; Miyazaki, S. *Shape Memory alloys*. CISM Courses and Lectures No 351. International Centre for Mechanical Sciences. Springer Wien New York. 147 p., 1996.

[18] [18] Bhattacharya, K. *Microstructure of martensite. Why it forms and how it gives rise to the shape-memory effect*. OXFORD University Press Inc., New York. 288 p., 2003.

[19] Delay, L; Walimont, H. *Martensitic Transformations in Copper-, Silver-and Gold-Based Alloys*. [Russian translation], Nauka, Moscow, 1980

[20] Fedotov, SG. *Investigations of Liquid and Solid Metals*. In: Kornilov II, editor. Moscow, Nauka Pub., 1982, 207-240.

[21] Collings, EW. *The Physical Metallurgy of Titanium Alloys*. 3rd Edition. Metals Park, Ohio; American Society for Metals, ASM; 1984.

[22] Nakata, Y; Tadaki, T; Shimizu, K. Thermal cycling effects in a Cu-Al-Ni shape memory alloy. *Trans. Japan. Inst. Metals,* 1985, 26-99, 646-652.

[23] Morin, M; Trivero, F. Influence of thermal cycling on the reversible martensitic transformation in Cu-Al-Ni shape memory alloy. *Materials Science and Engineering A*, 1995, 196, 177-181.

[24] Matlakhova, LA; Matlakhov, AN; Dias, FRM; Rodríguez, RS; Viahhi, I; Monteiro, LSG. Effect of thermal cycling treatment in the structure of the monocrystalline Cu-Al-Ni alloys with SME. In: *Annals of ABM*. Rio de Janeiro, RJ, Brazil. CD-ROOM, 2003, 3054-3061.

[25] Viahhi, I; Matlakhova, LA; Matlakhov; Dias, FRM; Monteiro, LSG; Pulnev, SA. Morphological, structure and phase alternations of single crystal Cu-Al-Ni Bulletin Of Tambov's University, 2003, v.8, n.4, c.661-664 [in Russo].

[26] Dias, FRM; Matlakhova, LA; Matlakhov, Carvalho, EA; Toledo, R; Rodríguez, RS. Behavior of monocrystalline Cu-Al-Ni alloys with SME under deformation by compression. In: *Proceedings of 59th Annual Meeting of the Brazilian Association for Metallurgical and Materials -ABM.* São Paulo, SP, Brazil. CD-ROM, 2004, 3087-3094.

[27] Matlakhova, LA; Dias, FRM; Matlakhov, AN; Rodríguez, RS; Castilló, TE; Toledo, R; Lima, JAP; Mirelles, LAM; Pereira, EC. Martensitic Transformation in monocrystalline Cu-Al-Ni alloys submitted the thermal cycling treatment. In: *Annals of CBECiMat.* Porto Alegre, RS, Brazil. CD-ROOM, 2004, 1355, 1-7.

[28] Matlakhova, LA; Matlakhov, AN; Pereira, EC; Rodríguez, RS; Viahhi, I; Pulnev, SA. Thermal cycling of shape memory Cu-Al-Ni single crystal. *Materials Sciences Transactions*. ISSN 1684-579X, 2005, 11-12-16 (in Russian).

[29] Pereira, EC; Matlakhova, LA; Matlakhov, AN; Toledo, R; Rodríguez, RS; Dias, FRM. Thermal hysterisis of reversible martensitic transformations in monocrystaline Cu-Al-Ni alloys. In: *Proceedings of 60th Annual Meeting of the Brazilian Association for Metallurgical and Materials - ABM.* Belo Horizonte, MG, Brazil. CD-ROM, 2005, 1-10.

[30] Silva, RJ; Matlakhova, LA; Pereira, EC; Matlakhov, AN; Monteiro, SN; Rodríguez, RS. Thermal cycling treatment and structural changes in Cu-Al-Ni monocrystalline alloys. *Materials Science Forum*, 2006, 514-516, 692-696.

[31] Pereira, EC. *Influence of thermal cycling treatment unloaded and under load in the structure and properties of the monocrystalline Cu–Al–Ni alloy with shape memory effect.* Dissertation submitted to the State University of the Northern Rio de Janeiro, Advanced Materials Laboratory, Campos dos Goytacazes, RJ, Brazil. CDD 669.96131, 131p., 15 February, 2006.

[32] Pereira EC; Matlakhova LA; Matlakhov NA; Monteiro SN; Carvalho EA. Fracture of a single crystal Cu-Al-Ni alloy. Brazilian Journal of Morphological Sciences 2005; 349-350.

[33] Matlakhova, LA; Pereira, EC; Matlakhov, AN; Monteiro, SN; Toledo, R. Mechanical behavior and fracture characterization of a monocrystalline Cu–Al–Ni subjected to thermal cycling treatments under load. *Materials Characterization*, 2008, 59, 1630-1637.

[34] Duerig T, Pelton A, Stöckel D. An overview of Nitinol medical applications. Materials Science and Engineering A 1999; 273-275:149-160.

[35] Auricchio F, Petrini L, Pietrabissa R, Sacco E. Shape-memory alloys in orthodontics: mechanical, experimental, modeling and clinical considerations. Proc Workshop on Shape Memory Alloy Materials. Experimental Investigation, Modeling, Metallurgy, Applications Warsaw, Poland, September 2001.

[36] Sakamoto H, Shimizu K. Experimental investigation on cyclic deformation and fatigue behavior of polycrystalline CuAlNi shape memory alloys above Ms. Trans. Japan Inst. Metals 1986; 27-8: 592-600.

[37] Van Humbeeck J, Stalmans R. Characteristics of shape memory alloy. In: Otsuka K, Wayman CM, editors. Shape Memory Materials. Cambridge: University Press; 1998, p.149-183.

[38] Tadaki T. Cu-based shape memory alloys. In: Otsuka K, Wayman CM, editors. Shape Memory Materials. Cambridge: University Press; 2000, p.97-116.

[39] Martynov V, Bokaie MD, Johnson AD, Gray GR. Superelastic and Shape Memory Single Crystal CuAlNi: Fabrication and Applications. Proc Int Conf on Shape Memory and Superelastic Technologies. California, USA, May 7-11; 2006.

[40] Suresh N, Ramamurty U. Effect of aging on mechanical behavior of single crystal Cu-Al-Ni shape memory alloys. Materials Science and Engineering A 2007; 454-455: 492-499.

[41] Friske, H; Anorg, Z. JCPDS 11-0010. *Allg. Chem.* 1949, 258-198.

[42] Karsson, J. JCPDS 07-0108. *Inst. Met.* 1951, 391-79.

[43] Warlimont, H; Wilkens, Z. JCPDS 28-0005.*Metallkd.* 1964, 382-55.

[44] Bown, M. JCPDS 28-0016. *Acta Crystallografica.* 1956, 70-9.

[45] Jouneau, PH; Stadelmann, P. Electron Microscopy Image Simulation [on line]. Centre Interdépartemenal de Microscopie Electrronique. EPEL, Lausanne, 1998. Capturado em 20 de junho de 2005 from http://cecm.insa-lyon.fr/CIOLS/crystal4.pl/.

[46] Gong CW, Wang YN, Yang DZ. Phase transformation and second phases in ternary Ni-Ti-Ta shape memory alloys. *Materials Chemistry and Physics* 96, 2006, 183-187

[47] Matlakhova, LA; Pereira, EC; Matlakhov, AN; Monteiro, SN; Rodríguez, RJS. Stress assisted Thermal cycling treatment of a monocrystalline Cu-Al-Ni alloy (to be published)

[48] Pereira, EC; Matlakhova, LA; Matlakhov, AN; Monteiro, SN; Rodríguez, RJS. Influence of the applied load during thermal cycling treatment on the structure and characteristics of RMT in a monocrystalline Cu-13.5Al-4Ni alloy. (to be published)

[49] Matlakhova, LA; Matlakhov, AN; Monteiro, SN; Fedotov, SG; Goncharenco, BA. Properties and structure characteristics of Ti-Nb-Al alloys. *Materials Science and Engineering A, 2005,* 393, 320-326.

[50] Pai Mizar, S. *Thermomechanical characterization of NiTiNOL and NiTiNOL based structures using ACES methodology.* Dissertation Submitted to the Faculty of the Worcester Polytechnic Institute in partial fulfillment of the requirement for the Doctor of Philosophy in Mechanical Engineering. 236 p., 2 December, 2005.

[51] Matlakhova, LA; Motta, AC; Matlakhov, NA; Kolmakov, AG; Sevostianov, MA. Modulus of elasticity of the Cu-Al-Ni monocrystalline alloy deformed by compression. In: *Annals of CBECiMat.* Porto de Galhinhas, PE, Brazil. 2008, 6463-6474

[52] Reed-Hill, RE; Abbaschian, R. *Physical Metallurgy Principles.* 3rd Edition. PWS Publishing Company. Boston. pp. 561-585, 1994.

In: Shape Memory Alloys
Editor: H. R. Chen, pp. 145-180
ISBN 978-1-60741-789-7

Chapter 5

INTERACTION BETWEEN MARTENSITIC TRANSFORMATIONS AND DEFECTS DURING THERMAL AND PSEUDOELASTIC CYCLING IN CUALNI SINGLE CRYSTALS

R. Gastien[a], M. Sade[b,c,d] and F. C. Lovey[b,d]

aDto. Ciencia y Técnica de Materiales, Instituto de Investigaciones Científicas y Técnicas para la Defensa (CITEFA),[1] Buenos Aires, Argentina
bCentro Atómico Bariloche, Comisión Nacional de Energía Atómica, [2] Río Negro, Argentina
cConsejo Nacional de Investigaciones Científicas y Técnicas (CONICET), Argentina.
dInstituto Balseiro,Universidad Nacional de Cuyo, Argentina

ABSTRACT

CuAlNi shape memory materials are promising for high temperature technological applications over other Cu based alloys, as they show low diffusion at temperatures close to room temperature or higher [1, 2]. For the development of actuator devices using CuAlNi alloys, it is necessary to study the thermal and mechanical response of the material due to cycling through the martensitic transformation. Additionally, the use of a single crystal material increases the life to fracture, and finally enhances the performance of future devices based on these materials.

In some CuAlNi alloys, a metastable β-bcc ordered structure obtained at room temperature can undergo martensitic transformations by cooling or applying stress (σ). Three types of martensitic transformations can be induced for chemical compositions between 13-15Al and 0-6Ni (wt.%): β↔γ', β↔β' and β↔β'+γ'.

In this chapter, the authors review the more recent research made on thermal and pseudoelastic tensile cycling in CuAlNi single crystals, and the interaction between martensitic transformations and defects which causes a novel effect in these alloys.

[1] J. B. de La Salle 4397, 1603 Villa Martelli, Buenos Aires, Argentina.
[2] 8400 S. C. de Bariloche, Río Negro, Argentina

Pseudoelastic $\beta\leftrightarrow\beta'$ cycling was performed in CuAlNi single crystals for chemical compositions around 14Al and 4Ni (wt.%). No evolution of the stress-strain curves is observed up to 3000 cycles. On the other hand, for the $\beta\leftrightarrow\beta'+\gamma'$ cycling, the martensitic transformation changes with the number of cycles, showing a clear inhibition of the γ' martensite as the cycling progresses [3]. The same effect was observed for thermal $\beta\leftrightarrow\beta'+\gamma'$ cycling, which had not been observed previously.

A proper estimation of the energy involved in this inhibition process was done [4] with experimental data [5] and a thermodynamical analysis through σ-T diagrams, using determinations of the entropy changes during $\beta\leftrightarrow\beta'$ and $\beta\leftrightarrow\gamma'$ transformations [6].

The interaction between dislocations generated during cycling and the twinned structure of the γ' martensite is the key to understand the origin of the inhibition effect. A first approach concerning the generation of stacking faults was introduced in [7].

The authors proposed a microscopic model to explain the γ' inhibition, related to the localized interaction between a dislocation array and the twinned γ' structure [8]. The model takes into account the interaction between the martensitic stress free transformation strains and the stress field created by the dislocation arrays. It is shown that the interaction is different for each twin related variant in the γ' martensite. The energy necessary to maintain the right volume relationship of the twinned γ' variants to produce an undistorted β/γ' habit plane is defined as the inhibition energy. The calculations give a value of around 12 J/mol, in reasonable agreement with experimental results. The studies and the model proposed to explain this effect are also useful to further understand how the defects interact with martensitic transformation.

INTRODUCTION

In some CuAlNi alloys, a metastable β-bcc ordered phase (also called parent phase) can be obtained at room temperature. This β phase can undergo martensitic transformations (MTs) by cooling (its product is called TIM: thermally induced martensite) or by means of an applied stress (SIM: stress induced martensite). These MTs are responsible for the shape memory and pseudoelastic effects that present some copper based alloys.

The type of TIM obtained after cooling the β phase below some critical temperature depends mainly on the chemical composition of the alloy [9]. Three types of MTs can be induced in CuAlNi alloys with chemical compositions between 13-15Al (wt.%) and 0-6Ni (wt.%): i) $\beta\leftrightarrow\gamma'$, ii) $\beta\leftrightarrow\beta'$and iii) $\beta\leftrightarrow\beta'+\gamma'$.

The crystal structure of the parent β phase is a DO_3 type ordered structure. All the martensitic structures are long period stacking order structures with the same basal plane, and the difference between them lies in their stacking sequence [10]. The β' martensite is also called 18R, as the period of the stacking sequence is 18 and it has rhombohedral symmetry. In the case of γ' martensite, also called 2H, the period is 2 and the symmetry is hexagonal.

The 2H martensite in Cu-Al-Ni alloys is generally composed by a sequence of two twin-related variants as a condition to have an undistorted habit plane with the β parent phase [1, 11]. Two possibilities have been found: the twin interface can be parallel to the {121} plane (type I twinning) with a twin thickness ratio very close to 1:2. This lattice invariant shear system gives a habit plane near to the $\{133\}_\beta$ plane [1], where the subscript β refers to the BCC type parent phase. Alternatively, another twin system has been observed where the twin interface is very closely parallel to the $\{231\}$ plane (Type II twinning), this twin system gives also a habit plane close to $\{133\}_\beta$ [11]. On the contrary, the β' martensite is single variant.

By cooling the β phase, the MT can be monitored by a physical property that changes significantly with temperature. One of the most used magnitudes is the electrical resistance of the sample. An example of the typical electrical resistance vs temperature curve for a β↔γ' MT is shown in Figure 1. On cooling the sample, the MT begins at a temperature called M_s, and finishes at a temperature M_f, at which the sample is in a complete martensitic state. The parent phase can be recovered on heating the sample. The retransformation begins at a temperature called A_s, and finishes at A_f, recovering the original β phase. This transformation/retransformation cycle has a certain hysteresis width, depending on the type of transition. The β→γ' MT has an hysteresis width of 20-40 K, and the β→β' around 5-10 K [9, 12].

On the other hand, if the MT is induced by loading, the type of SIM obtained depends on several factors, such as chemical composition, crystallographic tensile orientation, test temperature and type of loading [13-17]. A stress-strain curve for a β↔β' type MT is shown in Figure 2 (tensile stress is applied in this case). When a mechanical load is applied to the sample, after elastic deformation of the specimen, the MT begins at a critical stress called σ_c. Upon unloading, the sample recovers its original shape. This transformation/retransformation cycle also has a certain hysteresis width as the thermally induced one, depending on the type of transition. The β→β' MT has an hysteresis width of 1-5MPa (applied stress), and the β→γ' more than10MPa [3].

The critical stress (σ) to induce a MT has a linear relationship with test temperature (T) according to a Clausius-Clapeyron type law [18]. A σ-T diagram for a given alloy composition and crystallographic tensile orientation can be drawn using the experimental data of the critical stresses obtained at different temperatures, classifying each point (T,σ) with a particular stress induced MT. Determinations of the change of entropy associated with the MTs [6] and the transformation temperatures at zero stress are helpful data in order to construct the transformation and equilibrium lines in these diagrams. The relative position of these lines in the σ-T diagram is the key to understand the probability of nucleation of the involved martensitic structures. They are also helpful to know the precise σ-T coordinates of a SMA element for the design of actuator devices.

Several studies have been performed to analyze the effects of pseudoelastic cycling, i.e., of repeated stress induced martensitic transitions, on Cu-Al-Ni single crystals [19-22] and polycrystals [23-26]. Yang et al. [23, 24] published detailed reports about pseudoelastic cycling in polycrystalline β-CuAlNi. Lives to fracture were short, and failure occurred after nucleation of cracks at grain boundaries by a mechanism very similar to the one in conventional metallic systems. L. C. Brown [19] has studied the pseudoelastic cycling in single crystals. He noticed that the cycling life appeared to be largely independent of the amount of strain, crystal orientation and testing environment, but strongly dependent on surface preparation. Another important contribution has been made by H. Sakamoto [22], he has analyzed the cyclic response in single crystals under various deformation modes. He determined that the principal factor controlling N_f (number of cycles to fracture) was stress, and neither orientation nor strain rate dependence were observed.

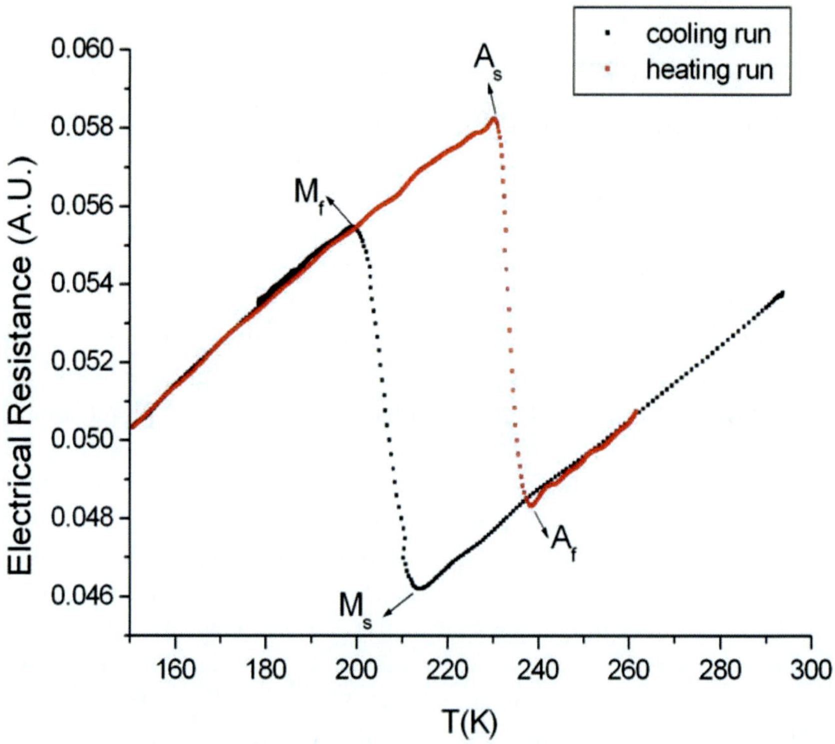

Figure 1. Electrical resistance vs. temperature curve of a Cu-14.3Al-4.1Ni(wt.%) single crystal sample. A β↔γ' type MT can be observed.

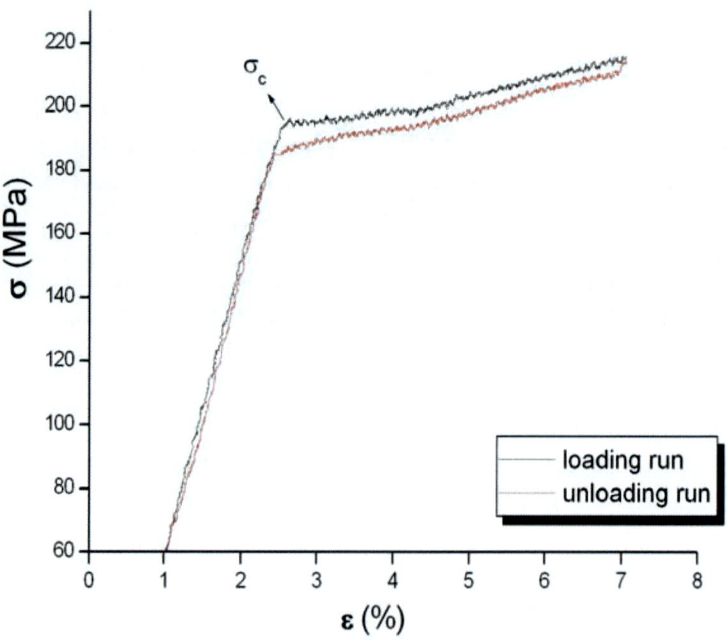

Figure 2. Applied tensile stress (σ) vs. strain (ε) curve of a β↔β' MT in a CuAlNi single crystal sample.

He compared these results with those previously reported in polycrystalline Cu-Al-Ni, and he found that the strain allowed for a given N_f in single crystals was larger than that in polycrystals. In a previous paper, Sakamoto et al. [21] suggested that the cause of low cycling life of Cu-Al-Ni single crystals is defect formation at martensite-matrix interfaces during transformation cycles.

As the pseudoelastic martensitic transformations in shape memory single crystals of chemical compositions around Cu-14Al-4Ni (wt.%) are well known [16, 17, 27, 28], the present authors focus their research on the mechanical evolution during pseudoelastic cycling of single crystals specimens and its relation with the formation of defects in the bulk of the material, for two different initial conditions: when β transforms to β' martensite, and the case where the matrix transforms to a mixture of β' and γ' martensites.

EXPERIMENTAL PROCEDURE

Most of the results that will be presented in the next Sections consist of experimental data that were previously reported by the present authors and other co-workers. The most significative experimental procedures are commented in the next paragraphs, although more details can be found in references [3, 5, 6].

CuAlNi alloys were prepared from 99.99% Cu and 99.99% Ni by melting in an arc furnace under argon atmosphere, adding 99.99% Al to the primary alloy by melting in a resistance furnace. The CuAlNi ingots were cut and the single crystals grown by the Bridgman method in sealed quartz tubes. The single crystals were solution treated at 1173 K for 1 h and then quenched into a mixture of ice and water at 273 K, to obtain an ordered β phase at room temperature. The crystallographic orientations of the single crystals were determined by the back-reflection Laue method.

The martensitic transformation temperatures (M_s, M_f, A_s and A_f) were obtained by electrical resistance vs. temperature measurements by the four-wire method.

Mechanical tensile tests were performed to follow the evolution of stress induced martensitic transformations. For this purpose, cylindrical specimens 23 mm long and 5 mm in diameter were machined from the as-grown crystals, and their central part, 10 mm long, were spark-machined to a diameter of 3 mm. The surface of this central part of the specimens was mechanically (SiC, 600 grit) and electrolytically polished (in 15% nitric acid in methanol at 293 K, 10 V). Tensile tests were carried out in an Instron 5567 machine, under well-controlled temperature conditions in an Instron 3119-005 temperature chamber. The load cell is connected to one of the tensile grips via an universal joint, which decreases the effect of possible torque. The measurement of elongation was determined by the crosshead displacement.

The surface of the cycled specimens was observed in an optical microscope Carl Zeiss Axiovert 100A. This microscope was also used to study the morphology of the stress induced martensitic structures, observing the surface of the sample while a special device (particularly developed for this microscope) applies tensile load to the specimen at different temperatures [29].

Discs belonging to the family of planes $\{001\}_\beta$ were cut from the cycled samples, and the observations of the bulk of these specimens were made in a Philips CM 200 Transmission Electron Microscope (TEM), operating at 200 kV.

Differential Scanning Calorimetry (DSC) measurements were carried out in a MDSC TA Instrument 2910. The heating and cooling rates were of 10 K/min, using empty aluminum sample pan as reference. Disc specimens of approximately 50 mg extracted from the single crystals were used.

RESULTS

1. Thermomechanical Characterization of Cualni Single Crystals

Thermally induced MTs (also called spontaneous MTs) of the CuAlNi single crystals samples were studied previously to the mechanical characterization. The characteristic martensitic transformations temperatures (M_s, M_f, A_s and A_f) were obtained by electrical resistance vs. temperature measurements. These curves also give information about the type of MT that is induced only by cooling the sample, when no stress is applied [10].

Tensile pseudoelastic cycles were performed at different temperatures, above the M_s temperature, in order to characterize the SIM transitions. One example is shown in Figure 3, where different behaviors can be observed for this Cu-14.3Al-4.1Ni(wt.%) single crystal sample. The curve at 294 K shows a maximum hysteresis width of about 4 MPa, which is typical for the $\beta \leftrightarrow \beta'$ transformation [13, 16]. From temperatures around 268 K, the hysteresis width increases gradually as the temperature is lowered. This behavior is attributed to the $\beta \leftrightarrow \beta' + \gamma'$ transformation, the larger the amount of γ' the wider the hysteresis on lowering the temperature [27]. It was shown earlier by Novak et al. [16] that the hysteresis of the γ' martensite depends strongly on the nucleation process. It is suggested that in most cases the β' martensite assists the nucleation of the γ' phase. For transformations taking place about and below 266 K a larger hysteresis is obtained and eventually no retransformation is observed on unloading, indicating that mainly the $\beta \rightarrow \gamma'$ transformation is induced [16], and the transformation stresses in this region are ill defined. It must be noticed that these experiments were performed from high to low temperatures. In cases in which martensite was retained after unloading, the sample was heated up to room temperature before the following test.

The authors studied the morphology of the SIMs, the martensite nucleation process and the order of nucleation when more than one martensitic structure is induced. These experiments were performed by *in situ* stress induced observations in an optical microscope with polarized light, using a special device which allows applying a tensile load to the sample at different temperatures [29]. Photographic sequences of the surface of the sample as the applied load is increased were performed at different temperatures. A sequence made at 298K is shown in Figs. 4(a) to (c), increasing the applied load on the sample. In the first photograph, fine plates of martensite can be observed as the applied stress reaches the critical value. When the load increases, new martensite plates appeared together with the coalescence of the first plates. This morphology and the way they appear correspond to a $\beta \rightarrow \beta'$ stress induced MT [30-32]. Another photographic sequence was made at 287 K, shown in Figs. 5(a) to (c). As the applied load reaches the critical value, fine plates of β' martensite appear firstly (Figure 5(a)).

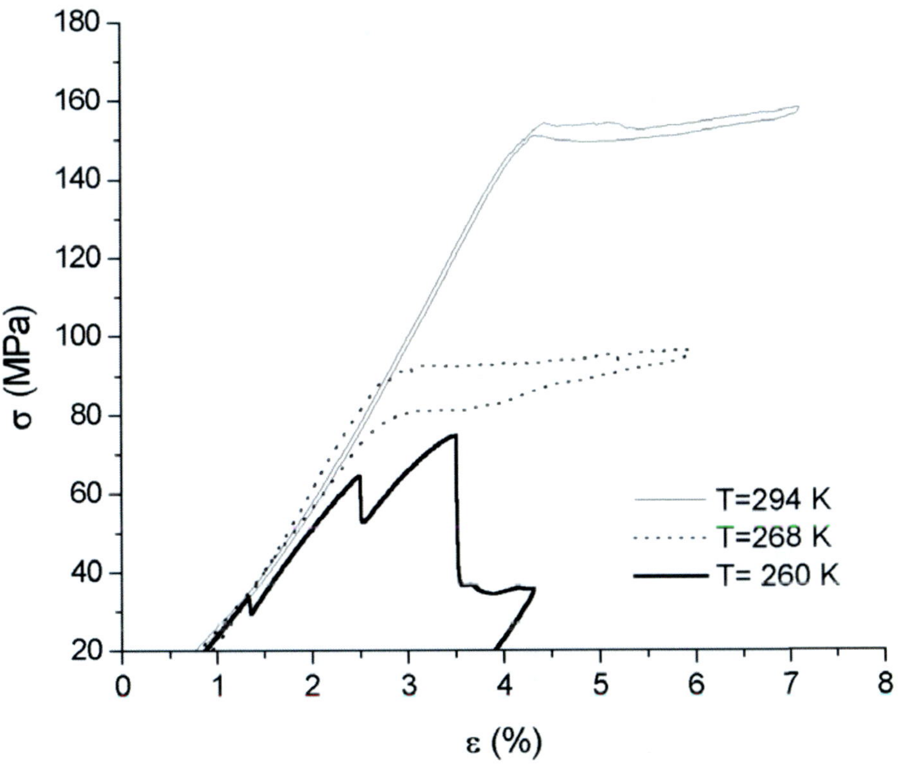

Figure 3. Pseudoelastic cycles for a Cu-14.3Al-4.1Ni(wt.%) single crystal sample, made at 294, 268 and 260 K. Strain rate ~ 0.0003 s-1. Crystallographic tensile orientation: [5 1 10]. Extracted from [6].

As the load increases, these plates disappeared and a coarse band appears abruptly (Figure 5(b)), corresponding to a γ' martensite nucleation [32]. This martensitic phase grows as the load increases (Figure 5(c)). This sequence corresponds to a β↔β'+γ' MT. This figure clearly shows that β' martensite nucleates firstly, and this structure assists the nucleation of the γ' phase.

After the characterization of the SIMs and TIMs, a σ-T diagram is constructed for a particular chemical composition and crystal orientation, determining values of the critical stress at different temperatures. The corresponding σ-T diagram for the sample of Figure 3 is shown in Figure 6. It can be observed that the points lie in a good linear relationship, with a slope $d\sigma/dT=(2.285\pm0.031)$ MPa/K (solid line). The open symbol indicates the case in which the β→β'+ γ' transition takes place: the corresponding transformation stress fits the linear behavior of the β→β' transition. The intersection of this line with the temperature axis gives the M_s for the spontaneous β→β' pure transformation. A value of M_s(18R)=227 K is obtained for this sample.

Tensile pseudoelastic cycles for a Cu-14.1Al-4.2Ni(wt.%) single crystal sample were performed at different temperatures, and the corresponding σ-T diagram obtained from the critical transformation stresses is presented in Figure 7. A general behavior similar to that of alloy Cu-14.3Al-4.1Ni(wt.%) can be observed.

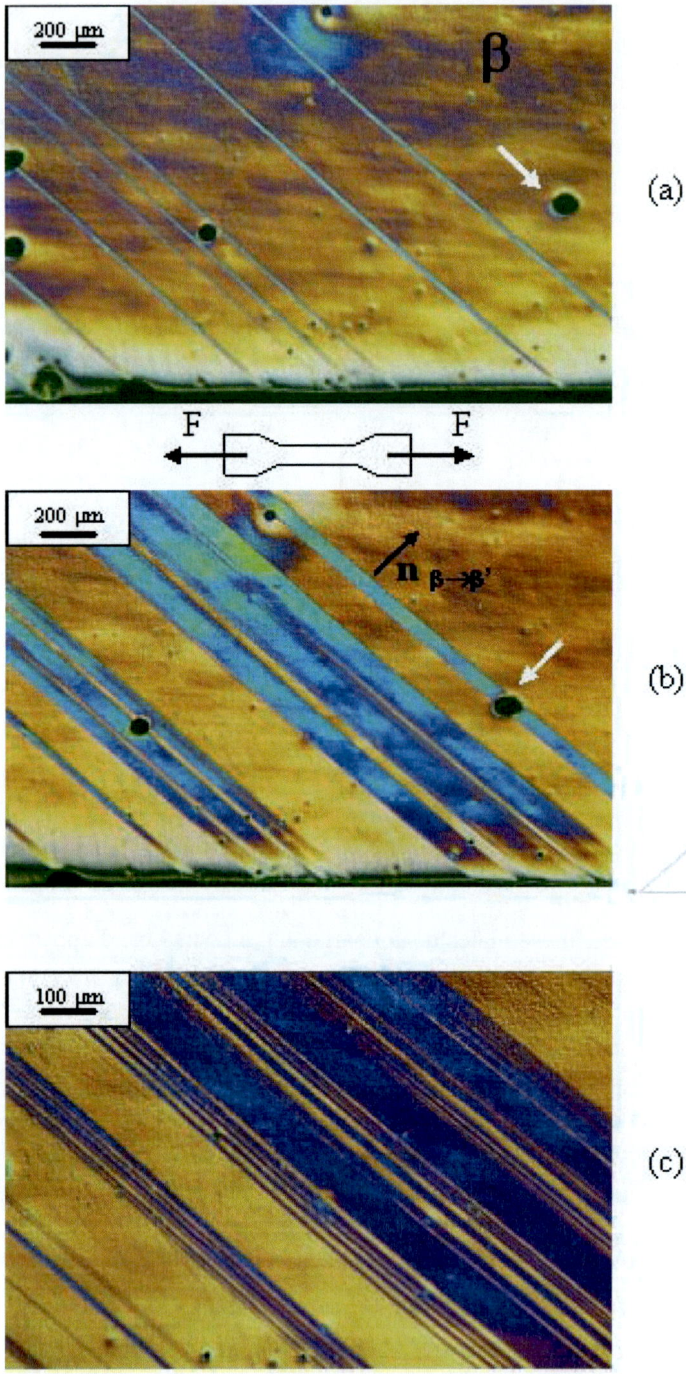

Figure 4. Photographic sequence made in the same zone of a Cu-14.1Al-4.2Ni(wt.%) single crystal sample, made at 298 K. The applied load is increased from (a) to (b). (a): the first β' martensite plates appear as the applied load reaches the critical value. (b): new β' martensite plates appear together with the coalescence of the first plates that appeared in (a). (c): a detail of the β' martensite plates of photograph (b). The stereographic projection of the tensile axis is shown in the triangle at the right of the photographs. The white arrow marks a defect of the surface used as a reference. Golden zone: β phase. nβ→β' : direction of the normal of the habit plane of the β→β' MT.

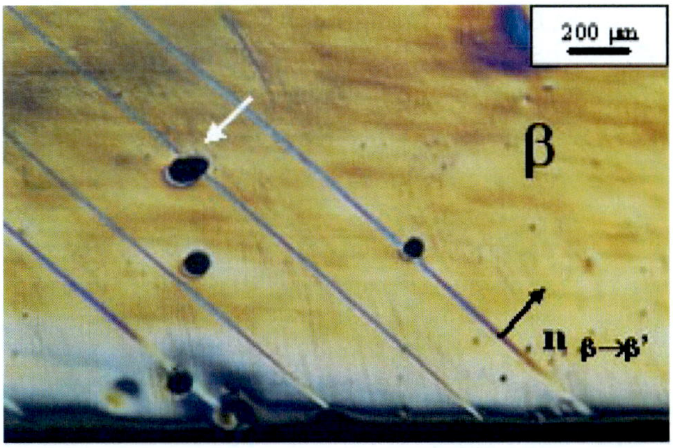

(a)

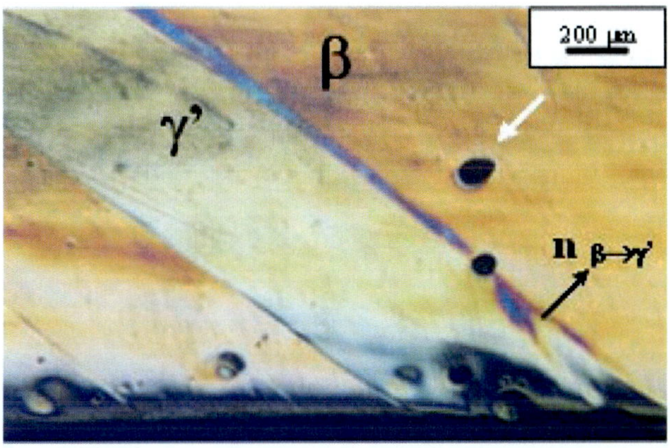

(b)

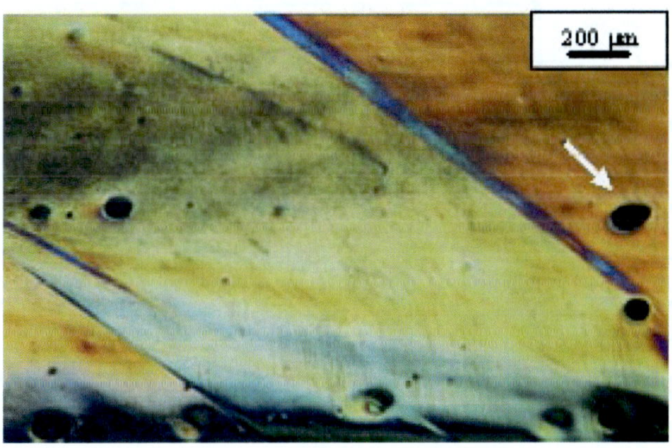

(c)

Figure 5. Photographic sequence made on the same sample as Figure 4, T=287 K. The applied load is increased from (a) to (c). (a): the first β' martensite plates appear as the applied load reaches the critical value. (b): a coarse γ' martensite band appears, which grows as the load is increased, as can be seen in (c).

Points corresponding to different kind of transformations, have a good linear relationship, with a slope of dσ/dT=(2.846±0.020) MPa/K (solid line). The extrapolated value of M_s at σ=0 for the β→β' transformation for this sample of Cu-14.1Al-4.2Ni(wt.%) is 250 K.

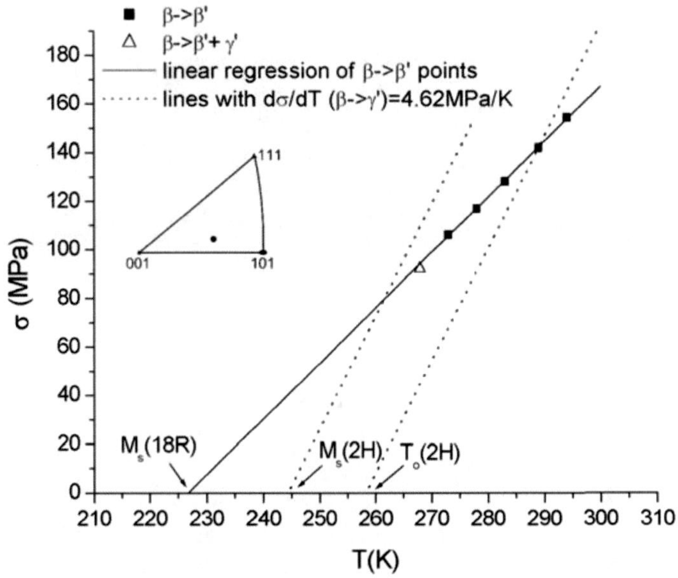

Figure 6. σ-T diagram for the sample of Figure 3. Triangle inside: crystallographic tensile orientation. Extracted from [6].

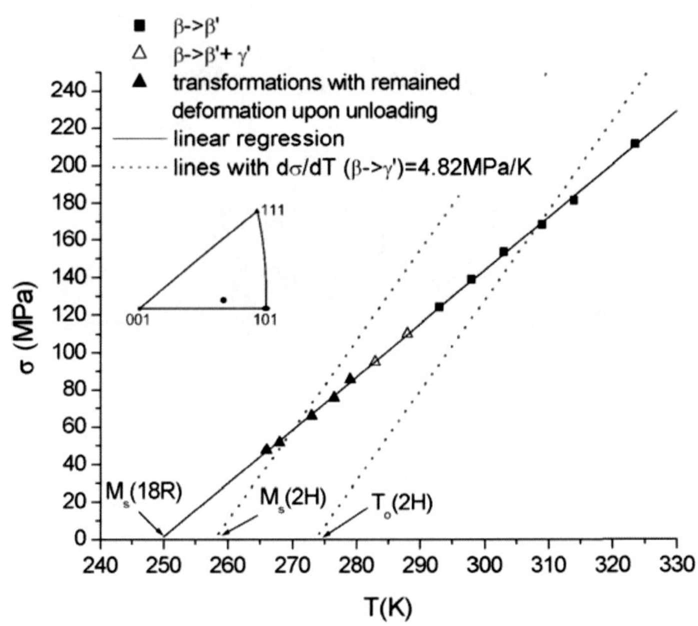

Figure 7. σ-T diagram for a Cu-14.1Al-4.2Ni(wt.%) single crystal sample. Triangle inside: crystallographic tensile orientation. Extracted from [6].

The entropy and enthalpy changes involved in phase transitions are important magnitudes in order to characterize an alloy. The entropy change during β→β' transformation can be obtained from Figs. 6 and 7 for both alloys, using a Clausius-Clapeyron type equation [10]:

$$\Delta S = \gamma \frac{d\tau}{dT} \quad (1)$$

where ΔS is the entropy change per unit of volume during the transformation, τ is the critical resolved shear stress ($\tau=\mu\sigma$, where μ is the Schmid factor) and γ is the transformation shear strain. A value of $\gamma=0.1686$ can be calculated from [33] using the lattice parameters obtained by Otsuka et al. [1]. Since both alloys are very close in composition, the same value of γ was used to calculate the entropy change $\Delta S^{\beta\to\beta'}$ for both alloys (Table I). For these chemical compositions, it is not possible to obtain $\Delta S^{\beta\to\beta'}$ by DSC measurements, as there is no β→β' thermally induced MT.

Assuming that $\Delta S^{\beta\to\beta'}$ remains constant down to M_s, it is possible to determine the equilibrium enthalpy change for the thermally induced β↔β' transition, as $\Delta H=T_0\Delta S$. The equilibrium temperature T_0 is obtained by adding to M_s half of the hysteresis width obtained from the isothermal mechanical tests, for several samples of both alloys. Average values are presented in Table I.

Table I. Average values of: ΔSβ→β' (from isothermal mechanical tests), ΔH β↔β'(calculated from ΔSβ→β'), ΔH β↔γ'(from DSC measurements) and ΔSβ↔γ'(calculated from ΔH β↔γ'), for Cu-14.3Al-4.1Ni(wt. %) and Cu-14.1Al-4.2Ni(wt. %). Values of ΔSβ→β' and ΔSβ↔γ' obtained by other authors for compositions around Cu-14Al-4Ni (wt.%) are also shown. Extracted from [6]

Author	**Alloy** (wt. %)	$\Delta S^{\beta\to\beta'}$ (J/molK)	$\Delta H^{\beta\leftrightarrow\beta'}$ (J/mol)	$\Delta H^{\beta\leftrightarrow\gamma'}$ (J/mol)	$\Delta S^{\beta\leftrightarrow\gamma'}$ (J/molK)
Gastien et al. [6]	Cu-14.3Al-4.1Ni	1.28 ± 0.08	267 ± 33	326 ± 9	1.41 ± 0.05
Gastien et al. [6]	Cu-14.1Al-4.2Ni	1.39 ± 0.08	317 ± 20	362 ± 7	1.44 ± 0.05
Friend [35]	Cu-14.1Al-4.2Ni	1.17			1.26
Ortín et al. [36]	Cu-14.1Al-4.2Ni	1.11 ± 0.10			1.54 ± 0.07
Friend et al. [38]	Cu-14.1Al-4.2Ni	-			1.54 ± 0.07 (p→m) 1.18 ± 0.07 (m→p)
Otsuka et al. [13]	Cu-14.0Al-4.2Ni	1.30			-
Horikawa et al. [37]	Cu-14.1Al-3.9Ni	1.35			
Friend et al. [34]	Cu-14.4Al-4.2Ni	-			1.61

Concerning the β↔γ' transition, the associated $\Delta H^{\beta\leftrightarrow\gamma'}$ enthalpy change can be obtained by differential scanning calorimetry (DSC) measurements from the heat exchanged in the stress free transition β↔γ' [34], by averaging values from cooling and heating runs, since the available data do not show differences in the heat capacities of the involved phases. The corresponding entropy change is calculated by dividing the enthalpy change by the equilibrium temperature T_o, taking $T_o=(M_s+A_f)/2$ [18]. Typical thermal power vs. temperature curves obtained for samples belonging to alloy Cu-14.3Al-4.1Ni(wt.%) can be seen in Figure 8. Average values of $\Delta H^{\beta\leftrightarrow\gamma'}$ and $\Delta S^{\beta\leftrightarrow\gamma'}$ for both alloys are given in Table I, where we compare our values with those stated by other authors for the same composition and for alloys around Cu-14Al-4Ni (wt.%). It can be observed that our values differ from the ones obtained by Friend [35] and Ortín [36], specially for the β→β' transition, but it must be taken into account that those determinations were done measuring thermal powers during stress induced tests, and on the other hand our values of $\Delta S^{\beta\rightarrow\beta'}$(and the ones made by Otsuka [13] and Horikawa [37]) were determined from isothermal mechanical tests by the use of the Clausius-Clapeyron equation.

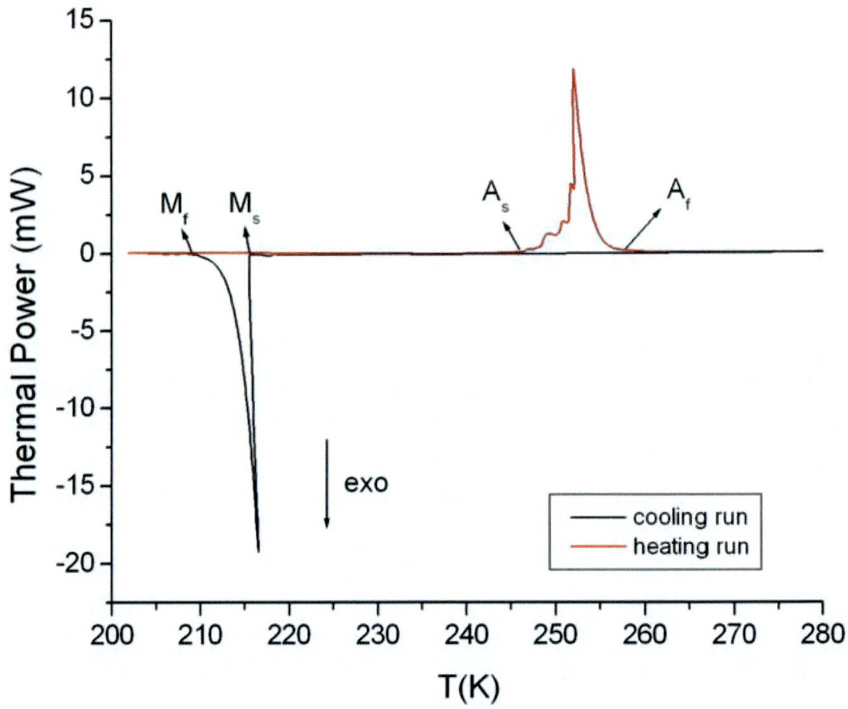

Figure 8. Thermal power vs. temperature measurement for a Cu-14.3Al-4.1Ni(wt.%) single crystal sample.

2. β↔β' Cycling

When dealing with shape memory alloys it is possible to perform mechanical cycles inducing a martensitic transition in each cycle. We will focus in this type of cycling. It has been shown that fracture can occur as a consequence of cumulative irreversible phenomena that are generated during the change of crystal structure that occurs in each cycle [39, 40]. These irreversible phenomena are mainly the generation of dislocations, which play an important role in the response of pseudoelastic cycling, through their interaction with the martensitic structure that is induced during each cycle.

Mechanical tensile cycling through the β↔β' transition were performed in single crystals specimens of various chemical compositions: Cu-14.3Al-4.1Ni (wt.%), Cu-14.1Al-4.2Ni (wt.%) and Cu-14.0Al-6.0Ni (wt.%). No evolution of the stress-strain curves was observed up to N=2700 in specimens oriented close to the [100] direction.

For example, in Figure 9, cycles obtained at 293 K for N=3 and N=2200 (N: number of cycle) for a Cu-14.3Al-4.1Ni (wt.%) single crystal sample are compared. The hysteresis width remains almost unchanged with cycling. The transformation stress was unaltered during cycling and small changes in the shape of the σ-ε curves can be observed. This repetitive pseudoelastic response is important to take into account for the design of shape memory actuator devices.

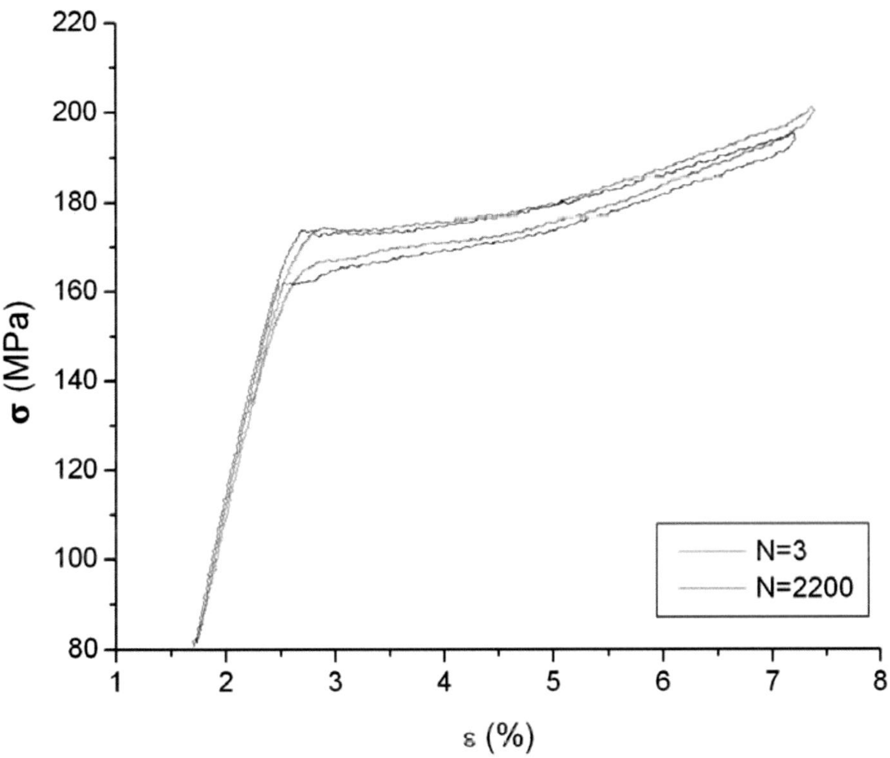

Figure 9. Pseudoelastic cycles at N=3 and 2200 for a Cu-14.3Al-4.1Ni (wt.%) single crystal sample, T=293 K, strain rate ~ 0.0003 s-1, tensile crystal orientation: [2 1 14].

3. β↔β'+γ' Cycling: The γ' Inhibition Effect

Pseudoelastic β↔β'+γ' cycling were performed on several CuAlNi single crystals samples, in order to study the mechanical evolution of the σ-ε curves with the number of cycles. Figure 10 shows the difference between N=2 and N=2000, performed at T=288K. A broad transformation hysteresis was observed for N=2, indicating a mixture of β' and γ' martensitic structures.

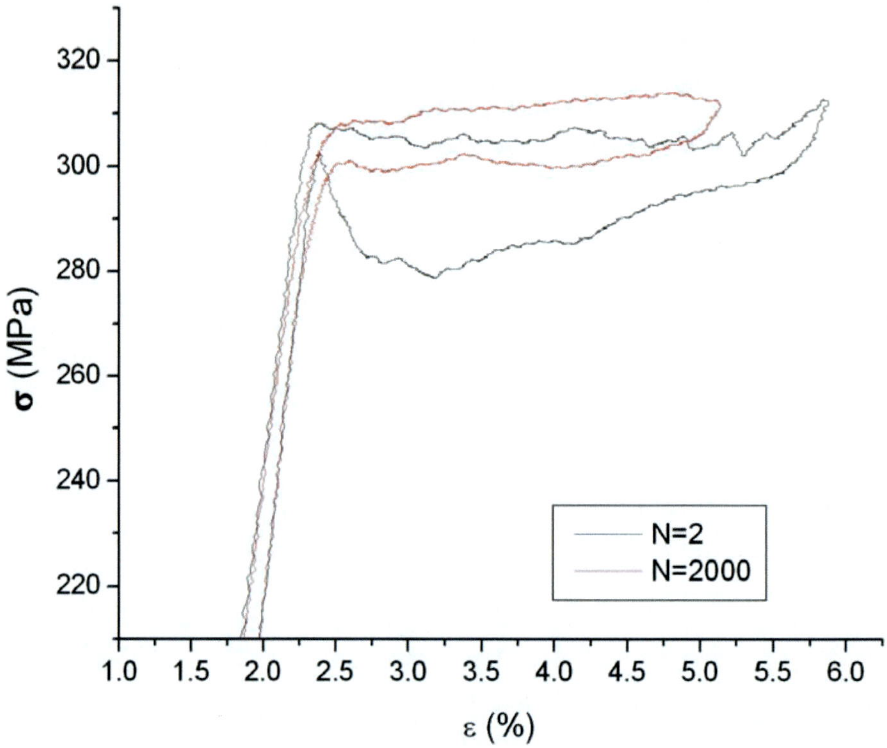

Figure 10. Pseudoelastic cycles at N=2 and 2000 for a Cu-14.3Al-4.1Ni (wt.%) single crystal sample, T=288 K, strain rate ~ 0.0003 s-1, tensile crystal orientation: [4 0 5].

The hysteresis reduces drastically for N=2000 showing a typical β↔β' transformation. It can also be noted that the transformation stress was unaltered regardless of the evolution of the transformation. This suggests that the β' martensite is present in the transformation since the first cycle. Besides, the broad hysteresis observed in the first cycles revealed the presence of γ'. The transition from β to a mixture of martensitic structures has been reported by other authors [16, 27]. The present results show an inhibition of the 2H structure as the pseudoelastic cycling evolves. The introduction of defects during cycling probably causes the disappearance of the 2H martensite, as it will be discussed later.

Another example of pseudoelastic β↔β'+γ' cycling for a Cu-14.1Al-4.2Ni (wt.%) single crystal sample is shown in Figure 11. This figure presents cycles 1, 500 and 1000, all of them performed at a strain rate of 0.0003 s^{-1}, at T=287 K. It is noticed that the morphology of the pseudoelastic cycle changes as the number of cycles increases, i. e., the transformation to γ' martensite is inhibited and finally the β→β' plays the main role in the stress induced transition.

This inhibition effect of the γ' structure as a β↔β'+γ' cycling progresses is also observed during thermal cycling through the mixed transition. A Cu-14.1Al-4.2Ni (wt.%) single crystal specimen was thermally cycled, while the electrical resistance response vs. temperature was registered. Figure 12 shows the change in the hysteresis curve as the number of cycles increases: the amount of γ' martensite decreases while the amount of β' increases, as the electrical resistance signal grows for the β↔β' transition and it diminishes for the β↔γ'.

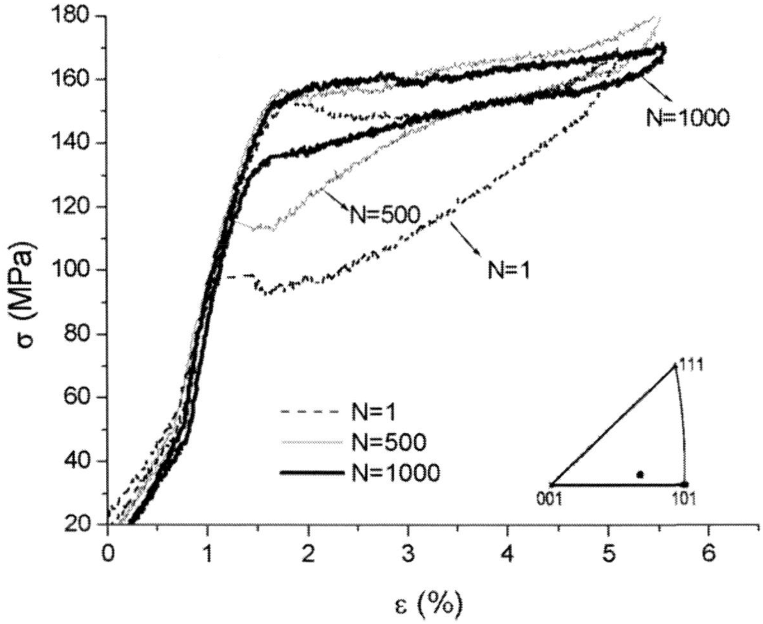

Figure 11. Pseudoelastic cycles at N=1, 500 and 1000 for a Cu-14.1Al-4.2Ni (wt.%) single crystal sample, T=287 K, strain rate ~ 0.0003 s-1. Triangle inside: crystallographic tensile orientation. Extracted from [5].

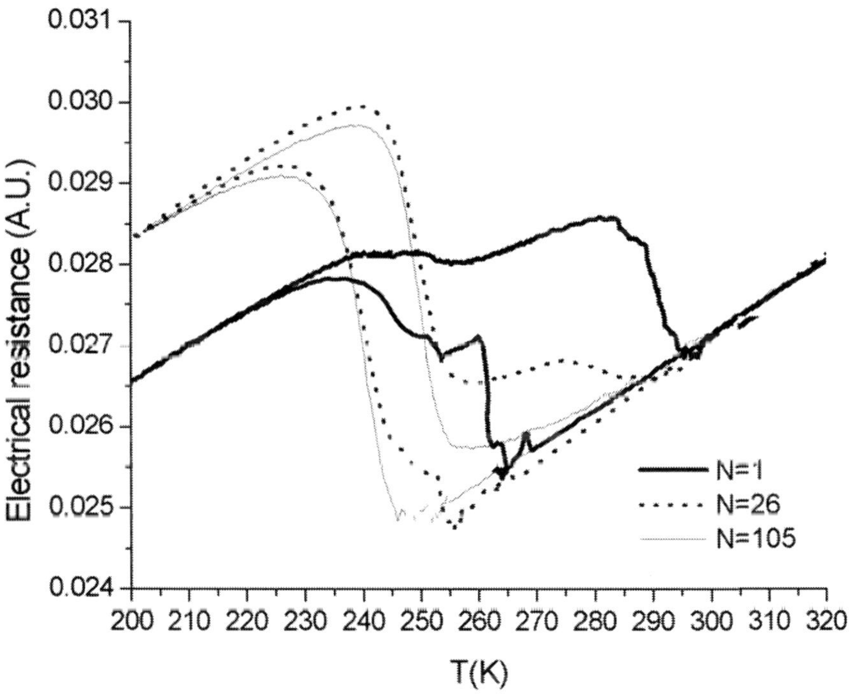

Figure 12. Electrical resistance vs. temperature for N=1, 26, 105. Cu-14.1Al-4.2Ni (wt.%) single crystal sample. Extracted from [5].

The curves show not only changes in the shape of the electrical response but in the value of the transformation temperatures as the cycling progresses. For the first cycle M_s(2H)=265 K and M_s(18R)=254 K, for N=105 both martensites nucleate nearly at the same temperature, with a value of M_s(2H)= M_s(18R)=246 K. This implies a shifting of the M_s(2H) temperature of about 20 K towards the low temperature region, and of about 8 K for the M_s(18R) temperature. In spite of this shifting, the hysteresis width of the martensitic transformation remains nearly constant.

Results obtained by Seguí et al. [41] and Recarte et al. [42] show that, when the thermally induced mixed transformation is involved, some plastic deformation is present due to an incomplete self-accommodation of both martensitic structures, and a stabilization of the β' martensite with respect to the γ' martensite is promoted during the reverse transformation. In this way, as the thermal cycling progresses, this plastic deformation would favor the transformation to the β' martensite, as it forms autoacomodant groups of variants easier than the γ', and the internal stresses are better accommodated. Consequently, the amount of γ' martensite would be reduced, as was observed in the thermal cycling presented in Figure 12.

4. Defects

4.1.In β↔β' cycled samples

The surface of the cycled specimens was analyzed by optical microscopy. No surface defects could be detected in this specimen, at variance with the behavior of Cu-Zn-Al alloys where an important density of surface defects have been observed for the same number of cycles and working temperature [43, 44].

Bulk defects generated during a β↔β' cycling were analyzed in a Transmission Electron Microscope (TEM) in samples subjected to different numbers of cycles, at 298 K (M_s=213 K). This is shown in Figure 13(a) and (b) for samples cycled 460 and 2200 times respectively. It can be remarked that the dislocations arrays are less dense compared with the well-known cycled Cu-Zn-Al single crystals [43, 44] for similar number of cycles and test temperatures.

4.2.In β↔β'+γ' Cycled Samples

Bulk defects generated by pseudoelastic β↔β'+γ' cycling were observed by TEM. Figs. 14(a) and (b) show the presence of dislocation bands, in a specimen subjected to 2400 cycles. The Burgers vector of the dislocations in the bands was analyzed following the same procedure as in Ref. [45]. The great majority of the dislocations have a mixed character, a Burgers vector direction corresponding to a $\langle 100\rangle_\beta$ direction and line direction of the $\langle 111\rangle_\beta$ type. In Figure 14(a), a band of approx. 20 dislocations separated by 0.09 μm can be observed. In another zone of the same sample (Figure 14(b)), a band of approx. 40 dislocations separated by 0.06 μm is shown.

Distances between bands were measured, and an average value of 21 μm was obtained for a sample cycled up to N=1370 (Figure 15). This sample shows longer and denser dislocations arrays than another sample cycled up to N=540 (Figure 16), which have an average distance between bands of 13 μm.

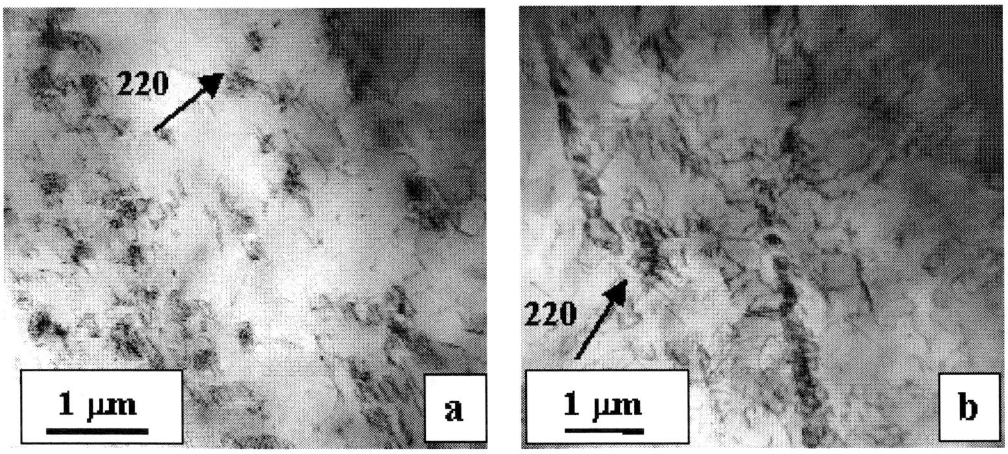

Figure 13. Transmission electron microscopy micrographs for specimens subjected to β↔β' pseudoelastic cycling. Dislocations partially arranged in bands are observed. a) after 460 cycles and b) after 2200 cycles. Extracted from [3].

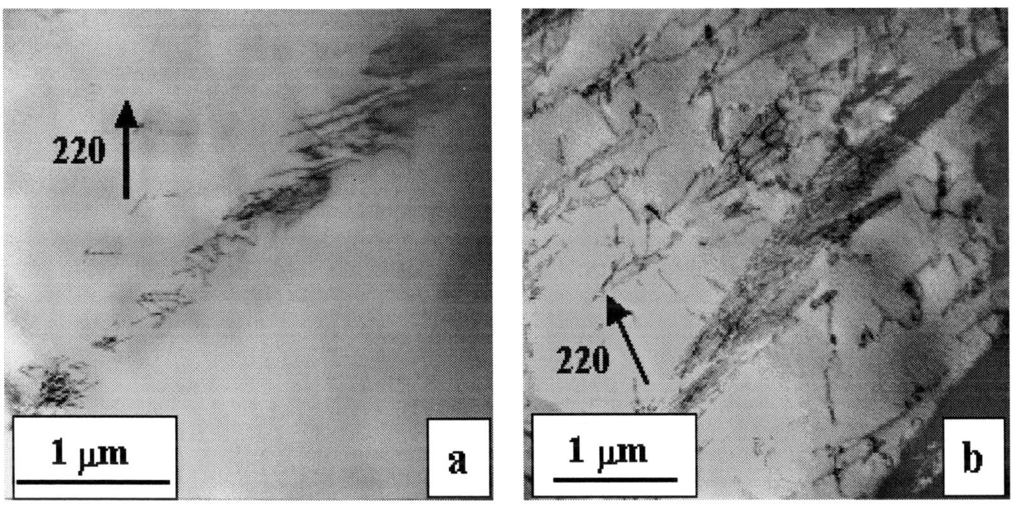

Figure 14. Transmission electron microscopy images of dislocation bands in a specimen after 2400 β↔β'+γ' cycles. (a) and (b) correspond to different area of the foil. Extracted from [3].

5. Estimation of The γ' Inhibition Energy

An experiment was performed in order to estimate the energy of inhibition of the γ' martensite after pseudoelastic β↔β'+γ' cycling. The order of magnitude of this energy is an important key to understand the origin of this phenomenon. A possible way to obtain the energy involved in the γ' inhibition is to determine a value of the energy necessary to induce the γ' structure again after the inhibition. This reasoning was used to design the experiment that will be described in the next paragraphs, although the complete development can be seen in reference [5].

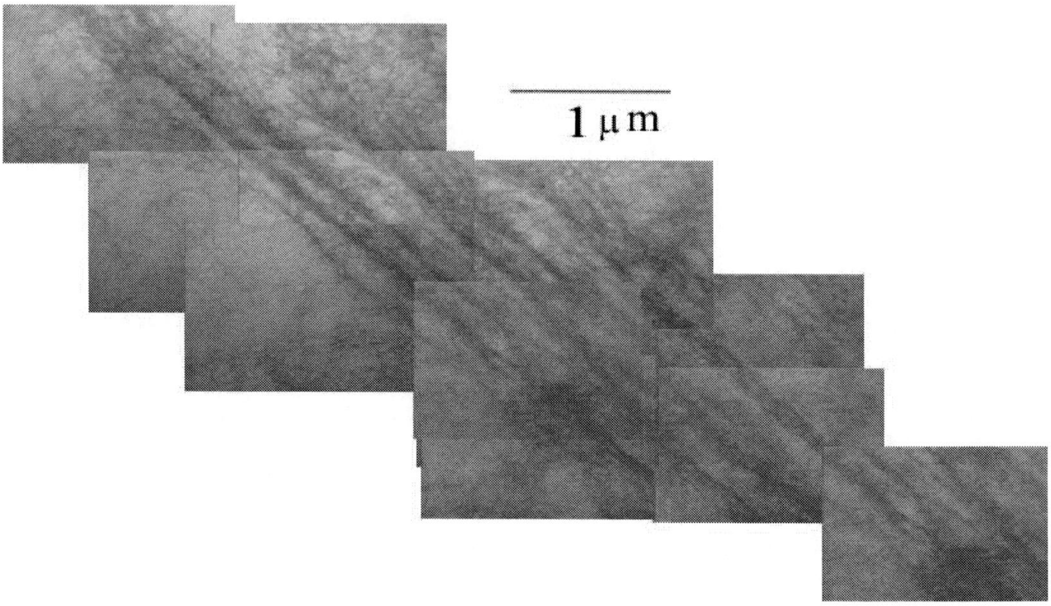

Figure 15. Transmission electron microscopy micrographs showing a dislocation array for a sample subjected to 1370 pseudoelastic $\beta\leftrightarrow\beta'+\gamma'$ cycles. Extracted from [5].

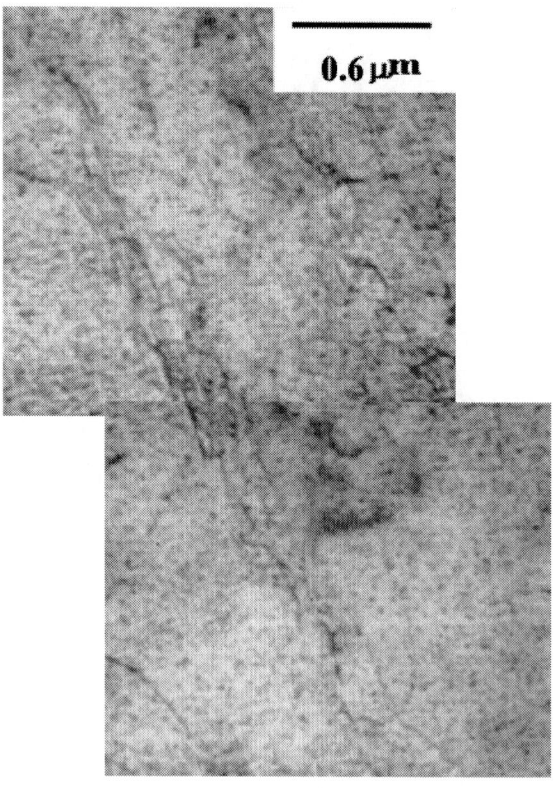

Figure 16. Transmission electron microscopy micrographs showing a dislocation array for a sample subjected to 540 pseudoelastic $\beta\leftrightarrow\beta'+\gamma'$ cycles. Extracted from [5].

First of all, a sample was subjected to a $\beta \leftrightarrow \beta' + \gamma'$ mechanical cycling at a temperature called T_{exp}, until the γ' structure is inhibited. After that, isothermal pseudoelastic cycles were performed descending the temperature from T_{exp} in intervals of 2 K. The hysteresis width of these cycles increases as the temperature is decreased, and at a temperature called T_{rec} the width obtained in the first cycle is recovered. The difference between T_{exp} and T_{rec} (called ΔT_{in}) gives a value of the undercooling necessary to obtain the γ' structure again. This procedure was conducted for several samples, obtaining an average value of $\Delta T_{in} = T_{exp} - T_{rec} \approx 10$ K.

Together with the value of ΔT_{in}, a decrease in the transformation stress is obtained from this experiment. With these data and a thermodynamical analysis through σ-T diagrams, a proper estimation of the energy of γ' inhibition involved in pseudoelastic $\beta \leftrightarrow \beta' + \gamma'$ cycling was obtained. The complete development of this analysis can be read in [4], and the most important steps will be shown in the forthcoming paragraphs.

Before calculating the inhibition energy, the Gibbs free energy difference between β phase and the martensitic structures has to be evaluated, and the possible contributions to this energy have to be considered.

As was stated in Sections 2 to 4, dislocations are created when a material is subjected to thermal or pseudoelastic cycling. The contributions of the dislocations to the Gibbs energy may come from the self-energies of the dislocations and generation of stacking faults. Any contribution of these defects to the configurational entropy can be neglected [46].

The Gibbs energy difference per mol between β and any of the other j phases (j=β'and γ') has the general form:

$$\Delta G^{\beta\to j}(T,\sigma) = \Delta H^{\beta\to j} - T\,\Delta S^{\beta\to j} - V\sigma\,\varepsilon_t^{\beta\to j} + \Delta G_{el}^{\beta\to j} + V\,\Delta E_d^{\beta\to j} \qquad (2)$$

$\Delta H^{\beta\to j}$: change of enthalpy of the β→j transition
$\Delta S^{\beta\to j}$: change of entropy of the β→j transition
V: molar volume
$\varepsilon_t^{\beta\to j}$: deformation associated with the β→j transition
$\Delta G_{el}^{\beta\to j}$: change of elastic energy associated with the β→j transition
$\Delta E_d^{\beta\to j}$: average energy difference (per unit volume) of the dislocations

If the difference between specific heats of the two involved phases is neglected [18], $\Delta H^{\beta\to j}$ and $\Delta S^{\beta\to j}$ can be considered independent on temperature and stress.

On the other hand, as the difference of elastic deformation between β and j phases is negligible compared with the deformation associated with the β→j phase transformation, the term $\Delta G_{el}^{\beta\to j}$ in eq. (2) can be neglected [47].

Besides, $\Delta E_d^{\beta\to j}$ in eq. (2) is the average energy difference per unit volume of the dislocations in both phases, which can be written as:

$$\Delta E_d^{\beta\to j} = \Delta E_d^{self} + E_d^{fault} \qquad (3)$$

ΔE_d^{self} is the self-energy difference of the dislocations embedded in the β and martensite phases. It has been shown [48] that the self-energies are nearly equal for β and β' phases, thus the term ΔE_d^{self} in eq. (3) can be neglected. We shall assume that such difference between β and γ' phases can also be neglected. $\overline{E}_d^{fault}$ represents the energy of stacking faults per unit volume averaged in the whole sample. The faults are created while the martensitic structure crosses the dislocation arrays, due to the loss of the Burgers vector translation symmetry [49]. This contribution to the Gibbs energy can be neglected in eq. (2) (see Ref. [4)], comparing with typical values of the change of enthalpy of the MTs obtained for these alloys (around 300 J/mol [6]).

In the following, the experience reported in [5] will be analyzed through the σ-T diagram shown in Figure 17. In this figure, the β→β' and β→γ' transformation lines are plotted together with the equilibrium β↔β' and β↔γ' lines. It will be assumed that the retransformation line (for the (β'+γ') martensitic states) lies parallel and at the right of the β↔γ' line. Along the equilibrium lines, eq. (2) becomes:

$$\Delta G^{\beta \to j}(T, \sigma_o) = 0 \ \text{(for j=β'and γ')} \tag{4}$$

σ_o : equilibrium stress

Taking into account the small hysteresis of the β→β' transition compared with the β→γ' [9, 12], the stress required to transform to β' martensite (called $\sigma_{T_{exp}}^{\beta \to \beta'}$) is nearly equal to the equilibrium stress $\sigma_{o,T_{exp}}^{\beta \to \beta'}$, this means that:

$$\Delta G^{\beta \to \beta'}\left(T_{exp}, \sigma_{T_{exp}}^{\beta \to \beta'}\right) \approx \Delta G^{\beta \to \beta'}\left(T_{exp}, \sigma_{o,T_{exp}}^{\beta \to \beta'}\right) = 0 \tag{5}$$

According to Figure 17 and taking into account that in the β→β'+γ' MT the γ' phase nucleates from the β', it could be assumed that the driving force to induce the γ' martensite at the temperature T_{exp} (called DF(T_{exp})), would be given by:

$$DF(T_{exp}) = \Delta G^{\beta \to \gamma'}\left(T_{exp}, \sigma_{o,T_{exp}}^{\beta \to \beta'}\right) = \Delta H^{\beta \to \gamma'} - T_{exp}\, \Delta S^{\beta \to \gamma'} - V\, \sigma_{o,T_{exp}}^{\beta \to \beta'}\, \varepsilon_t^{\beta \to \gamma'} \tag{6}$$

where $\sigma_{o,T_{exp}}^{\beta \to \beta'}$ means the equilibrium stress between β and β' at T= T_{exp}.

Analogously, the driving force to nucleate the γ' from the already formed β' at a temperature T_{rec} (called DF(T_{rec})), would be:

$$DF(T_{rec}) = \Delta G^{\beta \to \gamma'}\left(T_{rec}, \sigma_{o,T_{rec}}^{\beta \to \beta'}\right) = \Delta H^{\beta \to \gamma'} - T_{rec}\, \Delta S^{\beta \to \gamma'} - V\, \sigma_{o,T_{rec}}^{\beta \to \beta'}\, \varepsilon_t^{\beta \to \gamma'} \tag{7}$$

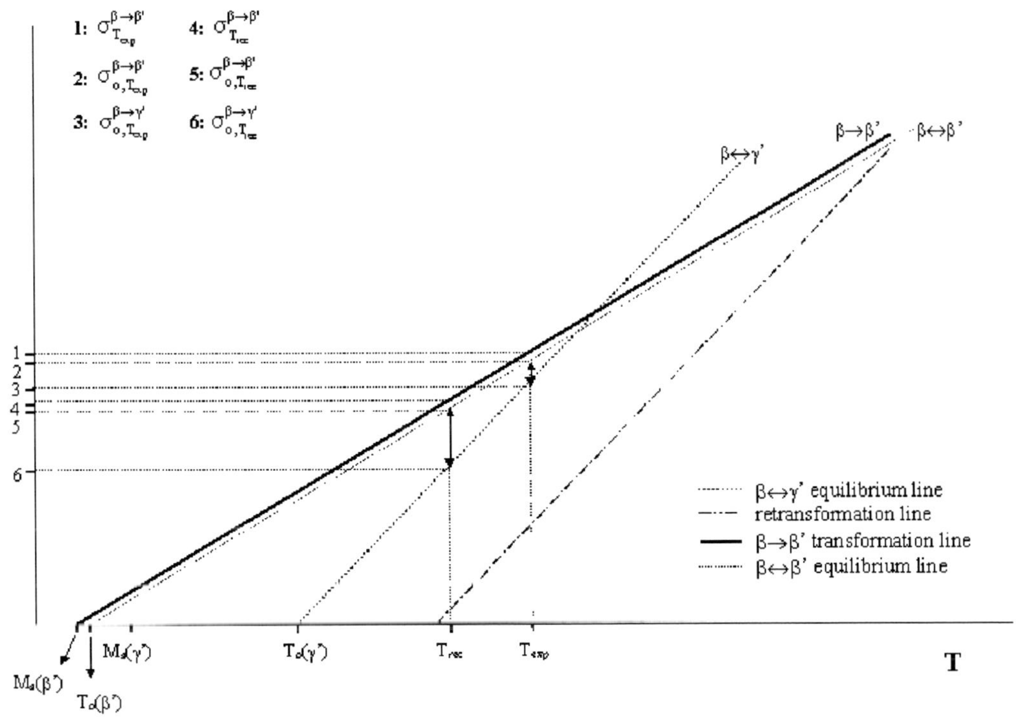

Figure 17. A schematic σ-T diagram for analyzing the experiment reported in [5], in order to estimate the energy of γ' inhibition involved in pseudoelastic β↔β'+γ' cycling. To means the equilibrium temperature. Extracted from [4].

Taking into account the equilibrium condition of eq. (4), the driving force to induce the γ' phase at T_{exp} and T_{rec} will be:

$$DF(T_{exp}) = \Delta G^{\beta\to\gamma'}\left(T_{exp}, \sigma_{0,T_{exp}}^{\beta\to\beta'}\right) = V\left(\sigma_{0,T_{exp}}^{\beta\to\gamma'} - \sigma_{0,T_{exp}}^{\beta\to\beta'}\right)\varepsilon_t^{\beta\to\gamma'} \tag{8}$$

$$DF(T_{rec}) = \Delta G^{\beta\to\gamma'}\left(T_{rec}, \sigma_{0,T_{rec}}^{\beta\to\beta'}\right) = V\left(\sigma_{0,T_{rec}}^{\beta\to\gamma'} - \sigma_{0,T_{rec}}^{\beta\to\beta'}\right)\varepsilon_t^{\beta\to\gamma'} \tag{9}$$

Since any contribution of the dislocations to the free energy difference between the phases can be neglected, the equilibrium lines plotted in Figure 17 remain the same after cycling. Thus, the inhibition of the γ' phase by cycling is not due to a change in the relative stability between β, β' and γ' phases. Instead it is attributed to an increase of the barrier for the growth of a γ' nucleus. When the driving force to obtain the γ' martensite (from the already formed β' martensite) is increased by cooling from T_{exp} to T_{rec}, then the γ' martensite appears again. Therefore, the energy associated with the inhibition of γ' (ΔG_{in}) is related to the difference between the driving force to nucleate γ' at T_{exp} and T_{rec}. Using eqs. (8) and (9):

$$\Delta G_{in} = DF(T_{rec}) - DF(T_{exp}) = \Delta G^{\beta\to\gamma'}\left(T_{rec}, \sigma_{o,T_{rec}}^{\beta\to\beta'}\right) - \Delta G^{\beta\to\gamma'}\left(T_{exp}, \sigma_{o,T_{exp}}^{\beta\to\beta'}\right)$$
$$= V\left[\left(\sigma_{o,T_{rec}}^{\beta\to\gamma'} - \sigma_{o,T_{exp}}^{\beta\to\gamma'}\right) - \left(\sigma_{o,T_{rec}}^{\beta\to\beta'} - \sigma_{o,T_{exp}}^{\beta\to\beta'}\right)\right]\cdot \varepsilon_t^{\beta\to\gamma'} \tag{10}$$

The difference $\left(\sigma_{o,T_{rec}}^{\beta\to\gamma'} - \sigma_{o,T_{exp}}^{\beta\to\gamma'}\right)$ can be evaluated through the Clausius-Clapeyron law on the β→γ' equilibrium line:

$$\frac{\left(\sigma_{o,T_{rec}}^{\beta\to\gamma'} - \sigma_{o,T_{exp}}^{\beta\to\gamma'}\right)}{\left(T_{rec} - T_{exp}\right)} = -\frac{\Delta S^{\beta\to\gamma'}}{\varepsilon_t^{\beta\to\gamma'}} \tag{11}$$

and $\left(\sigma_{o,T_{rec}}^{\beta\to\beta'} - \sigma_{o,T_{exp}}^{\beta\to\beta'}\right)$ on the β→β' transformation line:

$$\frac{\left(\sigma_{o,T_{rec}}^{\beta\to\beta'} - \sigma_{o,T_{exp}}^{\beta\to\beta'}\right)}{\left(T_{rec} - T_{exp}\right)} = -\frac{\Delta S^{\beta\to\beta'}}{\varepsilon_t^{\beta\to\beta'}} \tag{12}$$

In the experiment performed in [5], an average value for $(T_{exp}-T_{rec})=10$ K was obtained, values of $\Delta S^{\beta\to\beta'}$ and $\Delta S^{\beta\to\gamma'}$ were reported by the present authors in a previous paper [6] (also see Table I, Section 1 of Results), and values for $\varepsilon_t^{\beta\to\beta'}$ and $\varepsilon_t^{\beta\to\gamma'}$ can be taken from the literature [11]. Using these data in eqs. (11) and (12) and replacing in (10), the energy associated with the inhibition of γ' martensite in pseudoelastic β↔β'+γ' cycling is:

$$\Delta G_{in} \approx 7 \text{ J/mol} \tag{13}$$

The origin of this energy is most probably related to the localized interaction between the twinned γ' structure and the dislocations arrays generated during cycling. A microscopic model in order to explain the origin of this inhibition energy will be treated in the next Section.

6. Model for The Inhibition of γ' Martensite

This microscopic model is proposed by the present authors to explain the γ' inhibition in β↔β'+γ' cycling. It is related to the localized interaction between a dislocation array and the twinned γ' structure. Dislocations with Burgers vector $[100]_\beta$ and line direction $[111]_\beta$ (accordingly to the microscopic observations) in an isotropic β matrix are considered.

The generation of stacking faults is considered as one of the contributions to this interaction. These faults are created every time the Burgers vector of a dislocation in the β phase lies out of the basal plane of the martensite [3]. This will be treated in Section 6.1.

The other contribution arises from the interaction between the martensitic stress free transformation strains and the stress field created by the dislocation arrays (see Section 6.2). It is shown that the interaction is different for each twin related variant in the γ' martensite.

Taking into account both contributions, a value for the γ' inhibition energy is obtained and compared with the one estimated in Section 5.

6.1. Generation Of Stacking Faults

In this Section, the localized contribution of the stacking faults to the inhibition energy is analyzed in the transformed region of the material.

The first step is to analyze the origin of these stacking faults that are generated during β↔β'+γ' cycling. These faults appear when the twinned γ' structure crosses the dislocations generated by cycling.

The Burgers vectors of the dislocations are of the $<100>_{\beta}$ type, and these dislocations are embedded in the β phase, as was stated in Section 4. A view of one (010) atomic plane of the β phase is shown in Figure 18(i). Two possible Burgers vectors in this projection are $\frac{a_{\beta}}{2}[001]$ and $\frac{a_{\beta}}{2}[100]$. Thus an antiphase boundary is associated to each dislocation in the β phase. When these dislocations are absorbed by the growing 2H martensite, their Burgers vectors change into the $\mathbf{b}_1$ and $\mathbf{b}_2$ vectors respectively in the 2H structure (see Figure 18(ii)). We can deduce from Figure 18(ii) that $\mathbf{b}_1$ and $\mathbf{b}_2$ can be expressed as:

$$\mathbf{b}_{1,2} = \frac{1}{2}\left(\mathbf{a}_{2H} \pm \mathbf{c}_{2H}\right) \text{ (both signs)} \quad (14)$$

The small orthorhombic distortion of the basal plane is neglected in this analysis. The $\mathbf{b}_{1,2}$ vectors have no translation symmetry of the basic 2H structure (when ordering is disregarded), thus a stacking sequence fault will be created. The vectors given by Eq.(14) can be decomposed into:

$$\mathbf{b}_{1,2} = \mathbf{t}_{1,2} + \frac{1}{6}\mathbf{a}_{2H} \quad (15)$$

where $\mathbf{t}_{1,2} = \frac{1}{3}\mathbf{a}_{2H} \pm \frac{1}{2}\mathbf{c}_{2H}$

The vectors $\mathbf{t}_{1,2}$ are translation vectors of the stacking faults commonly observed in the 2H martensites when ordering is disregarded [50]. Hence, it is expected that stacking faults with the $\mathbf{t}_{1,2}$ translation vectors could be formed from the interaction of the martensite with the dislocations. The small difference of $\frac{1}{6}\mathbf{a}_{2H}$ given by Eq.(15) would be absorbed by the creation of a dislocation with this Burgers vector at the starting point of the stacking faults.

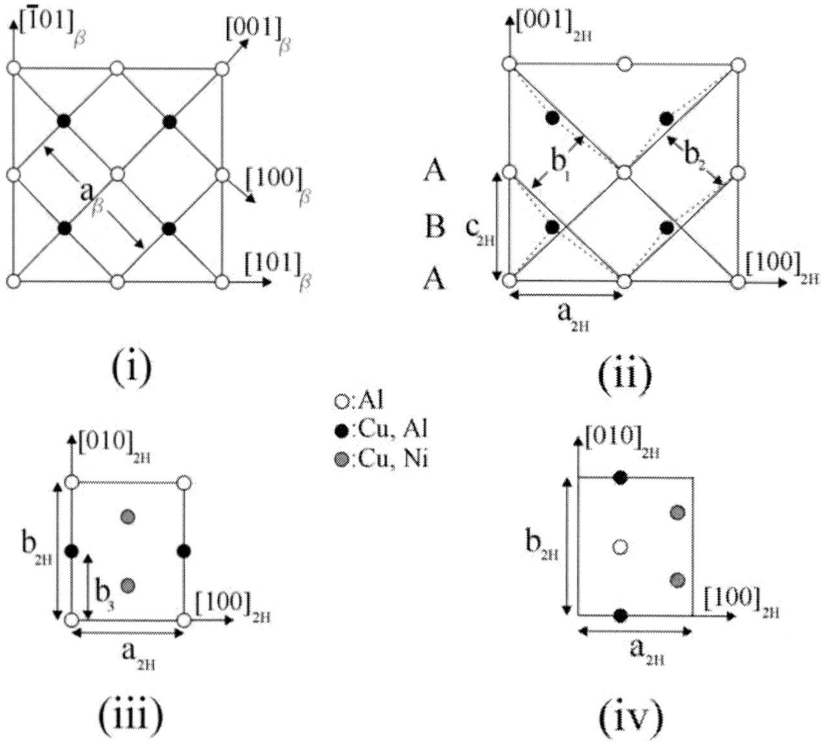

Figure 18. Lattice correspondence between β (i) and 2H (ii) structures. From (i) to (ii) the $\frac{a_\beta}{2}$ type Burgers vectors change into b1 and b2, losing their translation symmetry. Only one atomic plane parallel to the paper is drawn. Figures (iii) and (iv) show the A and B types of basal planes of the 2H martensite respectively. Lattice parameters and atomic distribution were taken from reported data [1]. Extracted from [3].

On the other hand, if the Burgers vector belongs to the basal plane of the martensite (vector $\mathbf{b}_3$ in Figure 18 (iii)), the same translation property as in the β is preserved and no stacking fault will be created for these particular dislocations.

The experimental observations show that the dislocations are grouped in arrangements having in majority the same Burgers vector direction. Thus when a twinned 2H plate crosses such arrangements it may happen that:

I. The Burgers vector belongs to the basal plane of one variant but it lies out of the basal plane of the twin related variant.

II. The Burgers vector lies out of the basal planes of both twin related variants.

When the case (i) takes place, stacking faults will be dragged only by the second twinned variant (having the Burgers vector out of the basal plane). Since introducing stacking faults will increase the free energy of the second variant. A higher density of dislocations leads to an increased difficulty in the formation of the undistorted habit plane.

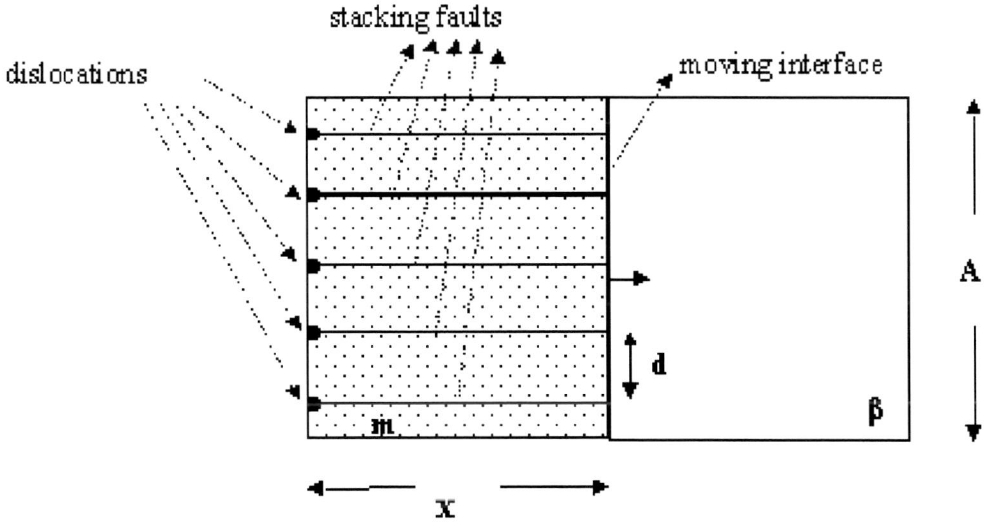

Figure 19. Simplified scheme concerning the interaction between the martensitic structure and a dislocation array (dislocation line perpendicular to this figure). When the interface β/martensite moves during the martensitic transformation, from each dislocation in the β phase the growing martensite drags a stacking fault. Dotted zone: martensitic phase, white zone: β phase. Extracted from [7].

A simplified scheme concerning the interaction between a martensitic structure and a dislocation array embedded in the β phase, which will be used to calculate the total stacking fault energy in a transformed volume of the sample is shown in Figure 19. Dislocation arrays observed in cycled CuAlNi single crystals samples are formed by approximately n=40 dislocations separated by d≈0.08 μm (see Section 4). In Figure 19, dislocations are represented by black dots (dislocation line perpendicular to the figure), with a distance d between them, and A is the length of the array. When the interface β/martensite moves during the martensitic transformation, from each dislocation in the β phase (with ***b*** out of the basal plane of the martensite) the growing martensite drags a stacking fault. The dislocation lines and the fault planes (L: thickness) are perpendicular to this figure, and the distance x in Figure 19 will be considered as the twin unit width (x=0.1 μm) observed in copper based alloys [51]. The total stacking fault energy in this transformed volume is obtained as:

$$E_{sf}^{T} = n\left(x \cdot L\right)\gamma_{sf} \tag{16}$$

with $n = A/d$, which is the quantity of faults with area equal to ($x.L$), and γ_{sf} is the stacking fault energy per unit area. The stacking fault energy per transformed volume is:

$$\mathrm{E}_{\mathrm{sf}} = \frac{E_{\mathrm{sf}}^{\mathrm{T}}}{x \cdot L \cdot A} = \frac{\gamma_{\mathrm{sf}}}{d} \tag{17}$$

Using an average distance between dislocations obtained by transmission electron microscopy observations, d= 0.08μm [3], and a value for γ_{sf} = 5 x 10^{-2} J/m^2 according to reported data about the generation of stacking faults in copper based alloys [52, 53], the stacking fault energy per transformed volume is:

$$E_{sf} = \frac{\gamma_{sf}}{d} \approx 6.3 \cdot 10^5 \text{ J/m}^3 = 4 \text{ J/mol} \tag{18}$$

The contribution of this stacking fault energy to increase the energy of the γ' martensite can be evaluated taking into account that the γ' martensite is a twinned system. For the following reasoning the type II twinning will be considered, but the results are analogous for type I twinning. The condition of having an undistorted habit plane between the type II twinning and the β matrix imposes a volume relationship between the twinned related variants. If V is the volume occupied by the twinned γ' martensite, and V_1 and V_3 the volumes that occupy variants 1 and 3 respectively:

$$V=V_1+V_3 \text{ with } V_1=\tfrac{2}{3}V \text{ and } V_3=\tfrac{1}{3}V \tag{19}$$

Thus the stacking fault energy in the twinned γ' martensite would depend on the crystallographic orientation between the Burgers vectors of the dislocations and the basal plane of variants 1 and 3. Since the dislocations with Burgers vectors $\boldsymbol{b}=[100]_\beta$ are predominant in the cycled CuAlNi samples, we shall restrict the analysis to this case.

The Burgers vector $\boldsymbol{b}=[100]_\beta$ lies in the basal plane of variant 1 of γ' martensite, but is out of the basal plane of the other variant. So, stacking faults will be dragged only by variant 3, increasing the free energy of this variant. According to eq. (19) the volume occupied by variant 3 is equal to ($\tfrac{1}{3}V$), thus the increase of energy, per volume unit, of the twinned system of volume V will be:

$$E^{sf(\gamma')}_{b=[100]} = \tfrac{1}{3} E_{sf} \tag{20}$$

Concerning the β' martensite, since there is only one variant in the stress induced transformation, and the $[100]_\beta$ Burgers vector of the dislocations lies in the basal plane of this variant, no stacking fault energy will be created in the β' phase:

$$E^{sf(\beta')}_{b=[100]} = 0 \tag{21}$$

We shall evaluate the inhibition energy as the difference of energy between the β' and the γ' martensites when these phases cross the dislocation arrangements. Since the stacking faults in the γ' martensite are absorbed by the twin interface [54], the interaction will be localized around the dislocations to a distance of the order of the twin width. In this respect the inhibition energy is considered as the difference in the barrier to overcome a dislocation arrangement between β' and γ', but no stacking fault energy is created in the β' phase.

Therefore, the contribution of the generation of stacking faults to the inhibition energy (E_{inh}^{sf}) will be simply given by eq. (20). Using eq. (18) in (20) we obtain:

$$E_{inh}^{sf} = E_{b=[100]}^{sf(\gamma')} = 1.33 \text{ J/mol} \quad (22)$$

This result can be considered as a rough contribution to the inhibition energy, it is in fact somewhat low compared to the experimental value of 7 J/mol estimated in Eq.(13) (Section 5) [4]. Nevertheless, this is not the only interaction mechanism acting between martensite and dislocations, since another contribution may arise from the localized interaction between the stress field of the dislocation arrays and the twinned related variants of γ' martensite. This will be treated in the next Section.

6.2.Interaction Between Dislocation Stress Fields and Martensitic Deformation Strain

This part of the model takes into account the interaction between the martensitic stress free transformation strains and the stress field created by the dislocation arrays. The most important steps of the calculus are shown in the next paragraphs, although the complete analysis can be read in [8].

An expression for the Gibbs free energy involved in this type of interaction was first given by Eshelby [55] as:

$$\Delta G_{d}^{int} = -\int \sigma_{ij}^{d} \cdot \varepsilon_{ij}^{m,tr} \, dV \quad (23)$$

where ΔG_{d}^{int} is the interaction of the martensite nucleus with the stress field of the defects, σ_{ij}^{d} is the stress field generated by the defects, and $\varepsilon_{ij}^{m,tr}$ the stress-free martensitic transformation strain. This expression was obtained taking into account that the strain $\varepsilon_{ij}^{m,tr}$ is higher than the strain field of the β phase and the elastic strain of the martensite [56]. Other approximations were applied, since the heat capacity coefficients as well as the thermal expansion coefficients of both phases are nearly equal, so the internal energy difference and the entropy difference will be nearly independent on temperature and stress [18, 56].

In order to evaluate Eq.(23) we have to analyze the differences between the interaction of the defects with β' and γ' martensites. The β' martensite is a single variant, therefore a β' plate will have positive and negative energy interactions when crossing the dislocations bands depending on the sign of the stress fields around the dislocation arrangements. On the average, the interaction energies from these regions will cancel to each other. Thus, at the end a very weak interaction of the β' phase with the dislocation stress fields is predicted, as it is indeed experimentally observed.

Concerning the γ' martensite we should remark that a very important characteristic of the MT is the presence of a macroscopically undistorted habit plane [57]. In the γ' martensite this strong condition for the habit plane can only be fulfilled by twinning. The twinning crystallography of the γ' martensite in CuAlNi alloys has been analyzed by Okamoto et al [11]. As in the analysis of the previous Section, type II twinning is considered for this treatment, which is formed by the twinned related variants 1 and 3 [11]. As the position of each variant within the volume V is unknown, the interaction energy with each variant will be evaluated in the whole volume V. Then the average interaction energy per unit volume will be taken as:

$$\widetilde{E}^{int}_{d,v} = -\frac{1}{V}\int \sigma^{d}_{ij}\cdot\varepsilon^{v}_{ij}\,dV \quad \text{for variants v=1,3} \tag{24}$$

ε^{v}_{ij} : components of the strain tensor $\boldsymbol{\varepsilon}^{\mathbf{m,tr}}$ for variant v of the γ' martensite.

The lower the value of $\widetilde{E}^{int}_{d,i}$ the more stable variant i will be in the volume V. As the total interaction energy to create variants 1 and 3 will be equal to ($\widetilde{E}^{int}_{d,1}.V_1+\widetilde{E}^{int}_{d,3}.V_3$), two possibilities arise:

- If $\widetilde{E}^{int}_{d,1}<\widetilde{E}^{int}_{d,3}$, it would be energetically more favorable to have the volume V fully transformed to variant 1, with an energy of interaction in this volume equal to ($\widetilde{E}^{int}_{d,1}.V$). Let us assume now that part of the variant 1 in volume V is replaced by variant 3, occupying the latter a volume V_3. The energy in V will be increased by an amount equal to $[(\widetilde{E}^{int}_{d,3}-\widetilde{E}^{int}_{d,1}).V_3]$. Using the volume fractions of V_1 and V_3 given in Eq.(19), the increase of energy in the volume V reads:

$$E^{1<3}_{inh} = [(\widetilde{E}^{int}_{d,3}-\widetilde{E}^{int}_{d,1}).V_3]/V = \tfrac{1}{3}(\widetilde{E}^{int}_{d,3}-\widetilde{E}^{int}_{d,1}) \tag{25}$$

This expression gives the increase of energy, with respect to the possible state of minimum energy in the volume V, when the right volume ratio between the twinned variants is created in order to obtain the habit plane. Since the habit plane is essential for the twinned γ' martensite to grow, $E^{1<3}_{inh}$ represents a barrier for the γ' transformation, and it is considered as a contribution to the γ' inhibition energy.

- If $\widetilde{E}^{int}_{d,3}<\widetilde{E}^{int}_{d,1}$, it is more favorable to have the volume V fully transformed to variant 3, with an energy of interaction in this volume equal to ($\widetilde{E}^{int}_{d,3}.V$). Here, the energy necessary to replace a volume V_1 with variant 1 will be equal to $[(\widetilde{E}^{int}_{d,1}-\widetilde{E}^{int}_{d,3}).V_1]$. Therefore, the energy of inhibition for this case is:

$$E^{3<1}_{inh} = [(\widetilde{E}^{int}_{d,1}-\widetilde{E}^{int}_{d,3}).V_1]/V = \tfrac{2}{3}(\widetilde{E}^{int}_{d,1}-\widetilde{E}^{int}_{d,3}) \tag{26}$$

Since both possibilities can equally appear around a dislocation band the inhibition energy may be taken as an average value from expressions (25) and (26) as:

$$\widetilde{E}_{inh} = \frac{1}{2}\left(E_{inh}^{1<3} + E_{inh}^{3<1}\right) \tag{27}$$

Before evaluating $\widetilde{E}_{d,v}^{int}$ (Eq. (24)), some assumptions have to be made. For simplicity the β-CuAlNi phase will be treated as isotropic. With respect to the bulk defects observed in CuAlNi cycled samples, dislocation arrays formed by N=40 dislocations of a mixed character, separated by 0.08 μm, each of them having **b**=$[100]_\beta$ and **u**= $[111]_\beta$ are considered for this analysis. This corresponds to an average configuration of the dislocation bands, according to the observations made by transmission electron microscopy [3].

An adequate coordinate system has to be chosen in order to separate the screw and edge contributions of the stress tensor **σ** of the dislocation array. Taking the direction of the dislocation line parallel to the z axis and the direction of the Burgers vector parallel to the x axis, the edge and the screw components of the stress tensor of the dislocation array can be properly expressed as in reference [58].

The components σ_{ij} of the stress tensor of a dislocation array, composed of N parallel dislocations separated by 0.08 μm, aligned along the x axis and with the first dislocation located in (x, y)=(0,0) are [58]:

$$\sigma_{xx} = -\sum_{k=1}^{N} \frac{G\,b}{2\pi(1-\nu)} \cdot \frac{y\left\{3[x-0.08(k-1)]^2 + y^2\right\}}{\left\{[x-0.08(k-1)]^2 + y^2\right\}^2} \tag{28}$$

$$\sigma_{yy} = \sum_{k=1}^{N} \frac{G\,b}{2\pi(1-\nu)} \cdot \frac{y\left\{[x-0.08(k-1)]^2 - y^2\right\}}{\left\{[x-0.08(k-1)]^2 + y^2\right\}^2} \tag{29}$$

$$\sigma_{xy} = \sum_{k=1}^{N} \frac{G\,b}{2\pi(1-\nu)} \cdot \frac{[x-0.08(k-1)]\left\{[x-0.08(k-1)]^2 - y^2\right\}}{\left\{[x-0.08(k-1)]^2 + y^2\right\}^2} \tag{30}$$

$$\sigma_{zz} = -\sum_{k=1}^{N} \frac{G\,b\,\nu}{\pi(1-\nu)} \cdot \frac{y}{\left\{[x-0.08(k-1)]^2 + y^2\right\}} \tag{31}$$

$$\sigma_{xz} = -\sum_{k=1}^{N} \frac{G\,b}{2\pi} \cdot \frac{y}{\left\{[x-0.08(k-1)]^2 + y^2\right\}} \tag{32}$$

$$\sigma_{yz} = \sum_{k=1}^{N} \frac{G\,b}{2\pi} \cdot \frac{[x-0.08(k-1)]}{\left\{[x-0.08(k-1)]^2 + y^2\right\}} \tag{33}$$

where G is the shear modulus, ν is the Poisson modulus, b is the Burgers vector length, and k is an index used to indicate the dislocation number. Micrometers are used to measure distances in the x and y directions. Values for G and ν were calculated using reference data of the elastic constants C_{11} and C_{12} [59]. The Burgers vector length was estimated as $b \approx a_\beta/2$ from lattice parameters given in reference [1].

On the other hand, the strain tensor **ε** associated to the martensitic transformation can be written as a diagonal tensor in an appropriated coordinate system [33] (one coordinate system for each variant, for more details see [8]), using lattice parameters of the β and γ' phases, taken from reference [1].

To investigate the behavior of this energy of interaction (Eq. (24)) around a dislocation array the width of the integration volume, along the x direction, is taken as the width of the twin unit, w_0 , i.e., the width of variant 1 plus the width of variant 3 measured in the direction normal to the twin boundary. The same value is taken for the height of the volume along the y direction. Integration volumes of (0.1 μm x 0.1 μm x 1 μm) are considered according to the reported twinning width of the γ' martensite [51]. As the array of dislocations is chosen to be along the x axis (in the dislocation system d), a distance of around 2b (b: length of the Burgers vector) from y=0 was considered for the position of the lower integration limit [58].

Calculations were performed for several positions of the integration volume along the x axis. Such positions were chosen to have the integration volume located symmetrically with respect to a given k^{th} dislocation. Table II resumes some of the results, and for this part of the analysis, integration volumes for y>0 (above the dislocation array) and y<0 (below the array, symmetric to x=0) were considered. It can be noted that for y>0, the condition $\widetilde{E}^{int}_{d,3} < \widetilde{E}^{int}_{d,1}$ is fulfill, which means that the dislocation array interacts differently with each martensite variant, and this difference would cause the final inhibition of the γ' structure. For these integration volumes, values of $E^{3<1}_{inh}$ were calculated following the described procedure (Eq. (26)). On the other hand, for y<0, the condition $\widetilde{E}^{int}_{d,1} < \widetilde{E}^{int}_{d,3}$ is fulfill, and $E^{1<3}_{inh}$ were determined using Eq. (25).

An average of these values can be taken as the contribution to the inhibition energy of the γ' martensite (Eq.(27)):

$$\widetilde{E}_{inh} = 10.7 \text{ J/mol} \tag{34}$$

Since the strength of the elastic fields decreases with the distance from the dislocation arrays, the value of the inhibition energy given by Eq.(34) depends on the choice of the size and position of the integration rectangles. It was verified that $\widetilde{E}_{inh}$ varies from 15 J/mol to 6 J/mol if the height of the integration volume is varied from 2b to 3 μm respectively (fixing the width of the volume). It was also found that the interaction energy decreases very quickly as the integration rectangle is located at positions outside the interval $0 < x < 3.12\,\mu m$ [8]. Thus the value given in Eq.(34) remains as a good representative value for the inhibition energy.

Following the same procedure, the energy of inhibition was evaluated for dislocations having other line direction of the type $<111>_\beta$, obtaining values of the same order of magnitude as the obtained for **u**=$[111]_\beta$ [8].

It is interesting to study how the strength of this inhibition energy varies with the distance to the dislocation array.

Table II. Values of the energy of interaction between a dislocation array and variants 1 and 3 of γ' martensite ($\widetilde{E}^{int}_{d,1}$ and $\widetilde{E}^{int}_{d,3}$), $E^{1<3}_{inh}$ and $E^{3<1}_{inh}$ (calculated using Eqs. (25) and (26), see text), for b=[100]β and line direction u=[111]β, for some integration volumes for y>0 ($V^{k}_{y>0}$) and y<0 ($V^{k}_{y<0}$). Extracted from [8]

$V^{k}_{y>0}$	$\widetilde{E}^{int}_{d,1}$ (J/mol)	$\widetilde{E}^{int}_{d,3}$ (J/mol)	$E^{3<1}_{inh}$ (J/mol)
$V^{1}_{y>0}$	10.40	-5.02	10.28
$V^{3}_{y>0}$	13.79	-7.63	14.28
$V^{7}_{y>0}$	14.89	-8.12	15.34
$V^{11}_{y>0}$	15.15	-8.14	15.53
$V^{16}_{y>0}$	15.27	-8.07	15.56
$V^{20}_{y>0}$	15.30	-8.00	15.53
$V^{21}_{y>0}$	15.30	-7.98	15.52
$V^{26}_{y>0}$	15.26	-7.85	15.41
$V^{31}_{y>0}$	15.11	-7.65	15.17
$V^{38}_{y>0}$	13.79	-6.80	13.73
$V^{40}_{y>0}$	10.40	-5.87	10.85

$V^{k}_{y<0}$	$\widetilde{E}^{int}_{d,1}$ (J/mol)	$\widetilde{E}^{int}_{d,3}$ (J/mol)	$E^{1<3}_{inh}$ (J/mol)
$V^{1}_{y<0}$	-10.40	5.87	5.42
$V^{3}_{y<0}$	-13.79	6.80	6.86
$V^{7}_{y<0}$	-14.89	7.44	7.44
$V^{11}_{y<0}$	-15.15	7.65	7.60
$V^{16}_{y<0}$	-15.27	7.80	7.69
$V^{20}_{y<0}$	-15.30	7.98	7.76
$V^{21}_{y<0}$	-15.30	7.98	7.76
$V^{26}_{y<0}$	15.26	8.09	7.78
$V^{31}_{y<0}$	-15.11	8.14	7.75
$V^{38}_{y<0}$	-13.79	7.63	7.14
$V^{40}_{y<0}$	-10.40	5.02	5.14

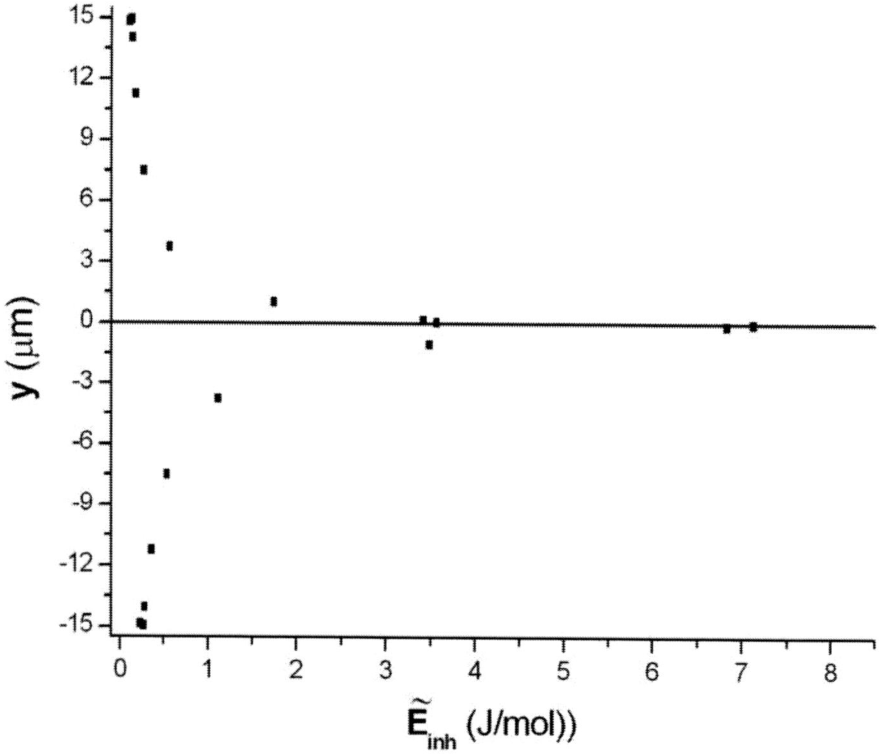

Figure 20. Behavior of $\widetilde{E}_{inh}$ as a function of the distance (along the y axis) to the next dislocation array. Y=0 indicates the position of the dislocation array. Extracted from [7].

For this analysis, the y position (for y>0 and y<0) of the integration volumes were varied up to the next dislocation array (we assume that the cycled sample is a network formed by dislocation arrays separated by 15 μm, as observed in CuAlNi cycled samples [3]). In Figure 20 the values of $\widetilde{E}_{inh}$ obtained for each y position are plotted, and it can be seen that the inhibition energy decreases rapidly with the y distance to the dislocation array, confirming that the interaction that causes the inhibition is strongly localized around the array.

Finally, assuming the contributions to the inhibition energy given by the generation of stacking faults and the interaction between stress and strain fields, the total inhibition energy can be calculated as the difference between the interaction of the defects with the two martensitic structures β' and γ'. So, the total inhibition energy can be written as:

$$E_{inh}^{T} = E_{inh}^{\gamma'} - E_{inh}^{\beta'} \tag{35}$$

As was stated previously, the interaction between defects and the β' phase will be in average nearly null, and no stacking faults are generated in the β' phase (as the Burgers vector of the dislocations lies in the basal plane of the β' structure). Therefore, $E_{inh}^{\beta'} \approx 0$, and the inhibition energy of the γ' martensite becomes:

$$E^{\gamma'}_{inh} = E^{sf}_{inh} + \widetilde{E}_{inh} \tag{36}$$

Using the results given by Eq. (34) and the value obtained for the stacking fault energy Eq.(22), the total inhibition energy is:

$$E^{T}_{inh} \approx 12 \text{ J/mol} \tag{37}$$

The performed analysis showed that the twinned γ' structure would find a barrier to maintain the right twin ratio for the undistorted habit plane. This essentially happens when the γ'martensite intersects an array of dislocations. It should be considered that an average distance of 15 μm [3] separates the dislocation arrays. Thus, the γ' phase could still grow in between the dislocation bands. However, in such a case high energy β/γ' interfaces would be needed around the dislocation bands in order to surround the obstacle. Moreover, even if part of the front achieves to grow in between the dislocation arrays, it will be stopped in a next dislocation band, and so on. In this sense the dislocation bands would act as an effective barrier for the γ' growth, while produce almost no effect on the β' martensite. By lowering the temperature, the stability of the γ' with respect to the β' martensite increases [4]. When the decrease of the free energy of the γ' martensite balances the inhibition energy the obstacles can be surmounted and the twinned γ' martensite would grow at the expense of the β'.

This mechanism is supported by the reasonable agreement observed between the total inhibition energy calculated in Eq. (37) and the inhibition energy estimated in Section 5, Eq.(13) [4], using experimental data. Nevertheless, we have to point out that the material was treated as elastically isotropic to simplify the involved calculations. Thus, the result given in Eq.(37) must be considered as an approximated evaluation giving the right order of magnitude.

It was also verified that the inhibition energy changes slightly with the number of dislocations forming the arrays. This can be explained in terms of the localized character of the interaction. It must be remarked that the proposed model can also be used to explain the inhibition of γ' observed after thermal β↔β'+γ' cycling [5], as the dislocations generated during this type of cycling are similar to those observed in mechanically cycled samples [60-64].

CONCLUSION

A thermomechanical characterization of CuAlNi single crystals samples (chemical compositions around Cu-14Al-4Ni (wt.%)) were done in order to study their response after thermal and mechanical cycling. This involves the characterization of the stress induced martensites and the thermal induced ones, as well as other physical magnitudes like enthalpy and entropy changes during MTs.

No evolution of the stress-strain curves was observed up to N=2700 for the β↔β' cycling in a specimen oriented close to the [100] direction. Moreover, no defects were observed at the specimen surface. This is at variance with the characteristics of pseudoelastically cycled Cu-Zn-Al [43, 44]. In Cu-Zn-Al alloys, the basal plane type defects are formed by dense arrangements of dislocations with retained martensite inside, being the dislocations generated by local plastic deformation of the β' martensite. The lack of this type of basal plane defects in the cycled Cu-Al-Ni alloy suggests that β' in this alloy is harder for plastic deformation than the 18R martensite in Cu-Zn-Al. Besides, there is evidence that the intrusion-extrusion defects formed at the surface in pseudoelastically cycled Cu-Zn-Al alloys are somehow related to diffusion processes [44]. The lack of such defects in Cu-Al-Ni could be explained from the fact that the atomic diffusion in this alloy is much lower that in Cu-Zn-Al at around room temperature [2]. Bulk defects observed in β↔β' cycled CuAlNi single crystals are dislocations arrays less dense compared with the observed in cycled CuZnAl single crystals, for similar number of cycles and test temperatures.

A novel effect was observed after thermal and pseudoelastic β↔β'+γ' cycling in CuAlNi single crystals, the inhibition of γ' structure as the cycling progresses. A suitable experiment and a thermodynamical analysis through σ-T diagrams give an estimation of the energy involved in this effect.

The authors analyzed the origin of this γ' inhibition effect through the interaction between MTs and defects produced during cycling. A microscopic model is proposed, based on the observations of the bulk defects observed after β↔β'+γ' cycling. It consists of two contributions, the first one deals with the generation of stacking faults in the γ' phase when crossing the dislocations embedded in the β phase. These faults increase the free energy of the variant for which the Burgers vector of the dislocations lies out of its basal plane. The other contribution is related to the localized interaction between the stress field generated by the dislocations and the twinned structure of γ' martensite. It was found that the defects interact differently with each twinned γ' variant, creating an energetic unbalance that makes difficult the creation of an undistorted habit plane, and finally inhibits the formation of this martensitic structure. It was shown that the dislocations produce no effect on the β' martensite, whose transformation is not affected by cycling. Taking into account both contributions, the inhibition energy is evaluated giving a value in reasonable agreement with the value estimated from a thermodynamical treatment of the experimental data. Although some approximations were made for obtaining these results, we can conclude that this model gives a good description of the microscopic physical mechanisms responsible of this inhibition effect, together with a reasonable estimation for the energy involved. This model is suitable to explain the γ' inhibition after thermal β↔β'+γ' cycling, as the defects observed in thermal cycled samples are of the same type as the one observed in pseudoelastic cycled specimens. Moreover, this model could be applied in other materials whenever a phase transition interacts with defects.

Finally, the present chapter gives a broad picture concerning thermal and mechanical cycling phenomena through martensitic transitions in shape memory CuAlNi single crystals.

REFERENCES

[1] Otsuka, K.; Shimizu, K. *Trans. JIM* 1974, *vol 15*, 103.

[2] Cingolani, E.; Van Humbeeck, J.; Ahlers, M. *Metall. Mater. Trans. A* 1999, *vol 30*, 493.

[3] Gastien, R.; Corbellani, C.E.; Álvarez Villar, H.N.; Sade, M.; Lovey, F.C. *Mater. Sci. Eng. A* 2003, *vol 349*, 191.

[4] Gastien, R.; Corbellani, C.E.; Sade, M.; Lovey, F.C. *Scripta Mater.* 2006, *vol 54*, 1451.

[5] Gastien, R.; Corbellani, C.E.; Sade, M.; Lovey, F.C. *Acta Mater.* 2005, *vol 53*, 1685.

[6] Gastien, R.; Corbellani, C.E.; Sade, M.; Lovey, F.C. *Scripta Mater.* 2004, *vol 50*, 1103.

[7] Gastien, R.; Sade, M.; Lovey, F.C. *Mat. Sci. Eng. A* 2008, *vol 481-482*, 518.

[8] Gastien, R.; Sade, M.; Lovey, F.C. *Acta Mater.* 2008, *vol 56*, 1570.

[9] Recarte, V.; Pérez-Sáez R.B.; Bocanegra E.H.; Nó M.L; San Juan J. *Mater. Sci. Eng. A* 1999, *vol 273-275*, 380.

[10] Otsuka, K.; Wayman, C.M. *Shape Memory Materials*; Cambridge University Press: Cambridge, UK, 1998.

[11] Okamoto, K.; Ichinose, S.;Morii, K; Otsuka, K.; Shimizu, K. *Acta Metall.* 1986, *vol 34*, 2065.

[12] Van Humbeeck, J.; Van Hulle, D.; Delaey, L.; Ortín, J.; Seguí, C.; Torra, V. *Trans. JIM* 1987, *vol 28*, 383.

[13] Otsuka, K.; Sakamoto, H.; Shimizu, K. *Acta Metall.* 1979, *vol 27*, 585.

[14] Sakamoto, H.; Shimizu, K.; Otsuka, K. *Trans. JIM* 1985, *vol 26*, 638.

[15] Novák, V.; Malimánek, J.; Zárubová, N. *Key Eng. Mat.* 1994, *vol 97-98*, 419.

[16] Novák, V.; Malimánek, J.; Zárubová, N. *Mater. Sci. Eng. A* 1995, *vol 191*, 193.

[17] Novák, V.; Sittner, P.; Vokoun, D.; Zárubová, N. *Mater. Sci. Eng. A* 1999, *vol 273-275*, 280.

[18] Wollants, P.; Roos, J.R.; Delaey, L. *Prog. Mat. Sci.* 1993, *vol 37*, 227.

[19] Brown, L. C. *Metall. Trans. A* 1979, *vol 10*, 217.

[20] Sakamoto, H.; Kijima, Y.; Shimizu, K.; Otsuka, K. *Scripta Metall.* 1981, *vol 15*, 281.

[21] Sakamoto, H.; Shimizu, K.; Otsuka, K. *Trans. JIM* 1981, *vol 22*, 579.

[22] Sakamoto, H. *Trans. JIM* 1983, *vol 24*, 665.

[23] Yang, N. Y. C.; Laird, C.; Pope, D.P. *Metall. Trans. A* 1977, *vol 8*, 955.

[24] Ritter, A.; Yang, N.Y.; Pope, D.P.; Laird, C. *Metall. Trans. A* 1979, *vol 10*, 667.

[25] Sakamoto, H.; Kijima, Y.; Shimizu, K. *Trans. JIM* 1982, *vol 23*, 585.

[26] Sakamoto, H.; Shimizu, K. *Trans. JIM* 1986, *vol 27*, 592.

[27]] Novák, V.; Malimánek, J.; Zárubová, N. *J. Phys. IV C8 Suppl. J. Phys. III* 1995, *vol 5*, 997.

[28] Novák, V.; Sittner, P.; Zárubová, N. *Mater. Sci. Eng. A* 1997, *vol 234-236*, 414.

[29] Gastien, R. *PhD Thesis* 2005, Universidad de Buenos Aires, Argentina.

[30] tsuka, K.; Shimizu, K. *Int. Metals Reviews* 1986, *vol 31*, 93.

[31] Otsuka, K.; Wayman, C.M.; Nakai, K.; Sakamoto, H.; Shimizu, K. *Acta Metall.* 1976, *vol. 24*, 207,

[32] Novák, V.; Landa, M.; Sittner, P. *J Phys. IV France* 2003, *vol. 112*, 593

[33] De Vos, J.; Aernoudt, E.; Delaey, L. *Z. Metallkd.* 1978, *vol 69*, 438.

[34] Friend, C.M.; Ortín, J.; Planes, A.; Mañosa, Ll.; Yoshikawa, M. *Scripta Metall.* 1990, *vol 24*, 1641.

[35] Friend, C.M. *Scripta Metall.* 1989, *vol 23*, 1817.

[36] Ortín, J.; Mañosa, Ll.; Friend, C.M.; Planes, A.; Yoshikawa, M. *Phil. Mag. A* 1992, *vol 65*, 461.

[37] Horikawa, H.; Ichinose, S.; Morii, K.; Miyazaki, S.; Otsuka, K. *Metall. Trans. A* 1988, *vol 19*, 915.

[38] Friend, C.M.; Mañosa, Ll.; Ortín, J.; Planes, A. *J. Phys. IV C4* 1991, *vol 1*, 71.

[39] Sade, M. *PhD Thesis* 1985, Instituto Balseiro, U.N. Cuyo, Argentina.

[40] Malarría, J. *PhD Thesis* 1996, Instituto Balseiro, U.N. Cuyo, Argentina.

[41] Seguí, C.; Cesari, E.; Van Humbeeck, J. *Mat. Trans. JIM* 1990, *vol 31*, 375.
[42] Recarte, V.; Pérez-Landazábal, J.I.; Rodríguez, P.P.; Bocanegra, E.H.; Nó, M.L.; San Juan, J. *Acta Mater.* 2004, *vol 52*, 3941.
[43] Sade, M.; Malarría, J.; Yawny, A.; Lovey, F.C. *Proceedings of the XVIII Conference on Applied Crystallography 2000*, Wisla, Poland. World Scientific Publishing Co., Pte. Ltd., Singapore, 2001, 153.
[44] Malarría, J.; Sade, M.; Lovey, F.C. *Mater. Sci. Eng. A* 2001, *vol 308*, 88.
[45] Sade, M.; Uribarri, A.; Lovey, F.C. *Phil. Mag. A* 1987, *vol 55*, 445.
[46] Weertman, J.; Weertman, J.R. *Elementary dislocations theory*. Oxford University Press: NY, 1992, 197.
[47] Wollants, P.; Bonte, M.D.; Roos, J.R. *Z. Metallk.* 1979, *vol 70*, 113.
[48] Rodriguez, P.L.; Condó, A.M.; Lovey, F.C. *Phys. Stat. Sol. (B)* 1996, *vol 197*, 279.
[49] Lovey, F.C.; Torra, V. *Prog. Mat. Sci.* 1999, *vol 44*, 189.
[50] Condó, A.M.; Lovey, F.C. *Philos. Mag. A* 1999, *vol 79*, 511.
[51] Condó, A.M.; Lovey, F.C.; Torra, V. *Phil. Mag. A* 2003, *vol 83*, 1479-1493.
[52] Ahlers, M. *Prog. Mater. Sci.* 1986, *vol 30*, 135-186.
[53] Lovey, F.C.; Hazarabedian, A.; Garces, J.E. *Acta Metall.* 1989, *vol 37*, 2321-2327.
[54] Lovey, F.C.; Torra, V.; Isalgué, A.; Condó, A.M. *J. Phys. IV France* 2003, *vol 112*, 479-482.
[55] Eshelby, J.D. *Prog. Solid Mech.* 1961, *vol 2*, 89.
[56] Kato, M.; Pak, H. *Phys. Stat. Sol. (b)* 1984, *vol 123*, 415.
[57] Wechsler, M.S.; Lieberman, D.S.; Read, T.A. *Trans. AIME* 1953, *vol 197*, 1503.
[58] Hirth, J.P.; Lothe, J. *Theory of Dislocations*. Krieger Publishing Company: Malabar, Florida, 1982.
[59] Recarte, V.; Pérez-Landazábal, J.I.; Nó, M.L.; San Juan, J. *Mat Sci Eng A* 2004, *vol 370*, 488.
[60] Kajiwara, S.; Kikuchi, T. *Acta Metall.* 1982, *vol 30*, 589.
[61] Kajiwara, S.; Kikuchi, T. *Phil. Mag. A* 1983, *vol 48*, 509.
[62] Ríos-Jara, D.; Morin, M.; Esnouf, C.; Guenin, G. *Scripta Metall.* 1985, *vol 19*, 441.
[63] Perkins, J.; Muesing, W.E. *Metall. Trans.* 1983, *vol 14A,* 33.
[64] Marukawa, K.; Kajiwara, S. *Phil. Mag. A* 1987, *vol 55*, 85.

In: Shape Memory Alloys
Editor: H. R. Chen, pp. 181-230
ISBN 978-1-60741-789-7

Chapter 6

IMPORTANCE OF THE STUDY OF THERMAL PROPERTIES IN THE RESEARCH OF NI-MN-GA SHAPE MEMORY ALLOYS

Alexandra Rudajevova
Charles University, Faculty of Mathematics and Physics, Department of Condensed Matter Physics[1], Praha , Czech Republic

ABSTRACT

The thermal properties of Ni-Mn-Ga based shape memory alloys were studied in the temperature range from room temperature up to 650 K. Both studied polycrystalline alloys, $Ni_{53.6}Mn_{27.1}Ga_{19.3}$ and $Ni_{54.2}Mn_{29.4}Ga_{16.4}$, are high-temperature shape memory alloys.

The thermal expansion of the $Ni_{53.6}Mn_{27.1}Ga_{19.3}$ alloy was investigated for samples with equiaxed grains and columnar grains. The thermal expansion characteristics of both structural types differ in the transient temperature range. Preferential orientations of the martensitic variants in the columnar grains make positive change on the temperature dependencies of the thermal expansion characteristics in the transformation temperature range. Only very small changes of the dilatation characteristics were found in the alloy with equiaxed grains. The thermal expansion characteristics of the $Ni_{53.6}Mn_{27.1}Ga_{19.3}$ alloy were studied also after compression deformation. Two types of deformation can be distinguished by the dilatation measurements. Thermal arrest memory effect was also studied by the dilatation technique for both Ni-Mn-Ga alloys. Analysis of these results shows that the thermal arrest memory effect is a consequence of both the shape memory effect and the stress focusing affect in the temperature range where the phase transformation occurs.

The temperature dependence of the thermal diffusivity, specific heat, and density were determined experimentally. The thermal conductivity was obtained as a product of these parameters for the shape memory of the $Ni_{53.6}Mn_{27.1}Ga_{19.3}$ alloy. The thermal diffusivity and thermal conductivity of the martensite was about 40% lower than for the austenite. Thermal diffusivity was measured by the flash method, where the time dependence of the temperature rise after irradiation was also determined. The latent heat of the phase

[1] Ke Karlovu 5, 121 16 Praha 2, Czech Republic.

transformation substantially lowers the temperature rise after irradiation in the transient temperature range.

Martensitic phase transformation in Ni-Mn-Ga-based shape memory alloys is an athermal phase transformation that starts practically immediately after reaching a certain transient temperature. This temperature is given in each point of the sample by two processes: the thermal conductivity and the phase transformation. Both processes take place in cooperation and their dynamics determines properties of the shape memory material in the transient temperature range. The complex process of the propagation of the phase transformation in the sample is analyzed. The influence of the heat transfer in the sample as well as in the measuring system on the transformation temperatures is also discussed.

INTRODUCTION

Martensitic transformation is a diffusionless structural transformation that occurs between a high-temperature (in most cases) cubic phase and a low-temperature phase with a lower symmetry. The lattice strain is a transformation parameter that determines the physical states of an initial phase and a product phase as well as local intermediate states of the alloy during the phase transformation. The free energy, ΔF, due to formation of a crystal of a product phase in the thermoelastic equilibrium is given by relation [1]:

$$\Delta F = -\Delta f V + E_{el} + \Gamma S; \qquad (1)$$

where V is the volume, S is the surface area, E_{el} is the elastic energy, and Γ is the specific interface energy of a martensite crystal. The gradual relaxation of the elastic energy during the phase transformation leads to a complex polydomain structure of the martensite. The martensitic transformation for most shape memory alloys is referred to as an athermal transformation [2]. This transformation appears to be instantaneous at any temperature and the amount of transformed material depends only on the temperature. The hysteresis in this case is not kinetic in nature. In some cases, the martensitic phase transformation can also be isothermal. Martensite may appear locally at a certain temperature and then the size of the martensitic regions increases continuously with time.

The alloys whose martensitic transformation temperatures are above room temperature are of potential technical interest because they have shown phase transformations at useful operational temperatures. The Ni-Mn-Ga alloy is a promising high-temperature shape memory alloy since it has an excellent shape memory effect and a high thermal stability. Although single crystals have been reported to yield giant magnetic field induced strains in previous experiments [3, 4], oriented polycrystals should be developed for possible applications. Various methods were employed for preparation of preferred crystallographic columnar polycrystals [5-7]. Bhattacharya and Kohn have shown [8] that the set of recoverable strains of a polycrystal depends not only on the texture of the polycrystal, but critically on the change of symmetry during the underlying transformation. The greater the change in symmetry, the greater the set of the recoverable strain.

The martensitic structure in Ni-Mn-Ga based alloys in stoichiometric as well as non-stoichiometric compounds is formed by the tetragonal distortion of the initial cubic lattice (c/a<1) [9]. For stoichiometric Ni_2MnGa, martensitic transformation temperature (~220 K) is below the Curie temperature (~380 K). Both transformation temperatures are very sensitive to the composition. For example, substitution of Ni in place Mn decreases the Curie temperature and the martensitic transformation temperature increases [10, 11]. The martensitic phase transformation is a first-order phase transformation, which is a transformation that involves latent heat. They are characterized by a discontinuous change in volume at a constant temperature and pressure.

DSC studies are the most frequently presented thermal property from all thermal properties of the Ni-Mn-Ga alloy [for example 12-15]. Figure 1 [14] shows typical DSC curves of $Ni_{54}Mn_{25}Ga_{21}$ during heating and cooling. The austenite transformation starting temperature, A_s, peaking temperature, A_p, and finishing temperature, A_f, are 536 K, 548 K, and 561 K, respectively. The martensitic transformation starting temperature, M_s, peaking temperature, M_p, and finishing temperature, M_f, are 533 K, 524 K, and 510 K, respectively. Heating and cooling rates in the measurement were 10 K/min. It is important to note that the difference between the starting temperatures of the martensite and the austenite transformations is only 3 K.

DSC measurements were also used to study the phase transformation, M→A, on Ti-Ni and Cu-Zn-Al shape memory alloys. In these measurements, the phase transformation, M→A, was interrupted and the sample was cooled. In the following the thermal cycle endothermic peak splits into two. This effect was first shown by Johnson, Dominique, and Reichman [16] on the NiTi alloy and it was called the thermal arrest memory effect. The thermal arrest memory effect has been the subject of intense investigation [17-21]; however, no clear explanation has been accepted.

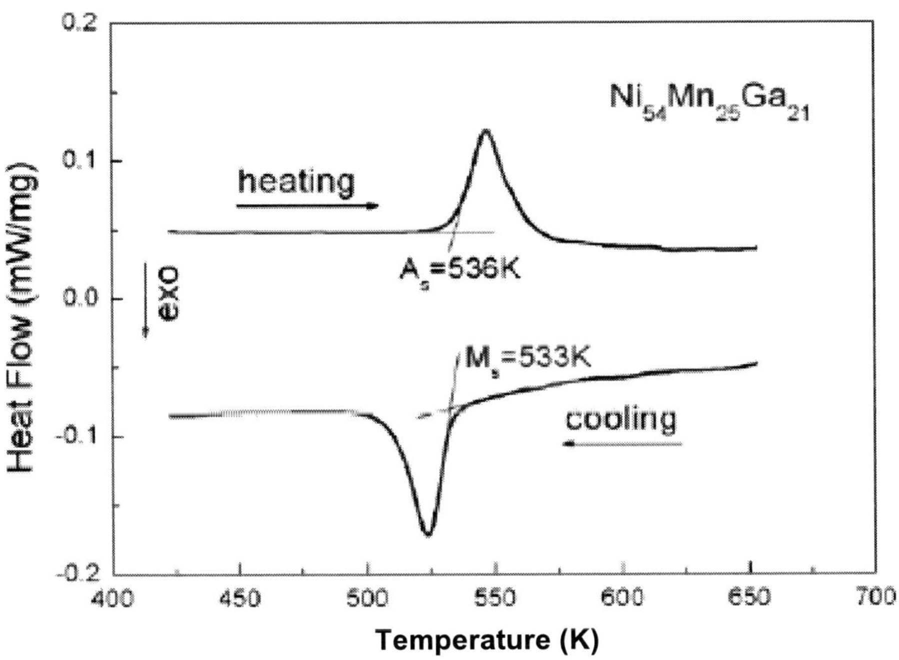

Figure 1. DSC curves of the $Ni_{54}Mn_{25}Ga_{21}$ alloy [14].

Specific heat is also an often studied thermal property of the shape memory alloys. The specific heat of two non-stoichiometric single crystalline Ni-Mn-Ga alloys was determined by E. Ceasari et al. [22] by the AC technique. They found no changes in the specific heat during the phase transformation. Small peaks in the temperature dependence of the specific heat measured by an AC calorimeter were found in [23]. The lambda-type peak for Ni_2MnGa is presented in [24]. Fraile-Rodrígues et al. [25] have made an analysis of the measurement of the specific heat for the Cu-Al-Ni alloy. They found a remarkable dependence of the calorimetric data on the experimental conditions. The results obtained by the quasi-isothermal method were different than the results obtained by the dynamic method, which were dependent on the heating and cooling rates used. The temperature dependence of the specific heat was studied for a heating/cooling rate from 5 K/h up to 80 K/h. The peak height and width increased with the heating/cooling rate. An explanation of the observed experimental behavior was found in the hypothesis that other heat sources could be responsible for this effect such as the frictional work from domain wall motion. The latent heat associated with the first-order phase transformation cannot be measured by the AC technique.

Equation (1) shows that the martensitic transformation is connected with the change of the elastic energy. This implies that various changes of the shape or volume occur during phase transformation. Dilatometry, which measures all deformation changes during thermal loading of materials, has been utilized by many researchers to investigate the martensitic phase transformations. Glavatska et al. [26, 27] studied the $Ni_{50.0}Mn_{29.6}Ga_{21.4}$ single crystal and found that the thermal expansion in the direction of the a-cell parameter decreases and in the c-cell direction increases. Furthermore, the tetragonality (c/a ratio) increases with temperature. Single crystals of near-stoichiometric Ni_2MnGa were studied by Wu at al. [28] with capacitance dilatometry. The thermal strain accompanying the martensitic transformation was not reproducible, but could be made more so with the application of external stress.

During measurement of the temperature dependencies of a given physical parameter, continual heating/cooling is usually used. The temperature dependencies of various physical parameters presented in literature are often dependent on the heating/cooling rate. Kostov et al. [29] studied the transformation temperature dependence of Cu-Zn-Al shape memory alloys on the heating/cooling rate by dilatometry. They used a Netzsch 402 E dilatometer with a protective gas atmosphere of nitrogen and hydrogen and their experiments were carried out at various heating/cooling rates of 3, 5, 10, 15, and 20 K/min. Characteristic temperatures decreased/increased with a decreasing heating/cooling rate. These dependencies were then used for determination of the activation energy of the martensitic phase transformation. Influence of the heating/cooling rate on the transformation temperatures may also be found in [30], where this effect is studied by DSC on a TiNiCu shape memory alloy. The authors demonstrate that M_f and A_f depend strongly on the heating/cooling rates. The thermal expansion measurement is also used as a control for the stress-induced martensitic transformation after cold-rolling as in Cu-Al-Mn-based alloys [31].

Dilatometry can also be used for the study of pre-deformed materials during heat treatment. If deformed material is heated, then part of the deformation energy is released. The dilatation characteristics of the compression pre-deformed $Ni_{53.6}Mn_{27.1}Ga_{19.3}$ alloy are presented in our previous works [32, 33]. During the first thermal cycle after deformation, two types of deformations were described: irreversible, which releases only during heating, and reversible, which releases during heating and once more during cooling. The first type is known as the one-way shape memory effect (OWSME) and the second is the two-way shape memory effect (TWSME) [34-37].

Detailed study of the thermal properties of $Ni_{2+x}Mn_{1-x}Ga$ (x=0, 0.04, 0.06, 0.10, 0.14, 0.18 and 0.24) alloys are presented in [23]. Apart from the thermal conductivity, the electrical resistivity, specific heat, and Seebeck coefficient were also studied in this work. The thermal conductivity was measured by a direct heat pulse technique. The temperature difference on the sample was determined by an E-type differential thermocouple with a junction thermally attached to two well-separated positions along the longest axis of the sample. The sample space was maintained in a good vacuum. The technique of the measurement is described in [38]. Figure 2 shows the temperature dependence of the thermal conductivity. In all temperature dependencies of the thermal conductivity, the martensitic transformation is connected with a step-like jump accompanied by a remarkable spike-shaped peak. Similar peaks were found on CDW (Charge-Density-Wave-Systems) materials, which undergo the Pierls transition at low temperatures [39, 40]. The appearance of the Kohn anomaly and the anomalous heat carried by the soft phonons were suggested as a possible origin for such peaks in the thermal conductivity.

From further works we can average the results of Faulkner et al. [41] where the thermal conductivity of a NiTi wire was measured for the austenite and martensite. It was found that the thermal conductivity of the austenite is twofold higher than the martensite. The NiTi shape memory alloy was also a subject of study by Zanotti et al. [42].

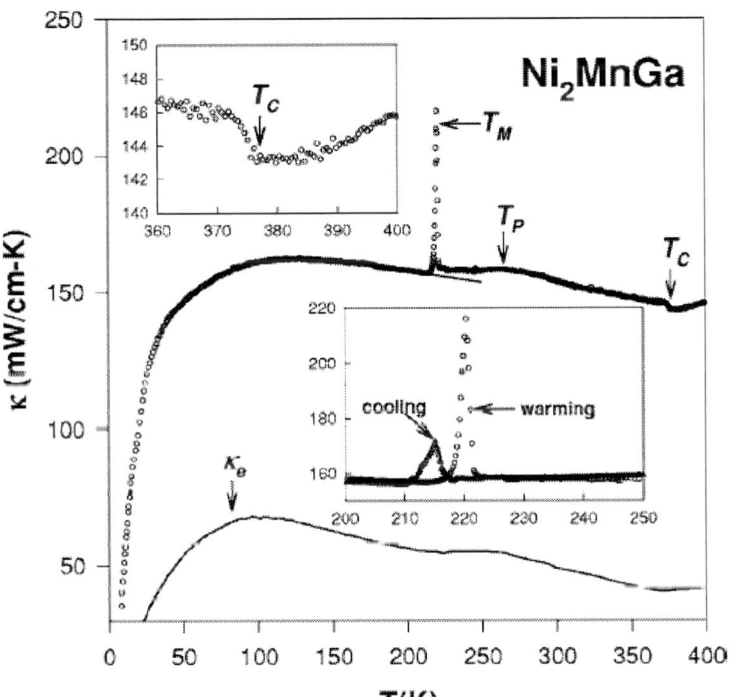

Figure 2. Thermal conductivity as a function of temperature for Ni_2MnGa. The estimated electronic (κ_e) contribution to the total conductivity is given as a solid line [23]

In their paper, the thermal diffusivity obtained by the experimental-numerical approach shows drastic changes during the martensitic phase transformation. The authors assume that a negative peak on the temperature dependence of the thermal diffusivity is a consequence of the heat capacity change. In the work of Zening and Yakuphanoglu [43], the dependence of the thermal conductivity was investigated on the hydrostatic pressure for Cu-13 wt% Al-4 wt% Ni. It was found that the thermal conductivity decreases with increasing pressure and the transformation temperature of M→A transformation increases with increasing pressure.

The martensitic phase transformation, as an athermal phase transformation, starts at the temperature which is known as the transformation temperature. Determination of this temperature is the aim of many experimental studies of shape memory materials. Knowledge of this transformation temperature is important for the practical use of these materials. It is therefore very important to determine the actual transformation temperature that is not influenced by the experimental method. In most of the experimental studies, the sample temperature is identified with the temperature measured outside the sample. The temperature of the sample and of the thermometer for every time and location is determined by the heat transport inside the experimental device. If the physical property is related to the temperature measured outside the sample we must assume that these temperatures will not be the same. In addition, the sample temperature is influenced by the absorption or release of the latent heat. Analysis of heat transport in the material studied as well as in the measuring system is therefore very important in the study of shape memory materials.

It can be seen that there are several problems where the thermal properties play an important role. The first problem is the dependence of the specific heat and thermal expansion coefficient (CTE) on the heating/cooling rate. This dependence cannot have a kinetic cause because the martensitic phase transformation is an athermal phase transformation that depends only on the temperature. To find the cause of this effect a detailed analysis of the experimental system is necessary. The question that occurs from this is whether the measured temperature represents the actual sample temperature? Further, how does the latent heat influence the sample temperature? The second question is crucial for all methods where the physical parameter is determined from the sample temperature. It can be important for the case where the thermal conductivity or specific heat parameters are calculated from the measured sample temperature.

To investigate these problems, a study of the dilatation characteristics is used for the temperature range from 290–650 K. Dilatometry determines not only deformations connected to thermal vibrations of the material, but also all deformations occurring in the material during thermal loading. Dilatometry can therefore give important information about the development of elastic energy as well as plastic energy in the shape memory alloys during heating and cooling. Dilatation results presented in this work are divided into three parts dealing with the influence of compression pre-deformation, and heating and cooling rates on the dilatation characteristics as well as the thermal arrest memory effect. The thermal conductivity that is studied in this work is determined from the thermal diffusivity, the specific heat, and the density. The thermal diffusivity is determined by the flash method in the temperature range from the room temperature to 650 K. The thermal expansion, the thermal diffusivity, and the thermal conductivity were investigated and analyzed for two shape memory alloys: $Ni_{53.6}Mn_{27.1}Ga_{19.3}$ and $Ni_{54.2}Mn_{29.4}Ga_{16.4}$.

EXPERIMENT

Materials

Polycrystalline ingots of Ni-Mn-Ga alloys were prepared by arc melting of the element under argon atmosphere. The microstructure of two ingots is shown in Figs. 3 and 4. Figure 3 reveals the cross-section of a polycrystalline ingot of the $Ni_{53.6}Mn_{27.1}Ga_{19.3}$ alloy (diameter 40 mm, length 100 mm). Three regions with different microstructures were found in this ingot. The central zone and the chill zone are composed from equiaxed grains and the rest of the ingot consists of columnar grains. The samples were prepared from the center of the ingot (with equiaxed grains as in Figure 5) and from the region with the columnar grains (Figure 6). The longitudinal axes of the samples were parallel to the longitudinal axes of the ingot. An ingot of the $Ni_{54.2}Mn_{29.4}Ga_{16.4}$ alloy (Figure 4) had a quadratic shape with dimensions of 25mm x 50 mm x 80 mm. The longitudinal axes of the prepared samples were perpendicular to the columnar grains. It can be seen from Figure 4 that only columnar grains were found in this ingot. The samples prepared from the ingots were annealed for four days at 1120 K.

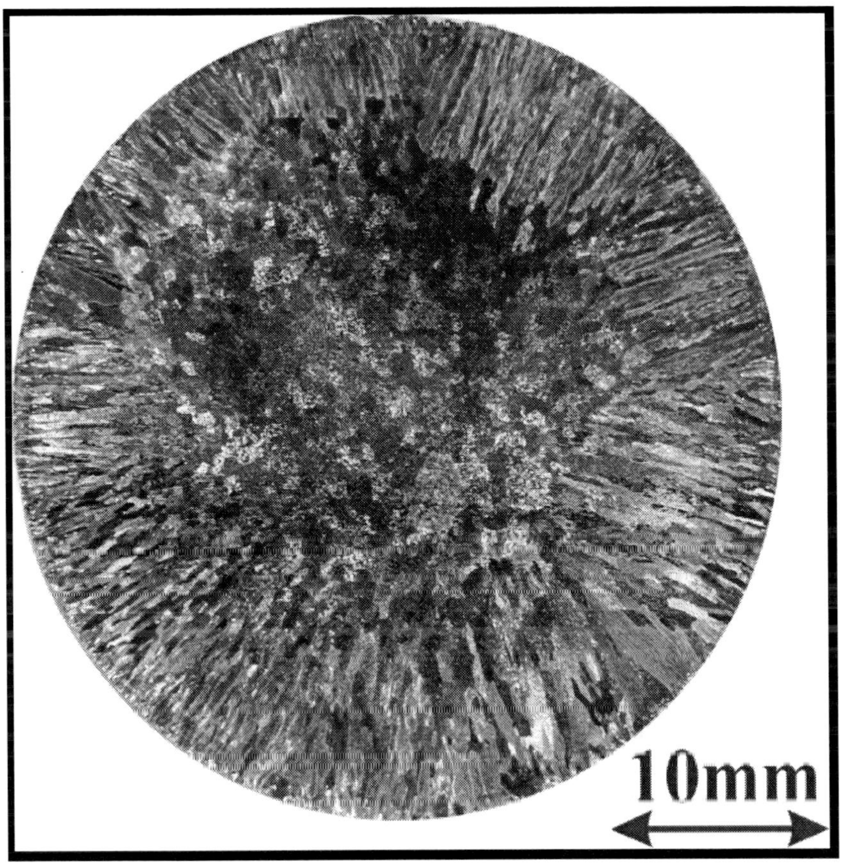

Figure 3. Microstructure of the cross-section of the polycrystalline ingot of the $Ni_{53.6}Mn_{27.1}Ga_{19.3}$ alloy.

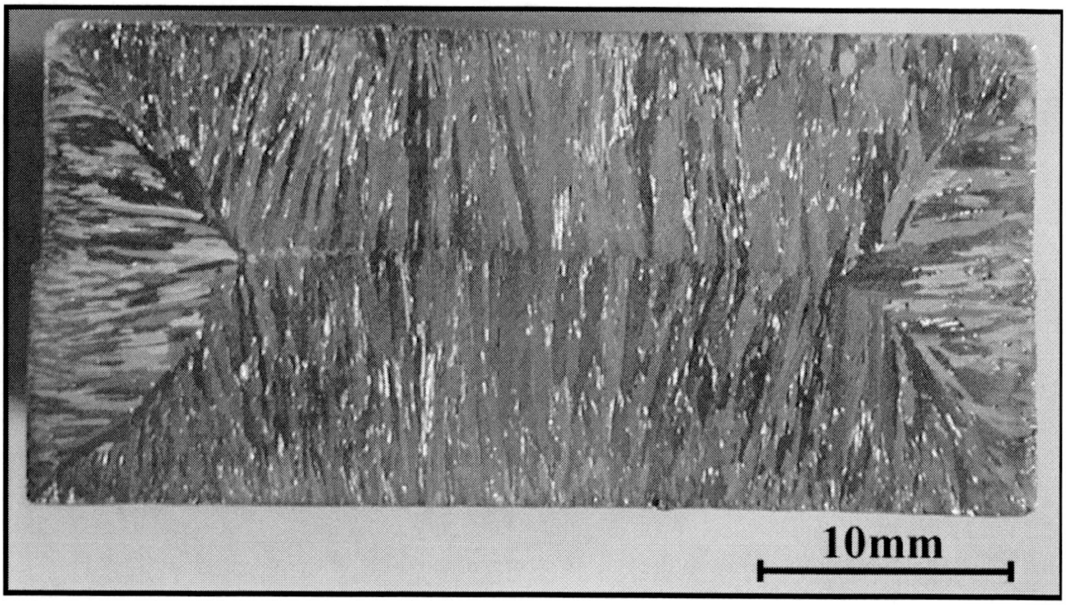

Figure 4. Microstructure of the cross-section of the polycrystalline ingot of the $Ni_{54.2}Mn_{29.4}Ga_{16.4}$ alloy.

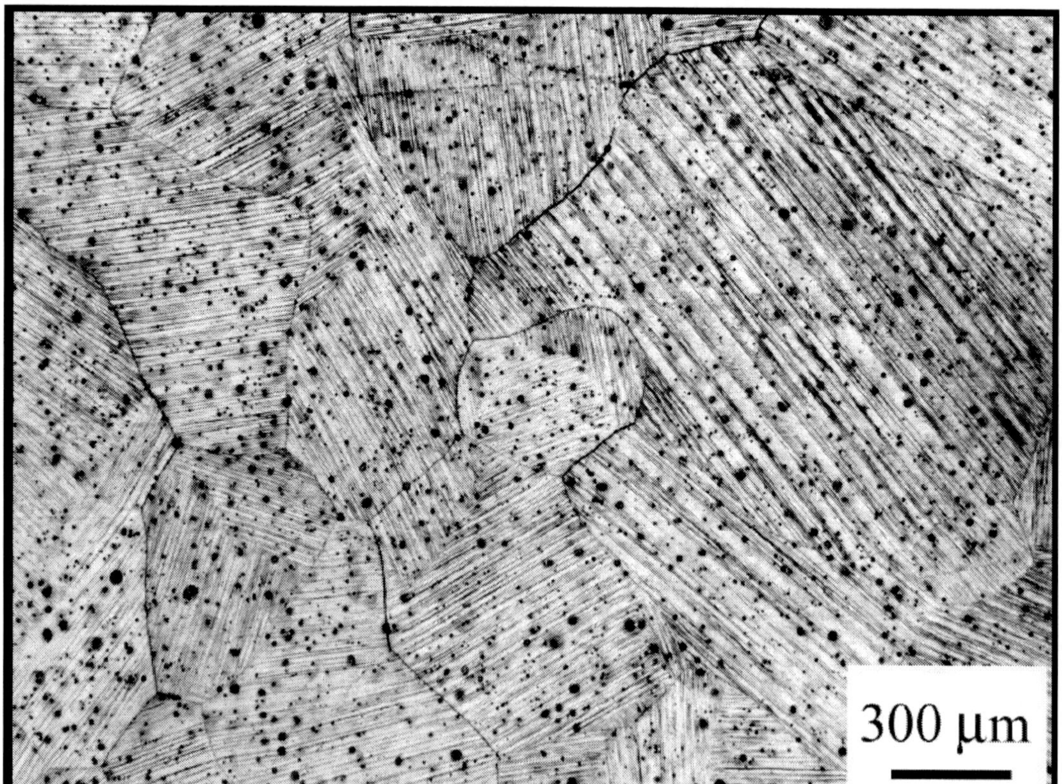

Figure 5. Microstructure of the $Ni_{53.6}Mn_{27.1}Ga_{19.3}$ alloy with equiaxed grains.

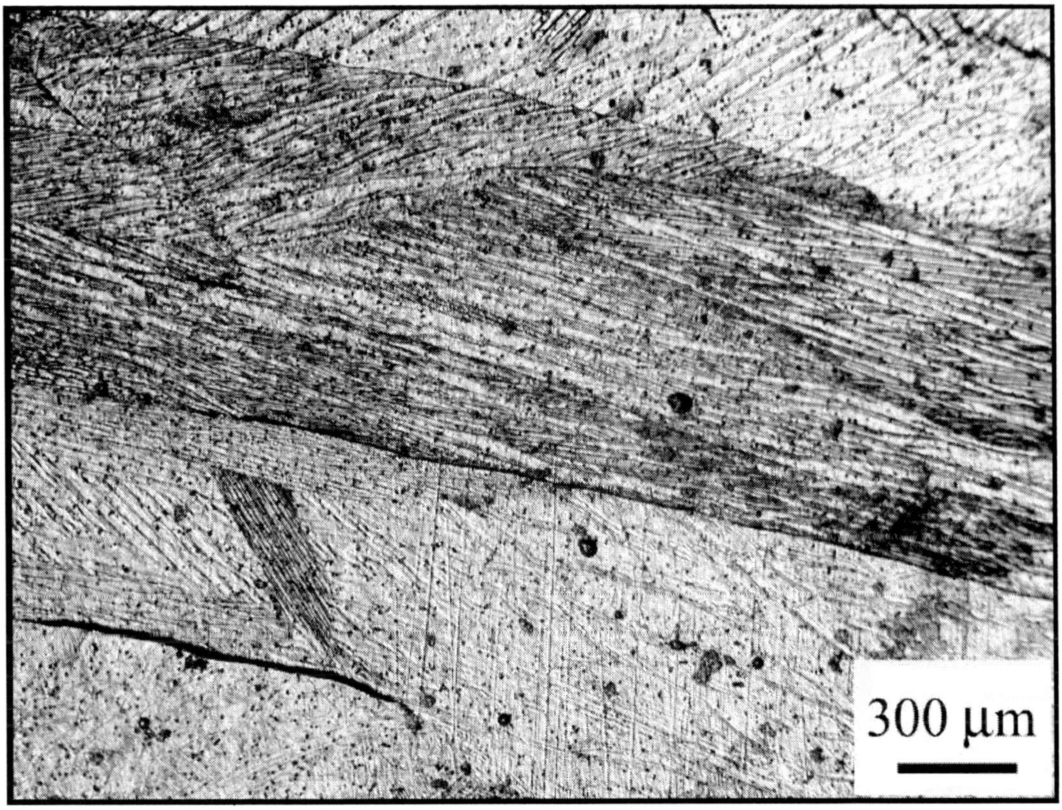

Figure 6. Microstructure of the $Ni_{53.6}Mn_{27.1}Ga_{19.3}$ alloy with columnar grains.

Thermal Expansion

During dilatation measurements, the relative elongation ($\Delta l/l_o$) is determined; where Δl is the difference in the sample length between two temperatures and l_o, which is the sample length at room temperature. The thermal expansion coefficient was obtained by the derivation of $\Delta l/l_o$ by the temperature.

The relative elongation of the samples was measured in an argon and helium atmosphere using a Netzsch 402E dilatometer over the range from room temperature to 650 K. Heating and cooling rates were 0.5, 2, 5, and 7 K/min. The samples were 6 mm in diameter and 25 mm in length. The accuracy of the measuring apparatus was checked by determining the CTE for pure Mg and comparing it with the data available from literature. The agreement between the measured values and the values in literature is in the range of ±1%. The thermal expansion curves were studied during the minimum three thermal (heating and cooling) cycles (runs). The dilatation characteristics in the first thermal cycle after preparation, after pre-deformation, or after interrupting the thermal cycle during the phase transformation were different from those that were obtained during the repeated thermal cycle. The dilatation characteristics were dependent upon the thermal and mechanical history of the material.

Figure 7 displays the experimental arrangement of the dilatometer interior. Presentation of this figure is important for discussion of the heat transfer during measurement of the dilatation characteristics.

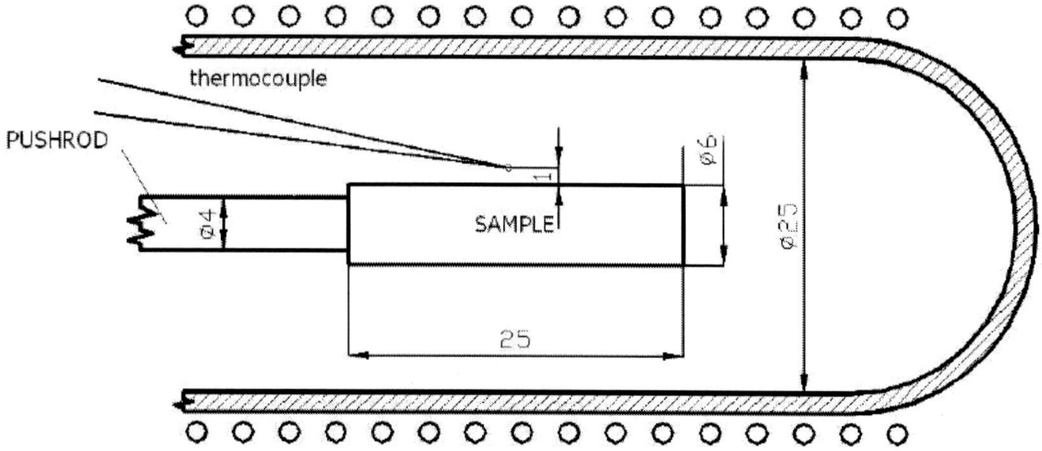

Figure 7. Schema of the interior of the dilatometer.

The sample is placed in the center of the furnace and is heated or cooled from the surface to the center. The heat is transported into the sample by the conduction from the corundum parts of the dilatometer by conduction in the protective gas and by irradiation. A Pt, Pt-Rh thermocouple is placed 1 mm above the sample.

The samples of the $Ni_{53.6}Mn_{27.1}Ga_{19.3}$ alloy were pre-deformed by compression up to 1%. Deformation was performed with an Instron-1186 type deformation machine and the cross-head displacement speed was 0.1 mm/s .

Thermal Diffusivity

Measurement of the thermal diffusivity was carried out in the temperature range from room temperature to 650 K in a helium atmosphere using the flash method. Flash methods are those in which a flash of thermal energy is supplied to a sample surface within a time interval that is short compared to the time required for the resulting thermal transient to propagate through a thin sample. Experimentally, they are characterized by their great simplicity, requiring only small amounts of material, and by being readily extended to high temperatures. Freedom from the heat loss problem stems largely from the short timescale aspect of the method. Parker et al. [44] were the first who used the flash method to measure the thermal diffusivity over a wide range of temperatures. Their results demonstrated that the method yielded data that were in excellent agreement with those obtained by other methods. The flash method is applicable to materials with a wide range of thermal diffusivities.

A xenon flash lamp was used for irradiation of the sample. After irradiation of a surface of the sample, the resultant temperature excursions at the rear face of the sample were recorded and the thermal diffusivity was computed from the resultant temperature rise versus time data. Transient times were usually less than 1 s. The thermal diffusivity, *a,* was calculated using the relation:

$$a = \frac{0.1388L^2}{t_{1/2}}; \tag{2}$$

where L is the specimen thickness and $t_{1/2}$ is the elapsed time needed for the rear face temperature to reach one-half of its maximum rise. A schema of the interior of the apparatus is shown in Figure 8.

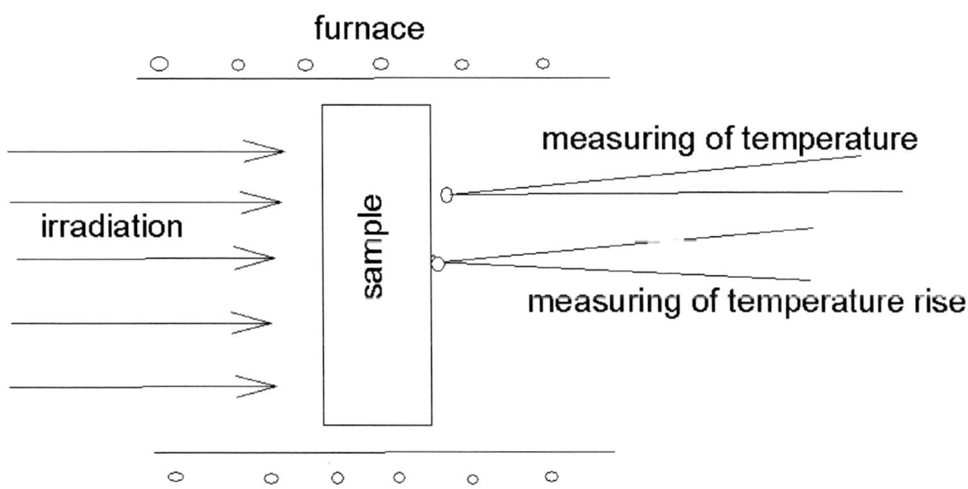

Figure 8. Schema of the experimental arrangement of the interior of the apparatus for the measurement of the thermal diffusivity.

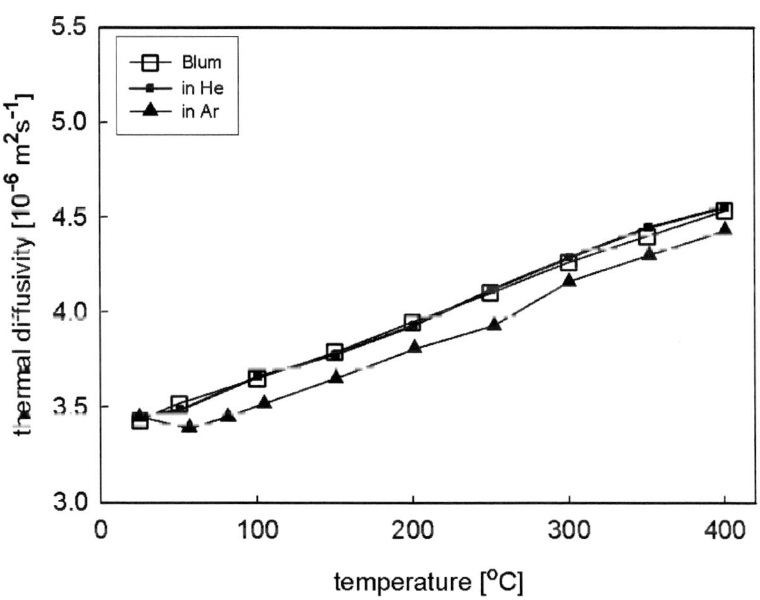

Figure 9. Temperature dependence of the thermal diffusivity for Inconel 600.

The thermal diffusivity of a metal is generally considered to be independent of its thickness. During the measurement of the thermal diffusivity by the flash method it was, however, found that there is a thickness effect. The origin of this effect may be a consequence of more than one phenomenon [45]. In each case, an optimum thickness may be found. This thickness depends on the thermal diffusivity of the measured material. It was found that the optimum thickness of the sample for our Ni-Mn-Ga alloys was about 2 mm. The diameter of the sample was 16 mm. The resultant temperature rise at the rear face of the specimen was 1-2 K. The sample temperature was changed by the distance of the flash lamp from the sample. Four distances were used: 180, 190, 200 and 210 mm. The measurement of each experimental point of the temperature dependence was performed during heating after 10 minutes at the test temperature; the time of the measurement was also 10 minutes. During the measurement the surface of the sample was blackened by graphite. The accuracy of the measurement was checked by measuring the thermal diffusivity of an Inconel 600 sample (Inconel 600 is the standard material for the measurement of thermal diffusivity and conductivity [46]). The results of this measurement are displayed in Figure 9. The measurement of the thermal diffusivity was made using the experimental arrangement described elsewhere [47, 48].

After irradiation of the sample, the front surface temperature may initially rise to a very high value because the energy is delivered in a very short time. Parker et al. [44] made a model for the estimation of the maximum front surface temperature. By this model the maximum temperature rise, ΔT_f, can be estimated by the following equation:

$$\Delta T_f = \frac{38 L \Delta T_M}{a^{\frac{1}{2}}}; \tag{3}$$

where ΔT_M is the maximum temperature rise measured on the back side of the sample. Using this relation for our case, the maximum temperature rise on the front of the sample after irradiation was determined to be 50 K for the $Ni_{53.6}Mn_{27.1}Ga_{19.3}$ alloy at room temperature.

Thermal Conductivity

The thermal conductivity was determined according to the following equation:

$$\lambda = \rho c_p a; \tag{4}$$

where λ is thermal conductivity, ρ is density, and c_p is the specific heat. The temperature dependence of the density was determined by the relation:

$$\rho = \rho_o \left(1 + \frac{\Delta l}{l_o}\right)^{-3}; \tag{5}$$

where ρ_o is the density of the material at room temperature and $\Delta l/l_o$ is the relative elongation of the material. The value of ρ_o was obtained by weighing the sample in rectified and dried toluene at room temperature. The temperature dependence of the relative elongation used for the calculation of the density was obtained by measuring the sample with equiaxed grains and the results of this measurement are shown in Figure 10. Determination of the temperature dependence of the specific heat will be described in connection with the thermal diffusivity results.

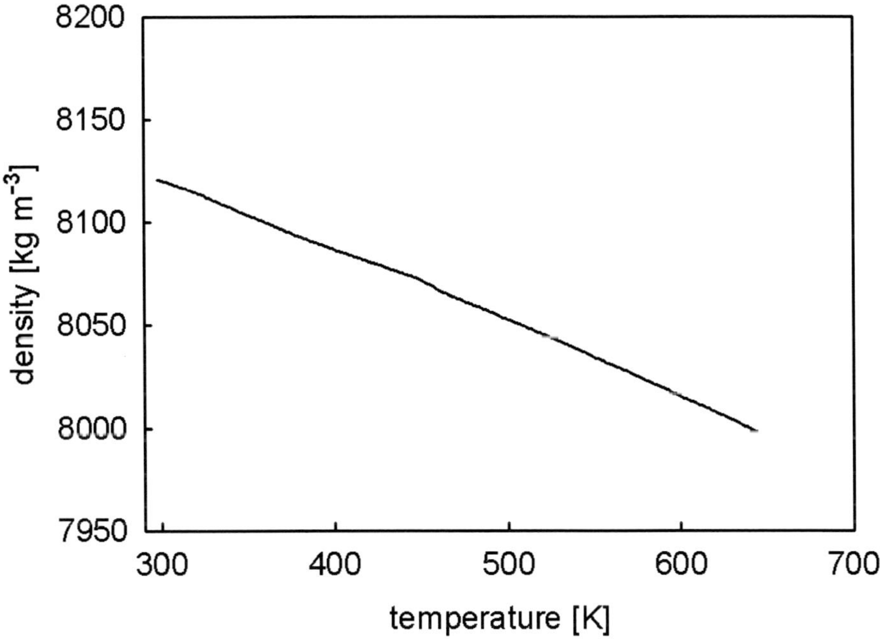

Figure 10. Temperature dependence of the density for the $Ni_{53.6}Mn_{27.1}Ga_{19.3}$ alloy.

Results and Discussion

1. Thermal Expansion

The atoms of crystalline materials at any temperature are in equilibrium between the attractive and repulsive forces from neighboring atoms. By increasing the temperature the amplitude of vibration about the equilibrium position increases; however, the attractive and repulsive forces are not symmetrical and since the repulsive force increases more rapidly than the attractive force with displacement, thermal expansion occurs as the temperature (and vibration amplitude) increases.

Most practical problems concerned with thermal expansion are related to the expansion of polycrystalline materials. Only in the case of isometric materials will the expansion of polycrystalline materials be the same as for a single crystal. For anisotropic materials, expansion coefficients vary with crystallographic orientation and stresses arise on heating and cooling because restrains of the individual crystallites. These stresses affect the observed thermal expansion. If the difference between the expansion coefficients of two crystallographic directions is sufficiently large, substantial stresses are built up. These stresses can lead to plastic flow in metals or grain boundary fractures in non-metallic materials.

Dilatometry detects not only change due to the lattice vibrations of the material, but also all volume changes in a material. It should be noted that all of these changes are additive. First-order transformations in single-phase systems are characterized by a discontinuous change in volume at a constant temperature and pressure. This phase transformation is connected with a discontinuous reconstruction of the lattice, including the change from one crystal structure to another. The changes in volume may be large or very small and positive or negative. In multiphase systems such as alloys, the transformation temperature range may be large and transformation temperatures on heating are higher than on cooling, with hysteresis as a result [49]. Second-order transformations are generally characterized by the disappearance of ordered structures that are connected with a discontinuous change of the coefficient of thermal expansion at constant temperature and pressure. The changes in the coefficient of thermal expansion may be large or very small and positive or negative [49].

1. 1. Dilatation Characteristics of Non-Deformed and Pre-Deformed Alloys

The characteristic strain connected with the transition, austenite→martensite, is called self-strain (self-stress) [50, 51] or self-accommodation [52, 53]. The tendency of this elastic energy to decrease has a dominant role on the structure, morphology, and crystallography of the transformation products. Experimental and theoretical studies of the internal structure of the martensite phase show [50] that the elastic energy of internal stresses is minimized when the orientational variants of the martensitic phases are ordered in the elastic domains and a special internal plate microstructure develops. The planes have a coherent boundary. Further decrease in the elastic energy leads to the formation of polydomain regions. A typical polydomain region of a martensite phase is a plate-like pack consisting of parallel planar domains, which are related to each other as twins. The martensitic transformation is always connected with large amounts of shear strain [54]. As a consequence of this effect, internal stresses occur in the alloy. They are reduced by intrinsic defects due to lattice invariant deformation. These defects can annihilate during the reverse transformation and this transformation is then fully reversible. If they cannot annihilate then they stay in the austenite where they accumulate during subsequent thermal cycles [55, 56].

If the dilatation characteristics were measured after manufacturing or heat treatment, then various non-equilibrium processes took place in the alloy and after these thermal cycles, permanent length changes of the sample were found. These results are not shown here. Figures 11 and 12 show the temperature dependencies of the relative elongation and the CTE for the $Ni_{53.6}Mn_{27.1}Ga_{19.3}$ alloy in the state when repeated measurement gives the same results. The results shown in Figure 11 belong to the sample comprised of the equiaxed grains. Figure 12 then displays the dilatation characteristics obtained for the sample with the columnar grains. The chief difference between both structures is in the transient temperature range. While the relative elongation of the sample with the equiaxed grains shows only small differences from a linear dependence, the relative elongation for the sample with the columnar grains is drastically changing. Figure 13 reveals the temperature dependence of the dilatation characteristics for the $Ni_{54.2}Mn_{29.4}Ga_{16.4}$ alloy. The samples were obtained from the ingot with the microstructure shown in Figure 4. The dilatation characteristics have practically the same shape as those for the $Ni_{53.6}Mn_{27.1}Ga_{19.3}$ alloy with the columnar grains. The vast jumps in the temperature dependence of the relative elongation are a consequence of the existence of preferential martensitic variants in the textured sample.

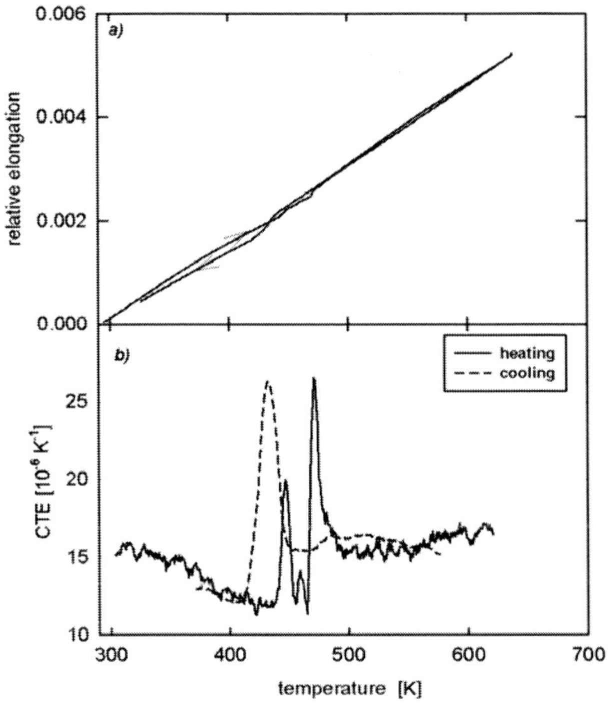

Figure 11. Temperature dependence of (a) the relative elongation and (b) the CTE for the $Ni_{53.6}Mn_{27.1}Ga_{19.3}$ alloy with equiaxed grains.

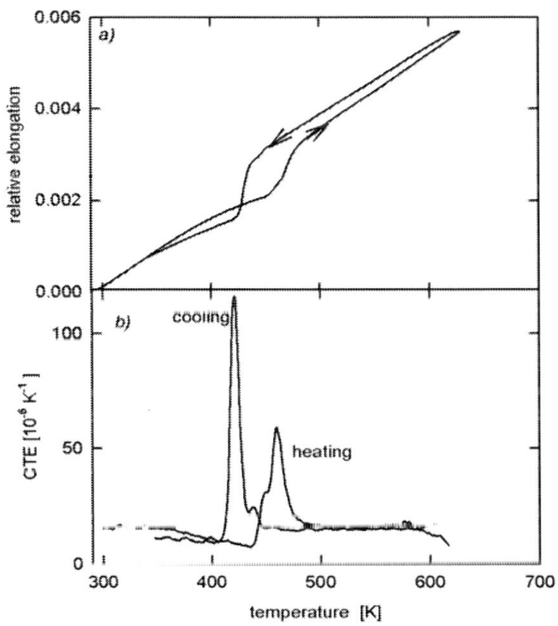

Figure 12. Temperature dependence of (a) the relative elongation and (b) the CTE for the $Ni_{53.6}Mn_{27.1}Ga_{19.3}$ alloy with columnar grains.

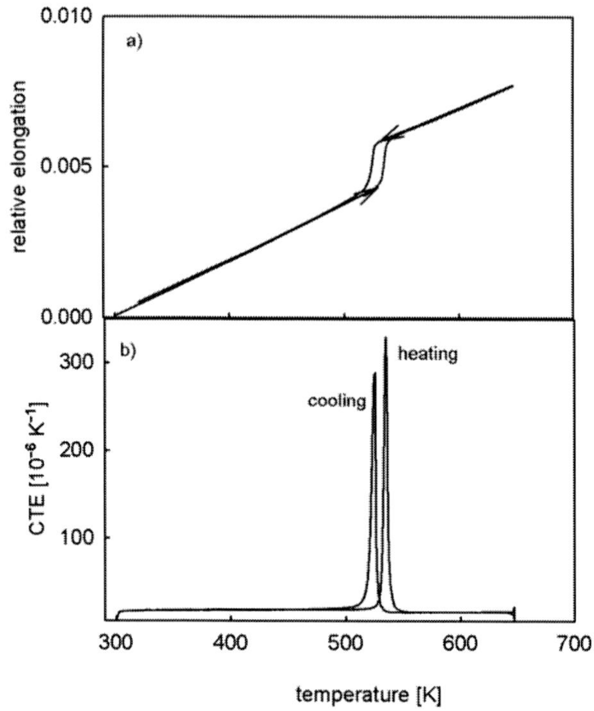

Figure 13. Temperature dependence of (a) the relative elongation and (b) the CTE for the $Ni_{54.2}Mn_{29.4}Ga_{16.4}$ alloy.

These structures occur during the manufacturing cooling and the transformation deformation cannot be removed from the material by any thermal cycling used in this work.

From all dilatation characteristics of NiMnGa alloys, it is clear that the alloy exists in three temperature ranges with distinctly different behavior. They are:

a) Martensite range;
b) Transient range;
c) Austenite range.

(A) Martensite Range

Figure 14 shows the temperature dependencies of the CTE for martensite in non-deformed as well as deformed $Ni_{53.6}Mn_{27.1}Ga_{19.3}$ martensite. The pre-deformed martensite has a larger CTE than the non-deformed martensite. Two regions may be distinguished on all dependencies. In the region from room temperature up to 378 K, the CTE remains nearly constant with increasing temperature. Above this temperature, the CTE decreases with increasing temperature. This behavior is anomalous as compared with typical alloys where the CTE usually increases with increasing temperature. The $Ni_{53.6}Mn_{27.1}Ga_{19.3}$ alloy is ferromagnetic up to 378 K and beyond this temperature is paramagnetic (the magnetization data were obtained on a PPMS device from Quantum Design Company (USA)).

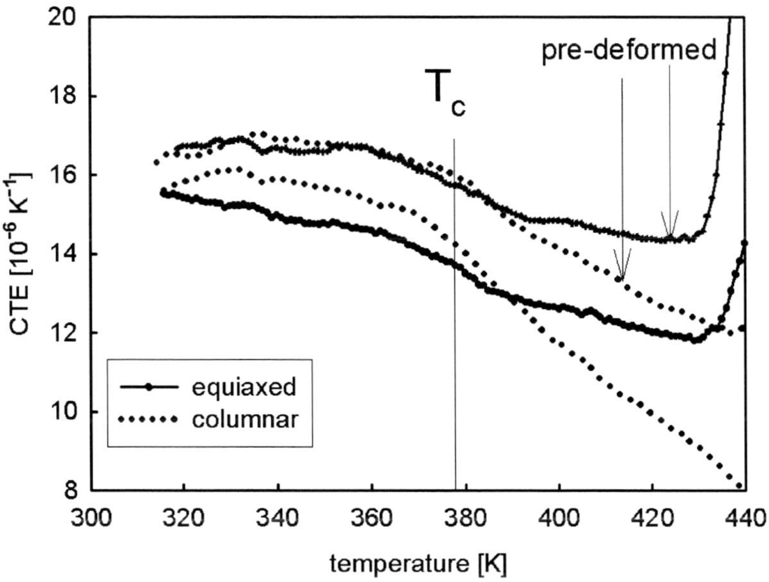

Figure 14. Temperature dependence of the CTE for the martensite structure in $Ni_{53.6}Mn_{27.1}Ga_{19.3}$.

The direction of the temperature dependence of the CTE is lower in the ferromagnetic temperature range than in paramagnetic range, which implies that there is magneto-volume effect that influences the dilatation characteristics. The influence of the magneto-volume effect can be seen in Figure 15 where the temperature dependence of the CTE is shown for the $Ni_{54.2}Mn_{29.4}Ga_{16.4}$ alloy and for the $Ni_{49.4}Mn_{23.6}Ga_{27.0}$ alloy. The second alloy is in an austenite state above room temperature. It can be seen that the temperature dependence of the CTE in martensite $Ni_{54.2}Mn_{29.4}Ga_{16.4}$ and $Ni_{53.6}Mn_{27.1}Ga_{19.3}$ are different. Comparison of both characteristics shows that the CTE is more influenced by the internal stresses in the $Ni_{53.6}Mn_{27.1}Ga_{19.3}$ alloy than in the $Ni_{54.2}Mn_{29.4}Ga_{16.4}$ alloy. We assume that the chief cause of this difference lies in the shape of the ingot from which the samples were prepared. The sample of the first alloy was prepared from the cylinder ingot (Figure 3) and the second sample was prepared from the quadratic ingot (Figure 4). Various shapes of ingots can be a cause of the various manners of cooling of the alloy inside the ingot during manufacturing. Both alloys were prepared by the same method, but their structure is different. The martensite structure remembers its mechanical and thermal history.

The relative elongation in the martensitic temperature range (Figure 14) is given by three contributions:

$$\frac{\Delta l}{l_o} = \left(\frac{\Delta l}{l_o}\right)_v + \left(\frac{\Delta l}{l_o}\right)_M + \left(\frac{\Delta l}{l_o}\right)_I; \qquad (6)$$

where $(\Delta l/l_o)_v$ is the strain as a result of the lattice vibration, $(\Delta l/l_o)_M$ is strain due to the magneto-volume effect, and $(\Delta l/l_o)_I$ is the strain as a result of the internal stresses.

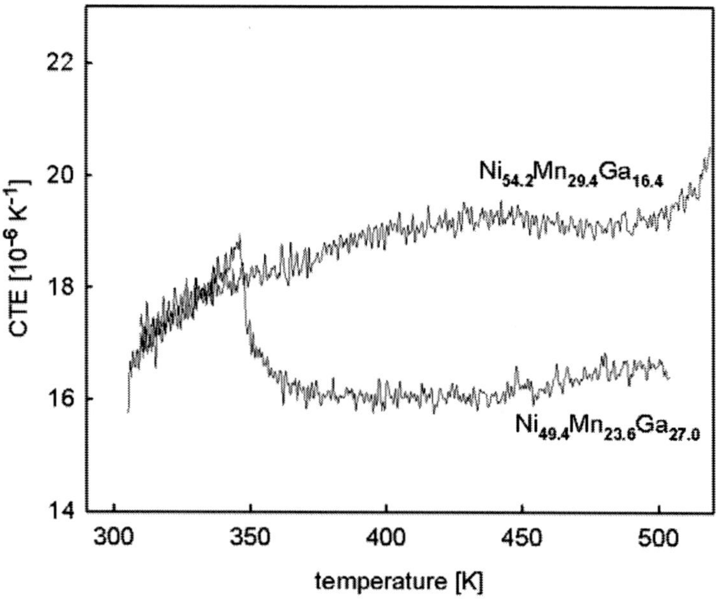

Figure 15. Temperature dependence of the CTE for the $Ni_{54.2}Mn_{29.4}Ga_{16.4}$ and $Ni_{49.4}Mn_{23.6}Ga_{27}$ alloys.

The first two members in Eq. (4) are positive and the third term is negative (it decreases dilatation characteristics). From a macroscopic view, the sample can be in tension in the direction of the longest axes at room temperature. With increasing temperature the tensile stresses decrease and the sample length decreases. From a microscopic view, the anisotropy of the CTE can be seen as the cause of this decrease. NiMnGa is an alloy where the CTE along the a-axis is much different from that along the c-axis [26]. In these materials, thermal stresses build in the alloy upon cooling from a high temperature to a low temperature. At the high temperature, in our case just below the completed austenite→martensite transformation, this type of stress, which is called the thermal stress, is assumed to be minimized. With decreasing temperature it increases. It is assumed that the thermal strain is expressed as: $\sim\Delta(\text{CTE})\Delta T$; where Δ(CTE) is the difference between the CTE values along different crystallographic orientations and ΔT is the difference between the temperatures of the stress-free state and the actual temperature. The thermal strain is then maximized at room temperature since it is the lowest temperature studied. When heating this alloy, the thermal strain decreases to reach the stress-free state. This effect decreases a relative elongation of the alloy during heating and reduces the CTE. The influence of the thermal stresses on the dilatation characteristics is also observed in composites where the metal matrix is reinforced by ceramic particles and where the difference between the CTE of both components is large.

(B) Transient Range

In all cases, jumps in the temperature dependence of the relative elongation on the transformation temperature range were found. These jumps correspond to transformation deformation that is reversible. On the temperature dependence of the CTE the transformation deformation is connected with a peak. In Figure 11 we can see three peaks. Several peaks or splitting peaks are often presented in literature [56, 57] in the transformation temperature region. These peaks are frequently explained by inter-martensitic transformations.

NiMnGa alloys have an electron concentration, *e*/a, between 7.5 and 7.7 [58]. This value lowers the stability of the temperature-induced martensite structure and gives rise to inter-martensitic phase transitions [59, 60]. Segui et al. [61] studied inter-martensitic transitions of the single crystal, $Ni_{52.6}Mn_{23.6}Ga_{23.8}$, by means of the use of a number of experimental techniques. Two thermally induced forward and reverse inter-martensitic transformations have been found to occur in that alloy. The difference in the transformation temperatures obtained by various methods of the inter-martensite transformation temperature was 11 – 16 K for the martenisite→austenite transition and 33-59 K for the reverse transition. The work by Segui et al. shows that each experimental technique used has a specific sensitivity to the phase transformation and that the internal stresses have a significant role in the transformation sequences.

A further cause of several peaks in the temperature dependence of the CTE may be non-homogeneous composition of the sample. For this reason an analysis of the chemical composition of the sample was made at three points of the sample. The composition ranged from 52.4-Ni, 28.8-Mn, 18.8-Ga (top of sample) to 53.9-Ni, 26.4-Mn, 19.7-Ga (center of sample) and 54.4-Ni, 26.1-Mn, 19.5-Ga (bottom of sample). It is assumed that these differences in composition are not likely the cause of the three peaks in Figure 11.

C) Austenite Range

The standard behavior of the dilatation characteristics is similar to those alloys where thermal vibrations are the sole source of the deformation.

Both $Ni_{53.6}Mn_{27.1}Ga_{19.3}$ alloys whose behavior was described in previous paragraphs were then pre-deformed in compression up to 1 %. Figures 16 and 17 show the dilatation characteristics of the pre-deformed alloys. We can see that permanent elongation of the samples after the first thermal cycle after deformation occurs in both cases and part of the compression deformation is recoverable. Figures 18 and 19 reveal the temperature dependence on the dilatation characteristics in the second thermal cycle after pre-deformation. In each figure, the dilatation characteristics obtained before deformation are also revealed. It was mentioned that dilatometry measures not only deformations connected with the lattice vibrations but also all deformations in the material during its thermal loading. For a better understanding of the processes occurring in shape memory alloys we can divide the measured relative elongation into several contributions (deformation are additive, Eq. (6)). By subtracting the relative elongation obtained in the second thermal cycle after pre-deformation from that obtained in the first thermal cycle after pre-deformation we obtain the temperature dependence of strain that is known as the residual strain. Relaxation of this residual strain occurs only during heating and it is a permanent process. Relaxation of this type of strain is generally known as OWSME. The temperature dependence of the deformation that was obtained as the difference between the relative elongation of the second thermal cycle of the pre-deformed sample and the one before deformation (0th run) is shown as the memory strain. This strain is reversible; it releases during heating and once more occurs during cooling (TWSME). Figures 20 and 21 show the temperature dependencies of both types of released strains. Permanent elongation after the first thermal cycle is higher for the sample with equiaxed grains than for the sample with columnar grains. This can be a consequence of different structures in both samples.

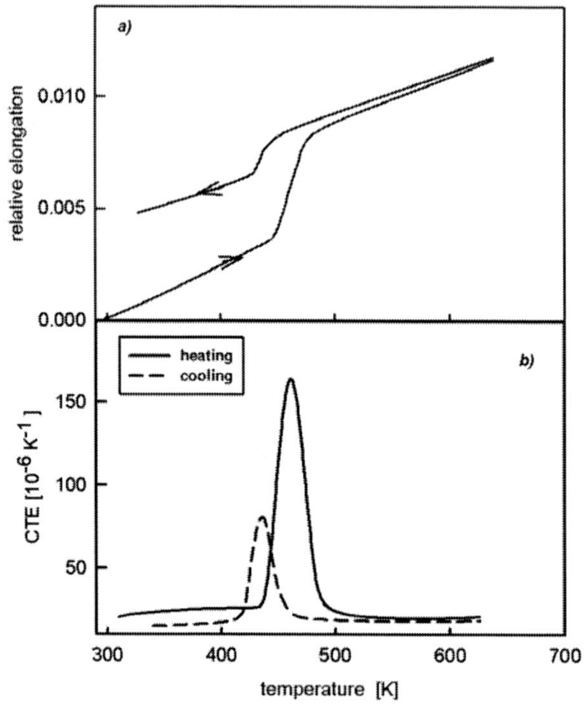

Figure 16. Temperature dependence of the dilatation characteristics for the compression pre-deformed $Ni_{53.6}Mn_{27.1}Ga_{19.3}$ alloy with equiaxed grains.

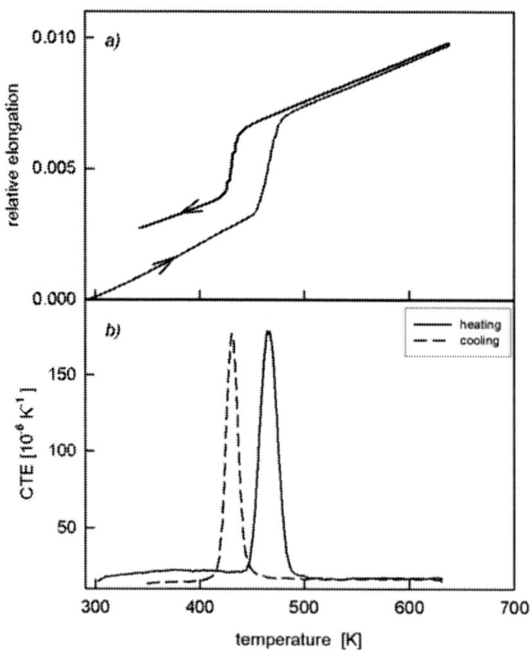

Figure 17. Temperature dependence of the dilatation characteristics for the compression pre-deformed $Ni_{53.6}Mn_{27.1}Ga_{19.3}$ alloy with columnar grains.

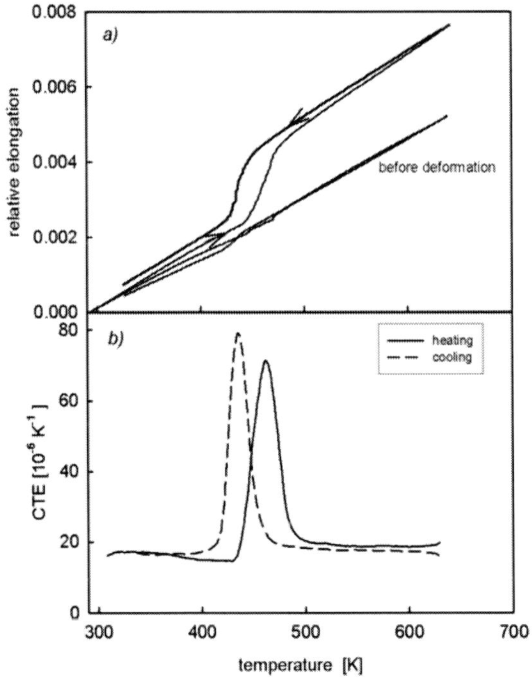

Figure 18. Temperature dependence of the dilatation characteristics for the compression pre-deformed $Ni_{53.6}Mn_{27.1}Ga_{19.3}$ alloy with equiaxed grains in the second thermal cycle after pre-deformation.

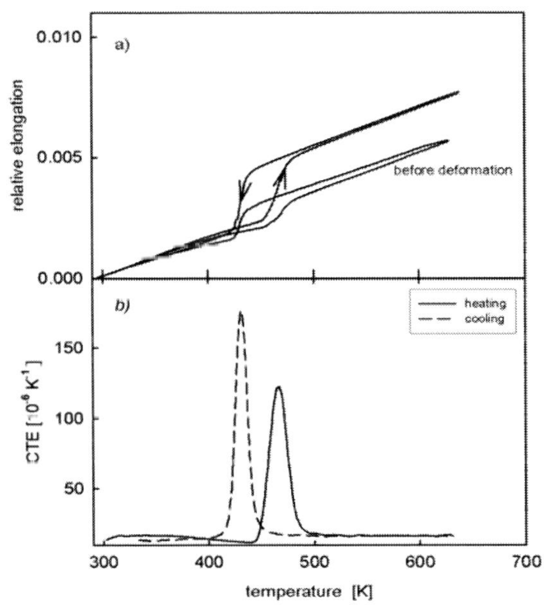

Figure 19. Temperature dependence of the dilatation characteristics for the compression pre-deformed $Ni_{53.6}Mn_{27.1}Ga_{19.3}$ alloy with columnar grains in the second thermal cycle after pre-deformation.

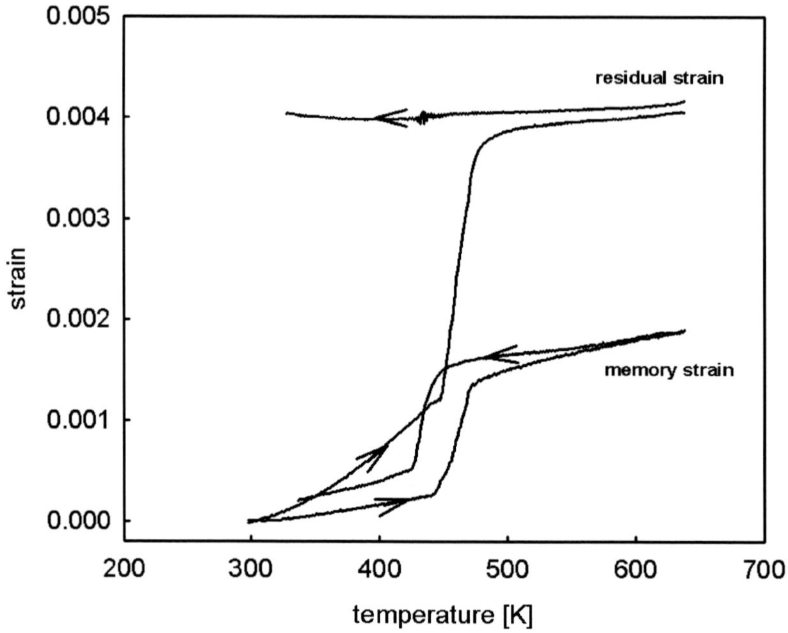

Figure 20. Temperature dependence of the residual and the memory strain for the $Ni_{53.6}Mn_{27.1}Ga_{19.3}$ alloy with the equiaxed grains.

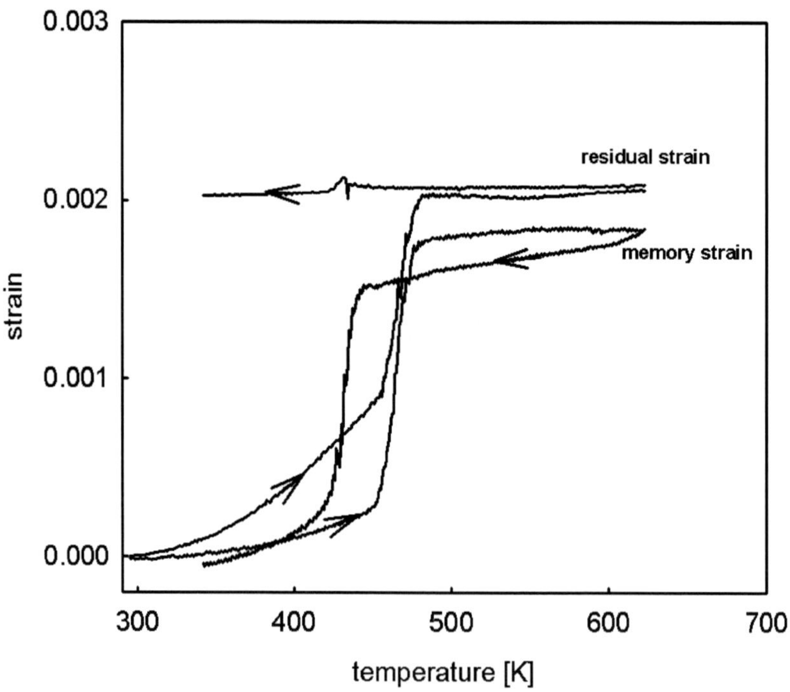

Figure 21. Temperature dependence of the residual and the memory strain for the $Ni_{53.6}Mn_{27.1}Ga_{19.3}$ alloy with the columnar grains.

The memory strain is practically the same for both alloys. Liu et al. [62] studied single crystals of the $Ni_{1.95}Mn_{1.42}Ga_{0.63}$ alloy. The samples were pre-compressed up to 1.6 %, 2.9 %, 4.2 %, and 6.5 %. The total strains were completely recovered after heating up to 625 K except for the sample that was pre-deformed to 6.5 % where the recovery was incomplete. Van Caenegem et al. [63] have established a shape recovery ratio, η:

$$\eta = \frac{l_1 - l_2}{l_1 - l_o}; \tag{7}$$

where l_1-l_2 is the change in length during heating (measured on the dilatation curve) and l_1-l_o is change in length after deformation. They studied Fe-Mn-Si-based shape memory alloys deformed in compression. The shape recovery ratio decreased with increasing deformation from 41 % for a 1.6 % deformation up to 8 % for a 17.5 % deformation. They assume that after the high strain, recovery reverse transformations are more complex because of the intersection of the martensitic plates and dislocation slip. No TWSME was found for this case. The grain refined $Ni_{54}Mn_{25}Ga_{21}$ alloy with a high transformation temperature was studied after compression deformation by X-ray diffraction. It was found that at a strain of 2% and stress 120 MPa a linear dependence of the stress-strain curve was obtained. This deformation corresponds to the elastic strain and martensitic reorientation

Liu [64] studied the detwinning process and its anisotropy in the NiTi alloy. If the alloy was deformed in a martensite structure containing 100% lattice twins then the deformation occurs through a variant reorientation/detwinning process. The residual strain stays in the martensite structure until the transformation temperature, A_s, is reached. The residual strain can then be recovered through a reverse phase transformation [64]. In this work anisotropy of the martensite deformation process between tension and compression is also investigated. Figure 22 shows the results presented in this work. It shows that tension deformation leads to martensite detwinning while compression deformation results in the formation of a high density of dislocations and no significant detwinning process took place at low deformations. These dislocations occur in martensitic twin bands. The first thermal cycle after tension deformation is connected with the reduction of the sample length after compression deformation elongation of the sample occurs. A higher dislocation generation occurs in the sample with a poor two-way memory effect. Dislocations also influence the hysteresis and shift of the transformation temperatures. We can see that there are large differences between the deformed alloys even when they have a similar chemical composition and are deformed up to similar deformations.

Generally, it was observed that the deformation of the martensite structure consists of two mechanisms in the microstructural scale.

- The reorientation and detwinning process is not connected with permanent changes during a single thermal cycle
- Dislocations generation (annihilation) leads to permanent changes of the sample length.

Figures 20 and 21 show two types of strains releasing during the first thermal cycle after deformation: residual (recoverable) strain and memory (non-recoverable) strain.

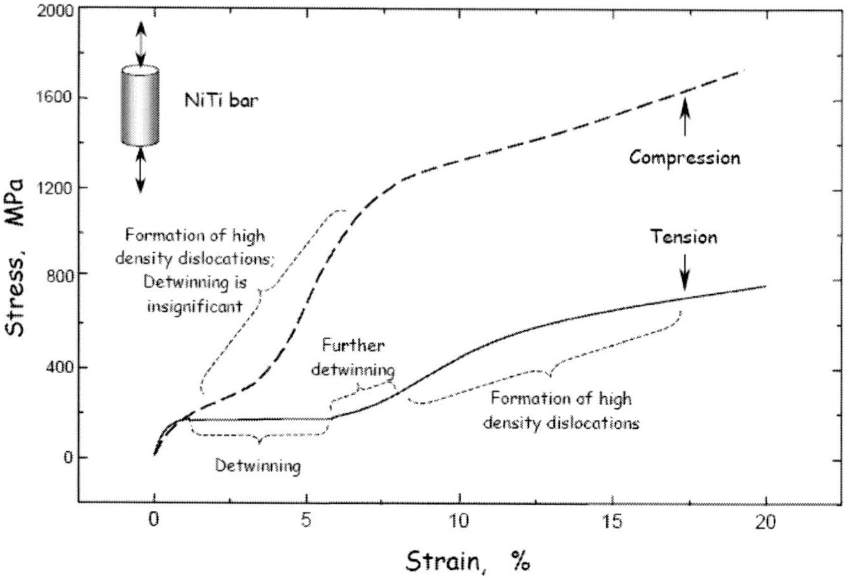

Figure 22. Anisotropy of the martensite deformation process between tension and compression. Microstructure study shows that tension deformation leads to martensite detwinning while compression deformation results in the formation of a high density of dislocations with no significant detwinning process.

Release of the residual strain occurs only during heating, which implies that it is a thermally activated process. Releasing of the memory strain is not a thermally activated process as it occurs as a consequence of phase transformation. The memory strain releases during heating and once more occurs during cooling. This is the primary difference between both types of strains or the OWME (permanent changes) and the TWME (reversible changes). The cause of both effects, therefore, cannot be the same. A further important difference between the OWSME and TWSME is the temperature at which both processes start. While the OWSME starts practically immediately when the heating starts, the TWSME is localized only in the transformation temperature range (Figs. 20 and 21).

Hornbogen [54] analyzed the case of accumulation defects in shape memory alloys. He describes two types of defects:

a) dislocations (b_α) produced by lattice invariant shear in the martensite structure (α) at $T<M_f$. These dislocations can transform into a lattice vector of the austenite structure (β) during heating of the temperature $T>A_f$:

$$b_\alpha \to b_\beta . \tag{8a}$$

N-times repeated thermal cycles can lead to an accumulation of tangles and persistent bands, or intrusions and extrusions in the surface [65].

b) weaker defects:

$$b_{\alpha}\rangle b_{\alpha}^{\times} \rightarrow b_{\beta} = 0\,; \tag{8b}$$

that are formed in the martensite structure if it is exposed to one small strain amplitude. Dislocations, $b_{\alpha}^{\times}$, are localized inside the martensitic domain boundaries or habit planes [48]. These defects can then annihilate during the first thermal cycle in the transformation temperature range. This is known as the effect of the first cycle (EFC). If these dislocations do not annihilate, they stay in the austenite structure as ghost dislocations in an arrangement that resembles the interior structure of the martensite structure. Repeated deformations accumulate plastic unrecoverable strain in the sample and increase the TWME [66].

The sample with equiaxed grains was compression pre-deformed up to 1%. In the first thermal cycle 40% of the deformation was removed, approximately 20% was bound as memory strain, and 40% of the deformation, in the form of dislocations, stays in the alloy. We assume that this type of dislocation cannot annihilate and does not change during the phase transformation. The architecture of these dislocations in the austenite structure can determine the arrangement of the martensite structure during the phase transformation (TWSME). By our assumption, the first type of dislocation (recoverable dislocations) can annihilate immediately after the heating of the sample (in contrast to the Hornbogen hypothesis). The shape memory strain is a consequence exclusively of deformation reorientation and detwinnning. Our results show that preferential martensite variants occurring during compression pre-deformation (in the direction of longitudinal sample axis) are the same as preferential martensitic variants in the columnar grains. The structure of the ingot can give some information about the history of the solidification process during preparation of this shape memory alloy. The alloy remembers its mechanical and thermal history and this problem will be further studied.

1.2. Dilatation Characteristics and Heating/Cooling Rates

The martensitic transformations in shape memory alloys have traditionally been regard as athermal. The experimental results studying these transformations would not be dependent on the heating/cooling rates. Numerous experimental results presented in the literature, however, contradict this fact. Many discussions were made on this theme [67-69]; however, no final conclusion was accepted. It was shown in the introduction that DSC or dilatation results are dependent on the heating/cooling rates. In all these works, it is assumed that it is a consequence of only the phase transformation of the material and the temperature to which the given physical parameter is related is identified with the temperature of the inside of the sample even though this temperature is measured outside of the sample. Various terms, such as an incubation time, or various thermal fluctuations have been used to explain the time dependence of the experimental results. The influence of the thermal transport on physical parameters during the phase transformation still has not been considered.

The thermal dependencies of the physical properties of materials, especially metals, are measured in an atmosphere of inert gases to prevent oxidation. The measurement atmosphere is not always presented in the papers, but argon is used in most cases [23, 70]. If there is any distance between the thermometer and the sample, it becomes very important to know the type of atmosphere used. The thermal conductivity of this media influences the difference in the temperatures between the thermometer and the sample. Two important questions occur in these cases: (a) Does the temperature difference between the thermometer and the sample depend on the type of atmosphere in the experimental device? (b) Does the temperature difference between the thermometer and the sample depend on the heating/cooling rate for the various atmospheres?

To obtain answers to these questions, detailed analysis of the measurement conditions is necessary. Figure7 shows the experimental arrangement of the interior of the dilatometer. The sample is placed in the center of the furnace and the thermocouple is placed 1 mm above the sample and the sample is heated from the surface to its center. The influence of the protective atmosphere on the CTE is revealed in Figure 23, where this parameter was determined in an argon and a helium atmosphere. This figure shows the results obtained during measurement of the CTE for four heating rates (0.5, 2, 5, and 7 K min^{-1}). We can see that the influence of the protective atmosphere on the dilatation characteristics is considerable. To characterize the influence of the heating/cooling rates on the CTE we select a maximum peak temperature (T_{MP}). The dependence of this temperature on the heating/cooling rates is displayed in Figure 24. The influence of the heating rate is more important in argon than in helium. Hysteresis depends on the protective atmosphere and is not a consequence of the kinetics of the phase transformation, but depends on the thermal transport in the experimental device during the measurement. The differences between the transient temperatures are a consequence of the different thermal conductivity of argon and helium. The thermal conductivity of helium gas at room temperature and a pressure of 1 atm. is 0.152 $Wm^{-1}K^{-1}$, whereas the thermal conductivity of argon for the same conditions is 0.0177 $Wm^{-1}K^{-1}$. The higher thermal conductivity of helium secures better heat transfer between the thermocouple and the sample. The influence of the thermal transport on the measured results is thus larger for an increased heating/cooling rate.

We have made a simplified, one-dimensional model [71] of the heat transfer during measurement of the thermal expansion. In considering the boundary conditions for the model, the sample surface is subjected to radiation from the furnace wall at a given temperature and perfect thermal contact was considered between the furnace wall and the gas and between the gas and the sample surface. Heat conduction throughout the gas was computed with temperature-dependent material properties of the gas. Heat conduction throughout the solid sample was computed with respect to the temperature-dependent thermal conductivity and heat capacity [72-74]. The thermocouple was not simulated as an additional part of the model; its temperature is evaluated as equal to the temperature of the gas at a selected distance from the sample surface. Figure 25 shows the calculated dependence of the temperature difference on the thermocouple (T_c) and the sample surface ($T_{surface}$), $T_c - T_{surface}$, on the heating rate. These differences were obtained in a manner such that $T_{surface}$ was always 437 K (the temperature below the start of the phase transformation). Figure 25 also shows the measured temperature differences between $T_s-(T_s)_{0.5}$; where T_s is the temperature when the phase transformation starts and $(T_s)_{0.5}$ is the phase transformation temperature for a heating rate of 0.5 K/min. It can be seen that both dependencies have the same characteristics. These differences are dependent on the thermal conductivity of the protective atmosphere and on the heating rate. Hence, we can assume that when the given transient temperature is reached on the sample surface, the temperature of the thermocouple is higher than the surface temperature for all cases presented here. These calculations were made in the temperature range outside of the phase transformation. The difference between the thermocouple temperature and the surface temperature will be higher in the temperature range where the phase transformation occurs due to latent heat.

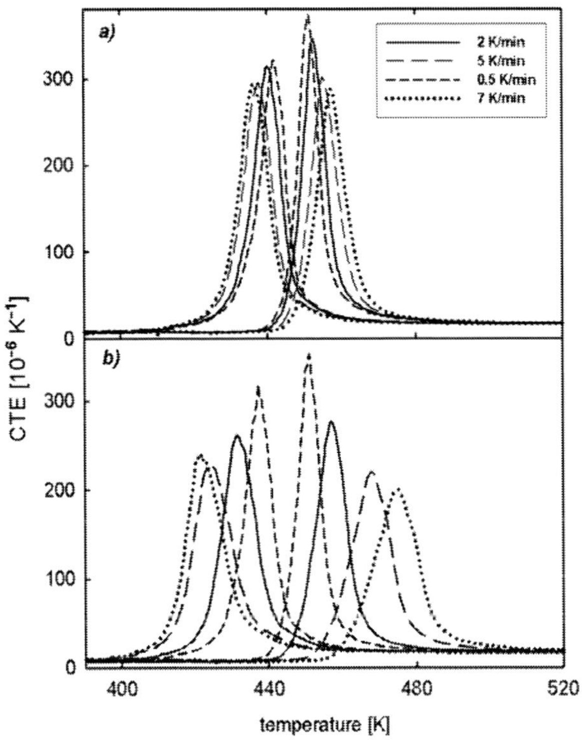

Figure 23. Temperature dependence of the CTE in a) an He and b) an Ar atmosphere.

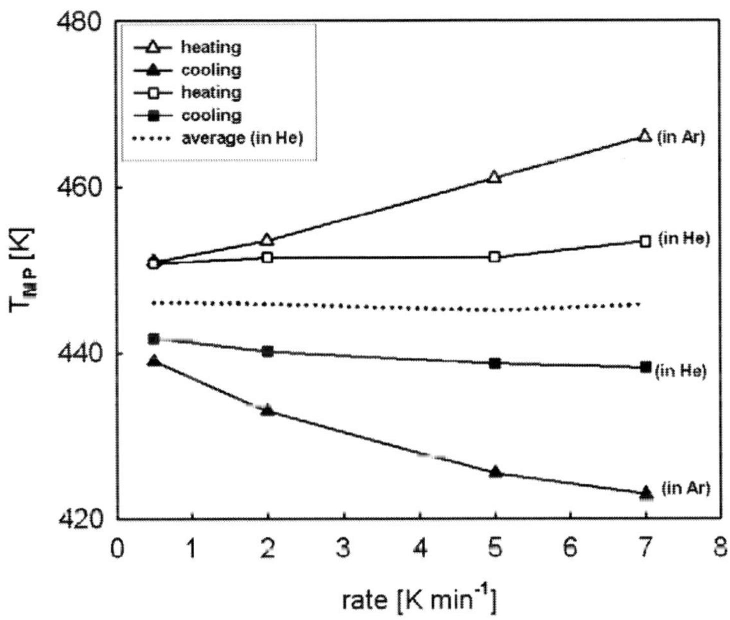

Figure 24. Dependence of the maximum peak temperatures T_{MP} on the heating/cooling rates in Ar and He for $Ni_{53.6}Mn_{27.1}Ga_{19.3}$ alloy.

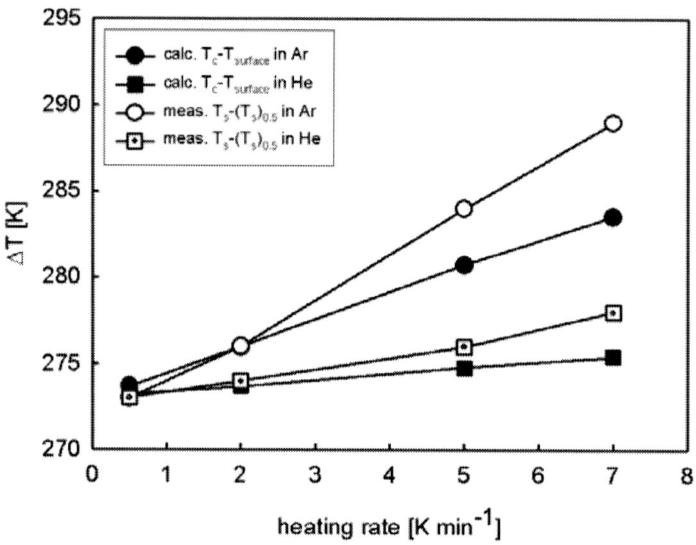

Figure 25. Dependence of the temperature difference on the thermocouple (T_c) and on the sample ($T_{surface}$), T_c-$T_{surface}$, on the heating rate. These differences were obtained in a manner such that $T_{surface}$ was always 437 K. This figure also shows the measured temperature differences between T_s-$(T_s)_{0.5}$; where T_s is the temperature at which the phase transformation starts and $(T_s)_{0.5}$ is phase transformation start temperature for a heating rate 0.5K/min.

When the transformation temperature is reached on the sample surface and the phase transformation starts, the latent heat is consumed and the surface temperature increases more slowly than in the case without phase transformation.

The latent heat of the phase transformation of our Ni-Mn-Ga alloys is large. This statement is based on fact that the thermocouple temperature (above sample) is influenced by the latent heat. Figures 26 and 27 show the time dependencies of the thermocouple temperature for heating/cooling rates of 0.5, 2, 5, and 7 K/min. These dependencies were obtained with the experimental arrangement when the thermocouple was located directly on the sample in a helium atmosphere. The influence of the latent heat is obvious; during heating/cooling the latent heat decreases/increases the sample temperature and therefore also the thermocouple temperature. This effect is maximized for the highest heating/cooling rate and it is minimized for the rate of 0.5 K/min. The temperature at which the time dependence of the temperature starts to deviate from a linear dependence is defined as the transformation temperature. This temperature was determined to be 445±2 K and is not dependent on the heating/cooling rate.

In our experimental arrangement, when the continual heated/cooled sample is placed in the furnace center, the highest/lowest temperature is reached always initially on the surface of the sample. The phase boundary propagates into the sample as the temperature increases or decreases. The rate of the propagation of the phase boundary is determined by the diffusion of the temperature through the sample (athermal phase transformation). The temperature inside of the sample is then given by the heat flow, the thermal conductivity, and the latent heat. The heat conduction and the phase transformation thus propagate in cooperation. This is a very complex process that will be described in more detail in connection with the thermal conductivity. The temperatures, A_s and M_s, are the temperatures connected with the start of the austenite and the martensite phase transformation, respectively. It was previously mentioned that for the sample placed in the furnace, these temperatures are connected with the surface changes of the given physical property. For example, for the cylinder sample in the dilatometer changes of the front and end of the sample (between push rod) will be the most important for the dilatation characteristics during the first state of the transformation. On the other hand, the phase transformation in the center of the sample will determine the final temperatures, A_f and M_f. The phase transformation lasts from 4 min for a heating/cooling rate of 7K/min up to 40 min for a heating/cooling rate of 0.5 K/min. This sufficiently long time in our experimental system allows us to localize the development of various transformation thermoelastic strains to a given part of the sample. If, for example, we find splitting of the CTE peak in the second portion of the transformation temperature range, we can assume that development of this thermoelastic strain occurs in the center of the sample rather than on the surface.

The temperatures A_s, M_s, A_f, and M_f do not give information about the sample temperature, but rather they determine the start and end of the phase transformation identified by the thermocouple outside the sample. The initial temperatures, A_s and M_s, are closest to the real transformation temperatures, whereas the temperatures, A_f and M_f, are the temperatures that provide information about the thermocouple temperature after the time that is necessary for the transformation of the entire sample. These temperatures depend on the experimental conditions of the measurement (distance of the thermocouple from the sample, sample size and shape, heating/cooling rate, thermal properties of surroundings of the sample, etc.). Nurveren et al. [75] studied the evolution of the transformation characteristics with respect to the heating/cooling rate in a NiTi alloy. While the A_s temperature was nearly the same for all studied heating rates (1, 5, 10, 15, and 20 K/min), the A_f temperature increased at about 10 K between the heating rate of 1 to 20 K/min.

Dunne and Wayman [76, 77] have recently classified the thermoelastic phase transformations into two categories: Class I (for example NiTi [78]), where $A_s > M_s$ and Class II where $M_s > A_s$. The temperature region between M_s and M_f is small in Class I and large in Class II. For Class II, Kaufman and Cohen [79] defined the characteristic temperature where the chemical free energies of the austenite and martensite phases are equal. This temperature, T_o, is known as the metastable equilibrium temperature for the A↔M phase transformation. The authors suggested that T_o can be determined by bracketing it between M_s and A_s according to:

$$T_o = \frac{1}{2}\left(M_s + A_s\right). \tag{9}$$

They have found it useful to consider M_s as the forming temperature of a single isolated plate of martensite during cooling and the reversion of this plate at A_f upon heating. It is evident that this cannot be used for our experimental arrangement. The sample is cooled from the surface so that the first martensite plate occurs at M_s on the sample surface and the last martensitic plate occurs at M_f in the sample center. During heating, the transformation, M→A, starts on the sample surface because this is the area with the highest temperature on the sample, which implies that, in our arrangement, the first martensitic plate arisen during cooling transforms between the first plates during heating at A_s. The end of the phase transformation of the sample is always connected with the transformation in the sample center.

Ni-Mn-Ga alloys studied in this chapter belong to Class II because $M_s > A_s$. The characteristic temperature, T_o, was determined for all heating/cooling rates as: 443 K for 0.5

K/min, 442 K for 2 K/min, 445 for 5 K/min, and 445 K for 7 K/min. It can be seen that this temperature is practically the same (444 ±1 K) as our transformation temperature determined from the time dependence of the temperature (Figs. 26 and 27) and is not dependent on the heating/cooling rate. Therefore, we assume that the phase transformation in Ni-Mn-Ga alloys is a typical first-order phase transformation with respect to the latent heat where the phase transformation temperatures for heating and cooling are the same. Similar results can be found in [14] and Figure 1, where the starting temperatures of the M↔A phase transformation differ by a value of only 3 K. Any hysteresis of the temperature dependence of physical parameters is not kinetic in nature. Hysteresis occurs as a natural consequence of the fact that the temperature measured outside of the sample is identified with the temperature in the sample interior.

This conclusion does not hold true for Ni_3Ta alloys that we are currently studying in our laboratory. Initial results show that this alloy behaves as a shape memory alloy. The martensitic structure at room temperature is visible in Figure 28 and the dilatation characteristics are shown in Figure 29, where it can be seen that this alloy belongs to Class I. Further differences in the comparison with Ni-Mn-Ga lies in the non-symmetry of the transformation strain obtained during heating and cooling. No latent heat was found on the time dependence of the temperature. This alloy was studied by Firstov et al. [80] by DSC, dilatometry, and X-ray diffraction. DSC measurement has shown the remarkable stability in the characteristic temperatures, while the transformation temperatures are different from those obtained by dilatometry. X-ray diffraction shows that the martensitic transformation occurs from the tetragonal austenite structure into the monoclinic martensite structure. The dependence of the transformation temperatures exhibits a much smaller dependence on the heating/cooling rates than Ni-Mn-Ga alloys. The difference between the maximum peak temperature for 0.5 K/min and 7 K/min is only 4 K in contrast to Ni-Mn-Ga where this difference is about 20 K. This difference is evidently a consequence of the latent heat and this alloy needs to be further investigated.

1.3. Thermal Arrest Memory Effect

The thermal arrest memory effect is undoubtedly a very interesting phenomenon that, with further investigation, can provide very important information about the behavior of shape memory alloys. Johnson et al. [16] were the first to show the influence of the interruption of the phase transformation on behavior of the alloy in subsequent thermal cycle. When arresting the M→P phase transformation in the Ni-Ti alloy with subsequent cooling, the endothermic peak in following thermal cycles splits into two peaks. This implies that the M→P transformation in the second cycle remembers the arrest in the first thermal cycle and this effect has been called the thermal arrest memory effect (TAME) by Madangopal [81]. The authors of the cited work assume that the TAME is a consequence of the locked-in transformation strain energy in the self-accommodation martensitic structure. The TAME effect has been the subject of intense investigation [17-21, 82]. In contrast to Madangopal [81], Airoldi and co-workers [18] (who called this effect SMART – stepwise martensite to austenite reversible transformation) suggested that the TAME phenomenon is related to the hysteresis associated with the martensitic transformation. Both hypotheses are based on DSC studies performed on Ti-Ni and Cu-Zn-Al shape memory alloys.

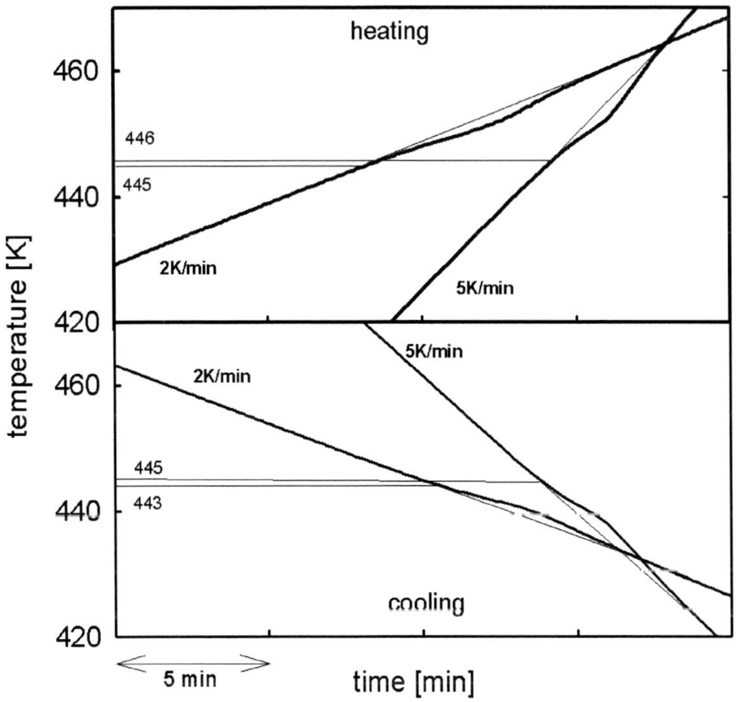

Figure 26. Time dependence of the temperature for a heating/cooling rate of 2 and 5 K min^{-1} (in He).

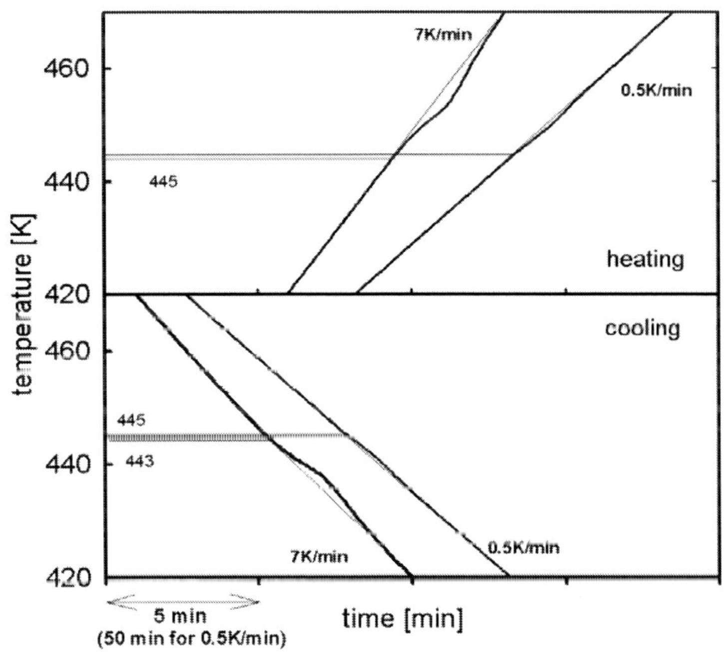

Figure 27. Time dependence of the temperature for a heating/cooling rate of 0.5 and 7 K min^{-1} (in He).

Figure 28. The microstructure of Ni_3Ta.

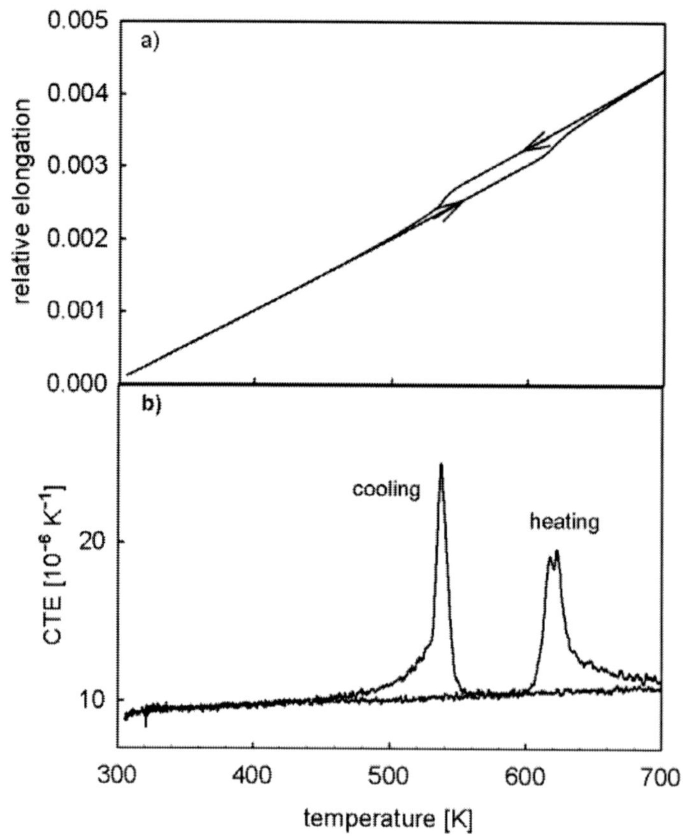

Figure 29. Dilatation characteristics of the Ni_3Ta alloy.

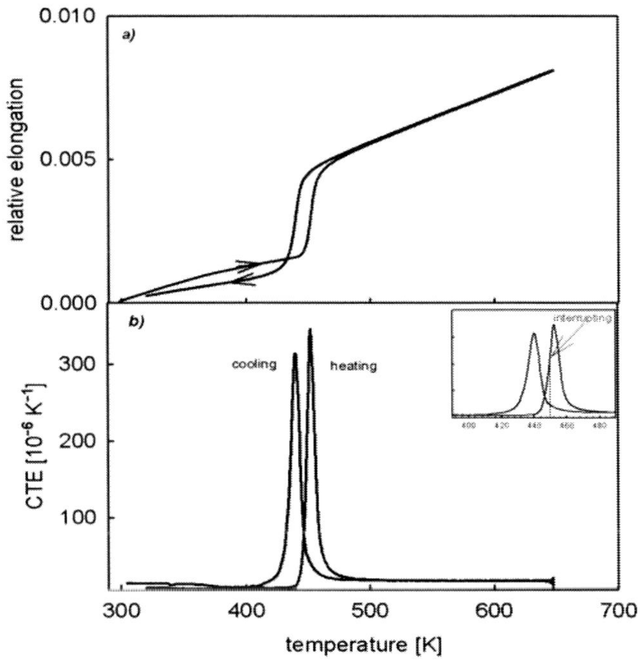

Figure 30. Temperature dependence of (a) the relative elongation and (b) the CTE for the $Ni_{53.6}Mn_{27.1}Ga_{19.3}$ alloy for a complete cycle.

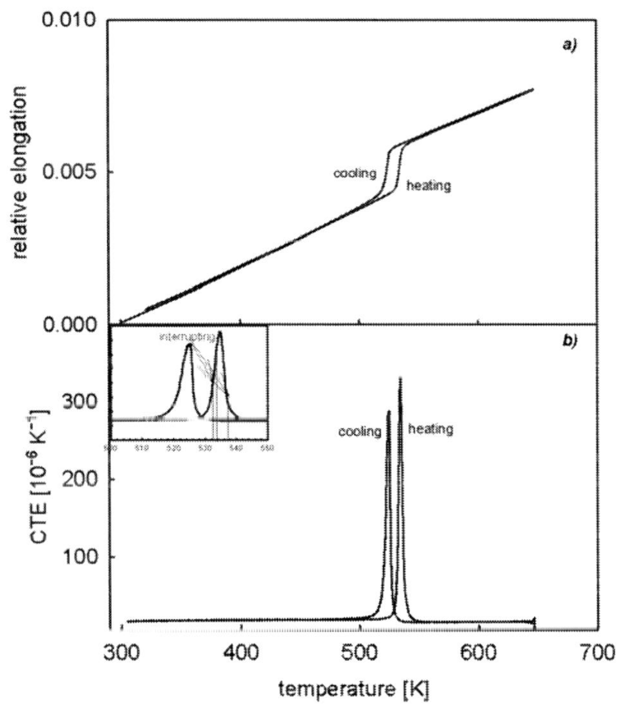

Figure 31. Temperature dependence of (a) the relative elongation and (b) the CTE for the $Ni_{54.2}Mn_{29.4}Ga_{16.4}$ alloy for a complete cycle.

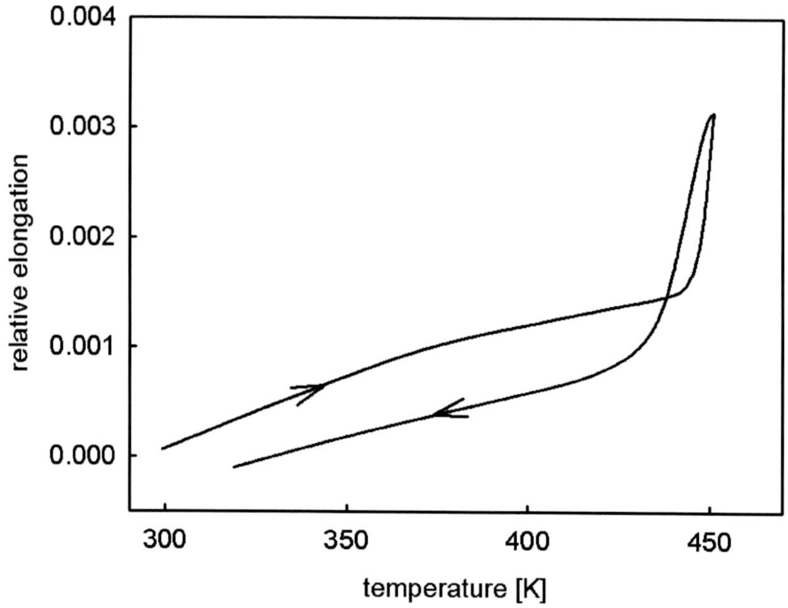

Figure 32. Temperature dependence of the relative elongation for the $Ni_{53.6}Mn_{27.1}Ga_{19.3}$ alloy during an incomplete cycle with a stop at T= 450 K.

Dilatometry is a method that has not been used for studying this effect. TAME effect in the $Ni_{53.6}Mn_{27.1}Ga_{19.3}$ and $Ni_{54.2}Mn_{29.4}Ga_{16.4}$ alloys was studied over the temperature range from room temperature to 650 K.

The dilatation characteristics of both materials before interruption are presented in Figs. 30 and 31. The interruption temperatures are marked in these figures. Figure 32 reveals the temperature dependence of the relative elongation for the $Ni_{53.6}Mn_{27.1}Ga_{19.3}$ alloy (it was pre-deformed and several thermal cycles were performed after pre-deformation) in the experiment where the temperature was increased up to 450 K and then stabilized for one hour prior to being cooled to room temperature. With this temperature sequence, the phase transformation takes place only partially and it can be seen that shortening of the sample by about 20 μm occurs after this thermal cycle. The results presented in Figure 33 show the temperature dependence of the relative elongation obtained in the following thermal cycle. It can be seen that permanent elongation of the sample occurs after this thermal cycle. Figure 34 displays the splitting of the peaks of the CTE during the phase transformation, M→A, when the total thermal cycle was made after interruption. If additional complete (without interruption) thermal cycles were then made, the dilatation characteristics would be the same as before interruption. The same experiments were performed with the $Ni_{54.2}Mn_{29.4}Ga_{16.4}$ alloy and demonstrated similar results; however, splitting of the CTE peaks is more visible than in the first alloy. Figure 35 shows the temperature dependence of the CTE after interruption when the interruption temperatures were 533, 534, and538 K. The temperature dependence of the CTE without interruption is also shown in this figure.

It was previously mentioned in the experimental section that our samples have a cylindrical shape. If a cylindrical sample is suddenly subjected to an impact cooling or a phase transformation, thermal stress waves occur at the surface of the sample.

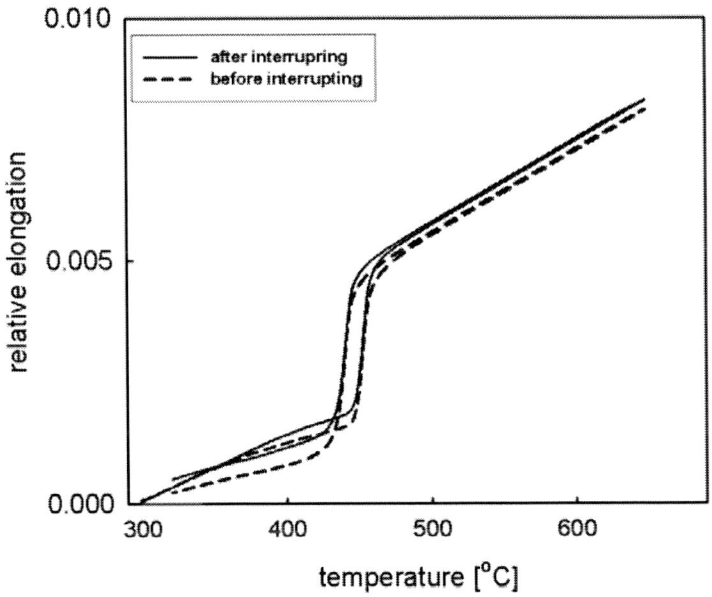

Figure 33. Temperature dependence of the relative elongation for the $Ni_{53.6}Mn_{27.1}Ga_{19.3}$ alloy in the thermal cycle following after incomplete thermal cycle.

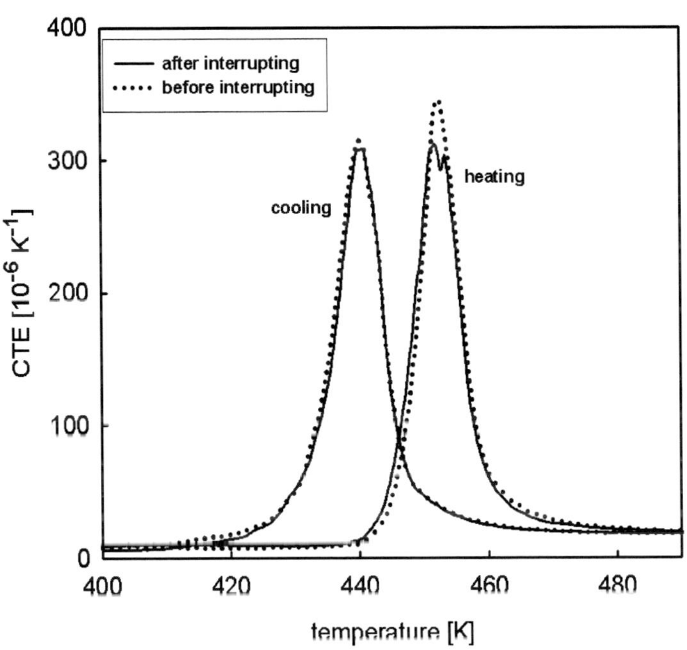

Figure 34. Temperature dependence of the CTE for the $Ni_{53.6}Mn_{27.1}Ga_{19.3}$ alloy after an incomplete cycle.

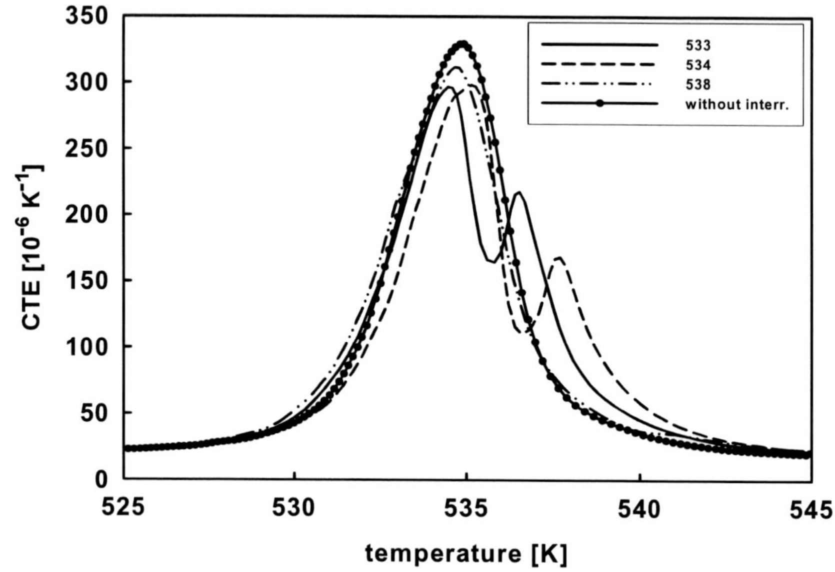

Figure 35. Temperature dependence of the CTE for the $Ni_{54.2}Mn_{29.4}Ga_{16.4}$ alloy after a single stop at T= 533, 534, and 538 K.

Stress waves in a cylinder with a focusing point proceed radially inward to the center of the cylinder sample. The waves may accumulate at the focusing point and give rise to very large magnitude stresses. This phenomenon is known in the literature as the stress focusing effect [83, 84]. Theoretical analysis of the stress focusing effect induced by thermal expansion and phase transformation is presented in the work of Hata [85]. This author has shown that when a cylindrical metal sample is cooled instantaneously below the phase transformation temperature, the stress focusing effect occurs via expansion due to phase transformation and thermal contraction.

When the transformation temperature is reached at the sample surface, the phase interface propagates into the sample and divides the sample into two ranges that have different structures and sizes. During the phase transformation, M→A, the austenite layer occurs first on the surface of the cylinder. As the austenite volume increases in the cylindrical sample, the stress focusing effect occurs at the center where the martensite structure exists. This martensite structure then experiences compressive stress. If we interrupt the phase transformation in a certain state, the sample will stay in this state as long as the temperature does not change (Figure 32). Decreasing the temperature causes the austenite structure to transform to the martensite structure (only the outside part of the sample). As the martensite structure in the center of the sample is deformed by compression due to the focusing effect at the end of the interrupted thermal cycle, the sample length is reduced (Figure 32). The sample returns to its initial size when the total thermal cycle including the phase transformation is applied (Figure 33). It can be seen that the influence of the interruption on the CTE is higher in the non-deformed $Ni_{54.2}Mn_{29.4}Ga_{16.4}$ alloy than in the pre-deformed $Ni_{53.6}Mn_{27.1}Ga_{19.3}$ alloy. Figure 35 shows the influence of the interruption temperature on the splitting of the CTE. Splitting of the transformation peak occurs only in the case when the interruption was applied in the first part of the phase transformation (the temperature region or time). It can be seen from Figure 35 that the second peak shifts to a higher temperature with increase in the interruption temperature. A higher interruption temperature corresponds to a higher compression stress in the center of the sample. Release of the compression deformation is a thermally activated process. We assume that the compression deformed center of the sample transforms at a higher temperature than the non-deformed outside of the sample. The difference between these temperatures is 2 K for an interruption temperature of 533 K, and 2.5 K for an interruption temperature of 534 K. This shift in the transient temperature results in a splitting of the CTE transient peak, which does not occur when the phase transformation is interrupted before the end of the phase transformation (537 K) because nearly the whole sample is transformed. Our results and conclusions are confirmed by the latest results of Rodríguez-Aseguinolaza et al. [86] where the TAME effect was studied by adiabatic calorimetry. The TAME effect is shown in the temperature dependence of the specific heat. With an increase of the interruption temperature, a new peak on the temperature dependence of the specific heat increases. If the interruption was made in the second part of the phase transformation no effect was found.

The evolution of the thermoelastic deformations during the phase transformation is schematically shown in Figure 36. In this schema we consider only heating because all deformation changes occur during heating. The schema of the phase transformation, M→A, without interruption is shown in the first part of Figure 36. In this schema we divide the duration of the phase transformation into two parts. First, the transformation occurs in the external part of cylinder. This transformation is connected with the compression of the internal part of the sample due to the stress focusing effect. In the second cycle of the transformation, the central part of the sample transforms and stresses that occurred in the first cycle of the transformation release.

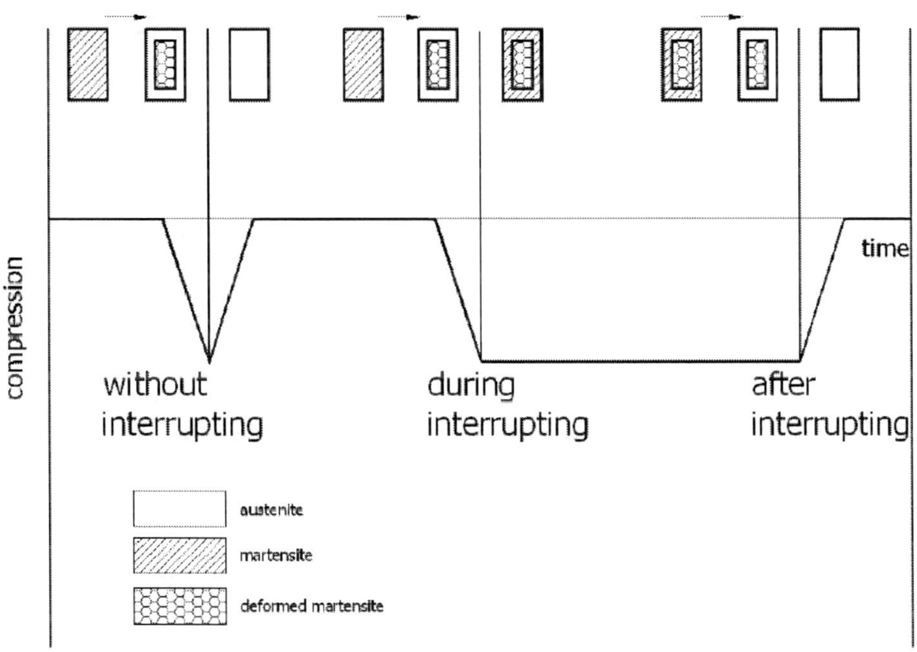

Figure 36. Schema of the evolution of the thermoelastic deformations in the sample during the phase transformation.

If we interrupt the M→A phase transformation the sample stays deformed in compression (the second part of Figure 36). The sample is then comprised of two parts: the center of the sample, which is the deformed martensite structure, and the outside of the sample, which is the austenite structure. During the following A→M transformation, only the austenite part of the sample transforms and the martensite center of the sample stays unchanged. After this thermal cycle, the sample stays macroscopically deformed in compression at room temperature resulting in a shortening of the sample. The compression deformation is released only after the total M→A transformation resulting in elongation of the sample (OWSME).

Our experiments with pre-deformed $Ni_{53.6}Mn_{27.1}Ga_{19.3}$ show that the transformation temperatures of the pre-deformed alloy are higher than for the non-deformed alloy. The same results are often presented in literature [87-89]. Relaxation of two species of strain took place in the thermal cycle after pre-deformation. First, the memory strain could not be removed from the sample, and then the residual strain is released only during heating as a thermally activated process. The residual strain is released in two temperature regions, during heating of the martensite structure below the transformation temperature and during the phase transformation. Release of the compression deformation after interruption takes place in a similar manner. Figure 33 shows that the strain after interruption releases from the room temperature. We therefore assume that both types of these strains are the same despite the fact that in the first case the entire sample was deformed in the axial direction and in the second case only a part of the sample was deformed. The authors of [42] studied the effects of hydrostatic pressure on the martensitic phase transformation, thermal conductivity, and thermodynamic parameters of Cu-13 wt%, Al-4 wt% Ni. They found that all these parameters are strongly affected by the applied pressure. The transformation temperatures increased and the thermal conductivity decreased with applied stress. The effect of the applied stress is connected with an increase in the dislocation density.

2. Thermal Diffusivity and Thermal Conductivity

The thermal conductivity may be determined by many methods. All methods are based on the solution of the general heat conduction equation that is given as:

$$\frac{dT}{dt} = \frac{\lambda}{\rho c_p}\left(\frac{\partial^2 T}{\partial x^2} + \frac{\partial^2 T}{\partial y^2} + \frac{\partial^2 T}{\partial z^2}\right); \qquad (10)$$

where T is the temperature and t is the time. The term, $\lambda/\rho c_p$, was defined as the thermal diffusivity by Lord Kelvin. This equation may be solved only for given initial and boundary conditions. Solution of the temperature distributions for a large number of relatively simple shapes and boundary conditions can be found in the book of Carslaw and Jaeger [90]. Many methods of measurement of the thermal diffusivity and thermal conductivity are based on solutions presented in this book. The accuracy of the measurement of the thermal diffusivity and thermal conductivity then depends on the experimental realization of the initial and boundary conditions of the solution of the heat conduction equation. One of the basic conditions for the derivation and solution of the heat conduction equation is that all thermophysical parameters in Eq. (10) are not dependent upon temperature. Generally, they depend on the temperature, but to a good approximation it is assumed that they are a function only of the local material and state. However, in the temperature range where phase transformations occur the thermophysical parameters dramatically change with the temperature and non-filling of the boundary conditions may lead to wrong conclusions during the interpretation of the thermal conductivity results. Careful analysis of the heat conduction in such systems is necessary before physical interpretation of the results is conducted.

The temperature dependence of the thermal diffusivity for the $Ni_{53.6}Mn_{27.1}Ga_{19.3}$ alloy is shown in Figure 37. This figure clearly shows an increase of the thermal diffusivity at about 450 K where the phase transformation, M↔A, occurs. The thermal diffusivity of the austenite is about 40% higher than that of the martensite (the values before and after phase transformation were compared) and it increases with increasing temperature faster than that for the martensite. A small peak below 370 K demonstrates a phase transformation of the ferromagnetic martensite to the paramagnetic martensite. This magnetic phase transformation temperature was found with magnetic measurements using a PPMS device from Quantum Design Company (USA). The temperature dependence of the thermal diffusivity has a reverse peak characteristic for this type of phase transformation (see Fe in [91]). The temperature dependence of the maximum temperature rise after irradiation of the sample is shown in Figure 38 (the distance of the flash lamp from the sample was 190 mm). It can be seen from both figures that the thermal diffusivity as well as the temperature rise changes significantly over the transformation temperature range. When the transformation temperature is reached in a sample after irradiation, the phase boundary between the austenite and the martensite structures propagates through the sample. The transformation temperature obtained from the temperature dependence of the thermal diffusivity was 450 K (from derivation of the temperature dependence of the thermal diffusivity) and from the temperature dependence of ΔT_M was 449 K (temperature of the maximum). The transformation temperature determined from the temperature dependence of the CTE (start of the phase transformation during heating) was 445 K. It is necessary to note that all of these temperatures were obtained by thermocouple measurements when the thermocouple was placed outside the sample.

For the determination of the thermal conductivity from the thermal diffusivity, knowledge of the specific heat and the density is necessary. In the literature we can find several studies concerning the measurement of the specific heat of Ni-Mn-Ga-based alloys. Kreissl et al. [92] have shown that the specific heat of the polycrystalline Ni_2MnGa alloy depends on the thermal history of the alloy. The peak in the transition temperature range was found only in the alloy quenched from 1273 K but not after quenching from 1073 K. Cesari at al. [22] present no changes of the specific heat during the phase transformation for two single crystals of the non-stoichiometric Ni-Mn-Ga alloys.

The specific heat for our purposes was obtained by the flash method because we find it important that the sample be heated in the same manner as during the thermal diffusivity measurement.

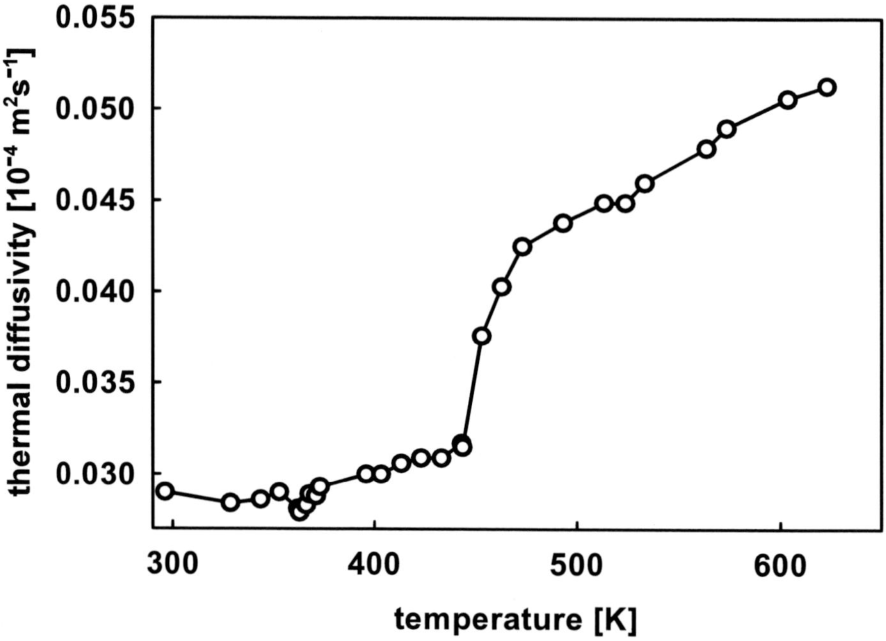

Figure 37. Temperature dependence of the thermal diffusivity (distance of the flash lamp from the sample was 190 mm).

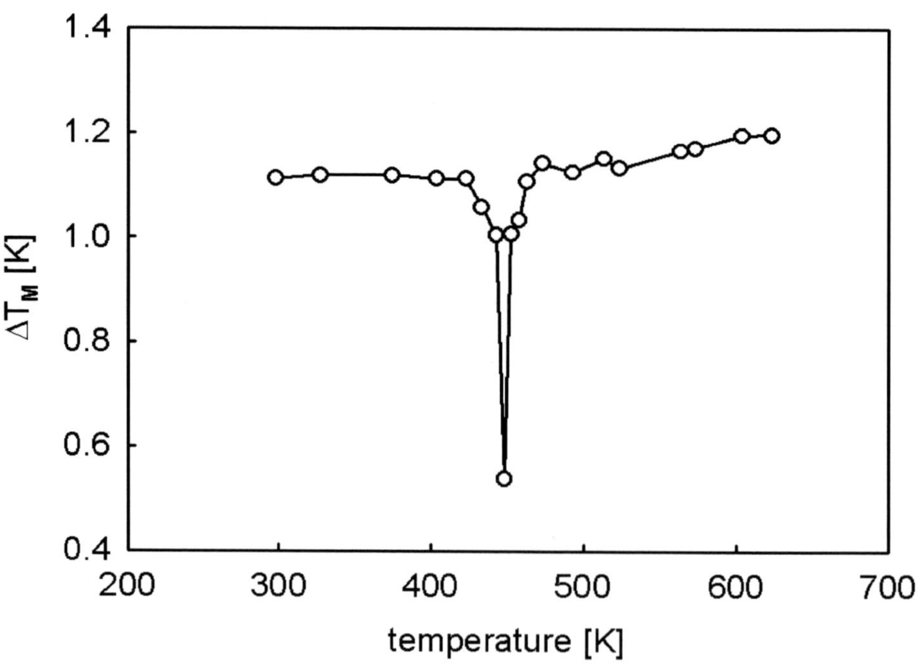

Figure 38. Temperature dependence of the temperature rise, ΔT_M, after irradiation (distance of the flash lamp from the sample was 190 mm).

For a determination of the specific heat by this method the following relation is used [93]:

$$c_p = \frac{Q}{\rho L \Delta T_M}; \qquad (11)$$

where Q is the energy input per unit area, ΔT_M is the maximum temperature rise after irradiation, and L is the sample thickness. Knowledge of Q is not necessary for the determination of the thermal diffusivity by the flash method [43]. The specific heat at room temperature was determined with a Differential scanning calorimeter from Perkin Elmer Pyris Diamond DSC. This value was used for the calculation Q by Eq. (11). Temperature dependence of the specific heat was then obtained using Eq. (11) where the maximum temperature rise, ΔT_M, (distance of the flash lamp from the sample was 190 mm) was measured with the flash apparatus. No correction for the radiation losses was made and the result is shown in Figure 39. Good agreement between these values of the specific heat and the specific heat determined in DSC for the austenite (above the transition temperature) can be seen from this figure. The changes of the specific heat during the magnetic transformation were not found by this method. The density is a further parameter necessary for the determination of the thermal conductivity. The temperature dependence of the density is shown in Figure 10 and it can be seen that only a small deformation change occurs over the transformation temperature range because, for the calculation of the density, we used the relative elongation of the sample with equiaxed grains.

The thermal conductivity determined as a product of the thermal diffusivity, density, and the specific heat is shown in Figure 40. The austenite has a higher thermal conductivity than martensite. Three temperature regions can be seen: the region of the martensite structure, the transient region, and the austenite structure region. In the transition temperature region a peak occurs, which is a consequence of the peak of the specific heat. This peak, however, has no physical meaning as the heat conduction equation was solved with the boundary condition that no phase transformation occurs in the material (c_p, ρ, and λ are not dependent on T). It was mentioned in the introduction that peaks in the temperature dependence of the thermal conductivity in the transient temperature region were found for various materials; however, these peaks, which were obtained by methods other than the flash method, have found their physical interpretation. For example, in [23] the peak found in Ni_2MnGa by the direct heat-pulse technique was connected with the additional heat carried by soft phonons.

When the phase transformation with the latent heat propagates through the sample, absorption of the latent heat leads to a decrease of the temperature, which is in accordance with [94] where the phase transformation in the Cu-Al-Ni shape memory alloy was studied by an infrared camera. The measurement of the thermal conductivity by a direct heat-pulse technique is based on the determination of the temperature gradient in the sample. If the phase transformation propagates through the sample in the location of the thermometers, their temperature is reduced by absorption of the latent heat, which naturally influences the thermal conductivity value. In our opinion, various peaks in the temperature dependencies of the thermal conductivity over the transformation temperature range are a natural consequence of neglect of the boundary conditions used during the solution of the heat conduction equation.

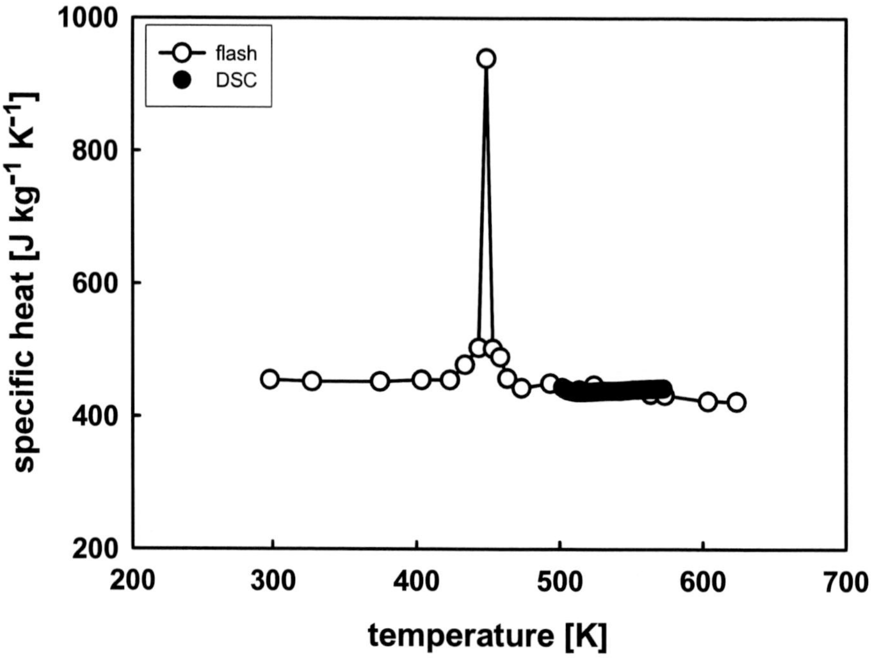

Figure 39. Temperature dependence of the specific heat.

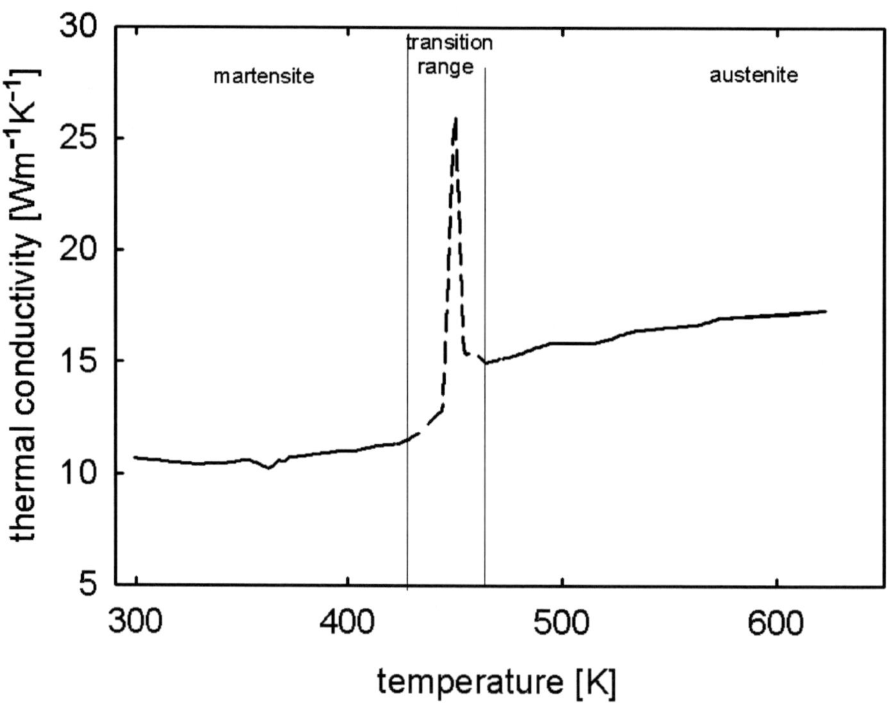

Figure 40. Temperature dependence of the thermal conductivity.

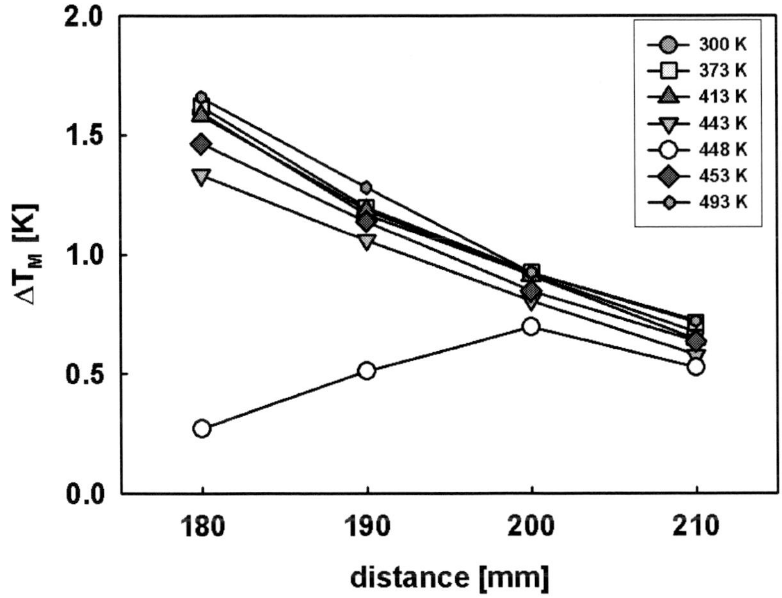

Figure 41. Dependence of the temperature rise, ΔT_M, on distance of the Xe-flash lamp from the sample.

Analysis of this problem should be made before these peaks are connected with material processes.

All of our thermal diffusivity measurements start at room temperature where heating of the martensite structure begins. In the temperature range outside of the transformation temperature range the heat conduction is a single process occurring in the sample. When the surface temperature of the sample is sufficiently close to the transformation temperature and the sample is irradiated so that the sample temperature exceeds the transformation temperature the phase interface propagates into the sample. Two processes determine the temperature of this interface as well as the velocity of the movement of this interface: thermal conductivity and phase transformation. Furthermore, both processes propagate together. The phase transformation influences the energy balance, temperature, and the rate of movement of the interface by latent heat. The velocity of the transient process (u) thus depends on the thermal conductivity of both phases (K_M, K_A), the latent heat (H), the density (ρ), the heating rate, and time. For a given time, u can be determined by the following relation [95]:

$$u = \frac{1}{\rho H}\left(K_M \left(\frac{\partial T}{\partial x} \right) - K_A \left(\frac{\partial T}{\partial x} \right) \right). \qquad (12)$$

This relation was modified from a relation that was originally derived for the propagation of a melting front in a conducting region where it was assumed that the density of both phases is the same. The velocity of the propagation of the phase interface changes with time. As a consequence of moving interface, with conditions such as in Eq. (12), the formula is nonlinear. This is discussed in Carslaw and Jaeger [90] and demonstrated by Lunardini [96].

The velocities of moving interfaces were determined experimentally in [97,98]. In the first of these works, the stress-induced martensitic phase transformation was studied on Cu-Al-Ni-based single crystals. The velocities of the moving interfaces were on the order of 10^1 m s^{-1}. In the second work, the velocity of moving interface (10^{-3} m s^{-1}) was determined for the thermally induced stress-free phase transformation in the Cu-Zn-Al alloy. The large difference between both values can be found in the various initiation of the phase transformation process. The process of the thermal conductivity in the sample can have a retarding effect on the propagation of the martensitic phase transformation in comparison with the stress-induced phase transformation. Analysis of the heat conduction and the martensitic phase transformation in [99] take into account the latent heat. They have used the enthalpy method [100] to describe the martensitic transformation of a Ni-Ti ring. Results show that the actuation time decreases with temperature and is larger when latent heat is taken in account than when the effect of latent heat is neglected. Generally, it is assumed [95] that the temperature of the phase interface does not change during its propagation through the material; however, Seiner et al. [94] have found that the temperature of the interface changes with time.

To study the influence of the phase transformation on the heat conduction we have measured the maximum temperature rise on the opposite side of the sample at various temperatures and for various input energy. The change of the input energy was obtained by a change of the distance of the flash lamp from the sample. The results are shown in Figure 41. Practically the same dependencies were obtained for all temperatures except for a temperature of 448 K. The substantial decrease of the temperature rise for higher input energies is an unambiguous influence of the phase transformation on the heat flow in the sample after irradiation. Figure 42 shows the time dependence of the temperature rise for selected temperatures with the input energy corresponding to a flash lamp distance of 190 mm. A region of the quasi-stationary behavior can be found on each curve in Figure 42. For this state we have proposed the schema presented in Figure 43 where it is assumed that the heat flows only in one dimension and that the starting temperature (T_o) is close, but less than, the transient temperature. The absorbed energy, Q_1 or Q_2, (Q_2>Q_1) rises with the temperature at x=0. If no phase transformation occurs in the material then the temperature would reach T_{11} or T_{21}. The temperature in the sample would decrease as the dashed lines indicate in the schema. If, however, the temperatures T_{11} and T_{21} are higher than T_{transf} and a phase transformation takes place in the sample, then the temperature decreases on T_{12} and T_{22} due to the storage of energy in the form of latent heat. The temperature decreases in the sample as x increases. When the temperature reaches T_{transf}, the reverse phase transformation takes place, the latent heat is released, and the temperature abruptly increases. For the case of lower input energy, Q_1, this process occurs inside the sample and the resultant temperature measured at x=L does not detect a phase transformation. If the input energy is sufficiently high (Q_2) then T_{trans} is not reached in the sample and the measured temperature rise at x=L is lower than for the case without a phase transformation ($\Delta T_2 < \Delta T$). The reverse phase transformation occurs in the sample after a longer time than the time where ΔT_M was determined. Figure 42 shows that the sample temperature decreases after achievement of the maximum value more slowly at 448 K than at other temperatures. The results clearly show that the phase transformation, martensite↔austenite, and the heat conduction take place in the material in mutual cooperation. This is a very complicated non-equilibrium process where the thermal conductivity is a crucial parameter.

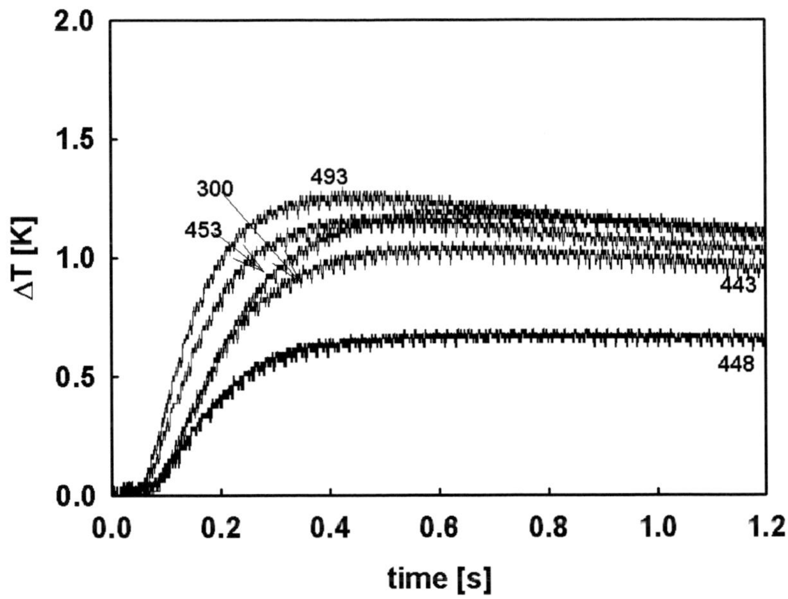

Figure 42. Time dependence of the temperature rise, ΔT, (distance of the flash lamp from the sample was 190 mm).

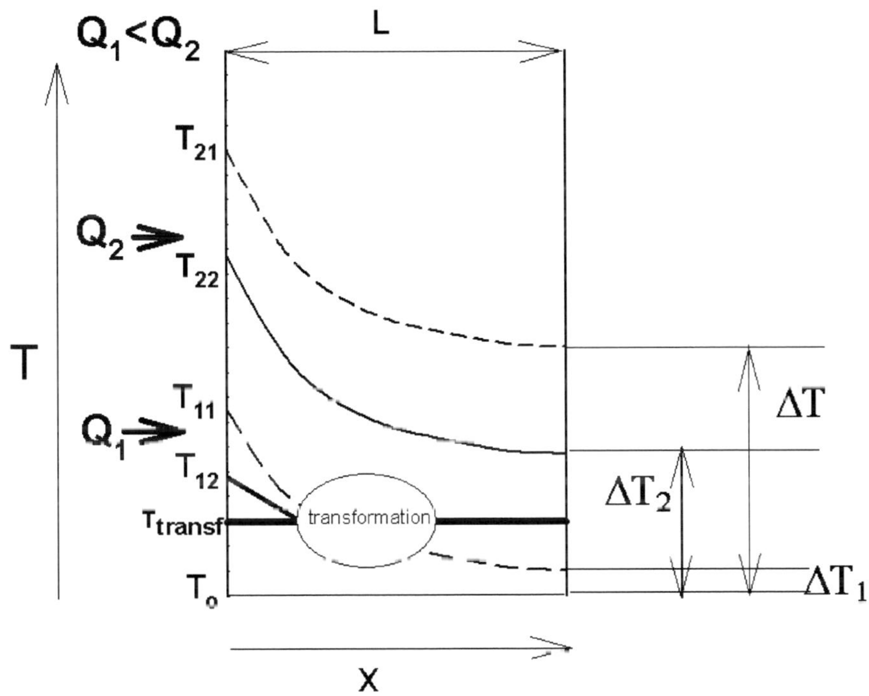

Figure 43. Schema of the heat conduction in the sample thickness, L, for a given time; where Q_1 and Q_2 are the input energy, T is temperature, and ΔT is the maximum temperature rise at x=L.

Conclusion

The results presented in this chapter show that if the sample of the shape memory alloy is heated/cooled from outside, then the phase transformation starts on the surface and propagates into the sample center. The phase boundary moves with a velocity that is determined by the heat flow, the latent heat, and the thermal conductivity of both phases. The velocity of the phase boundary then determines the time between the start and end of the phase transformation, which is given by the size and shape of the sample. The characteristic transformation temperature is temperature when the phase transformation starts. Further temperatures, such as the maximum peak temperature or the finish temperature, are not material properties because they are influenced by the experimental conditions of the measurement. The dilatation characteristics are strongly influenced by the thermal conductivity of the protective atmosphere in the dilatometer. The influence of the heating/cooling rates on the dilatation characteristics is higher in an argon atmosphere then in a helium atmosphere. Analysis of the thermal transport in the dilatometer shows that the temperature difference between the sample surface and the thermocouple is higher in argon than in a helium atmosphere. If the transformation temperature (start of the phase transformation) was determined from the time dependence of the thermocouple temperature when the thermocouple was placed on the sample, then the transformation temperature was the same for all heating/cooling rates and it was the same for heating as well as cooling.

The shape memory alloy remembers its mechanical and thermal history. This memory is based on the reorientation and detwinning of the low temperature martensitic phase during mechanical loading. The dilatation results presented in this chapter show that the compression pre-deformation of the studied Ni-Mn-Ga alloys results in three types of the strains:

1) The residual strain (OWSME), which may be removed by heating. Its removal is connected with the permanent length (shape) changes. It releases not only during the M→A phase transformation but also over the entire temperature range of the existence of the martensite phase.
2) The memory strain (TWME), which cannot be removed from the sample and is a consequence exclusively of reorientation and detwinning of the martensitic variants. Its release and formation are bound on only the transformation temperature range.
3) The dislocation structures, which stay in the martensite as well as in the austenite phases after the phase transformation and cannot be removed from the material by thermal cycling.

The thermal arrest memory effect is a consequence of the shape memory effect and the stress focusing effect occurring in the cylindrical sample. The thermal expansion results obtained in the thermal cycle, which was interrupted during the first state of the phase transformation, clearly show that the phase boundary propagates from the surface to the center of the sample. The sample center stays compression deformed if its phase transformation is interrupted. When the total thermal cycle is then applied on the sample this deformation is released (OWSME). The compression deformed region of the sample transforms at a higher temperature than the non-deformed outside part and splitting of the CTE peak occurs.

The thermal diffusivity and the thermal conductivity of the austenite structure are about 40% higher than the martensite structure. No peaks occur in the temperature dependence of the thermal diffusivity. The relation, $\lambda = \rho c_p a$, cannot be used for the calculation of the thermal conductivity in the transformation temperature range because the general heat conduction equation was derived on the condition that no phase transformations occur in the material. It was shown that during the heating of the sample in the transformation temperature range, the sample temperature rise is lower than for the case without the phase transformation. This is a consequence of the absorption of latent heat. If this fact is not respected in methods where the determination of the physical parameter is based on the measured temperature then misleading conclusions can be accepted.

AKNOWLEDGMENT

This work is a part of the research program MSM 0021620834 that is financed by the Ministry of Education of the Czech Republic. The authors acknowledge financial support from the Grant Agency of the Czech Republic under Grant 106/06/0368.

REFERENCES

[1] Roytburd, A. L. *Mater. Sci. Eng.* A 1999, 273-275, 1-10.
[2] Cao, W.; Krumhansl, J. A.; Gooding, R. *J. Phys. Rev.* B 1990, 4, 11319-11327.
[3] O`Handley R. C.; Murray, S. J.; Marioni, M.; Nembach, H.; Allen, S. M. *J. Appl. Phys.* 2000, 87, 4712-4717.
[4] Murray, S. J.; Marioni, M.; Tello, P. G.; Allen, S. M.; O`Handley, R. C. *Magn. Magn. Mater.* 2001, 226-230, 945-947.
[5] Murray, S. J.; Marioni, M.; Allen, S. M.; O`Handley, R. C.; Lograsso, T. A. *Appl. Phys. Lett.* 2000, 77, 886-888.
[6] Liang, T.; Jiang, C. B.; Xu, H. B.; Liu, Z. H.; Zhang, M.; Cui, Y. T. *J. Magn. Magn. Mater.* 2004, 268, 29-32.
[7] Oikawa, K.; Wulff, L.; Lijima, T.; Gejima, F.; Ohmori, T.; Fujita, A. *Appl. Phys. Lett.* 2001, 79, 3290-3292.
[8] Bhattacharya, K.; Kohn, R. V. *Acta Mater.* 1996, 44, 529-542.
[9] Pons, J.; Chernenko, V. A.; Santamarta, N.; Cesari E. *Acta Mater.* 2000, 48, 3027-3038.
[10] Khovailo, V. V.; Oikawa, K.; Abe, T.; Takagi, T. *J. Appl. Phys. Lett.* 2003, 93, 8483-8485.
[11] Vasilev, A. N.; Bozhko, A. D.; Khovailo, V. V.; Dieshtein, I. E.; Shavrov, V. G., Buchelnikov, V. D.; Matsumoto, M.; Suzuki, S.; Takagi, T.; Tani, J. *Phys. Rev.* B 1999, 59, 1113-1120.
[12] He, X. M.; Rong, L. J.; Yan, D. S.; Li, Y. Y. *Scripta Mater.* 2005, 53, 1411-1415.
[13] Ma, Y.; Jiang, Cha.; Li, Y.; Xu, H.; Wang, C.; Liu, X. *Acta Mater.* 2007, 55, 1533-1541.
[14] Xu, H. B.; Li, Y.; Jiang, C. B. *Mater. Sci. Eng.* A 2006, 438-440, 1065-1070.
[15] Besseghini, S.; Pasquale, M.; Passaretti, F.; Sciacca, A.; Villa, E. *Scripta Mat.* 2001, 44, 2681-2681.

[16] Johnson, W. A.; Dominique, J. A.; Reichman, S. H. *J. Phys.* IV 1982, 12, 43-48.
[17] Airoldi, G.; Corsi, A.; Riva, G. *Scripta Mater.* 1997, 36, 1273-1278.
[18] Airoldi, G.; Besseghini, S.; Riva, G. *J. Phys.* IV 1995, 5, 483-488.
[19] Wang, Z. G.; Zu, X. T.; Zhu, S.; Wang, L. M. *Mater. Lett.* 2005, 59, 491-494.
[20] Krishnan, M. *Scripta Mater.* 2005, 53, 875-879.
[21] Wang, Z.G.; Zu, X.T.; Yu, H. J.; Fu, H. *Scripta Mater.* 2006, 55, 497-500.
[22] Cesari, E.; Chernenko, V. A.; Font, J.; Muntasell, J. *Therm. Acta* 2005, 43, 153-156.
[23] Kuo, Y. K.; Sivakumar, K. M.; Chen, H. C. *Phys. Rev.* B 2005, 72, 054116-1-054116-10.
[24] Brown, P. J.; Bargawi, A. Y.; Crangle, J.; Neumann, K. U.; Ziebeck, K. R. A. *J. Phys. Condens. Matter.* 1999, 11, 4715-4722.
[25] Fraile-Rodríguez, A.; Ruiz-Larrea, I.; Lopéz-Echarri, A.; San Chuan, J. *Scripta Mater.* 2006, 54, 1199-1203.
[26] Glavatska, A.; Mogylny, G.; Glavatskiy, I.; Gavriljuk, V. *Scripta Mater.* 2002, 46, 605-610.
[27] Cherepin, V. T.; Glavatska, N. I.; Glavatsky, I. N.; Gavriljuk, V. G. *Meas. Sci. Technol.* 2002, 13, 174-178.
[28] Wu, X. D.; Finlayson, T. R. *J. Phys. Condens. Matter.* 2007, 19, 026218-026227.
[29] Kostov, A. J.; Živkovič, Ž. D. *Therm. Acta* 1997, 291, 51-57.
[30] Wang, Z. G.; Zu, X. T.; Huo, Y. *Therm. Acta* 2005, 436, 153-155.
[31] Sutou, Y.; Omori, T.; Wang, J. J.; Kainuma, R.: Ishida, K. *Mater. Sci. Eng.* A 2004, 378, 278-282.
[32] Rudajevova, A. *J. Alloys Comp.* 2007, 430, 153-157.
[33] Rudajevova, A.; Frost, M.; Jäger, A. *Mater. Sci. Technol.* 2007, 23, 542-546.
[34] Nagasawa, A.; Enami, K.; Yshino, Y.; Abe, Y.; Nenno, S. *Scripta Metall.* 1974, 8, 1055-1060.
[35] Perkins, J. *Scripta Metall.* 1974, 8, 1469- 1476.
[36] Saburi, T.; Nenno, S. *Scripta Metall.* 1974, 8, 1363- 1368.
[37] Stalmans, R.; van Humbeeck, J.; Delaey, L. *Acta Metall. Mater.* 1992, 40, 501-511.
[38] Kuo, Y. K.; Lue, C. S.; Hsu, F. H.; Li, H. H.; Yang, H. D. *Phys. Rev.* B 2001, 64, 125124-125130.
[39] Smontara, A.; Biljakovic, K.; Artemenko, S. N. *Phys. Rev.* B 1993, 48, 4329-4334.
[40] Lue, C. S.; Kuo, Y. K.; Hsu, F. H.; Li, H. H.; Yang, H. D.; Fodor, P. S.; Wenger, L. E. *Phys. Rev.* B 2002, 66, 033101-033105.
[41] Faukner, M. G.; Amalraj, J. J.; Bhattacharya, A. *Smart Mater. Struct.* 2000, 9, 632-639.
[42] Zanotti, C.; Giuliani, P.; Riva, G.; Tuissi, A.; Chrysanthou, A. *J. Alloys Comp.* 2008 in press
[43] Zengin, R.; Yakuphanoglu, F. *Mater. Lett.* 2003, 57, 3107-3110.
[44] Parker, W. J.; Jenkins, R. J.; Butler, C. P.; Abort, G. L. *J. Appl. Phys.* 1961, 32, 1679-1684.
[45] Walter, A. J.; Gilchrist, K. E.; Taylor, R. *High Temp-High Press* 1972, 4, 439-436.
[46] Blumm, J.; Lindemann, A.; Niedrig, B. ECTP 17th European Conference on Thermophysical Properties, Bratislava 2005.
[47] Rudajevova, A. *Silikaty* 1989, 33, 61-67.
[48] Rudajevova, A. *Mater. Res. Bull.* 1991, 26, 1363-1369.

[49] Touloukian, Y.; Kirby, R. K.; Taylor, R. E.; Desai, P.D. Thermal expansion, Metallic Elements and Alloys, IFI/ Plenum, New York and Washington 1977.
[50] Roitburd, A. L. *Phys. Stat.. Sol.* (a) 1977, 40, 333-342.
[51] Roitburd, A. L.; Kurdjumov, G. V. *Mater. Sci. Eng.* 1979, 39, 141-167.
[52] Madangopal, K. *Acta Mater.* 1997, 45, 5347-5365.
[53] Wayman, C. M. *Scripta Metall.* 1971, 5, 489- 495.
[54] Hornbogen, E. *J. Mater. Sci.* 2004, 39, 385-399.
[55] Kajiwara, S.; Kikuchi, T. *Acta Metall.* 1982, 30, 589-598.
[56] Segui, G.; Chernenko, V. A.; Pons, J.; Cesari E. *J. Phys.* IV 2003, 112, 903-906.
[57] Chernenko, V. A.; Pons, J.; Cesari, E.; Ishikawa, K. *Acta Mater.* 2005, 53, 5071-5077.
[58] Chernenko, V. A. *Scripta Mater.* 1999, 40, 523-527.
[59] Wang, W. H.; Liu, Z. H.; Zhang, J.; Chen, J. L.; Wu, G. H.; Zhan, W. S. *Phys. Rev.* B 2002, 66, 052411-052415.
[60] Soolshenko, V.; Lanska, N.; Ullako, K. *J. Phys.* IV 2003, 112, 947-950.
[61] Segui, G.; Chernenko, V. A.; Pons, J.; Cesari, E.; Khovalio, V.; Takagi, T. *Acta Mater.* 2005, 53, 111-120.
[62] Liu, Y.; Ai, F.; Jiang, B.; Liu, Y.; Qui, X. *Mater. Sci. Eng.* A 2006, 438-440, 999-1002.
[63] Van Caenegem, N.; Duprez, L.; Verbeken, K.; Segers, D.; Houbaert, Y. *Mater. Sci. Eng.* A 2008, 481-482, 183-189.
[64] Liu, Y. *Smart Materials Proceedings of SPIE*, 2001, 4234, 82-93.
[65] Sade, M.; Hornbogen, E. *Z. Metallkunde* 1988, 79, 782-789.
[66] Airoldi, G.; Ranucci T. *Materials Smart Systems II* 1997, 459, 323-328.
[67] Planes, A.; Prez-Reche, F. J.; Vives, E.; Mañosa, L. *Scripta Mater.* 2004, 50, 181-186.
[68] Kakeshita, T.; Takeguchi, T.; Fukuda, T.; Saburi, T. S. *Mater. Trans. JIM* 1996, 37, 229-235.
[69] Aspelmeyer, M.; Klemradt, U.; Wood, L. T.; Moss, S. C.; Peisel, J. *Phys. Stat. Sol.* (a) 1999, 174, R9.
[70] Friman, M.; Hämäläinen, H. *J. Phys.* IV 2003, 112, 929-934.
[71] Rudajevova, A.; Lang, V.; Šroub J. *Int. J. Thermophys.* 2009, 30, 969-975.
[72] Touloukian, Y. S.; Powel, R. W.; Ho, Y. Q.; Klemens, P. G. The TPRC Data Series, Volume 1: Thermal Conductivity- Metallic Elements and Alloys, IFI/Plenum, New York-Washington, 1970.
[73] Touloukian, Y. S.; Powel, R. W.; Ho, Y. Q.; Klemens, P. G. The TPRC Data Series, Volume 2: Thermal Conductivity-Nonmetallic Solids, IFI/Plenum, New York-Washington, 1970.
[74] Touloukian, Y. S.; Buyco, E. H. The TPRC Data Series, Volume 5: Specific Heat Nonmetallic solids, IFI/Plenum, New York-Washington, 1970.
[75] Nurveren, K.; Akdogan, A.; Huang, W. M. *J. Mat. Process. Technol.* 2008, 196, 129-134.
[76] Dunne, D. P.; Wayman, C. M. *Met. Trans.* 1973, 4, 137-141.
[77] Dunne, D. P.; Wayman, C. M. *Met. Trans.* 1973, 4, 147-151.
[78] Uchil, J.; Mohanchandra, K. P.; Kumara, K. G.; Mahesh, K. K.; Murali, T. P. *Physica B* 1999, 270, 289-297.
[79] Kaufman, L.; Kohen, M. *Prog. Met. Phys.* 1958, 7, 165-179.
[80] Firstov, G. S.; Koval, Yu. N.; Van Humbeeck, J.; Ochin, P. *Mater. Sci. Eng.* A 2008, 481-482, 590-593.

[81] Madangopal, K.; Banerjee, S.; Lele, S.. *Acta Metall. Mater.* 1994, 42, 1875-1885.
[82] He, X. M.; Rong, L. J.; Yan, D. S.; Li, Y. Y. *Scripta Mater.* 2005, 53, 1411-1415.
[83] Ho, Ch. ASME *J. Appl. Mech.* 1976, 43, 464-468.
[84] Hata, T. ASME *J. Appl. Mech.* 1994, 61, 676-680.
[85] Hata, T.; Sumi, N. *Acta Mech.* 2008, 195, 69-80.
[86] Rodríguez-Aseguinolaza, J.; Ruiz-Larrea, I.; Nó, M. L.; López-Echarri, A.M.; San Juan, J. M. *Acta Mater.* 2008, 56, 3711-3722.
[87] Airoldi, G.; Corsi, A.; Riva, G. *Mater. Sci. Eng.* A 1998, 241, 233-240.
[88] Patel, J. R.; Chen, M. *Acta Metall.* 1953, 1, 531-538.
[89] Tatar, C.; Yakuphanoglu, F. *Termochim. Acta* 2005, 430, 129-133.
[90] Carslaw, F. R., Jaeger J. C. Conduction of heat in solids 2nd ed. Oxford Univ. Press London 1959.
[91] Shanks, H. R.; Klein, A. A. H.; Danielson G. C. *J. Appl. Phys.* 1967, 38, 2885-2892.
[92] Kreissl, M.; Neumann, K. U.; Stephens, T.; Ziebeck, K. R. A. *J. Phys. Condens. Matter.* 2003, 15, 3831-3839
[93] Xue, J.; Taylor, R. *Inter. J. Thermophys.* 1993, 14, 313-320.
[94] Seiner, H.; Landa, M.; Sedlak, P. *Proc. Estonian Acad. Sci. Phys. Math.* 2007, 56, 218-225.
[95] Gebhart, B. Heat conduction and mass diffusion McGraw-Hill, Inc. New York 1993.
[96] Lunardini, V. J. Heat Transfer in Cold Climates, Van Nostrand Reinhold, New York 1981.
[97] Escobar, J. C.; Clifton, R. J. *SPIE* 1995, 2427, 186-197.
[98] Pelegrina, A. J.; Ahlers, M. A*cta Metall. Mater* .1990, 38, 293-299.
[99] Elhadrouz, M.; Ben Zineb, T.; Pator E. *J. Phys.* IV 2003, 112, 183-186.
[100] Necati Ozisik, M. Finite Difference Methods in Heat transfer, CRC Pr I LlC 1994.

In: Shape Memory Alloys
Editor: H. R. Chen, pp. 231-246

ISBN 978-1-60741-789-7

Chapter 7

CHRACTERISTIC DEFORMATION AND STRUCTURAL CHANGES OF NI-TI ALLOYS BY MOLECULAR DYNAMICS: PHASE TRANSFORMATION AND AMORPHIZATION

Tomohiro Sato, Ken-Ichi Saitoh, Keisuke Kubota and Noboru Shinke

Kurimoto Center for Innovative Technology, Kurimoto. Ltd.
2-8-45, Shibatani, Suminoe-ku, Osaka-shi, Osaka, Japan
and Faculty of Engineering Science, Kansai University
3-3-35, Yamate-cho, Suita-Shi, Osaka, Japan

ABSTRACT

In this study, molecular dynamics simulation (MDS) is conducted by using embedded atom method (EAM), which is an interatomic potential function suitable for Ni-Ti system. The EAM potential reproduces the experimental results of the energetic for stable Ni-Ti phases (B2, B19') and the Ti precipitates which are consisted of hct lattices. In the simulation model, we put disc-shaped Ti precipitates inside Ni-Ti parent alloy. Firstly, in the model with the disc-shaped precipitates, martensite is maintained around the interfaces because inevitable local residual strain can be balanced, when the whole simulation model has no strain (completely relaxed). Under tensile loading, in the case with disc precipitates, martensite region propagates from the interface plane. In the propagation, the unit lattices show structurally two pathways to complete transformation. The split into these pathways is strongly depending on the angle between tensile direction and the orientation of the martensitic unitcell. The results suggest that the interface is a preferred site for nucleation of martensite and that there is a certain crystallographic condition for the next variants to form smoothly. Secondly, stress-induced amorphization is calculated by MDS with a simple potential function derived from modified EAM (MEAM) framework. Relation between deformation (elastic and/or plastic) and structural changes into amorphous phase is characterized by the method of common neighbor analysis (CNA).It is found that the amorphization caused by shear deformation is well enhanced especially when slip motion is restricted due to the choice of shear direction. The obtained amorphous is found energetically meta-stable. Besides, it is found that martensite is observed around the amorphous region.

1. Introduction

In the recent years, it has been discovered that Ni-Ti alloys manufactured under extraordinary conditions show various unique functions. For example, by sputtering technique, Ni-Ti alloys become thin films which can be used for micro actuators[1-3]. By peculiar intense plastic forming as HPT (high-pressure torsion), the alloys show superior mechanical properties[4][5].

Usually, these unique properties are researched by a lot of experimental methods. However, computational methods have abilities to analyze unknown micro phenomena, i.e. nano-scale materials' dynamics.

There are several studies on phase transformation of Ni-Ti alloys or shape memory alloys by molecular dynamics simulation(MDS). For example, for Ni-Al alloy, being utilized as another kind of shape memory alloy like Ni-Ti alloy, [6] phase transformation is analyzed by MDS. In Ni-Ti alloy itself, MDS of phase transformation using Lennard-Jones potential [7,8] is investigated by focusing on relation between elastic constant and phase transformation. These researches have unique results because their martensite unitcell is in $L1_0$ structure (fcc system) not in B19'. Besides, for radiation-induced amorphization in Cu-Ti system, MDS is performed[9].

In this study, molecular dynamics simulation (MDS) is conducted for some specific models. First, we construct a model of nano-sized Ni and Ti system including disc-shape precipitates. This model reveals a nucleation behavior of martensite which leads to propagation of variant structures.

Secondly, amorphization of Ni-Ti alloys under shear stress is simulated. It is found that the amorphization is a kind of stacking fault. Slipping and rotation of crystalline occur and they induce the amorphization. Around the amorphous structure, there exist some martensite regions because of local stress and strain.

2. Theory and Procedure

2-1. Interatomic potentials for MDS

For MDS of the Ni-Ti system to reproduce the precipitates models, EAM potentials developed by Lai et al.[10] are adopted. The total energy of the system is given by

$$E = \sum_i \left[\sum_{j \neq i} A_{\alpha\beta} \exp\left\{ -P_{\alpha\beta} \left(\frac{r_{ij}}{d_{\alpha\beta}} - 1 \right) \right\} - \sqrt{ \sum_{j \neq i} \xi_{\alpha\beta}^2 \exp\left\{ -2q_{\alpha\beta} \left(\frac{r_{ij}}{d_{\alpha\beta}} - 1 \right) \right\} } \right] \tag{1}$$

Table 1. Potential parameters

i-j	Ni-Ni	Ti-Ti	Ni-Ti
Ec[eV/atom]	4.44	4.85	5.02
d[Å]	2.49	2.95	2.607
A [eV/atom]	0.104	0.153	0.3
P	11.198	9.253	7.9
ξ [eV/atom]	1.591	1.879	2.48
Q	2.413	2.513	3.002

In this EAM expression, α and β indicate atomic species, r_{ij} is the distance between atoms i and j while the other parameters are listed in Table 1. These potential parameters are derived from experimental data or first-principles calculations, such as cohesive energy (E_c), elastic constants, nearest-neighbor distances and lattice parameters.

For amorphization model, modified EAM (MEAM) potential[11-17] is referred. The basic form of atomic energy in MEAM is as bellow.

$$E_i = \left[F_i(\bar{\rho}_i / Z_i) + \frac{1}{2} \sum_{j(\neq i)} \phi(R_{ij}) \right] \qquad (2)$$

In the expression, parameters are included for embedding function F, electrical density ρ, the number of neighbor atoms Z, and interatomic distance R. Especially, details of MEAM parameters for Ni-Ni bonding and Ti-Ti bonding are well researched by Lee et al.[15] and Kim et al.[16] and their parameters are available. On the other hand, Ni-Ti bonding is not general at present and it has to be derived by experimental data.

2.2. Simulation Models

At first, MDS model of uniaxial tensile loading is prepared as shown in Figure1., in which 19,404 atoms are simulated. In the center of the simulation cell on the plane parallel to the xy plane, the plate precipitate is introduced. The interatomic distance in z direction within this precipitate is chosen so that it guarantees a lower energy compared to the periodic model defined in the previous study[18]. The size of this model is 4.857 nm × 10.14 nm × 5.010 nm ($x \times y \times z$). The size of disc precipitate is chosen as φ 3.0 nm (on xz plane) ×0.81 nm(in y direction). The z direction corresponds to the [0 0 1] direction of the B2 structure. For the tensile simulation, atoms at the end region in y direction are constrained as shown in Figure1. As an initial structure, the B2 structure at 0 K is chosen in this simulation. Most of the atoms in the xy and yz surfaces are forming free surfaces. Before performing the tensile simulation, the initial structure has been sufficiently relaxed without constraining atoms at the end-region. In this process, in order to avoid inessential lattice vibrations at finite temperature, the temperature of the system is always brought back to 0 K by using the velocity-scaling scheme.

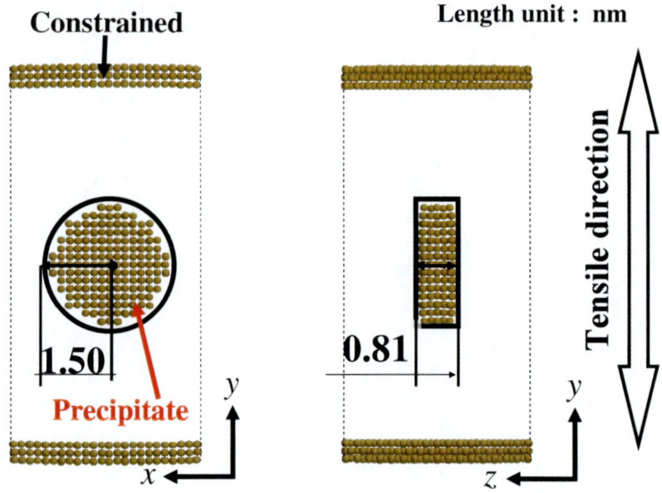

Figure 1. Simulation model with disc precipitate.

On the other hand, the temperature is not controlled in the simulation of tensile stress because velocity-scaling is artificial and is not suitable for observing the genuine atomic dynamics in the phase transformation. In the uniaxial tensile loading, a tensile speed corresponding to the strain rate (= 5.92×10^{8} 1/s) is added to the constrained atoms.

Other MDS models are constructed as well. Figure2 shows one of the models for studying stress-induced amorphization. There are three models: model (100), model (110) and model (111).

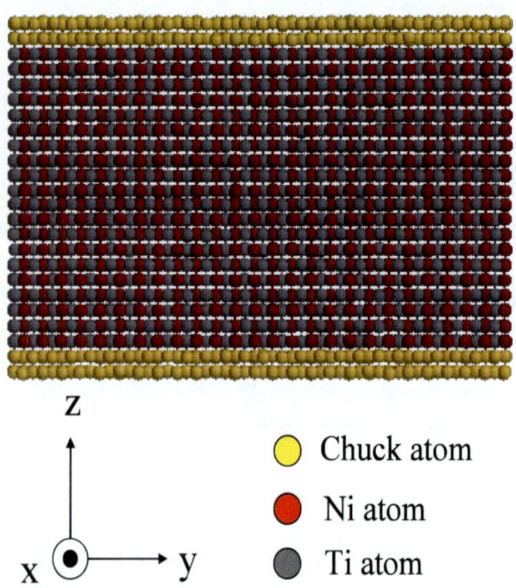

Figure 2. Simulation model for amorphization.

Table 2. Size of amorphous models

	Cell x [nm]	Cell y [nm]	Cell z [nm]
(111)model	6.81	11.80	8.34
(110)model	6.81	12.04	8.51
(100)model	6.62	12.04	9.03

Each of the three models has a specific relation between shear direction and slip plane. The size of models is shown in Table 2. These models are imposed periodic boundary conditions only in x and y direction. After enough structural relaxation, shear loading is applied by moving the chucking regions indicated in Figure2 (strain rate = 1.00×10^{10} 1/s). In order to avoid the burst of phase transformation, the temperature of the system is kept at 80K by using the velocity-scaling scheme. After all shearing processes, relaxation is performed for 40 ps at 80 K.

2.3. Method for Detecting the Martensite and Amorphous Structures

Figure 3 shows the schematic unitcells of B2 structure and B19' structure in Ni-Ti system and relation between these two unitcells are also shown. The present B19' unitcell is consisted of four pieces of B2 unitcells. Each B2 unitcell is cut on {110} plane. As a result, the B19' unitcell has four facets of {110} plane as well as two facets of {100} plane. Lattice parameters a, b, c and angle γ are indicated in Figure3.

From the instantaneous coordinates of atoms in the simulation, we can identify the part of martensite [19]. For Ni–Ti alloys, the martensite structure exhibits B19' structure in the unitcell. In our scheme, the lattice lengths of the B19' unit cell (a, b, c) and the angle γ between a and b are searched for from positions of neighbor atoms which are forming triplets. By setting criteria for these lengths and angle with certain ranges, the B19' unit cell is found. Conditions for the search are: 0.286 nm $\leq a \leq$ 0.296 nm, 0.440 nm $\leq b \leq$ 0.470 nm, 1.50 $\leq a / b \leq$ 1.62 and 93.0 degrees $\leq \gamma \leq$ 98.0 degrees.

On the other hand, conditions for the martensitic research are 0.279 nm $\leq a \leq$ 0.296 nm, 0.450 nm $\leq b \leq$ 0.476 nm, 1.51$\leq b/a \leq$1.66, 96.3 degrees $\leq \gamma \leq$ 98.6 degrees, which are used in the amorphous model.

For distinguishing martensite and amorphous regions in the amorphous model, another range for detection of maretensite is assumed in the amorphous model. This is because MEAM potential and EAM potential for the amorphous model have their own stable structures in both B2 and B19' unitcells.

Thus, B19' unit cells, emerging in an instantaneous structure, can be identified during the MDS. Interrelation between deformation (elastic and/or plastic) and structural changes into amorphous phase is characterized by the method of conventional common neighbor analysis (CNA)[20][21]. The CNA method specifies four indices for one atom, seeing the relation between neighbor atoms with regard to their atomic bondings.

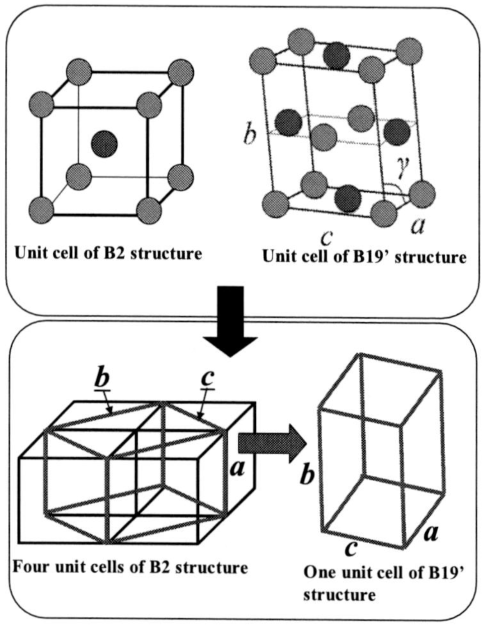

Figure 3. Schematic of the relation between B2 and B19' structures.

3. Results and Discussion

3.1. Precipitates Model

Figure 4 shows changes of martensite ratio ($R_{\rm m}$) which is the ratio of triplets detected as martensite to all the triplets in the system. The ratio increases drastically at the strain range from 0.085 to 0.090. Here, we would like to focus on the phase transformation around the precipitates. To do so, the change of the martensite ratio with regard to nucleation and propagation is analyzed. In Figure5, increase of R_m drastically changes at a couple of points of applied strains. Generally, this tendency is not observed in a model without precipitates. At the first point of increase, martensite starts to nucleate near the interface between precipitates and parent matrix structure of B2.

Figure6. shows the atomic configuration of martensite part found at strain 0.085. The nucleation of martensite is likely to have two pathways (x and y) to propagate.

A feature that martensite region has two directions of growing (pathways for propagation), especially near interfaces between parent and martensite, is effectively analyzed by an angle φ defined as a new parameter described in the following part and shown in Figure7.

The angle φ is measured as an angle between the orientation of unitcell and the tensile direction. That is, if the [0 0 1] direction of unitcell is heading to y direction (tensile direction), the angle φ becomes zero.

In the previous study [9], spontaneous division into multiple pathways of phase transformation in unitcells from B2 to B19' is clarified.

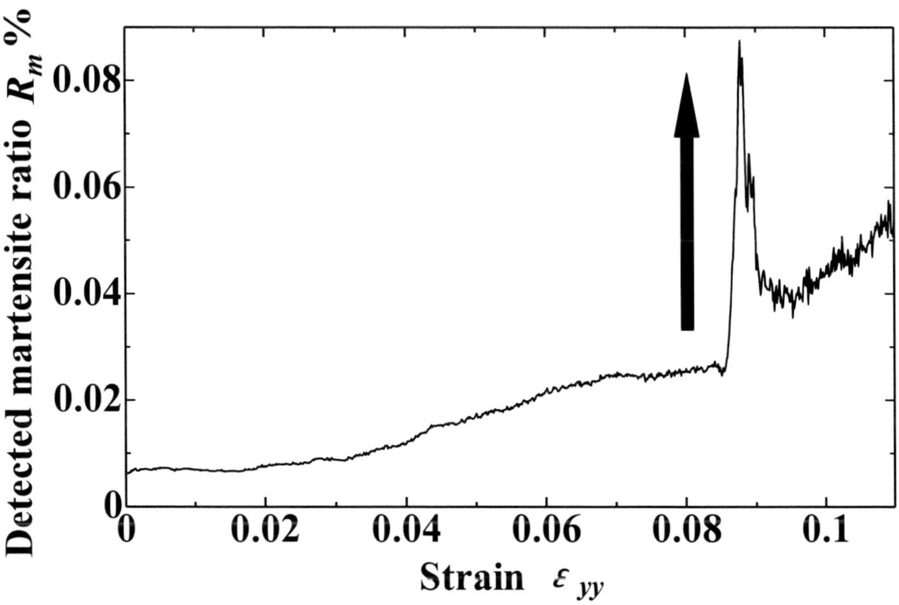

Figure 4. Changes of detected martensite ratio (The strain range is for whole the tensile simulation).

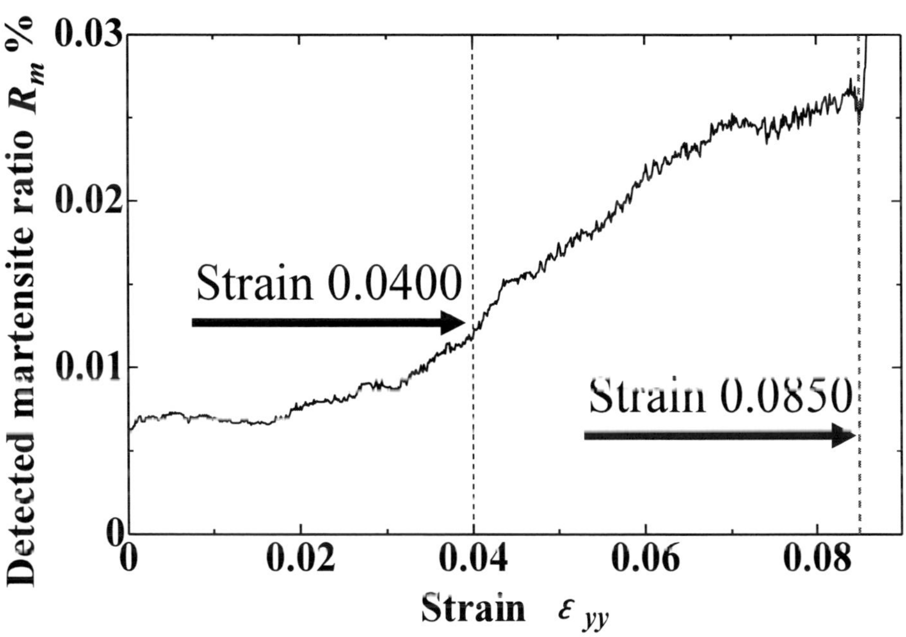

Figure 5. Changes of detected martensite ratio (Vertical lines show strain of drastic changes. The strain range is for a part of the tensile simulation).

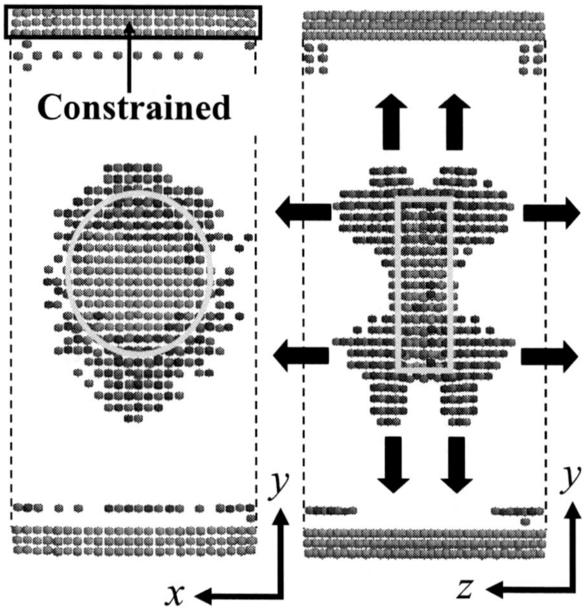

Figure 6. Martensite structure appearing around the precipitate at strain 0.085.

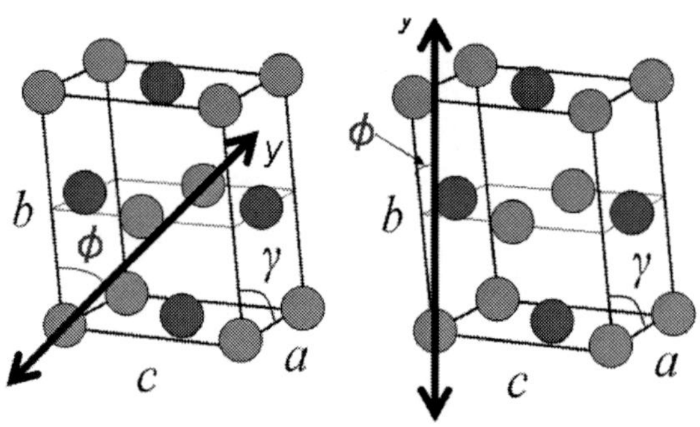

Figure 7. Schematic drawings showing the relation between the tensile direction (arrows: y direction) and the orientation of B19' (e.g. edges with length b).

There are mainly two pathways. During loading, one group of unitcells shows that angle γ increases linearly as strain increases (categorized as G1), while another one shows that lattice lengths increase linearly (categorized as G2). At the point where B19' unitcell begins to be detected, lattice lengths change instantaneously in the group G1, while angle γ changes instantaneously in G2. Since the angle φ introduced here relates the orientation of unitcells to the direction of applied stress, the group G2, in which the lengths change gradually, will be further categorized by the angle φ.

In G2, γ and φ has a certain correlation, where especially $\varphi$is lower than 10 degrees. So, G2 is re-divided by using the angle φ. The result will show whether particular orientation of martensite unitcells is observed or not.

We try to compare change of φ in each unitcell with its own γ (bond angle). Figure8. shows the relation between φ and γ, where Figure8(a) shows only those with decreasing φ with increasing strain and Figure8(b) does only those with increasing φ. Here, they are called φ_{in} and φ_{de} , respectively. At no strain, both φ_{in} and φ_{de} are almost proportional to the bonding angle γ. On the other hand, at strain 0.085, γ is scattered over the range from 93 degrees to 98 degrees, with relatively low φ value.

Figure9 shows changes of lattice lengths and angles in unitcells. Fig9(a)(b) are those for φ_{de} and φ_{in}, respectively. During tensile loading, lattice lengths increase linearly and then the angleγ changes instantaneously on a vertical dot line, just when B19' unitcells are detected, as shown in both Fig9(a) and Figure9(b). By comparing φ_{in} and φ_{de} in Figure9, their equilibrium angles after the onset of B19' unitcells are slightly different. Although they are both categorized as the G2 group with regards to unitcell shape, it is guessed that two pathways with regards to another parameter angle φ exist.

Figure11 shows the relation between changes of φ_{in} and φ_{de} and γ during the loading simulation. The plots in Figure11 are obtained by calculating each average of the plots in Figure8. In this figure, the leftmost plot means that obtained at no strain, and the rightmost one means that at strain 0.085. The vertical broken line in the figure indicates γ=93 degrees which has been one of criteria in our detecting method. That is, martensite is detected when γ is beyond 93 degrees. At no strain, averaged value of γ for φ_{in} and φ_{de} are estimated at about 91.7 degrees and 92.0 degrees, respectively. Up to the critical angle γ=93 degrees, φ_{in} increases proportionally with the bonding angle γ. However φ_{de} decreases in inverse proportional manner.

Figure12 shows schematic of unitcells viewing on *y-z* plane. For the unitcell with φ_{de}, φ decreases, which means that the orientation of lattice edge *b* and tensile direction tends to be parallel each other. In this case, next martensite would be formed from the interface of {110} plane in B2, because another stacking on {100} plane which is parallel to edge *b* would require much atomic movement. On the other hand, for the unitcells with φ_{in}, φ increases, which means that the direction of the edge *b* tends to be inclined. In this case, it is easy for the next martensite to stack up on {100} plane of B2 structure (see in Figure3). The forming of martensite can relax the local strain. In other words, there appears self-adjusting mechanism of variant structures.

3.2. Amorphous Model

Figure 13 shows the changes of amorphous ratio, martensite ratio and bcc ratio in three models (their bottom planes correspond to (100), (110) and (111) planes of B2 structure, respectively) subjected to applied shear strain. Each ratio is detected basically by using the CNA method.

In the model (100), when bcc ratio decreases, martensite ratio increases. After that event, amorphous ratio begins to increase with bcc ratio recovering.

In the model (110), both increase of martensite ratio and decrease of bcc ratio occur at almost the same value of shear strain. At this point, amorphous ratio has accumulated only a little.

a)

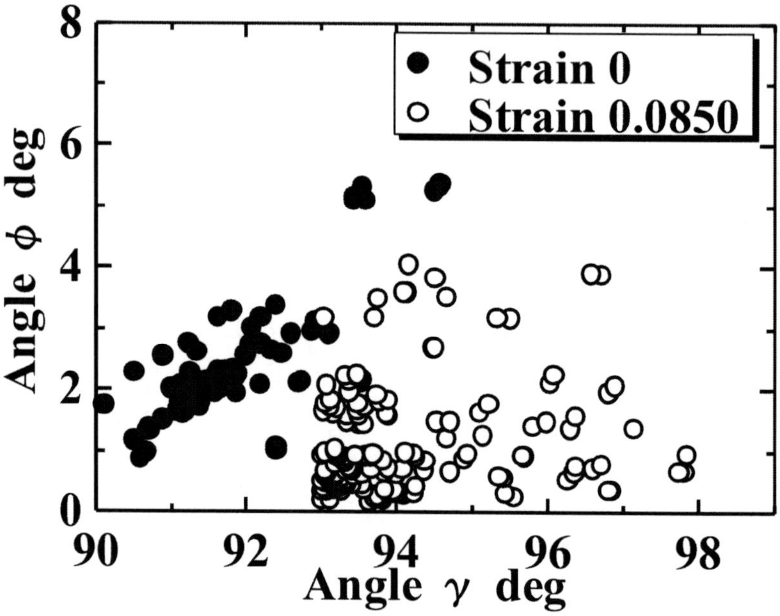

b)

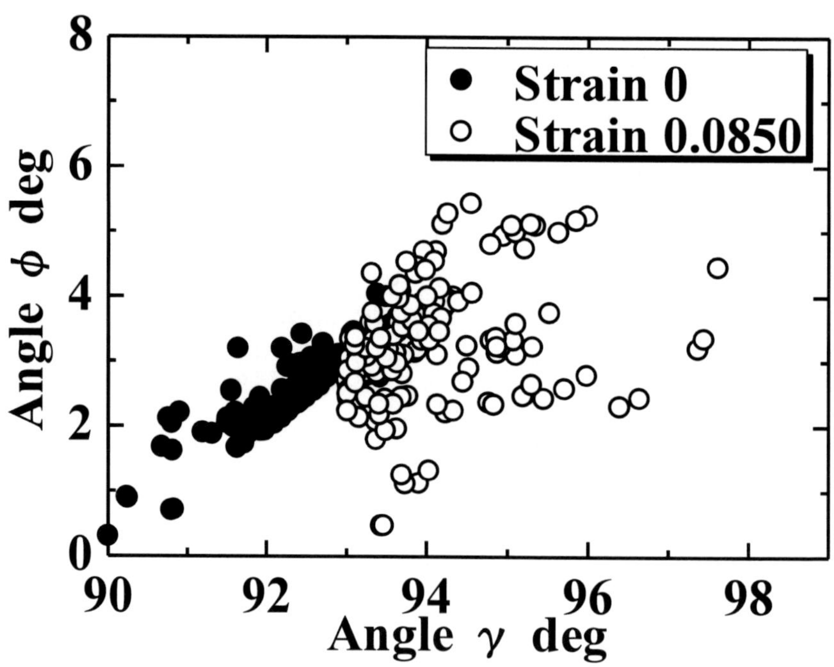

Figure 8. Relation between bond angle γ andφ ((a)φ de, (b)φ in. The definition of φ and the way of categorization is shown in the text.).

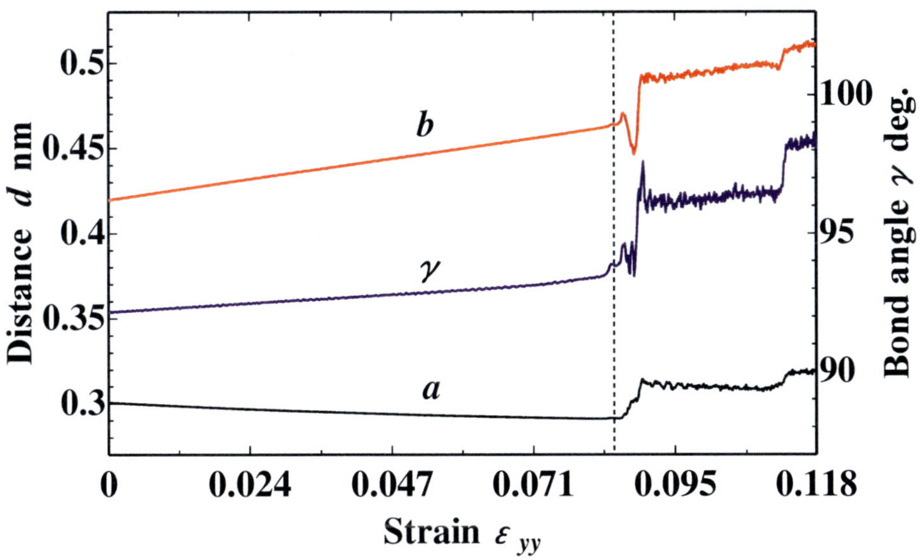

(b) ϕ_{in}

Figure 9. Changes of lattice lengths and bond angle during loading.

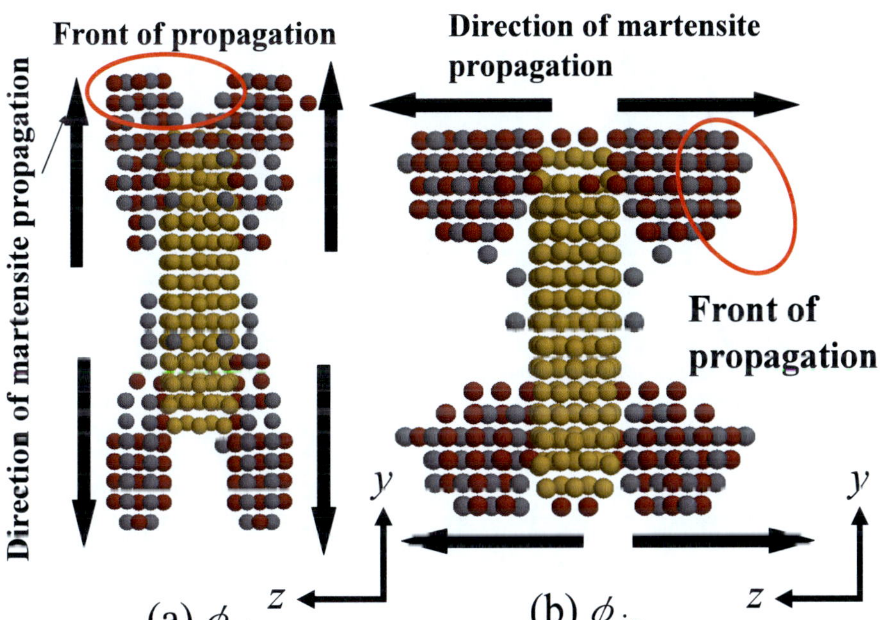

(a) ϕ_{de} (b) ϕ_{in}

Figure 10. Martensite divided into two groups around the precipitates ((a) forφ de, (b) for φ in).

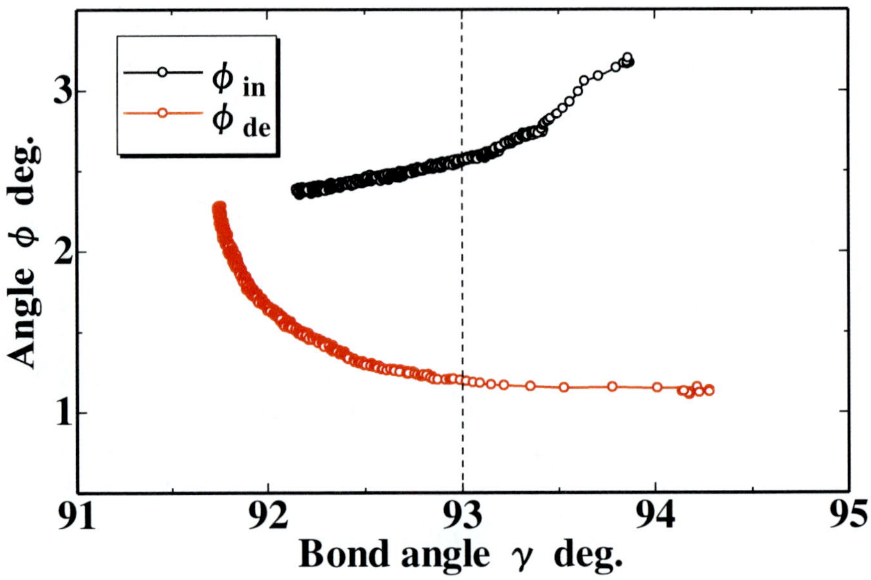

Figure 11. Changes of φ with regard to bond angle γ (each plot shows the average value).

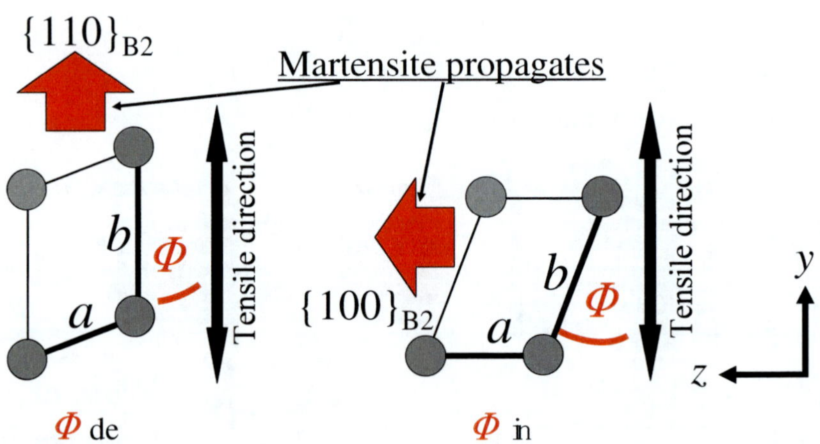

Figure 12. Explanation of the relation between orientation of martensite lattice and the defined angleφ .

In the model (111), when bcc ratio decreases, martensite ratio increases. After that, these two ratios together display minimum at shear strain around 0.25, then, with some delay, amorphous ratio shows maximal peak. At shear strain of 0.3, three ratios all start to increase.

When bcc ratio decreases, indices of CNA showing local atomic bondings enormously change. These changes mean nucleation of stacking fault which is closely related to dislocation slip. These changes also mean that amorphous is detected. From these results, it is confirmed that slip after decreasing bcc ratio causes the amorphous ratio to increase.

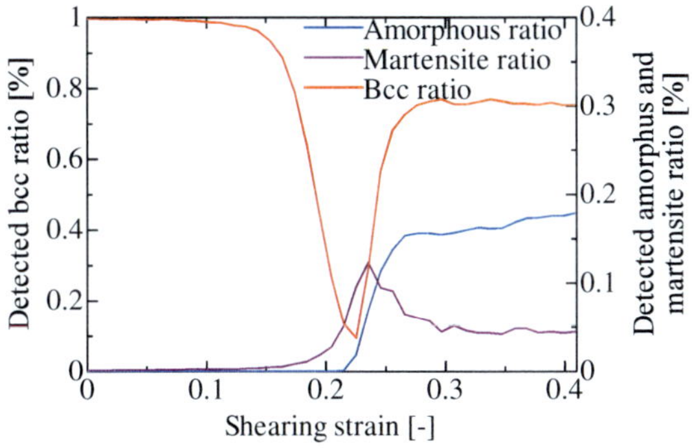

(a) model(100)

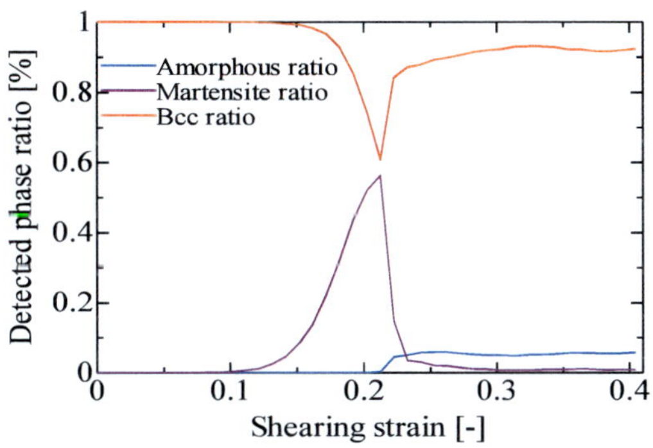

(b) model(110)

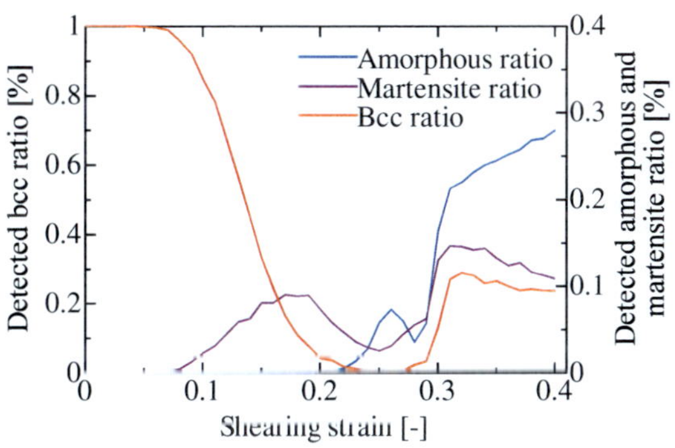

(c) model(111)

Figure 13. Changes of the amount of crystalline categories during shear simulation.

Generally, it is hard to slip in model (111) because slip system in this model prohibits slipping in the present shear direction. Therefore, a part of crystals may rotate under plastic deformation during decreasing bcc ratio.

As a result of difficulty for slip deformation, the (111) model shows more nucleation of amorphous than other two models and it exhibits much amorphous.

Figure14 shows changes of crystalline categories (bcc, martensite or amorphous) monitored in the structural relaxation.

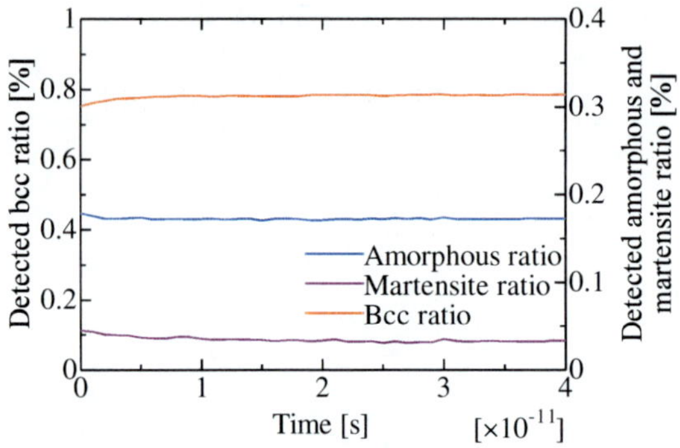

(a) model(100)

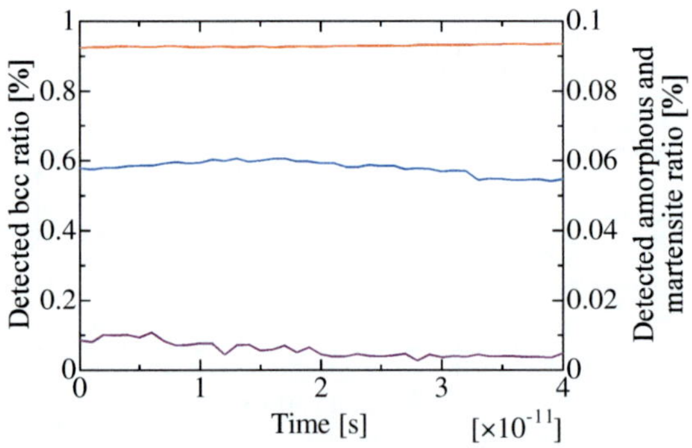

(b) model(110)

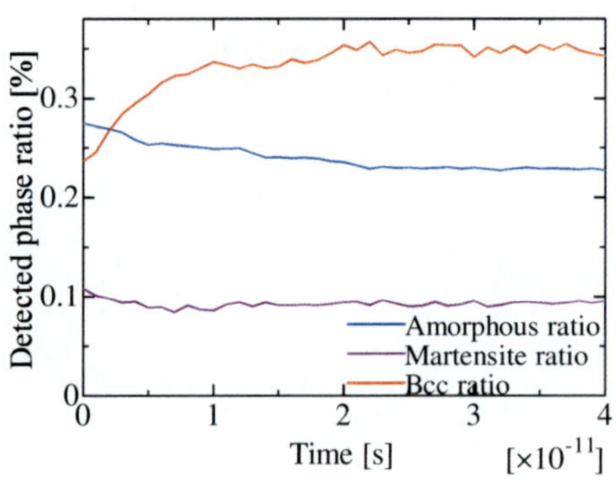

(c) model(111)

Figure 14. Changes of the amount of crystalline categories during relaxation simulation.

The amount of each structure does not change in both (100) and (110) models. However, their composition changes in (111) model so as to increase amorphous structure. Together with this phase transformation, some remained crystalline orientations are likely to rotate.

In spite of structural change, potential energy remains the same in the structural relaxation of all three models as shown in Figure15. It can be judged that produced amorphous structures are, at least energetically, meta-stable. During structural relaxation after shear loading, some groups of amorphous are detected around martensite.

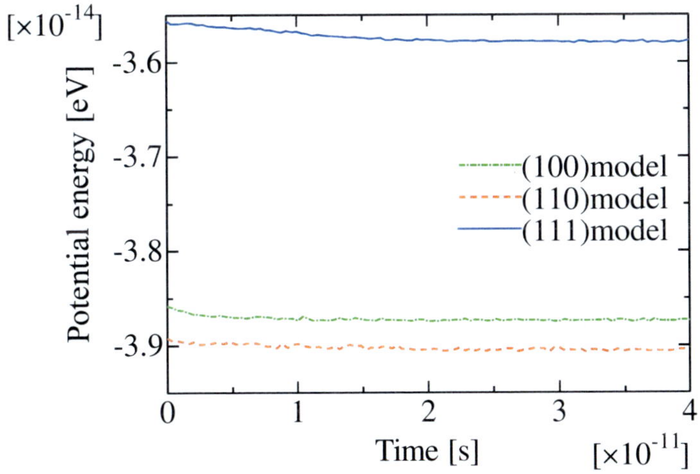

Figure 15. Changes of system energy under relaxation simulation.

Positions of amorphous and martensite are almost the same. Local stress or strain around the interface between amorphous and parent crystal is likely to cause martensite structure as an intermediate structure, the situation of which is alike in the model of the precipitates shown in section 3.1 above.

4. Conclusion

In order to study the microscopic dynamic behavior of Ni-Ti alloys under martensite and amorphous transformation, molecular dynamics simulation (MDS) is performed.

One of the models includes disc precipitates and is applied uniaxial tensile loading. The resulted martensite structure and etc. are investigated. Results are as follows:

- In the Ni-Ti MDS model including the disc precipitates, martensite phase arises around the interface. This is because local strain between Ni-Ti parent matrix and precipitates can be structurally balanced and then the whole model has no strain.
- The Ni-Ti model including the disc precipitates shows that the martensite around the precipitates are categorized by two growing directions (pathways). This feature strongly depends on the geometrical relationship between the orientation of unitcell and the tensile direction.

Other models are to study amorphization of Ni-Ti induced by shear deformation. For these models, results are summed up as follows.

- Amorphization occurs by shear deformation. Not only slip motion inside crystals, but also the rotation of them is observed. Especially, under the condition where slip is hard to be activated, nucleation of amorphous is enhanced.

- Energy-level of amorphous is determined as meta-stable. Local stress or strain around the interface between amorphous and parent crystal is likely to cause martensite.

REFERENCES

[1] Y.Fu.; D.Hejun.; *J.Trans. Mater. Res. Soc.* , 2004, *29*, 2993-2996.

[2] T.Kikuchi.;K.Ogawa.;S.Kajiwara.;T.Matsunaga.;S.Miyazaki.;Y.Tomota.; *Phil. Magn. A.*,1998,*78*,467-489.

[3] S.Kajiwara.; T.Kikuchi.; K.Ogawa.; T.Matsunaga.; S.Miyazaki.; *Phil. Magn. Lett.*,1996,*74*,137-144.

[4] M. Peterlechner.;T. Waitz.;H.P. Karnthaler.; *Scr. Mater.*, 2008,*59*, 566-569.

[5] T. Waitz.; W. Pranger.; T. Antretter.; F.D. Fischer.; H.P. Karnthaler.; *Mater. Sci. Eng. A.*, 2008, *481-482*, 479-483.

[6] S.Rubini.; P.Ballone.;*Phys. Rev. B*,1993,*48*-1,99-105.

[7] T.Suzuki.;M.Shimono.;X.Ren.;M.Wuttig.;*J. Alloys. Comp.*,2003,*355*,183-187.

[8] T.Suzuki.;M.Shimono.;*J. Phys. IV France*, 2003, *112*, 129-132.

[9] M.J.Sabochik.;N.Q.Lam.;*Phys. Rev. B*, 1991, *43*-7, 5243-5252.

[10] W.S.Lai.;B.X.Liu.;*J.Phys.Condens.Matter.*,2000,*12*,L53-L60.

[11] M.I.Baskes.;*Phys. Rev. B*, 1992, *46*,2727-2742.

[12] M.I.Baskes.;S.P.Chen.;F.J.Cherne.;Phys.Rev.B,2002,*66*,104107 1-9.

[13] B.J.Lee.;B.D.Wirth.;J.H.Shim.;J.K.Sang.;C.Kwon.;J.H.Hong.;*Phys.Rev.B*, 2005,*71* , 184205 1-15.

[14] M.I.Baskes.;S.G.Srinivasan.;S.M.Valone.;R.G.Hoagland.;*Phys.Rev.B*, 2007,*75* ,094113 1-16.

[15] B.J.Lee.;J.H.Shim.;M.I.Baskes.;*Phys.Rev.B*,2003,*68*,144112 1-11.

[16] Y.M.Kim.;B.J.Lee.;M.I.Baskes.;*Phys.Rev.B*,2006,*74*,014101 1-12.

[17] B.J.Lee.;M.I.Baskes.;H.Kim.;Y.K.Cho.;*Phy. Rev.B*,2001,*64*,184102 1-10.

[18] T.Sato.;K.Saitoh.;N.Shinke.; *Mater. Sci. Eng. A.*, 2008, *481-482*, 250-253.

[19] T.Sato.;K.Saitoh.;N.Shinke.;*Modell.Simul.Mater.Sci.Eng.*,2006,*14*,S39-S46.

[20] J.D. Honeycutt.; H.C. Andersen.; *J. Phys. Chem.*,1987,*91*, 4950–4963.

[21] H.Tsuzuki.; P.S.Branicio.;J.P.Rino.; *Comput.Phys.Commun.*, 2007,*177*, 518-523.

In: Shape Memory Alloys
Editor: H. R. Chen, pp. 247-265
ISBN 978-1-60741-789-7

Chapter 8

EVOLUTION OF THE STRESS-INDUCED MARTENSITIC TRANSFORMATION IN A CUALBE SHAPE MEMORY ALLOY UNDER UNIAXIAL TENSION

***F. M. Sánchez-Arévalo*[1], *G. Pulos*[2] *and R. Pérez-Cabeza de Vaca*[3]**

1 CCADET – Universidad Nacional Autónoma de México, A.P. 70-186, México D.F., C.P. 04510

2 Instituto de Investigaciones en Materiales, Universidad Nacional Autónoma de México, Apdo. Postal 70-186, Cd. Universitaria, México, D. F. C.P. 04510

3 Facultad de Ingeniería, Universidad Nacional Autónoma de México Apdo. Postal 70-186, Cd. Universitaria, México, D. F. C.P. 04510

ABSTRACT

The evolution of the stress-induced martensitic transformation was determined for a Cu-Al 11.2 wt%-Be 0.6 wt% shape memory alloy by digital image analysis. The CuAlBe tensile specimens were subjected to uniaxial tension; during the tension test, digital images of the polished and chemically etched surface's specimens were taken. From the tension test, the stress-strain curve representing the macromechanical behavior of the material was obtained. From the acquired images, the stress-induced martensitic transformation was observed and the number of grains, the area of the grains and the martensitic plates in the grains were estimated. Additionally, the orientation of the martensitic plates was determined for each image associated to the stress-strain curve.

Image analysis showed that, during the tension test, the martensitic variants appeared and disappeared due to the interaction between a grain and its neighbors, which is caused by the stress state of the grain, the original variant or variants, and the state of transformation of the neighboring grains. With the results from the micromechanical behavior, the strain measured with the strain gage could be resolved into its elastic and transformational components.

INTRODUCTION

Shape memory alloys (SMA's) are materials that can remember their original shape, after being elastically or pseudo-plastically deformed, by increasing their temperature. This behavior is useful in replacing conventional materials and develop new applications in science and daily life. The superelastic effect is one of the reasons there has been a continuous effort to understand, predict and explode the shape memory behavior of these materials.

The change of shape in shape memory alloys is due to a martensitic transformation. The martensitic transformation is defined as a first-order displacive process, where a body center cubic parent phase (austenitic phase) transforms by a shearing mechanism into a monoclinic or orthorhombic martensitic phase [1]. With this change of phase in the solid state other associated effects are present like the single shape memory effect, the double shape memory effect and the superelastic effect. All these effects are well known and they have been reported in literature [2-6].

The martensitic transformation has been observed in several classes of materials like metals, ceramics, polymers and in a biological structure [7]. Although the shape memory effect is present in almost all classes of materials, the metallic alloys with shape memory effects are preferred for developing new applications to replace conventional materials.

In this chapter, the behavior of a CuAlBe shape memory alloy under uniaxial tension is presented; in addition, a brief historical review of the research with the CuAlBe system is mentioned. The CuAlBe system appeared in 1982 and the first studies revealed that the system presented shape memory effects [8-10]. Since then, theoretical and experimental studies of the crystallography, microstructure, phase stability, thermomechanical and mechanical behavior as well as the micromechanical behavior have been carried out.

In 1997, Jurado *et al.*, worked on the order-disorder phase transition of the CuAlBe system [11]. Also in 1997, Hautcoeur *et al.* [12] determined the stress-temperature diagram for the CuAlBe system; in this diagram, they show the stability regions for the different types of martensite and determined the slope of transformation line. In 2000, Siredey and Eberhardt [13] studied the fatigue behavior of CuAlBe and showed that cyclic loading reduces the critical stress considerably. Balo, Ceylan and Aksoy studied, in 2001, the effect of deformation on the microstructure of CuAlBe [14, 15]; they reported that the addition of small quantities of Beryllium does not change the DO3 super-cell of the austenitic phase nor the martensitic 18R phase.

Trying to understand the mechanical behavior of SMA, Bouvet, Calloch and Lexcellent carried out a series of experiments using uniaxial and biaxial loading and determined the critical stress for two different compositions [16]; they used digital image correlation to check the boundary conditions for biaxial compression [17]. In 2004, Kaouache *et al.* studied the martensitic transformation in the superelastic regime for both monocrystal and polycrystal CuAlBe [18]; later on, in 2006 [19], they measured the crystalline orientation of three grains, determined the 24 variants present in CuAlBe and proposed a transformation criterion.

While in 1999, Siridey *et al.* proposed constitutive equations for SMAs at the grain level, no one had determined experimental strain fields in SMAs at the grain level. In 2006, Sanchez and Pulos [21] studied the stress-induced martensitic transformation and obtained displacement vector fields; recently, in 2008 [22], they obtained strain measurements averaged over several grains. Although several studies have been carried out to understand the CuAlBe system, nobody had studied the evolution of the stress-induced martensitic transformation in a 2-D confined CuAlBe shape memory alloy using digital image analysis until now.

The objective of this chapter is to study the evolution of the stress-induced martensitic transformation in a CuAlBe shape memory alloy undergoing uniaxial tension. The grain size is comparable to the thickness of the sample so that the measured behavior on the surface is representative of the behavior throughout the thickness of the sample; on average, one grain spans the thickness of the sample and thus confinement of the grains occurs in two dimensions only. The behavior of the CuAlBe shape memory alloy is studied at the macro and micromechanical level. To understand the macromechanical behavior of CuAlBe, a uniaxial tension test was carried out using an MTS load frame; additionally, an optical microscope was attached to the load frame to reveal the micromechanical behavior of the material. The complete details are explained in the next section.

EXPERIMENT

To study the supereslastic effect, an austenitic alloy was elaborated. The composition was Cu-Al 11.2 wt%-Be 0.6 wt%, which is close to the eutectoidal composition [23]. An induction furnace (Leybold-Heraeus) was employed to elaborate the alloy by a melting process; this furnace has a controlled atmosphere and in our case argon gas was used. From the castings, suitable slices were cut and then hot rolled to obtain thin sheets. The length, width and thickness of the sheets were 280, 57 and 0.8 millimeters, respectively. These dimensions were reached after a 191 % hot-rolled (800 °C) deformation process. The hot-rolled process was carried out in an oven (Sola Basic-Lindberg model 847) and a roll machine (Fenn Amca International). Subsequently, the sheets were subjected to a heat treatment, called betatization, to reveal the shape memory effects; the sheets were heated at 750 °C during 15 minutes and then water-quenched to 95 °C during 20 minutes [21-25]. To determine the transformation temperatures, a differential calorimeter (TA Instrument model DSC 2910; USA) was used. This measurement was the first evidence that the shape memory effect in the CuAlBe alloy used in this study was present. From the sheets the uniaxial tension samples were cut in a numerically-controlled machine center; tensile specimens were obtained having a width of 3 mm, a thickness of 0.7 mm and an equivalent length[1] of 25.88 mm. The sample was polished and chemically etched with Ferric chloride ($FeCl_3$) to reveal the grains on the material's surface. The polished and chemically etched samples had a final thickness of 0.68 mm and were subsequently tested under uniaxial tension.

Tensile tests were carried out on a servohydraulic loading device (MTS 858 MiniBionix axial). To characterize the micromechanical behavior, an optical microscope was adapted to a high resolution digital camera –Nikon D2X 4288 × 2848 pixels– as shown in figure 2. The modular microscope works as an infinity-corrected compound microscope with magnifications of 10X. To control the MiniBionix MTS a 407 MTS controller was used while data and images were acquired with a National Instruments PXI-1002 system –including a 6281, 8331 and 4220 PXI-boards– and a PC. A virtual instrument (VI) was programmed in LabVIEW in order to synchronize and save the data and images. The virtual instrument was able to save the

[1] Length of a rectangular strip with the same width and thickness that would exhibit the same force-displacement behavior as the tensile specimen [26].

following parameters: time, displacement, force, strain and digital images. With the displacement, strain and force data, the macroscopic stress-strain curve was obtained and the acquired images were tagged to points in the stress-strain curve. Hence the micro and macro-mechanical behavior was determined simultaneously.

Image analysis was accomplished using a matlab scrip. The scrip consists in labeling objects or grains per image, the calculation of grains areas, (x, y) centroids of grains, the number of martensitic plates by grain and their orientation with respect to the cross-section area. From these results it was possible to determine the variants –groups of martensitic plates with the same orientation– that were induced by the applied loading; in addition, regions where martensitic plate reorientation occur were determined. The martensitic reorientation regions were analyzed in order to obtain the displacement vector field, strain measurements and the influence that neighboring grains have on the phase transformation.

With the series of images acquired during the tension test, displacement vector fields were calculated from pairs of images for each illumination type. The Willert and Gharib algorithm [28] was used to calculate $u_k(x_k, y_k)$ and $v_k(x_k, y_k)$, where u and v represent the displacements of an analysis object in the x and y directions respectively [26, 27]. The x and y represent the position coordinates of the analysis object in every image; subindex k indicates the corresponding object, which is defined as a 256x256 pixels area. The in-plane strains were determined by minimizing the errors of a six parameters linear model (typically used in linear elasticity theory) given by:

$$u_k(x_k, y_k) = A_1x_k + B_1y_k + C_1 + \delta_u(x_k, y_k) \qquad \text{Eq. (1a)}$$

$$v_k(x_k, y_k) = A_2x_k + B_2y_k + C_2 + \delta_v(x_k, y_k) \qquad \text{Eq. (1b)}$$

The minimization returns values for the A's, B's and C's where $A_1=\varepsilon_x$ (strain in x direction), C_1= translation in x direction, $B_2= \varepsilon_y$ (strain in y direction), C_2= translation in y direction. B_1 and A_2 are linear combinations of shear strain and rotation where the shear strain $\varepsilon_{xy}= (A_2+B_1)/2$ and the rotation $\theta= (A_2-B_1)/2$; $\delta_u(x_k, y_k)$ and $\delta_v(x_k, y_k)$ represent the (small) errors to be minimized

From the displacement vector field specific regions were selected to get the in-plane strains corresponding to those grains were the reorientation process occurred. The same analysis was done with the neighboring grains to observe the influence of neighbors on the phase transformation. Through image analysis the micro-mechanical behavior was determined separately from its macro-mechanical counterpart.

RESULTS

The Cu-Al 11.2 wt%-Be 0.6 wt% shape memory alloy had the following transformation temperatures M_s= -95 °C, M_f=-103 °C, A_s=-91 °C and A_f=-76 °C, where M_s and M_f indicate the start and end of the direct transformation (austenite to martensite); A_s and A_f indicate the start and end of the inverse transformation (martensite to austenite).

The most important transformation temperature, for the superelastic effect (SE), is M_s because this temperature is related to the transformation (or critical) stress –usually denoted by

σ_c. The transformation stress denotes the point where the slope changes in the stress vs elongation ratio curve. The transformation stress and M_s are related by the Claussius Clapeyron equation as shown in equation 2 where T is the test temperature.

$$\sigma_c = 1.97MPa(T - M_s) \quad \text{Eq. (2)}$$

Taking into account an M_s= -95 °C and employing equation 2, the transformation stress is around 230 MPa for the alloy's composition.

From the uniaxial tension test the superelastic effect can be observed in the stress vs elongation ratio curve, as shown in Figure 1. This effect is present when a material, in austenitic phase, is stressed until the martensitic phase appears in the material. The stress-induced martensitic transformation causes a reduction in the slope in the stress vs elongation ratio curve and represents the point when martensitic plates begin to appear. The slope in the martensitic region was not equal to zero, as expected, since the sample is a polycrystal [5].

In this case, an elastic deformation close to 0.23 % could be accommodated before the phase change started and due to the superelastic effect the deformation could be increased to 0.58 % without very little plastic deformation[2]. In addition, this curve showed the slope change starting around 240 MPa; furthermore, the theoretical transformation stress calculated from equation 2 and using M_s temperature is very close to the critical stress obtained from the uniaxial tension test. Using the first five data points –labeled A through E– and using a linear least-squares fit, the elastic modulus was estimated to be 114 ± 1 GPa.

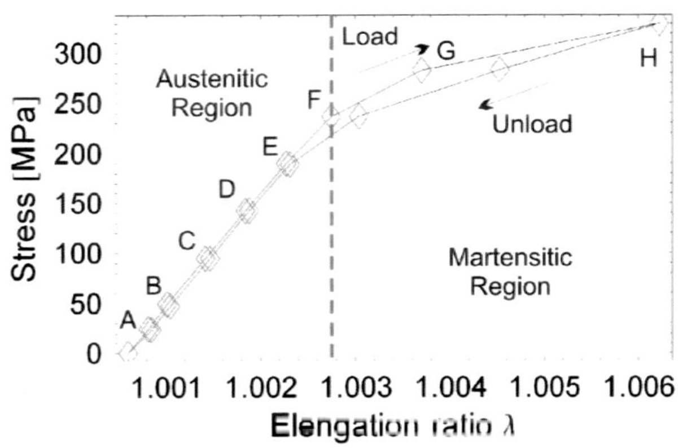

Figure 1. Stress vs elongation ratio curve.

The letters A to H in Figure 1, represent the state of stress and elongation ratio for each mosaic showed in Figure 2. The mosaics were obtained from an assembly of images that were acquired during the tensile test; each mosaic was constructed from six photographs and is 3890 x 12811 pixels. The mosaics in figure 2(a) to figure 2(h) show close to 100 grains in an area of 2.193 x 7.222 mm, which is the central part of the 3.01 x 14.00 mm test section; the area of the mosaic is roughly 38% of the test section –and considering an equivalent length of 25.9 mm, the mosaic represents 20.3% of the test sample. This number will later be used as a scaling factor to estimate the total number of plates in the test sample. The crystalline structure of the

[2] The accumulated plastic deformation can be estimated from the shift in the elastic return.

sample can be observed in these mosaics and the grain size could be determined; it was around 500 microns. As was mentioned earlier, from the grain size and the thickness of the sample, one can assume that the grains are confined only in the plane of the sample and that the measurements on the surface are representative of the through-the-thickness averages. Figures 2(a) to 2(e) show the sample in the austenitic phase. The austenitic phase presents a smooth (scratch-free) surface while the martensitic phase presents a roughness due to the growth of martensitic plates –which show up as dark lines or scratches due to the bright field illumination used– as it is shown in figures 2(f) to 2(h). From the mosaics, it is clear that the martensitic phase does not appear in all the grains in the region of study since the phase transformation depends on several factors such as stress level, crystalline orientation, test temperature, grain size and granular interaction, among others. Thus a more detailed and local analysis of a grain and its neighbors is needed to determine the micro-mechanics of the phase transformation.

As it was mentioned, the mosaics represent different snapshots of the phases present in the grains at different levels of stress in the sample. From figure 2(a) it is clear that the sample has the grains in the austenitic phase at 25 MPa. When the stress was increased, the martensitic phase began to appear at the top of the mosaics as shown the figures 2(b) trough 2(e) between 50 MPa to 190 MPa; nevertheless, the vast majority of grains have only the austenitic phase. From figure (2f) it is evident that the phase transformation has started since the martensitic plates show up in several grains.

In order to track the phase change and understand the micromechanical behavior, the grains were labeled as shown in Figure 3. This figure was used as a map of the transformation state during image analysis. As can be observed from Figure 3, the sample has 95 grains in the region of study, which is about the minimum number of grains needed to capture the macroscopic (averaged) mechanical behavior [29]; thus, the mosaic shows the minimum area required to do a representative study of the material.

A) 25 MPa B) 50 MPa C) 100 MPa D) 140 MPa

E) 190 MPa F) 240 MPa G) 280 MPa H) 330 MPa

Figure 2. Mosaics at different stress levels.

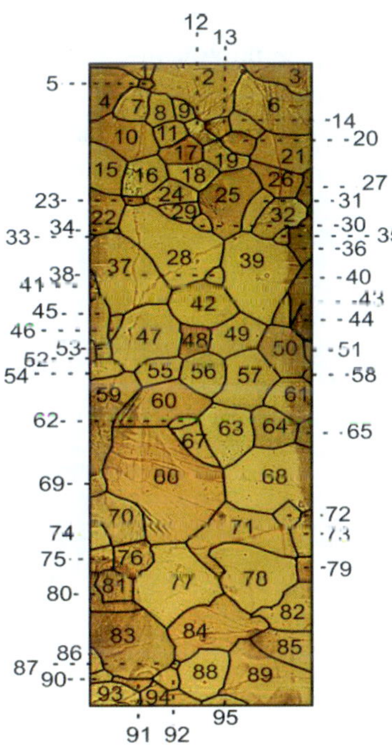

Figure 3. Labeled grains in the mosaics.

The labeling process of grains was done for all mosaics and the grain label was kept the same for each mosaic throughout the sequence. This analysis was able to reveal the transformational level of each grain. Figure 4 represents the evolution of the stress-induced martensitic transformation in each grain; the grain number, the number of martensitic plates and the stress level are represented in the x, y and z axis respectively. According to figure 3 the grain number goes from the top to the bottom of the mosaic; this grain number is plotted in the x axis in figure 4 and goes from left to right. From figure 4 it is also obvious that some grains at the top and bottom of the mosaic present martensitic plates at the initial stress level around 25 MPa; nevertheless, the central part of the mosaic does not present any grains with martensitic plates.

When the stress is increased, the number of martensitic plates increases including grains that had plates initially; the central part of the mosaic, that was free of the martensitic phase, begins to be covered with new plates.

The curves in red indicate that almost all grains are in austenitic phase and the curves in blue indicate that the previous relation has changed and now the majority of grains have undergone the change of phase (Austenite to Martensite). Although the number of plates increases as the stress is increased, there are cases in which plates disappear as the stress is increased. One such case is the grains 39-55 when the stress is increased from 50 to 100 MPa (last two orange curves). In this case, the peak –number of plates– at grain number 39 is reduced from three to two plates while the two plates in grain 47 disappear; in grain 55, the number of plates is reduced from four to three while the number of plates in grain 42 remains constant. There are other cases in which this happens but it serves as an example to point out the following: since the strain field must be compatible and a grain is surrounded by neighbors that may have different orientations and stiffness in the loading direction, the number of plates is not necessarily a monotonically increasing function of the stress. Plates may disappear or reappear in different orientations as the stress is increased even though, as it will be pointed out later, one can associate an increase in the transformational strain with each new plate. The increase in the transformational strain is the reason for the apparent plastic behavior –past the critical stress– of a shape memory alloy; the disappearance of the martensite plates, together with their associated reduction in transformational strain is the reason for superelastic unloading and recovery of the apparent plastic strains.

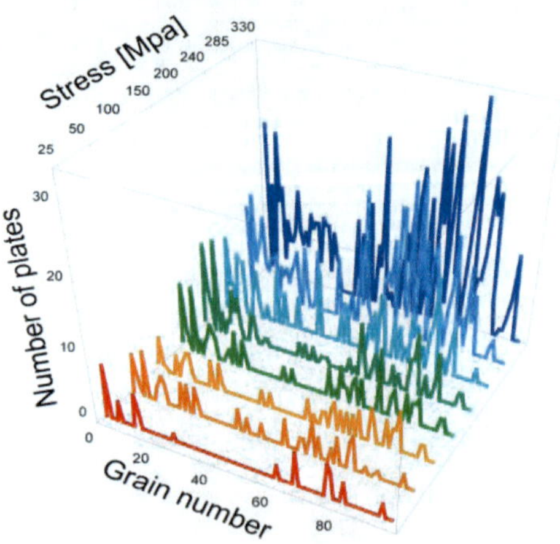

Figure 4. Evolution of the stress-induced martensitic transformation.

The stress vs. elongation ratio curve, presented in figure 1, shows a slope change due to the martensitic phase transformation. Now it is clear that there must be a representative percentage of the martensitic phase to observe that slope change. This change in figure 1 is also associated to the martensitic evolution curve marked in light blue in figure 4 –occurring at a stress level of 240 MPa. Thus observing, in the same figure, all the peaks in the curves at 240, 280 and 330 MPa, it could be assumed that the material presents a considerable percentage of the martensitic phase.

To quantify this percentage, all the grains with martensitic plates were taken into account from each mosaic and using the scaling factor that was defined earlier, the total number of plates in the sample can be estimated; in this case there were a maximum of 779 plates detected in the mosaic associated to 330 MPa. With a scaling factor of 1/0.203 = 4.93, the estimated maximum number of plates is 3840; the 41 plates detected in the mosaic at 25 MPa scale to an estimated 202 plates for the whole sample. The (total) number of plates as a function of stress is shown in figure 5.

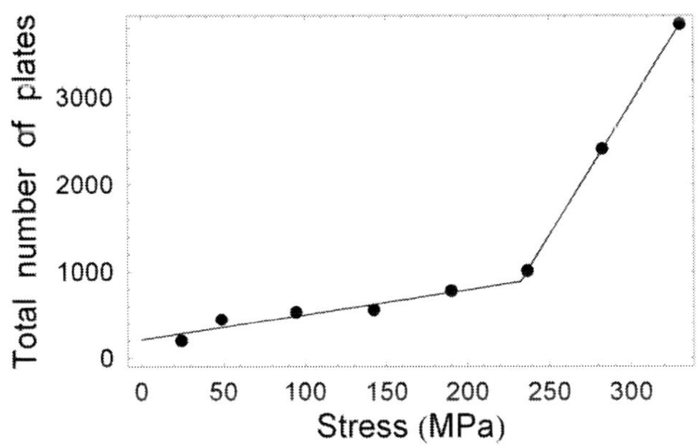

Figure 5. Total number of plates in the sample at different stress levels.

It should be noted that the (total) number of plates is a monotonically increasing function of stress –at least with the data that has been examined. Figure 5 shows an exponential growth of the martensitic plates where it is clear that the plates have grown suddenly, after 240 MPa, in different grains –and at different orientations. To characterize the number of plates as a function of stress, a (nonlinear) least squares fit of a hyperbola was used; initially, a five parameter model was adjusted but the parameter associated to the radius of curvature of the hyperbola at its vertex[3] had a larger uncertainty than value and was eliminated. The four-parameter hyperbola is the union of its asymptotes, which are shown in the graph; the results of the fit are shown in table 1.

[3] The intersection of the asymptotes

Table 1. Four parameter hyperbolic fit

parameter	value	uncertainty	units
σ_c	233	4	MPa
plates @ σ_c	900	100	plates
slope before	2.9	0.5	plates/MPa
slope after	30	1	plates/MPa

The critical stress, σ_c, and the number of plates at the critical stress are the coordinates of the intersection of the asymptotes; the slopes represent the generation of plates before and after the critical point. Thus, martensite plates are generated at a rate of three for an increment of one MPa and this rate increases by a factor of ten after the critical stress is reached; two things should be noted: (1) martensitic plates are generated at stress well below the critical –or transition– stress and (2) the tenfold increase in the plate generation rate is has a direct effect in the sharp reduction of the (macroscopic) tangent modulus. In addition to the calculated parameters, the model was used to estimate the number of plates at zero load; the (residual) number of plates is 200 ± 70. This number matches well with the one obtained from a residual stress calculation that will be shown later.

From the image analysis the orientation of each martensitic plate was measured and arranged into groups or families using a simple threshold criterion; the criterion to define a group was a difference of 12 degrees between orientations. Using this criterion, the minimum and maximum number of groups was 3 and 210 at the lowest and highest stress level respectively (figure 6a and figure 6f). It should be pointed out that this criterion is somewhat arbitrary since the number of groups depends on the threshold –especially at the higher stresses; nevertheless, it is useful for presenting the data and it serves to detect the two main orientations present.

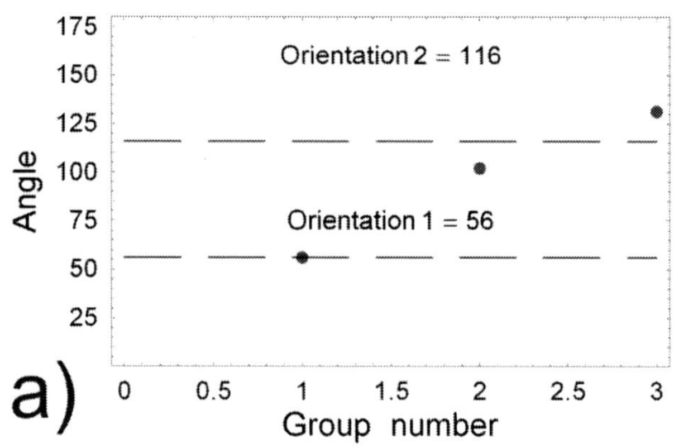

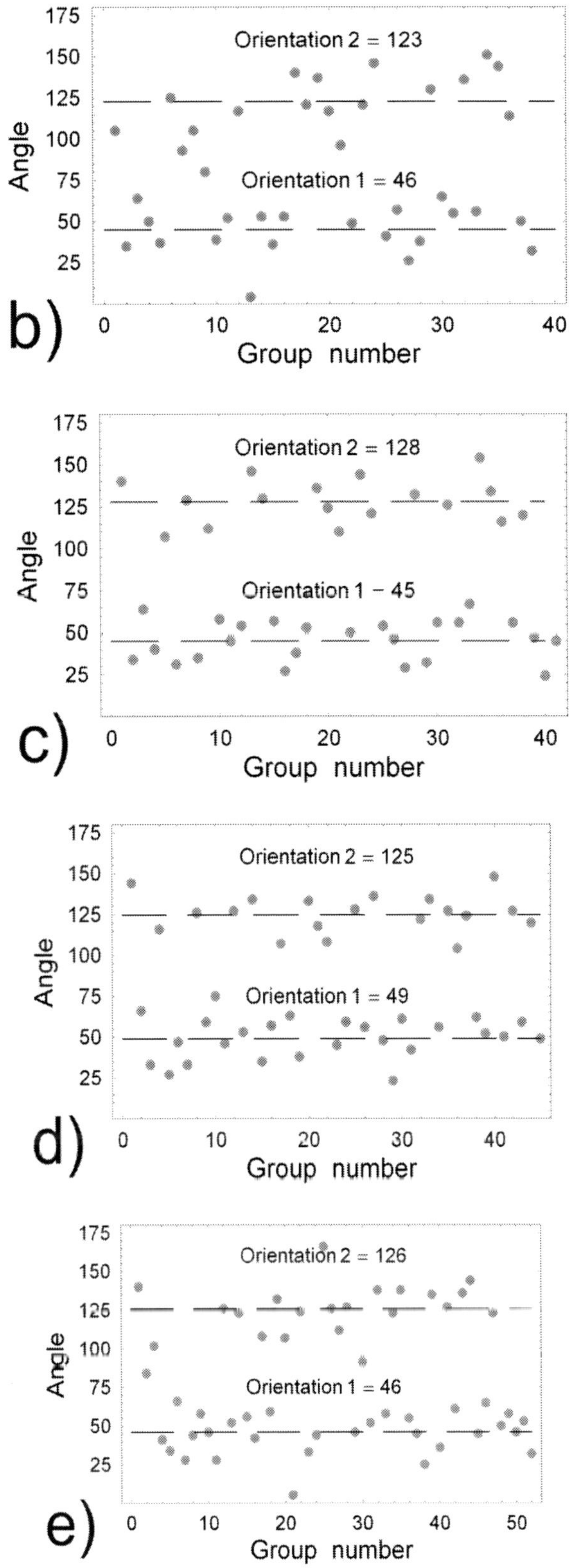
b)
Orientation 2 = 123
Orientation 1 = 46
Angle
Group number
c)
Orientation 2 = 128
Orientation 1 − 45
Angle
Group number
d)
Orientation 2 = 125
Orientation 1 = 49
Angle
Group number
e)
Orientation 2 = 126
Orientation 1 = 46
Angle
Group number

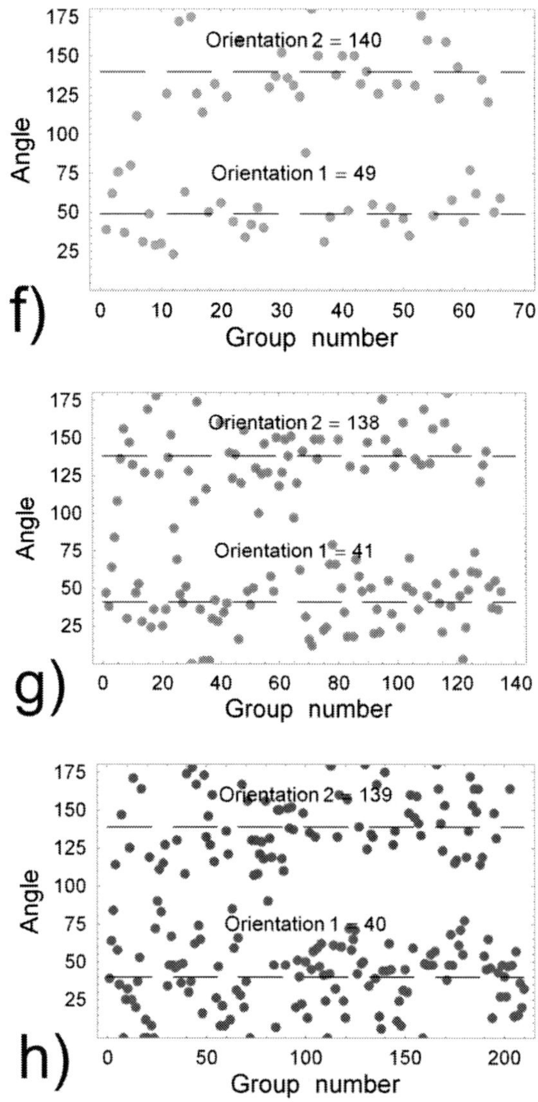

Figure 6. Stress-Induced orientation distribution due to the applied force direction and grain geometry.

Figure 6 shows the angle vs. the number of groups for all mosaics; two main orientations can be detected. The first one corresponds to a range between 25 and 75 degrees while the second one to a range between 10 and 160 degrees. The clustering of the groups is centered around the angles of maximum shear stress, which in the case of uniaxial tension are 45 and 135 degrees. The averages of the clusters are shown in the figures. The evolution of orientations presented in figure 6 shows that clustering is somewhat dynamic since it is easier to detect it for some of the stress levels but seem to disappear for some others. Clustering at 100 MPa is easier to detect than the clustering at 50 MPa or at the higher stress levels; for the last two stress levels, clustering occurs for the larger group numbers which are groups associated with larger grain numbers that correspond to grains at the bottom of the mosaics.

As it was pointed out earlier, plate formation depends, among other things, on the stress level, the grain geometry and the neighboring grains; the orientation also depends on which of variants in the cell is activated. CuAlBe has the D03 cell, which has 24 variants; grain orientation and loading direction are critical factors to determine which of the variants is activated but at higher stress levels, it may be possible to activate many of the variants. The end effect of this complex problem is the evolution of orientations presented in figure 6.

Although predicting or analyzing the orientation of martensite plates is a very complex problem, tracking the number of plates generated can be used to split the total strain into an elastic component and a transformational one. If one assumes that each plate generates a given amount of transformational strain, the macroscopically measured total strain can be divided into the elastic response of the austenite phase and the additional (transformational) strain that comes from the phase change. Thus,

$$\sigma = E_A \varepsilon_e + \sigma_0 \qquad \text{Eq. (3a)}$$

$$\varepsilon_T = \varepsilon_e + \varepsilon_{AxM} = \varepsilon_e + \alpha \text{plates}[\sigma] \qquad \text{Eq. (3b)}$$

where σ is the stress, E_A is the elastic modulus of the austenite phase, ε_e is the elastic strain, σ_0 is the residual stress, ε_T is the total strain, ε_{AxM} is the transformational strain, α is the constant of proportionality and the number of plates as a function of stress is the fitted hyperbola found before. The results of the analysis are shown in table 2.

Table 2. Elastic and transformational strain fit

parameter	value	uncertainty	units
E_A	176	3	GPa
σ_0	39	1	MPa
α	1.03	0.01	□□/plate

The elastic modulus is higher than the one obtained from the strain gage measurements (114 ± 1 GPa); it is clear that this should be the case since there is some plate generation occurring before the critical stress is reached. The residual stress calculated is equivalent to 220 ± 10 $\mu\varepsilon$, which can be converted into some 210 ±10 residual plates; this measurement is comparable to the previous estimate of 200 ± 70 plates.

The proportionality constant of 1.03 ± 0.01 $\mu\varepsilon$/plate has a small uncertainty since the calculated plate function, shown in figure five, has the same form as the macroscopic behavior shown in figure 1; the only data point which does not fit the analysis is the one labeled "G", which has a smaller strain than the expected value. In this case, the difference in strain is about 770 $\mu\varepsilon$; this strain is very large compared to the standard deviation of the analysis and was not used in the calculations. The results are summarized in figure 7, which shows the elastic behavior (blue) that characterizes the austenite phase and the transformational strain (green) that is 1.03 times the plate function that was shown in figure 5. The addition of these two strain models the total strain vs stress behavior that was shown in figure 1; in the plot, the datum that was labeled "G" is shown together with the predicted one "G'."

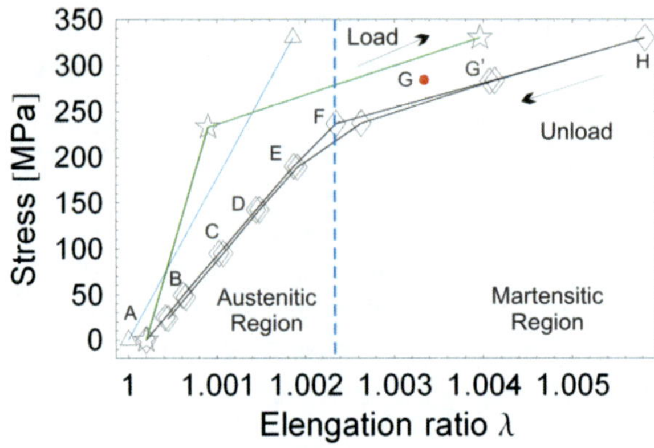

Figure 7. Elastic, transformational and total strain as a function of stress

Using these results, one can compare the elastic, transformational and total strains with the ones obtained from local image analysis. Since granular interaction plays an important role in the evolution of the transformation phase, three different regions of the mosaic were selected at different stress levels. To study the granular interaction of these regions two images per region were selected as shown in figure 8a and 8b for the first region, figure 8d and 7e for the second one and figure 8g and 8h for the third one. The arrow next to figure 8a indicates the loading direction.

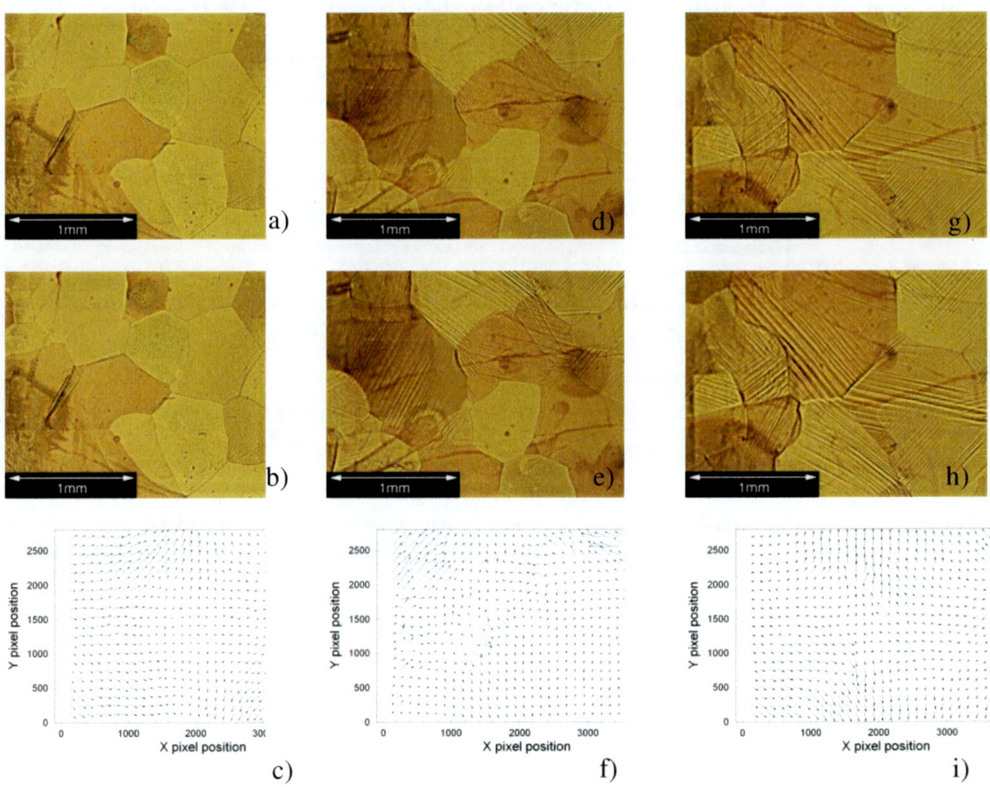

Figure 8. Granular interaction observed with displacement vector fields

One of the three cases selected comes from data before the critical stress is reached (austenite loading range) while the other two are cases taken after the critical stress was reached; of the two later cases, one is from a region that has lower plate density while the last one come from a region that ends up with a high plate density. From each pair of images the displacement vector field was calculated using the Willert and Gharib algorithm [28]. The images in figure 8a and 8b correspond to the central part of the mosaic. These images showed that almost all grains, except the grain labeled with the number 60 in figure 3, were in austenitic phase; a smooth hyperbolic vector field was obtained. The last pair of images showed that all grains were in martensitic phase as it is shown in figure 7g and 7h; in this case, the displacement vector field obtained from them was not a smooth hyperbolic field. For the second and third cases, figures 8f and 8i showed disturbances in some sectors of the displacement field. Although one may argue that the disturbances are due to the granular interaction caused by the growth of martensitic plates, the main assumption in the calculation of the displacement vector field is violated. As it was pointed out before, to obtain a displacement vector field, two images –from two different loads– are compared; the second image is assumed to be the first one except that it has been stretched or deformed. When new plates appear in the second image, which are not in the first one, the main assumption in the calculation is violated and the results become unreliable. This fact can be illustrated when the strains calculated from the displacement field are compared with the macroscopic strain calculations; these results are shown in table 3.

Table 3. Local and global strain calculations

Case	Stress increment (MPa)	Total strain from strain gage ($\mu\varepsilon$)	Local strain from DIC ($\mu\varepsilon$)
C to D	40 ± 10	300 ± 100	350 ± 10 + ??
G to H	50 ± 10	1800 ± 400	2300 ± 100 + ??
G to H	50 ± 10	1800 ± 400	$10, 100 \pm 200$ + ??

For the first case (C to D) both calculations coincide; the actual reading from the strain gage was 346 $\mu\varepsilon$ but it was rounded off to 300 due to the large uncertainty. The large uncertainty comes mostly from the uncertainty in the stress increment, which is large since the stress increment is small although the uncertainties in the stresses are about 5%. The strain from the DIC calculation was 354 $\mu\varepsilon$, which was rounded to 350; there is a missing uncertainty that comes from the calculation of the displacement field but the uncertainties could not be extracted from the software that was used.

For the second case, which has stresses larger than σ_c, the strains are much larger since (martensite) plates are generated at ten times the rate below σ_c; within the give uncertainties, both results coincide. This case was taken from a region that was relatively free of plates. In the last case, which has images full of plates, the calculated strain from DIC is unrealistically high; the reason may be that the main assumption –comparing the same image– is clearly violated since there are more plates in image 8h than there are in image 8g.

To determine if the DIC strain calculation could be extended to the grain level, the best case (C to D) was selected and the image that was shown in figure 8b was superposed with the displacement vector field (figure 8c). The grain boundaries were used to select the displacement vectors located at those grains; thus, individual grains were isolated and processed. The grains selected were labeled in figure 3 with the following numbers 55, 56, 57, 60, 61, 62, 63 and 64. The colored displacement field vector is shown in figure 9.\

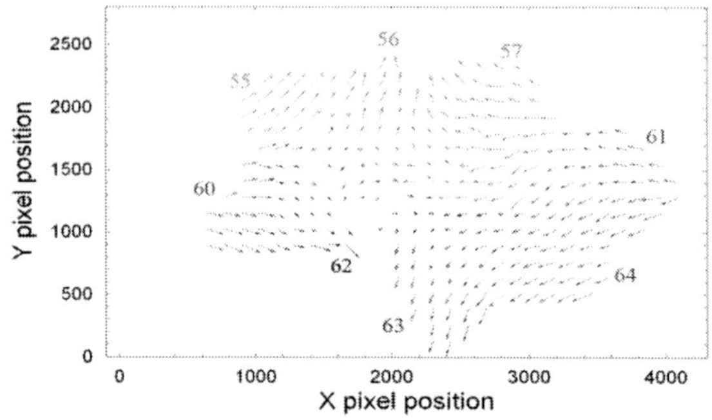

Figure 9. Displacement vectors for each grain by colors.

Using this data, the longitudinal strain for each grain was calculated with equations 1a and 1b. The resulting strains –named longitudinal strain to indicate that it is the strain in the loading direction– are plotted in figure 10; the levels of the elastic (blue), transformational (green) and total strains are also shown in figure 10.

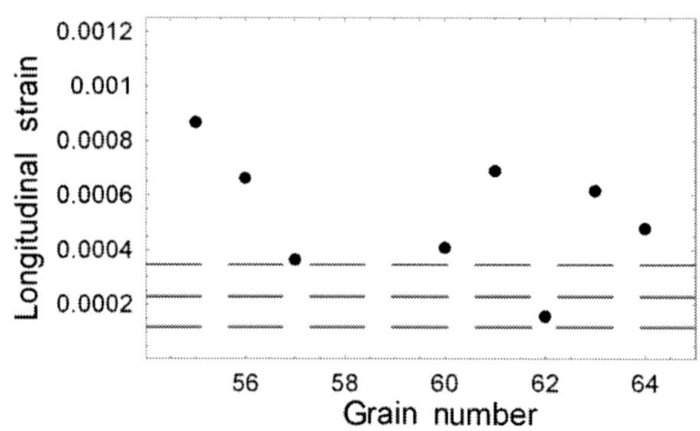

Figure 10. Longitudinal strain per grain.

Uncertainties are not shown in figure 10 since they could not be obtained for the displacement field calculation; narrowing down the selection of vectors corresponding to a single grain show that there is a tendency towards inhomogeneous strain at the grain level. These microscopic calculations yield similar results that the (macroscopically) calculated total strain. Clearly, higher resolution and the uncertainties from DIC are needed to characterize further the strain field at the grain level.

CONCLUSION

The evolution of the stress induced martensitic transformation was characterized by tracking the number of martensite plates that appeared –and some times disappeared– during a uniaxial test of a CuAlBe SMA alloy. A critical stress of 233 ± 4 MPa was obtained and the rates of plate generation before and after the transition were also determined.

It was observed that martensite plates are generated at stresses well below the critical one, although the generation rate is only 10% of the rate of plate generation past the critical stress. The sharpness of the transition could not be determined since the uncertainty in the parameter related to the radius of curvature had an uncertainty much larger than its value; in this case the hyperbola mapped onto its asymptotes, which was the function used to determine (martensite) plate generation as a function of stress. Having this function similar in shape to the macroscopic stress-strain behavior allowed us to make the key assumption that the transformational strain is proportional to the number of plates, which in turn allowed us to separate the total strain into its elastic and transformational components. With this analysis, the elastic modulus of the austenite phase and the constant of proportionality were determined. A better assumption may be that the transformational strain is proportional to the volume fraction of the transformed phase; however, counting the number of plates in each of the mosaics may be useful to estimate the transformational strain if the area examined is comparable to the area of the sample.

Comparison between macroscopic and microscopic strain calculations was in agreement for cases where there were a small number of plates in the images; in the case where the two images have a large –and different– number of plates, the results from the microscopic calculation yield unrealistically high strain. Extending the calculation to individual grains showed inhomogeneous strains but more resolution and known uncertainties are needed to determine the strain field down to the grain level.

Acknowledments

This work was developed with financial support from the PAPIIT DGAPA-UNAM program through grants IN100706, IX121504 and from CONACyT through grants P47758212 and NC204. UNAM CTIC-DGAPA provided a postdoctoral fellowship for F.M.S.A. and additional resources were provided by the ICyTDF program (Instituto de Ciencia y Tecnologia del Distrito Federal). The authors are grateful to Luis A. Ferrer and Mario Acosta for the use of the electrical strain gage equipment and to Gabriel Lara for his technical support.

References

[1] Olson, M.; Cohen, G. B.; Clapp, P.C. On the classification of displacive phase transformations. *Proceedings of the international conference on martensitic transformation. ICOMAT 79*, Cambridge-Massachussets U.S.A. 1979 pp. 1-11.

[2] Patoor, E.; Berveiller, M. *Les alliages à mémoire de forme. Technologies de pointe*; Hermes, PARIS, 1990 ; pp 09-63.

[3] Wayman, C. M.; Duerig, T. M. *An introduction to martensite and shape memory, Engineering Aspects of Shape Memory Alloys.* Butterworth-Heinemann, London (UK), pp. 3-20.

[4] Otzuka, K.; Wayman, C. M.; Nakay, K.; Sakamoto, H,; Shumizu K. *Acta Metallurgica*, 1976. Vol.24, pp. 207-226.

[5] Yang, J. H.; Wayman, C. M. *Mater. Charact.* 1992. vol 28, pp. 23-35.

[6] Yang, J. H.; Wayman, C. M. *Mater. Charact.* 1992.vol. 28, pp. 37-47.

[7] Czichos, H. "Adolf Martens and the research of martensite." *Proceedings of the european conference on martensitic transformation in science and technology*. Bochum, Alemania, 1989; pp. 3-14.

[8] Higuchi, A.; Suzuki, K.; Matsumoto, Y.; Sugimoto, K.; Komatsu, S.; Nakamura, N. "Shape memory effect in Cu-Al-Be ternary alloy." In: *Proceedings of the international conference on martensitic transformations*, ICOMAT 1892. Leuven, Belgica. pp. 767-772.

[9] Higuchi, A.; Suzuki, K.; Sugimoto, K.; Nakamura, N. "Thermal stability of Cu-Al-Be shape memory alloy." In: *Proceedings of the international conference on martensitic transformations,* ICOMAT 1896. Nara, Japan. pp. 886-890.

[10] Belkahla S.; Flores H.; Guenin G. *Mater. Sci. Eng. A-Struct. Mater. Prop. Microstruct. Process.* 1993, A169. p. 119-124.

[11] Jurado, M.; Castan, T.; Mañosa, L.; Planes, A.; Bassas, J.; Alcobe, X; Morin, M. *Philos. Mag.* 1997, Vol. 5. pp. 1237-1250.

[12] Hautcoeur, A.; Eberhardy, A.; Patoor, E.; Berveiller M. "Thermomechanical behavior of monocrystalline Cu-Al-Be shape memory alloys and determination of the metastable phase diagram." *Journal de physique IV. Colloque C2, suplément au Journal de Physique III.*1995,Vol. 5. pp. c2-459 to c2-464.

[13] Siredey, N.; Eberhardt, A. *Mater. Sci. Eng. A-Struct. Mater. Prop. Microstruct. Process.*. 2000, A290, pp. 171-179.

[14] Balo, N.; Ceylan, M.; Aksoy, M. *Mater. Sci. Eng. A-Struct. Mater. Prop. Microstruct. Process.* 2001, A311, pp. 151-156.

[15] Balo, N.; Ceylan M. *J. Mater. Process. Technol.* 2002 Vol. 124. pp. 200-208.

[16] Bouvet C.; Calloch S.; Lexcellent, C. *Transactions of ASME.* 2002,Vol 124. pp. 112-124.

[17] Chevalier, L.; Calloch, S.; Hild, F.; Marco, Y. Eur. *J. Mech. A/Solids.* 2001, vol. 20. pp. 169-187.

[18] Kaouache, B.; Berveiller, S.; Inal, K.; Eberhardt, A.; Patoor, E. *Mater. Sci. Eng. A-Struct. Mater. Prop. Microstruct. Process.* 2004, A378. pp. 232-237.

[19] Kaouache, B.; Inal, K.; Berveiller, S.; Eberhardt, A.; Patoor, E. *Mater. Sci. Eng. A-Struct. Mater. Prop. Microstruct. Process.* 2006, A 438-444. pp. 773-778.

[20] Siridey, N.; Patoor, E.; Berveiller, M.; Eberhardt, A. Int. J. Solids Struct. 1999, Vol. 36 pp. 4289-4315.

[21] Sánchez, F. M.; Pulos, G. *Materials science Forum.* 2006,Vol. 509. pp. 87-92.

[22] Sánchez-Arévalo, F. M.; Pulos, G. *Mater. Charact.* 2008,Vol. 59. Issue 11. pp. 1572-1579.

[23] Belkahla, S. "Elaboration et caracterisation de nouveaux allieages á mémoire de forme basse température type Cu- Al-Be." Ph. D. Thesis.1990, Inst. Nat. Sc. Appl., Lyon France. Pp 132.

[24] Belkahla, S. Dégration thermique de l'effet mémoire de forme d'un alliage ternaire: Cu-Zn-Al *These de magister en phisyque des matérriaux.* Université d'Annaba (Algérie) 1985, pp. 87.

[25] Flores, H.. "Stabilite thermique de la phase _ et de l'effet memoire double sens d'un alliage a memoire de forme du type Cu-Al-Be." Ph. D. Thesis.1993, Inst. Nat. Sc. Appl., Lyon France. Pp 150.

[26] Sánchez-Arévalo, F. M. "Estudio experimental del comportamiento mecánico de un material con memoria de forma." Ph. D. Thesis. 2007, Universidad Nacional Autónoma de México. México. pp 90.
[27] Chu, T. C.; Ranson, W. F.; Sutton, M.A.; Peters, W.H. *Exp. Mech.*. 1985, 25(3). pp. 232-244.
[28] Willert, C. E.; Gharib, M. *Exp. Fluids*. 1991,Vol. 10. pp. 181-193.
[29] Nygards, M. *Mech. Mater*. 2003. Vol. 35. pp. 1049-1057.

In: Shape Memory Alloys
Editor: H. R. Chen, pp. 267-285
ISBN 978-1-60741-789-7

Chapter 9

FORMATION OF TRANSITORY BAINITE AS A PRECURSOR OF α-PHASE DURING TEMPERING OF MARTENSITIC CU-ZN-AL SMAS

Leandru-Gheorghe Bujoreanu
Faculty of Materials Science and Engineering, The "Gh.Asachi" Technical University Iaşi, Romania

ABSTRACT

The paper assumes that, in the case of a martensitic Cu-Zn-Al shape memory alloy subjected to heating above 770 K, the endothermic peaks revealed by differential scanning calorimetry (DSC) and differential thermal analysis (DTA) and located around 630 K would correspond to the transitory formation of α_1 bainite. This endothermic phenomenon has been regarded as transitory since heating proceeding causes exothermic precipitation of equilibrium α-phase which is copper-rich and therefore its formation augments both Zn and Al amounts in β_2 austenitic matrix.

Considering that the formation of tempering bainite has been completed at 670 K, specimens which were water-quenched from this temperature were analyzed by optical and transmission electron microscopy (OM and TEM, respectively) and bainite formation was ascertained by its plate-like morphology with irregularly distributed internal striations having microtwin-appearance and by its selected-area diffraction (SAD) pattern with orthorhombic unit cell having 9R atomic ordering.

In order to reveal the irreversible character of tempering-martensite formation, induced by its diffusion controlled growth, specimens water-quenched from 770 K were subjected to DSC-scans during heating. In this case no martensite reversion to austenite was detected and the accumulation of irreversibly formed α-phase caused a gradual change of chemical composition of β_2 matrix accompanied by a diminution in the intensity of its order-disorder transition.

INTRODUCTION

The discovery of bainite in eutectoid steels, under the form of "acicular, dark etching aggregate" which was "tougher for the same hardness than tempered martensite", is owed to the pioneering studies on the isothermal transformation of austenite performed by Davenport and Bain, in 1930, the metallographic constituent, itself, being named in the honor of the latter, in 1934 [1].

The nature of the mechanism of bainitic transformations in Fe-Cr-C and Fe-Si-C steels, was revealed by Ogawa and Kajiwara by means of high resolution electron microscopy (HREM) observations which have shown that, in contrast to martensite transformation, which is of shear-type (diffusionless), bainite transformation is of diffusion-controlled type because no intermediate, transition lattice was observed even in the very tip of the bainite plate [2].

During almost eight decades, the importance of fundamental research on bainite transformation has gradually increased in such a way that distinctive sections were assigned to this phenomenon within International Conferences on Martensitic Transformations (ICOMAT), ever since 1986, as well as special issues of some cutting edge journals such as Scripta Materialia, where it was considered as "the phenomenon that has caused most controversy in the field of physical metallurgy" [3].

As a microstructural constituent of steels, bainite is defined as "the grain boundary allotriomorph morphology of ferrite containing a non-lamellar distribution of carbides formed during eutectoid decomposition" [4] or more general as "the nonlamellar two-phase product of eutectoid decomposition in which diffusional and sequential precipitation and noncooperative growth of the low-temperature phases occurs [5].

On the time-temperature-transformation (TTT) diagrams, the isothermal transformation of austenite to bainite occurs at temperatures located below the maximum of the "C" curves, corresponding to the shortest time for bainite (B) formation, which is also called "bainite nose". On continuous cooling, bainite is typically formed at cooling rates illustrated with ③ which must range between the critical minimum (Min) and maximum (Max) values, illustrated with grey, in **Figure 1** corresponding to plain carbon steel [5, 6].

Bainite formation has also been associated with martensite-like surface relief described as "multiple and complex" for upper bainite and as "single and uniform „for lower bainite which may also present "tent-shaped" and "inverse-tent-shaped" tilts [4, 5, 7]. However, besides martensite, surface relief accompanies the formation of both bainite and Widmanstätten ferrite, which was described as "carbide-free bainite", yet bainite growth must be neither rapid nor accompanied by high supersaturation of carbide [8] although in some cases it may be accelerated by stress, as an effect of diffusivity increase of solute atoms [9].

In spite of the intensive effort spent for fundamental research of the transformation mechanism of bainite formation and growth in steels, various aspects are still not fully understood, such as bainite formation from plastically deformed austenite and its large growth rate and autocatalytic nucleation, identified by Bhadeshia [10].

Nevertheless, as an effect of accepting the occurrence of martensite transformation in various other materials, beside ferrous alloys [11], bainite transformation was also reported in nonferrous materials such as Shape Memory Alloys (SMAs, which include Ni-Ti-, Cu-Zn-Al-, Cu-Al-Ni- and Fe-Mn-Si-based alloys) the governing mechanism of which relays on stress-induced reorientation of martensite plate variants and their reversion to austenite on heating

[12], even if there have been some opinions that the term is misused and "that bainite is no longer used as the name for a decomposition product of β brass" [13].

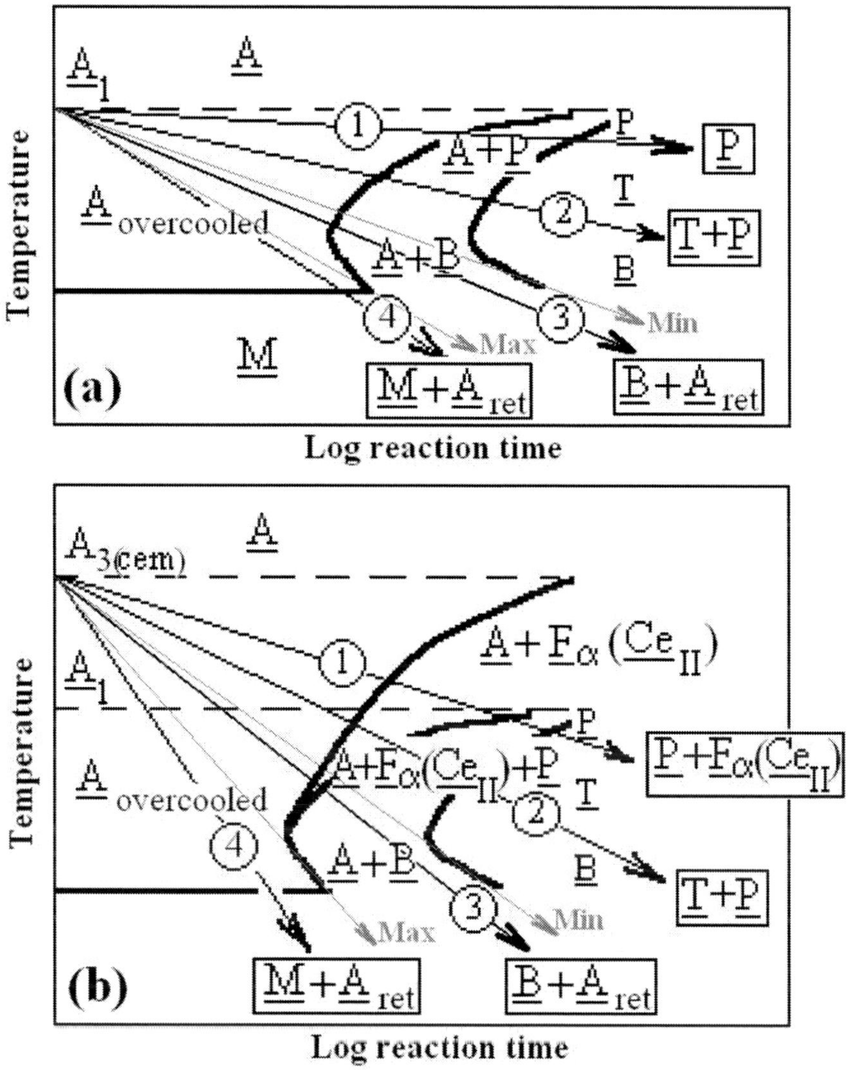

Figure 1. Schematic TTT diagrams of plain carbon steels illustrating the transformation of austenite (A) into equilibrium constituents, such as pearlite (P) and the proeutectoid phases ferrite (F_{α}) and cementite (Ce_{II}); into transitory phase such as troostite (T) and bainite (B) as well as into metastable martensite (M): (a) eutectoid steel; (b) hipo(hyper)eutectoid steel

The isothermal formation of α_1 bainite was initially observed in Cu-Zn alloys [14] and a specific "C-curve" was associated with bainitic transformation on the time-temperature diagrams on both isothermal or continuous cooling (TTT or CCT, respectively) [15].

Later, the studies performed by Takezawa *et al.* revealed that, during stress induced formation of α_1 bainite in Cu-Zn-Al, nucleation occurred after an incubation period evaluated to about 6 minutes, irrespective to stress magnitude and orientation, and that growth was controlled by Zn-atoms diffusion [16]. Consequently, up to 3 wt.% Zn fluctuations were reported between chemical compositions of α_1 plates and β matrix [17], representing parent phase which, for simplicity reasons, will be hereinafter termed "austenite" since it undergoes either a thermally or a stress-induced martensitic transformation [18]. Further studies of Takezawa and Sato, on Cu-Zn-Al alloys, emphasized the mechanism [19] and morphology [20] of bainite transformation during isothermal ageing and considered that bainite plates nucleate as an effect of releasing considerable amounts of stress concentrations developed in the crossed regions of $\alpha_{\{111\}}$ phase boundary facets [21]. On further increases of time and temperature, during aging heat treatment of martensitic Cu-Zn-Al SMAs, the preferred formation of α-phase according to a specific orientation [22] and morphology [23] was noticed.

Excepting for some isolated cases, where diffusion-controlled nucleation of bainite at 523 K was found responsible for the formation of a solute-depleted zone in Cu-28 Zn-4 Al (mass. %) [24], it is generally accepted that bainitic transformation has a diffusionless nucleation and a diffusion-controlled growth, even if its formation is accompanied by a geometry change designated as “reverse shape memory effect” [25]. Therefore, in contrast to martensitic transformation, bainitic transformation is irreversible and must be retarded in order to avoid degradation of Shape Memory Effect (SME) [26].

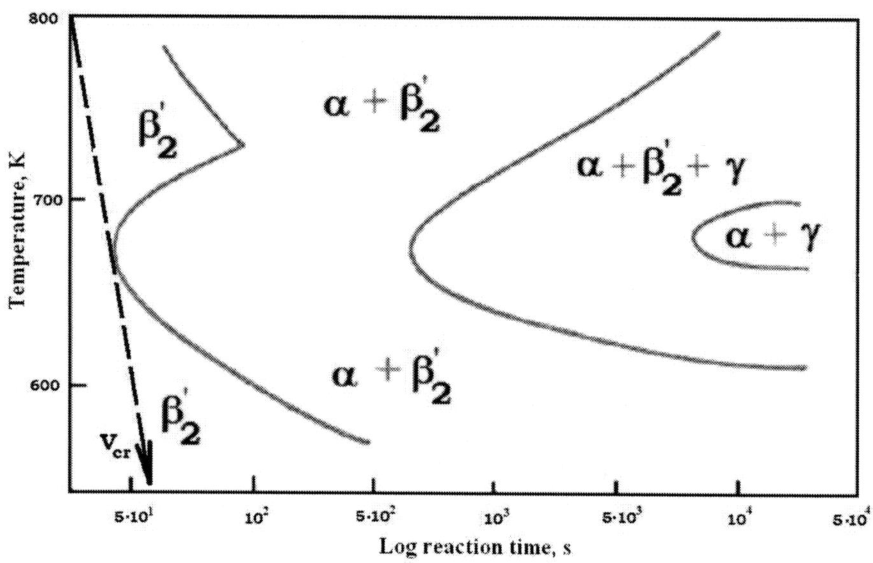

Figure 2. TTT diagram of a Cu-20 Zn-5 Al (mass. %) illustrating minimum critical cooling rate for martensite formation without any accompanying precipitates [27]

From a kinetic point of view, bainite formation in Cu-Zn-Al SMAs is represented, in the same way as in steels, by a “bainite nose” on the TTT curves, as illustrated in Figure 2, corresponding to Cu-20 Zn-5 Al (mass. %) [27]. During isothermal ageing between 453-573 K of martensitic Cu-Zn-Al SMAs with up to 5 mass. % Al, the formation of transitory α_1 bainite (9R, orthorhombic) was reported that became disordered (3R) and finally transformed into equilibrium α-phase (fcc) [28].

In the case of a Cu-Zn-Al SMA with 7 mass. % Al, stable bainite was observed only after homogenization and tempering at 573 K while in other heat treatment states, of both air-cooled hot-forged specimens and water quenched austenitized specimens, bainite was considered as a precursor of equilibrium α-phase formation during the heating of martensitic alloys [29]. By means of X-ray diffraction (XRD), differential thermal analysis (DTA) and differential scanning calorimetry (DTA) α_1 bainite was identified on both optical and transmission electron micrographs (OM and TEM, respectively) [30], under the form of a plate-like product with internal striations [31] which seem to be irregularly distributed and have the appearance of microtwins [32].

In the present chapter new evidence is brought on the transitory formation of α_1 bainite as a precursor of α-phase during tempering of martensitic Cu-Zn-Al SMAs, on the microstructural morphological characteristics of bainite and on the irreversibility of high-temperature phase transitions, occurring in this alloy system.

2. EXPERIMENTAL

After casting, hot pressing (40 % thickness reduction/ 1023 K/ air cooling), homogenizing (1073 K/ 18 ksec/ water) and tempering (300 sec/ water) at three temperatures (573, 673 and 773 K), the specimens of a Cu-21.6 Zn-7 Al (mass. %) SMA were prepared for: (i) DTA; (ii) DSC; (iii) dilatometry; (iv) OM and (v) TEM studies. The experiments were performed on: (i) a MOM Q-01500D derivatograph with a heating rate of 5 K/min [22]; (ii) a 2920 Modulated DSC TA INSTRUMENTS unit, with a temperature variation rate of 10 K/min and isothermal maintaining for 3 min. at the end of each ramp [29]; (iii) on a uni axial vacuum dilatometer, comprising a Quartz tube connected to a displacement transducer (HBM, type W1E) with computer interface, using Ø5 x 6.77 mm specimens, cut by means of a low rate-saw in order to preserve their thermomechanical history, which were heated with 2 K/min, up to 773K, followed by free air cooling [33]; (iv) a LEICA optical line comprising Aristomet light microscope and Orthomat E exponometer, using metallographically prepared specimens [29] and (v) a JEM 2000FX transmission electron microscope with double tilt sample holder, operating at 200kV and using thin foils for TEM observations, prepared by twin-jet electropolishing [29].

3. RESULTS AND DISCUSSION

When a homogenized (1073 K/ 18 ksec/ water) Cu-21.6 Zn-7 Al (mass. %) SMA was heated from room temperature (RT) to 800 K similar results were obtained by DTA and DSC, as illustrated in **Figure 3**. Considering that the specimen is in martensitic state, as an effect of water-quenching, the first endothermic peak would correspond to reverse martensitic transformation, the exothermic peak would represent α-phase precipitation and the second endothermic peak would reveal an order-disorder transition of austenite [34]. As previously reported [29] it is assumed that the endothermic peak marked by a circle and located at above 600 K would correspond to transitory bainite formation [30]. The two thermograms from Figure 3 have similar aspects and locations of the three high-temperature transformations. However, even if this detail is unimportant for the present chapter, they reveal different locations for reverse martensitic transformation, around 470 K on DTA chart and around 330

K on DSC chart. This could be an effect of the difference between heating rates, since DSC heating was performed with 10 K/ min which is the double of DTA heating rate. Since reverse martensitic transformation is favored by high cooling rates, this could be the reason why this reaction occurred at higher temperatures on DTA chart. The presumed peaks of transitory bainite formation look different on DTA and DSC charts: more prominent on DTA and flatter on DSC but this could be another effect of different heating rates, as well.

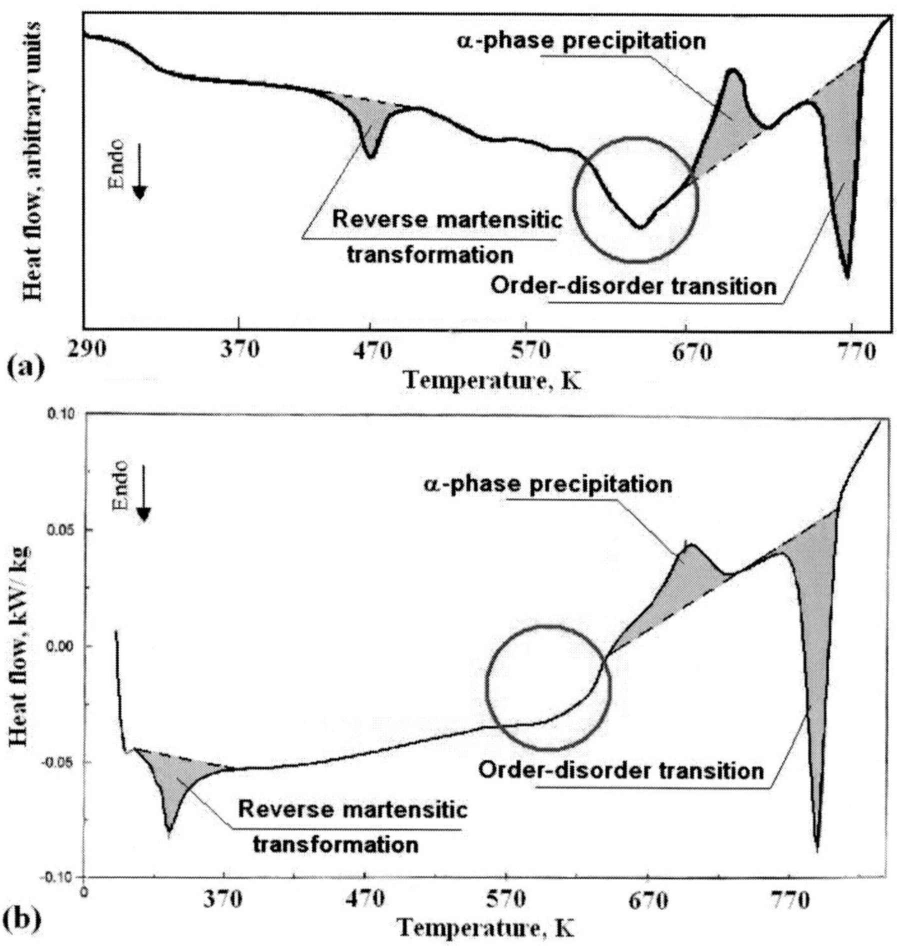

Figure 3. Phase transitions occurring during the heating of a Cu-21.6 Zn-7 Al (mass. %) SMA homogenized (1073 K/ 18 ksec/ water): (a) DTA chart; (b) DSC chart. The circle indicates an endothermic peak assumed to correspond to transitory bainite formation

In order to corroborate DSC results with other thermal analysis techniques, **Figure 4** compares DSC with dilatometry results in the case of an hot pressed (40 % thickness reduction/ 1023 K/ air cooling) Cu-21.6 Zn-7 Al (mass. %) SMA. It is obvious that, in spite of the difference between the two austenitization states, hot pressed-air cooled and homogenized-water cooled, the DSC thermograms displayed in Figure 3(b) and 4(a) are almost identical. Therefore baselines were drawn, in Figure 4(a), in order to emphasize reverse martensite transformation and transitory bainite formation along with the corresponding start and finish temperatures for phase transformation. The results were, for reverse martensitic transformation, A_s = 308 K and A_f = 365 K and for bainite formation, B_s = 537 K and B_f = 643 K. On the other hand, the dilatogram from Figure 4(b) also reveals two marked deviations from linear thermal expansion, which could fairly be connected with martensite reversion and transitory bainite formation. The former is associated with a marked contraction [35] while the latter could be the effect of a slight extension. Accordingly, critical transformation temperatures were approximated as A_s = 347 K and A_f = 371 K, for reverse martensitic transformation and B_s = 610 K and B_f = 657 K for bainite transitory formation. It is noticeable that critical temperatures determined by dilatometry, with a heating rate of 2 K/min, were higher than the corresponding values determined by DSC with a heating rate of 10 K/min, in accordance with previous comment regarding the differences between DTA and DSC. In addition, thermal hysteresis has been less than half, for both phase transitions, in Figure 4(b) as compared to Figure 4(a), which suggests that dilatometric techniques are more useful than calorimetric techniques in regards of the study of bainitic transformation [36].

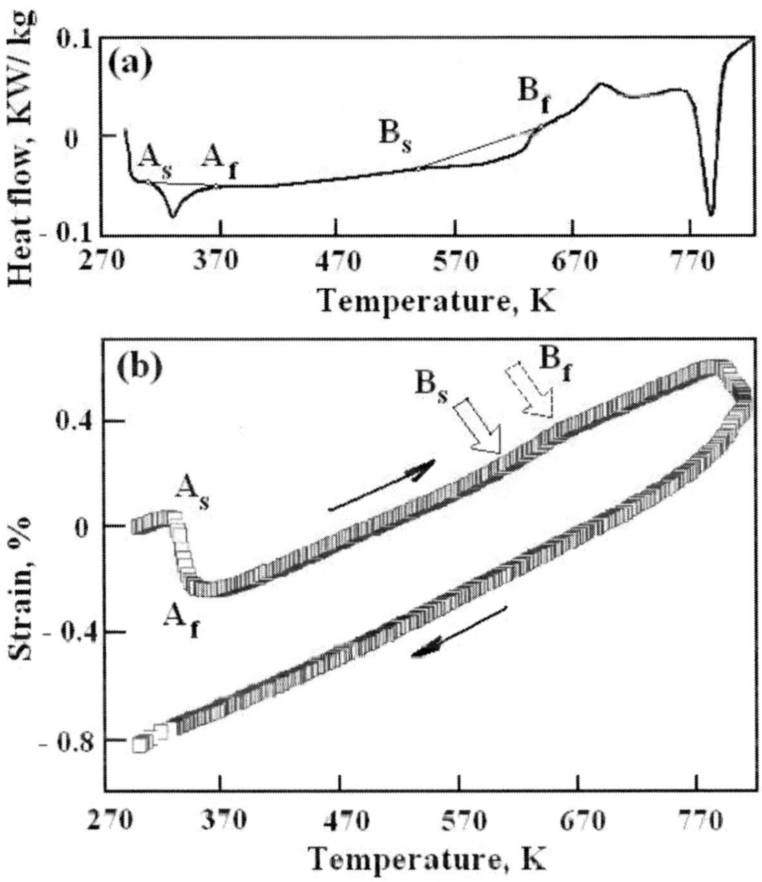

Figure 4. Bainite formation effects in a Cu-21.6 Zn-7 Al (mass. %) SMA hot pressed (40 % thickness reduction/ 1023 K/ air cooling): (a) DSC chart on heating; (b) dilatogram during heating-cooling

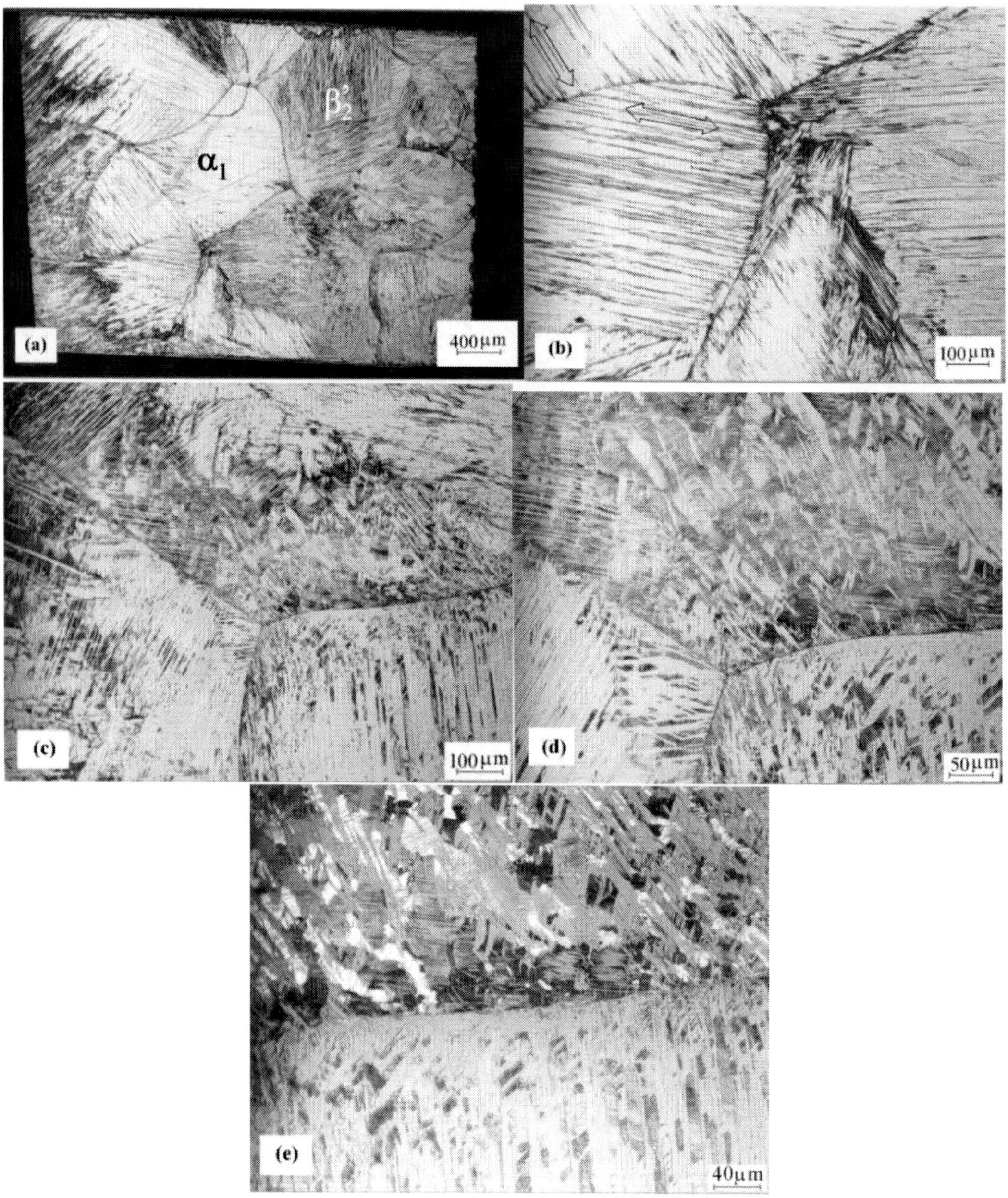

Figure 5. OM micrographs of a Cu-21.6 Zn-7 Al (mass. %) SMA homogenized (1073 K/ 18 ksec/ water) and tempered (573 K/ 300 s/ water): (a) general aspect; (b) aspect of bainite plates at the intersection of three grains; (c) internal striations of bainite plates in the intersection area of three grains; (d) detail of internal striation of bainite plates; (e) substructure of bainite plates at a grain boundary without visible allotriomorphic precipitation

The microstructural morphological characteristics of bainite are further revealed by OM and TEM on three tempering states (573, 673 and 773 K) on a homogenized (1073 K/ 18 ksec/ water) Cu-21.6 Zn-7 Al (mass. %) SMA.

Five OM micrographs of a specimen tempered at 573 K are shown in **Figure 5**. Figure 5(a) shows the general aspect of the specimen, revealing an average grain size of 0.8 mm.

Some of the right-side grains have areas of β'_2 martensite with diamond-like structure. However, there are at least six grains which will be further analyzed since they reveal plate-like morphology, corresponding to α_1 bainite description [31]. Figure 5(b) illustrates the aspect of bainite plates at the intersection of three grains. It is noticeable that most of the long bainite plates in upper and left-side grains possess single orientations, within the same grain and none of them cross grain boundaries which are rather narrow, without allotriomorphic precipitates. A more detailed view of the structure of bainite plates, in another intersection area of three grains, is given in Figure 5(c). Thus, left-side grain reveals even thickness bainite plates while upper grain displays plates with various orientations, some of them possessing internal striations. The enlarged detail of this intersection area of three grains is shown in Figure 5(d) which enables the observation of internal striations in most of the bainite plates. This micrograph allows noticing that the striations are equally spaced and preserve the same orientation within the same plate. At higher magnification, Figure 5(e) shows that not all of the bainite plates, in upper grain, have internal striations yet some of the plates have extremely fine striations under the form of sub-plates, which could be caused by large dislocations densities [19], as it will shown later.

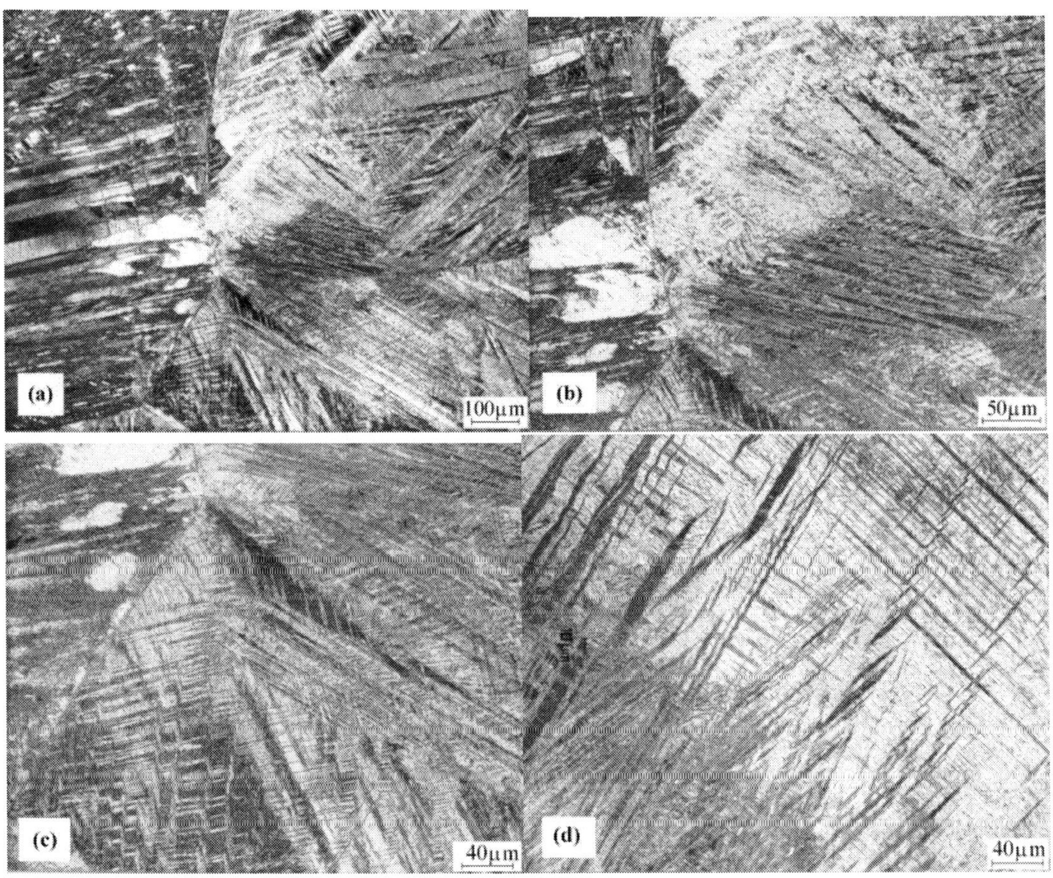

Figure 6. OM micrographs of a Cu-21.6 Zn-7 Al (mass. %) SMA homogenized (1073 K/ 18 ksec/ water) and tempered (673 K/ 300 s/ water): (a)-(c) progressively enlarged images of an intersection of three grains revealing bainite-plate refining, while preserving their V-formation; (d) substructure of bainite plates comprising white parallel equal-width bands

Figure 6 displays four OM micrographs of the specimen homogenized and tempered at 673 K (300 sec/ water). In accordance with Figure 3, after being tempered at this temperature, this specimen should comprise, besides bainite, a small amount of α-phase the precipitation of which should be triggered at this temperature. In three details, shown in Figure 6(a)-(c), there are progressively enlarged images of an intersection of three grains revealing bainite-plate refining as an effect of increasing tempering temperature, while preserving their V-formation. The internal striations become again visible in Figure 6(c). However there are differences between these striations and those shown in Figure 5, since they are no longer confined to the bainite plates. For instance, the lower grain from Figure 6(c) displays mostly horizontal striations, on different bainite plates. At larger magnifications, the substructure of bainite plates becomes visible in Figure 6(d), being formed by white parallel equal-width bands.

After further increasing tempering temperature to 773 K, it is expected that bainite transformed to α-phase which would represent most of the structure of the specimen, as illustrate OM micrographs of **Figure 7**. The general aspect reveals large rods-like α-phase crystallites, the length of which was estimated to 1 mm, together with other crystallites with various sizes and shapes. This variety is shown in Figure 7(a) which displays idiomorphic white crystallites with different widths and length. At a closer look, it seems that the long thin parallel rods and the short crystallites, with two orientation tendencies at about 70^0, occupy different depth levels on the surface, since shorter crystallites are visible under the long thin ones. Figure 7(b) shows the intersection of three grains among which the left-side one contains both long and short rods. Typical Widmanstätten [37] structures are visible inside the grains The structure of a grain boundary and the α-phase rods located on each side of it are shown in Figure 7(c). In the central area, there is a colony of crystallites, some of which seem to be able to cross the grain boundary with a slight deviation. Two intersecting tendencies of α-phase rods, at approx. 20^0 and at 90^0, are noticeable in Figure 7(d) and in the detail shown in Figure 7(e). In this case, as well, the shorter crystallites are located bellow the longer ones. The coexistence of short disoriented crystallites with thin long parallel evenly distributed rods of α-phase is illustrated in Figure 7(f). It is assumed that the periodicity of α-phase rods is due to the nucleation dislocations in periodically arranged slip planes [38].

The expected presence of bainite in the specimen tempered at 573 K is confirmed by means of the TEM micrographs shown in Figures 8 and 9. According to data found in literature the plate-shaped product and its selected-area diffraction (SAD) pattern, from **Figure 8** could be ascribed to α_1 bainite [17]. The width of the observed bainite plates is typically about 200 nm. SAD pattern matches 9R orthorhombic structure since no arrangement of spots, indicating the inclination of *c*-axis, as well as no superlattice-spots are noticeable [20].As noticeable from the micrograph, bainite plates, which were initially free of stacking faults after their formation at 573 K [39], are composed of subplates which at their turn comprise subunits [40] corresponding to parallel slip dislocation lines [19].

The presence of high densities of slip dislocations, on each side of a boundary between two variant plates of bainite [20] is shown in **Figure 9**. It is noticeable that most of the dislocation lines have counterparts on the other side of the boundary [19].

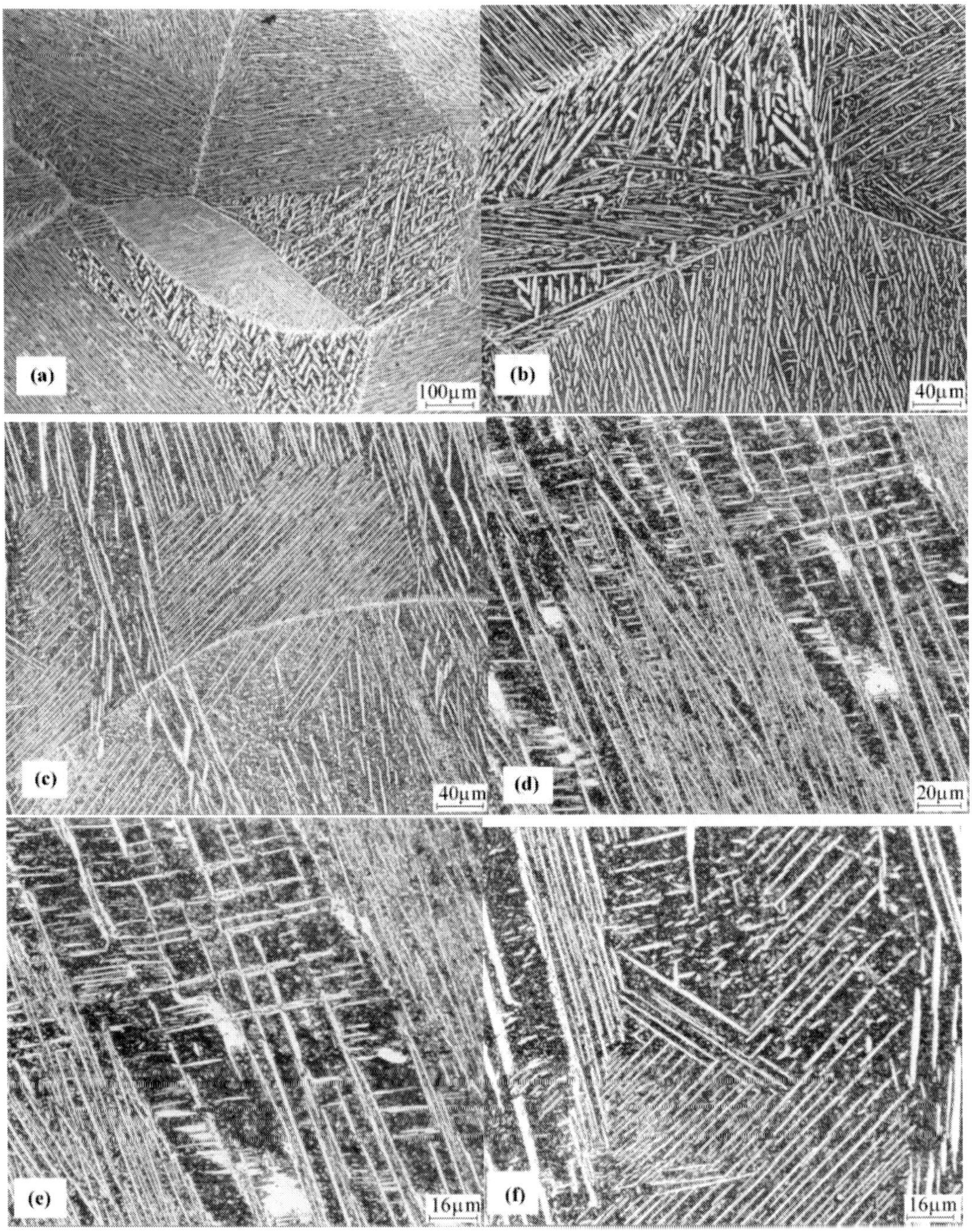

Figure 7. OM micrographs of a Cu-21.6 Zn-7 Al (mass. %) SMA homogenized (1073 K/ 18 ksec/ water) and tempered (773 K/ 300 s/ water): (a) general aspect of idiomorphic α-phase rods of different size and orientation tendencies in different grains; (b) aspect of α-phase rods at the intersection of three grains; (c) orientation tendencies of α-phase rods at a grain boundary which some of them seem to cross; (d) two intersecting tendencies of α-phase rods; (e) detail of the area with perpendicular α-phase rods; (f) coexistence of thin long parallel evenly distributed rods with short disoriented crystallites of α-phase

Figure 8. TEM micrographs of a Cu-21.6 Zn-7 Al (mass. %) SMA homogenized (1073 K/ 18 ksec/ water) and tempered (673 K/ 300 s/ water) illustrating bainite like plates and corresponding SAD pattern

Figure 9. TEM micrographs of a Cu 21.6 Zn 7 Al (mass. %) SMA homogenized (1073 K/ 18 ksec/ water) and tempered (673 K/ 300 s/ water) illustrating high densities of dislocations on each side of a boundary between a pair of bainite plates

The influence of transitory bainite formation on the other two high temperature phase transitions, α-phase precipitation and order-disorder transition of β_2 austenite has been revealed by means of three consecutive heating-cooling cycles applied up to 873 K on the DSC device, as shown in **Figure 10**. Figure 10(a) displays a diminution of both transformations with increasing the number of cycles, numbered from 1 to 3. An enlarged image of the three thermographs is illustrated in Figure 10(b). Considering that α-phase precipitation occurs between 670 and 770 K, it is assumed that after the first heating cycle most of this phase has already formed. If heating had been applied up to above 1023 K, the specimen would become entirely austenitic since all α-phase would decompose to austenite. However, in the present case, heating stopped at 873 K; therefore α-phase did not dissolve. During cooling its amount remains unchanged and during subsequent heating a small additional amount of α-phase is added to pre-existing one. After each α-phase precipitation β_2 austenite becomes more and more depleted in copper (since α-phase is richer in Cu than austenite which is based on CuZn intermetallic compound). Consequently the matrix becomes enriched in zinc which causes a diminution of critical transformation temperatures in such a way that no martensite formation occurs during subsequent cooling [29] and no reverse martensitic transformation is observed during next cycle.

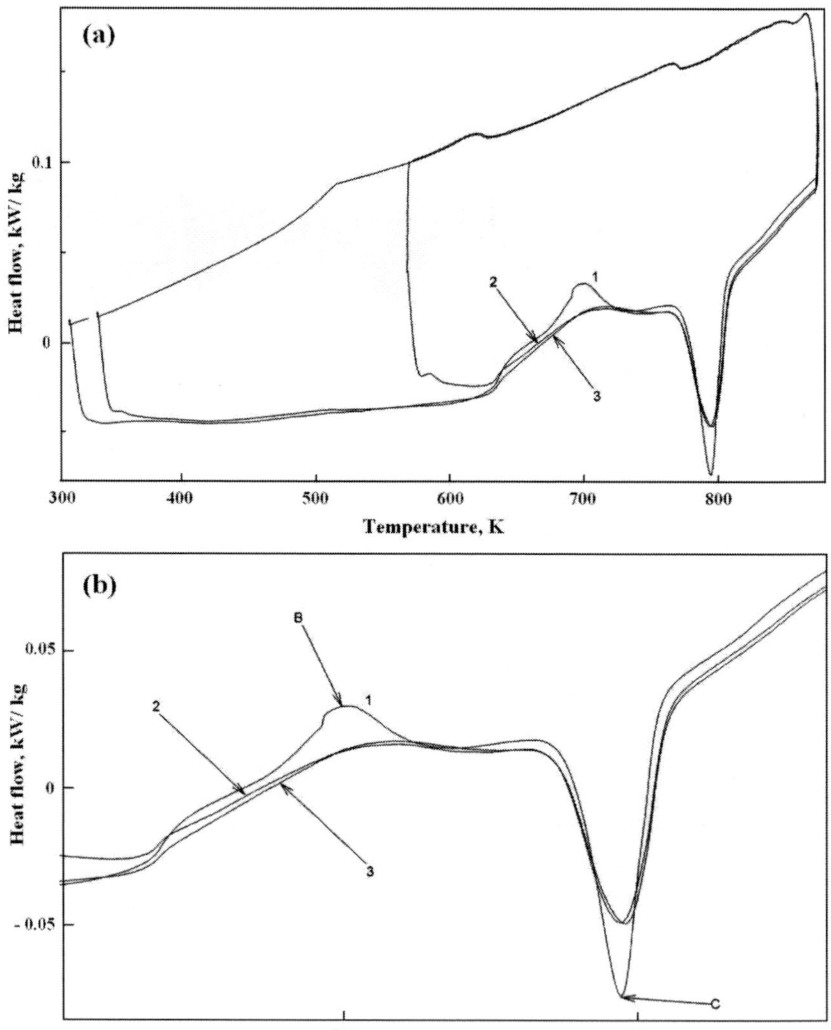

Figure 10. DSC chart of a Cu-21.6 Zn-7 Al (mass. %) SMA hot pressed (40 % thickness reduction/ 1023 K/ air cooling) subjected to two heating-cooling cycles from RT to 873 K and to a third cycle between 573 and 873 K: (a) general aspect; (b) detail of the transformation area

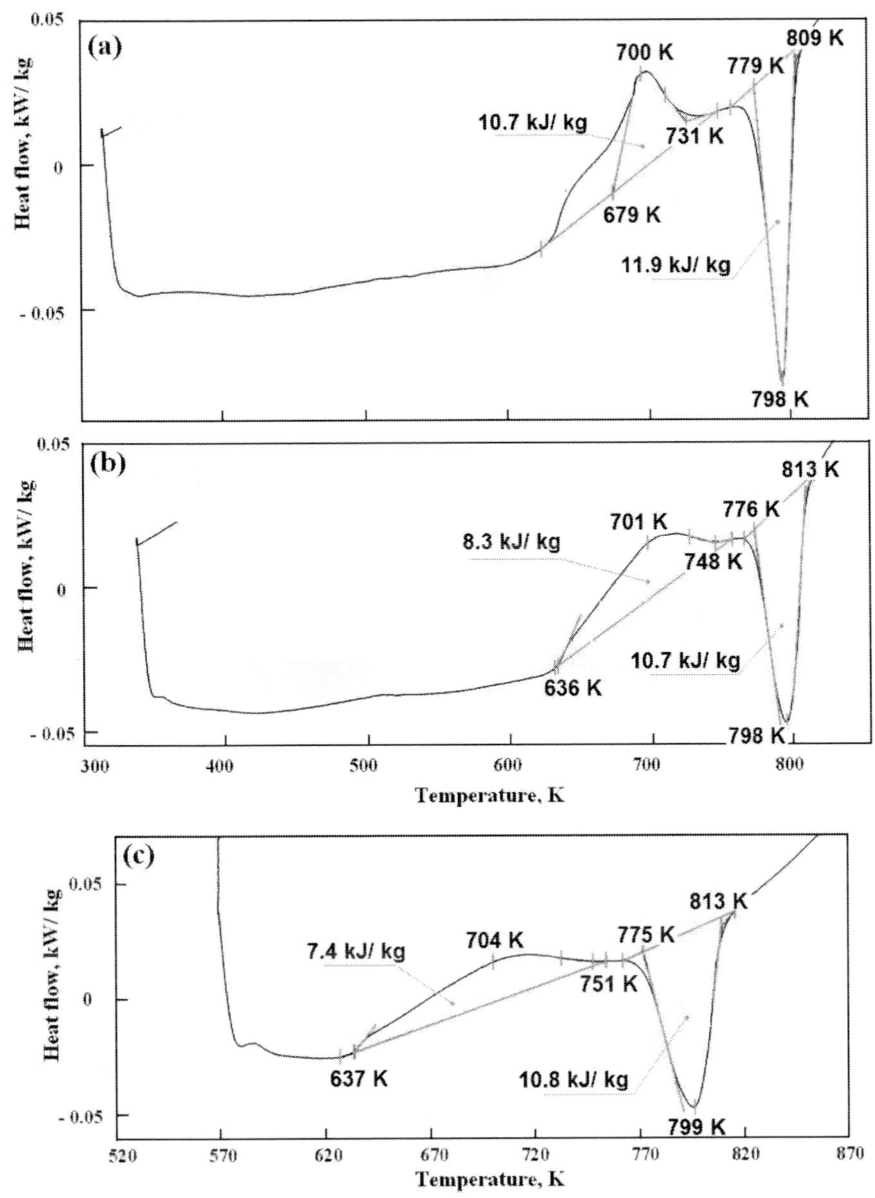

Figure 11. Individual DSC charts recorded during thermal cycling in Figure 10: (a) first cycle; (b) second cycle; (c) third cycle

For a better insight of the evolution of α-phase precipitation and order-disorder transition of β_2 austenite, during thermal cycling, **Figure 11** separately illustrates the DSC thermograms for each cycle. It is noticeable that, during the first heating cycle, α-phase precipitation occurs in two stages. On each of the three charts auxiliary lines were drawn in a different color by device software in order to allow determining critical start, average and finish temperature as well as specific transformation heat. These values are shown on each thermograph. When analyzing the variation tendencies of average transformation temperatures, from the first to the third cycle, it is noticeable that the temperatures tend to increase for both α-phase precipitation and order-disorder transition of austenite. The increase is more obvious for α-phase precipitation (from 700 to 704 K) and much slighter for order-disorder transition (from 798 to 799 K), which is one of the effects of increasing zinc content [41]. On the other hand, as expected from the noticeable decrease of the magnitude of corresponding peaks, it is obvious that both phase transitions tend to become less intense with increasing the number of thermal cycles, as illustrated by the variation of specific transformation heat. In this respect, as well, the variation tendency is more obvious in α-phase precipitation (the specific transformation heat of which decreased from 10.7 to 7.4 kJ/ kg) than in order-disorder transition of austenite (where the decrease occurred from 11.9 to 10.8 kJ/ kg). In regards to bainite formation, with increasing the number of heating cycles, it appears that the corresponding endothermic maxima become less and less prominent. This could be a logical consequence of increasing the amount of α-phase, from one cycle to the other, since it has been argued that, if this phase is already present in the microstructure, no bainite formation is expected to occur [29].

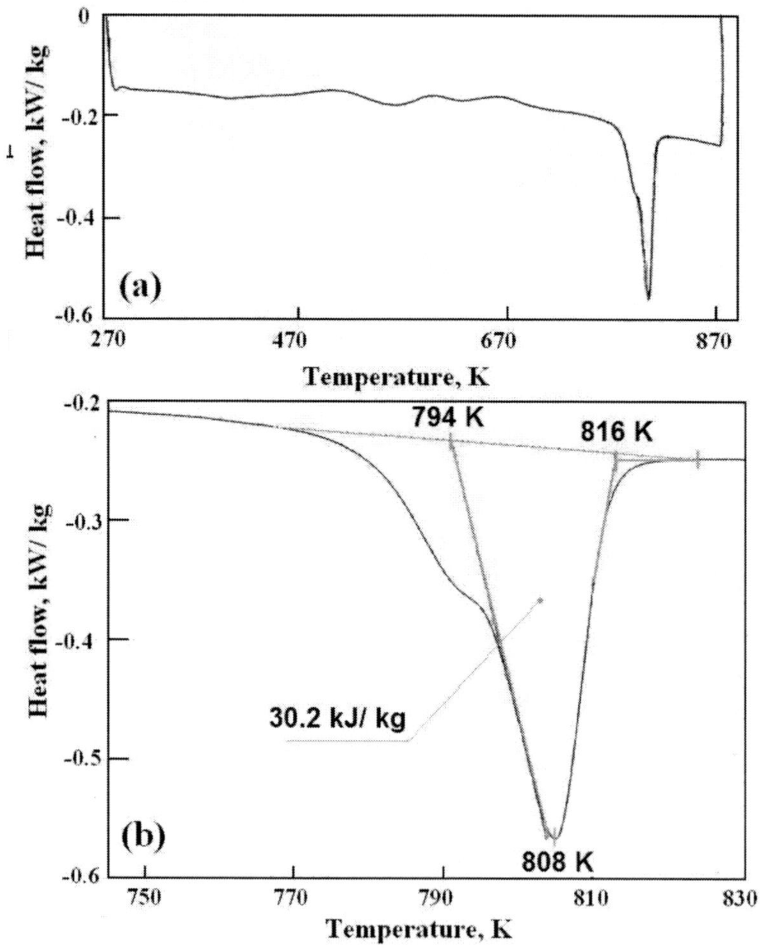

Figure 12. DSC chart recorded during the heating to 873 K of a Cu-21.6 Zn-7 Al (mass. %) SMA homogenized (1073 K/ 18 ksec/ water) and tempered (773 K/ 300 s/ water): (a) general aspect; (b) detail of the endothermic peak

The effects of α-phase pre-existence, on the behavior of a specimen tempered at 773 K during heating up to 873 K, are noticeable in **Figure 12**. Figure 12(a) displays the general aspect of the DSC thermograph. Obviously, there is no evidence of reverse martensitic transformation, bainite formation and α-phase precipitation. This suggests that, during 300 seconds maintaining at 773 K, α-phase precipitation was complete. This assumption is sustained by the OM micrographs shown in Figure 7, which suggest a complete development of α-phase rods on the entire diameter of the grains. Such a structure, with fully developed α-phase rods, would not undergo any bainitic transformation, because tempering bainite can only be regarded as a precursor and not a successor of α-phase. From Figure 12(b) it is noticeable that there are in fact two order-disorder transitions of austenite, probably in the succession $L2_1 \rightarrow B2$ and $B2 \rightarrow A2$ [42].

4. Conclusions

The transitory formation of α_1 bainite, during the heating of a martensitic Cu-21.6 Zn-7 Al (mass. %) SMA was associated to an endothermic peak, detected by both DSC and DTA techniques. Being located between approx. 537 and 643 K, on DSC charts, transitory bainite formation was accompanied by a sensible thermal expansion, between 610 and 657 K, on the dilatograms. Bainite was formed as a precursor of equilibrium α-phase, being characterized by specific enthalpy absorption of approx. 14 kJ/ kg.

Assuming that, after subjecting homogenized specimens (with almost fully martensitic structure) to tempering (573K/ 300 sec/ water), most of their martensitic structure transformed to austenite, it is further assumed that a small part of martensite turned to α_1 bainite, while the rest of martensite became stabilized and remained untransformed [43, 44]. On these grounds, the presence on bainite, predicted by DSC, DTA and dilatometry, was revealed by OM on the specimens tempered at 573 K. On OM micrographs α_1 bainite revealed plate-like morphology comprising subplates with internal striations. TEM micrographs confirmed the preservation of plate-shaped form and characteristic SAD pattern for 9R orthorhombic α_1 bainite in Cu-Zn-Al SMAs and revealed the presence of high density of slip dislocations on both parts of bainite subplate pairs. Bainite plates had typical widths of 20 nm.

The occurrence of α-phase together with parallel equal-width bands of bainite was revealed by OM, as an effect of increasing tempering temperature to 673 K. Bainite plates had a V-formation characteristic for Widmanstätten structures also observed at higher tempering temperatures.

OM micrographs of the specimens tempered at 773 K confirmed the completion of α-phase precipitation and revealed rod-like morphologies with preferred orientation tendencies according to (111) and (200) close-pack planes families, characterized by intersections at about 70^0 and 90^0, respectively.

The role of undissolved α-phase, on the order-disorder transition of austenite, consisted in increasing transformation temperatures and diminishing phase-transition intensity. When subjected to heating up to 873 K, a specimen tempered at 773 K underwent only order-disorder transition of austenite which, in this case occurred in two steps.

These results suggest that, in the case of martensitic Cu-Zn-Al SMAs subjected to tempering, bainite is formed only as a precursor and never as a predecessor of α-phase. In other words, during heating bainite plates transform to α-phase rods but no bainite will form if α-phase is already present in microstructure.

ACKNOWLEDGMENTS

The author wishes to thank Assoc. Prof. S. Stanciu for valuable help in casting and specimen preparation. Special thanks are brought to Prof. G. Eggeler for support in performing DSC measurements and to Dr. T. Kikuchi for his expertise in TEM observations. Dr.T. Sawaguchi is gratefully acknowledged for the fruitful discussions and perpetual encouragement.

Financial support was offered by UEFISCU by means of the research grant PN II - ID 301 - PCE - 2007 - 1, contract no. 279/ 01.10.2007

REFERENCES

[1] Bhadeshia, H. K. D. H. *Bainite in Steels*. Second edition, IOM Communications Ltd, London, 2001; pp 2-4

[2] Ogawa, K.; Kajiwara S. *Mater Sci Eng A* 2006, *438–440*, 90–94

[3] Hillert, M. *Scripta Mater* 2002, *47*, 137-138

[4] Aarson, H. I.; Spanos, G.; Reynolds, W.T.Jr. *Scripta Mater* 2002, *47*, 139-144

[5] Sinha, A. K. *Ferrous Physical Metallurgy*, Butterworth, Boston, 1989; pp. 282-293

[6] Anderson, J. C.; Leaver, K. D.; Rawlings, R. D.; Alexander, J. M. *Materials Science, fourth* edition, Chapman and Hall, 1990; pp 265-267

[7] Bhadeshia, H. K. D. H.; Sir Honeycombe, R Steels. *Microstructure and Properties.* Third edition, Butterworth- Heinemann, London; 2006; pp 129-154

[8] Hillert, M. *Scripta Mater* 2002, *47*, 175-180

[9] Hsu, T.Y. *Curr Opin Solid St M* 2005, *9(6)*, 256-268

[10] Bhadeshia, H. K. D. H. *Mater Sci Eng A* 1999, 273–275, 58–66

[11] Wayman, C. M. In *Shape Memory Effects in Alloys*; Perkins, J. (Ed.); Plenum Press, New York 1975; pp. 1-27

[12] Wayman, C. M.; Duerig, T. W. In *Engineering Aspects of Shape Memory Alloys*, Duerig, T.W.; Melton, K.N.; Stöckel, D.; Wayman, C. M., (Eds.), Butterworth-Heinemann, London, 1990; pp. 3-20

[13] Hillert, M.; Purdy, G. R. *Scripta Mater* 2000, *43*, 831-833

[14] Cornelis, I.; Wayman, C. M. *Acta Metall* 1974, *22*, 301-311

[15] Hornbogen, E. *Mater Sci Forum* 2000, *329-330*, 57-64

[16] Takezawa, K.; Imamura, H.; Tanikzaki, K.; Sato, S. *Proceedings of the ICOMAT-82*, *J. Phys.* 1982, *43*, C4 741-746

[17] Wu, M. H., Wayman, C. M. In *Proceedings of the ICOMAT-86*, The Japan Institute of Metals, Nara, JP, 1986; pp. 619-624

[18] Lovey, C.; Torra, V. *Progr Mater Sci* 1999, *44*, 189-289

[19] Takezawa, K.; Sato, S. In *Proceedings of the ICOMAT-86*, The Japan Institute of Metals, Nara, JP, 1986; pp. 625-630

[20] Takezawa, K.; Sato, S. In *Proceedings of the ICOMAT-86*, The Japan Institute of Metals, Nara, JP, 1986; pp. 631-636
[21] Takezawa, K.; Sato, S. *Mater Trans JIM* 1991, *32(8)*, 766-773
[22] Bujoreanu, L.-G.; Craus, M. L.; Rusu, I.; Stanciu, S.; Sutiman, D. *J Alloys Compd* 1998, *278*, 190-193
[23] Bujoreanu, L.-G.; Craus, M. L.; Stanciu, S.; Dia, V. *Mater Sci Technol* 2000, *16*, 612-616
[24] Zhang, X.; Jin, X. J.; Xu, Z. Y. *J Mater Sci Technol* 2002, *18(1)*, 1-3
[25] Pops, H. In *Shape Memory Effects in Alloys*; Perkins, J. (Ed.); Plenum Press, New York 1975; pp. 525-536
[26] Hsu, T. Y. *Mater Sci Forum* 2001, *394-395*, 369-374
[27] Suzuki, Y. In *Shape Memory Materials*, Otsuka, K.; Wayman, C.M. (Eds.), Cambridge University Press, 1998, pp 133-148
[28] Lee, E.-S.; Kim, Y. G.. *Acta Metall Mater* 1990, *38*, 1669-1676
[29] Bujoreanu, L. G. *Mater. Sci. Eng. A* 2008, *481–482*, 395–403
[30] Bujoreanu, L.-G.; Stanciu, S.; Barsanascu, P.; Lohan, M. N. In *Advanced Topics in Optoelectronics, Microelectronics, and Nanotechnologies IV*, edited by Schiopu, P., Panait, C., Caruntu, G., Manea, A., Proc. of SPIE 2008, Vol. 7297, 72970B
[31] Cornelis, I.; Wayman, C. M. *Acta metall* 1974, *22*, 301-311
[32] Garwood, R.D., In *Martensite. Fundamentals and Technology*, Petty, E. R. (Ed.), Longman, Bristol, 1970, pp. 95-117
[33] Bujoreanu L.-G.; Schürhoff, J.; Stanciu, S.; Roman C. *Metall New Mater Res* 1997, *4*, 12-24
[34] Spielfield, J. *Mater Sci Eng A* 1999, *273-275*, 639-643
[35] Bujoreanu, L.-G.; Stanciu, S.; Ioniță, I.; Munteanu, C.; Găluşcă, D. G. *Metalurgia Int* 2005, *X(2)* 3-10
[36] Zel'dovich, V.; Khomskaya, I.; Frolova, N.; Novoselova, T. V. *J Phys IV* 1997, *7(C5)*, 179-184
[37] Guilemany, J. M.; Fernández, J. *J Mater Sci* 1996, *31,* 4981-4984
[38] Khomskaya, I.; Zel'dovich, V.; Frolova, N. *J Phys IV* 2003, *112*, 529-532
[39] Li, C. M.; Yang, Z. G.; Wang, J. J.; Zheng, Y. K.; Fang, H. S. *Metall Mater Trans A* 1997, *28(8)*, 1617-1623
[40] Li, C. M.; Fang, H. S.; Yang, Z. G.; Bo, X. Z.; Wang, J. J.; Zheng, Y. K. *J Mater Sci* 1989, *33(2)*, 487-496
[41] Ahlers, M. *Progr. Mater. Sci.* 1986, *30*, 135-186
[42] Janssen, J.; Van Humbeeck, J.; Chandrasekaran, M.; Mwamba, N.; Delaey, L. *Proceedings of the ICOMAT-82, J. Phys.* 1982, *43*, C4, 715-720
[43] Kustov, S.; Golyandin, S.; Sapozhnikov, K.; Cesari, E.; Van Humbeeck, J.; De Batist R. *Acta Mater* 2002, *50*, 3023-3044
[44] Kustov, S.; Pons, J.; Cesari, E.; Van Humbeeck, J.; Morin, M. *Mat. Sci. Eng. A* 2004, *378*, 283–288

In: Shape Memory Alloys
Editor: H. R. Chen, pp. 287-315
ISBN 978-1-60741-789-7

Chapter 10

THE SHAPE MEMORY EFFECT IN ANTIFERROMAGNETIC ALLOYS

J. H. Zhang
School of Materials Science and Engineering, Shanghai Jiao Tong University,
Shanghai 200240, China

There are shape memory effect (SME) in Mn-based antiferromagnetic alloys same as MnCu, MnNi and MnFe etc. There exist both antiferromagnetic transition (AT) and martensitic transformation (MT) in the alloys. The MT with thermoelastic effect is following AT. When Ms of MT temperature is close to T_N of AT each other, the coupling of transformation between MT and AT is formed, which resulted in hysteresis-free characteristics of two-way SME associated with temperature. There also are magnetic field induced strain (MFIS) and magnetic shape memory effect (MSME) in the antiferromagnetic alloys. Twin boundaries of *fct* martensitie and walls of antiferromagnetic domain are consistent in the alloys. With the help of movement of (011) twin boundaries in *fct* martensite, the reorientation of martensite is produced and a larger MFIS is obtained under an applied field and showed SME property.

1. ANTIFEROMAGNETIC TRANSITION, DOMAIN WALLS AND ITS MOVEMENT UNDER APPLIED FIELD AND EXTERNAL STRESS, MAGNETOSTRICTION

1.1. The Characteristics of Antiferromagnet

The antiferromagnetic materials are also a magnetic order on magnetic structure. And the magnetic structure is composed of two ferromagnetic subcrystalline structures, which possess equivalent magnetic moment and opposite direction. So they do not show any magnetism and spontaneous magnetization. A weak magnetism can be obtained under an applied magnetic field and the susceptibility is about an order of 10^{-2}~10^{-5} magnitudes.

1) The critical temperature and the susceptibility of antiferromagnetic materials

There is a critical temperature of antiferromagnetic transition, T_N, which is named as Néel point. When $T > T_N$, materials are paramagnetic and $T < T_N$, that show antiferromagnetic property.

At $T > T_N$ the change of the susceptibility with temperature obeys Curie-Weiss law[1]

$$\chi = \frac{C}{T+\theta} \tag{1}$$

where C is constant, T is absolute temperature, θ is Néel point.

At $T < T_N$ the change of the susceptibility with temperature in single crystal and polycrystal is shown in Figure1. Where the relation of χ between single crystal and polycrystal is

$$\chi_{poly} = \frac{1}{3}(\chi_{//} + 2\chi_{\perp}) \tag{2}$$

where $\chi_{//}$ and $\chi_{\perp}$ are the susceptibility parallel to and perpendicular to applied field, respectively, in single crystal. $\chi_{\perp} > \chi_{//}$.

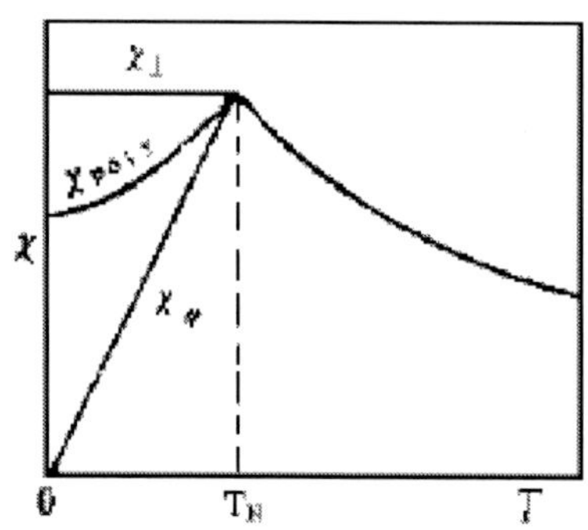

Figure 1. The curve of χ-T in single crystal and polycrystal

2) The order of antiferromagnetic transition

Antiferromagnetic transition belongs to a second order transition[2]. At transition point chemical potential and its first derivative of two phases are equivalent, but the second derivative of the chemical potential shows no equivalent. Expansion coefficient(α), volume compressibility(β) and specific heat(C_p) are also not equivalent.

$$\mu_1 = \mu_2, \left.\frac{\partial \mu_1}{\partial P}\right|_T = V_1 \neq \left.\frac{\partial \mu_2}{\partial P}\right|_T = V_2, \left.\frac{\partial \mu_1}{\partial T}\right|_p = -S_1 \neq \left.\frac{\partial \mu_2}{\partial T}\right|_p = -S_2 \tag{3}$$

$$\alpha_1 \neq \alpha_2,\ \frac{1}{V_1}\left(\frac{\partial}{\partial T}\left(\frac{\partial \mu_1}{\partial p}\right)_T\right)_p \neq \frac{1}{V_2}\left(\frac{\partial}{\partial T}\left(\frac{\partial \mu_2}{\partial p}\right)_T\right)_p \tag{4}$$

$$\beta_1 \neq \beta_2,\ -\frac{1}{V_1}\left(\frac{\partial^2 \mu_1}{\partial p^2}\right)_T \neq -\frac{1}{V_2}\left(\frac{\partial^2 \mu_2}{\partial p^2}\right)_T \tag{5}$$

$$C_{p1} \neq C_{p2},\ -T\left(\frac{\partial^2 \mu_1}{\partial T^2}\right)_p \neq -T\left(\frac{\partial^2 \mu_2}{\partial T^2}\right)_p \tag{6}$$

1.2. The Displacement of Antiferromagnetic Domain Wall

1) The formation and observation of antiferromagnetic domain walls

The magnetic domains and domain walls are formed in antiferromagnetic transition. With the help of the local fluctuation of antiferromagnetic order parameter, antiferromagnetic transition leads to form some macroscopic magnetic domains with different antiferromagnetic vectors in the antiferromagnet. The domains are divided with walls each other. Because the vector of antiferromagnetic spontaneous magnetization is almost close to zero, so some method for observing ferromagnetic domains can be not used to antiferromagnetic domains. Generally antiferromagnetic resonance, Neutron diffraction, ultrasonic damping[3], neutron topography or X-ray topography[4,5] etc. can be used to measure antiferromagnetic domains. When the size of the domains is larger, a polarized-light microscope can also be used to observe some antiferromagnetic domains.

Roth [6] observed the movement of T-domain walls with a polarized-light microscope in NiO single crystal, the different morphology of ant ferromagnetic domains for applied magnetic field on [$\bar{1}\,\bar{1}\,2$] and [$\bar{1}\,10$] directions shows in Figure2.

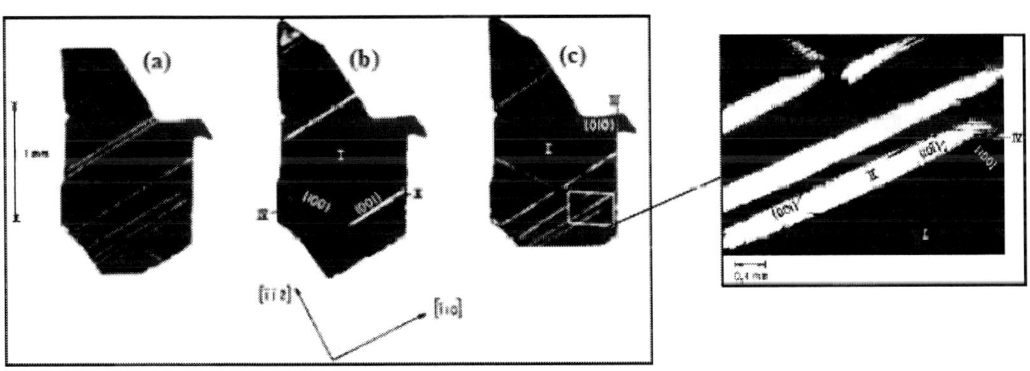

*Detail of the facet in (c) is above right image.

Figure 2. The displacement of NiO domain walls with the change of magnetic field direction (a) without any applied magnetic field, (b) with $[\bar{1}\,\bar{1}\,2]$ and (c) with $[\bar{1}\,10]$.

Antiferromagnetic domain walls can be moved under applied magnetic field and external stress. Safa and Tanner[7] studied directly domain wall movement in the cubic antiferromagnets $KNiF_3$ and $KCoF_3$ with X-ray topography in field of 1.5T at 4.2K. Clark and Tanner[8] reported antiferromagnetic domain walls motion on (101) and ($\bar{1}01$) planes under external stress on [001] direction under increasing uniaxial stress with synchrotron X-ray topography in $KCoF_3$ at 77K. The wall displacement is linearly with stress above σ_{th}(threshold).

2) The thermodynamics theory of antiferomagnetic domain wall motion[7]

In a domain with spin directions having direction cosines α, β and γ with the cube axes, the free energy per unit volume due to the interaction with an applied magnetic field, the direction cosines of which are a, b and c with respect to the cube axes, is

$$E_H = -\frac{1}{2}\chi_{//}H_{//}^2 - \frac{1}{2}\chi_{\perp}H_{\perp}^2 = \frac{1}{2}(\chi_{\perp} - \chi_{//})H^2(\alpha a + \beta b + \gamma c)^2 - \frac{1}{2}\chi_{\perp}H^2 \quad (7)$$

If a wall between two domains with direction cosines α, β, γ and α', β' and γ' respectively, is displaced by a distance x, the change of free energy (per unit wall area) is

$$\Delta E_H = \frac{1}{2}s(\chi_{\perp} - \chi_{//})H^2[(\alpha a + \beta b + \gamma c)^2 - (\alpha' a + \beta' b + \gamma' c)^2] \quad (8)$$

So an effective force on the wall can be written

$$F_H = \frac{1}{2}(\chi_{\perp} - \chi_{//})H^2[(\alpha a + \beta b + \gamma c)^2 - (\alpha' a + \beta' b + \gamma' c)^2] \quad (9)$$

In 1954, Néel[9] postulated a restoring force, -Cx, proportional to the displacement, which leads to a contribution to the free energy given by $\Delta E_R = Cx_2/2$. That implies at least over certain ranges of displacement, antiferromagnetic domain wall motion is reversible.

In order to explain a threshold field, H_{th}, a constant coercive force $-B$ acting on the wall is assumed and leading to a free energy contribution $\Delta E_C = Bx$.

Consider the displacement of a wall between [001] and [100] (or [010]) domain in a field applied to [001]. The total change in free energy is

$$\Delta E = -\frac{1}{2}x(\chi_{\perp} - \chi_{//})H^2 \sin^2\theta + \frac{1}{2}Cx^2 + Bx + x(\frac{K_1}{4})\sin^2 2\theta \quad (10)$$

where θ is the angle to which spins turn with respect to [001] in [100] (or [010]) domain. $K1$ is the first anisotropy constant in the cubic system. According to equilibrium condition, the partial derivative of ΔE with respect to θ and x equal to zero. We obtained $\theta = n\pi/2$ (n is integral) and

$$x = \frac{\left[\frac{1}{2}(\chi_{\perp} - \chi_{//})H^2 - B\right]}{C} \tag{11}$$

From Eq.(11), when $x > 0$, we obtained a threshold field H_{th} given by

$$H_{th} = \left[\frac{2B}{(\chi_{\perp} - \chi_{//})}\right]^{1/2} \tag{12}$$

At a critical field, H_{cr}, the displacement will equal the spacing L between walls and wall motion stops. Thus

$$H_{cr}^2 - H_{th}^2 = \frac{2CL}{(\chi_{\perp} - \chi_{//})} \tag{13}$$

From Eqs. (11), (12) and (13) we obtained

$$x = L\left[\frac{H^2 - H_{th}^2}{H_{cr}^2 - H_{th}^2}\right] \tag{14}$$

Under an external stress σ whose direction cosines with the cube axis are β_1, β_2, β_3, when for stresses above σ_{th} walls are also expected to be displaced linearly with stress until a critical stress σ_{cr} is reached where the displacement equals original wall separation L. for a stress parallel to [001] and the angle of the spins θ_1, θ_2 with [001] in [001] and [100] (or [010]) domains. Then

$$\sigma_{th} = -\frac{2B}{3\lambda_{100}} \tag{15}$$

When a critical stress σ_{cr} reached where the displacement equals the original wall separation L. Similar to Eq.(13),

$$CL = \frac{3}{2}\lambda_{100}(\sigma_{cr} - \sigma_{th}) \tag{16}$$

where λ_{100} is strain on [100] direction.

1.3. The Magnetostriction of Antiferromagnet

There is distortion of crystal lattice in antiferromagnetic transition, an applied magnetic field can result in the magnetostriction of antiferromagnet. Alberts[10] obtained the magnetostriction of 2×10^{-6} on $[\bar{1}10]$ direction at 2.1T in a single crystal of NiO, the strain is proportional to H^2 before saturation. After compressing on [111] direction, the strains of 33×10^{-6} on $[\bar{1}10]$ and 23×10^{-6} on $[11\bar{2}]$ were respectively shown[11].

Large magnetostrictive effects were observed in dysprosium and holmium single crystals. The maximum linear strains of 7500×10^{-6} for dysprosium at 20K and 3500×10^{-6} for holmium at 45K were reported[12].

Magnetostrictive effect is reversible, the magnetostriction of polycrystalline $Pr_{0.5}Sr_{0.5}MnO_3$ at T=25K and H=14.7T with an order of 10^{-3} is obtained. λ_{pall} parallel to the direction of applied magnetic field is positive and λ_{per} perpendicular to that is negative. With decrease of temperature the both of magnetic strain is increased [13]. Seen Figure3.

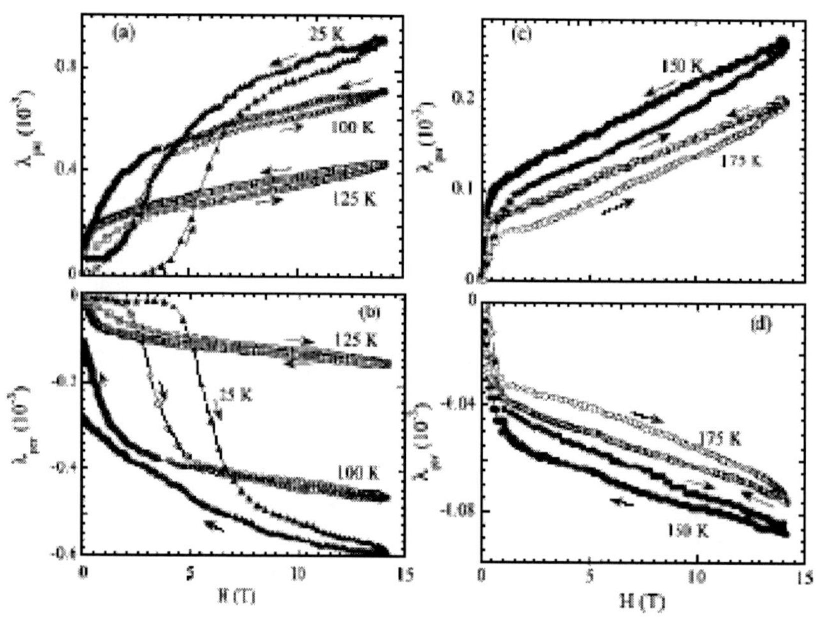

Figure 3. Parallel (λ pall) and perpendicular (λ per) magnetostriction isotherms of $Pr_{0.5}Sr_{0.5}MnO_3$.

The motion of antiferromagnetic domain wall under applied stress or applied magnetic field and magnetostriction are reversible and show shape memory effect. But though the distortion increases with the decrease of temperature, the magnetic induced strain is lower due to the distortion of crystal lattice in antiferromagnet is a little, specially in polycrystalline condition. And that must be at very low temperature.

A comparatively large magnetostriction in antiferromagnets can be obtained, only in antiferromagnetic compounds, at cryogenic temperature or with a relatively strong magnetic field.

2. Antiferromagnetic Transition and Martensitic Transformation, Magnetic Structure, Lattice Distortion, Transformation Softening, the Transformation Coupling in γMn-based Alloys.

2.1. The Antiferromagnetism of γMn-Based Alloys

1) The magnetic structure of γMn

Mn belongs to transition metal, there is four allotropes, α, β, γ, δ. It is an antiferromagnetic metal with magnetic moment of 2.3 μB. Antiferromagnetic structure suggested by Meneghetti[14] and after that by Umehayshi etc[15] is show in Figure4. The spin directions are on [001], [110] and [111], to reach lowest energy state. The magnetic structures of [001] or [110] directions posses tetragonal symmetry and [111] forms cubic symmetry. Generally, [001] structure is named as collinear spin structure or single-Q and [111] structure as non collinear spin structure or triple-Q.

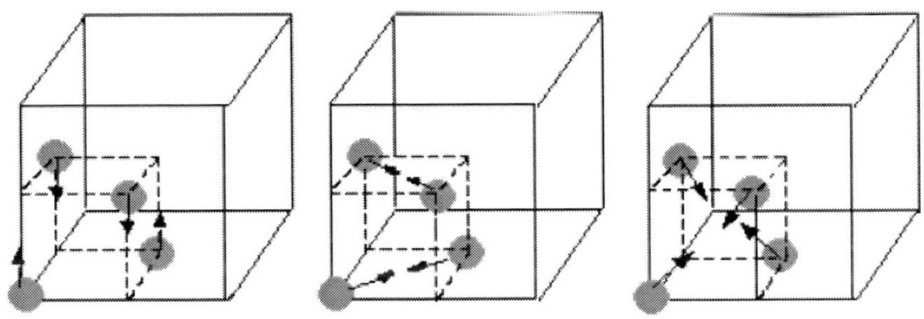

(a) (b) (c)

Figure 4. Antiferromagnetic structure of γMn (a) [001] (b) [110] and (c) [111].

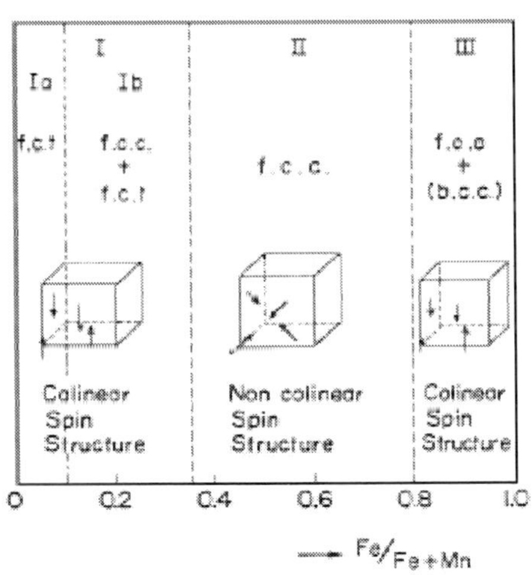

Figure 5. Magnetic phases in Mn-Fe alloys[16].

2) The magnetic structure of γMn-based alloys In γMn-based alloys the magnetic structure is related with composition of the alloys. For example,the antiferromagnetism in $Mn_{1-x}Fe_x$ alloys can be classified into three groups: i) $0 \leq x \leq 0.3$, ii) $0.4 \leq x \leq 0.8$, iii) $0.8 \leq x \leq 1.0$, as shown in Figure5.

But the accurate concentration of their boundaries, has not yet been determined. It is seen in Figure5 that both i) and iii) regions belong to collinear spin structure and ii) is with non spin structure.

The magnetic properties in the first region are almost the same as the one for γMn. The Néel temperature and the magnetic moment of the alloys dec- rease gradually with increasing iron concentration.

The different magnetic structure may be effect their mechanism of magnetic field induced strain.

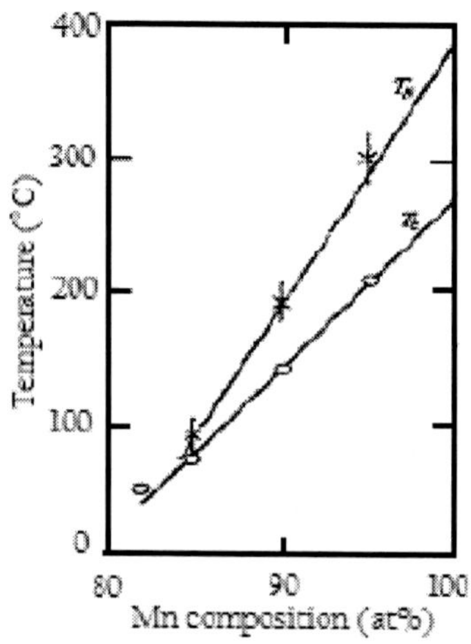

Figure 6. Change of T_N and Ms[17].

2.2. Martensitic Transformation and Antiferromagnetic Transition

1) Martensitic transformation in γMn-based alloys In γMn-based alloys martensitic transformations take place which are associated with antiferromagnetic ordering of Mn atoms. Generally the martensitic transformation is following the antiferromagnetic transition.

In Mn-Cu alloys, once the Mn content is higher than 69 at%, *fcc*→*fct* martensitic transformation comes to heel antiferromagnetic transition. For Mn>80at% alloys, the relationship between T_N and Ms is shown in Figure6. When Mn<82at%, TN and Ms are almost consistent and can dot be differentiated.

According to tight relationship of the both transformations, Shimizu[18] suggested that antiferromagnetic transition leads to a tetragonal distortion of *fcc* crystal (an order of 10^{-6}) which touches off *fcc*→*fct* martensitic transformation with distortion of 10^{-2} and forms a lot of (110) twins of martensite.

Similar to Mn-Cu alloy, there is the *fcc→fct* martensitic transformation following antiferromagnetic transition in Mn-Ni, Mn-Pd and Mn-Fe(Cu) alloys, as shown in Figure7 and Figure8.

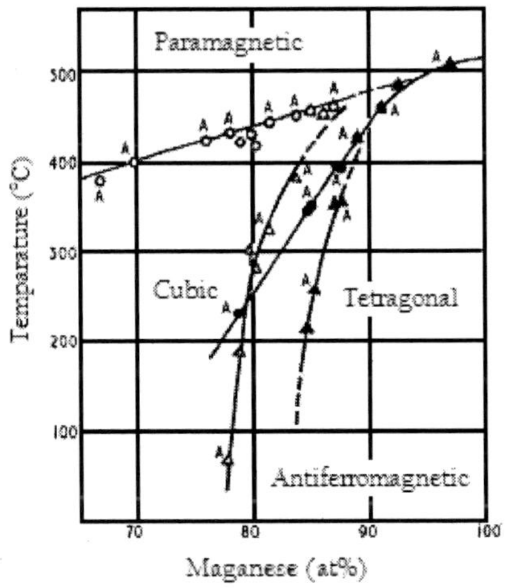

Figure 7. The magnetic and structural phase diagram Mn-Ni of (hollow)and Mn-Pd (solid)○ ● (T_N)△ ▲(Ms)[19]

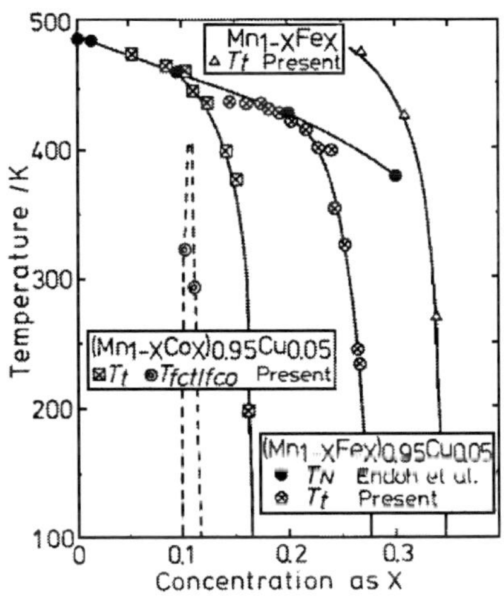

Figure8 The magnetic and structural phase diagram of Mn-Fe (Cu) alloys[20].

The martensitic transformation in γMn-based alloys may be divided in three groups. The first group is characterized by coincidence of of T_N (Néel point) and M_S (the temperature of martensitic start transformation) at which *fcc* structure transforms into *fct* structure with c/a<1. For the alloys belonging to the second group Ms<T_N and the *fct* structure is characterized by c/a>1. In the intermediate narrow concentration fields alloys exist where rhombic structure (R) is formed and the chain of martensitic transformation can take place with *fcc→fct*(c/a<1)→*fco* (R).

For Mn-Fe alloys, when Fe content <25at%Fe *fct* martensitic structure belongs first group. Fe content of 27~30at% is second [21], as shown in Figure9(a). In Mn-Ni alloys[22], similar to Mn-Fe alloys when Ni<15at%, Ms≈TN and c/a<1 the region belong first group. Ni content of 15~22at%, Ms<TN, the region is second and c/a>1. For Ni content of 13~17at% may take place with *fcc*→*fct*(c/a<1)→*fco* (R) and that belongs to third. As shown in Figure9(b).

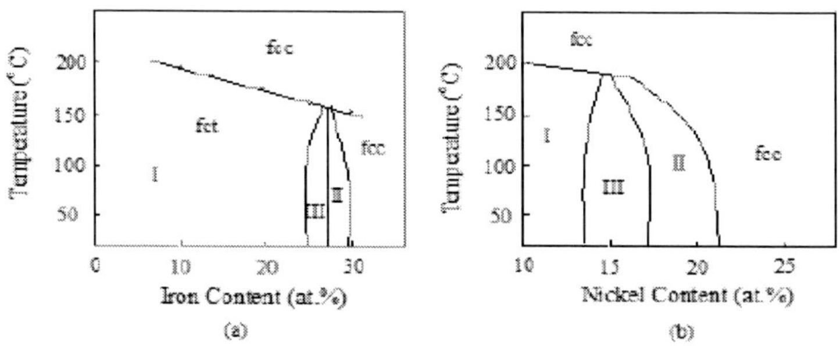

Figure 9. The phase diagram for Mn-Fe(a) and Mn-Ni(b) alloys.

2) The soften model of antiferromagnetic transition and martensitic transformation

In γMn-based alloys there is soften model of modulus of elasticity at transformation temperature for both antiferromagnetic transition and martensitic tran2sformation. But they are very different each other in different alloys. TN and Ms are almost consistent of Mn-Cu alloy and *fct* martensitic transformation formed (110) twins. At transformation temperature elastic modulus shows soften, as show in Figure10. Sugimono[24] suggested to be C′ soften. However it is discovered with neutron diffraction that soften phonon was not [110]T (shear moduli) but [100]L(vertical) in Ms temperature, it is not soften of $(C_{11}\text{-}C_{12})$ but $(C_{11}+2C_{12})\rightarrow 0$. That may be related with the coupling of antiferromagnetic transition.

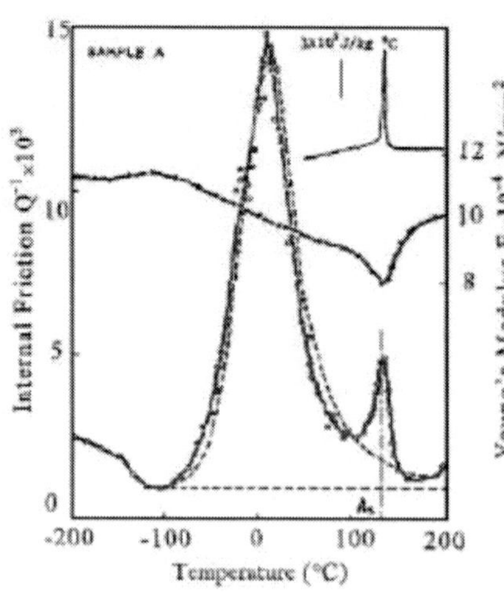

Figure 10. Variation if internal friction and Young' modulus in 88.26wt%Mn-Cu alloy[23].

For Mn-Ni-C alloys there also is *fcc→fct* martensitic transformation Figue11 is related with change of elastic moduli in Mn-9.2at%Ni-6.1at%C alloy.

Here first C11 decreases with temperature, then $C'=(C_{11}-C_{12})/2$ decreases. At Ms(125K) C_{11} reaches minimum (0.017) and the C', (primary soften model: [110]T) is not zero but 5Gpa. It shows the effect of antiferromagnetic transition on *fct* martensitic transformation.

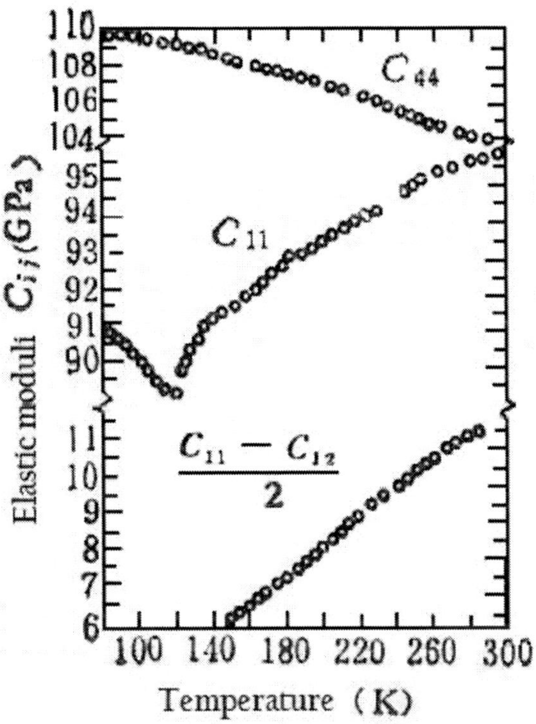

Figure11. The change of elastic moduli with temperature in Mn-9.2Ni-6.1C alloy modulas[25]

Figure12 is Temperature dependence of Young's in Mn-Ni alloy with different composition. Compare with Figure7, it is clear that there is soften in both antiferromagnetic transition and martensitic transformation in γ-Mn based alloys and shows effect each other. From Figure12 it is can be seen that the closer is the relation between Ms temperature and T_N point, the rapider is the soften of antiferromagnetic transition. It shows that the effect of martensitic transformation soften on antiferomagnetic transition one. Ms of Mn-6.6at%Ni alloy is almost equals to T_N, because of the coupling of both softens, that leas to a maximum decrease of the alloy with this composition. When $T_N << Ms$, the soften of antiferromagnetic modulus is very little.

Figure13 is Temperature dependence of the elastic Moduli in Mn-61.5at%Fe alloy. Also compare with Figure8, soften temperature in Figure13 is corresponding to antiferromagnetic transition T_N and so the antiferromagnetic transition in Mn-Fe alloys has soften of [110]T. Namely, there is the same soften for antifer- rmagnetic transition and martensitic transformation in Mn-Fe alloys.

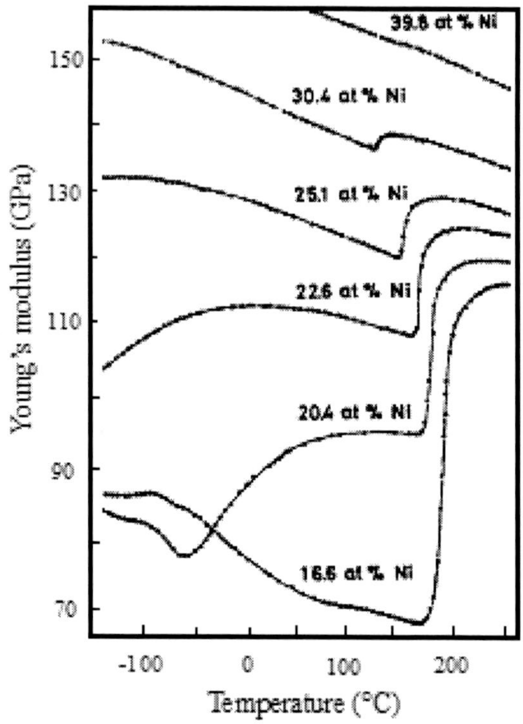

Figure12 The change of Young's modulus in Mn-Ni alloy with different composition[26].

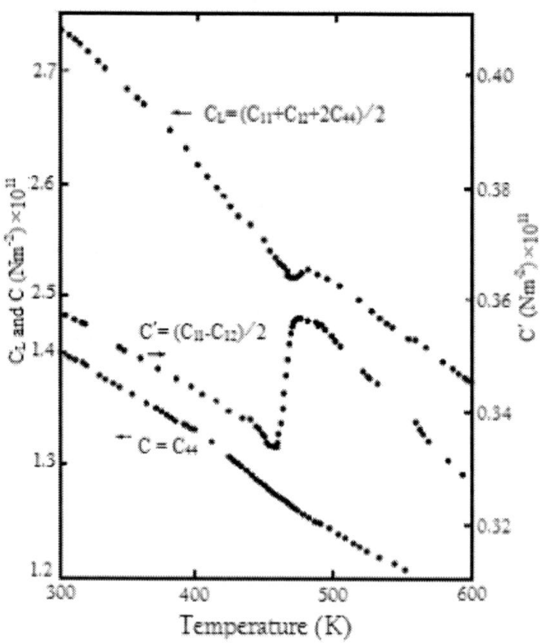

Figure13. is Temperature dependence of the elastic Moduli in Mn-61.5at%Fe alloy[27].

2.3. The Coupling between Antiferromagnetic Transition and Martensitic Transformation

Coupling between antiferromagnetic transition and martensitic transformation leads to a continual change of lattice structure during direct martensitic transformation and its reversible one in cooled and heated. Figure14 is the change of both Mn-Ni-C alloys with temperature. The *fct* martensite belongs to the tetragonal structure with c > a .The lattice constant is continually changed with temperature for the B alloy. The martensitic transformation is very like as a second order transformation. It can be attributed to its Ms be close to T_N[25].

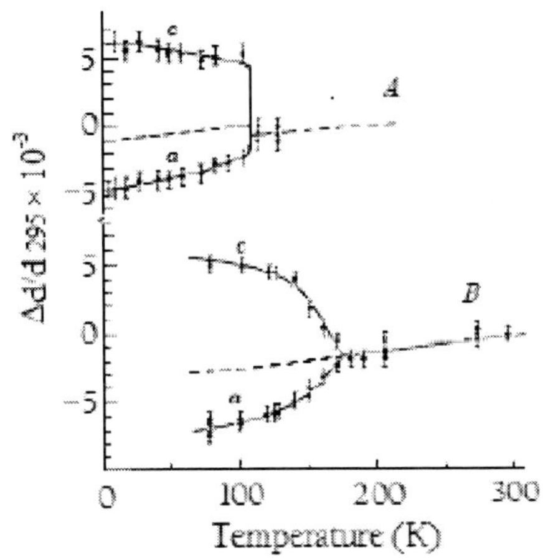

Figure14. Change of lattice measured with neutron diffraction(A) Mn-9.2±0.2at%Ni-6.1±0.2at%C and (B) Mn-8.8±0.2at%Ni-5.9±0.3at%C alloys.

In Mn-Fe and Mn-Fe-Cu alloys there also is the coupling between antiferromagnetic transition and martensitic transformation, as shown in Figure15(a) and Figure15(b) [28]. Lattice parameters are c < a in the alloys.

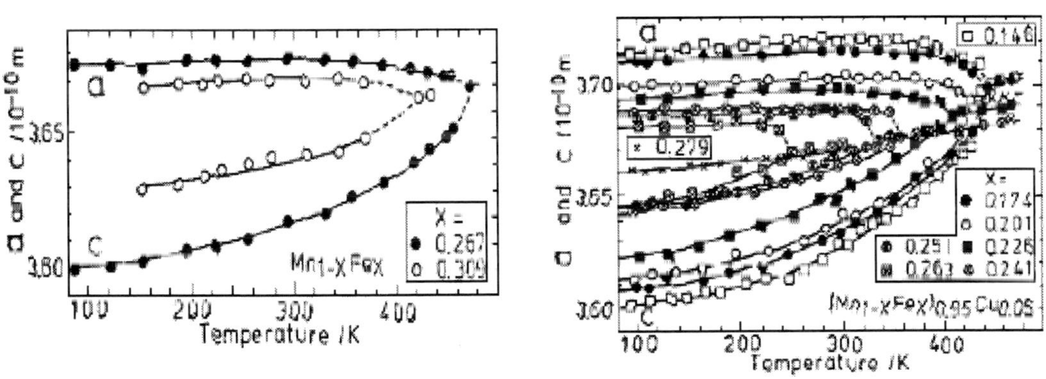

a) b)

Figure 15. Temperature vs. lattice parameters of (a) Mn-Fe alloys (b) Mn-Fe-Cu alloys.

From some results it can be obtained that at Ms=T_N the coupling between first order martensitic transformation and second order antiferromagnetic transition leads to the martensitic transformation to have the characteristics of second order transition.

3. Shape Memory Effect Associated with Temperature with Hysteresis-Free Characteristics

3.1. Improper Martensitic Transformation

Martensitic transformation can be classed with proper or improper martensitic transformation. The characteristics of proper martensitic transformation are i) first order nucleation-grown transformation ii) there is no diffusion and with typed-shear iii) there is a invariable plane strain and surface relief.

Rich-Mn γMn-based alloys belong to improper martensitic transformation at Ms ~T_N. i) Martensitic transformation and antiferromagnetic transition are continuous or close to one another in transformation process ii) the product of transformation shows twin structure iii) there is lattice soften with shear-typed property. As shown in chapter 2.

3.2. Hysteresis-Free Shape Memory Effect

In 1978 Vintaykin etc[29] reported two-way shape memory effect in Mn-Ge and Mn-Ga alloys. With a pre-bending deformation of 5.0% it is shown in Figure16.

In Figure16 the curve 1 is Mn-10.3at%Ge; 2 is Mn-13.3at%Ge; 3 is Mn-15.5at%Ge; 4 is Mn-24.1at%Ga. In the alloys the change of bending deformation with temperature shown two-way shape memory effect and keep their stability in heating and cooling cycles.

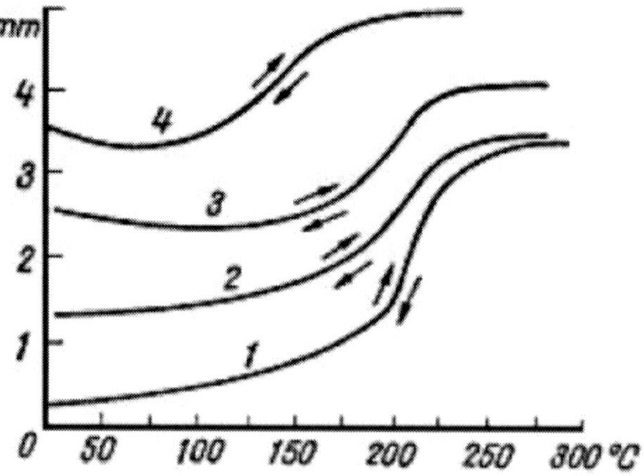

Figure 16 Two-way shape memory effects.

1) The first class alloys (Ms=TN, c/a<1)

Ms temperature of Mn-19at%Cu-1at%Ni alloy[30] quenched from 900°C and then Spindal decomposed at 450°C for 2h is at 170°C. Ms and Af are practically consistent with T_N. The crystal lattice parameter of the alloy changed with the temperature from face centered cubic to tetragonal i.e. *fcc→fct*. The change is gradual and continuous, almost a second order transformation. That shows the coupling between antiferromagnetic transition with second order characteristics and martensitic transformation with first order one. As shown in Figure17.

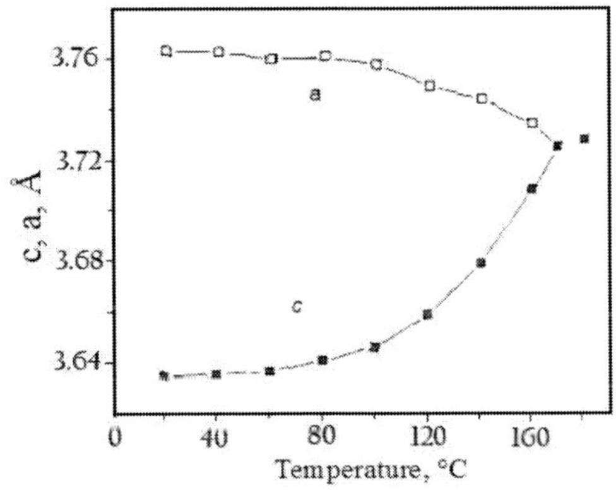

Figure 17. The variation of lattice parameters with temperature of Mn-19at%Cu-1at%Ni alloy

Figure18 illustrates the change of shape of Mn-19at%Cu-1at%Ni alloy in the form of an Archimedean spiral manufactured by the twist of a strip at the room temperature. When the temperature is raised the spiral untwists in the direction opposed to the sign of the preliminary plastic deformation (curve 1). The process is completed near the temperature of reverse martensitic finishing point Af. The shape recovery is approximately 0.2~0.5 of pre-deformation value depending on degree of preliminary deformation. With subsequent decrease of temperature the spiral is deformed along curve 2. With continuation of temperature cycling it keep the change along curve 2. The result shows there is a two-way shape memory effect with hysteresis-free characteristics of the Mn-19%Cu-1%Ni alloy.

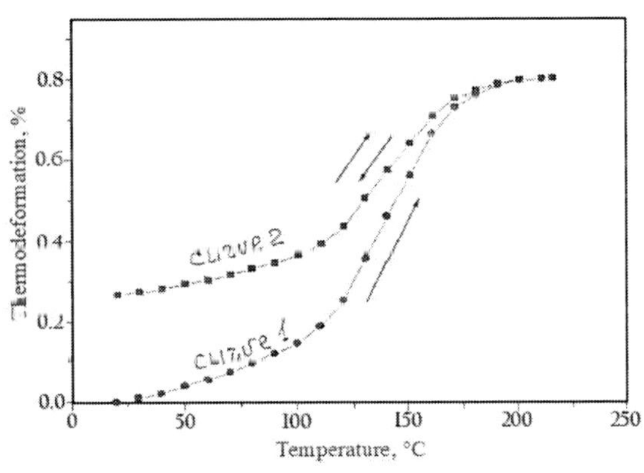

Figure18 Strain recovery upon reversible shape memory effect of Mn-19at%Cu-1at%Ni alloy.

In Mn-15.8at%Cu alloy belonged also to first class after bending strain at room temperature the sample with heating and cooling cycles shows also an analogous two way hysteresis-free shape memory effect, see Figure19[21].

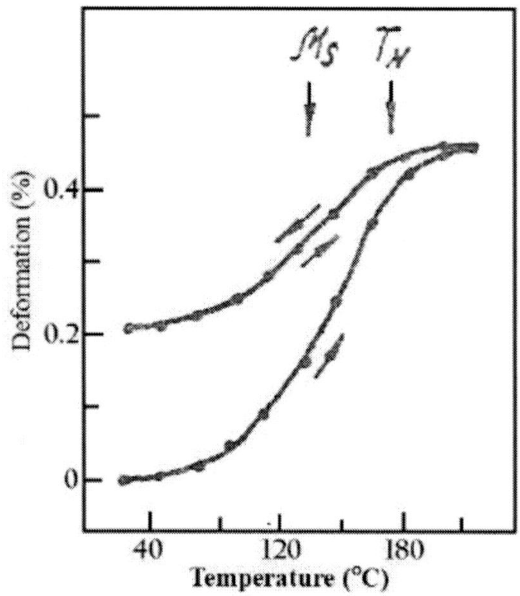

Figure 19.The shape memory effect in Mn-15at%Cu alloy.

Figure20 is anther analogous example with two way shape memory effect in Mn-9.5at%Fe-5.0at%cu alloy [31].

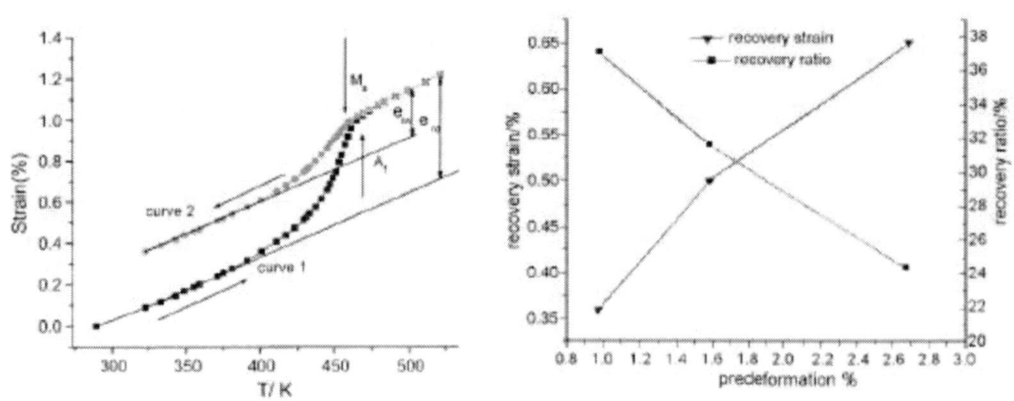

a) b)

Figure 20 The tow way shape memory effect in Mn-9.5at%Fe-5.0at%cu alloy (a) Dilatometer curve, (b)Effect of pre-strain ere and η_0.

Two-way temperature shape memory effect is measured on a dilatometer. A pre-compressive strain can result in both twins reorientation and plastic strain by dislocations at the same time. When sample compressed at room temperature in martensitic state was reheated above Af (the completed temperature of inverse martensitic transformation), the strain formed by dislocation can recover but the plastic strain by dislocation slipping can remain, as shown in Figure20(a). Here one-way reversible recovery ratio is determined from $\eta_0 = e_{re}/e_{pre}$, where e_{pre} is the pre-strain and e_{re} is the one-way recover strain, etw is two-way recovery strain, see the curve 1 in Figure20(a). The effect of pre-strain values on ere and η0 is listed in Figure20(b). ere increases with the increase of pre-strain and η_0 decreases with that. After first heating, by further heating and cooling the change of the strain follows curve 2. Curve 2 remain invariable for 50 cycles.

2) The second and third class alloys[21]

Ms ~ T_N and c/a >1 of second class alloy in γ-Mn based alloys, for example Mn-17at%Ni alloy, (see Figure9). The character of the shape memory effect is some differentiated for first class with Ms = T_N. However under the condition of Ms ~ T_N which is higher then Ms, also there is coupling between antiferromagnetic transition and martensitic transformation. That results in to show two-way shape memory effect similar to first class alloys. Figure21 is the result after bending strain. We can here observed a recovering and a subsequent reversible hysteresis-free deformation.

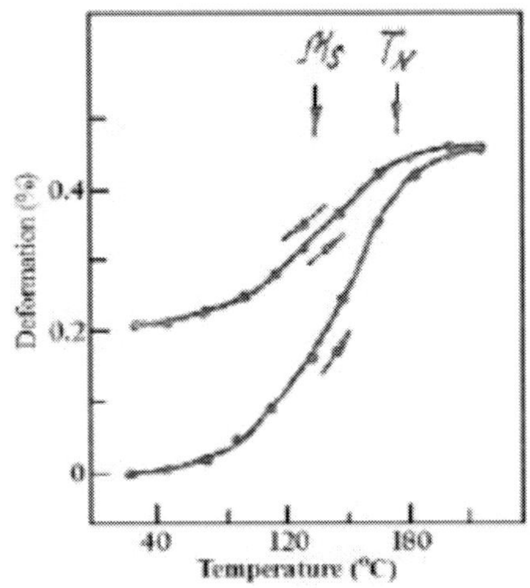

Figure 21. The shape memory effect in Mn-17at%Ni alloy (II class).

Mn-15at%Ni alloy belongs third class in γ-Mn based alloys. Ms ~ T_N, but c/a < 1. In the intermediate narrow concentration fields the alloys exist where rhombic structure (R-phase) is formed. With decrease of temperature its transformation approach is antiferromagnetic transition at T_N, next martensitic transformation (*fcc* →*fct*) and rhombic structure.

The change of lattice parameters is continuous and gradual, as shown in Figure22(a). In the third class alloys there is "a sign-variable reversible deformation". During the first heating usually an incomplete restoration of the shape is seen to take place, whereas in the subsequent cooling and heating cycles a hysteresis-free deformation occurs, a change in the effect sign taking place. The temperature of the sign variation corresponds to a transition temperature of *fct* →an orthorhombic phase, as shown in Figure22(b).

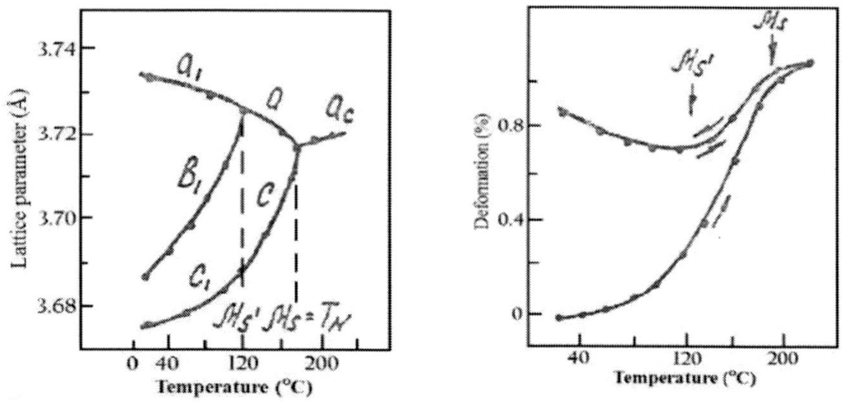

a) b)

Figure 22. The shape memory effect in Mn-15at%Ni alloy (III class) (a) the temperature dependency of lattice parameters (b) the shape memory effect.

3.3. Effect of Transformation Coupling on the Two-Way Shape Memory Effect[32,33]

From the viewpoint of thermodynamics, the driving force of a first order martensitic transformation can be written as

$$\Delta G^{p\rightarrow M} = \Delta G_c^{p\rightarrow M} + \Delta G_e^{p\rightarrow M} + \Delta G_s^{p\rightarrow M} \tag{17}$$

where $\Delta G_c^{p\rightarrow M}$ is a chemical free energy term originating in the difference of volume free energy between parent and martensite. The value of this term is negative, which provides for the increase of an elastic strain energy, $\Delta G_e^{p\rightarrow M}$, and an interfacial energy, $\Delta G_s^{p\rightarrow M}$, between parent and martensite during martensitic transformation. The value of ($\Delta G_e^{p\rightarrow M} + \Delta G_s^{p\rightarrow M}$) is positive, when the value of the ($\Delta G_e^{p\rightarrow M} + \Delta G_s^{p\rightarrow M}$) is larger that $\left|\Delta G_c^{p\rightarrow M}\right|$, a diving force of martensitic transformation, $\Delta G^{p\rightarrow M}$, is necessary. The larger the difference between $\left|\Delta G_c^{p\rightarrow M}\right|$ and ($\Delta G_e^{p\rightarrow M} + \Delta G_s^{p\rightarrow M}$), the larger are the driving force and thermo-hysterisis. As above shown for γ-Mn based alloys with Ms = TN (I class alloys) or Ms ~ TN (II and III class alloys), there is no obvious shear *M*s, and the lattice parameters change continuously as the temperature decreases below *M*s. Accordingly, it can be deduced that $\Delta G_s^{p\rightarrow M}$ is very small.

It is known that antiferromagnetic ordering is often accompanied by a tetragonal lattice distortion. There is an acoustic soft mode in first-order *fcc→fct* martensitic transformation. The acoustic soft mode can be coupling with the softening of the second-order antiferromagnetic transition (seen in above Section 2.2), so $\Delta G_e^{p\rightarrow M}$ is also very small.

Therefore, $\Delta G_e^{p\to M} + \Delta G_s^{p\to M}$ may be close to $\left|\Delta G_c^{p\to M}\right|$ and the driving force of martensitic transformation , $\Delta G^{p\to M}$, tends to become zero. That results in the *fcc→fct* martensitic transformation being hysterisis-free or with little thermal hysterisis.

With the increase of temperature, there also is exist the coupling between antiferromagnetic continuously disorder process and inverse martensitic transformation, which results in hysterisis-free or vary little thermo-hysterisis during inverse martensitic transformation.

4. MFIS and MSME of Antiferromagnets

4.1. The Correction of Domain Walls and Twin Boundaries

1) The motion of domain walls and magnetostriction in antiferromagnets

In low-anisotropy cubic antiferromagnets, a spin-flop transition is not observed and Néel (1954) suggested that the reorientation occurred via antiferromagnetic domain wall motion. Because susceptibility $\chi_\perp > \chi_{//}$ of antiferromagnets under an applied field, when the difference in magnetic energy $(\chi_\perp > \chi_{//})H2/2$ equals the anisotropy energy *K*, a spin reorientation can take place. Here there is a threshold value, *H*t, (Seen above Section 1.2)

$$H_t = \sqrt{\frac{2K}{(\chi_\perp - \chi_{//})}} \qquad (18)$$

The motion of domain walls results in magnetostriction of antiferromagnets. The most of magnetostriction obtained an order magnitude of 10^{-6}, due to little lattice distortion brought by antiferromagnetic transition, for example in NiO[10]. And that is at very low temperature, owing to low T_N, 4.2K for $KNiF_3$ and $KCoF_3$[7], and 20K for dysprosium and 45K for holmium (though they have more large strain value with an order magnitude of 10^{-3}[12]. In polycrystalline $Pr_{0.5}Sr_{0.5}MnO_3$ the magnetostriction with an order of 10^{-3} is obtained but the same at a vary low temperature of T=25K and under a very high applied field of H=14.7T[13].

Generally magnetic field induced strain with magnetostriction is very little, because there not is twin boundaries formed by martensitic transformation.

2) The motion of twin boundaries (domain walls) in antiferromagnets

When there are twin boundaries in antiferromagnet, the motion of twin boundaries caused under an applied field may be lead to more strain. Generally, the twin boundaries are consistent with antiferromagnetic domain walls[34,35] because of the coupling between antiferromagnetic transition and martensitic transformation.

The domain structure of an antiferromagnet has not been studied by now due to experimental problems arising in the investigation. The spontaneous magnetization of an antiferromagnetic domain equals to zero. Therefore, ordinary methods of ferromagnetic domains investigation cannot be applied to the antiferromagnetic domains. When the twins formed by a structural transformation are consistent with antiferromagnetic domains. With the help of observation for the twins the antiferromagnetic domains can be investigated with ordinary methods. Lavrov etc[36] studied magnetic shape-memory effect and the motion of twin boundaries (domain walls) in an antiferromagnetic $La_{2-x}Sr_4CuO_4$ crystal with high-temperature superconductivity at room temperature under an applied field of 14T by metallographic polarized-light.

When $La_{2-x}Sr_4CuO_4$ crystal is cooled, the tetragonal-orthorhombic structure transformation take place and form a domain structure (known as twins), where the orthorhombic distortion alters sign upon crossing domain walls, as shown in Figure23(a). Applying an applied of 14T, bc (red) domains dominate under an applied magnetic field along parallel to the plane of observation or ac(blue) domains along perpendicular to the plane of observation, as shown in Figure23.

It show clearly when there is coupling between antiferromagnetic transition and structural transformation, the twin boundaries formed by structural transformation are consistent with the domain walls of the antiferromagnet and the twin boundaries (domain walls) can be moved with the reorientation of magnetic order under an applied magnetic field.

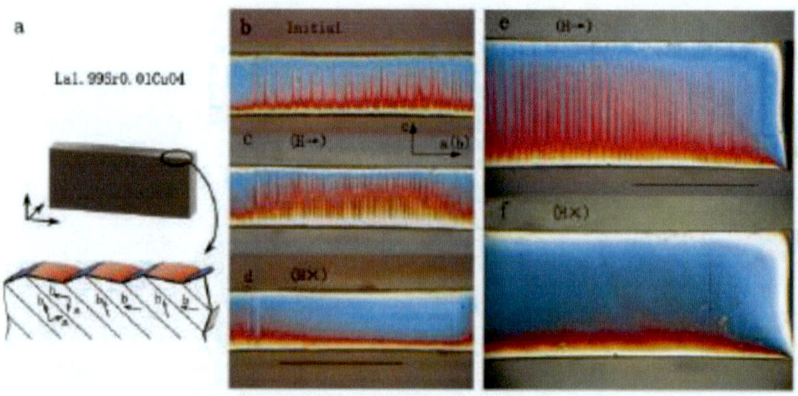

Figure 23. The motion of twin boundaries of La2-xSr4CuO4 crystal under an applied field (a) the orthorhombic-twins (b) initial twins with a larger fraction ac(blue) domains (c) bc (red) domains dominate under 14T along parallel to the plane of observation (d) ac (blue) domains dominate along perpendicular to the plane of observation (e) and (f) are magnified view of (c) and (d), respectively. Scale bars, 1mm (b-d) and 0.5mm (e,f).

4.2. The Mechanism Moved Twin Boundaries (Domain Walls) under an Applied Field

1) The motion of the dislocations on twin boundaries

Twin dislocation is suggested to play a key role for the rearrangement of twins (domains) in ferroelectric and ferromagnetic materials[37,38]. A simple model of dislocation motion on twin boundary is shown in Figure24[39].

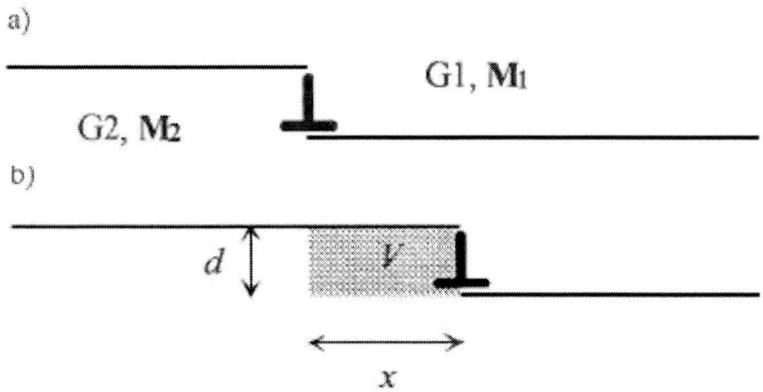

Figure 24. The motion on a twin boundary a)Dislocation in a twinning plane b)By moving the distance x.

In Figure24 the grains on either side are labeled G1 and G2. The corresponding magnetization vectors are **M1** and **M2**, respectively. Under an applied magnetic **H** the twinning dislocation with the line length L move the distance x, the work done is $\Delta W{-}FA{-}FxL$. Due to the motion of the twinning dislocation, the volume V transforms from G1 orientation into G2 orientation and, therefore, ΔE equals the difference of the magnetic energy densities of G1 and G2 times V.

$$\Delta E = V(\mathbf{M_2} - \mathbf{M_1}) \bullet \mathbf{H} = dxL\Delta\mathbf{M} \bullet \mathbf{H} \qquad (19)$$

where $\mathbf{M}=\chi\mathbf{H}$, in the case of antiferromagnet $\chi_{\perp} > \chi_{//}$ and under an applied field, **H**, there is a difference of energy at the both sides of the dislocation in Figure24, the difference of energy can also drive the motion of the dislocations on twin boundaries in antiferromagnet.

2) Twin of *fct* martensite in γMn-based alloys

There is the structural transformation, which is *fcc*→*fct* martensitic transformation with (011) twins following antiferromagnetic transition in rich-Mn γMn-based alloys.

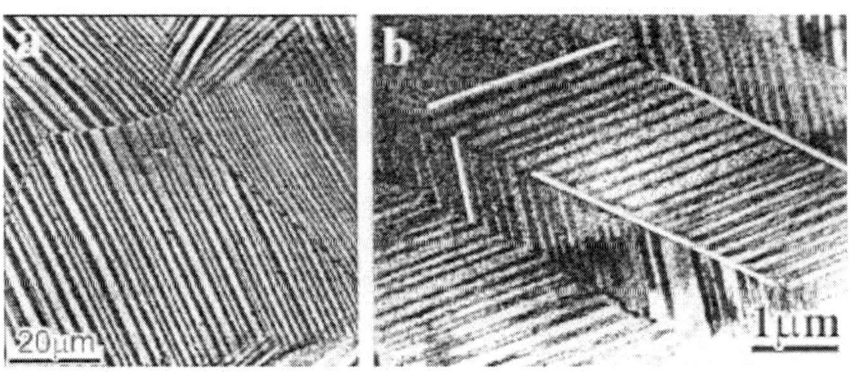

Figure 25. Orientation imaging microscope (a) and TEM micrograph(b) of twins in Mn-30at%Cu alloy[40].

Figure25 shows the orientation imaging microscope(OIM) and TEM micrograph at rich-Mn region (90.6at%Mn). The twinning bands with similar widths and observed and regularly oriented in every crystalline grain, as shown in Figure25(a). In Figure25(b) TEM micrograph revealed that there is the secondary twinning bands within the twinning bands shown by (OIM) analysis.

There also are (011) twins of *fcc→fct* martensitic transformation, which is following antiferromagnetic transition in MnNi and MnFe(Cu) alloys. T_N and Ms of Mn82.2Ni17.8 alloy are 180°C and 88°C, respectively, by using DSC measurement. Its twin image is shown in Figure26[41]. Figure27 is a dark field image taken from two spots, respectively, under [111] two-beam condition in $Mn_{85.5}Fe_{9.5}Cu_{5.0}$ alloy Ms of which is close to T_N (~450K)[42].

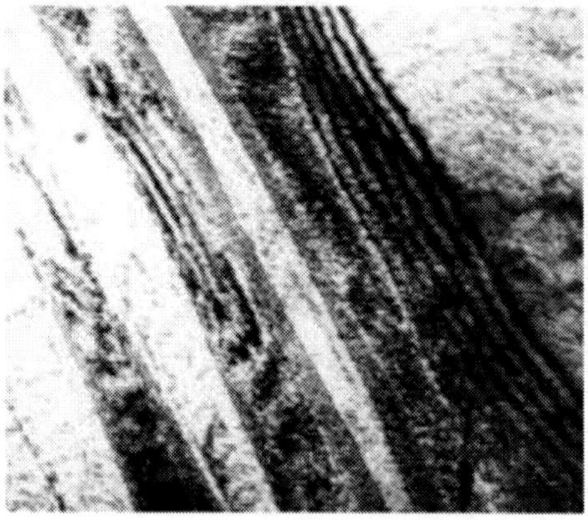

Figure 26. Twin structure of Mn82.2Ni17.8 alloy.

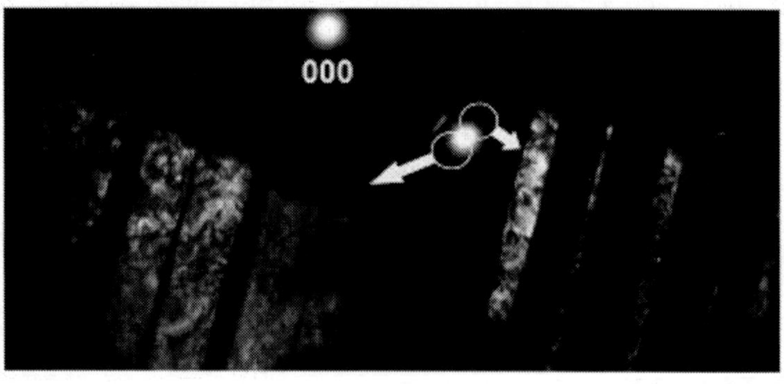

Figure 27. The dark field images using the two split beams, respectively, in $Mn_{85.5}Fe_{9.5}Cu_{5.0}$ alloy.

3) The dislocation on twin boundaries in antiferromagnet[43]

The results recently researched with HRTEM have revealed that there are twin boundaries in antiferromagnet. Figure28 shows the HRTEM image taken across the twin boundary with [100] zone axis in $Mn_{85.5}Fe_{9.5}Cu_{5.0}$ alloy. Due to the small tetragonality (c/a=0.97), in Figure28(a) by drawing two white reference lines along certain direction, the location of the twin boundary is determined from where the lines begin to deviate from the lattice. The location of twin dislocation is circled and the space between the two white arrows is the transition region between the both twin variants.

The plane pointed out by black arrows is (011) twin boundary. Figure28(b) is transformed into a glancing-angle view of (a), where the periodicity perpendicular to the glancing direction is exaggerated and the crystallographic orientation change of the lattice near the twin boundary can be observed clearly. In the positions indicated by arrow there are two lattice fringe misfits of opposite direction.

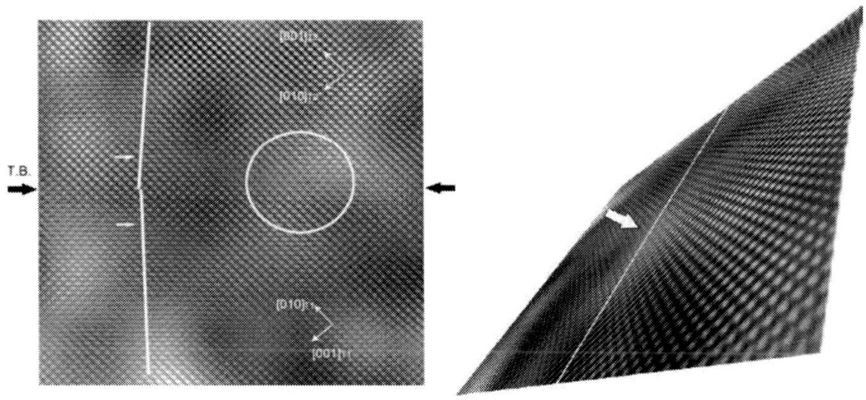

(a) (b)

Figure 28. HRTEM image across twin boundary(a) and distorted glancing view of (a).

According to the results of HRTEW, an atomic model of the twin dislocation in a tetragonal lattice can be shown in Figure29. In most γ-Mn based alloys with (011) twin martensite, the tetragonality, $|1\text{-}c/a|$, is close to or much less than 0.05. In $Mn_{85.5}Fe_{9.5}Cu_{5.0}$ alloy the angle is given by

$$\alpha = 90^o - 2arctg\frac{c}{a} \approx 1.8^o \quad (20)$$

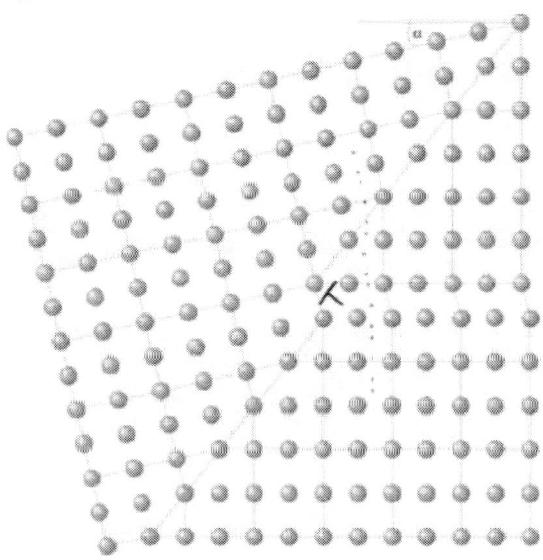

Figure 29. Atomic model of a twinning dislocation in tetragonal lattice.

Compare with Figure24, it is clear when the twin dislocation is moved under an applied stress or magnetic field along different directions, the change between the both variants take place, macro-strain with different directions may be shown.

4.3. MFIS and MSME in Polycrystalline γMn-Based Alloys

1) MFIS in the case without applied stress (or pre-strain)

The resistance measurement of $Mn_{82.2}Ni_{17.8}$ alloy shows that T_N point of antiferromagnetic transition is at 453K. Ms (start temperature of martensitic transformation), Mf (final temperature), As (start temperature of reverse martensite) and Af (final temperature) are 361, 278, 348 and 413K, respectively, as shown in Figure30. The microstructure at room temperature is twin martensite (Figure26). Figure31 shows the change of MFIS at parallel to an applied field within a range of 0 to ±10kOe and maximum strain reaches to $+30\times10^{-6}$. Due to the tetragonality (1-c/a) of the alloy is only 0.014, so the strain is more lower. And there is a threshold field Hth which is 2.2kOe when direct field direction and –1.6kOe while opposite field direction. The MFIS are reversible and shows MSME[41].

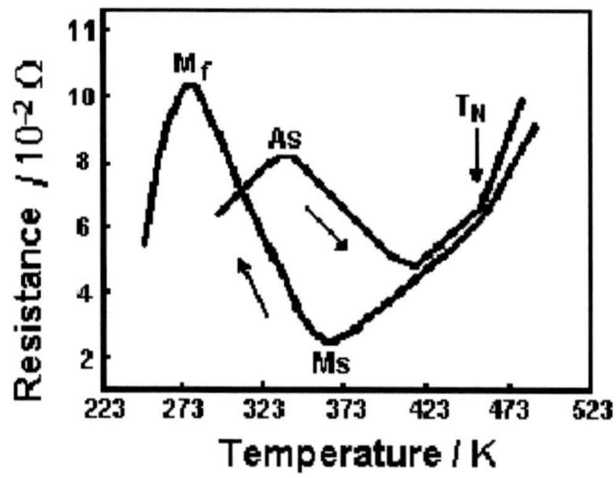

Figure 30. Resistance-temperature curve of $Mn_{82.2}Ni_{17.8}$

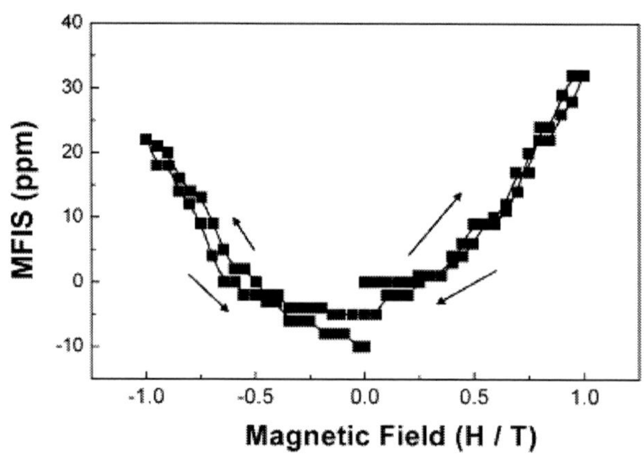

Figure31 MFIS under an applied field in $Mn_{82.2}$Ni17.8.

In $Mn_{85.5}Fe_{9.5}Cu_{5.0}$ alloy two-way temperature shape memory effect shows hysteresis-free charac- teristics (Figure20) and the microstructure at room temperature is (011) twin martensite (Figure27). Ms(~450K) is close to T_N, there is a very good coupling between antiferromagnetic transition and martensitic transformation. Due to the tetragonality (1-c/a) of the alloy reaches to 0.031, a very great and reversible MFIS parallel to H which reaches to $+1.6\times10^{-3}$ is observed at an applied field of 4T, as shown is Figure32. Also there is a threshold field Hth of ~1T.

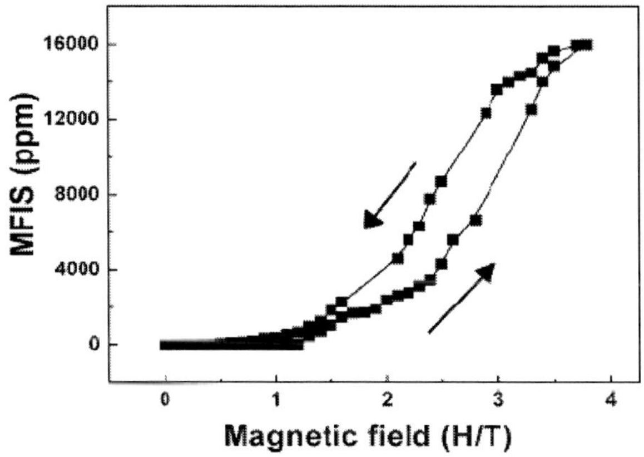

Figure 32. MFIS in $Mn_{85.5}Fe_{9.5}Cu_{5.0}$ alloy[44].

Martensitic transformation does not take place in $Mn_{42}Fe_{58}$ alloy cooled to room temperature and the lattice belongs to face-centered-cubic structure with lattice constant, a, of 0.3733nm, as seen in Figure 9(a). Though there is not martensitic twins in micro-structue of the alloy, but there is the domains of anti-ferromagnetic transition. Also in the polycrystalline alloy MFIS is reversible. Figure33 shows that MFIS ($\varepsilon_{//}$) parallel to the applied magnetic field reaches to 169×10^{-6} and that ($\varepsilon_{\perp}$) perpendicular to the applied field is only -5×10^{-6} at 1T. Especially the area of its hysterias loop is very small. The MFIS may be the motion of antiferromagnetic domains or the related with magnetostriction. But no matter how the MFIS is also reversible to possess the MSME characteristics.

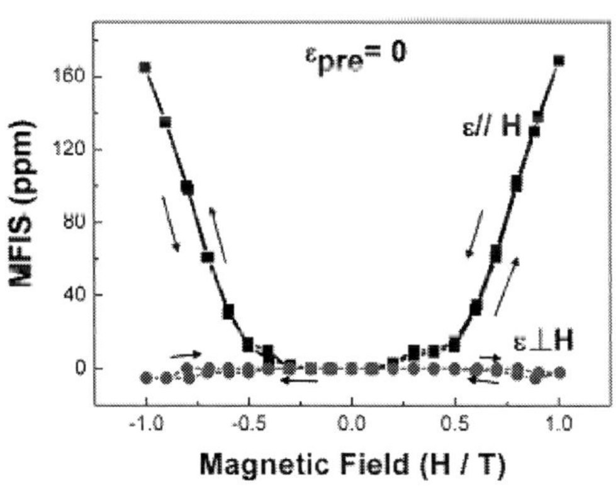

Figue33 MFIS in $Mn_{42}Fe_{58}$ alloy[45]

2) MFIS with applied stress (or pre-strain)

After a pre-compressive strain (i.e pre-stress) of –200ppm with the same direction parallel to following applied magnetic field on the specimen before an applied magnetic field was applied, MFIS direction were changed from a positive direction to negative under following applied magnetic field, as shown in Figure34.

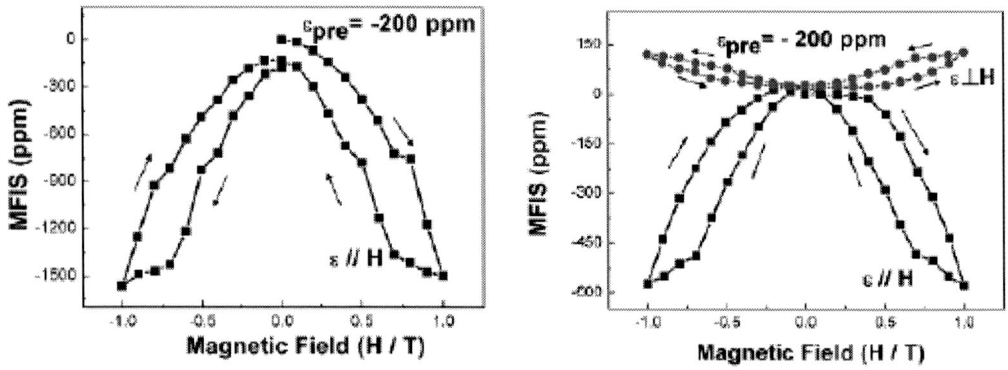

(a) (b)

Figure 34. MFIS with a pre-compressive strain of –200ppm(a) $Mn_{85.5}Fe_{9.5}Cu_{5.0}$ alloy[44](b) $Mn_{42}Fe_{58}$ alloy[45].

Figure34(a) is a result of negative strain at the direction parallel to applied magnetic field in $Mn_{85.5}Fe_{9.5}Cu_{5.0}$ alloy and Figure34(b) is that in $Mn_{42}Fe_{58}$ alloy. Comparing with Figure32 and Figure33, it is considered that internal microstructure of the antiferromagnetic crystal, that include spin vector of antiferromagnet, the orientation of twin or magnetic domain, produced the change of different direction.

From the macroscopic scale, the strain (positive or negative) is observed under an applied magnetic field (without applied stress or with an applied pre-stress). On the mesoscopic scale, this strain is related to the motion of twin boundaries of martensite or domain walls of antiferromagnet, that results is the reorientation of part lattice. According to microscopic scale, the reorientation is carried out by the motion of different direction of the dislocations on twin boundaries or domain walls. However, because of the lack of some quantitative data, such as anisotropic constant, etc, for γ-Mn based alloys the mechanism of MFIS is only qualitatively discussed and further study on theoretical analysis and experimental measurement must be carried out.

POSTSCRIPT

Up to now the attention for antiferromagnetic materials is very limited, especially on practical application when mass of ferromagnetic materials are widely applied in industrial and daily area, due to there is difficulty on experiment.

In 1948, a high damping property of Mn-12(wt)Cu alloy with *fct* martensitic (011) twin was discovered, the alloy taken on both antiferromagnetic transition and martensitic transformation, its high damping properties deal with stress-induced twin boundaries movement, which there only is a example in antiferromagnetic alloys.

The antiferromagnetic γ-Mn based alloys have not magnetism, they can be used to the place required without magnetism. And they belong to the category of austenite stainless steels, which have corrosion and oxidation stability. Anther the γ-Mn based alloy, Mn-Fe[46] and Mn-Ni[47], show also high damping characteristics. From some recent results the γ-Mn based alloys can considered a type-new antiferromagnetic multifunction materials[48]. So we may be pay attention to function properties and practical application for the new field of antiferromagnetic materials, which is not clearly known and opened out.

ACKNOWLEDGMENTS

The some works in the paper were supported by the Natural Science Foundation of China (Grant No.50171041 and No.50571065).

REFERENCES

[1] L. Néel, Ann. De Phys, 5(1936), 160.

[2] P. Ethernets, Proc. Amsterdam Acad., 36(1933), 153.

[3] A.A. Minakov, I.V. Shvets, and V.G. Veselago, Magnetostriciton and antiferromagnetic domains dynamics in helical antiferromagnets, *J. Magn. Magn. Mater.*, 88(1990), 121-133.

[4] M. Schlenker and J. Baruchek, Neutron techniques for the observation of ferro- and antiferromagentic domains, *J. Appl. Phys.*, 49(1978), 1996-2001.

[5] M. Ando and S. Hosoya, Size and behavior of antiferromagnetic domains in Cr directly observed with x-ray and neutron topography, *J. Appl. Phys.*, 49(1978), 6045-6051.

[6] W. L. Roth, Neutron and optical studied of domain in NiO, *J. Appl. Phys.*, 31(11)(1960), 2000-2011

[7] M. Safa and B. K. Tanner, Antiferromagnetic domain wall motion in KNiF3 and KCoF3 observed by X-ray synchrotron topography, *Philosophical Magazine* B, 37(6)(1978), 739-750.

[8] G. F. Clark and B. K. Tanner, Antiferromagnetic domain wall motion under external stress, *Phys. Stat. Sol.* (a) 59(1980),241-249.

[9] L. Néel, Proceedings of the Conference on Theoretical Phusics, Kyoto, Science Council of Japan, (1954), 701.

[10] L. Alberts and E.W. Lee, Magnetostriciton in antiferromagnetic nickel oxide, *Proc. Phys. Soc.*, 78(1961), 728-733.

[11] P. De V. Plessis, S.J. Van Tonder and L.Alberts, Magnetostriction of a NiO single crystal II, *J. Phys. C: Solid St. Phys.*, 4(1971), 2565-2571.

[12] S. Legvold, J. Alstad, and J. Rhyne, Giant magnetostriction in dysprosium and holmium single crystals, *Phys. Rev. Lett.*, 10(12)(1963), 510-512.

[13] R. Mahendiran, M.R. Ibarra, C. Marquina, B. Garcia-Landa, L. Morellon, A. Maignan, B. Raveau, A. Arulraj and C.N. Rao, Giant anisotropic magnetostriction in Pr0.5Sr0.5MnO3, *Appl. Phys. Lett.*, 2003, 82: 242-243.

[14] D. Meneghetti, S. Sidhu, Magnetic structure in Cu-Mn alloy, *Phys. Rev.*, 105(1957), 130-134.

[15] H. Umebayashi, Y. Ishikawa, Antiferromagnetism of γFe-Mn alloys, *J. Phys. Soc. Jap.*, 21(7)(1966), 1281-1294.

[16] Y. Enboh and Y. Ishikawa, Antiferromagnetism of γ Iron Manganes Alloys, *J. Phys. Soc.* Japan, 30(6)(1971), 1614-1627.

[17] G. E. Bacon, I. W. Dunmur, J. H. Smith, R. Street, The antiferromagnitism of manganese copper allolys, *Proc. Roy. Soc.*, 214A(1957), 223-238.

[18] K. Shimizu, Y. Okumura, H. Kubo, Crystallographic and morphological studies on the fcc to fct transformation in Mn-Cu alloy. Transactions of Japan Institute of Metals, 23(2)(1982), 53-59.

[19] T. J. Hichs, A. R. Pepper, J. H. Smoth, Antiferromagnetism in γ phase manganese-palladium and manganese-nichel alloys, *J. Phys.* C !(1968), 1683-1689.

[20] Y. Endoh and Y. Ishikawa, J. Phys. Soc. Japan, 30(1971), 1614.

[21] E.Z.Vintaikin, D.F. Litvin, V. A. Udovenko, G. V. Shcherbedinskij, Martensitic transformation in γ-Mn alloy and shape memory effect, Proceeding of ICOMAT'79, Cambrige: MA, USA, MIT-press, 673-678.

[22] N.Honda, Y.Tanjji, Y.Nakagawa, The Orthorhombic distortion in γ manganese alloys containing Nickel. *J. Phys. Soc. Japan.* 38(1975), 589.

[23] K. Sugimoto, T. Mori, S. Shiode, Effect of composition on the internal friction and Young's modulus in γ-phase Mn-Cu alloys, Metal Science Journal, 7(1973), 2085-2090.

[24] K. Sugimoto, K. Ohta, Proc. 6th, ICFUAS, Tokyo, (1977).

[25] R. D. Lowde, R. T. Harley, G. A. Saunders, M. Sato, R.Scherm, G. Underhill, Proc. *R. Soc.*, A374(1981), 765.

[26] N.Honda, Y.Tanjji, Y.Nakagawa, Lattice distortion and elastic properties of antiferromagnetic γ Mn-Ni alloys, *J. Phys. Society Japan*, 41(1976), 1931-1937.

[27] J. T. Lenkkeri, Measurement of elastic moduli of face-centred cubic alloys of transition metals, *J. Phys. F: Metal Phys.*, 11(1981), 1991-1996.

[28] Kunio Ito, Makoto Tsukishima and Masanobu Kobayashi, Temperature Dependence of Lattice Parameters and *fcc/fct/fco* Transformation in Some Mn-Bbase Metastable γ-Phase Alloys, *Transactions of Japan Institute of Metals,* 24(7)(1983), 487-490.

[29] Y.Z. Vintaykin, V.M. Sakhno and V.A. Udovnko, Reversibile shape memory effect in alloys of manganese with germanium and gallium, *Phys. Met. Metall.*, 46, 1978: 161-163.

[30] G. Nosova, E. Vintaikin, Investigation of nature of two-way shape memory effect in γ Mn based alloys, *Scripta Mater.*, 40(3)(1999), 347-351.

[31] J. H. Zhang, W. Y. Peng, J. J. Zhang, Xu Zuyao (T. H. Hsu), Shape memory effct of an antiferromagnetic Mn-9.5at%Fe-5.0at%cu alloy, Met. Sci. Eng. A, 481-482(2008), 326-329.

[32] Jihua Zhang, Wenyi Peng, Ping Lu, T.Y.Hsu (Xu Zuyao), The coupling between antiferromagnetic transition and martensitic transformation in γ-MnFe based alloys, *Phys. Stat. Sol.* (C) 1(7), 1772 (2004).

[33] Zhang Jihua, Peng Wenyi, Gu Suyi, Deng Huanming, Chen Shuchuan, Damping mechanism associated with coupling between antiferromagnetic transition and martensitic transforma -tion, *Mat. Sci. and Eng.,* A442(2006), 229-232.

[34] H. Gomonay and M.Loktev, Magnetostriciton and magnetoelastic domians in antiferromagets, *J.Phys.:Condens. Matter.,* 14(2002), 3959-3971.

[35] U.Hocke and H.Warlimont, Structural changes associated with antiferromagnetic ordering in Mn-rich Mn-Ni alloys, J.Phys. F:Metal phys., 7(1977), 1145-1155.

[36] A. N. Lavrov, Seiki Komiya and Yoichi Ando, Magnetic shape-memory effect in a crystal, *Nature,* 418(2002), 385.

[37] K. Ullakko, Magnetically controlled shape memory alloys: A new class of actuator materials, *J. Mater. Eng.* Perf, 5(1996), 405-409.

[38] K. Ullakko, J. K. Huang, C. Antner, R. C. O'Handly, and V. V. Kokorin, Large magnetic-field-induced strains in Ni2MnGa single crystal, *Appl. Phys. Lett.*, 69(1966), 1966-1968.

[39] P. Müllner and K. Ullakko, The force of magnetic/electric field on a twinning dislocation, *Phys. Stat. Sol.* (b), 208(1998), R1.

[40] Fuxing Yin, Takuya Sakaguchi, Qingchao Tian, Atsuko Sakurai and Kotobu Nagai, The twinning microstructure and damping behavior in Mn-30at%Cu alloy, Materials Transactions, The japan Institute of Metals, 46(2005), 2164-2168.

[41] W. Y. Peng, X. Y. Wang, J. J. Zhang, and J. H. Zhang, Magnetic-field-induced strain in an antiferromagnetic γ-MnNi alloy, *J. App. Phys.*, 99,033908 (2006).

[42] X. Y. Wang, W. Y. Peng, J. H. Zhang, Martensitic twin and antiferromagnetic domains in γ-MnFe(Cu) alloy, *Mater. Sci. Eng.* A 438-440(2006), 194-197.

[43] Xiaoyu Wang, Jihua Zhang, Structure of twin boundaries in Mn-based shape memory alloy: A HRTEM study and strain driving force, Acta Materialia, 55(2007), 5169-5176.

[44] J. H. Zhang, W. Y. Peng, T. Y. Hsu(Xu Zuyao), The magnetic foeld induced strain without prestress and with stress in a polycrystalline Mn-Fe-Cu antiferromagnetic alloy, *Appl. Phys. Lett.*, 93(2008), 1-3.

[45] W. Y. Peng and J. H. Zhang, Magnetostriction studies in an antiferromagnetic polycrystalline Mn42Fe58 alloy, *Appl. Phys. Lett.*, 89, 262501 (2006).

[46] [46] K. Ito, M. Kobayashi, H. Abe, Damping capacity of metastable γ-phase in manganese rich Mn-Fe alloys, *Met Sci.*, 11(1979), 502-504.

[47] G. Hausch, A.Schmolz, E.Torok, H.Warlimont, Internal friction and elastic anomalies antiferromagnetic Mn-Ni alloy, *J. De Physique C*, 9(1983), 471

[48] J. H. Zhang, Typed-new multifunctional Mn-rich antiferromagnetic alloys, ScienceDirect, *Current Opinion in Solid State and Materials Science*, 9(2005), 326-330.

In: Shape Memory Alloys
Editor: H. R. Chen, pp. 317-338
ISBN 978-1-60741-789-7

Chapter 11

PHASE TRANSFORMATION IN NITI SHAPE MEMORY ALLOY UNDER THERMOMECHANICAL CONDITIONS

Manjunatha Pattabi*[1] *and K. Ramakrishna
Department of Materials Science, Mangalore University,
Mangalagangotri -574 199, India

ABSTRACT

The effects of thermal cycling and mechanical operations like cutting and polishing on the phase transformation behavior of heat-treated NiTi shape memory alloy (SMA) have been investigated by means of Powder X-Ray Diffraction (XRD) and Differential Scanning Calorimetry (DSC). The results of studies carried out on the effect of thermal cycling on SMA heat-treated at 660 °C and effect of mechanical cutting and polishing on SMA heat-treated at 560 and 660 °C, are presented. In NiTi alloy heat-treated at 660 °C, the integrated peak area (integrated intensity) corresponding to $(110)_A$ plane of the austenitic phase increases and the area corresponding to martensitic phase decreases with increasing temperature. Austenitic start and finish temperatures change during the first few thermal cycles. The former decreases by ≈15 °C and the latter decreases by ≈35 °C during the first cycle, whereas martensitic start and finish temperatures do not get affected by thermal cycling. The transformation characteristics determined through XRD are compared with those obtained from DSC. The thermal width observed from XRD studies is larger than that from DSC measurements due to the better sensitivity of the former in detecting smaller volume fractions of the phases in the martensitic transformations as compared to the thermal methods like DSC. Mechanically cut and polished samples show broad thermal width and lesser heat flow associated with the M↔A transformation compared to chemically etched samples due to the internal stresses generated during mechanical operation. Among the samples heat-treated at 560 °C, the sample subjected to mechanical cutting and polishing (HT560MC) shows a single step M→A on reverse phase transformation and two step A→R→M on forward transformation whereas chemically etched sample (HT560CE) shows the intermediate R-phase both during forward and reverse transformations. In the case of HT560MC, the dislocations created by the

[1] Corresponding author. Tel.: +91 824 2287249; Fax: +91 824 2287367; e-mail: manjupattabi@yahoo.com (M. Pattabi).

mechanical cutting and polishing may be responsible for the non-appearance of R-phase in the reverse transformation even up to 150 cycles, possibly due to the entanglement of dislocations at high density of dislocations. Heat treatment at a higher temperature of 660 °C does not facilitate the formation of R-phase in the reverse transformation due to the reduced dislocation density. However, dislocations created due to mechanical operations (in HT660MC) help the formation of R-phase to a very small extent during forward transformation. Both the heat-treated samples show the effect of mechanical cutting and polishing even at 150th thermal cycle.

1. BACKGROUND

Smart materials are materials that have one or more properties that can be significantly changed in a controlled fashion by external stimuli such as stress, temperature, moisture, pH, electric or magnetic fields etc. Shape Memory Alloys (SMAs) are one of the smart materials (since they function as sensors as well as actuators) that can recover from large strains through the application of heat. A shape memory alloy undergoes thermoelastic martensitic transformation, which gives rise to the shape memory effect (SME) and pseudoelasticity (PE). SME is one of the most important and interesting property exhibited by these materials. SMAs have attracted tremendous interest both in scientific research and in practical utilization owing to their novel thermomechanical behavior. It is established that the novel thermomechanical properties of these alloys are associated with thermoelastic martensitic transformations and it can be induced either by temperature (thermally induced martensitic transformation) or by applied stress (stress-induced martensitic transformation). This transformation and its reversion are accompanied by the variation of a certain physical properties [1] and they are listed in Table 1.

These varying physical properties are used to characterize the transformation behavior of SMAs. Therefore, specific macroscopic changes occur simultaneously with variation in physical properties associated with the microscopic evolutions. Stability of the phase transformation behavior (phase transformation temperatures, number of phases, thermal width etc.) is very much essential for any applications.

Table 1. Physical Properties that change during SME Transformation

Yield Strength	Acoustic Emission
Hardness	Magnetic Properties
Young's Modulus	Internal Friction
Expansion	Lattice Spacing
Roughness	Hall Coefficient
Color	Electromotive Force
Ring	Electron Density Waves
Bend	Heat Capacity
Feel	Latent Heat of Transformation
Damping	Thermoelectric Power
Velocity of Sound	Thermal Conductivity
Electrical Resistance	

In principle, shape memory effect is associated with thermoelastic martensitic transformation and reverse phase transformation. However, some irreversible processes cannot be avoided in the transformation. Irreversible processes are repeated during thermal cycling with a cumulative effect, causing a change in transformation characteristics such as transformation temperatures. When phase transformation takes place degradation of the material could be accelerated in every cycle and the performance of any actuator, which is designed to operate in a narrow temperature range, can be affected.

Due to the good shape memory effect, in addition to excellent mechanical properties, NiTi alloy and its ternary alloys are most important among many shape memory alloys, which make use of either SME or PE, with practical applications in areas like military equipment, industry and medicine as connectors, aerials, reflectors, safety devices, valves etc. Hence, the influence of cyclic reversible martensitic transformation on the stability of properties of SMA, including shape recovery and characteristic temperatures of transformations, is of particular significance [2,3]. However, the martensitic transformation in this system is least understood compared with other systems due to the complexity of the transformation and sensitivity to composition [4-6], alloying addition [7-11] and thermomechanical treatments [2-6,12-20]. Many studies on the effect of traditional material processing procedures such as cold working and aging on the thermomechanical response of SMAs have been carried out. The transformation temperatures in NiTi SMAs are related to the presence of lattice defects introduced by cold working [20-22]. For a better understanding of the transformation behavior of NiTi, the investigation of the effect of deformation is important because it is predicted that defects such as dislocation, introduced by plastic deformation, influence the stability of each phase [22]. High dislocation density restricts growth of the martensitic phase and development of stress-induced reoriented martensite resulting in increased transformation hardening in the stress-strain curve [23]. Due to thermomechanical treatments, the single-step austenite-to-martensite (A→M) transformation on cooling is replaced by a two-step transformation from austenite to the R-phase and then to martensite (A→R→M). In this case, the reverse transformation on heating may be either a single-step M→A transformation or a two-step M→R→A transformation [24]. The R-phase transformation occurring in some NiTi SMAs prior to Austinite to Martensite (B2→B19') transition due to thermomechanical effect has attracted much attention because of its unusual behaviour and small thermal hysteresis that make them attractive for applications. Most of the studies on NiTi binary alloys have been carried out on Ni-rich [4-6,12-15] or equiatomic [2,5,15-21] NiTi alloys. However, there are also a few reports on Ti-rich NiTi alloys [3,4,17,22,25-29]. It is known that Ti-rich NiTi alloys contain a second phase, Ti_2Ni in the matrix and have higher transformation temperatures than those of the Ni-rich or equiatomic NiTi alloys [27].

Mayazaki et al. [5] showed that in solution-treated NiTi specimens with various Ni-contents, the transformation temperatures change with thermal cycling. However, in aged Ni-rich specimen and those annealed at a temperature lower than the recrystallization temperature after cold working, they were found to be constant even after thermal cycling. Transmission electron microscopy reveled that dislocations were introduced by thermal cycling in the solution-treated specimens and their density increased with increasing number of thermal cycles. However, in the specimens with constant transformation temperatures, no substantial change in the microstructures was observed with thermal cycling. On the basis of these results, they concluded that the shift of the transformation temperature with thermal cycling is due to the introduction of dislocations, but neither due to any aging effect nor due to a change in the degree of order during thermal cycling. SMA properties are influenced not only by the cyclic mode but also by the cycling temperature range. Depending on the temperature range, there is 'complete' and 'incomplete' cycling [30]. The cyclic modes for SMA can be classified in five groups: Stress-free thermal cycling, Thermal cycling at constant strain, Thermal cycling at constant load, Thermal cycling combined with stress/strain cycling and Stress or strain cycling at constant temperature. For each cycling mode the variation of properties as a function of the number of cycles has been evaluated [31] based on the existing data of NiTi alloys from the literature. According to this, for simple thermal cycling without external stress, the M_s temperature generally decreases with increasing cycle number. Special treatments, such as annealing, heavy cold work, etc., can suppress this effect. M_s is almost constant during thermal cycling if (1) a R-phase transition is involved and (2) M_s is below 0°C. In thermal cycling with constant strain the recovery force deteriorates with cycling. Below a critical strain a linear relation between the recovery force and the logarithm of the number of cycles exists for incomplete cycling. For two-way actuators the cycling life can be estimated by cycling at maximum strain. Thermal cycling at constant load is one training method to obtain two-way memory (TWM) performance. The R-phase should be avoided in order to develop significant TWM strain. Improving critical slip stress for SMA is a key point to stabilize superelastic behavior. Superelastic (SE) cycling itself could be applied to achieve stable characteristics before service. Batches with lower M_s correspond to higher critical slip stress and higher energy density stored as well as more stable SE behavior against stress cycling. Regardless of the cycling modes, the variation in shape memory properties caused by incomplete cycling is far less than for complete cycling, which means that the transformation sequence involved plays a critical role. Among the transformation temperatures, M_s is very sensitive to cycling while R-phase transformation temperature T_R is almost stable. Batches with a lower M_s temperature and R-phase transition (accordingly, $M_s << T_R$) are more stable to thermal [6] and stress cycling.

2. Thermal Cycling Effect

Depending on the physical property [1] that changes during phase transformation, various techniques like Differential Scanning Calorimetry (DSC) [3,11-14,17-22,32-37], Electrical Resistivity (ER) measurements [5,6,38-42], Internal Friction [26,35], Magnetic Susceptibility measurements [42] and Thermomechanical Analysis (TMA) [32-34] have been commonly employed to study the transformation behavior of different shape memory alloys. X-Ray diffraction (XRD) has been used to characterize SMAs [36-38,43]. X-ray diffraction studies as a function of specimen temperature have not been found in literature, which would yield information on the transformation characteristics as a particular phase evolves with temperature. A new attempt has been made to study the phase transformation behavior of SMA under thermal cycling by determining the austenite peak area i.e. integrated area under the X-ray diffraction peak with different temperatures [44,45].

Investigations have been carried out on samples prepared from Ti -rich NiTi (≈51 at % Ti) SMA alloy sheet of nominal dimension 88.9 x 406.4 x 0.5 mm, as-rolled having an oxide surface, obtained from Shape Memory Applications Inc., USA. Local composition measurements were made on a Karl Zeiss Scanning Electron Microscopy (SEM) model

LEO 440 fitted with an EDAX system. All the specimens were heat-treated at 660 °C in an electric muffle furnace. The specimens were heated to the required pre-set temperature and held at that temperature for 20 minutes before quenching them in a water bath at room temperature (RT). The samples were mechanically polished using fine-grained emery paper under cold-water circulation to remove the oxide layer formed during heat-treatment as well as to trim the edges to fit into the sample stage. Thermal cycling was carried out by immersing the samples first in a liquid paraffin bath maintained at +160 °C for two minutes followed by immersion in another liquid paraffin bath maintained at room temperature (≈28 °C) for two minutes. The samples were cleaned with isopropyl alcohol to remove the oil.

X – ray diffraction measurements were made on a Bruker D8 Advance X – ray diffractometer, using the (θ,θ) geometry at a scan speed of 1°/min. Parallel beam Cu-K_α radiation with Ni filter was used by setting the aperture slit, anti-scatter slit and receiving slit widths at 2 mm, 2 mm and 0.2 mm, in that order. For a given sample, the XRD spectra were recorded in the range +25 °C to +160 °C at intervals of 10 °C, during heating and cooling, using a TTK-450 attachment combined with Anton Paar TCU 100 temperature control unit. The samples were maintained at a particular set temperature for 5 minutes before recording the XRD. A vacuum of 10^{-3} mbar was maintained in the sample stage to prevent any possible oxidation at elevated temperatures. The (hkl) planes corresponding to the lines in the XRD spectra were identified from the ICDD database (Diffrac Plus Release 2001) using the software Eva provided by Bruker. The lines were also indexed using PowderX software [46] by providing lattice parameters as a_o = 3.015 Å for austenitic phase, a = 2.889 Å, b = 4.120 Å, and c = 4.622 Å with β =96.8° for martensitic phase [38] for confirmation. DSC was carried out on the specimen using Shimadzu DSC-50 in the temperature range from RT to +160 °C at a scanning rate of 10 °C/min. The phase transformation temperatures are determined by determining the onset point of slope change in the DSC curves.

A typical X-ray diffractogram recorded at RT for the Ti-rich NiTi SMA sheets heat-treated at 660 °C, prior to any thermal cycling, is shown in Figure 1. Three peaks are observed at 2θ values of 39°, 40.945° and 44.837°, which correspond to (002), (-111) and (012) planes of martensitic phase, designated as $(002)_M$, $(-111)_M$ and $(012)_M$, in that order. A fourth peak corresponding to (110) reflection of the austenitic phase is observed at a 2θ value of 41.975°, which is designated as $(110)_A$. The line intensity corresponding to $(110)_A$ peak is relatively less than the peak $(-111)_M$ corresponding to martensitic phase, indicating the presence of austenite phase to a lesser extent at RT. The XRD patterns at different temperatures, during heating and cooling, for N = 0, 1, 5 and 50, are shown in figures 2(a) - 2(h). The $(110)_A$ peak intensity increases rapidly up to a certain temperature beyond which the rate of increase comes down. It is to be noticed that the intensity of the peaks corresponding to the martensitic phase, $(002)_M$, $(-111)_M$ and $(012)_M$, decreases with increasing temperature, with $(-111)_M$ and $(012)_M$ peaks almost disappearing beyond 140 °C. These peaks show up again while cooling below 100 °C, indicating that the austenitic phase grows at the expense of martensitic the phase, as is expected.

The integrated area under the $(110)_A$ peak (peak area) is considered in the ensuing discussions. Figure 3 shows a plot of peak area corresponding to $(110)_A$ with temperature for a sample which is not subjected to any thermal cycling. A thermal hysteresis can be seen. During heating, the peak area increases slowly in the lower temperature range with a slope change being observed at around 90 °C.

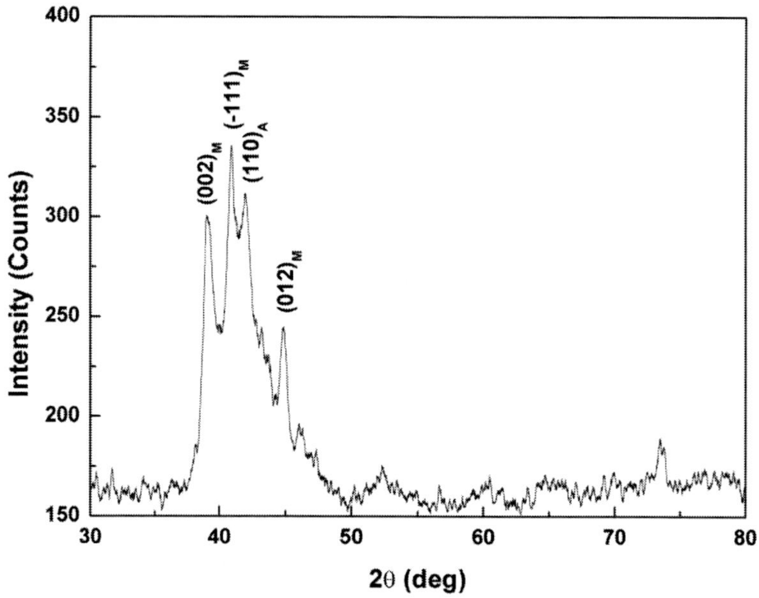

Figure 1. A typical X-ray diffractogram of the heat-treated NiTi SMA at room temperature.

With further increase in temperature, the extent of increase in the peak area decreases again at around 135 °C. While cooling, the peak area remains fairly constant up to 72 °C with a large decrease in the peak area in the temperature range 72 to 47 °C. The sharp change in the integrated peak area, around 90 °C while heating and 72 °C while cooling, is attributed to martensitic-to-austenitic (M→A) and austenitic-to-martensitic (A→M) transformations, respectively. Transformation temperatures were determined by noting the intersection of the tangents drawn to the part of the curves on either side of the point where a sudden change in integrated peak area takes place. This is similar to the method used for ER measurements by Uchil et al [39].

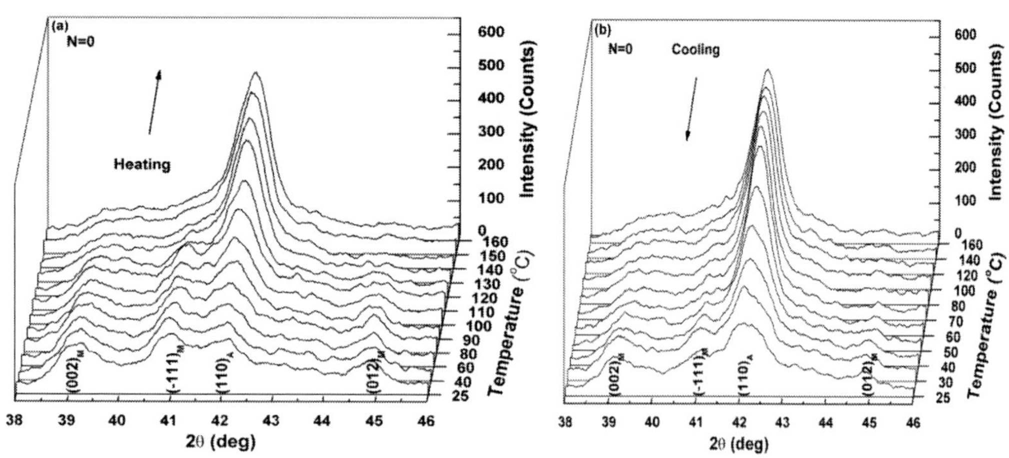

a b

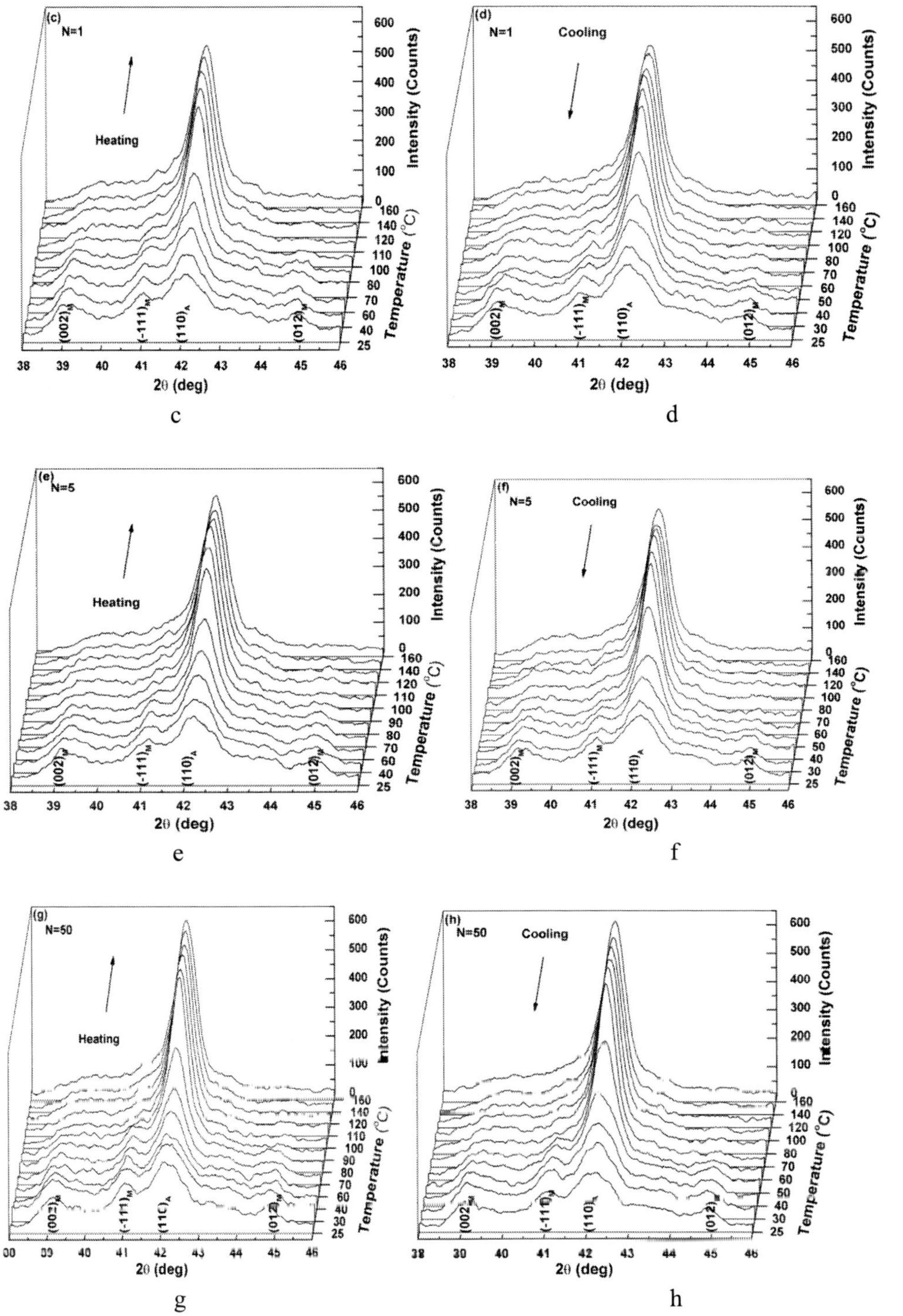

Figure 2(a)-2(h). XRD patterns of the heat-treated NiTi SMA at different temperatures, during heating and cooling, for N = 0, 1, 5 and 50.

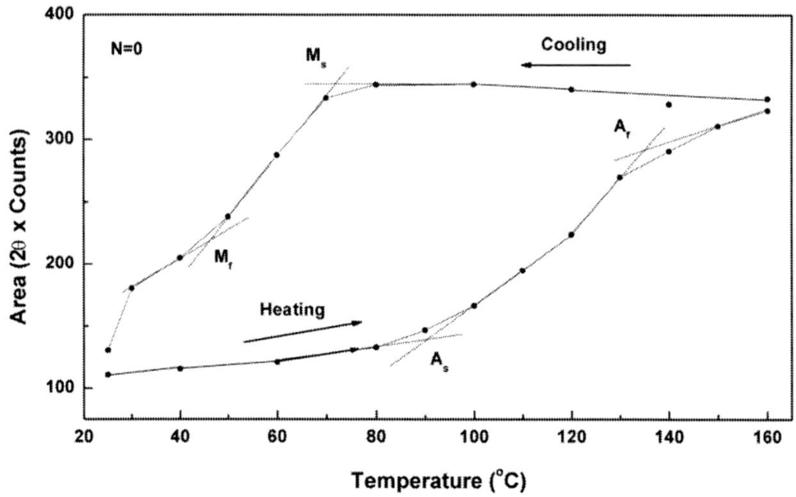

Figure 3. A plot of the integrated area of $(110)_A$ peak with temperature for the heat-treated NiTi SMA sample.

Figure 4 shows the DSC thermogram for a specimen without any cycling. Table 2 gives the transformation temperatures determined from DSC and from the peak area under $(110)_A$ versus temperature plot for the sample without any cycling (N = 0) for comparison. The thermal width observed from XRD studies is larger than that from DSC measurements. This is attributed to the sensitivity of X–ray diffraction to the phases present, as even a small volume fraction of the phases can be detected by XRD when compared to thermal techniques. The different transformation temperatures obtained from XRD and DSC may be due to the different masses of the sample used. To check this possibility DSC was carried out on three samples of different masses (Figure 5). The transformation temperatures for these samples were almost the same, ruling out such a possibility.

The thermal hysteresis corresponding to different numbers of thermal cycling, N = 0, 1, 5, 25 and 50, are shown in Figure 6. The thermal hysteresis associated with N = 0 is very large compared to that for N = 1, 5, 25 and 50. There is a drastic change in transformation temperatures A_s and A_f due to the first thermal cycling. The decrease in A_s and A_f during thermal cycling is significant only up to a few cycles. Subsequently, it is fairly constant. Thermally stabilized dislocations suppress the effect of thermal cycling [5]. X-ray diffractograms at 100 °C during cooling cycle, for the samples subjected to different number of thermal cycling are shown in figure 7 (a). With increasing number of thermal cycles, the intensity of the $(110)_A$ peak increases without much change in the intensity of peaks corresponding to martensitic phase. The variation of integrated area under $(110)_A$ peak (at 100 °C) with the number of thermal cycles is shown in figure 7 (b). Initially, the integrated area increases rapidly and the increase is slower at higher number of cycles. The increase in the austenitic peak intensity or the peak area can be due to the growth of austenitic phase or due to a better orientation of this phase. Figure 8 shows the DSC thermograms of the samples subjected to different thermal cycles. The latent heat of austenitic phase transformation increases only up to N = 5 after which it starts decreasing. The transformation temperature decreases with increase in the number of cycles. On the other hand, integrated peak area increases rapidly up to N = 5 and continues to increase even up to N = 50. Therefore, the initial large increase in the integrated peak area with thermal cycling (up to N = 5) is attributed to the increased austenitic phase.

Table 2. Transformation temperatures determined from the integrated peak area of $(110)_A$ reflections versus temperature plot and DSC for a sample without any thermal cycling (N = 0).

Techniques	Transformation Temperature (° C)			
	A_s	A_f	M_s	M_f
XRD	90	135	72	47
DSC	98	123	78	62

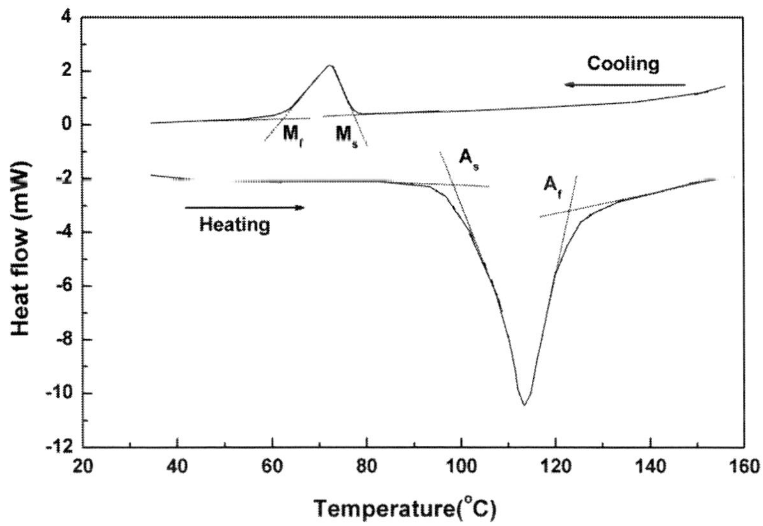

Figure 4. A DSC thermogram for the heat-treated NiTi SMA.

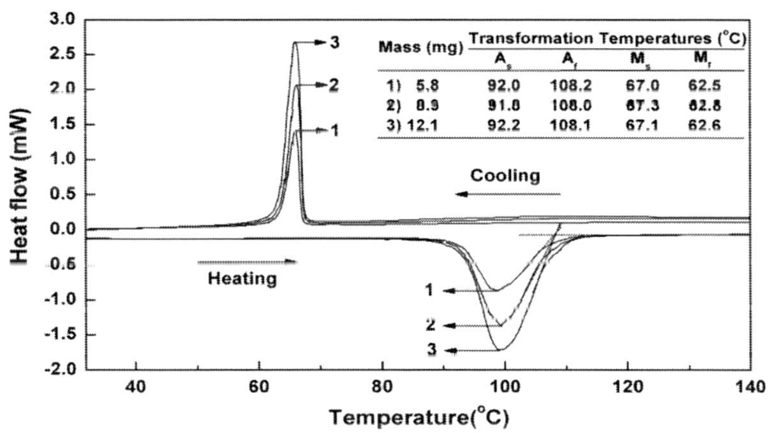

Figure 5. DSC thermograms for the heat-treated NiTi SMA of different masses.

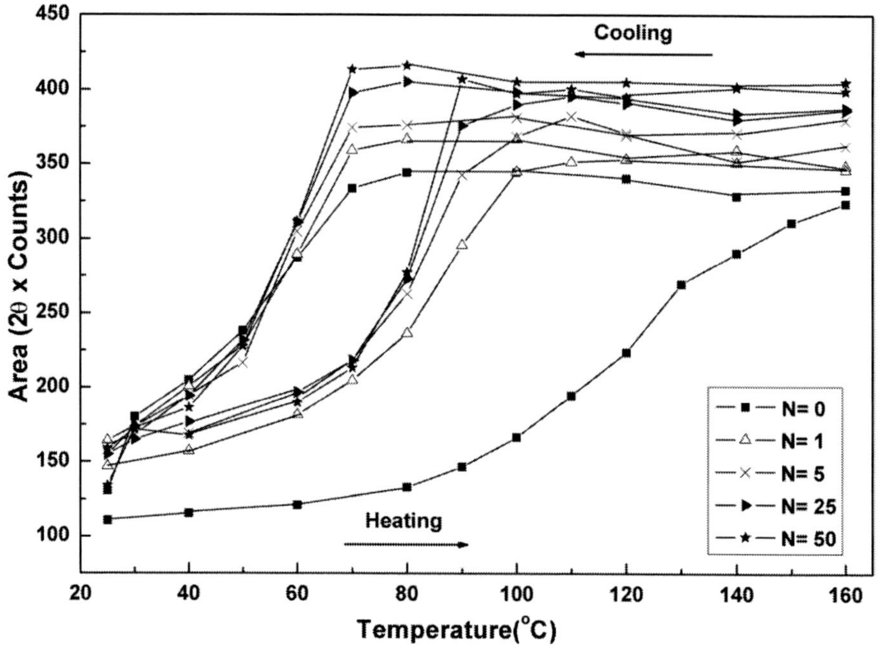

Figure 6. Phase transformation profiles obtained through XRD studies for NiTi SMA samples subjected to different thermal cycles (N = 0, 1, 5, 25 and 50).

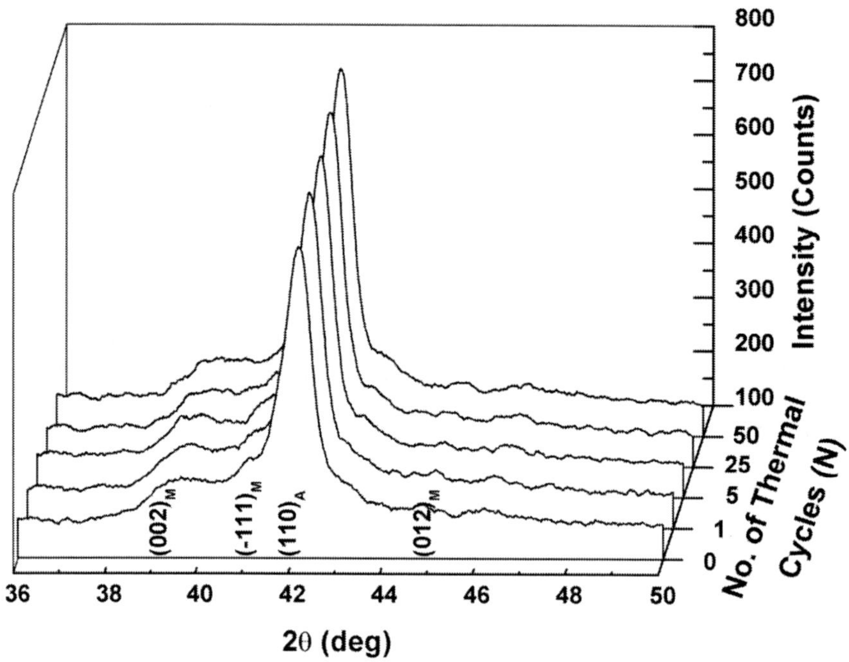

a

Figure 7(a). X-Ray diffractograms obtained at 100 °C, while cooling, for samples subjected to thermal cycling.

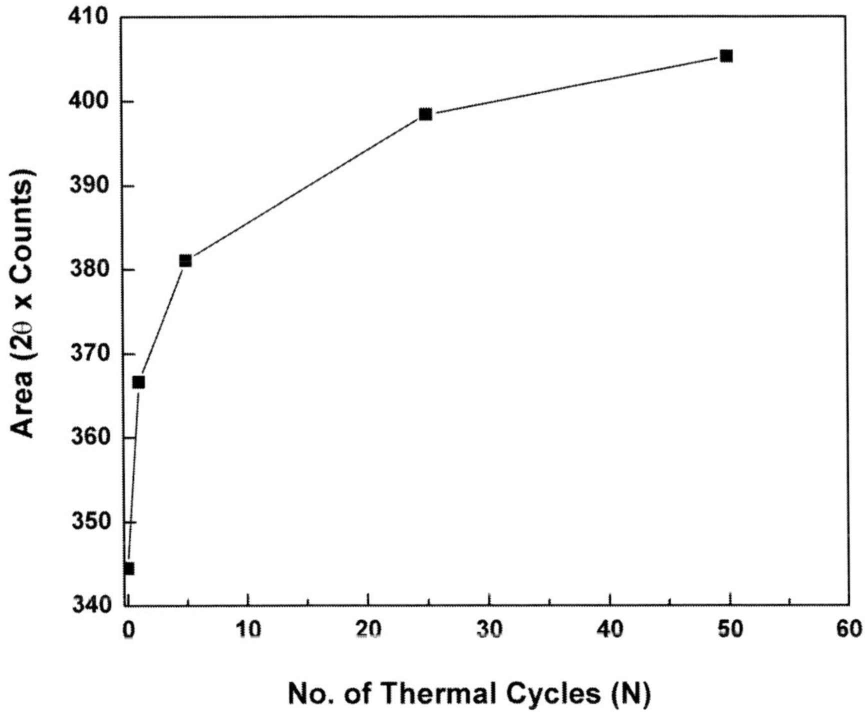

b

Figure 7(b). The variation of integrated area under $(110)_A$ peak recorded at 100 °C during cooling, after different thermal cycles.

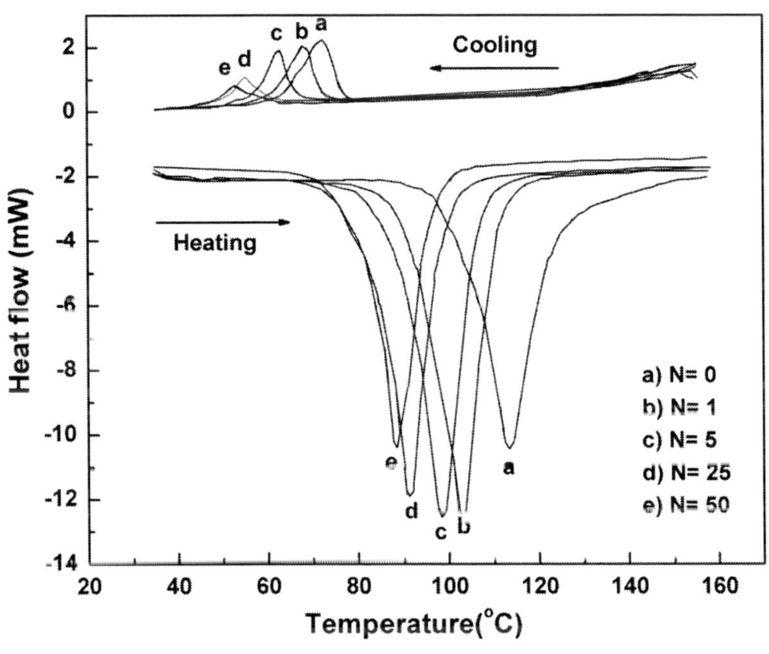

Figure 8. DSC thermograms of the NiTi SMA subjected to different thermal cycles (N = 0, 1, 5, 25 and 50).

At larger number of cycles, the increase may be due to a better orientation of the grains. The decrease in A_s and A_f during initial thermal cycling and the rapid increase in the intensity of the austenitic phase indicate an easier M→A phase transformation which may be related to the presence of a dislocation structure that serves as a network of nucleation sites for austenite phase [43]. This is supported by the fact that the NiTi alloy shows thermoelastic martensitic transformations only after heat-treatment to produce an optimum number of defects. M_s and M_f determined from the temperature-dependent XRD studies are almost constant where as DSC shows a decrease in M_s and M_f with increasing number of thermal cycling as is expected due to the increased defects.

3. Mechanical Polishing Effect

Shape memory alloys, like any other alloys, experience oxidation during hot working or intermediate annealing during cold working process. Unlike other alloys, shape memory alloys need heat treatment for shape setting at final process of fabrication. Since Ti is highly reactive with oxygen, oxidation during heat treatment is unavoidable. Surface modification makes the alloy react with mainly oxygen, carbon or nitrogen. Ti reacts preferentially with the elements, thus the chemical composition changes near the surface [47]. Drastic change in transformation temperature may be expected as it depends strongly on the composition [48]. Kim et al [49] studied the effect of composition change near the surface of a Ni-Ti alloy on transformation and deformation behavior. Preferential reaction of Ti with oxygen results in compounds such as Ni_3Ti and Ni_4Ti_3 and a Ni-rich region between the bulk and the surface oxide layer. Martensitic transformation did not take place in the Ni-rich region due to the high Ni content while R-phase transformation temperature decreased significantly. Oxidation deteriorated the superelasticity due to the formation of the Ni-rich compounds as well as the Ni-rich region in which slip deformation occurs instead of induction of martensite. Hence, it is clear that presence of oxide layer affect the transformation behaviors of SMAs and to avoid these effects the oxide layer needs to be removed. The approach adopted for removing oxidation may be of significance from the thermomechanical considerations. Further, the SMA sheets or wires need to be cut to size and shape, for any particular application. Therefore, we made an attempt to investigate if mechanical cutting and polishing have any effect on the shape memory transformation behavior of NiTi alloy [50].

The specimens (same as those mentioned earlier) were subjected to heat treatment by annealing at a pre-set temperature for 20 minutes before quenching them in a water bath at RT. Two sets of samples were studied, one heat-treated (HT) at 560 °C and another at 660 °C. In each set, one sample is prepared by cutting a small piece from the larger SMA sheet using a mechanical cutter followed by a mechanical polish using a fine-grained silicon carbide emery papers of grit P320, P600 and P800 (with average particle diameter of 46.2, 25.8 and 21.8 μm) under cold-water circulation to remove the oxide layer, after the completion of the heat treatment and designated as “HT560MC” or “HT660MC”. The other sample is prepared by cutting the SMA sheet using mechanical cutter prior to heat treatment and the oxide layer is removed by chemical etching (CE) using 45% vol HNO_3 + 45% vol H_2O + 10% vol HF and designated as “HT560CE” or “HT660CE”. Both the samples in each set were of equal mass and heat treatment was carried out together to eliminate experimental errors associated with

HT. Final mass of the samples in first set was 15.7 mg each and in second set 19.7 mg each. The SMAs were subjected to thermal cycling as explained earlier. The phase transformation behaviour of all the samples was studied using a Shimadzu DSC-50 Differential Scanning Calorimetry. The measurements were made in the temperature range from RT to +160 °C at a scanning rate of 3 °C/min under N_2 atmosphere. Contact mode Nanosurf easyScan Atomic Force Microscope (AFM) with scanning speed of 1024 lines at 1.2 sec/line was used to observe the surface topography of the samples.

Figure 9 shows the surface topography of heat-treated (560 °C) NiTi SMA samples with oxide layer and without oxide layer (chemically etched and mechanically polished) for zero thermal cycling. Straight dark lines on the surface of the mechanically polished sample indicate the presences of deformed layers (scratch) due to mechanical polishing. Figures 10 and 11 show the DSC thermograms of HT560MC and HT560CE subjected to various numbers of thermal cycling (N) up to 150, during heating and cooling respectively. Table 3 gives the phase transformation temperatures of HT560MC and HT560CE determined form DSC. In case of HT560MC at N=0, reverse phase transformation starts (A_s) at 97.6 °C and austenitic phase formation continues beyond 140 °C without any clear finishing temperature A_f. Corresponding thermogram for HT560CE shows a sharper change in heat flow at A_f and hence, smaller thermal width compared to HT560MC. During cooling also, HT560CE exhibits sharper thermal width compared to that of HT560MC. The internal stress generated due to plastic deformation is known to restrict the martensitic transformation, keeping the martensitic phase pinned in the microstructure [20,51]. The mechanical work carried out on HT560MC must be responsible for the broad thermal width as well as smaller heat flow, associated with the M↔A transformation.

The specimen subjected to thermal cycling exhibit different transformation characteristics. It is seen that the reverse M→A transformation peak shifts progressively to lower temperatures with increase in the number of thermal cycling for both HT560MC and HT560CE. Forward transformation on cooling also exhibits a similar trend. Further, the heat flow of austenite phase transformation shows rapid increase up to N=10 beyond which it start decreasing gradually for HT560MC. On the other hand, for the HT560CE the increase is up to N = 5 followed by a decrease which is more rapid than for HT560MC. The sudden decrease in A_s and A_f during initial thermal cycling and the rapid increase in the heat flow of the austenitic phase indicate an easier M → A phase transformation which may be related to the presence of a dislocation structure introduced by thermal cycling that serves as a network of nucleation sites for austenite phase [43]. The change in transformation heats, particularly the decrease, as determined by thermal measurement, has been suggested to be related to the stabilization of martensite, which reduces the amount of the material participating in the transformation [52-54]. The decrease in heat flow starts at an earlier cycle (N = 5) in case of HT560CE and the decrease is more rapid as compared to the case of HT560MC. This may be due to the formation of R-phase, as the single-step M →A transformation changes to a two-step transformation M→R→A after 5th cycle in the case of HT560CE (Figure 10b). However, HT560MC continues to be single step transformation even up to N = 150. The residual dislocations due to the cold working not favoring martensite-austenite transformation by interacting with the dislocations created during thermal cycling has been reported by Turenne et al [43]. In the case of HT560MC, the dislocations created by the mechanical cutting and polishing may be responsible for the non-appearance of R-phase even up to 150 cycles, possibly due to the entanglement of dislocations at high dislocation density.

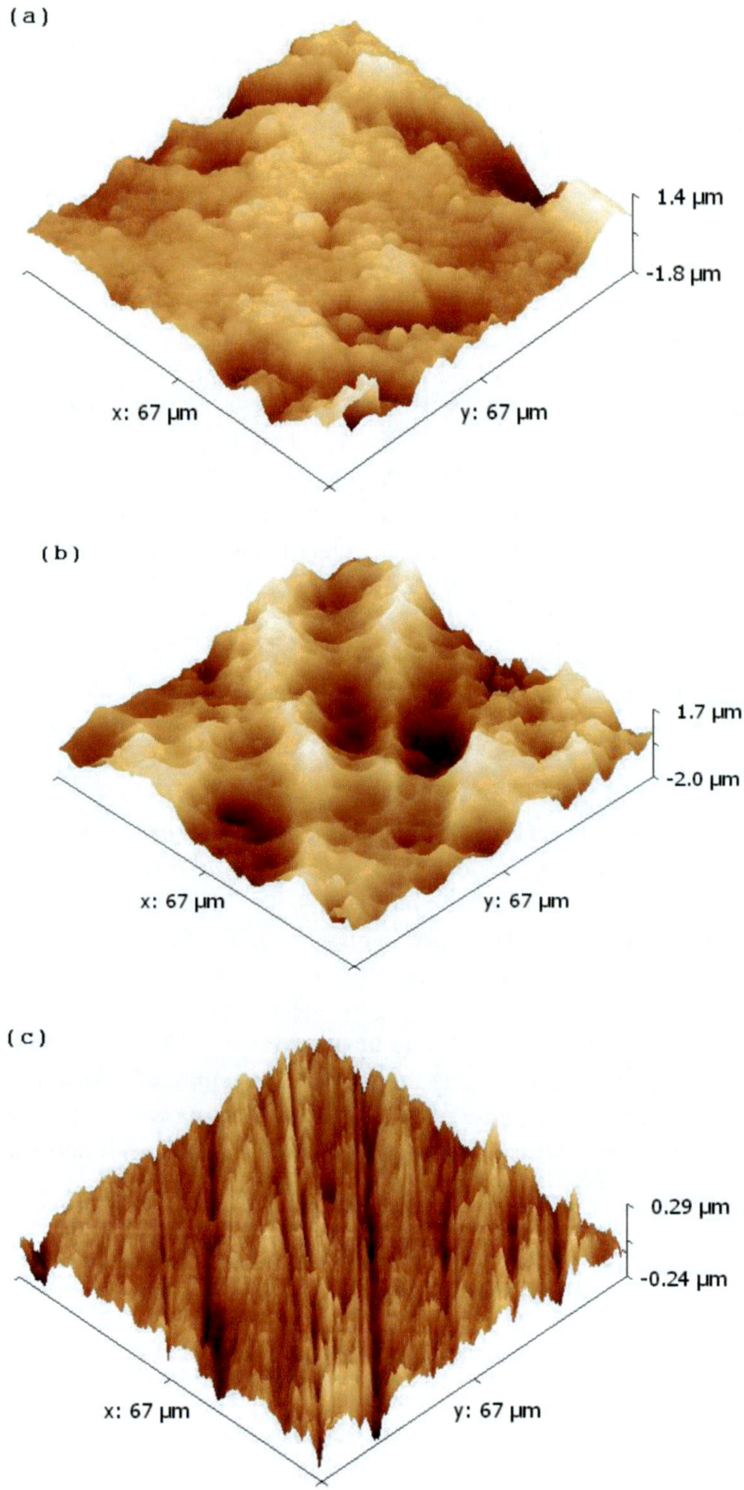

Figure 9. AFM image of heat-treated NiTi SMA Surface a) with oxide layer, b) chemically etched, c) mechanically polished..

Table 3. Phase Transformation temperatures of HT560MC and HT560CE at different number of cycles

No. of Thermal Cycles (N)	Phase Transformation Temperatures (°C)												
	HT560MC					HT560CE							
	A_s	A_f	R_s	M_s	M_f	A_s	A_f	R_s'	A_f'	R_s	R_f	M_s	M_f
0	97.6	NC[a]	--	73.2	61.6	93.8	112.7	--	--	--	--	70.2	64.8
1	88.5	104.3	--	71.1	60.5	91.3	105.5	--	--	--	--	66.4	61.6
5	85.7	98.9	--	65.2	58.3	84.7	100.2	--	--	--	--	61.6	58.3
10	83.9	95.5	--	63.8	54.4	--	--	79.8	95.9	62.8	--	--	54.1
15	83.4	94.1	63.9	--	51.4	--	--	77.8	94.3	62.7	56.8	56.2	51.8
25	81.4	92.4	63.1	--	49.0	--	--	76.1	92.6	62.4	55.2	53.9	49.5
50	79.8	92.0	62.4	--	47.7	--	--	74.2	91.1	62.2	54.0	52.0	48.4
100	78.7	92.9	61.5	--	45.7	--	--	72.9	89.6	62.0	52.8	50.4	46.9
150	77.2	92.2	61.2	--	45.1	--	--	71.6	88.3	61.2	52.2	49.5	44.8

[a] Not Clear

Table 4. Phase Transformation temperatures of HT660MC and HT660CE at different number of thermal cycles

No. of Thermal Cycles (N)	Phase Transformation Temperatures (°C)								
	HT660MC					HT660CE			
	A_s	A_f	R_s	M_s	M_f	A_s	A_f	M_s	M_f
0	103.2	NC[a]	--	77.7	62.1	93.3	114.5	77.172	65.4
1	88.2	105.9	--	73.0	60.4	90.1	105.9	.2	62.4
5	85.3	100.9	--	68.0	57.5	87.4	100.2	66.5	58.0
10	83.1	97.7	--	65.7	55.3	84.0	96.9	62.5	55.1
50	78.1	93.4	63.6	--	48.5	78.5	91.2	56.8	49.2
75	78.2	93.9	63.2	--	47.2	77.7	90.2	56.2	47.7
100	76.4	93.9	62.6	--	46.7	76.6	89.9	55.8	47.1
150	75.9	94.0	61.9	--	45.7	76.2	89.4	55.5	46.3

[a] Not Clear.

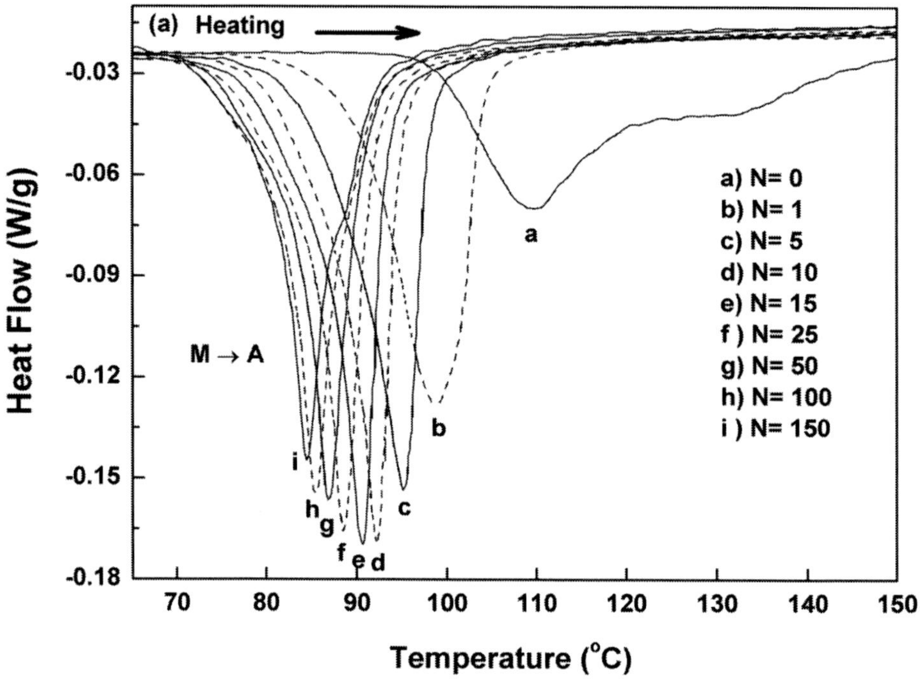

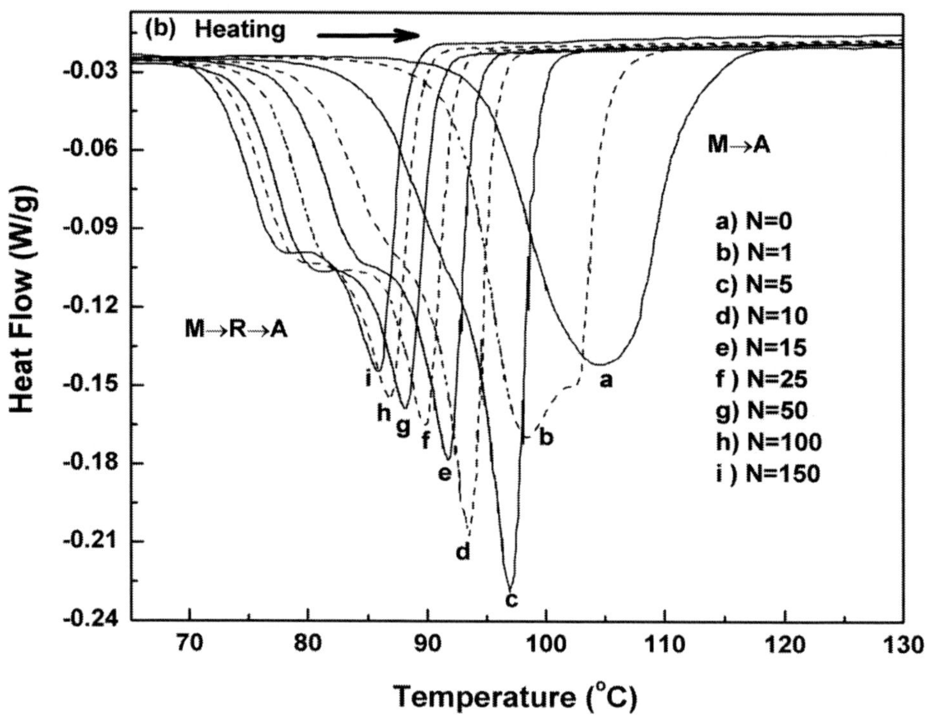

Figure 10. DSC thermograms corresponding to N = 0, 1, 5, 10, 15, 25, 50, 100 and 150 of (a) HT560MC and (b) HT560CE during reverse transformation.

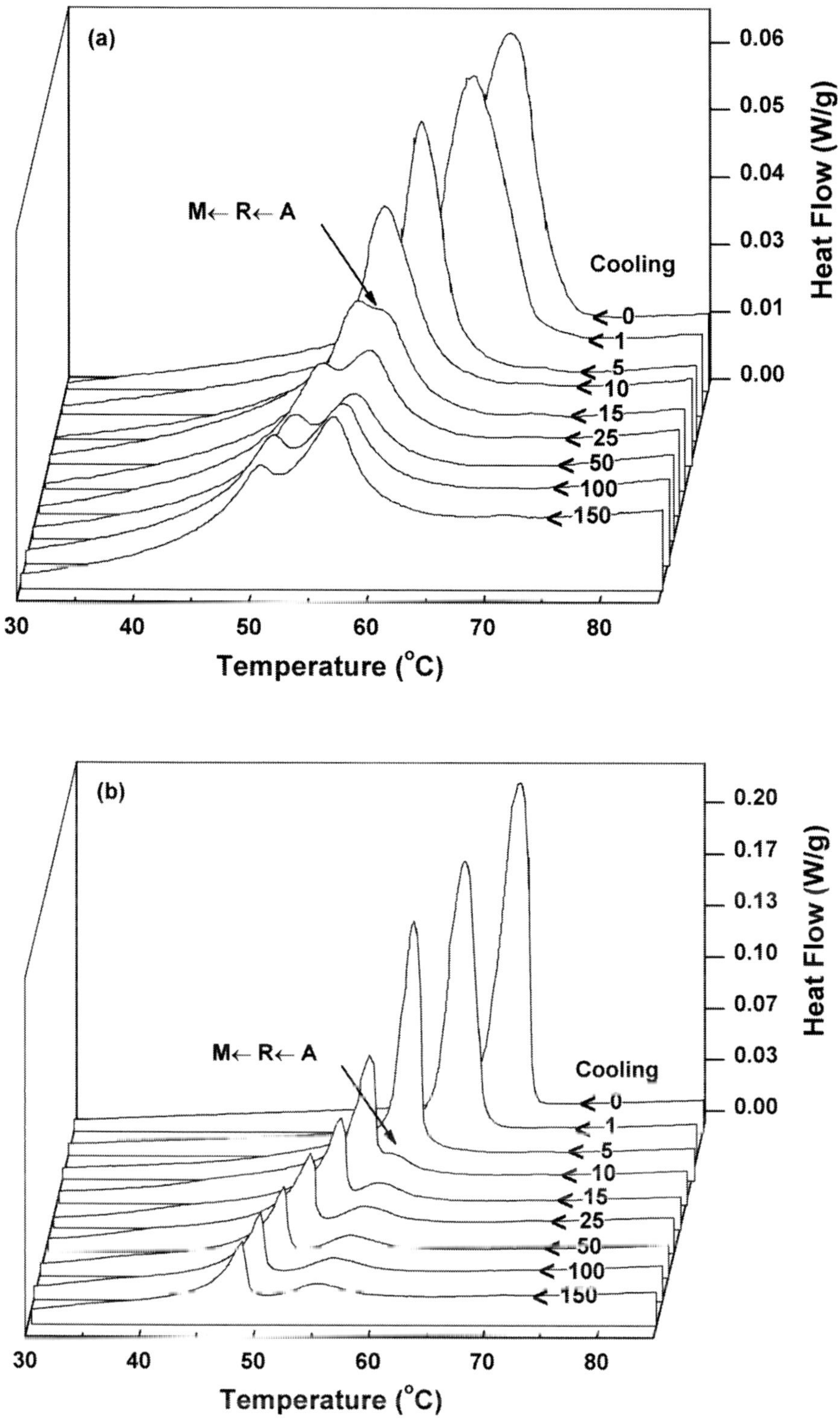

Figure 11. DSC thermograms corresponding to N = 0, 1, 5, 10, 15, 25, 50, 100 and 150 of (a) HT560MC and (b) HT560CE during forward transformation.

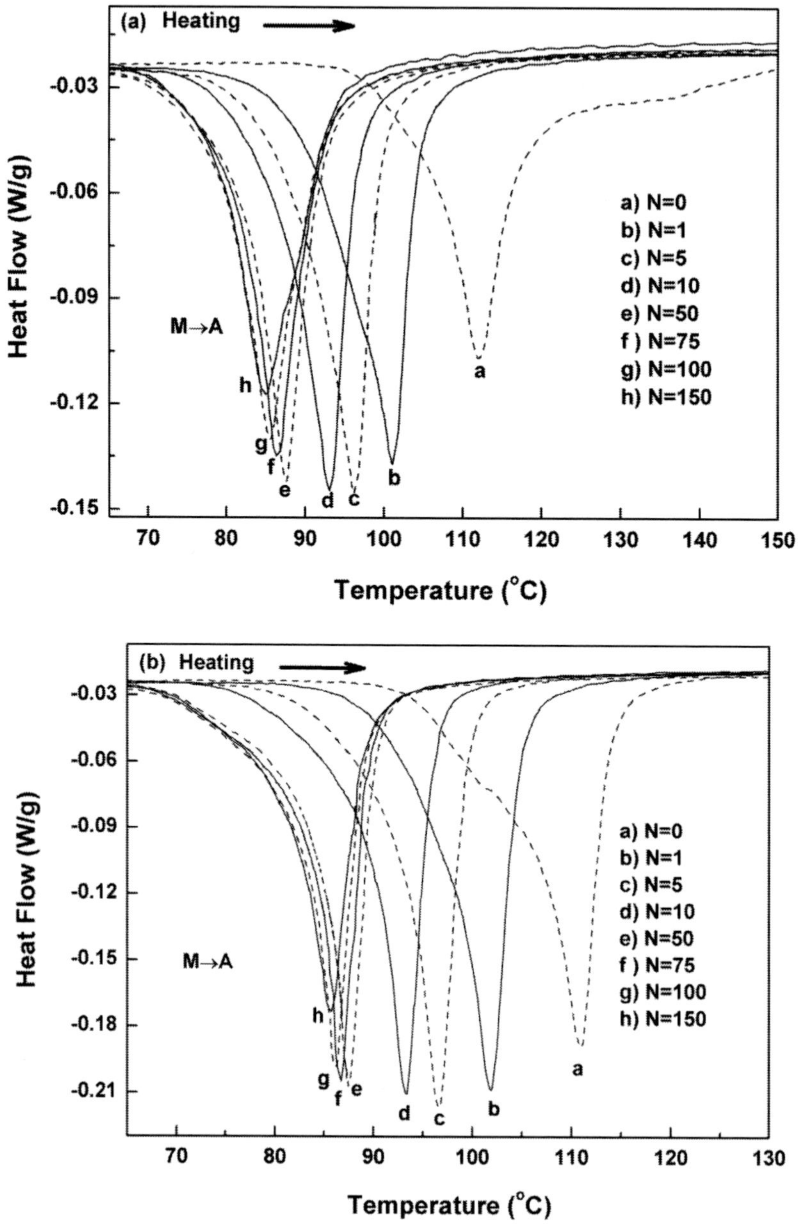

Figure 12. DSC thermograms corresponding to N = 0, 1, 5, 10, 50, 75, 100 and 150 of (a) HT660MC and (b) HT660CE during reverse transformation.

During forward transformation both HT560MC and HT560CE show decrease in M_s and M_f with increasing N, which may be attributed to the effect of dislocation stress fields in opposing the transformation [5,24]. Both the samples show single-step phase transformation A→M initially and after a certain number of cycles, exhibit two-step phase transformation A→R→M. However, the way in which R-phase grows is different in HT560MC and HT560CE. R-phase shows up in case of HT560MC by splitting up of M-phase during the 15th thermal cycle. With increase in the number of cycles, separation continues even up to N = 150. A shoulder is formed on the high-temperature side of the exothermic peak which corresponds to the R-phase for HT560CE at the 10th cycle with the well-separated peak clearly visible at N=15.

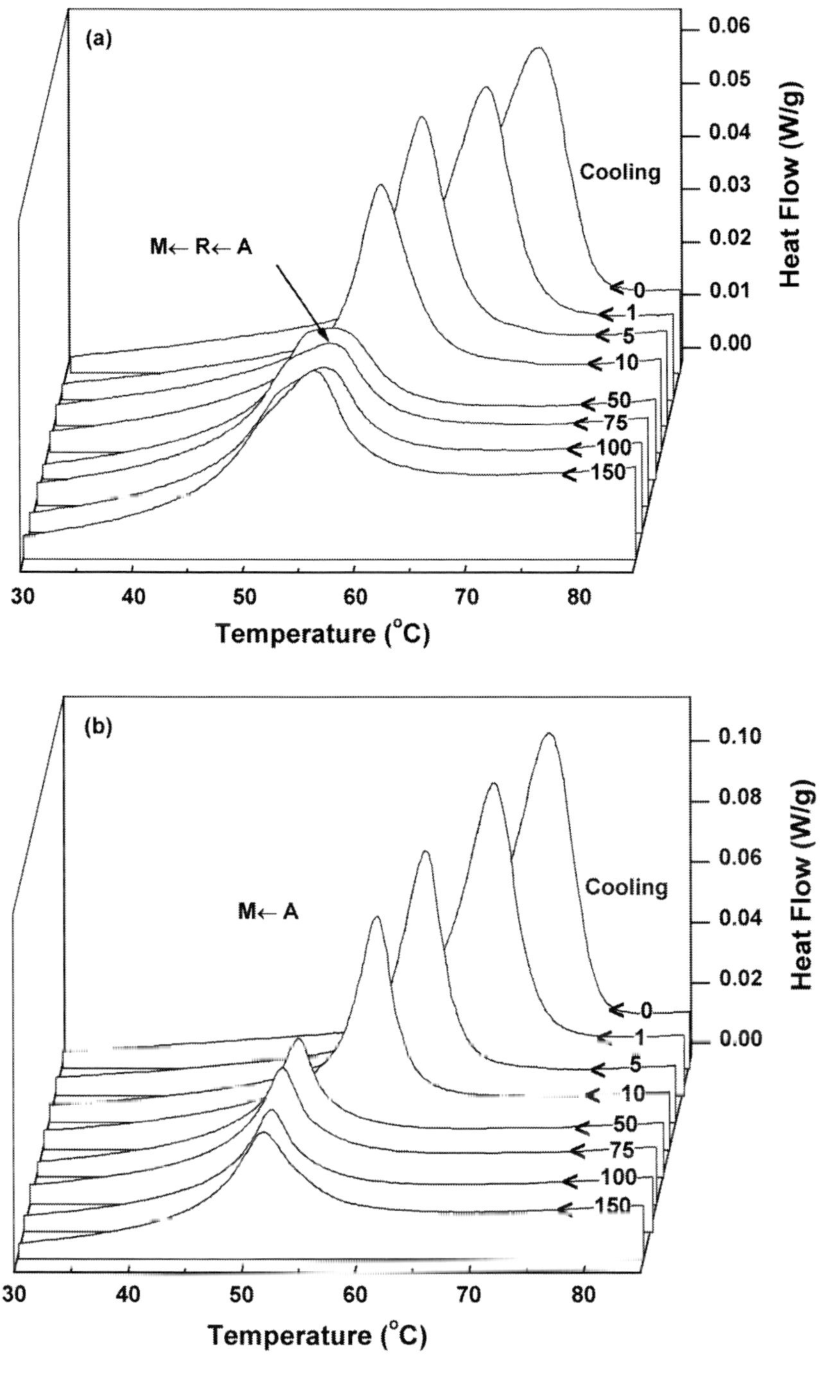

Figure 13. DSC thermograms corresponding to N = 0, 1, 5, 10, 50, 75, 100 and 150 of (a) HT660MC and (b) HT660CE during forward transformation.

Figures 12 and 13 show the DSC thermograms of HT660MC and HT660CE subjected to various numbers of thermal cycling (N) up to 150, during heating and cooling respectively. Table 4 gives the phase transformation temperatures of HT660MC and HT660CE determined from DSC. Most of the transformation characteristics are similar to those of corresponding samples heat-treated at 560 °C. The difference is mainly in the appearance of intermediate R-phase. Here both the samples show single step reverse phase transformation, even up to 150 cycles. While cooling, HT660MC shows (Figure 13a) single step phase transformation A→M up to 10 cycles beyond which a two-step phase transformation (A→R→M) takes place. HT660MC shows a broader peak when the specimen is subjected to cycles more than 10 indicating the appearance of a R-phase. However, in the case of HT660CE, the R-phase seems to be completely absent in the complete transformation cycle in contrast to the HT560CE sample, where R-phase showed up at a lesser number of cycles in comparison with HT560MC. The heat treatment at higher temperature results in smaller dislocation density, which in turn does not facilitate the formation of R-phase. The dislocations created due to the mechanical cutting and polishing may be responsible for the R-phase being exhibited by HT660MC.

4. SUMMARY

1. XRD technique used for characterization of SMAs were found to be very much useful and promising.
2. The thermal hysteresis width corresponding to the M↔A phase transformation observed through the XRD technique is large when compared to that measured from DSC.
3. A_s and A_f decrease during the initial few thermal cycles, after which they remain fairly constant, whereas M_s and M_f do not change significantly due to thermal cycling, as observed from the temperature dependent XRD studies.
4. Integrated area under the austenitic peak increases rapidly with the number of thermal cycling up to N = 5 and continues to increase even up to N = 50. The initial large increase in the integrated peak area with thermal cycling (up to N = 5) is attributed to the increased austenitic phase, whereas at larger number of cycles, the increase may be due to a better orientation of grains as indicated by DSC.
5. Mechanically cut and polished samples show broad thermal width and lesser heat flow associated with the M↔A transformation compared to chemically etched samples due to the internal stresses generated during mechanical operation.
6. The dislocations created by the mechanical cutting and polishing may be responsible for the absence of R-phase in HT560MC during reverse transformation, in samples subjected to thermal cycling. Heat treatment at a higher temperature of 660 °C does not facilitate the formation of R-phase in the reverse transformation and helps the formation of R-phase to a very small extent, that too only in HT660MC, during forward transformation due to the reduced dislocation density.
7. Both the heat-treated samples show the effect of mechanical cutting and polishing even at 150th thermal cycle.

REFERENCES

[1] Harrison, J. D. In *Engineering Aspects of Shape Memory Alloys;* Duerig, T. W.; Melton, K. N.; Stöckel, D.; Wayman, C. M. Ed.; Butterworth-Heinemann Ltd.: London, UK, 1990; pp 106-111.
[2] Kwarciak, J.; Lekston, Z.; Morawiec, H. *J. Mater. Sci.* 1987, *22*, 2341-2345.
[3] Paula, A. S.; Canejo, J. P. H. G.; Silva, R. J. C.; Fernandes, F. M. B. *Mater. Sci. Forum* 2004, *455-456*, 346-350.
[4] Filip, P.; Mazanec, K. *Scr. Metall. Mater.* 1994, *30*, 67-72.
[5] Miyazaki, S.; Igo, Y.; Otsuka, K. *Acta Metall.* 1986, *34*, 2045-2051.
[6] Airoldi, G.; Bellini, G.; Francesco, C. D. *J.Phys. F: Met. Phys.* 1984, *14*, 1983-1987.
[7] Eckelmeyer, K. H. *Scr. Metall.* 1976, *10*, 667-672.
[8] Matsumoto, H. *J. Mater. Sci. Lett.* 1989, *8*, 232-233.
[9] Huisman-Kleinherenbrink, P. M., Jian, Y., Beyer, J. *Mater. Lett.* 1991, *11*, 145-150.
[10] Piao, M.; Miyazaki, S.; Otsuka, K.; Nishida, N. *Mater. Trans. JIM* 1992, *33*, 337- 345.
[11] Ma, J. L.; Wu, K. H. *Mater. Sci. Technol.* 2000, *16*, 716- 719.
[12] Miyazaki, S.; Otsuka, K. *Metall. Trans. A* 1986, *17A*, 53- 63.
[13] Stroz, D.; Kwarciak J.; Morawiec H. *J. Mater. Sci.* 1988, *23*, 4127-4131.
[14] Huang, X.; Liu, Y. *Scr. Mater.* 2001, *45*, 153-160.
[15] Todoroki, T.; Tamura, H. *Trans. Jpn. Inst. Met.* 1987, *28*, 83-94.
[16] Wang, Z. G.; Zu, X. T. *Scr. Mater.* 2005, *53*, 335-339.
[17] Matsumoto, H. *J. Alloys Compd.* 2003, *350*, 213-217.
[18] Ren-Der, J.; Jong-Cheng, T. *Scr. Metall. Mater.* 1994, *30*, 1027-1030.
[19] Liu, Y.; Chen, X.; McCormick, P. G. *J. Mater. Sci.* 1997, *32*, 5979-5984.
[20] Miller, D. A.; Lagoudas, D. C. *Mater. Sci. Eng. A* 2001, *308*, 161-175.
[21] Matsumoto, H. *Physica B* 1993, *190*, 115-120.
[22] Matsumoto, H. *Mater. Lett.* 1991, *11*, 40- 42.
[23] Filip, P.; Mazanec, K. *Scr. Metall. Mater.* 1995, *32*, 1375-1380.
[24] McCormick, P. G.; Liu, Y. *Acta Metall. Mater.* 1994, *42*, 2407-2413.
[25] Kudoh, Y.; Tokonami, M.; Miyazaki, S.; Otsuka, K. *Acta Metall.* 1985, *33*, 2049- 2056.
[26] Pelosin, V.; Riviere, A. *Metall. Mater. Trans. A* 1998, *29A*, 1175-1180.
[27] Lin, H. C.; Wu, S. K.; Lin, J. C. *Mater. Chem. Phys.* 1994, *37*, 184-190.
[28] Lòpez, H. F. *Mater. Lett.* 2001, *51*, 144-150.
[29] Paula, A. S.; Canejo, J. P. H. G.; Schell, N.; Fernandes, F. M. B. *NIM B* 2005, *238*, 111-114.
[30] Wayman, C. M.; Cornelis, I.; Shimizu, K. *Scr. Metall.* 1972, *6*, 115-122.
[31] Tang, W.; Sandström, R. *Mater. Des.* 1993, *14*, 103-113.
[32] Yinong, L.; Yong, L.; Humbeeck, J. V. *Acta Mater.* 1999, *47*, 199-209.
[33] Uchil, J.; Mohanchandra, K. P.; Ganesh Kumara, K., Mahesh, K. K. *Physica B* 1999, *270*, 289-297.
[34] Liu, Y.; Xie, Z. *Mater. Sci. Eng. A* 2003, *361*, 185-190.
[35] Liu, Y.; Xie, Z.; Humbeeck, J. V. *Mater. Sci. Eng. A* 1999, *273-275*, 673-678.
[36] Zhao, X. K.; Cai, W.; Tian, Y.; Zhao, L. *Mater. Sci. Forum* 1999, *394-395,* 153-156.
[37] McNaney, J. M.; Imbeni, V.; Jung, Y.; Papadopoulos, P.; Ritchie, R. O. *Mech. Mater.* 2003, *35*, 969-986.

[38] Matsumoto, O.; Miyazaki, S.; Otsuka, K.; Tamura, H. *Acta Metall.* 1987, *35*, 2137-2144.
[39] Uchil, J.; Mohanchandra, K. P.; Ganesh Kumara, K.; Mahesh, K. K. *Mater. Sci.Eng. A* 1998, *251*, 58-63.
[40] Matsumoto, H. *J. Alloys Compd.* 2004, *368*, 182-186.
[41] Li, Z. C.; Zhao, X. K.; Zhang, H.; Liu, L.; Xu, Y. B. *Mater. Lett.* 2003, *57*, 1086- 1090.
[42] Wang, F. E.; DeSavage, B. F.; Buehler, W. J.; Hosler, W. R. *J. Appl. Phys.* 1968, *39*, 2166- 2175.
[43] Turenne, S.; Prokoshkin, S.; Brailovski, V.; Sacepe, N. *Can. Metall. Quart.* 2000, *39*, 217-224.
[44] Pattabi, M., Ramakrishna, K., Mahesh, K. K. *Mater. Sci. Eng. A* 2007, *448*, 33-38.
[45] Pattabi, M., Ramakrishna, K., Mahesh, K. K. *Mater. Sci. Eng. A* 2009, *505*, 191.
[46] Dong, C. *J. Appl. Crystallogr.* 1999, *32*, 838.
[47] Shabalovskaya, S.; Ryhanen, J.; Yahia, L. *Mater. Sci. Forum* 2002, *394-395*, 131- 138.
[48] Tang, W. *Metall. Mater. Trans. A* 1997, *28A*, 537-544.
[49] Kim, K. S.; Jee, K. K.; Kim, Y. B.; Jang, W. Y.; Han, S. H. *Eur. Phys. J. Special Topics* 2008, *158*, 67– 71.
[50] Pattabi, M.; Ramakrishna, K. *Mater. Sci.Eng. A* 2008, *486*, 14-18.
[51] Wang, Z. G.; Zu, X. T.; Feng, X. D.; Zhu, S.; Zhou, J. M.; Wang, L. M. *Physica B* 2004, *353*, 9-14.
[52] Stalmans, R.; Humbeeck, J. V.; Delaey, L. *Acta Metall. Mater.* 1992, *40*, 501-511.
[53] Liu, Y.; McCormick, P. G. *Mater. Sci. Forum* 1990, *56-58*, 585-590.
[54] Favier, D.; Liu, Y. *J. Alloys Compd.* 2000, *297*, 114-121.

In: Shape Memory Alloys
Editor: H. R. Chen, pp. 339-360
ISBN 978-1-60741-789-7

Chapter 12

DEPENDENCE OF TRANSFORMATION TEMPERATURES OF SHAPE MEMORY ALLOYS ON THE NUMBER AND CONCENTRATION OF VALENCE ELECTRONS

***Mehrdad Zarinejad and Yong Liu*[1]**
School of Mechanical and Aerospace Engineering
Nanyang Technological University[2], Singapore

ABSTRACT

Dependence of transformation temperatures of shape memory alloys on the number (e_v/a) and concentration (c_v) of valence electrons is investigated. The alloys are categorized into low (e_v/a<5), medium (5≤e_v/a≤7.50), and high (e_v/a>7.50) valence electron groups. For majority of alloys in medium and high valence electron group, clear correlations between transformation temperatures and c_v are found. M_s and A_s both decrease from 1100 and 1150˚C to as low as -206˚C and -153˚C respectively, with increasing c_v from 0.10 to around 0.30. When e_v/a>7.50, the transformation temperatures show much less dependence on c_v and can be lower than the medium group temperatures even at medium or low c_v values. High electron NiMnX (X=Ga, In, Al) Heusler alloys containing Mn (>30 at%) show high transformation temperatures at high c_v values due, most likely, to anti ferromagnetism of excess Mn which reduces the elastic moduli. The low valence electron group alloys show significantly lower transformation temperatures for their c_v values compared to the main group. The influence of c_v is discussed on the basis of the variations of elastic constants due to composition change.

Keywords: metals and alloys; shape memory; phase transformation; electronic structure.

---

[1] Please send all correspondences to: Prof. Yong Liu, School of Mechanical and Aerospace Engineering, Nanyang Technological University, Singapore 639798, Tel/Fax: +65-6790 4951, Email: mliuy@ntu.edu.sg.
[2] Nanyang Avenue, Singapore 639798.

1. Introduction

Transformation temperatures of shape memory alloys are of paramount importance from both fundamental and technological points of view [1-10]. Despite of many efforts to relate transformation temperatures to chemical composition, the dependence of M_s and A_s on alloy composition has remained unclear, and the design of alloys, has been rather empirical, as an in depth understanding about the effect of composition change has not been achieved. The key influential factors that have been referred to are elastic constants of the matrix crystal before martensitic transformation and microstructural features such as precipitates [8]. Very recently, in order to understand the mechanisms that control the transformation temperatures in NiTi-based shape memory alloys, we suggested the necessity of acquiring an in-depth knowledge of how the atomic bonding strength of the matrix crystal is influenced by chemical composition of the alloys [10]. We introduced the dependence of transformation temperatures on the number and concentration of valence electrons in NiTi-based shape memory alloys. It was shown that in these alloys the transformation temperature was distinctly related to whether the number of valence electrons per atom of the alloy is $(e_v/a)=7$ or $(e_v/a)\neq 7$. More importantly, it was demonstrated that by increasing the valence electron concentration (c_v) of the NiTi-based alloys, the transformation temperatures decrease as a result of higher elastic moduli of the crystal which influence the resistance against shear, responsible for martensitic transformation [10]. This understanding was achieved based on the knowledge that prior to martensitic transformation a softening of elastic moduli occurs with lowering temperature [8,11,12]. Experimental works, in Cu-based and NiTi-based shape memory alloys [4-7] are indicative of existence of a critical value of elastic constant at which transformation occurs which is not sensitive to alloy compositions [1] and is only slightly dependant on temperature [8]. Based on this, the elastic moduli of the alloy, which are dependant on both composition and temperature are the key parameters influencing the transformation temperatures [8,9]. Since in metallic materials the delocalized valence electrons dominate the strengths of bonds and elastic properties [13], it is of importance to study the electronic structure of alloys.

In the present work, the investigation is extended to nearly all existing transition-metal-based shape memory alloys in order to reveal a more in depth understanding of the dependence of transformation temperatures (M_S, A_S) on the electronic structure of transition metal based shape memory alloys. The electronic structures of nearly all the alloy systems are examined. Two parameters are considered: 1) the number of valence electron per atom, and 2) the valence electron concentration of the alloy. Furthermore, the possible influence of other factors such as, Mn antiferromagnetism, or electron orbital occupancy, on the transformation temperatures of shape memory alloys is introduced. The findings of this study take important steps towards understanding the influence of chemical composition on the transformation temperatures of shape memory alloys and will enable more effective alloy design.

2. Analysis

A survey of the literature reveals a number of transition-metal shape memory alloy systems with a wide range of martensitic transformation temperatures [8,10,14-113]. There are many alloy systems with or without magnetic effects. These include NiTi, NiMn CuAlNi,

AuCd, FeMn, ZrCu, NbRu, TaRu, CoNi, CoAl, Ni_3Ta, CuZn, MnCu, TiPt, TiPd, TiNb, ZrNb, Fe_3Pt, Fe_3Ge, etc. Tables 1-3 present more than 200 alloy compositions which comprise nearly all the transition-metal-based shape memory alloys, together with their transformation temperatures (M_S, A_S) in ascending order. The transformation temperature data were extracted from the relevant literature as listed in tables or presented in graphs. The majority of these data correspond to solution treated alloys to minimize the effects of precipitates and mechanical work. Nonetheless, in few cases where solution-treated data were not available in the literature, the data pertaining to annealed conditions have been used. Thus, the possible effect of precipitates was not considered. Also, the possible influence of different crystal structures both in austenite and martensite for different alloy systems has not been considered in this work.

2.1. Dependence on the Number of Valence Electrons

The number of valence electrons are usually considered as the number of d and s electrons for an atom of transition metals, and p and s electrons for an atom for non-transition ones. The valence electrons per atom of an alloy can be calculated based on the atomic fraction of the elements in the alloy by equation (1):

$$\frac{e_v}{a} = f_A e_v^A + f_B e_v^B + f_C e_v^C + f_D e_v^D + \ldots \qquad (1)$$

where f_A, f_B, f_C and f_D represent the atomic fractions of A, B, C, and D elements in the alloy and, e_v^A e_v^B, e_v^C and e_v^D are the corresponding number of valence electrons of elements A, B, C, D, respectively. The e_v/a values of the alloys examined in this study are within a wide range of 4.20-11.50 (Tables 1-3, and Figure 1). The M_S (A_S) temperatures as high as 1100°C (1150°C) down to -206°C (-153°C) are examined. For analysis in this study, the alloys are categorized into three groups with low (e_v/a<5), medium (5≤e_v/a≤7.50) and high (e_v/a>7.50) number of valence electrons per atom as tabulated in Tables 1, 2, and 3, respectively. A small group of shape memory alloys have e_v/a<5. These alloys are mostly of TiNb-based compositions. A considerable number of non-magnetic alloys happen to belong to the medium range. Majority of the NiTi-based and many of high temperature alloys are within this group The highest M_s is observed in this group when e_v/a is around 6.50 to 7.00, and tends to lower when e_v/a deviates from this range. While most of the non-magnetic alloys have e_v/a values in the medium range, the majority of magnetic alloys belong to the high valence electron category (e_v/a>7.50). Nearly all NiMn-based Heusler compositions Co-based, Fe-based, and Cu-based alloys fall into this category. An overall view indicates that M_S temperatures of these alloys are comparatively lower compared to those of the alloys in the medium e_v/a group (Figure 1). A similar variation with the number of valence electrons is observed for A_S temperature. The number of reported A_S data is smaller than M_S only due to unavailability of the A_S data in the literature.

Table 1. Shape memory alloys with e_v/a<5

Alloy Composition [at.%]	M_s [°C]	A_s [°C]	e_v/a [e atom^{-1}]	c_v	Ref.
$Ti_{72}Nb_{28}$	-93	-	4.28	0.16	[14]
$Ti_{72}Nb_{22}Ta_6$	-33	-	4.28	0.15	[15]
$Ti_{74}Nb_{26}$	0	-	4.26	0.16	[16]
$Ti_{78}Nb_{22}$	137	-	4.22	0.16	[15]
$Ti_{80}Nb_{20}$	217	-	4.20	0.16	[14]
$Ti_{80}Nb_{18.40}Pd_{1.60}$	288	405	4.28	0.17	[17]
$Ti_{67}Nb_{30}Pd_3$	300	-	4.48	0.16	[17]

Table 2. Shape memory alloys with 5≤e_v/a≤7.50

Alloy Composition [at.%]	M_s [°C]	A_s [°C]	e_v/a [e atom^{-1}]	c_v	Ref.
$Ni_{47.25}Mn_{23}Ga_{24.5}Fe_{5.25}$	-173	-	7.47	0.27	[18]
$Fe_{60.4}Mn_{8.5}Si_{13}Cr_{10.2}Ni_{7.9}$ ($Fe_{65}Mn_9Si_7Cr_{10}Ni_9$ (wt))	-168	-58	7.35	0.30	[19]
$Fe_{58.6}Mn_{15.4}Si_{9.4}Cr_{12.2}Ni_{4.4}$ ($Fe_{62}Mn_{16}Si_5Cr_{12}Ni_5$ (wt))	-153	-53	7.31	0.30	[19]
$Fe_{58.7}Mn_{19.2}Si_{9.4}Cr_{8.1}Ni_{4.6}$ ($Fe_{62}Mn_{20}Si_5Cr_8Ni_5$ (wt))	-133	-33	7.36	0.30	[19]
$Ni_{43}Ti_{50}Co_7$	-100	-	6.93	0.28	[8]
$Ni_{44}Ti_{50}Fe_6$	-100	-	6.88	0.28	[20]
$Ni_{47}Ti_{50}Fe_3$	-98	-	6.94	0.28	[8]
$Ni_{48.5}Ti_{50}Cr_{1.5}$	-95	-	6.94	0.28	[8]
$Ni_{48}Ti_{50}Mn_2$	-95	-	6.94	0.28	[8]
$Ni_{47.5}Ti_{50}Fe_2Mo_{0.5}$	-90	5	6.94	0.28	[21]
$Ni_{45}Ti_{50}V_5$	-80	-	6.75	0.27	[8]
$Ni_{50}Mn_{25}Ga_{25}$	-71	-	7.50	0.27	[22]
$Ni_{46}Ti_{50}V_4$	-50	-	6.80	0.27	[8]
$Ni_{47.6}Ti_{46.4}Nb_6$	-50	-29	6.92	0.27	[23]
$Ni_{49}Mn_{23.5}Ga_{27.5}$	-46	-	7.37	0.26	[24]
$Ni_{50}Ti_{48}W_2$	-37	-26	7.04	0.27	[25]
$Ni_{46}Ti_{50}Cu_3Mo_1$	-30	-	6.99	0.28	[26]

Alloy Composition [at.%]	M_s [°C]	A_s [°C]	e_v/a [e atom^{-1}]	c_v	Ref.
$Ni_{50}Ti_{49}W_1$	-28	-12	7.02	0.27	[25]
$Ni_{48.3}Ti_{51}Mo_{0.7}$	-13	20	6.91	0.27	[27]
$Ni_{49}Ti_{50}Mo_1$	10	20	6.96	0.28	[27]
$Ni_{45}Ti_{50}Pd_5$	20	-	7.00	0.27	[8]
$Fe_{58.92}Mn_{14.59}Si_{12.99}Cr_{9.09}Ni_{4.41}$ ($Fe_{63.43}Mn_{15.45}Si_{7.03}Cr_{9.11}Ni_{4.98}$ (wt))	24	112	7.24	0.30	[28]
$Ni_{39.5}Ti_{49.5}Cu_{10}Cr_1$	25	30	7.08	0.28	[29]
$(Co_{45}Ni_{25}Ga_{30})_{0.95}Ta_5$	27	37	7.33	0.24	[30]
$Fe_{59.85}Mn_{13.99}Si_{12.98}Cr_{8.90}Ni_{4.28}$ ($Fe_{64.44}Mn_{14.82}Si_{7.03}Cr_{8.87}Ni_{4.84}$ (wt))	28	97	7.25	0.30	[28]
$Fe_{58.2}Mn_{30.6}Si_{11.2}$ ($Fe_{62}Mn_{32}Si_6$ (wt))	37	-	7.25	0.29	[19]
$Ni_{52}Ti_{46.74}Re_{1.26}$	37	87	7.16	0.28	[31]
$Ni_{47.9}Mn_{26.1}Ga_{26}$	39	52	7.10	0.27	[32]
$Ni_{50}Ti_{45}Ta_5$	44	76	7.05	0.26	[33]
$Ti_{32}Hf_{18}Ni_{49.8}Re_{0.2}$	52	101.5	6.99	0.21	[34]
$Ni_{48}Ti_{50}W_2$	55	70	6.92	0.27	[25]
$Ni_{50}Ti_{47}Ta_3$	58	90	7.03	0.26	[33]
$Ni_{30}Ti_{50}Cu_{20}$	58	60	7.20	0.28	[35]
$Ni_{50}Ti_{50}$	60	80	7.00	0.28	[8]
$Ni_{25}Ti_{50}Cu_{25}$	61	62	7.25	0.29	[36]
$Ni_{42}Ti_{49}Ag_9$	61	20	7.15	0.27	[37]
$Ni_{44}Ti_{49}Ag_7$	63	23	7.13	0.27	[37]
$Fe_{85}Mn_{15}$	67	197	6.81	0.26	[38]
$Ni_{49.5}Ti_{45.5}Zr_5$	70	110	6.97	0.27	[39]
$Ni_{44}Ti_{40}Hf_{10}Cu_6$	72	90	7.06	0.24	[40]
$Ni_{48}Ti_{50}Au_2$	75	-	7.02	0.27	[41]
$Co_{45}Ni_{25}Ga_{30}$	77	105	7.45	0.26	[30]
$Ni_{29}Ti_{51}Pd_{20}$	85	-	6.94	0.24	[42]
$Ni_{49.5}Ti_{40.5}Zr_{10}$	90	170	6.97	0.26	[40]
$Ni_{30}Ti_{50}Pd_{20}$	90	-	7.00	0.24	[8]
$Ni_{40}Ti_{50}Pt_{10}$	90	100	7.00	0.23	[43]
$Ni_{26.5}Ti_{51}Pd_{22.5}$	95	-	6.94	0.24	[42]

Table 2. (Continued)

Alloy Composition [at.%]	M_s [°C]	A_s [°C]	e_v/a [e atom^{-1}]	c_v	Ref.
$Zr_{48}Cu_{50}Ti_2$	110		7.5	0.22	[44]
$Ni_{50}Ti_{40}Hf_{10}$	120	165	7.00	0.23	[40]
$Ti_{50}Pd_{43.5}Cr_{6.5}$	127	-	6.74	0.21	[45]
$Zr_{50}Cu_{50}$	140	230	7.50	0.22	[46]
$Ni_{40}Ti_{50}Hf_{10}$	140	185	6.40	0.22	[10]
$Zr_{50}Cu_{50}$	150		7.50	0.22	[44]
$Zr_{42.3}Cu_{29.9}Ni_{11}Co_{10.2}Ti_{6.6}$	177	137	7.26	0.22	[47]
$Mn_{85.5}Fe_{9.5}Cu_5$	182	180	7.29	0.29	[48]
$Ni_{49.5}Ti_{35.5}Zr_{15}$	190	235	6.97	0.25	[39]
$Ni_{49.5}Ti_{35.5}Hf_{7.5}Zr_{7.5}$	195	230	6.97	0.23	[49]
$Ti_{50}Rh_{40}Ni_{10}$	210	-	6.60	0.21	[50]
$Ni_{15}Ti_{50}Pd_{35}$	225	-	7.00	0.22	[42]
$Ni_{36}Ti_{49}Hf_{15}$	230	-	6.16	0.20	[8]
$Ni_{35}Ti_{50}Hf_{15}$	235	256	6.10	0.19	[10]
$Zr_{49.6}Cu_{28.2}Ni_{6.8}Co_{15.4}$	253	212	7.15	0.21	[47]
$Ni_{30}Ti_{50}Pt_{20}$	266	271	7.00	0.20	[43,51]
$Ni_{49.5}Ti_{30.5}Hf_{10}Zr_{10}$	290	315	6.97	0.22	[49]
$Ni_{50}Ti_{30}Hf_{20}$	290	317	7.00	0.20	[10]
$Zr_{50}Cu_{40}Ni_{10}$	300	-	7.40	0.21	[44]
$Ti_{50}Rh_{45}Ni_5$	300	100	6.55	0.20	[50]
$Ti_{50}Rh_{46}Co_4$	300	150	6.50	0.20	[50]
$Ti_{50}Rh_{50}$	345	140	6.50	0.19	[50]
$Ni_{31}Ti_{49}Hf_{20}$	350	-	5.86	0.17	[8]
$Ti_{43.33}Pd_{50}Hf_{6.67}$	360	-	7.00	0.19	[45]
$Ni_{25}Ti_{50}Pt_{25}$	430	450	7.00	0.19	[42]
$Zr_{50}Rh_{50}$	480	570	6.50	0.15	[53]
$Zr_{50.59}Cu_{29.56}Ni_{19.85}$	497	597	7.26	0.21	[47]
$Ti_{50}Pd_{50}$	537	577	7.00	0.20	[54]
$Ti_{50}Pd_{50}$	544.7	574	7.00	0.20	[55]
$Ni_{20}Ti_{50}Pt_{30}$	560	594	7.00	0.18	[51]

Alloy Composition [at.%]	M_s [°C]	A_s [°C]	e_v/a [e atom^{-1}]	c_v	Ref.
$Ru_{43}Nb_{50}Fe_7$	650	-	6.50	0.16	[56]
$Ni_{15}Ti_{50}Pt_{35}$	680	750	7.00	0.16	[43]
$Zr_{50}Ir_{50}$	740	880	6.50	0.11	[52]
$Nb_{50}Ru_{36}Fe_{14}$	795	797	6.50	0.16	[56]
$Ni_{10}Ti_{50}Pt_{40}$	810	850	7.00	0.16	[43]
$Nb_{50}Ru_{43}Fe_7$	841	830	6.50	0.16	[56]
$Ru_{50}Nb_{50}$	865	-	6.50	0.15	[57]
$Nb_{50}Ru_{50}$	887	890	6.50	0.15	[56]
$Ni_5Ti_{50}Pt_{45}$	900	950	7.00	0.14	[43]
$Ti_{50}Pt_{50}$	1020	1040	7.00	0.14	[58]
$Ru_{50}Ta_{50}$	1050	1030	6.00	0.10	[57,59]
$Ru_{46}Ta_{54}$	1100	1150	6.38	0.11	[60]

The alloy compositions in parentheses () are the weight percent formula of the alloys as presented in the original references.

Table 3. Shape memory alloys with e_v/a>7.50

Alloy Composition [at.%]	M_s [°C]	A_s [°C]	e_v/a [e atom^{-1}]	c_v	Ref.
$Au_{50}Zn_{50}$	**-206**	**-**	**11.5**	**0.21**	[61]
$Fe_{75}Pt_{25}$	**-188**	**-**	**8.50**	**0.22**	[62]
$Ni_{52}Mn_6Fe_{15}Ga_{27}$	-153	-153	7.63	0.27	[63]
$Ni_{51}Fe_{22}Ga_{27}$	-132	-	7.67	0.30	[64]
$Co_{38.5}Ni_{32.5}Al_{29}$	131	-113	7.59	0.33	[65]
$Ni_{49.6}Mn_{27.7}Ga_{20.6}Si_{2.1}$	-128	-111	7.60	0.27	[32]
$Ni_{52}Mn_{16}Fe_5Ga_{27}$	-123	-98	7.53	0.27	[63]
$Ni_{49.5}Mn_{23}Ga_{24.5}Fe_3$	-108	-	7.54	0.27	[18]
$Ni_{51.7}Mn_{28.5}Ga_{15.6}In_{4.2}$	-95	-73	7.76	0.27	[32]
$Cu_{54.74}Al_{29.75}Zn_{15.51}$ ($Cu_{72.3}Al_{6.66}Zn_{21.04}$(wt))	-94	-108	8.77	0.36	[66]
$Co_{38}Ni_{33}Al_{29}$	-91	-73	7.59	0.33	[66]
$Fe_{49.8}Ni_{28.11}Co_{17.35}Ti_{4.74}$ ($Fe_{50}Ni_{33}Co_{12}Ti_5$ (wt))	-90	10	8.53	0.32	[67]
$Ni_{50}Mn_{35}Sb_{15}$	-88	-128	8.2	0.27	[68]

Table 3. (Continued)

Alloy Composition [at.%]	M_s [°C]	A_s [°C]	e_v/a [e atom^{-1}]	c_v	Ref.
$Ni_{42}Co_2Mn_{45}Sn_{11}$	-81	-84	8.04	0.27	[69]
$Ni_{43}Mn_{46}Sn_{11}$	-73	-78	7.96	0.27	[70]
$Fe_{70}Pd_{24}Pt_6$	-73	-78	8.60	0.25	[71]
$Ni_{52}Mn_{11}Fe_5Co_5Ga_{27}$	-68	27	7.63	0.27	[63]
$Co_{41}Ni_{32}Al_{27}$	-54	-85	7.70	0.33	[72]
$Ni_{40}Mn_{30}Ga_{20}Fe_{10}$	-67	-82	7.74	0.28	[73]
$Pd_{29.5}Fe_{64.25}Co_{6.25}$	-53	-	8.65	0.27	[74]
$Co_{37}Ni_{34}Al_{29}$	-58	-43	7.60	0.33	[65]
$Ni_{43}Co_1Mn_{45}Sn_{11}$	-53	-52	8.05	0.27	[69]
$Fe_{65.9}Mn_{12.9}Si_{4.5}Ni_{6.7}Cr_{9.5}Nb_{0.5}$	-50	83	7.62	0.30	[75]
$Fe_{71.4}Mn_{12.5}Ni_{6.2}Cr_{9.4}Ti_{0.5}$	-50	69	7.79	0.30	[75]
$Ni_{50}Mn_{36}Sb_{14}$	-38	-58	8.22	0.27	[76]
$Ni_{42.5}Ti_{42.5}Cu_{15}$	-34	-30	7.60	0.30	[12]
$Cu_{68.0}Al_{28.2}Ni_{3.8}$ ($Cu_{81.7}Al_{14.1}Ni_{4.2}$ (wt))	-33	-	8.71	0.35	[77]
$Ni_{51}Mn_{23}Ga_{24.5}Fe_{1.5}$	-33	-	7.56	0.27	[19]
$Ni_{41.75}Mn_{44.65}Sn_{10.65}B_{2.95}$	-31	-33	7.81	0.27	[70]
$Pd_{56.8}In_{25.4}Fe_{17.8}$	**-29**	**-21**	**7.87**	**0.18**	[78]
$Cu_{70.6}Al_{25.3}Mn_{4.1}$	-25	-18	8.81	0.35	[79]
$Fe_{62.40}Mn_{12.76}Si_{9.50}Cr_{8.84}Ni_{6.37}C_{0.13}$ ($Fe_{65.92}Mn_{13.25}Si_{5.05}Cr_{8.70}Ni_{7.05}C_{0.03}$ (wt))	-22	113	7.43	0.30	[75]
$Fe_{70}Pd_{28}Pt_2$	-20	-24	8.60	0.26	[71]
$Co_{39}Ni_{33}Al_{28}$	-5	17	7.65	0.33	[65]
$Ni_{50}Mn_{29}Ga_{21}$	-5.3	-4.4	7.66	0.28	[80]
$Ni_{40..95}Mn_{43.81}Sn_{10.47}B_{4.77}$	2	-1	7.72	0.28	[70]
$Cu_{70.9}Al_{20.4}Mn_{8.7}$	3	7	9.02	0.35	[79]
$Pd_{29.43}Fe_{68.55}Co_{2.02}$	3	-	8.61	0.27	[74]
$Ni_{54}Fe_{19}Ga_{27}$	3	7	7.73	0.27	[81]
$Pd_{29.23}Fe_{66.67}Co_{4.10}$	6	-	8.63	0.27	[24]
$Ni_{52}Mn_{24.5}Ga_{23.5}$	12	16	7.62	0.27	[82]
$Cu_{69}Al_{27.5}Ni_{3.5}$	15	40	8.76	0.35	[83]

Alloy Composition [at.%]	M_s [°C]	A_s [°C]	e_v/a [e atom^{-1}]	c_v	Ref.
$Fe_{70}Pd_{30}$	15	14	8.60	0.27	[71]
$Ni_{50}Mn_{37}Sb_{13}$	17	2	8.24	0.28	[76]
$Ni_{52}Mn_{1}Fe_{15}Co_{5}Ga_{27}$	17	297	7.73	0.27	[63]
$Fe_{70}Pd_{30}$	17	14	8.60	0.27	[84]
$Au_{43.3}Cu_{31.8}Al_{24.9}$	**17**	**70**	**9.00**	**0.19**	[85]
$Fe_{70.1}Pd_{29.9}$	18	6	8.60	0.27	[86]
$Ni_{50}Mn_{28}Ga_{22}$	18	19	7.62	0.27	[87]
$Co_{66.66}Ni_{28.34}Ga_{5}$	20	-	8.98	0.32	[88]
$Co_{41}Ni_{32}Al_{24}Sb_{3}$	20	24	7.76	0.31	[72]
$Fe_{72.6}Mn_{25.6}Mo_{1.8}$ ($Fe_{72}Mn_{25}Mo_{3}$ (wt))	21	-	7.71	0.30	[89]
$Ni_{50.4}Mn_{28.8}Ga_{19.8}Si_{1}$	25	-3	7.69	0.28	[32]
$Ni_{49}Fe_{18}Ga_{27}Co_{6}$	27	-	7.69	0.27	[90]
$Au_{43.75}Cu_{31.25}Al_{25}$	**27**	**-**	**9.00**	**0.19**	[91]
$Au_{50.5}Cd_{49.5}$	**31**	**-**	**11.50**	**0.18**	[92]
$Ni_{49.7}Mn_{28.7}Ga_{21.6}$	32	38	7.63	0.27	[32]
$Fe_{80}Mn_{20}$	33	154	7.80	0.30	[38]
$Cu_{73.7}Sn_{26.3}$ ($Cu_{60}Sn_{40}$ (wt))	37	-	9.16	0.27	[93]
$Ni_{50.5}Mn_{28.2}Ga_{21.3}$	38	41	7.66	0.27	[32]
$Ni_{45}Co_{5}Mn_{36.7}In_{13.3}$	39	37	7.92	0.27	[94]
$Co_{60.7}Ni_{29.9}Si_{9.4}$ ($Co_{63.9}Ni_{31.4}Si_{4.7}$ (wt))	39.6	264.8	8.83	0.34	[95]
$Ni_{50}Mn_{38}Sb_{12}$	42	17	8.26	0.28	[76]
$Ni_{50}Mn_{34}In_{16}$	47	-	7.86	0.26	[96]
$Pd_{57}In_{25}Fe_{18}$	**50**	**-**	**7.89**	**0.18**	[78]
$Fe_{61.96}Mn_{30.47}Co_{7.57}$ ($Fe_{62}Mn_{30}Co_{8}$ (wt))	52	162	7.77	0.30	[97]
$Fe_{73}Mn_{24.52}Ge_{2.48}$ ($Fe_{72}Mn_{24}Ge_{4}$ (wt))	52	-	7.66	0.29	[98]
$Ni_{55}Fe_{18}Ga_{27}$	54	27	7.75	0.27	[81]
$Ni_{52}Fe_{18}Ga_{27}Co_{3}$	58	-	7.72	0.27	[90]
$Co_{50}Ni_{23.5}Ga_{26.5}$	60	-	7.64	0.27	[88]
$Ni_{50}Mn_{28}Ga_{21}Y_{1}$	60	62	7.62	0.27	[87]
$Ni_{50}Mn_{30}Ga_{15}Al_{5}$	62	-1	7.68	0.28	[73]
$Co_{50}Ni_{22}Ga_{28}$	64	-	7.54	0.27	[99]

Table 3. (Continued)

Alloy Composition [at.%]	M_s [°C]	A_s [°C]	e_v/a [e atom^{-1}]	c_v	Ref.
$Cu_{67.71}Zn_{19.45}Al_{12.84}$ ([$Cu_{72.65}Zn_{21.5}Al_{5.85}$ (wt))	69	70	10.17	0.37	[100]
$Ni_{50}Mn_{29}Ga_{20}Gd_1$	70	78	7.74	0.27	[80]
$Ni_{50}Mn_{29}Ga_{20}Dy_1$	75.9	76.3	7.66	0.27	[80]
$Cu_{68.5}Zn_{18.8}Al_{12.7}$	76	-	10.2	0.37	[101]
$Co_{50}Ni_{22}Ga_{27}Al_1$	77	-	7.54	0.27	[99]
$Ni_{51.4}Mn_{27.2}Ga_{21.4}$	78	80	7.69	0.28	[31]
$Mn_{85}Cu_{15}$	80	-	7.60	0.30	[102]
$Pt_{10}Ni_{40}Mn_{25}Ga_{25}$	82	77	8.5	0.34	[103]
$Cu_{68.3}Zn_{20.4}Al_{11.3}$	85	-	10.3	0.37	[101]
$Ni_{51.3}Mn_{14.4}Ga_{26.3}Fe_8$	86	88	7.57	0.27	[32]
$Ni_{40}Mn_{30}Ga_{20}Co_{10}$	88	91	7.64	0.27	[73]
$Fe_{69.9}Mn_{25.37}Co_{4.73}$ ($Fe_{70}Mn_{25}Co_5$ (wt))	97	177	7.79	0.30	[97]
$Fe_{67}Mn_{25.41}Co_{7.59}$ ($Fe_{67}Mn_{25}Co_8$(wt))	100	175	7.82	0.30	[97]
$Cu_{58.51}Al_{32.43}Zn_{9.06}$ ($Cu_{78.9}Al_{8.54}Zn_{12.56}$ (wt))	102	48	8.50	0.35	[66]
$N_{50}Mn_{29}Ga_{19}Gd_2$	105	115	7.82	0.27	[84]
$Ni_{54}Mn_{23}Ga_{23}$	110	115	7.70	0.27	[104]
$Co_{86}Al_{14}$	112	236	8.16	0.32	[99]
$Co_{50}Ni_{22}Ga_{26}Al_2$	115	-	7.54	0.27	[99]
$Ni_{50}Mn_{29}Ga_{19}Dy_2$	126.2	123.7	7.66	0.27	[80]
$Ni_{50}Mn_{28}Ga_{19}Y_3$	130	132	7.62	0.27	[87]
$Co_{50}Ni_{22}Ga_{23}Al_5$	150	-	7.54	0.27	[99]
$Ni_{55.6}Mn_{11.4}Fe_{7.4}Ga_{25.6}$	166	180	7.72	0.27	[105]
$Co_{35}Ni_{40}Al_{25}$	167	-	7.90	0.33	[106]
$Ni_{55}Mn_{22.5}Ga_{22.5}$	175	200	7.75	0.27	[104]
$Ni_{48.75}Mn_{35.5}Ga_{15.75}$	175.1	178.7	7.83	0.28	[107]
$Ni_{50}Mn_{40}In_{10}$	177	-	8.10	0.28	[96]
$Ni_{50}Mn_{29}Ga_{18.8}Co_{2.2}$	223	224	7.79	0.28	[31]
$Ni_{75}Ta_{25}$	250	-	8.75	0.22	[108]
$Cu_{72.12}Al_{26.45}Pd_{1.43}$ ($Cu_{84.13}Al_{13.08}Pd_{2.79}$ (wt))	262.8	262.8	8.87	0.35	[109]

Alloy Composition [at.%]	M_s [°C]	A_s [°C]	e_v/a [e atom^{-1}]	c_v	Ref.
$Cu_{79.23}Al_{20.77}$ ($Cu_{90}Al_{10}$ (wt))	265	-	9.34	0.36	[110]
$Ni_{50.7}Mn_{34.6}Ga_{14.7}$	311	319	7.93	0.29	[111]
$Ni_{50}Mn_{40}Ga_{10}$	347	-	8.10	0.30	[112]
$Cu_{73.1}Al_{21.5}Ag_{5.4}$ ($Cu_{80}Al_{10}Ag_{10}$(wt))	350	-	9.28	0.35	[110]
$Ni_{50}Mn_{37.5}Al_{12.5}$	390	-	8.00	0.32	[112]
$Ni_{50}Mn_{45}Sn_5$	437	-	8.35	0.30	[112]
$Ni_{49.6}Mn_{45.5}In_{4.9}$	487	509	8.29	0.30	[113]
$Ni_{50}Mn_{50}$	627	-	8.50	0.32	[113]

The alloy compositions in parentheses () are the weight percent formula of the alloys as presented in the original references.

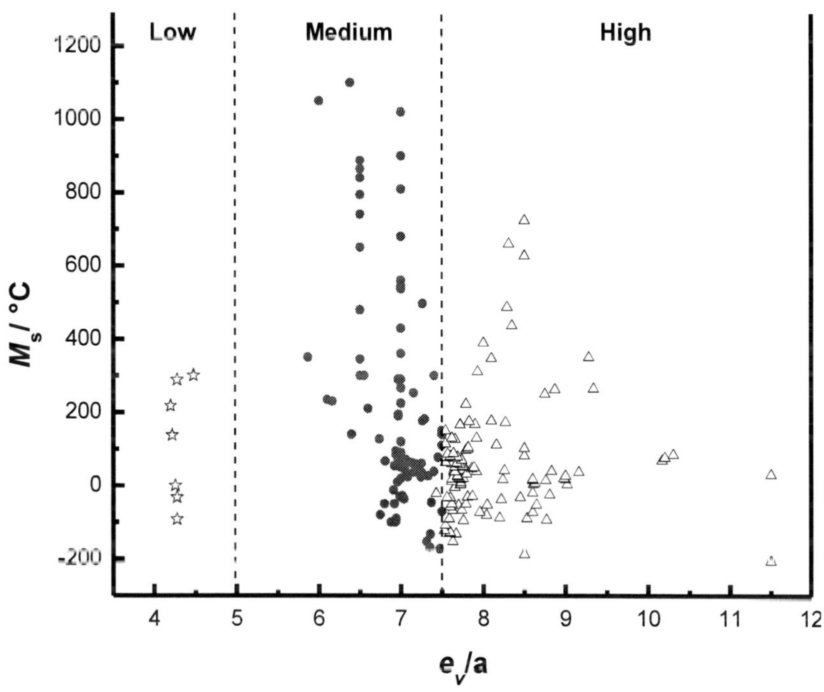

(a)

Figure 1. (Continues)

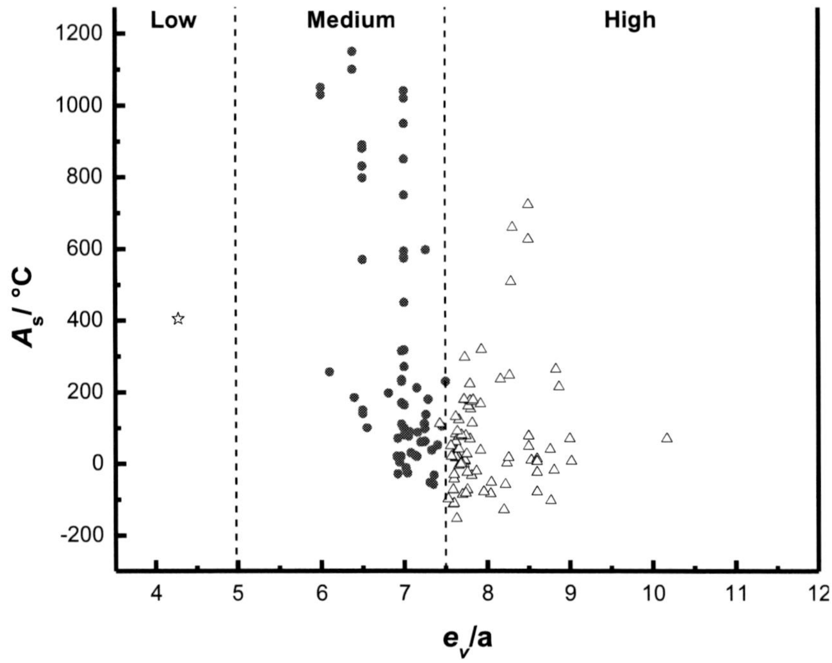

(b)

Figure 1. Variations of (a) M_s and (b) A_s with the number of valence electrons.

2.2. Dependence on Valence Electron Concentration

The average concentration of valence electrons c_v of an alloy can be defined as the ratio of number of valence electrons to the total number of electrons of the alloy, $c_v = (e_v/e_t)$ [10], which can be simply calculated as follows (equation (2)):

$$c_v = \frac{e_v}{e_t} = \frac{f_A e_v^A + f_B e_v^B + f_C e_v^C + f_C e_v^D}{f_A Z_A + f_B Z_B + f_C Z_C + f_D Z_D} \quad (2)$$

where $Z_{A,}$ $Z_{B,}$ $Z_{C,}$ Z_D etc., represent the atomic numbers of elements A, B, C, D, etc., respectively. Similarly f_A, f_B, f_C, f_D etc., represent their corresponding atomic fractions in the alloy. Depending on the alloying elements, atomic fractions of elements comprising an alloy, different c_v values are resulted. These values for the alloys examined are tabulated in Tables 1-3. The variations of M_s and A_s temperatures versus valence electron concentration are plotted in Figure 2. In both cases, a main trend can be highlighted with increasing the electron concentration. For alloys within the indicated bands both M_s and A_s decrease consistently from temperatures as high as 1100°C (1150°C) down to as low as -206°C (-153°C) for alloys with a c_v of within 0.10-0.27. The decreasing trend seems to stop around $c_v \approx 0.27$. By further increase of c_v, M_s and A_s remain in a relatively constant range or start to increase when the electron concentration is above 0.34-0.35. The general trends of variation of M_s and A_s seem to be similar.

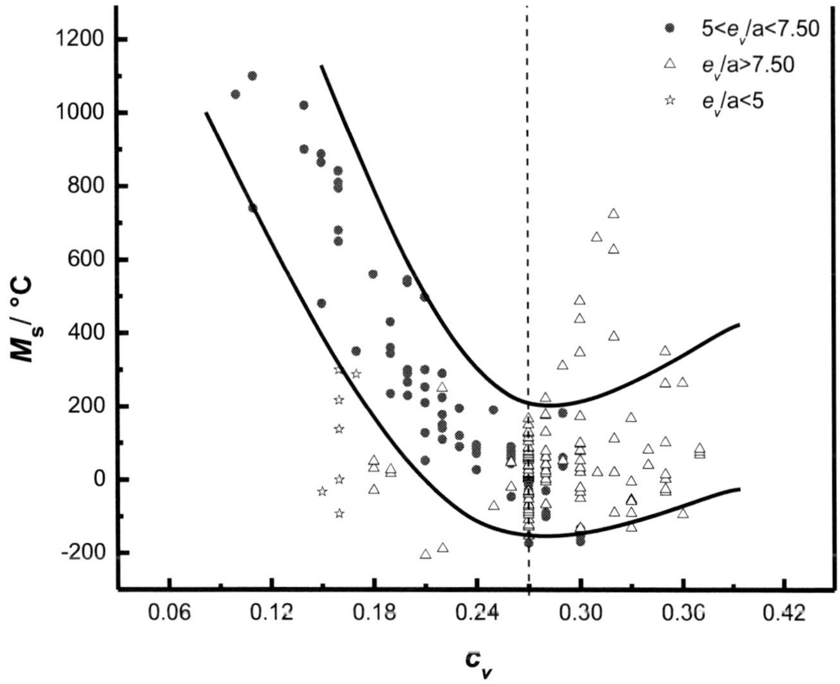

(a)

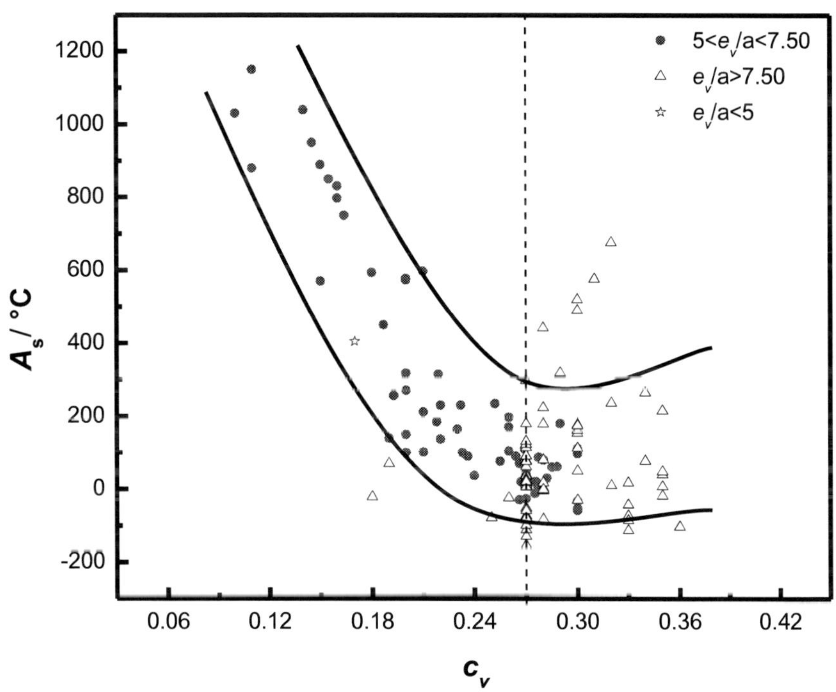

(b)

Figure 2. Variations of (a) M_s and (b) A_s with the valence electron concentration.

2.3. Deviations from the Main Trend

Apart from the main trend which involves the majority of alloys, three other minor groups can be addressed which show some degree of deviation from the behavior of the main group. These alloys are introduced as follows.

2.3.1. Mn-Rich NiMnX Heusler Alloys

When the electron concentration of the alloy is sufficiently high (c_v>0.27) in NiMnX (X=Ga, In, Al) Heusler alloys, by increasing the Mn content beyond a certain limit (Mn at%>30) M_s and A_s temperatures both increase considerably and deviate from the general trend in alloys (Figure 2). Figure 3 is a plot of transformation temperature of Mn containing Heusler alloys versus the Mn atomic fraction in the alloy. By increasing the Mn content to ~30 at.%, the M_s temperatures increase.

2.3.2. High Valence Electron Alloys (e_v/a >7.50)

There are limited data points at medium or low c_v values with relatively low M_s temperatures as indicated by (triangle) in Figure 2. These data points which show some deviation from the main trend pertain to alloys with high e_v/a. Despite high number of valence electrons (e_v/a>7.50), their c_v values are low or medium (0.17-0.22). The corresponding data of these alloys have been highlighted in Table 3 in bold format.

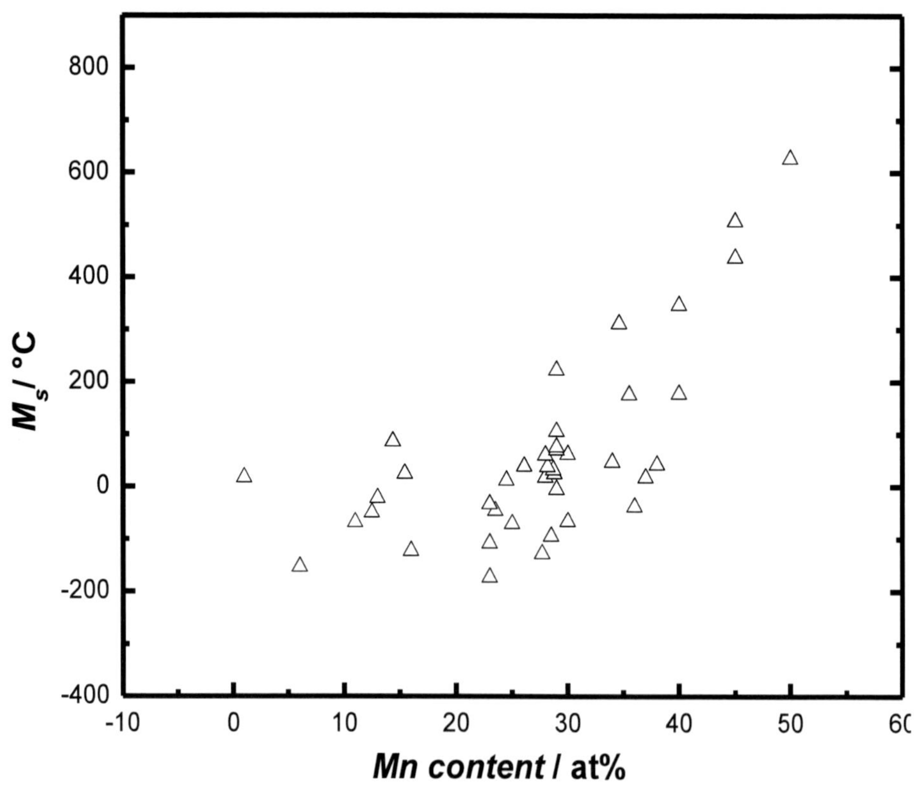

Figure 3. The effect of Mn content in Heusler alloys on the transformation temperatures.

2.3.3. Low Valence Electron Alloys ($e_v/a<5$)

There are a limited number of TiNb-based alloys with low electron concentration (on average 0.15), and low number of valence electrons ($e_v/a<5$) which have M_s and A_s values lower than what is observed for ordinary alloys with similar electron concentrations. These alloys tend to have an average atomic number of around 25. These data points have been shown by stars outside the main trend in Figure 2.

3. DISCUSSION

3.1. Valence Electron Concentration, Elastic Properties, and Transformation Temperatures

In order to understand the fundamental reason behind the influence of the main factor, c_v, on martensitic transformation temperature of shape memory alloys, the relationship between valence electron concentration, elastic properties and transformation temperatures of the alloys are discussed here. In the metallic bonding the valence electrons act like 'glue', bonding non-valence electrons and ion kernels together; [10] whereas non-valence electrons contribute to the total atomic volume of the alloy. The way the strengths of particular bonds in the alloys are influenced by the change of c_v remains to be investigated in detail. However, one possible explanation may be provided. Increasing valence electron concentration could mean thickening of the 'glue' bonding the ionic kernels together. In fact, an established relationship between the valence electron concentration and bulk modulus of metallic materials and intermetallic compounds is known. Higher valence electron concentration usually results in higher bulk and therefore shear moduli [114,115]. Bulk modulus is a measure of resistance of a solid to volume change and shear modulus is a measure of resistance to shape change [10,114]. Increasing valence electron concentration of the alloys results in higher elastic and shear moduli. It is thus logical to think that change of valence electron concentration is accompanied by a change in the elastic constant as the interatomic bonding is affected [10,114]. The key factor controlling the bonding is the electronic structure of the alloy. Before martensitic transformation the elastic moduli decrease during cooling and reach a critical value [8]. If the elastic moduli of austenite is enhanced as a result of higher c_v, the cooling should continue to lower temperatures before a critical elastic constant is reached as a result of pre-martensitic softening of bulk and/or shear moduli, hence M_s is decreased. On the contrary, if the alloy has lower elastic bulk and shear moduli in austenite phase, the critical elastic constant can be reached at higher temperatures and M_s is increased. The decreasing trend continues until the electron concentration reaches around 0.27 and remains in a relatively constant range for most alloys beyond this range. A possible explanation is that the 'glue' that was referred to, becomes too thick, and further increase in its thickness does not fortify the bonds, rather at very high c_v>0.34-0.35, high valence electron concentration may even cause antibonding and slight increase of the temperatures (Figure 2). M_s is influenced based on the elastic constants of austenite and A_s follows its variation on the basis of the degree of reversibility of transformation for each specific shape memory alloy. These explain why some of the shape memory alloy compositions have high and some others have low transformation temperatures.

3.2. Effect of Mn

NiMnX (X=Ga, In, Al) Hesuler alloys with excess Mn and relatively high concentration of valence electrons show increased transformation temperatures (Figure 3) and therefore deviate from the main trend. As discussed, the concentration and number of valence electrons both play critical roles in elastic stiffness of the transition metal crystals but they do not usually act alone. The d-orbitals may become hybridized with p- and s-orbitals. Therefore the occupancy behavior of the orbitals can also be influential. As one proceeds across one of the long rows of transition metals in the periodic table of elements, the atomic volume changes relatively little but the elastic stiffness changes markedly, reaching a maximum near the middle of each row [114]. Mn is an exception in which the bonding is strongly affected by its antiferromagnetism [114]. The calculations of Moruzzi and Marcus indicate that the elastic moduli of Mn is considerably reduced due to anti-ferromagnetism [116]. In $Ni_{50}Mn_{25}Ga_{25}$, the Ni magnetic moment is much smaller than Mn moment. However, the Ni 3d state has a larger contribution to the electronic state near Fermi energy than Mn 3d state [117,118]. Therefore both elements play important roles in determining the properties of the alloy system [119]. When the Mn content of this Heusler alloy is high, it is possible that some of the anti-ferromagnetism of Mn can be maintained. This results in considerable reduction of the elastic moduli and consequently the elevation of the transformation temperatures (Figure 2). Clear evidences of anti-ferromagnetism and/or negative magnetic moments on Mn atoms in NiMnGa alloys have been shown in some recent works [120, 121]. It is possible that Mn-rich NiMnIn and NiMnAl alloys which also exhibit increased M_s temperatures, are similarly affected by Mn antiferromagnetism. In depth clarification of the influence of Mn requires further study.

3.3. High Valence Electron Alloys with Medium or Low Electron Concentration

In the alloys with e_v/a well above 7.50, the sensitivity of transformation temperature to the concentration of valence electrons is considerably reduced. This fact is evident from the plot of M_s-c_v for these alloys (Figure 4). This implies that when the number of valence electrons is sufficiently high, the strength of the bonds and therefore elastic properties can be high even at medium c_v(0.17-0.24). This originates from the fact that higher number of valence electrons can provide sufficiently strong bonding for larger or more dense ion kernels in the metallic bonds [115]. The net result is lower transformation temperature for these alloys and some degree of deviation from the main trend (Figure 1).

3.4. Low Valence Electron Alloys

The existing low valence electron alloys show c_v values of 0.14-0.17, for which alloys in the main group have M_s of usually more than 400°C. In low electron alloys the M_s values are significantly lower (-93-300°C) than the main trend. These alloys contain considerable amount of Nb, and their average atomic number is nearly 25-26. The elastic moduli of low valence electron Ti-based alloys have been studied by some researchers [122-124].

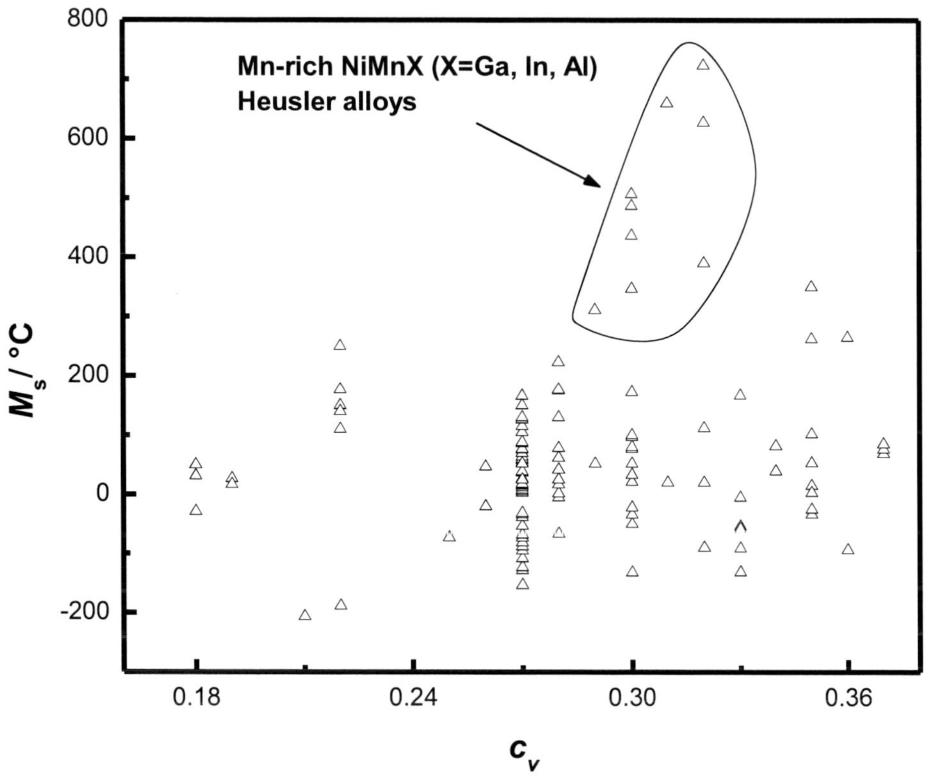

Figure 4. Variation of M_s with valence electron concentration for high valence electron alloys.

In Ti-based alloys the stability of austenite phase increases by increasing e_v/a from nearly 4 by the solid solution strengthening with the addition of alloying elements near Ti group in periodic table [122]. There are reports of maximum elastic modulus of the alloy when the e_v/a is in around 4.15 [123,124].

In terms of electronic structure of the alloys it is not clear why the austenite stability can be extended to lower temperatures. If an alloy can be viewed, as an element with average atomic number of the alloy (in this case 25-26), the electronic configuration can be written as, $1s^2 2s^2 2p^6 3s^2 3p^6 4s^2 3d^5$, for which the d-orbital is almost half filled. Hence we are faced with a condition in which nearly half way through a row in the table of transition metals. In this condition usually the maximum elastic moduli and strength in the row is reached [114]. Higher elastic moduli translate to lower transformation temperatures. When the alloy deviates from this condition (higher or lower e_v/a), at low electron concentration and number of electrons the transformation temperatures is expected to be higher. This may be indicative of the considerable influence of orbital occupancy behavior on the elastic properties and transformation temperature of the shape memory alloys with low valence electrons.

4. Conclusions

The main correlations shown in this work introduce probably the most influencing factors, i.e. e_v/a and c_v, and antiferromagnetic effect of Mn, responsible for the changes of transformation temperatures in transition metal shape memory alloys. The following conclusions can be drawn:

- Shape memory alloys can be divided into low ($e_v/a<5$), medium ($5\leq e_v/a\leq 7.50$), and high ($e_v/a>7.50$) electron groups with majority of the alloys in these groups following a relatively similar tendency of transformation temperature change.
- The valence electron concentration (c_v) is found to be an important parameter influencing the transformation temperatures for the majority of the shape memory alloys (main trend). M_s and A_s both decrease with increasing c_v from 0.10 to around 0.27, remain almost constant beyond that and start to increase beyond 0.34-0.35. Increasing c_v translate into increasing elastic constants of the alloys which explains the transformation at lower temperatures.
- Three minor groups of alloys deviate from the main trend, 1) NiMnX (X=Ga, In, Al) Heusler alloys containing high Mn content (Mn$\geq$ 30 at.%) show higher transformation temperatures most likely due to anti-ferromagnetism effect of this element and the consequent reduction of elastic properties, 2) high valence electron alloys having low or medium c_v as a result of sufficient bonding provided by high number of valence electrons, and 3) low valence electron alloys with lower transformation temperatures possibly due to electron occupancy behavior of the d orbital.

References

[1] K. Otsuka, C. M. Wayman, Shape Memory Materials, Cambridge University Press, Cambridge, UK 1998.

[2] C.S. Barrett, T.B. Massalski, Structure of Metals, Pergamon, Oxford, UK 1987.

[3] K. Bhattacharya, S. Conti, G. Zanzotto, J. Zimmer, *Nature* 428 (2004) 55-59.

[4] A. Planes, L. Manosa, D. Rioa-Jara, J. Ortin, *Phys. Rev.* B 45 (1992) 7633-7639.

[5] L. Manosa, J. Zarestky, T. Lograsso, D.W. Dalaney, C. Stassis, *Phys. Rev.* B 48 (1993) 15708-15712.

[6] L. Manosa, M. Jurado, A. Planes, J. Zarestky, T. Lograsso, C. Stassis, *Phys. Rev.* B 49 (1994) 9969-9972.

[7] X. Ren, N. Miura, J. Zhang, K. Otsuka, K. Tanaka, M. A. Koiwa, *Mater. Sci. Eng.* A 312 (2001) 196-206.

[8] K. Ostuka, X. Ren, *Prog. Mater. Sci.* 50 (2005) 511-678.

[9] X. Ren, K. Taniwaki, K. Otsuka, T. Suzuki, K. Tanaki, Y. I. Chumlyakov, T. Ueiki, *Phil. Mag.* A 79 (1999) 31-41.

[10] M. Zarinejad, Y. Liu, *Adv. Func. Mater.* 18 (2008) 2789-2794.

[11] P. C. Clapp, *Phys. Stat. Sol.* B 57 (1973) 561-570.

[12] P. C. Clapp, *Met. Trans.* A 12 (1981) 589-595.

[13] W.D. Callister, Materials Science and Engineering: An Introduction, John Wiley and Sons, Asia 2007.

[14] S. Miyazaki, H.Y. Kim, H. Hosoda, *Mater. Sci. Eng.* A 438-440 (2006) 18-24.

[15] H.Y. Kim, S. Hashimoto, J.I. Kim, T. Inamura, H. Hosoda, S. Miyazaki, *Mater. Sci. Eng.* A 417 (2006) 120-128.

[16] H.Y. Kim, Y. Ohmatsu, J.I. Kim, H. Hosoda, S. Miyazaki, *Mater. Trans.* 45 (2004) 1090-1095.

[17] D.H. Ping, Y. Mitarai, F.X. Yin, *Scripta Mater.* 52 (2005) 1287-1291.

[18] D.E. Soto-Parra, F. Alvarado-Hernandez, O. Ayala, R.A. Ochoa-Gamboa, H. Flores-Zuniga, D. Rios-para, *J. Alloys Compd.* 464 (2008) 288-291.

[19] M. Valeanu, G. Filoti, V. Kuncser, F. Tolea, B. Popescu, A. Galatanu, G. Schinteie, A. D. Jianu, I. Mitelea, D. Schinle, C. M. Craciunescu, *J. Magne. Magne. Mater.* 320 (2008) e164-e167.

[20] S. Miyazaki, K. Otsuka, *Phil. Mag. A* 50 (1984) 393-408.

[21] F.S. Liu, Y. Li, Y. Li, H. Xu, *Mater. Sci. Eng. A* 438-440 (2006) 896-899.

[22] P.J. Webster, K.R.A. Ziebeck, S.L. Town, M. S. Peak, *Phil. Mag. B* 1984, 49, 295–310 .

[23] T. Takagi, Y. Suto, R. Kainuma, K. Yamauchi, K. Ishida, *Appl. Biomater.* B 76 (2006) 179-183.

[24] P. Zhao, L. Dai, J. Cullen, M. Wuttig, *Metall. Mater. Trans* A 38 (2007) 745-750.

[25] S.F. Hsieh, S.K. Wu, H.C. Lin, C.H. Yang, *J. Alloy. Compd.* 387 (2005) 121-127.

[26] G.B. Cho, T.Y. Kim, C.A. Yu, Y. Liu, T.H. Nam, *J. Alloy. Compd.* 449 (2008) 129-133.

[27] T.H. Nam, D.W. Chung, J. S. Kim, S.B. Kang, *Mater. Lett.* 52 (2002) 234-239.

[28] B.C. Maji, C.M. Das, M. Krishnan, R.K. Ray, *Corrosion Sci.* 48 (2006) 937-949.

[29] H.C. Lin, C.H. Yang, K.M. Lin, G.K. Hsu, *J. Alloy. Compd.* 386 (2005) 157-164.

[30] J. Liu, H.X. Zheng, M.X. Xia, J.G. Li, *Scripta Mater.* 52 (2005) 955-958.

[31] H. Steckmann, V. I. Kolomytsev, A. V. Kozlov, *Ultrasonics* 37 (1999) 59-62.

[32] I. Glavatskyy, N. Glavatska, O. Söderberg, S.-P. Hannula, J.-U. Hoffmann, *Scripta Mater.* 54 (2006) 1891-1895.

[33] C.W. Gong, Y.N. Wang, D.Z. Yang, *J. Alloy. Compd.* 419 (2006) 61-65.

[34] F. Dalle, V. Kolomytsev, P. Ochin, R. Portier, *Scripta Mater.* 44 (2001) 929-934.

[35] Y.H. Li, L.J. Rong, Z.T. Wang, G.X. Qi, C.Z. Wang, *J. Alloy. Compd.* 400 (2005) 112-115.

[36] Y. Tong, Y. Liu, Z. Xie, *J. Alloy. Compd.* 456 (2008) 170-177.

[37] C. Zamponi, M. Wuttig, E. Quandta, *Scripta Mater.* 56 (2007) 1075-1079.

[38] A. Sato E. Chishima, Y. Yamaji, T. Mori, *Acta Metall.* 32 (1984) 539-547.

[39] S. F. Hsieh, S. K. Wu, J. *Alloy. Compd.* 270 (1998) 237-241.

[40] K. Chastaing, P. Vermaut, P. Ochin C. Segui J. Y. Laval, R. Portier, *Mater. Sci. Eng. A* 438-440 (2006) 661-665.

[41] K.H. Eckelmeyer, *Scripta Metall. Mater.* 10 (1976) 667-672.

[42] R. Zarnetta, A. Savan, S. Thienhaus, A. Ludwig, *Appl. Surf. Sci.* 254 (2007) 743-748.

[43] O. Rios, R. Noebe, T. Biles, A. Garg, A. Palczer, D. Scheiman, H. J. Seifert, M. Kaufman, in Proc. of SPIE Vol. 5761: *Smart Materials and Structures* (Ed: W.D. Armstrong) 2005, SPIE, USA, 376

[44] Yu.N. Koval, G.S. Firstov, L. Delaey, J.V. Humbeeck, *Scripta Metall. Mater.* 31 (1994) 799-802.

[45] T. Hoshiya, K. Inoue, K. Enami, In the Proc. of mechanical properties and phase transformations of multiphase intermetallic alloys 1994, TMS, USA, 99-106.

[46] Yu.N. Koval, G.S. Firstov, A.V. Kotko, *Scripta Metall. Mater.* 27 (1992) 1611-1616.

[47] G.S. Firstov, J. Van Humbeeck, Y.N. Koval, *Mater. Sci. Eng. A* 378 (2004) 2-10.

[48] J.H. Zhang, W.Y. Peng, J.J. Zhang, X. Zuyao (T. Y. Hsu), *Mater. Sci. Eng. A* 481-482 (2008) 326-329.

[49] S.F. Hsieh, S. K. Wu, *J. Alloy. Compd.* 312 (2000) 288-294.

[50] E.L. Semenova, V.M. Petyukh, Yu.V. Kurdryavstev, *J. Alloy. Compd.* 230 (1995) 115-119.
[51] R. Noebe, D. Gaydosh, S. Padua II, A. Garg, T. Biles, M. Nathal, in Proc. of SPIE Vol. 5761: Smart Materials and Structures (Ed: W.D. Armstrong) 2005, SPIE, USA, 364.
[52] E.L. Semenova, Yu.V. Kudryavstev, *J. Alloys Compd.* 203 (1994) 165.
[53] G.S. Firstov, Yu.N. Koval, J. Van Humbeeck, R. Portier, P. Vermaut, P. Ochin, *Mater. Sci. Eng. A* 438-440 (2006) 816-820.
[54] T. Yamamuro, Y. Morizono, J. Honjyo, M. Nishida, *Mater. Sci. Eng. A* 438-440 (2006) 327-331.
[55] K. Otsuka, K. Oda, Y. Ueno, M. Piao, Scripta Mater. 29 (1993) 1355-1358.
[56] X. Gao, W. Cai, Y.F. Zheng, L.C. Zhao, *Mater. Sci. Eng. A* 438-440 (2006) 862-864.
[57] R.W. Fonda, H.N. Jones, R.A. Vandermeer, *Scripta Mater.* 39 (1998) 1031-1037.
[58] T. Biggs, M.B. Cortie, M.J. Witcomb, L.A. Cornish, *Metall. Mater. Trans. A* 32 (2001) 1881-1886.
[59] R.W. Fonda, R.A. Vandermeer, *Phil. Mag. A* 76 (1997) 119-133.
[60] B.H. Chen, H.F. Franzen, J. Less-Common Metals 157 (1990) 37-45.
[61] R.D. McDonald, P.A. Goddard, J. Lashley, N. Harrison, C.H. Mielke, J. Singleton, H. Harima, M. Suzuki, *J. Phys. Chem. Solids* 67 (2006) 2100-2105.
[62] T. Sakomoto, T. Fukuda, T. Kakeshita, *Mater. Sci. Eng. A* 438-440 (2006) 1036-1040.
[63] K. Tsuchiya, Y. Sho, T. Kushima, Y. Todaka, M. Umemoto, *J. Magne. Magne. Mater.* 310 (2007) 2764-2766.
[64] Y. Murakami, *Appl. Phys. Lett.* 92 (2008) 102512-1-3.
[65] Z. Liu, S. Yu, H. Yang, G. Wu, Y. Liu, *Intermetallics* 16 (2008) 447-452.
[66] J.M. Guilemany, F.J. Gil, *Thermochimica Acta 1*67 (1990) 129-138.
[67] L. Jian C.C. Chou, C.M. Wayman, *Mater. Chem. Phys.* 34 (1993) 14-23.
[68] S. Chatterjee, S. Giri, S. Majumdar, A.K. Deb, S.K. De, V. Hardy, *J. Phys.: Condens. Matter* 19 (2007) 346213-1-10.
[69] H.S. Liu, C.L. Zhang, Z.D. Han, H.C. Xuan, D.H. Wang, Y.W. Du, *J. Alloys Compd.* 467 (2009) 27-30.
[70] H.C. Xuan, D.H. Wang, C.L. Zhang, Z.D. Han, B.X. Gu, Y.W. Du, *Appl. Phys. Lett.* 2008, 92, 102503 1-3.
[71] T. Wada, T. Tagawa, M. Taya, *Scripta Mater.* 48 (2003) 207-211.
[72] G.F. Xu, Z.M. Yin, S.Z. Mou, F.H. Lou, K. Oikawa, *Trans. Nonferrous Met. Soc.* China 16 (2006) 776-782.
[73] R.P. Mathur, R.K. Singh, V. Chandrasekaran, S. Ray, P. Ghosal, *Metall. Mater. Trans. A* 38 (2007) 2076-2084.
[74] D. Vokoun, Y.W. Wang, T. Goryczka, C.T. Hu, *Smart Mater. Struct.* 2005, 14, S261-S265.
[75] N. Stanford, D.P. Dunne, B.J. Monaghan, *J. Alloy. Comp.* 430 (2007) 107-115.
[76] J. Du, Q. Zheng, W. J. Ren, W. J. Feng, X. G. Liu and Z. D. Zhang, *J. Phys. D: Appl.* Phys. 40 (2007) 5523–5526.
[77] Y. Zhang, J. Gui, R. Wang, L. Gao, Y. Wu, Y. Tang, J. Phys.: Condens. Matter 5 (1993) 2719-2728.
[78] H. Ishikawa, Y. Sutou, T. Omori, K. Oikawa, K. Ishida, A. Yoshikawa, R.Y. Umetsu, Kainuma, *Appl. Phys. Lett.* 90 (2007) 261906-1-3.
[79] N. Zarubova, V. Novak, *Mater. Sci. Eng.* A 378 (2004) 216-221.
[80] L. Gao, Z.Y. Gao, W. Cai, L.C. Zhao, *Mater. Sci. Eng. A* 438-440 (2006) 1077-1080.
[81] C. Segui, J. Pons, E. Cesari, J. Dutkiewicz, *Mater. Sci. Eng.* A 438-440 (2006) 923-926.
[82] W.H. Wang, G.H. Wu, J.L. Chen, S.X. Gao, W.S. Zhan, *Appl. Phys. Lett.* 79 (2001) 1148-1150.

[83] M. Landa, P. Sedlak, P. Sittner, H. Seiner, V. Novak, *Mater. Sci. Eng.* A 462 (2007) 320-324.

[84] V. Recarte, C. Gomez-Polo, V. Sanchez-Alarcos, J. I. Perez-Landazabal, *J. Magne. Magne. Mater.* 316 (2007) e614-e617.

[85] L. Battezzati, G. Fiore, M. Massazza, *J. Alloys Compd.* 434-435 (2007) 264–267.

[86] D. Vokoun, T. Goryczka, C.T. Hu, *J. Alloy. Compd.* 372 (2004) 165-168.

[87] W. Cai, L. Gao, A.L. Liu, J.H. Sui, Z.Y. Gao, *Scripta Mater.* 57 (2007) 659-662.

[88] M. Wuttig, J. Li, C. Craciunescu, *Scripta Mater.* 44 (2001) 23-2397.

[89] K. Tsuzaki, Y. Natsume, Y. Tomota, T. Maki, *Scripta Met. Mater.* 33 (1995) 1087-1092.

[90] H. Morito, A. Fujita, K. Oikawa, K. Ishida, K. Fukamichi, R. Kainuma, *Appl. Phys. Lett.* 90 (2007) 062505-1-3.

[91] S. Urbano, A. Mancaa, S. Besseghinib, G. Airoldia, *Scripta Mater.* 52 (2005) 317-321.

[92] Y. Murakami, T. Ohba, K. Morii, S. Aoki, K. Otsuka, *Acta Mater.* 55 (2007) 3203-3211.

[93] M. Kamal, Radiation Effects and Defects in Solids 161 (2006) 189-191.

[94] Y. Murakami, T. Yano, D. Shindo, R. Kainuma, K. Oikawa, K. Ishida, *Scripta Mater.* 55 (2006) 683-686.

[95] W. Zhou, Y. Liu, B. Jiang, P. Zhou, *Mater. Sci. Eng. A* 438-440 (2006) 468-471.

[96] X. Moya, Ll. Manosa, A. Planes, S. Aksoy, M. Acet, F. Wassermann, T. Krenke, *Phys. Rev.* B 75 (2007) 184412-1-5.

[97] A.A.H. Hamers, C.M. Wayman, *Scripta Metall.* 25 (1991) 2723-2728.

[98] X. Lu, B. Zhang, Z. Qin, Y. Zhang, B. Ding, Z. Hu, *Mater. Sci. Eng.* A 347 (2003) 258-264.

[99] P.N. Zhang, J. Liu, J. Alloy Compd. 462 (2008) 225-228.

[100] Z.G. Wang, X. T. Zu, H. J. Yu, X. He, C. Peng, Y. Huo, *Thermochimica Acta* 448 (2006) 69-72.

[101] V. Asanovic, K. Delijic, N. Jaukovic, *Scripta Mater.* 58 (2008) 599-601.

[102] Q. Tian, F. Yin, T. Sakaguchi, K. Nagai, *Acta Mater.* 54 (2006) 54 1805-1813.

[103] Y. Kishi, Z. Yajima, K. Shimizu, M. Wuttig, *Mater. Sci. Eng. A* 378 (2004) 361-364.

[104] G. Feng, C. Jiang, T. Liang, H. Xu, *J. Magne. Magne. Mater.* 248 (2002) 312-317.

[105] H.B. Wang, C. Liu, Y.C. Lei, W. Cai, *J. Alloys Compd.* 465 (2008) 458-461.

[106] W. Maziarz, J. *Alloy Compd.* 448 (2008) 223-226.

[107] Y. Liu, F. Ai, B. Jiang, Y. Liu, X. Qi, *Mater. Sci. Eng.* A 438-440 (2006) 999-1002.

[108] G.S. Firstov, Yu.N. Koval, J. Van Humbeeck, P. Ochin, *Mater. Sci. Eng. A* 481-482 (2008) 590-593.

[109] Z.C. Lin, W. Yu, R.H. Zee, B.A. Chin, *Intermetallics* 8 (2000) 605-611.

[110] R.A.G. Silva, A. Cuniberti, M. Stipcich, A.T. Adorno, *Mater. Sci. Eng.* A 456 (2007) 5-10.

[111] Y. Ma, C. Jiang, Y. Li, H. Xu, C. Wang, X. Liu, *Acta Mater.* 55 (2007) 1533-1541.

[112] T. Krenke, X. Moya, S. Aksoy, M. Acet, P. Entel, Ll. Manosa, A. Planes, Y. Elerman, A. Yucel, E.F. Wassermann, *J. Magne. Magne. Mater.* 310 (2007) 2788-2789.

[113] X. Moya, L. Manosa, A. Planes, T. Krenke, M. Acet, E.F. Wassermann, *Mater. Sci. Eng.* A 438–440 (2006) 911–915.

[114] J. Gilman, *Electronic Basis of the Strength of Materials*, Cambridge University Press, Cambridge, UK 2003.

[115] J. Gilman, R.W. Cuberland, R.B. Kaner, *Int. J. Refract Met.* H. 24 (2006) 1-5.

[116] V.L. Moruzzi, P.M. Marcus, *Phys. Rev.* 48 (1993) 7665.

[117] P.J. Brown, *J. Phys. Condens. Matter* 11 (1999) 4715.

[118] S. Fujii, *J. Phys. Soc. Japan* 58 (1989) 3657.

[119] S. Imada, A. Yamasaki, T. Kanomata, T. Muro, A. Sekiyama, S. Suga, *J. Magne. Magne. Mater.* 310 (2007) 1857-1858.

[120] C. Jiang, Y. Muhammad, L. Deng, W. Wu, H. Xu, *Acta Mater.* 52 (2004) 2779–2785
[121] F. Albertini, L. Pareti, A. Paoluzi, L. Morellon, P.A. Algarabel, M.R. Ibarra, *Appl. Phys.* Lett. 81 (2002) 4032.
[122] E.W. Collings, H.L. Gegel, *Scripta Metall.* 7 (1973) 437-444.
[123] C.A. Luke, R. Taggart, D.H. Polonis, *Trans ASM* 57 (1964) 142–149.
[124] Y.L. Hao, S.J. Li, S.Y. Sun, C.Y. Zheng, R. Yang, *Acta Biomater.* 3 (2007) 277-286.

In: Shape Memory Alloys
Editor: H. R. Chen, pp. 361-369
ISBN 978-1-60741-789-7

Chapter 13

SOME FACTORS AFFECTING THE TRANSFORMATION HYSTERESIS IN SHAPE MEMORY ALLOYS

Yong Liu
School of Mechanical and Aerospace Engineering
Nanyang Technological University, Singapore 639798

Transformation hysteresis is an important property of shape memory alloys (SMAs) which requires careful consideration during the materials selection for targeted technical applications. When being used for repeated active actuation purposes (for example in MEMS, morphing wing of aerial vehicles, robotics, etc.) a small hysteresis is required since it strongly influences the operational frequency achievable and the power consumption needed for actuation. On the other hand, some applications may require very large transformation hysteresis in order to maintain a stable austenite phase within a large temperature range for retaining the predefined shape (for example in deployable space structures, pipe joining, etc). In order to obtain SMAs with either very low or very high hysteresis, a fundamental understanding on its influencing factors is ultimately important. This short communication highlights several recent results on the factors affecting the transformation hysteresis in shape memory alloys. This aricle is not aimed to provide a complete spectrum on the transformation hysteresis of shape memory alloys. It is hoped to encourage further investigation to understand the fundamental factors controlling the transformation hysteresis, thus enabling an effective alloy design.

It is well known that the thermoelastic phase transformation is associated with a local balance between chemical and non-chemical forces [1,2]. Chemical force arises from the difference in Gibbs free energies between austenite and martensite and acts as a driving force promoting the phase with lower energy at each temperature. Non-chemical force arises from two major contributions; one is the elastic energy to accommodate the shape and volume change during transformation and the other is the energy dissipated during phase transformation. The elastic energy stored during forward transformation is released upon reverse transformation acting as driving force, thus does not contribute to the transformation hysteresis. It can be demonstrated that the elastic energy is responsible for the transformation interval. The energy dissipation due to defect formation is irreversible and contributes to the transformation hysteresis. Thus, factors contributing to the frictional forces against structural phase transformation will contribute to the transformation hysteresis. In the following, several observations/investigations on the factors affecting the transformation hysteresis are briefly discussed namely, atomic radius of the alloying elements, lattice mismatch between martensite and austenite, substrate-induced stress in bimorph, and effect of precipitation.

1. Atomic Radius of Alloying Element

Alloying plays a major role in altering the transformation hysteresis. Hysteresis values of a number of NiTi-based alloys are listed in [3]. Several Ni-Ti-Hf and Ni-Ti-Zr alloys have hysteresis values in the range of 60-70°C while the hysteresis of $Ti_{50}Ni_{50-x}Cu_x$ ($x \geq 20$ at.%) is as low as 5°C. Analysis [3] further shows that the valence electron concentration plays an important role in determining the transformation temperatures. An overall decreasing trend in transformation temperatures with increasing valence electron concentration c_v (ratio of the number of valence electrons to the total number of electrons of the alloy) is observed. In addition, two distinct trends of transformation temperatures with respect to the number of valence electrons per atom (e_v/a) exist depending whether $e_v/a = 7$ or $e_v/a \neq 7$. A very recent work has extended the investigation to nearly all the existing transition-metal-based SMA systems and the results are presented in Chapter 16 of this book. If, however, choose the data from alloys with e_v/a of 7 or close to 7 ($6.97 \leq e_v/a \leq 7$), and consisting of 20 at% of ternary or ternary + quaternary alloying elements, it seems that the transformation hysteresis of the NiTi-based SMAs increases with increasing the atomic radius of the ternary and quaternary elements as shown in Figure 1 [3].

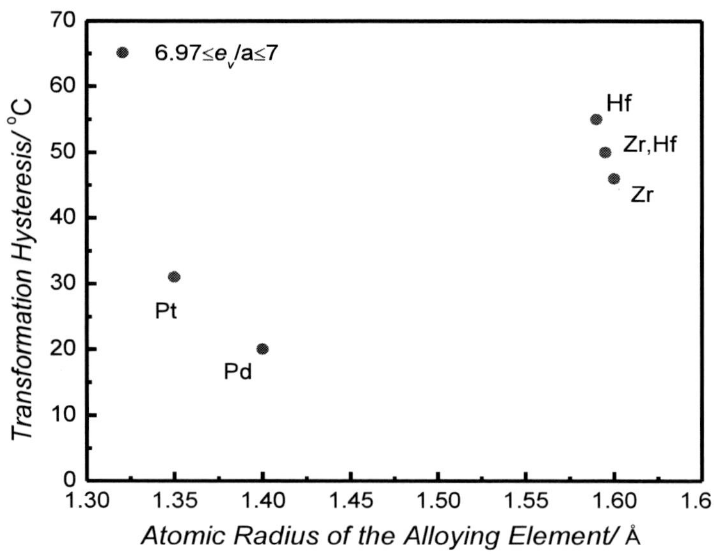

Figure 1. Dependence of transformation hysteresis of NiTi-based alloys on the atomic radius of the ternary and quaternary alloying elements. The alloys have $6.97 \leq e_v/a \leq 7$, and 20 at.% of ternary or ternary + quaternary alloying elements. Adapted from [3].

Energy dissipation associated with the frictional force against interatomic movements may be one of the factors contributing to the hysteresis. A larger atomic size of the additional alloying elements may result in more frictional force thus more energy dissipation leading to increased hysteresis. In addition, it is well established that lattice defects, particularly dislocations, also increase the frictional energy against interfacial movement during phase transformation, thus contributing to transformation hysteresis.

2. LATTICE COMPATIBILITY

It has been generally accepted that in martensitic transformation the geometric compatibility between martensite and austenite can be described by using three principal lattice distortions, namely λ_1, λ_2, λ_3 ($\lambda_1 < \lambda_2 < \lambda_3$). In the martensite coordinate system, the lattice distortion is given in the matrix form

$$B = \begin{pmatrix} \lambda_1 & 0 & 0 \\ 0 & \lambda_2 & 0 \\ 0 & 0 & \lambda_3 \end{pmatrix}$$

For B2 (cubic) to B19 (orthorhombic) transformation existed in $Ti_{50}Ni_{50-x}Cu_x$ ($x \geq 10$) SMAs, the principal lattice distortions can be calculated as [5]:

$$\lambda_1 = \frac{a_o}{a_c},\ \lambda_2 = \frac{b_o}{\sqrt{2}a_c},\ \lambda_3 = \frac{c_o}{\sqrt{2}a_c}$$

where a_o, b_o and c_o are lattice parameters of orthorhombic martensite, and a_c is the lattice parameter of B2 austenite.

There are six lattice correspondences between austenite and martensite. If taking one of the six correspondences i.e., $[001]_c//[100]_o$, $[110]_c//[010]_o$, $[\bar{1}10]_c//[001]_o$, the lattice distortion matrix in austenite coordinate system is given as:

$$U = \begin{bmatrix} \frac{\lambda_2+\lambda_3}{2} & \frac{\lambda_2-\lambda_3}{2} & 0 \\ \frac{\lambda_2-\lambda_3}{2} & \frac{\lambda_2+\lambda_3}{2} & 0 \\ 0 & 0 & \lambda_1 \end{bmatrix}$$

The determinants of both the matrices are equal, *i.e.*, $\det B = \det \mathbf{U} = \lambda_1\lambda_2\lambda_3$ which determines the volume change resulted from phase transformation. When there is no volume change, $\det B = \det \mathbf{U} = 1$. It is proposed [6] that in order for an SMA to have extremely low hysteresis, two conditions need to be satisfied, namely $\det \mathbf{U} = \lambda_1\lambda_2\lambda_3 = 1$ and $\lambda_2 = 1$.

The formation of martensite twins is to reduce the elastic strain energy arise from lattice incompatibility between martensite and austenite. As discussed in [6], if $\lambda_2 = 1$, austenite and martensite lattices are “crystallographically exactly compatible” and there is no need to form either “elastic transition layer” between martensite and austenite or the lattice twins, thus, a lowest hysteresis is expected. It was also proposed that elimination of the stored elastic energy arise from lattice incompatibility results in reduction of hysteresis. Since the value of λ_2 depends only on lattice parameters, it was thus hoped that this criterion can be used as a practical guide to develop superior SMAs with low hysteresis and improved fatigue life [6].

A combinatorial approach was taken to map the lattice parameters and hysteresis versus composition of NiTiCu and NiTiPd deposited thin films [6]. The results showed a decreasing tendency in hysteresis with λ_2 approaching 1 in both NiTiCu ($\lambda_2 < 1$) and NiTiPd ($\lambda_2 > 1$) alloys. The lowest hysteresis (~20°C) is located near a composition of 55 at.% Ti and 12 at.% Cu ($\lambda_2 \cong 1$). Any deviation from this composition results in an increase in hysteresis.

On the other hand, it has been widely reported that the transformation hysteresis in orthorhombic $Ti_{50}Ni_{50-x}Cu_x$ ($x \geq 10$) alloys can be as low as 5°C [7-12]. In order to further verify the effect of λ_2 and detU on hysteresis, we plotted some available data of NiTiCu alloys in Fig 2. All the data were taken from alloys undergoing only B2-to-B19 transformation. Results agree that the dependence of hysteresis on detU is not obvious. However, the hysteresis shows no clear decreasing tendency with λ_2. For most of the samples the hysteresis is between 5° and 9°C with λ_2 ranging from 0.975 to 1.0025. The discrepancy poses a fundamental question: what is the key factor dominating the transformation hysteresis? If λ_2 is the key factor, why the hysteresis was still as large as 20°C when λ_2 approaches 1 [6]? As discussed in [6], when $\lambda_2 = 1$, the material is assumed to have full lattice compatibility between martensite and austenite and no lattice defects (mismatch dislocations, accommodation twins) were expected in martensite. However, even in $Ti_{50}Ni_{25}Cu_{25}$ alloy which has a hysteresis as small as 5°C, martensite twins were still observed [9,12,13]. This might suggest that using value of λ_2 as a criterion to predict hysteresis needs reconsideration.

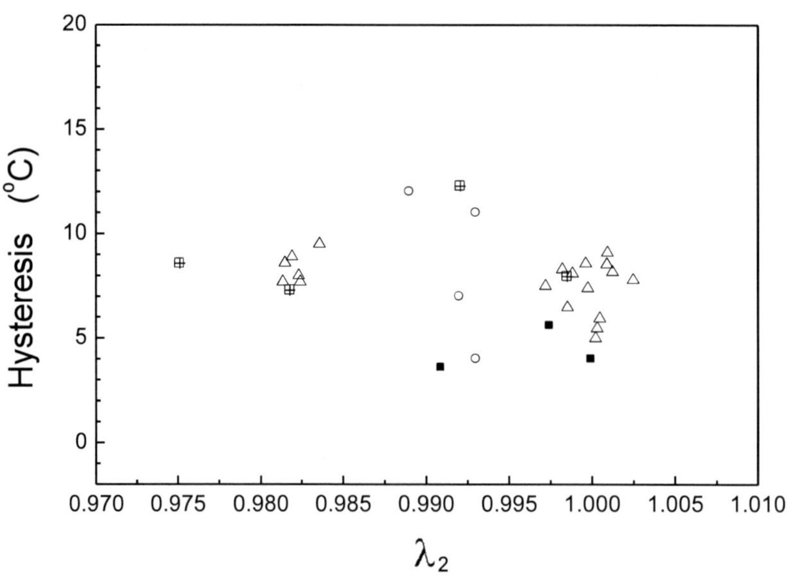

(a)

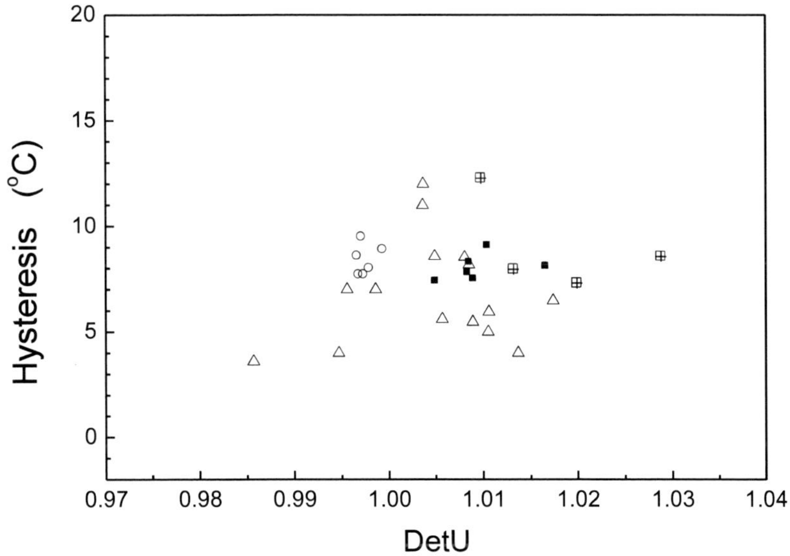

(b)

Figure 2. Hysteresis of $Ti_{50}Ni_{50-x}Cu_x$ bulk and Ti50Ni25Cu25 melt-spun ribbons shows no systematic change with λ_2 (a) and detU (b). Cu content varies between 10 and 30 at%. Data are taken from references [7,8,11,12,14]. All alloys involve B2-to-B19 transformation only.

In order to further understand the effect of the principal lattice distortions on the formation of lattice defects and hysteresis, we need to re-examine the lattice correspondences between martensite and austenite. Two types of twinning modes were reported in NiTiCu orthorhombic martensite, namely (111) type I twin and (011) compound twin [5,9]. Taking one lattice correspondence as an example, $[001]_c//[100]_o$, $[110]_c//[010]_o$, $[\bar{1}10]_c//[001]_o$, the orientation relation between B19 lattice and B2 lattice is shown in Figure 3, where the lattice is undistorted ($\lambda_1 = \lambda_2 = \lambda_3 = 1$). Two martensite twinning modes are illustrated schematically. The twinning shear is lying in both the twin plane and the habit plane [15] in order to minimize the transformation distortion on the habit plane. The directions of the principal lattice distortions (λ_1, λ_2, λ_3) due to phase transformation are also indicated. It can be seen that in order to have exact compatibility between martensite and austenite it is necessary to have: $\lambda_1 = \lambda_2 = \lambda_3 = 1$. Any deviation from this condition shall lead to atomic shear in lattice. In order to reduce the lattice distortion on habit plane, martensite twins are formed in SMAs. In B2-to-B19 transformation, $\lambda_1 < 1$, $\lambda_2 \cong 1$, $\lambda_3 > 1$ are reported. As can be seen from Figure 3, λ_2 is not parallel to any of the two twinning shear directions. Thus, even $\lambda_2 = 1$, since $\lambda_1 \neq 1$ and $\lambda_3 \neq 1$, atomic shear may still take place along the twinning shear direction leading to the formation of martensite twins. This may explain why martensite twins were still observed in orthorhombic NiTiCu alloy with λ_2 being very close to one (0.9996 and 1.0003, hysteresis is about 5°C) [9,12,13]. Thus, we may assume that the value of λ_2 can not be used alone as a criterion to evaluate the defect formation and transformation hysteresis.

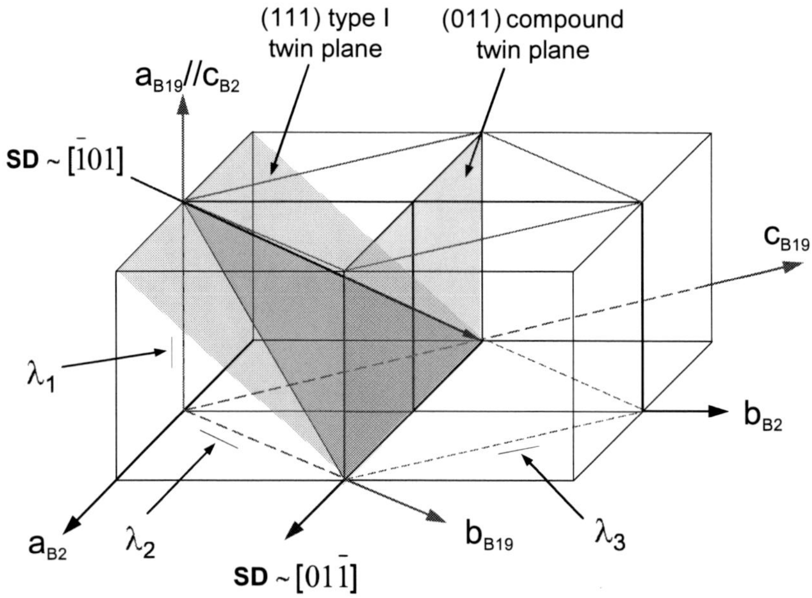

Figure 3. Orientation relations between B19 lattice and B2 lattice indicating martensite twin planes, twinning shear directions and directions of principal lattice distortions (λ_1, λ_2, λ_3).

Thus, we might consider that the principal lattice distortions between martensite and austenite are responsible for the formation of lattice defects which indirectly determine the hysteresis. In such case, the direct reason responsible for the hysteresis is the irrecoverable dissipated energy due to formation of martensite twins and other lattice defects. In order to relate the lattice mismatch to hysteresis, a link between the lattice mismatch and lattice defect formation needs to be established. If, however, in an extreme situation in which all the lattice distortions are accommodated elastically and no lattice defects are formed, one might not be able to expect transformation hysteresis.

3. Biasing Stress In Bimorph

When the deposited SMA thin films are used for MEMS actuators, the substrate-induced stress on the transformation hysteresis may also need consideration. It is known that the phase transformation in a constraint SMA thin film is a self-regulating process. The transformation process alters the stress magnitude in the thin film, meanwhile the stress variation also affects the transformation process. In a transforming constrained thin film, stress, strain and transformation temperatures are interdependent. Comparing with freestanding thin films, the transformation hysteresis of bimorph SMA thin films has been reported to decrease in NiTi, NiTiCu and NiTiPd thin films [16-18]. As shown in Figure 4, in the substrate-attached NiTiCu thin film, the transformation hysteresis is about 12 °C, which is a much smaller value than that of the same film in freestanding condition (26°C) [18].

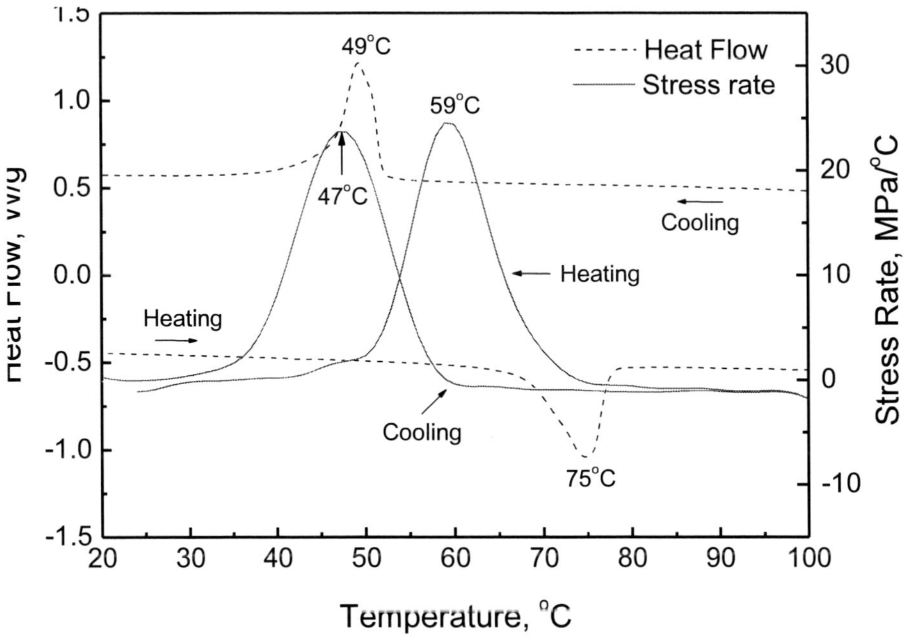

Figure 4. Differentiated substrate-induced stress as a function of temperature in $TiNi_{48.2}Cu_{1.8}$ thin film deposited on (100) Si substrate ($\Delta T = 12°C$). DSC curve of the same thin film in freestanding condition is also plotted for comparison ($\Delta T = 26°C$). Adapted from [18].

4. PRECIPITATION

It is interesting to note that the precipitation seems do not affect the transformation hysteresis of a NiTiCu melt-spun ribbon as shown in Figure 5. The results are obtained from a $Ti_{50}Ni_{25}Cu_{25}$ melt-spun ribbon after annealed at different temperatures for different durations. Precipitation, its distribution and volume fractions were determined for samples differently annealed and were reported in [19]. The precipitation processes are also highlighted in the figure. Three stages were identified, Stage I is the low temperature annealing and results in no precipitation, Stage II is within the intermediate temperature range and results in the formation of B11 precipitates, while Stage III is within the high temperature range whereby the transition from B11 precipitates to a new type of precipitate namely, $Ti_2(Cu, Ni)$, is observed. According to the data shown in Figure 5, the above mentioned precipitation processes do not have observable effect on the transformation hysteresis of the $Ti_{50}Ni_{25}Cu_{25}$ melt spun. The hysteresis value of 5-6°C is within the typical range reported in the literature.

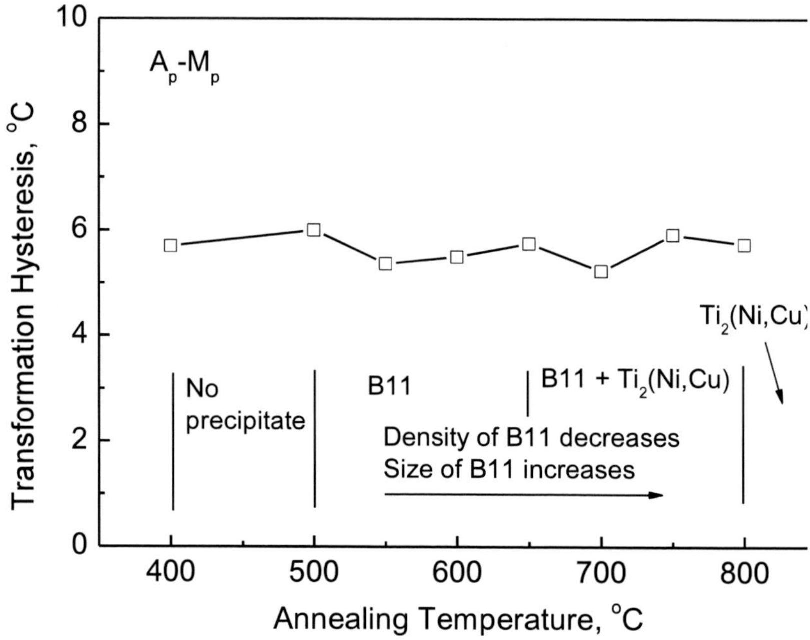

Figure 5. Transformation hysteresis values in a $Ti_{50}Ni_{25}Cu_{25}$ melt-spun ribbon after annealed at different temperatures shown. The formation of precipitates in the samples differently annealed is also indicated [20].

REFERENCES

[1] Olson G.B., M. Cohen. Thermoelastic behavior in martensitic transformation. *Scripta Metall.* 9 (1975) 1247-1254.

[2] Tong H.C. and C.M. Wayman. Thermodynamics of thermoelastic martensitic transformations. *Acta Metall.* 23 (1975) 209-215.

[3] Zarinejad M. and Y. Liu. Dependence of transformation temperatures of NiTi-based shape memory alloys on the number and concentration of valence electrons. *Advanced Functional Materials* 18/18 (2008) 2789-2794.

[4] Zarinejad M. and Y. Liu. Dependence of transformation temperatures of shape memory alloys on the number and concentration of valence electrons, submitted.

[5] Saburi T., Y. Watanabe, S. Nenno. Morphological characteristics of the orthorhombic martensite in a shape memory Ti-Ni-Cu alloy. *ISIJ International* 29 (1989) 405-411.

[6] Cui J., Y.S. Chu, O.O. Famodu, Y. Furuya, J. Hattrick-Simpers, R.D. James, A. Ludwig, S. Thienhaus, M. Wuttig, Z. Zhang, I. Takeuchi. Combinatorial search on thermoelastic shape-memory alloys with extremely small hysteresis width. *Nature Materials* 5 (2006) 286-290.

[7] Moberly W.J. and K.N. Melton. NiTiCu shape memory alloys, in Engineering Aspects of Shape Memory Alloys, edited by T.W. Duerig, K.N. Melton, D. Stöckel and C.M. Wayman (London: Butterworth-Heinemann), 1990, pp. 46-57.

[8] Nam T.H., T. Saburi, K. Shimizu. Cu-content dependence of shape memory characteristics in Ti-Ni-Cu Alloys. *Materials Transactions JIM* 31 (1990) 959-967.

[9] Xie Z.L., J. Van Humbeeck, Y. Liu and L. Delaey. TEM study of Ti50Ni25Cu25 melt spun ribbons. *Scripta Materialia* 37 (1997) 363-371.
[10] Liu Y. Mechanical and Thermomechanical Properties of Ti50Ni25Cu25 melt-spun ribbon. *Mater. Sci. Eng. A* 354 (2003) 286-291.
[11] Liu Y., Z.L. Xie, Y.X. Tong, C.W. Lim. Properties of rapidly annealed Ti50Ni25Cu25 melt spun ribbon. *Journal of Alloys and Compounds* 416/1-2 (2006) 188-193.
[12] Cheng G.P., Z.L. Xie, Y. Liu. Transformation characteristics of annealed Ti50Ni25Cu25 melt spun ribbon. *Journal of Alloys and Compounds* 415/1-2 (2006) 182-187.
[13] Xie Z.L., G.P. Cheng, Y. Liu. Microstructure and texture development in Ti50Ni25Cu25 melt-spun ribbon. *Acta Materialia* 55/1 (2007) 361-369.
[14] Cheng G.P., PhD thesis, Nanyang Technological University, 2006.
[15] Porter D.A. and KE Easterling. Phase transformations in metals and alloys, 2nd edition, Nelson Thornes Ltd 2001.
[16] Winzek B. and E. Quandt. Zero hysteresis in shape-memory Ti-Ni-X-films (X = Cu, Pd) under constrain. *Mat. Res. Soc. Symp. Proc.* 604 (2000) 117-122.
[17] Winzek B. and E. Quandt. Shape-memory Ti-Ni-X-films (X = Cu, Pd) under constraint. *Z. Metallk.*90 (1999) 769-802.
[18] Liu Y. and X. Huang. Substrate-induced stress and transformation characteristics of a deposited NiTiCu thin film. *Phil. Mag.* 84/19 (2004) 1919-1936.
[19] Tong Y.X., Y. Liu, Z.L. Xie, M. Zarinejad. Effect of precipitation on the shape memory effect of $Ti_{50}Ni_{25}Cu_{25}$ melt-spun ribbon. *Acta Materialia* 56/8 (2008) 1721-1732.
[20] Tong Y.X. and Y. Liu. Unpublished results.

In: Shape Memory Alloys
Editor: H. R. Chen, pp. 371-423
ISBN 978-1-60741-789-7

Chapter 14

MICROMECHANICAL MODELING OF SHAPE MEMORY ALLOY COMPOSITES

Jacob Aboudi[1,*] ***and Yuval Freed***[2,†]
[1]Faculty of Engineering, Tel-Aviv University
Ramat-Aviv 69978, Israel
[2]Department of Fracture Mechanics, Fatigue and Damage Tolerance
Engineering Devision, Israel Aerospace Industry, Israel

Abstract

This chapter reviews several approaches for the mathematical modeling of shape memory alloys to be embedded as constituents (e.g. fibers) thus forming shape memory alloy composites. To this end, shape memory alloy fibers that exhibit pseudoelasticity, one and two-way shape memory effects, rate dependence and transformation induced plasticity are considered. The behavior of the shape memory composites is determined by employing a micromechanical analysis which is capable of predicting the response of such composites that consist of polymer as well as metallic matrices. Since the shape memory alloy fibers behavior is dominated by the temperature, this micromechanical model accounts for the full thermomechanical coupling between the fibers and matrix in which the mechanical and thermal effects mutually affect each other. The micromechanical modeling is also applied to predict the behavior of shape memory alloy honeycombs, which form a promising new type of smart materials. The possibility of increasing the tensile strength of concrete by embedding activated shape memory wires is investigated. This is carried out by a micromechanical modeling in which both evolving damage effects as well as coupled damage-plasticity effects are incorporated in the prestressed concrete.

1. Introduction

Shape memory alloys (SMAs) are quite interesting materials, that are characterized by several unique properties which ordinary metals do not possess. It is common to separate the

*E-mail address: aboudi@eng.tau.ac.il
†E-mail address: yfreed@iai.co.il

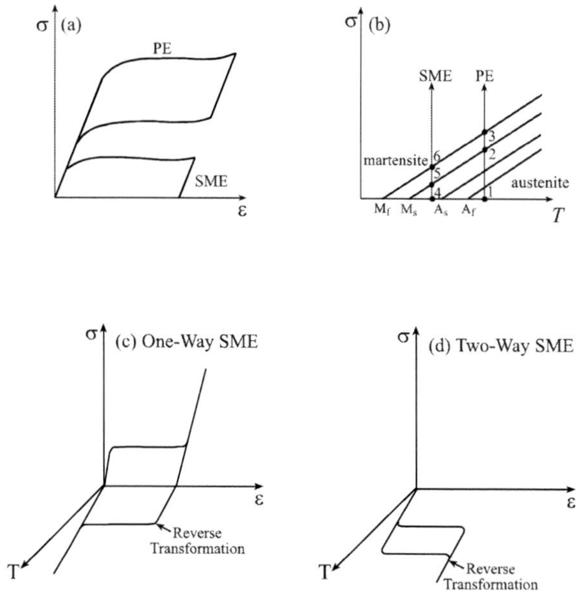

Figure 1. (a) Typical stress-strain curves of an SMA. (b) Typical stress-temperature curves. If $T > A_f$ (point 1), PE occurs as a result of a stress induced transformation (points 2 to 3). If $T < A_s$ (point 4), SME resulting from a stress induced transformation (points 5 to 6) is achieved. (c) Illustration of the one-way SME. (d) Illustration of the two-way SME.

observable behavior of shape memory alloys into two major phenomena. In the *pseudoelasticity* (PE) phenomenon, nonlinear elastic behavior is observed (see Fig. 1(a)). Here, very large strains upon loading occur, but full recovery is achieved in a hysteresis loop upon unloading. The second phenomenon occurs when a shape memory alloy experiences the *one-way shape memory effect* (SME), where it exhibits a large residual strain after loading and unloading cycle as illustrated in Fig. 1(a). This strain may be fully recovered by raising the temperature of the body.

The shape memory responses are usually separated into forward and backward responses. A *forward transformation* denotes that the alloy is transforming to its martensitic phase (during temperature decrease for example). Then, the temperatures corresponding to the start and finish of this transformation are called martensite start (M_s) and finish (M_f) temperatures, as shown in Fig. 1(b). In *reverse transformation*, when raising the temperature, the material is transforming from the martensite phase, back to the parent austenite phase. In this case, the temperatures associated with the transformation are called austenite start (A_s) and finish (A_f). However one may think of it differently; austenite is the only stable phase at temperatures above A_f, while martensite is the only stable phase at temperatures below M_f. In between, an austenite-martensite mixture is possible, and it depends upon the direction of the transformation. During the transformation, the amount of martensite in the material is called the *martensite fraction*. For instance, in a forward transformation, if the material is fully transformed, then the martensite fraction equals unity.

The relationship between the stress and temperature is presented in Fig. 1(b). It should be noted that above the curve denoted by M_f, the material is fully martensitic; whereas, below the curve denoted by A_f, it is fully austenitic. In addition to the phase transformation associated with a temperature change, the transformation may be stress induced. In this case, the phase changes from austenite to martensite upon tensile loading, with constant temperature. This temperature dictates whether shape memory effect or pseudoelasticity is taking place. If the temperature $T > A_f$ with $\sigma = 0$ (point 1 in Fig. 1(b)), PE occurs. Upon loading, the austenite transforms to martensite (points 2 to 3); upon unloading the body, full recovery is observed (see Fig. 1(a)), since austenite is the only stable phase at this temperature. If $T < A_s$ (point 4 in Fig. 1(b)), SME occurs. Similar to PE, the austenite transforms to martensite (points 5 to 6); since austenite is not a stable phase at this temperature, there is no reverse transformation upon unloading, and residual strains are observed (see Fig. 1(a)). If $A_s < T < A_f$, a reverse transformation occurs upon unloading, but still some residual strains are observed. For both stress or temperature induced martensite, a hysteresis loop is observed in the stress-strain curve. This loop may be explained using the material stress-temperature relations; for the case of stress induced martensite, the forward and reverse transformations occur in different stress ranges. Hence, recovery followed by a reverse transformation is on a path different from that of the loading path.

While both the pseudoelasticity and the one-way shape memory effect are intrinsic characteristics of the alloy, the *two-way shape memory effect* is an acquired characteristic. It can be obtained by applying a specific thermomechanical loading cycles (usually refer to as *training*) which results in a generation of permanent defects (see for example, Perkins [1], Stalmans et al. [2], Hebda and White [3], Miller and Lagoudas [4], Lahoz and Puertolas [5], Patoor et al. [6], Abel and Luo [7]). These defects are effectively considered as non-homogeneous plastic strains at the martensite or austenite phases. The outcome of the appearance of these strains is an alloy which exhibits two stable shapes instead of one: a high-temperature shape in the austenite phase and a low-temperature shape in the martensite phase; the transition between these shapes may be carried out by applying a temperature change without any mechanical loading, see Fig. 1(c) and (d). As a result, the two-way shape memory effect forms the basis of an actuation mechanism which may be utilized in industrial and surgical applications.

A special case of the two-way shape memory effect is that in which deformation due to martensite reorientation results in internal plastic strains. As discussed by Liu et al. [8], these strains are essential to compensate for the orientation mismatch that occurs between martensite variants in neighboring grains. From a macroscopic point of view, this mechanism results in a two-way shape memory effect which represents the directional dependency of the generated internal stress field due to applied thermal loading. In other words, if an external load is applied at a constant temperature which is lower than the martensite final temperature M_f, martensite reorientation is generated. At this state, while keeping the alloy loaded, two stable shapes can be identified: a low-temperature shape in the detwinned martensite phase and a high-temperature shape in the austenite phase.

Different applications of SMA may be found in various fields, such as civil engineering (see for example, Janke et al. [9], Song et al. [10]), aerospace, (Kudva [11]), medical devices (Machado and Savi [12], Morgan [13]), actuators (Luo and Salvia [14]), and in general engineering (Razov [15]). This results from relatively good mechanical properties (high stiffness, good resistance to corrosion), bio-compatibility, and a damping capacity which is far greater than that of standard materials. However, there are also several limitations. These alloys are still relatively expensive to manufacture with respect to ordinary materials such as steel and aluminum. Furthermore, these materials have poor fatigue properties, which are lower by two orders of magnitude than those of ordinary metals. Consequently, applications in which the pseudoelasticity or the shape memory effects are repeated may be designed for a relatively small number of transformation cycles. The fatigue behavior of several shape memory alloys was reviewed by Patoor et al. [6]. More details about the characteristics and applications of the SMA materials can be found in the recent book edited by Lagoudas [16].

In this chapter, modeling and several applications of shape memory alloy composites are explained and discussed. Currently, composite materials with SMA fibers have become a significant member of the family of smart materials. The use of embedded SMA fibers in different media is usually aimed at reducing the cost of monolithic SMAs. However, in many applications the use of SMA fibers may assist in controlling the behavior of the structure itself. There are numerous studies in which new concepts are employed to model the behavior of these composites. Among them, Wei et al. [17]-[18] who presented the main ideas in the design and fabrication of intelligent composites based on shape memory alloys. Lagoudas et al. [19] introduced a thermomechanical model that is capable of describing the response of active laminates with SMA strips. In their study, Lagoudas et al. [19] implemented a lamination theory in conjunction with the finite element method. Su et al. [20] suggested a constitutive law to describe the behavior of an SMA reinforced composite laminate using irreversible thermodynamics. Gilat and Aboudi [22] studied the dynamic response of active SMA composite plates whereas Thompson and Loughlan [21] and Gilat and Aboudi [23] focused on the control of the post-buckling response in composite plates. Other investigators focused on modeling of shape memory alloy fibers that are embedded in an elasto-plastic metallic matrix (Lee et al. [24], Wang et al. [25]), a polymeric matrix (Gall et al. [26]) and a carbon fiber reinforced plastic matrix (Taketa et al. [27]). In addition, several attempts were made to predict the macroscopic behavior of shape memory alloy composites by means of micromechanical models (Boyd and Lagoudas [28], Lagoudas et al. [29], Kawai et al. [30], Carvelli and Taliercio [31], Song et al. [32], Kawai [33], Marfia [34], Gilat and Aboudi [22]-[23]).

In this chapter, a micromechanical model, known as the high fidelity generalized method of cells (HFGMC), is employed. This micromechanical model, first presented by Aboudi [35] and reviewed by Aboudi [36], is capable of accurately predicting the macroscopic thermo-mechanical behavior of a composite material by properly taking into account the detailed interaction between its various constituents. Furthermore, shape memory alloys may exhibit large inelastic strains. Therefore, they induce full thermomechanical coupling (TMC). In other words, the temperature of the alloy changes upon applied mechanical loading, and the mechanical response of the alloy changes upon temperature change. This phenomenon occurs in many inelastic materials; however, in SMAs it is crucial, since the behavior of the alloys is dictated by the current temperature. In addition, this full TMC generates a rate-dependent material response. This is due to the coupling with the energy equation which includes rate-dependent terms.

In Section 2 of this chapter, the HFGMC micromechanical model with full thermomechanical coupling in general composite materials is established. In the next sections, this model is employed to predict the response of SMA composites in which the SMA fibers exhibit various types of behavior. Thus, in Section 3, the HFGMC method is applied to model the special case of thermomechanical coupling where unidirectional SMA fibers are embedded in either metallic or polymeric matrix, where both pseudoelasticity and one-way shape memory effects are considered (Aboudi and Freed [37]). This is followed by considering constitutive relations for SMA fibers undergoing transformation induced plasticity effects, that are embedded in either metallic or polymeric matrices (Freed and Aboudi [38]). Furthermore, the behavior of SMA honeycombs with various topologies are investigated (Freed et al. [39]). This Section is continued by presenting the mathematical background of the constitutive behavior of an SMA composite material which exhibits the two-way shape memory effect. The resulting constitutive equations are applied to investigate the behavior of a unidirectional composite with an embedded SMA fibers that exhibit the two-way shape memory effect (Freed and Aboudi [40]). Finally, specific constitutive relations of SMA fibers are employed to study the behavior of two types of plain concrete which are being prestressed by these fibers are given in Sections 4 and 5. In the first type of plain concrete which is described in Section 4, its constitutive behavior includes evolving damage (Freed et al. [41]). In Section 5, the second type of plain concrete, whose constitutive behavior includes coupled damage and plasticity effects, is considered (Freed and Aboudi [42]).

The present Chapter is confined to the modeling of SMAs and SMA composites in the framework of infinitesimal deformations. A finite strain model for SMAs which was presented by Auricchio [43], was employed by Aboudi [44] to micromechanically analyze the behavior of SMA composites undergoing large deformations.

2. Full Thermomechanically Coupled Micromechanical Analysis of Composites

The HFGMC micromechanical model was extended by Aboudi and Freed [37] and Aboudi [45] for the modeling of doubly periodic composites (e.g., long fibers embedded in a matrix) and triply periodic composites (e.g., short fibers or particles embedded in a matrix), respectively, with full TMC. In the present chapter, this HFGMC micromechanical model with full TMC is briefly outlined in the following. This model is based on a homogenization technique for composites with periodic microstructure as shown in Fig. 2(a) in terms of the global coordinates (x_2, x_3). The repeating unit cell, Fig. 2(b), defined with respect to local coordinates (y_2, y_3), of such a composite is divided into N_β and N_γ subcells in the y_2 and y_3 directions, respectively. Each subcell is labeled by the indices $(\beta\gamma)$ with $\beta = 1, ..., N_\beta$ and $\gamma = 1, ..., N_\gamma$, and may contain a distinct homogeneous material. The dimensions of subcell $(\beta\gamma)$ in the y_2 and y_3 directions are denoted by h_β and l_γ, respectively. A local coordinate system $(\bar{y}_2^{(\beta)}, \bar{y}_3^{(\gamma)})$ is introduced in each subcell whose origin is located at its center, see Fig. 2(c).

The local (subcell) constitutive equation of the material which, in general, is assumed to be thermoinelastic is given by

$$\boldsymbol{\sigma}^{(\beta\gamma)} = \mathbf{C}^{(\beta\gamma)} : \left(\boldsymbol{\epsilon}^{(\beta\gamma)} - \boldsymbol{\epsilon}^{I(\beta\gamma)}\right) - \boldsymbol{\Gamma}^{(\beta\gamma)} \Delta T^{(\beta\gamma)} \tag{1}$$

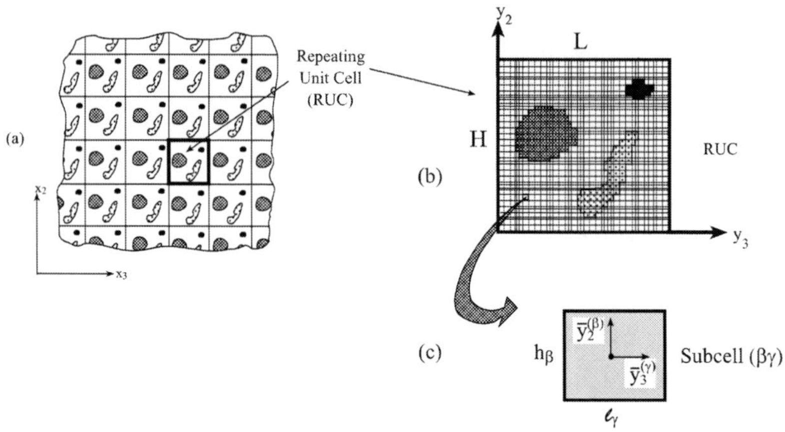

Figure 2. (a) A multiphase composite with doubly-periodic microstructures defined with respect to global coordinates (x_2, x_3). (b) The repeating unit cell is represented with respect to local coordinates (y_2, y_3). It is divided into N_β and N_γ subcells, in the y_2 and y_3 directions, respectively. (c) A characteristic subcell $(\beta\gamma)$ with local coordinates $\bar{y}_2^{(\beta)}$ and $\bar{y}_3^{(\gamma)}$ whose origin is located at its center.

where $\boldsymbol{\sigma}^{(\beta\gamma)}$, $\boldsymbol{\epsilon}^{(\beta\gamma)}$, $\boldsymbol{\epsilon}^{I(\beta\gamma)}$ and $\boldsymbol{\Gamma}^{(\beta\gamma)}$ are the stress, total strain, inelastic strain and thermal stress coefficients tensors, respectively, in subcell $(\beta\gamma)$. In Eq. (1), $\mathbf{C}^{(\beta\gamma)}$ is the stiffness tensor of the material in the subcell $(\beta\gamma)$, and ΔT denotes the temperature change from a reference temperature. The inelastic strain $\boldsymbol{\epsilon}^{I(\beta\gamma)}$ is governed by the appropriate flow rule of the inelastic material, that is given by the Prandtl-Reuss equations of the classical plasticity or by an appropriate viscoplastic flow rule. For a shape memory alloy constituent that fills this subcell, the flow rule is govern by the corresponding evolution equations that models this material. The basic assumption in HFGMC is that the displacement vector $\mathbf{u}^{(\beta\gamma)}$ in each subcell is expanded into quadratic forms in terms of its local coordinates $(\bar{y}_2^{(\beta)}, \bar{y}_3^{(\gamma)})$, as follows

$$\begin{aligned}\mathbf{u}^{(\beta\gamma)} &= \bar{\boldsymbol{\epsilon}} \cdot \mathbf{x} + \mathbf{W}_{(00)}^{(\beta\gamma)} + \bar{y}_2^{(\beta)} \mathbf{W}_{(10)}^{(\beta\gamma)} + \bar{y}_3^{(\gamma)} \mathbf{W}_{(01)}^{(\beta\gamma)} \\ &+ \frac{1}{2}\left(3\bar{y}_2^{(\beta)2} - \frac{h_\beta^2}{4}\right)\mathbf{W}_{(20)}^{(\beta\gamma)} + \frac{1}{2}\left(3\bar{y}_3^{(\gamma)2} - \frac{l_\gamma^2}{4}\right)\mathbf{W}_{(02)}^{(\beta\gamma)} \end{aligned} \qquad (2)$$

where $\bar{\epsilon}$ is the externally applied average strain, $\mathbf{W}_{(00)}^{(\beta\gamma)}$ is the volume-averaged displacement, and the higher-order terms $\mathbf{W}_{(mn)}^{(\beta\gamma)}$ must be determined as discussed below.

In the full thermomechanically coupled HFGMC model, the unknown temperature field change $\Delta T^{(\beta\gamma)}$ in the subcell is also expanded as follows

$$\begin{aligned}\Delta T^{(\beta\gamma)} &= \Delta T_{(00)}^{(\beta\gamma)} + \bar{y}_2^{(\beta)} \Delta T_{(10)}^{(\beta\gamma)} + \bar{y}_3^{(\gamma)} \Delta T_{(01)}^{(\beta\gamma)} \\ &+ \frac{1}{2}\left(3\bar{y}_2^{(\beta)2} - \frac{h_\beta^2}{4}\right)\Delta T_{(20)}^{(\beta\gamma)} + \frac{1}{2}\left(3\bar{y}_3^{(\gamma)2} - \frac{l_\gamma^2}{4}\right)\Delta T_{(02)}^{(\beta\gamma)} \end{aligned} \qquad (3)$$

where $\Delta T^{(\beta\gamma)}_{(00)}$ is the volume-averaged temperature and the higher-order terms $T^{(\beta\gamma)}_{(mn)}$ are additional unknowns.

The unknown terms $\mathbf{W}^{(\beta\gamma)}_{(mn)}$ and $\Delta T^{(\beta\gamma)}_{(mn)}$ are determined from the fulfillment of the coupled equilibrium and the energy equations as well as the periodic boundary conditions, and the interfacial continuity conditions of displacements, tractions, temperatures and heat fluxes between subcells.

It can be shown (Aboudi [35]) that in the framework of HFGMC, the equilibrium equations take the form

$$\frac{S^{(\beta\gamma)}_{2i(10)}}{h^2_\beta} + \frac{S^{(\beta\gamma)}_{3i(01)}}{l^2_\gamma} = 0, \qquad i = 1,2,3 \tag{4}$$

where $S^{(\beta\gamma)}_{2i(10)}$ and $S^{(\beta\gamma)}_{3i(01)}$ are the volume-averaged moments $\bar{y}^{(\beta)}_2 \sigma^{(\beta\gamma)}_{2i}$ and $\bar{y}^{(\gamma)}_3 \sigma^{(\beta\gamma)}_{3i}$, respectively.

For standard inelastic isotropic materials, the energy equation is given by

$$\left[\rho c_v \dot{T} + \nabla \cdot \boldsymbol{q}\right]^{(\beta\gamma)} = \left[\zeta \boldsymbol{\sigma} : \dot{\boldsymbol{\epsilon}}^I - T\boldsymbol{\Gamma} : \dot{\boldsymbol{\epsilon}}\right]^{(\beta\gamma)} \tag{5}$$

where dot denotes differentiation with respect to the time t, and $\mathbf{q}$, ρ, c_v are the heat flux vector, material mass density and the specific heat at constant deformation, respectively. The parameter ζ is a partition factor which indicates that only a portion (about 90 percent) of the inelastic work W_I is transformed into heat (the rate of the inelastic work is given by $\dot{W}_I = \boldsymbol{\sigma} : \dot{\boldsymbol{\epsilon}}^I$). In conjunction with the Fourier law, the heat flux is related to the temperature gradient by

$$\mathbf{q} = -k\nabla T \tag{6}$$

where k is the thermal conductivity. By employing Eq. (2)-(3), the spatial derivatives in Eq. (5) can be eliminated. As a result, this equation is reduced to an ordinary differential equation in time. Consequently, let us represent it in the following compact form

$$[M]^{(\beta\gamma)}\dot{T}^{(\beta\gamma)} = [S]^{(\beta\gamma)}T^{(\beta\gamma)} + [\dot{Q}]^{(\beta\gamma)} \tag{7}$$

where $[\dot{Q}]^{(\beta\gamma)}$ denotes the right-hand-side of Eq. (5). The implicit difference in time of Eq. (7) yields (Mitchell and Griffith [46], Chapt. 6)

$$\left\{[M]^{(\beta\gamma)} - \omega\Delta t[S]^{(\beta\gamma)}\right\}T^{(\beta\gamma)}_{n+1} = \left\{[M]^{(\beta\gamma)} + (1-\omega)\Delta t[S]^{(\beta\gamma)}\right\}T^{(\beta\gamma)}_{n} + [Q]^{(\beta\gamma)}_n - [Q]^{(\beta\gamma)}_{n-1} \tag{8}$$

where $\Delta t = t_{n+1} - t_n$ and ω is a parameter (for the Crank-Nicolson scheme: $\omega = 1/2$).

The periodic boundary conditions ensure that the displacements, tractions, temperatures and heat fluxes at opposite surfaces of the repeating unit cell (i.e., at $y_2 = 0$ and H as well as $y_3 = 0$ and L) are identical, see Aboudi [35]-[36] for more details pertaining the micromechanical analysis a partial TMC in which the temperature change is prescribed. A principal ingredient in the present micromechanical analysis is that all these conditions are imposed in the average (integral) sense. For example, the periodicity conditions for the displacement vector at the two opposite sides of the repeating unit cell of Fig. 2(b) are given by

$$\mathbf{u}^{(1\gamma)}|_{\bar{y}_2^{(1)}=-h_1/2} = \mathbf{u}^{(N_\beta\gamma)}|_{\bar{y}_2^{(N_\beta)}=h_{N_\beta}/2}, \quad \gamma = 1, ..., N_\gamma \tag{9}$$

In the framework of HFGMC, these conditions are replaced by

$$\int_{-l_\gamma/2}^{l_\gamma/2} \mathbf{u}^{(1\gamma)}|_{\bar{y}_2^{(1)}=-h_1/2}\, d\bar{y}_3^{(\gamma)} = \int_{-l_\gamma/2}^{l_\gamma/2} \mathbf{u}^{(N_\beta\gamma)}|_{\bar{y}_2^{(N_\beta)}=h_{N_\beta}/2}\, d\bar{y}_3^{(\gamma)}, \quad \gamma = 1, ..., N_\gamma \tag{10}$$

Similarly, the pointwise displacement continuity conditions at the interface between subcell $(\beta\gamma)$ and its neighboring one $(\beta\,\gamma + 1)$ that are given by

$$\mathbf{u}^{(\beta\gamma)}|_{\bar{y}_3^{(\gamma)}=l_\gamma/2} = \mathbf{u}^{(\beta\,\gamma+1)}|_{\bar{y}_3^{(\gamma)}=-l_\gamma/2} \tag{11}$$

for $\beta = 1, ..., N_\beta$ and $\gamma = 1, ..., N_\gamma - 1$ are replaced by

$$\int_{-h_\beta/2}^{h_\beta/2} \mathbf{u}^{(\beta\gamma)}|_{\bar{y}_3^{(\gamma)}=l_\gamma/2}\, d\bar{y}_2^{(\beta)} = \int_{-h_\beta/2}^{h_\beta/2} \mathbf{u}^{(\beta\,\gamma+1)}|_{\bar{y}_3^{(\gamma)}=-l_\gamma/2}\, d\bar{y}_2^{(\beta)} \tag{12}$$

for $\beta = 1, ..., N_\beta$ and $\gamma = 1, ..., N_\gamma - 1$

As a result of the imposition of the equilibrium, energy, continuity of tractions, displacements, temperatures and heat fluxes between the subcells, as well as the above mentioned periodic boundary conditions, a linear system of algebraic equations at the current time step is obtained which can be represented in the following form

$$\mathbf{KU} = \mathbf{f} + \mathbf{g} \tag{13}$$

where the matrix $\mathbf{K}$ contains information on the geometry and thermomechanical properties of the materials within the individual subcells $(\beta\gamma)$, and the displacement-temperature vector $\mathbf{U}$ contains the unknown displacement and temperature coefficients:

$$\mathbf{U} = [\mathbf{U}^{(11)}, ..., \mathbf{U}^{(N_\beta N_\gamma)}] \tag{14}$$

where in subcell $(\beta\gamma)$ these coefficients, which appear on the right-hand-side of Eq. (2)-(3), are

$$\mathbf{U}^{(\beta\gamma)} = \left[\mathbf{W}_{(00)}, \Delta T_{(00)}, \mathbf{W}_{(10)}, \Delta T_{(10)}, \mathbf{W}_{(01)}, \Delta T_{(01)}, \mathbf{W}_{(20)}, \Delta T_{(20)}, \mathbf{W}_{(02)}, \Delta T_{(02)}\right]^{(\beta\gamma)} \tag{15}$$

The mechanical vector $\mathbf{f}$ contains information on the applied average (far-field) strains ϵ. The inelastic force vector $\mathbf{g}$ appearing on the right-hand side of Eq. (13) contains the inelastic effects given in terms of the integrals of the inelastic strain distributions. These integrals depend implicitly on the elements of the displacement-temperature coefficient vector $\mathbf{U}$, requiring an incremental procedure of Eq. (13) at each point along the loading path, see Aboudi et al. [47]-[48] for more details.

Due to the dependence of the elements of $\mathbf{K}$ on the temperature, it is necessary to invert this matrix at every time step. The solution of Eq. (13) at a given time step enables the establishment of the following localization relation which expresses the average strain

$\bar{\epsilon}^{(\beta\gamma)}$ and temperature $\Delta\bar{T}^{(\beta\gamma)}$ in the subcell $(\beta\gamma)$ to the externally applied average strain $\bar{\epsilon}$ in the form:

$$\left\{ \begin{array}{c} \bar{\epsilon}^{(\beta\gamma)} \\ \Delta\bar{T}^{(\beta\gamma)} \end{array} \right\} = \left\{ \begin{array}{c} \mathbf{A}^{M(\beta\gamma)} \\ \mathbf{A}^{T(\beta\gamma)} \end{array} \right\} \bar{\epsilon} + \left\{ \begin{array}{c} \mathbf{V}^{T(\beta\gamma)} \\ \upsilon^{T(\beta\gamma)} \end{array} \right\} + \left\{ \begin{array}{c} \mathbf{V}^{I(\beta\gamma)} \\ \upsilon^{I(\beta\gamma)} \end{array} \right\} \tag{16}$$

where $\mathbf{A}^{M(\beta\gamma)}$ and $\mathbf{A}^{T(\beta\gamma)}$ are the mechanical and thermal concentration tensors of the subcell $(\beta\gamma)$, $\mathbf{V}^{T(\beta\gamma)}$ and $\mathbf{V}^{I(\beta\gamma)}$ are second-order tensors that involve thermal and inelastic effects in the subcell, and $\upsilon^{T(\beta\gamma)}$ and $\upsilon^{I(\beta\gamma)}$ are the corresponding scalars. These second-order thermal and inelastic tensors and the corresponding scalars arise due to the existence of $[Q]^{n-1}$ in Eq. (8) at the previous time step. It should be noted that in the present case of full TMC, the application of the far-field strain $\bar{\epsilon}$ induces a temperature change from the reference temperature $\Delta T^{(\beta\gamma)}$ in the subcell.

In order to establish the global (macroscopic) constitutive equation of the composite, we utilize the definition of the average stress in the composite in terms of average stress in the subcells:

$$\bar{\boldsymbol{\sigma}} = \frac{1}{HL} \sum_{\beta=1}^{N_\beta} \sum_{\gamma=1}^{N_\gamma} h_\beta l_\gamma \bar{\boldsymbol{\sigma}}^{(\beta\gamma)} \tag{17}$$

where $\bar{\boldsymbol{\sigma}}^{(\beta\gamma)}$ is the average stress in the subcell. By substituting Eq. (1) and (16) in (17), one obtains the final form of the effective constitutive law of the multiphase fully coupled thermo-inelastic composite, which relates the average stress $\bar{\boldsymbol{\sigma}}$, strain $\bar{\epsilon}$, thermal stress $\bar{\boldsymbol{\sigma}}^T$ and inelastic stress $\bar{\boldsymbol{\sigma}}^I$ tensors as follows

$$\bar{\boldsymbol{\sigma}} = \mathbf{C}^* : \bar{\epsilon} - (\bar{\boldsymbol{\sigma}}^T + \bar{\boldsymbol{\sigma}}^I) = \mathbf{C}^* : (\bar{\epsilon} - \bar{\epsilon}^I) - \boldsymbol{\Gamma}^* \Delta T \tag{18}$$

In this equation $\mathbf{C}^*$ is the effective stiffness tensor which is given by

$$\mathbf{C}^* = \frac{1}{HL} \sum_{\beta=1}^{N_\beta} \sum_{\gamma=1}^{N_\gamma} h_\beta l_\gamma \left[\mathbf{C}^{(\beta\gamma)} \mathbf{A}^{M(\beta\gamma)} - \boldsymbol{\Gamma}^{(\beta\gamma)} \mathbf{A}^{T(\beta\gamma)} \right] \tag{19}$$

The global thermal stress tensor $\bar{\boldsymbol{\sigma}}^T$ is determined from

$$\bar{\boldsymbol{\sigma}}^T = \frac{1}{HL} \sum_{\beta=1}^{N_\beta} \sum_{\gamma=1}^{N_\gamma} h_\beta l_\gamma \left[\mathbf{C}^{(\beta\gamma)} : \mathbf{V}^{T(\beta\gamma)} \quad \boldsymbol{\Gamma}^{(\beta\gamma)} \upsilon^{T(\beta\gamma)} \right] \tag{20}$$

from which the effective thermal stress tensor $\boldsymbol{\Gamma}^*$ can be detrmined. The global inelastic stress $\bar{\boldsymbol{\sigma}}^I$ is of the from

$$\bar{\boldsymbol{\sigma}}^I = -\frac{1}{HL} \sum_{\beta=1}^{N_\beta} \sum_{\gamma=1}^{N_\gamma} h_\beta l_\gamma \left[\mathbf{C}^{(\beta\gamma)} : \mathbf{V}^{I(\beta\gamma)} - \boldsymbol{\Gamma}^{(\beta\gamma)} \upsilon^{I(\beta\gamma)} - \bar{\boldsymbol{\sigma}}^{I(\beta\gamma)} \right] \tag{21}$$

where the inelastic stress tensor in the subcell is: $\bar{\boldsymbol{\sigma}}^{I(\beta\gamma)} = \mathbf{C}^{(\beta\gamma)} : \bar{\epsilon}^{I(\beta\gamma)}$. The global inelastic strain $\bar{\epsilon}^I$ is similarly detrmined as shown in Eq. (18). Finally, the global average temperature $\Delta\bar{T}$ in the composite can be determined from a simple averaging of all temperature change $\Delta\bar{T}^{(\beta\gamma)}$ in the subcells.

The average of the temperature change $\Delta\bar{T}$ over the the repeating unit cell, that is predicted from the micromechanics analysis, is given by

$$\Delta\bar{T} = \frac{1}{HL}\sum_{\beta=1}^{N_\beta}\sum_{\gamma=1}^{N_\gamma} h_\beta l_\gamma \Delta\bar{T}^{(\beta\gamma)} \tag{22}$$

where $\Delta\bar{T}^{(\beta\gamma)}$ is the average temperature change in the subcell.
In concluding this section it should be noted that the present full thermomechanically coupled HFGMC micromechanical model can be readily reduced to a the case of the HFGMC model with partial TMC by ignoring the effects of the energy equation (5).

3. Shape memory alloy composites

In this Section, the response of SMA composites is determined by the previously discussed full thermomechanically coupled HFGMC micromechanical model. Both metal (in which inelastic behavior is exhibited) and polymer matrices are considered.

3.1 Full TMC constitutive equations of the monolithic SMA

The fully coupled thermomechanical equations of the monolithic SMA material were presented by Auricchio and Petrini [49], [50], [51] who extended and improved the thermodynamical model of Souza et al. [52]. These equations are briefly presented in the following. The strain $\boldsymbol{\epsilon}$ is decoupled into a dilatation θ and deviatoric $\mathbf{e}$ parts as follows

$$\boldsymbol{\epsilon} = \frac{\theta}{3}\mathbf{I} + \boldsymbol{e} \tag{23}$$

where $\mathbf{I}$ is the unit tensor. The free energy function ψ, is a function of the dilatation strain θ, the deviatoric strain $\boldsymbol{e}$, the transformation strain $\boldsymbol{e}^{tr}$ and absolute temperature T, is given by

$$\begin{aligned} \rho\psi(\theta, \boldsymbol{e}, \boldsymbol{e}^{tr}, T) &= \frac{1}{2}K\theta^2 + G||\boldsymbol{e} - \boldsymbol{e}^{tr}||^2 - 3\alpha K\theta(T - T_0) + \beta < (T - M_f) > ||\boldsymbol{e}^{tr}|| \\ &+ \frac{h}{2}||\boldsymbol{e}^{tr}||^2 + (u_0 - T\eta_0) + \rho c_v\left[T - T_0 - T\log\frac{T}{T_0}\right] + \Upsilon(\boldsymbol{e}^{tr}) \end{aligned} \tag{24}$$

where ρ, K, G, α, T_0, β, M_f, h, and ρc_v are the mass density, bulk modulus, shear modulus, coefficient of thermal expansion, reference temperature, slope of the stress-temperature relation, martensite final temperature, slope of the stress-strain relation during the stress-induced phase transformation and the heat capacity, respectively. In addition, u_0 and η_0 are the internal energy and entropy at the reference state. In Eq. (24), $||.||$ and $< . >$ denote the Euclidean norm and the positive part of the argument respectively, and $\Upsilon(\boldsymbol{e}^{tr})$ is an indicator function defined as

$$\Upsilon(\boldsymbol{e}^{tr}) = \begin{cases} 0 & ||\boldsymbol{e}^{tr}|| \leq \epsilon_L \\ \infty & ||\boldsymbol{e}^{tr}|| > \epsilon_L \end{cases} \tag{25}$$

where ϵ_L is the norm of the maximum transformation strain reached at the end of the transformation during uniaxial test.

The resulting constitutive relations obtained from Eq. (24) are given by

$$p = \frac{\partial \psi}{\partial \theta} = K[\theta - 3\alpha(T - T_0)] \tag{26}$$

$$\boldsymbol{s} = \frac{\partial \psi}{\partial \boldsymbol{e}} = 2G(\boldsymbol{e} - \boldsymbol{e}^{tr}) \tag{27}$$

$$\eta = -\frac{\partial \psi}{\partial T} = \eta_0 + 3K\alpha\theta - \beta||\boldsymbol{e}^{tr}||\frac{< T - M_f >}{|T - M_f|} + \rho c_v \log\frac{T}{T_0} \tag{28}$$

$$\boldsymbol{X} = -\frac{\partial \psi}{\partial \boldsymbol{e}^{tr}} = \boldsymbol{s} - \left[\beta(T - M_f) + h||\boldsymbol{e}^{tr}|| + \frac{\partial \Upsilon(\boldsymbol{e}^{tr})}{\partial ||\boldsymbol{e}^{tr}||}\right]\frac{\partial ||\boldsymbol{e}^{tr}||}{\partial \boldsymbol{e}^{tr}} \tag{29}$$

where p and $\boldsymbol{s}$ are the volumetric and deviatoric parts of the stress $\boldsymbol{\sigma}$, η is the entropy and $\boldsymbol{X}$ is the transformation stress associated with $\boldsymbol{e}^{tr}$. It should be noted that the McAuley brackets $< \cdot >$ and the indicator function Υ in Eq. (24) are not smooth. In obtaining the entropy η and the transformation stress $\boldsymbol{X}$, derivatives of the free energy function with respect to the temperature T and the transformation strain $\boldsymbol{e}^{tr}$ are carried out. Following Aurrichio and Petrini [50], the derivative of the McAuley brackets $< T - M_f >$ with respect to T is taken as unity, while the derivative of the indicator function is defined as

$$\frac{\partial \Upsilon(\boldsymbol{e}^{tr})}{\partial \boldsymbol{e}^{tr}} \equiv \gamma \frac{\boldsymbol{e}^{tr}}{\| \boldsymbol{e}^{tr} \|} \tag{30}$$

where $\gamma = 0$ if $\| \boldsymbol{e}^{tr} \| < \epsilon_L$ and $\gamma \geq 0$ if $\| \boldsymbol{e}^{tr} \| = \epsilon_L$.

The evolution law of $\boldsymbol{e}^{tr}$ is

$$\dot{\boldsymbol{e}}^{tr} = \dot{\xi}\frac{\partial F(\boldsymbol{X})}{\partial \boldsymbol{\sigma}} \tag{31}$$

where the dot denotes a derivative with respect to time t, and F and $\dot{\xi}$ play the role of limit function and plastic consistent parameter, respectively, subject to the Kuhn-Tucker conditions:

$$\dot{\xi} \geq 0, \quad F \leq 0, \quad \dot{\xi}F = 0. \tag{32}$$

For symmetric SMA behavior, the function F is given by

$$F(\boldsymbol{X}) = ||\boldsymbol{X}|| - R \leq 0 \tag{33}$$

where R is the radius of the elastic domain.

The first law of thermodynamics can be written as

$$\rho\dot{u} = \rho(\dot{\psi} + T\dot{\eta} + \dot{T}\eta) = \boldsymbol{\sigma} : \dot{\epsilon} + r - \nabla \cdot \boldsymbol{q} \tag{34}$$

where u is the internal energy. The coupled energy equation is obtained by substituting Eq. (24) in Eq. (34), obtaining

$$\rho c_v \dot{T} + \nabla \cdot \boldsymbol{q} = D_{mech} + H_{tmc} \tag{35}$$

where D_{mech} is the heat production due the dissipative mechanical process which is given by

$$\begin{aligned} D_{mech} &= \boldsymbol{\sigma} : \dot{\boldsymbol{\epsilon}} - \rho\left(\frac{\partial \psi}{\partial \theta}\dot{\theta} + \frac{\partial \psi}{\partial \boldsymbol{e}}\dot{\boldsymbol{e}} + \frac{\partial \psi}{\partial \boldsymbol{e}^{tr}}\dot{\boldsymbol{e}}^{tr}\right) \\ &= [\boldsymbol{\sigma} : \dot{\boldsymbol{\epsilon}} - (p\dot{\theta} + \boldsymbol{s} : \dot{\boldsymbol{e}})] + \boldsymbol{X}\dot{\boldsymbol{e}}^{tr} \\ &= \boldsymbol{X}\dot{\boldsymbol{e}}^{tr} \end{aligned} \tag{36}$$

The heat production due to the thermomechanical coupling is

$$\begin{aligned} H_{tmc} &= \rho T\left(\frac{\partial^2 \psi}{\partial T \partial \theta}\dot{\theta} + \frac{\partial^2 \psi}{\partial T \partial \boldsymbol{e}}\dot{\boldsymbol{e}} + \frac{\partial^2 \psi}{\partial T \partial \boldsymbol{e}^{tr}}\dot{\boldsymbol{e}}^{tr}\right) \\ &= \left(\beta \frac{\partial ||\boldsymbol{e}^{tr}||}{\partial \boldsymbol{e}^{tr}} : \dot{\boldsymbol{e}}^{tr} - 3K\alpha\dot{\theta}\right)T \\ &= \left(\beta \frac{\boldsymbol{e}^{tr}}{||\boldsymbol{e}^{tr}||} : \dot{\boldsymbol{e}}^{tr} - 3K\alpha\dot{\theta}\right)T \end{aligned} \tag{37}$$

Hence the resulting coupled energy equation is given by

$$\rho c_v \dot{T} + \nabla \cdot \boldsymbol{q} = \left[\boldsymbol{X} + T\beta \frac{\boldsymbol{e}^{tr}}{||\boldsymbol{e}^{tr}||}\right] : \dot{\boldsymbol{e}}^{tr} - 3TK\alpha\dot{\theta} \tag{38}$$

together with the Fourier law given by Eq. (6).

The above nonlinear equations have been treated by Auricchio and Petrini [49], [50], [51] who employed for their solution a computational algorithm based on an implicit time procedure. This algorithm consists of integrating these equations over a time interval $[t_n, t_{n+1}]$ using a backward Euler scheme. Thus, assuming the knowledge of the solution at time t_n, as well as the strain ϵ at time t_{n+1}, the stresses are computed using the radial return mapping algorithm (Simo and Hughes [53]). It should be noted that in the case of standard isotropic materials, the energy equation is given by Eq. (5). A comparison between Eq. (38) and Eq. (5) shows that the term
$\left[\boldsymbol{X} + T\beta \frac{\boldsymbol{e}^{tr}}{||\boldsymbol{e}^{tr}||}\right] : \dot{\boldsymbol{e}}^{tr}$ replaces the rate of inelastic work: $\dot{W}_I = \boldsymbol{\sigma} : \dot{\boldsymbol{\epsilon}}^I$, where $\boldsymbol{\epsilon}^I$ is the inelastic strain which replaces the transformation strain $\boldsymbol{\epsilon}^{tr}$ of the SMA material.

3.1.1 Applications

In the present applications, SMA fibers that are oriented in the x_1−direction (see Fig. 2(a)) are considered. A set of nine subcells in a 3×3 configuration is used. The middle subcell (i.e. $\beta = \gamma = 2$) is filled with SMA fiber, while the rest of the subcells are taken as the matrix (aluminum or epoxy). In all cases, the volume fraction of the SMA fibers is taken as $v_f = 30\%$. The material properties of the SMA fibers, aluminum and epoxy matrices are given in Tables 1, 2 and 3, respectively. The response of the monolithic SMA material is shown in Figs. 2 and 3 of Ref. [37].

Table 1. Material properties of the SMA fibers (Auricchio and Petrini [49]).

Property	Value
E	70 GPa
ν	0.33
k	18 W/(mK)
α	1×10^{-6}/K
ρc_v	5.44 MJ/(m^3K)
h	500 MPa
R	45 MPa
β	7.5 MPa/K
ϵ_L	0.03
M_f	253.15K

E, ν denote the Young's modulus and Poisson's ratio. The other parameters were already defined in Section 2.

Table 2. Elastic, plastic and thermal parameters of the isotropic elastoplastic aluminum matrix.

E(GPa)	ν	α (10^{-6}/K)	σ_y(MPa)	E_s(GPa)	k(W/(mK))	ρc_v(MJ/(m^3K))
72.4	0.33	22.5	371.5	23	116.7	2.25

E, ν, α, σ_y, E_s, k and ρc_v denote the Young's modulus, Poisson's ratio, coefficient of thermal expansion, yield stress, secondary modulus, thermal conduction and heat capacity, respectively.

Table 3. Elastic and thermal parameters of the isotropic epoxy polymeric matrix.

E(GPa)	ν	α (10^{-6}/K)	k(W/(mK))	ρc_v(MJ/(m^3K))
3.45	0.35	54	0.18	1.28

E, ν, α, k and ρc_v denote the Young's modulus, Poisson's ratio, coefficient of thermal expansion, thermal conduction and heat capacity, respectively.

The induced temperature resulting from the full TMC coupling can be obtained by subjecting the SMA/aluminum composite to 5 cycles of uniaxial stress loading in the fiber direction at a reference temperature $T_0 = 253.15$ K. Note that this temperature induces the shape memory effect. The average axial stress response of the composite and the induced average temperature change are shown in Fig. 3. The corresponding response to 5 cyclic loads of the unreinforced aluminum matrix is quite similar (but with a higher amplitude as

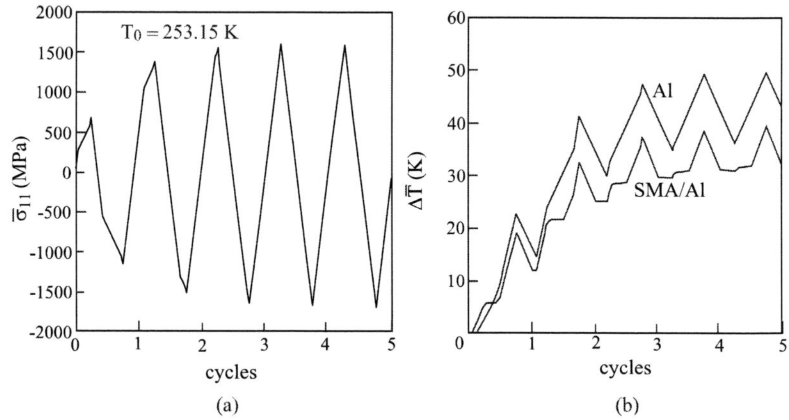

Figure 3. (a) The uniaxial response of the SMA/aluminum composite in 5 cycles loaded in the fibers direction and, (b) the resulting induced average temperature change at a reference temperature $T_0 = 253.15K$. Also shown is the corresponding induced temperature change in the unreinforced aluminum.

well as a higher maximum stress) to the composite response that is exhibited in Fig. 3(a) and is not shown. The induced temperatures of the SMA/aluminum and unreinforced aluminum are shown in Fig. 3(b). It may be observed that the maximum temperature induced by the thermomechanical coupling in the unreinforced aluminum is greater than that obtained in the SMA/aluminum composite. In other words, inclusion of shape memory alloy fibers results in a decrease in the global temperature that is being generated in the composite by the thermomechanical coupling.

It is interesting to observe a maximum temperature change of almost 50 K in the case of shape memory effect, which is induced by only five loading cycles. Consequently, a larger number of cycles may cause a significant temperature change, so that the current temperature will correspond to the pseudoelasticity temperature (it is understood that once the temperature of the composite increases, a partial recovery of the SMA fibers occurs). In other words, for a large number of cycles, the material behavior no longer exhibits the shape memory effect, but is rather pseudoelastic. For example, consider a NiTi alloy with 55.5 % Nickel, $A_s = 288$ K and $A_f = 308$ K. If an actuator is designed for a reference working temperature of $T = 280$ K, it exhibits the shape memory effect. However, in the absence of proper heat transfer, the temperature of the actuator may increase by several dozens of degrees upon cycling. Once the current temperature of the actuator reaches $T = A_f = 308$ K, it exhibits pseudoelasticity instead of shape memory effect, and the actuator is no longer usable.

It should be mentioned that in the above analyses, a repeating unit cell was investigated, and not a boundary value problem. Consequently, the boundary conditions were assumed to be periodic, and no heat transfer was considered. In the case of heat transfer, the temperature increase is expected to be less significant (Freed et al. [39]). In the case in which the

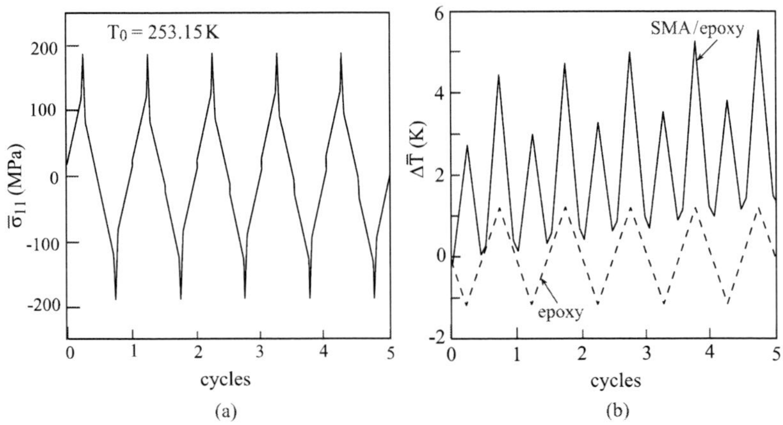

Figure 4. (a) The uniaxial response of the SMA/epoxy composite in 5 cycles loaded in the fibers direction and, (b) the resulting induced average temperature change at a reference temperature $T_0 = 253.15K$. Also shown is the corresponding induced temperature change in the unreinforced epoxy.

matrix is epoxy, it is understood that the thermomechanical coupling might be negligible. To this end, consider the epoxy matrix whose material parameters are given in Table 3. As a result of the absence of inelastic strains in this polymer, it is expected that the effect of the thermomechanical coupling will be weak and the corresponding induced temperature change insignificant. The thermomechanical coupling in this composite is induced only by the shape memory alloy fibers, and depends upon their volume fraction. The response of the composite to 5 cycles of uniaxial stress loading in the fiber direction at a reference temperature $T_0 = 253.15$ K is presented in Fig. 4. In Fig. 4(a), the macroscopic stress during the cycling process is illustrated. Of particular interest is the uniform periodic temperature change that is induced in the unreinforced epoxy caused by the thermomechanical coupling (in the absence of any irreversible effects). This is in contrast to the nonuniform and continuously increasing temperature in the unreinforced aluminum (see Fig. 3(b)) which is caused by the irreversible effects in this material (i.e., the effect of the rate of the inelastic strains). As a result, the induced temperature $\Delta\bar{T}$ caused by the thermomechanical coupling in the SMA/epoxy composite exhibits a slowly increasing pattern, as shown in Fig. 4(b).

To conclude this subsection, it was shown that the thermomechanical coupling effect may be significant for shape memory alloy composites. For a case of an aluminum matrix, the temperature of the composite changes rapidly. Although this phenomenon occurs in many hardening materials, it is extremely important in the case of shape memory alloy composites, since their constitutive behavior is governed by the current temperature. However, in the case of an epoxy that is reinforced by a shape memory alloy fiber, there is almost no thermomechanical coupling. It should be emphasized that the presented micromechanical analysis is rather complicated and costly. Therefore, whenever thermomechanical coupling may be neglected, the analysis can be reduced to the simpler partial TMC analysis, and the sought for a micromechanical prediction may be determined more quickly. From the results obtained in this subsection, it seems that it is possible to neglect the thermomechanical coupling effect in the case of shape memory alloy fibers that are embedded in a polymeric matrix. Further results and discussions can be found in Aboudi and Freed [37].

3.2 Full TMC constitutive equations of the monolithic SMA exhibiting transformation induced plasticity

In the following discussion, the transformation induced plasticity effects in the monolithic SMA material are included. To this end, the free-energy density function ψ of Auricchio et al. [54] which models this effect was generalized by Freed and Aboudi [38] to include the thermal effects as follows

$$\begin{aligned}
\rho\psi(\theta,\mathbf{e},T,\mathbf{e}^{tr},\boldsymbol{q}^{pr}) &= \frac{1}{2}K\theta^2 + G \parallel \mathbf{e}-\mathbf{e}^{tr} \parallel^2 + \beta < T - M_f > \parallel \mathbf{e}^{tr} - \boldsymbol{q}^{pr} \parallel + \frac{1}{2}h \parallel \mathbf{e}^{tr} \parallel^2 \\
&+ \frac{1}{2}H \parallel \boldsymbol{q}^{pr} \parallel^2 - A\mathbf{e}^{tr} : \boldsymbol{q}^{pr} + \Upsilon(\mathbf{e}^{tr}) + \rho c_v \left[(T-T_0) - T\ln\frac{T}{T_0} \right] \\
&- 3\alpha K\theta(T-T_0) + (u_0 - T\eta_0), \qquad (39)
\end{aligned}$$

where ρ is the material density, K and G are the bulk and shear modulus of the alloy, respectively, $\theta = \mathrm{tr}(\epsilon)$ is the trace of the strain tensor ϵ, $\mathbf{e}$ is the deviatoric strain tensor, $\parallel \bullet \parallel$ is an Euclidean norm, $\mathbf{e}^{tr}$ is the deviatoric part of the transformation strain tensor, β is a material parameter related to the transformation stress, T is the current temperature, M_f is the temperature which corresponds to fully martensitic transformation, $\boldsymbol{q}^{pr}$ is a permanent inelastic strain tensor, and h, H and A are material parameters which are related to the hardening of the phase transformation, the saturation of the permanent phase transformation, and the degradation of the model, respectively. In Eq. (39), α is the coefficient of thermal expansion of the shape memory alloy, ρc_v is the heat capacity, u_0 and η_0 are the reference internal energy and entropy, respectively, and T_0 is a reference (initial) temperature. In addition, $\Upsilon(\mathbf{e}^{tr})$ is an indicator function which is associated with the transformation strain and is given by Eq. (25).

The constitutive equations may be derived from the free-energy function $\psi(\theta,\mathbf{e},T,\mathbf{e}^{tr},\boldsymbol{q}^{pr})$ by differentiate it with respect to its internal variables,

$$\begin{aligned}
p &= \rho\frac{\partial\psi}{\partial\theta} = K\theta - 3\alpha K(T-T_0) \\
\mathbf{s} &= \rho\frac{\partial\psi}{\partial\mathbf{e}} = 2G(\mathbf{e}-\mathbf{e}^{tr}) \\
\mathbf{X} &= -\rho\frac{\partial\psi}{\partial\mathbf{e}^{tr}} = \mathbf{s} - \beta < T - M_f > \frac{\mathbf{e}^{tr}-\boldsymbol{q}^{pr}}{\parallel \mathbf{e}^{tr}-\boldsymbol{q}^{pr} \parallel} - h\mathbf{e}^{tr} + A\boldsymbol{q}^{pr} - \gamma\frac{\mathbf{e}^{tr}}{\parallel \mathbf{e}^{tr} \parallel} \\
\mathbf{Q} &= -\rho\frac{\partial\psi}{\partial\boldsymbol{q}^{pr}} = \beta < T - M_f > \frac{\mathbf{e}^{tr}-\boldsymbol{q}^{pr}}{\parallel \mathbf{e}^{tr}-\boldsymbol{q}^{pr} \parallel} - H\boldsymbol{q}^{pr} + A\mathbf{e}^{tr} \\
\rho\eta &= -\rho\frac{\partial\psi}{\partial T} = 3\alpha K\theta - \beta \parallel \mathbf{e}^{tr}-\boldsymbol{q}^{pr} \parallel + c\ln\frac{T}{T_0}, \qquad (40)
\end{aligned}$$

where p is the hydrostatic pressure, $\mathbf{s}$ is the deviatoric stress tensor, $\mathbf{X}$ is a stress-like tensor which is associated with the transformation strain $\mathbf{e}^{tr}$, $\mathbf{Q}$ is a stress-like tensor which is associated with the permanent inelastic strain tensor $\boldsymbol{q}^{pr}$ and η is the entropy. Observing Eq. $(40)_2$, it appears that the deviatoric stress tensor s does not depend on the permanent inelastic strain tensor $\boldsymbol{q}^{pr}$. However, the transformation strain $\mathbf{e}^{tr}$ *is coupled* with the permanent inelastic strain tensor $\boldsymbol{q}^{pr}$ as will be discussed in the following. It should be noted that the McAuley brackets $< \cdot >$ and the indicator function Φ_{ϵ_L} in eq. (39) are not smooth. In obtaining the entropy η and the transformation stress $\mathbf{X}$, derivatives of the free energy function with respect to the temperature T and the transformation strain $\mathbf{e}^{tr}$ are carried out. As indicated previously, the derivative of the McAuley brackets $< T - M_f >$ with respect to T is taken as unity, while the derivative of the indicator function is defined by Eq. (30).

The first law of thermodynamics is given by Eq. (34). Equation (34) yields after some manipulations the same form given by Eq. (35) but with

$$\begin{aligned} D_{mech} &= \boldsymbol{\sigma} : \dot{\boldsymbol{\epsilon}} - \rho \left(\frac{\partial \psi}{\partial \theta} \dot{\theta} + \frac{\partial \psi}{\partial \mathbf{e}} \dot{\mathbf{e}} + \frac{\partial \psi}{\partial \mathbf{e}^{tr}} \dot{\mathbf{e}}^{tr} + \frac{\partial \psi}{\partial \boldsymbol{q}^{pr}} \dot{\boldsymbol{q}}^{pr} \right) \\ &= \left[\boldsymbol{\sigma} : \dot{\boldsymbol{\epsilon}} - (p\,\dot{\theta} + \mathbf{s} : \dot{\mathbf{e}}) \right] + \mathbf{X} : \dot{\mathbf{e}}^{tr} + \mathbf{Q} : \dot{\boldsymbol{q}}^{pr} \\ &= \mathbf{X} : \dot{\mathbf{e}}^{tr} + \mathbf{Q} : \dot{\boldsymbol{q}}^{pr} \end{aligned} \tag{41}$$

and

$$\begin{aligned} H_{tmc} &= \rho T \left(\frac{\partial^2 \psi}{\partial T \partial \theta} \dot{\theta} + \frac{\partial^2 \psi}{\partial T \partial \mathbf{e}} \dot{\mathbf{e}} + \frac{\partial^2 \psi}{\partial T \partial \mathbf{e}^{tr}} \dot{\mathbf{e}}^{tr} + \frac{\partial^2 \psi}{\partial T \partial \boldsymbol{q}^{pr}} \dot{\boldsymbol{q}}^{pr} \right) \\ &= \left[\beta \frac{\mathbf{e}^{tr} - \boldsymbol{q}^{pr}}{\| \mathbf{e}^{tr} - \boldsymbol{q}^{pr} \|} : (\dot{\mathbf{e}}^{tr} - \dot{\boldsymbol{q}}^{pr}) - 3\alpha K \dot{\theta} \right] T. \end{aligned} \tag{42}$$

Hence, the energy equation in this case becomes

$$c\dot{T} + \nabla \cdot \boldsymbol{q} = \left[\mathbf{X} + T\beta \frac{\mathbf{e}^{tr} - \boldsymbol{q}^{pr}}{\| \mathbf{e}^{tr} - \boldsymbol{q}^{pr} \|} \right] : \dot{\mathbf{e}}^{tr} + \left[\mathbf{Q} - T\beta \frac{\mathbf{e}^{tr} - \boldsymbol{q}^{pr}}{\| \mathbf{e}^{tr} - \boldsymbol{q}^{pr} \|} \right] : \dot{\boldsymbol{q}}^{pr} - 3T\alpha K \dot{\theta}. \tag{43}$$

It is observed from Eqs. (42) and (43) that an appropriate flow rules for the evolution of the transformation strain $\mathbf{e}^{tr}$ and the permanent inelastic strain $\boldsymbol{q}^{pr}$ have to be determined. To this end, the associative flow rules are defined by

$$\dot{\mathbf{e}}^{tr} = \dot{\xi} \frac{\partial F}{\partial \mathbf{X}} = \dot{\xi} \frac{\mathbf{X}}{\| \mathbf{X} \|}, \tag{44}$$

and

$$\dot{\boldsymbol{q}}^{pr} - \dot{\xi} \frac{\partial F}{\partial \mathbf{Q}} - \kappa \dot{\xi} \frac{\mathbf{Q}}{\| \mathbf{Q} \|}, \tag{45}$$

where κ is a material parameter defining the amount of permanent inelastic strain and F is a limit function defined as

$$F(\mathbf{X}, \mathbf{Q}) = \| \mathbf{X} \| + \kappa \| \mathbf{Q} \| - R. \tag{46}$$

In Eq. (46), R is a material property which denotes the effective stress that induces the initial transformation, and has a similar meaning as the initial yield stress in plasticity. In addition, the classical Kuhn-Tucker conditions, Eq. (32), are employed to the model.

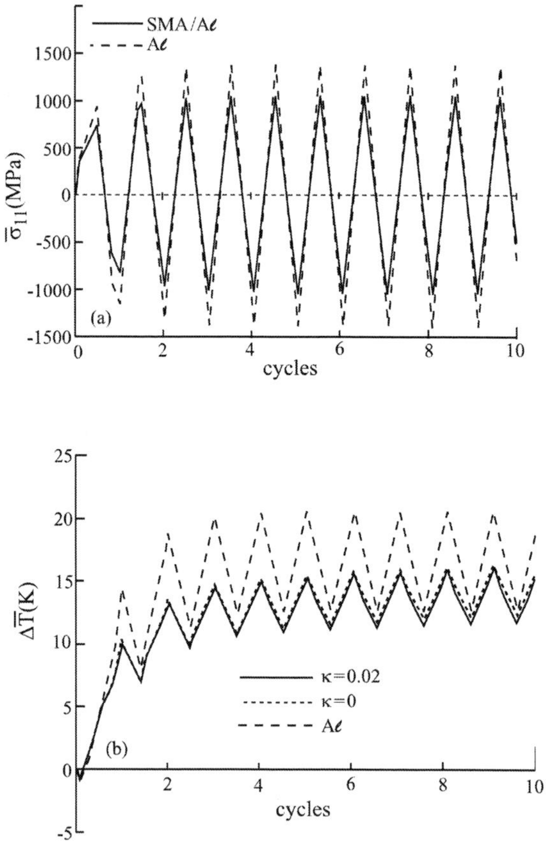

Figure 5. (a) The average stress responses of the SMA/aluminum composite and the unreinforced aluminum during 10 uniaxial loading-unloading cycles. (b) The resulting average temperature changes.

3.2.1 Applications

Here too, we consider unidirectional metal and polymeric matrix composites reinforced by SMA fibers that exhibit transformation induced plasticity. The aluminum and the epoxy properties of the metallic and polymeric matrices composites are given in Tables 2 and 3, respectively. The properties of the SMA fibers are described in Table 4. The response of the monolithic SMA material is exhibited in Figs. 2-6 of Ref. [38].

Table 4. Material properties of the SMA fibers (Auricchio et al. [54]).

Property	Value
E	50 GPa
ν	0.35
K^{th}	18 W/(mK)
α	1×10^{-6}/K
ρc	5.44 MJ/(m^{3}K)
h	1000 MPa
R	50 MPa
β	2 MPa/K
ϵ_L	0.04
M_f	223K
δ	10^{-8}

E, ν denote the Young's modulus and Poisson's ratio. The other parameters were already defined in the text.

In Fig. 5(a), the evolution of the stresses of the SMA/aluminum composite during 12 loading-unloading cycles are compared to that of unreinforced aluminum matrix. It should be noted that with the considered parameters of the shape memory alloy fibers, a degradation of the stresses is expected. However, the overall behavior of the composite does not exhibit a degradation. This can be expected since the shape memory alloy fibers do not reinforce the aluminum matrix [38]. Therefore, the effective stress of this metal matrix composite is governed mostly by the aluminum matrix and not by the shape memory alloy fibers. Consequently, the degradation of the shape memory alloy fibers is negligible as compared to the strain hardening effect of the aluminum matrix. In Fig. 5(b), the induced temperature of the shape memory alloy composite during 12 loading-unloading cycles is compared to that of the unreinforced aluminum matrix. The responses of two different composite materials were examined. In the first one, it was assumed that the transformation induced plasticity of the shape memory alloy fibers is negligible ($\kappa = 0$), while in the second case, $\kappa = 0.02$ was considered. It is observed that the there is a slight difference between these cases. Specifically, the temperature change of the composite without transformation induced plasticity ($\kappa = 0$) is greater than that of $\kappa = 0.02$. Recalling Eq. (42), it is observed that once $\dot{q}^{pr}$ vanishes (as in the case of $\kappa = 0$), the heat product due to thermomechanical coupling H_{tmc} is more significant. For both cases, the generated average temperature changes reach 15 K. The induced temperature change of aluminum matrix reaches a maximum level of 20 K.

The response of the SMA/epoxy unidirectional composite due to 15 loading-unloading cycles is exhibited in Fig. 6. In all cases, tension cycles and thermomechanical coupling were considered. For simplicity, only the responses of the 1 st, 5th, 10th and 15th cycles are illustrated in Fig. 6(b). In this figure, both saturation of the permanent inelastic strain, and a degradation of the stresses are observed. In Fig. 6(c), the evolution of the stress during the 15 loading-unloading cycles is shown. In this graph, two cases of thermomechanical coupling are examined. In the first case, it was assumed that the transformation induced plasticity is negligible ($\kappa = 0$), while in the second one, $\kappa = 0.02$ was considered. It is observed that the resulting stresses are greater in the case of transformation induced

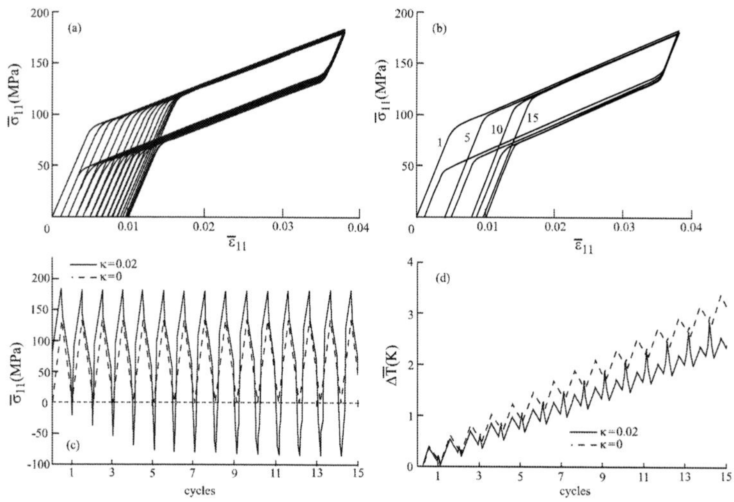

Figure 6. (a) The response of SMA/epoxy composite to 15 uniaxial loading-unloading cycles. (b) Cycles #1, 5, 10 and 15. (c) Evolution of average stresses during 15 uniaxial loading-unloading cycles. (d) The resulting average temperature changes.

plasticity, which is associated with a strain hardening. However, the generated temperature changes during this process exhibit an opposite behavior in the sense that the transformation induced plasticity reduces the resulting average temperature, see Fig. 6(d). This observation is due to the fact that once $\dot{q}^{pr}$ vanishes (as in the case of $\kappa = 0$), the heat product due to thermomechanical coupling is more significant (see Equation 42). For both cases, the induced average temperature change after 15 cycles is rather small, as compared to that of the SMA/aluminum composite material (cf. Fig. 5(b)).

3.2.2 Shape memory alloy honeycombs

The previously presented constitutive equations of shape memory alloy fibers that exhibit transformation induced plasticity was employed by Freed et al. [39], in conjunction with the HFGMC micromechanical model, to investigate the behavior of shape memory alloy honeycombs.

Hexagonal, circular, re-entrant (which generate auxetic materials) and triangles topologies of SMA honeycombs were investigated in [39]. The repeating unit cells of these topologies of honeycombs are shown in the insets to Fig. 7. The hexagonal and circular topologies were suggested by Grummon et al. [55]. In addition, two new topologies were examined in [39]; namely, re-entrant and triangles. The triangles topology is expected to exhibit a superior behavior, since it is very stiff, both for tension, compression and shear loading (Gibson and Ashby [56]). The re-entrant is a special structure which results in a negative effective in-plane Poisson's ratio (Gibson et al. [57], Almgren [58], Masters and Evans [59]).

The behavior of the different topologies is examined by comparing the macroscopic stress-strain responses and the corresponding transformation surfaces. In all cases, the material parameters are given by Table 4, and the constant $\kappa = 0.02$ in Eq. (45) was assumed.

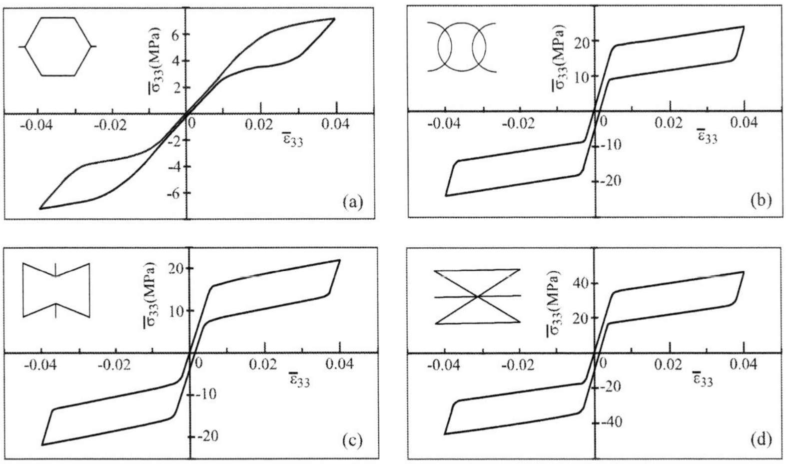

Figure 7. Pseudoelasticity ($T = 298$ K): macroscopic stress strain behavior of the (a) hexagonal, (b) circles, (c) re-entrant, and (d) triangles honeycomb topologies.

In Fig. 7, the responses in the pseudoelasticity region ($T = 298$ K) are illustrated. It is observed that the triangles topology results in a much greater transformation stress (approximately 38 MPa), while the transformation stress of the circles and re-entrant topologies is 18 MPa and 16 MPa, respectively. The transformation stress of the hexagonal topology is significantly smaller, and results in approximately 6 MPa. Furthermore, the elastic region of the stress-strain curve of the hexagonal topology is far greater than that of the other topology. This implies that the effective transverse Young's modulus of this structure is lower than that of the other topologies. In all cases, a small portion of the martensite does not transform back to austenite upon unloading, as was also observed by Shaw et al. [60].

A similar behavior is observed from Fig. 8 where the macroscopic stress-strain responses of the different topologies are obtained in the region of shape memory effect ($T = 223$ K). Here, the transformation stresses of the hexagonal, circles, re-entrant and triangles topologies are approximately 1.7 MPa, 5 MPa, 5 MPa and 10 MPa, respectively. As expected, a residual strain upon a full loading-unloading cycle is observed in all cases. Most of this strain may be recovered upon raising the temperature. However, a small portion of it remains permanent due to transformation induced plasticity (since κ in Eq. (45) is not taken as zero).

In Fig. 9, the transformation surfaces of the different topologies at the pseudoelasticity domain ($T = 298$ K) are illustrated. These surfaces resemble the initial yield surfaces of elastoplastic materials, but here they describe the initiation of the phase transformation in the shape memory alloy. It should be noted that the parameter R in Eq. (46) represents the effective stress which triggers a phase transformation of a monolithic shape memory alloy. Here however, the macroscopic transformation stress is much smaller, since the volume fraction of these structures is only 30%. The transformation surfaces were calculated by assuming that the effective local stress in a subcell $(\beta\gamma)$ which consists of a shape memory

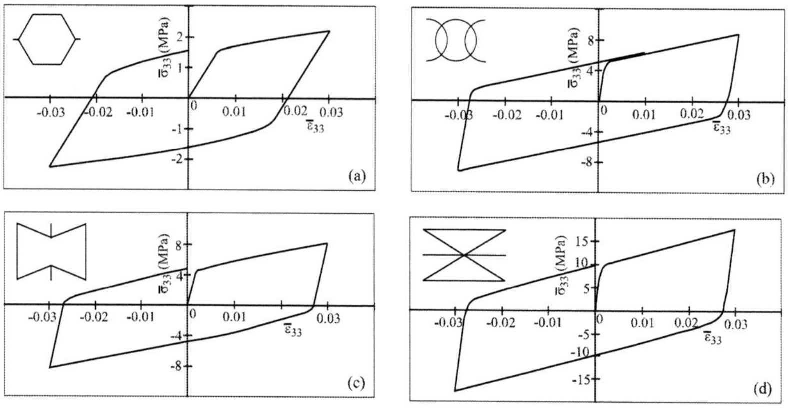

Figure 8. Shape memory effect ($T = 223$ K): macroscopic stress strain behavior of the (a) hexagonal, (b) circles, (c) re-entrant and (d) triangles honeycomb topologies.

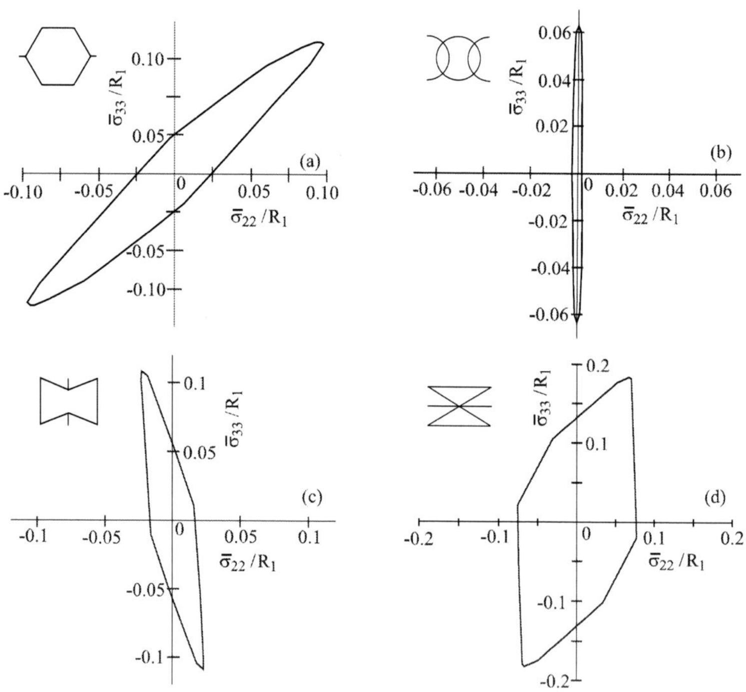

Figure 9. Transformation surfaces ($T = 298$ K) of the (a) hexagonal, (b) circles, (c) re-entrant and (d) triangles honeycomb topologies.

alloy reaches the transformation limit R such that

$$\| \mathbf{X}^{(\beta\gamma)} \| + \kappa \| \mathbf{Q}^{(\beta\gamma)} \| \equiv \| \mathbf{X}^{(\beta\gamma)} \| = R, \tag{47}$$

since $\| \mathbf{Q}^{(\beta\gamma)} \| = 0$ at the onset of the transformation. Observing Equation (40), the transformation stress $\mathbf{X}^{(\beta\gamma)}$ reduces to the deviatoric stress $\mathbf{s}^{(\beta\gamma)}$ at the onset of the transformation (since $\mathbf{e}^{tr} = \mathbf{q} = \mathbf{0}$ for subcell $(\beta\gamma)$). Consequently, Eq. (47) becomes

$$\left(\frac{3}{2} \mathbf{s}^{(\beta\gamma)} : \mathbf{s}^{(\beta\gamma)} \right)^{\frac{1}{2}} = \beta < T - M_f > + R \equiv R_1. \tag{48}$$

In the absence of thermomechanical coupling, Eq. (16) is reduced to

$$\bar{\boldsymbol{\epsilon}}^{(\beta\gamma)} = \mathbf{A}^{M(\beta\gamma)} \bar{\boldsymbol{\epsilon}}. \tag{49}$$

Similarly, it may be shown that

$$\bar{\boldsymbol{\sigma}}^{(\beta\gamma)} = \mathbf{B}^{(\beta\gamma)} \bar{\boldsymbol{\sigma}} \tag{50}$$

where the stress concentration tensor $\mathbf{B}^{(\beta\gamma)}$ is given by

$$\mathbf{B}^{(\beta\gamma)} = \mathbf{C}^{(\beta\gamma)} \mathbf{A}^{M(\beta\gamma)} [\mathbf{C}^*]^{-1}. \tag{51}$$

Consequently, for a given external loading $\bar{\boldsymbol{\sigma}}$ one can readily obtain the local stress tensor $\bar{\boldsymbol{\sigma}}^{(\beta\gamma)}$ from Eq. (50) which can be employed in Eq. (48) for the determination of the transformation surfaces. For all cases, the obtained transformation surfaces were found to be closed, convex, and anisotropic. Specifically, the transformation surface of the hexagonal topology, Fig. 9(a), is similar to that obtained by Wang and McDowell [61] for periodic metal honeycombs. The circles topology, Fig. 9(b), results in a preferred direction for the transformation surfaces. This implies a significant difference between the effective Young's moduli of the transverse x_2 and x_3-directions. In addition, vertices are observed at the intersections of the yield conditions for the re-entrant and triangles topologies. These vertices represents polyslip of crystals (see Wang and McDowell [61]). Finally, all transformation surfaces in Figs. 9(a)-(d) are asymmetric with respect to the $\bar{\sigma}_{22}/R$ and $\bar{\sigma}_{33}/R$ axes. As discussed by Wang and McDowell [61], this implies that these structures are expected to collapse in bending or stretching. It should be noted that a similar form of transformation surfaces are observed in the case of shape memory effect where the temperature is $T = 223$ K. This results from the fact that although the first term on the right-hand-side of Eq. (48) vanishes, the normalized curves in Fig. 9 remain unchanged. Further results and discussions can be found in [39].

3.3 Constitutive equations of monolithic shape memory alloys exhibiting the two-way shape memory effect

In this subsection, the three-dimensional constitutive model of shape memory alloys proposed by Helm and Haupt [62] which is capable to predict the two-way shape memory effect is adopted. A general inelastic framework for the derivation of three-dimensional thermomechanical constitutive equations for materials undergoing phase transformations is presented by this model. Presently, this model is briefly described with some minor changes. For a more complete description of the model, the reader is referred to Helm and Haupt [62].

As suggested in [62], the free-energy function ψ is decomposed into elastic ψ_e and inelastic ψ_s parts. The elastic part of the free-energy density function is given by

$$\begin{aligned} \psi_e &= \frac{\mu}{\rho}\boldsymbol{\epsilon}_e^D : \boldsymbol{\epsilon}_e^D + \frac{1}{2}\frac{\kappa}{\rho}\left[\boldsymbol{\epsilon}_e : \mathbf{I}\right]^2 - 3\frac{\kappa\alpha}{\rho}\left(T - T_0\right)\boldsymbol{\epsilon}_e : \mathbf{I} + \int_{T_0}^{T} c_{d_0}(\hat{\theta})d\hat{\theta} \\ &+ u_0^A + z\Delta u_0 - T\left[\int_{T_0}^{T}\frac{c_{d_0}(\hat{\theta})}{\hat{\theta}}d\hat{\theta} + \eta_0^A + z\Delta\eta_0\right] \end{aligned} \quad (52)$$

where a superscript D represents the deviatoric part of the tensor, μ and κ are the shear and bulk muduli of the alloy (same material properties for the austenite and martensite phases are assumed), ρ is the material mass density, α is the coefficient of thermal expansion, T and T_0 denote the current and initial temperature, c_{d_0} represents a part of the specific heat capacity, z is the martensite fraction and $\Delta u_0 = u_0^M - u_0^A$ and $\Delta\eta_0 = \eta_0^M - u_0^A$ are the initial values of the internal energy u_0 and entropy η_0 for the austenite and martensite phases (where superscripts A and M denote austenite and martensite phases, respectively). In Eq. (52), $\mathbf{I}$ is the 3×3 identity matrix and $\boldsymbol{\epsilon}_e$ is the elastic part of the total strain tensor $\boldsymbol{\epsilon}$ where the decomposition

$$\boldsymbol{\epsilon} = \boldsymbol{\epsilon}_e + \boldsymbol{\epsilon}_d \quad (53)$$

is adopted where $\boldsymbol{\epsilon}_d$ represents the inelastic strain tensor. An additional decomposition of the inelastic strain tensor is employed

$$\boldsymbol{\epsilon}_d = \mathbf{Y}_{e1} + \mathbf{Y}_{d1} \quad (54)$$

where $\mathbf{Y}_{e1}$ is used to represent the energy storage due to internal stress fields and $\mathbf{Y}_{d1}$ is introduced to model the dissipation effects during the evolution of internal stresses.

The inelastic part of the free energy function is given by

$$\psi_s = \frac{1}{2}\frac{c_1}{\rho}\mathbf{Y}_{e1} : \mathbf{Y}_{e1} \quad (55)$$

where c_1 is a material property.

Following classical arguments, the stress and internal stress tensors are derived from the free-energy components as follows

$$\begin{aligned} \boldsymbol{\sigma} &= \rho\frac{\partial\psi_e}{\partial\boldsymbol{\epsilon}_e} = 2\mu\boldsymbol{\epsilon}_e^D + \kappa\left(\boldsymbol{\epsilon}_e : \mathbf{I}\right)\mathbf{I} - 3\kappa\alpha\left(T - T_0\right)\mathbf{I} \\ \mathbf{X}_\epsilon &= \rho\frac{\partial\psi_s}{\partial\mathbf{Y}_{e1}} = c_1\left(\boldsymbol{\epsilon}_d - \mathbf{Y}_{d1}\right) \end{aligned} \quad (56)$$

where $\boldsymbol{\sigma}$ is the stress tensor, $\mathbf{X}_\epsilon$ is the internal stress field.

Let the internal variable $\mathbf{X} = \mathbf{X}_\epsilon + \mathbf{X}_T$ represents the overall residual stresses where

$$\mathbf{X}_T = \sqrt{\frac{2}{3}}\frac{\rho}{\gamma_d}\left\langle\frac{\partial\psi_e}{\partial z}\right\rangle\frac{\boldsymbol{\epsilon}_d}{\parallel\boldsymbol{\epsilon}_d\parallel} \quad (57)$$

with $< \bullet >$ as the McAuley brackets and γ_d is a material property. In Eq. (57), $\mathbf{X}_T$ represents the residual stresses which are associated with temperature induced phase transformation. The evolution of the inelastic strain tensor ϵ_d is governed by the flow rule

$$\dot{\epsilon}_d = \lambda_d \frac{\partial f}{\partial \boldsymbol{\sigma}} \tag{58}$$

where f is a yield function which is given by

$$f = \| \boldsymbol{\sigma}^D - \mathbf{X} \| - \sqrt{\frac{2}{3}} k \tag{59}$$

with k as an effective yield value. By combining Eqs. (59) and (58), it may be shown that

$$\frac{\partial f}{\partial \boldsymbol{\sigma}} = \frac{\boldsymbol{\sigma}^D - \mathbf{X}}{\| \boldsymbol{\sigma}^D - \mathbf{X} \|} = \mathbf{N}. \tag{60}$$

The inelastic multiplier λ_d may be explicitly determined as

$$\lambda_d = \begin{cases} \frac{1}{\eta_d}\left(\frac{f}{r_d}\right)^m & \begin{array}{lll} \mathrm{A}\rightarrow\mathrm{M} & \text{for} & \Delta\psi > 0,\ f > 0,\ z_{SIM} < 1, \\ & & \tau_{eff} > \| \mathbf{X}_T \|, \text{ and } \epsilon_d : \mathbf{N} \geq 0 \\ \mathrm{M}\rightarrow\mathrm{A} & \text{for} & \Delta\psi > 0,\ f > 0,\ z_{SIM} > 0, \\ & & \text{and } \epsilon_d : \mathbf{N} < 0 \\ \tilde{M}\rightarrow\mathrm{M} & \text{for} & f > 0,\ z_{SIM} < 1,\ z_{TIM} > 0, \\ & & \text{and } \epsilon_d : \mathbf{N} \geq 0 \\ \mathrm{M}\rightarrow\tilde{M} & \text{for} & f > 0,\ z_{SIM} > 0,\ z_{TIM} < 1, \\ & & \Delta\psi < 0 \text{ and } \epsilon_d : \mathbf{N} < 0 \end{array} \\ 0 & \text{elsewhere} \end{cases} \tag{61}$$

where η_d and r_d are material parameters, $\tau_{eff} = \| \boldsymbol{\sigma}^D - \mathbf{X} \|$, $\Delta\psi = \Delta u_0 - T\Delta\eta_0$, and z_{SIM} and z_{TIM} denote the stress and temperature induced martensite fractions, respectively, so that the overall martensite fraction is given by $0 \leq z = z_{SIM} + z_{TIM} \leq 1$. In Eq. (61), A and M represent the austenite and detwinned martensite phases, while $\tilde{M}$ represents the twinned martensite phase.

Next, the evolution terms of the stress induced and temperature induced martensite fractions are given by

$$\dot{z}_{SIM} = \sqrt{\frac{2}{3}} \frac{1}{\gamma_d} \frac{\epsilon_d : \dot{\epsilon}_d}{\| \epsilon_d \|} \tag{62}$$

and

$$
\dot{z}_{TIM} = \begin{cases}
-\dot{z}_{SIM} & \tilde{M} \rightarrow \text{M} \quad \text{for} \quad f > 0,\ z_{TIM} > 0,\ z_{SIM} < 1 \text{ and } \epsilon_d : \mathbf{N} \geq 0 \\
-\dot{z}_{SIM} & \text{M} \rightarrow \tilde{M} \quad \text{for} \quad \Delta\psi \leq 0,\ f > 0,\ z_{SIM} > 0,\ z_{TIM} < 1 \text{ and } \epsilon_d : \mathbf{N} < 0 \\
-\dfrac{|\dot{T}|}{M_s - M_f} \dfrac{\Delta\psi}{|\Delta\psi|} & \text{A} \rightarrow \tilde{M} \quad \text{for} \quad \dot{z}_{SIM} = 0,\ \dot{T} < 0,\ T \leq M_s,\ f < 0 \text{ and } z < 1 \\
-\dfrac{|\dot{T}|}{A_f - A_s} \dfrac{\Delta\psi}{|\Delta\psi|} & \tilde{M} \rightarrow \text{A} \quad \text{for} \quad \dot{z}_{SIM} = 0,\ \dot{T} > 0,\ T \geq A_s,\ \text{and } z > 0 \\
0 & \text{elsewhere}
\end{cases} \tag{63}
$$

where A_s, A_f, M_s and M_f are the austenite and martensite start and finish temperatures.

Finally, the evolution of the internal variable $\mathbf{Y}_{d1}$ is given by

$$
\dot{\mathbf{Y}}_{d1} = \xi \mathbf{X}_\epsilon \tag{64}
$$

with

$$
\xi = \sqrt{\frac{2}{3}} \left\{ \langle \dot{z}_{SIM} \rangle^0 \exp\left[-\beta_1 z_{SIM}^{\gamma_1} \right] + \langle -\dot{z}_{SIM} \rangle^0 \right\} \frac{b_1}{c_1} \parallel \dot{\epsilon}_d \parallel . \tag{65}
$$

It should be noted that Helm and Haupt [62] established the corresponding energy equation to describe the coupling between the current temperature and the generated inelastic strains in the shape memory alloy. This is similar to the studies of Aboudi and Freed [37] and Freed and Aboudi [38], in which the full TMC of shape memory alloys and shape memory alloy composites was extensively investigated. Here however, we focus on the two-way shape memory effect; hence, the effect of thermomechanical coupling is neglected.

3.3.1 Applications

Herein, some results for the behavior of a unidirectional SMA/epoxy composite are given. The SMA fibers and the polymeric epoxy matrix are characterized by Table 5 and 3, respectively. The behavior of the monolithic SMA materials is shown in Figs. 1 and 2 of Ref. [40]. In Figs. 10(a) and (b), the one-way shape memory effect is demonstrated for the SMA/epoxy composite material ($v_f = 0.3$) for loading in the axial and transverse directions, respectively. In both cases, the temperature of the composite material is initially set to $T = 200$ K (which corresponds to the shape memory effect). Thus, the shape memory alloy fiber is at a state of twinned martensite. Upon a subsequent applied external loading of the composite at the constant temperature of 200 K, a phase transformation between twinned and detwinned martensite is induced in the fiber. If the applied load is released to zero, macroscopic residual strains are observed in the composite material. As may be verified, the induced residual strains upon loading in the transverse directions are obviously greater than those generated by the axial loading. These strains may be fully recovered by increasing the temperature. For the case of axial applied loading (Fig. 10(a)), the reverse

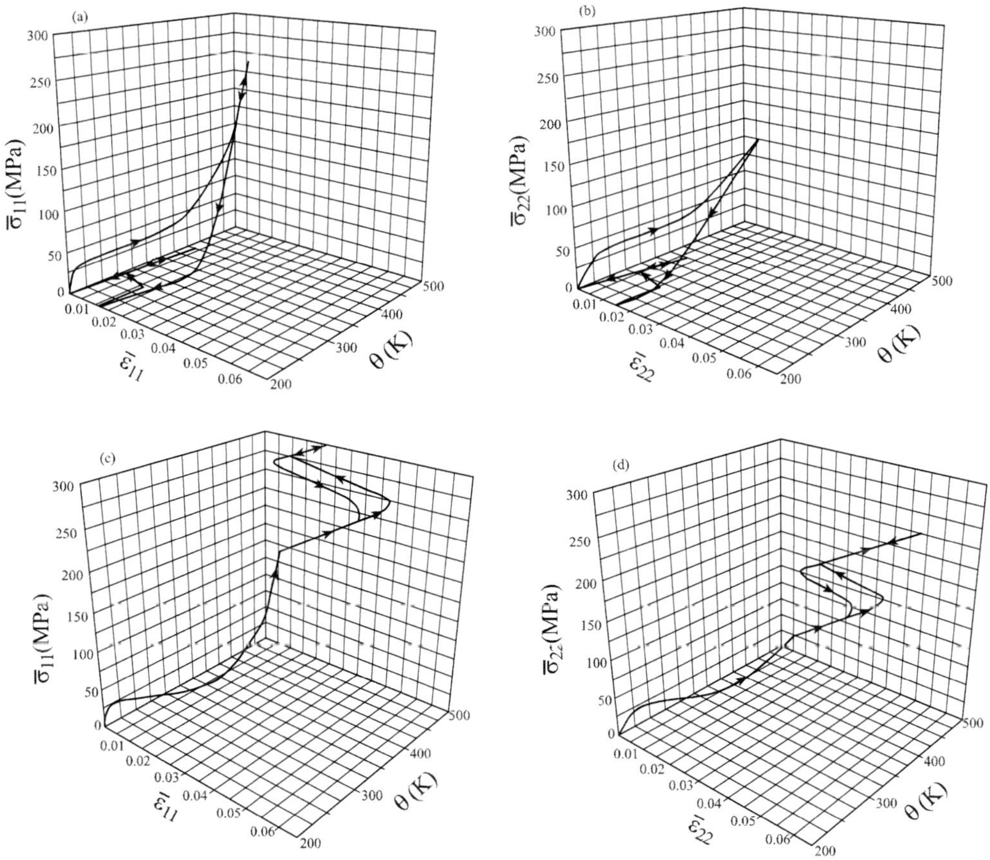

Figure 10. The SMA/epoxy composite material: (a) the one-way shape memory effect upon axial loading. (b) The one-way shape memory effect upon transverse loading. (c) The two-way shape memory effect upon axial loading. (d) The two-way shape memory effect upon transverse loading.

transformation to the austenite phase begins at $T = 268$ K and finishes at $T = 284$ K; recall that for monolithic shape memory alloys the reverse phase transformation occurs in the range 268 K$< T <$ 291 K. Hence, the epoxy constituent changes the transformation temperatures. In the case of the transverse loading (Fig. 10(b)), the reverse transformation occurs at a narrower range of 266 K$< T <$ 282 K. In other words, recalling that a larger amount of residual strains are being recovered, the evolution of the one-way shape memory effect is more rapid in the case of transverse loading. It should be noted that in both cases, the shape of the composite is not affected by an additional temperature change.

Table 5. Material properties of the shape memory alloy (Helm and Haupt [62]).

Property	Value
κ	43 GPa
μ	19.8 GPa
ρ	6400 kg/m^3
α	1.06×10^{-5}/K
η_d	2×10^7 s
m	3
r_d	1 MPa
c_1	14 GPa
b_1	700
β_1	10
γ_1	10
Δu_0	-12375 J/kg
$\Delta\eta_0$	-46.875 J/kg K
k	50 MPa
γ_d	0.05
M_s	320K
M_f	300K
A_s	350K
A_f	370K

The two-way shape memory effect of the SMA/epoxy composite is shown in Figs. 10(c) and (d) for loading in the axial and transverse directions, respectively. Here too, the temperature of the composite is initially set at a temperature of $T = 200$ K to induce a twinned martensite phase in the shape memory alloy fiber. Then, an externally applied loading is imposed on the composite as a result of which a stress induced transformation occurs to form a detwinned martensite in the fiber. Upon a subsequent temperature increase, the detwinned martensite phase in the fiber is transformed to austenite. In the case of axial loading (Fig. 10(c)), this process begins at $T = 339$ K, and the material is fully recovered at $T = 435$ K. Next, the temperature is decreased and a phase transformation between the austenite and the detwinned martensite phases takes place in the fiber. This transformation is initiated at $T = 407$ K and finishes at $T = 307$ K. In the case of transverse loading (Fig. 10(d)), the reverse transformation occurs in the range of 306 K$< T <$ 347 K while the forward transformation begins at $T = 318$ K and finishes at $T = 260$ K. In both axial and transverse loadings of the composite, two stable shapes can be readily observed in the shape memory alloy composite. The first shape is stable at low temperatures ($T < 307$ K and $T < 260$ K for the axial and the transverse loading, respectively) while the second shape is stable at high temperatures ($T > 435$ K and $T > 347$ K for the axial and the transverse loading, respectively). The transition between these shapes is carried out by applying the appropriate temperature change. It can be readily concluded that the embedded shape memory alloy fibers have a significant influence on the macroscopic response of the composite where both one-way and two-way shape memory effects can be detected and utilized.

The effect of the shape memory alloy fiber volume fraction of the SMA/epoxy composite on the two-way shape memory temperatures for applied axial and transverse loading is depicted in Figs. 11(a) and (b), respectively. To this end, fiber volume fractions of $v_f = 0.1, 0.3, 0.7$ together with the monolithic shape memory alloy ($v_f = 1$) are shown. Obviously, the axial and transverse responses of the latter are identical, which results from the isotropic behavior of the monolithic shape memory alloy; hence, its response is drawn only in Fig. 11(a). Let us define *stability temperatures* as the two temperatures which are associated with the stable shapes of the shape memory alloy composite. Between these two temperatures, phase transformation takes place in a hysteretic loop. Above and below these extreme temperatures, the high and low temperature shapes can be identified from the response that is exhibited by the monolithic shape memory alloy, the stability temperatures are 300 K and 366 K). Consequently, for both axial and transverse loading, it may be easily observed from Fig. 11 that the fiber volume fraction strongly affects the stability temperatures of the two-way shape memory process. To this end, two extreme cases can be identified: for $v_f = 1$, the range of stability temperatures is the smallest, while for $v_f = 0$ (i.e., pure epoxy), the two-way shape memory effect does not exist. For axial loading, the stability temperatures were found to be: $T = 317$ K and 494 K for $v_f = 0.1$, $T = 310$ K and 431 K for $v_f = 0.3$, $T = 303$ K and 376 K for $v_f = 0.7$, and $T = 300$ K and 366 K for the monolithic shape memory alloy ($v_f = 1$). In other words, the temperature change which is needed for transition from the low temperature shape to the high temperature shape decreases with the increase of the fiber volume fraction. In the case of transverse loading, similar behavior is observed. Specifically, the stability temperatures were found to be: $T = 258$ K and 356 K for $v_f = 0.1$, $T = 258$ K and 344 K for $v_f = 0.3$, and $T = 263$ K and 335 K for $v_f = 0.7$. In addition, it appears that the stable shapes are sensitive to the temperature change as the fiber volume fraction decreases (note that the strains outside the stability temperatures are not constant). This reflects the variations of the effective thermal strains of the composite material with the temperature change (it should be noted that the coefficient of thermal expansion of the epoxy matrix is much greater than that of the shape memory alloy fiber, resulting in a larger amount of macroscopic thermal strain with the decrease of fiber volume fraction). Further results are given by Freed and Aboudi [40] where the effect of the two-way shape memory effect on the metallic matrix in a metal matrix composite is studied.

4. SMA/Concrete Composite With Evolving Damage in the Concrete Phase

In this section, shape memory alloy fibers are embedded within a concrete matrix. In order to increase the tensile strength of the resulting reinforced concrete the SMA fibers are activated. To this end, the constitutive equations of the monolithic SMA fibers are described, followed by a brief derivation of the constitutive behavior of the unreinforced concrete with evolving damage effects. The macroscopic constitutive equations of the unidirectional SMA/concrete are determined by employing the HFGMC model. The activation procedure

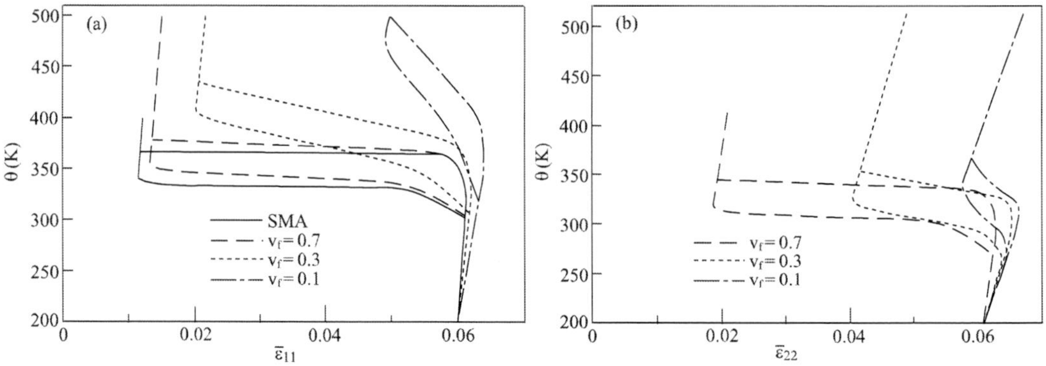

Figure 11. The effect of the fiber volume fraction on the two-way shape memory behavior of the SMA/epoxy composite material upon (a) axial and (b) transverse loading.

of the embedded SMA fibers is discussed.

4.1 Constitutive modeling of monolithic shape memory alloy

The modeling of the prestressed concrete with shape memory alloys fibers begins with choosing an appropriate constitutive model for the SMA fibers. To this end, the model proposed by Panoskaltsis et al. [63] is considered. This model performs a general inelastic framework for the derivation of general three-dimensional thermomechanical constitutive equations for materials undergoing phase transformations. Its framework is based on the generalized plasticity theory and on some basic elements from the theory of continuum damage mechanics. In this subsection, this model is briefly described, although it is presented in slightly different notations. At the end of this subsection, the numerical integration scheme of the constitutive equations of the model is derived and explained in details.

As in most of SMA constitutive equations, this model considers only one internal variable. The internal scalar variable is denoted here by ξ, and it describes the martensite fraction. The general form of the transformation surface is taken as a von-Mises type surface, and given by

$$F(\boldsymbol{\sigma}, T) = \sqrt{3J_2} - CT, \tag{66}$$

where $\boldsymbol{\sigma}$ is the Cauchy stress tensor, $J_2 = \boldsymbol{s} : \boldsymbol{s}/2$ is the second invariant of the deviatoric stress tensor $\boldsymbol{s}$, C is a material property, T is the current temperature and the loading surface $F(\boldsymbol{\sigma}, T)$ is constant. The austenite to martensite transformation surface is given by

$$F_M(\boldsymbol{\sigma}, T) = \sqrt{3J_2} - C_M T, \tag{67}$$

where C_M is the slope of the one-dimensional stress-temperature diagram for martensite behavior. Moreover, introducing F_{M_f} and F_{M_s} as the loading surfaces at the beginning and the end of the transformation yields

$$\begin{aligned} F_{M_s}(\boldsymbol{\sigma}, T) &= \sqrt{3J_2} - C_M(T - M_s), \\ F_{M_f}(\boldsymbol{\sigma}, T) &= \sqrt{3J_2} - C_M(T - M_f), \end{aligned} \tag{68}$$

where M_s and M_f represents the martensite start and finish temperatures, respectively.

It may be easily shown that the product $F_{M_s}F_{M_f}$ is negative only during the austenite to martensite phase transformation. Hence, the expression for evolution of the martensite fraction during the forward transformation is given by

$$\dot{\xi} = -\frac{\langle -F_{M_s}F_{M_f}\rangle}{|F_{M_s}F_{M_f}|}\frac{1-\xi}{F_{M_f}}\langle \dot{F}_M\rangle, \tag{69}$$

where the dot above a quantity represents its increment and $\langle\cdot\rangle$ is the McAuley bracket, defined as $\langle x\rangle = (|x|+x)/2$. Evolution of the transformation strain tensor (the flow rule) is given as

$$\dot{\boldsymbol{\epsilon}}^{tr} = \epsilon_L\dot{\xi}\boldsymbol{n}_{A\to M} \tag{70}$$

where ϵ_L is a material constant, which represents the maximum inelastic strain, and $\boldsymbol{n}_{A\to M}$ is the normal to the martensitic transformation surface, and defined by

$$\boldsymbol{n}_{A\to M} = \sqrt{\frac{3}{2}}\left(\frac{\boldsymbol{s}}{\bar{\sigma}}\right), \tag{71}$$

where the effective stress is given by

$$\bar{\sigma} = \sqrt{\frac{3}{2}\boldsymbol{s}:\boldsymbol{s}}\ . \tag{72}$$

Similarly, for the martensite to austenite transformation, the loading surfaces are defined as

$$\begin{aligned} F_A(\boldsymbol{\sigma},T) &= \sqrt{3J_2} - C_A T, \\ F_{A_s}(\boldsymbol{\sigma},T) &= \sqrt{3J_2} - C_A(T-A_s), \\ F_{A_f}(\boldsymbol{\sigma},T) &= \sqrt{3J_2} - C_A(T-A_f), \end{aligned} \tag{73}$$

where C_A denotes the slope of the one-dimensional stress-temperature diagram for austenite behavior. Similar to the forward transformation, the product $F_{A_s}F_{A_f}$ is negative only for an evolving backward transformation. Evolution of the martensite fraction during the backward transformation is given by

$$\dot{\xi} = -\frac{\langle -F_{A_s}F_{A_f}\rangle}{|F_{A_s}F_{A_f}|}\frac{\xi}{F_{A_f}}\langle \dot{F}_A\rangle, \tag{74}$$

the flow rule of the backward transformation is

$$\dot{\boldsymbol{\epsilon}}^{tr} = \epsilon_L\dot{\xi}\boldsymbol{n}_{M\to A} \tag{75}$$

where

$$\boldsymbol{n}_{M\to A} = \sqrt{\frac{3}{2}}\left(\frac{\boldsymbol{\epsilon}^{tr}}{\bar{\epsilon}^{tr}}\right), \tag{76}$$

and the effective strain is given by

$$\bar{\epsilon}^{tr} = \sqrt{\frac{3}{2}\,\epsilon^{tr} : \epsilon^{tr}} \quad . \tag{77}$$

Finally, The constitutive law of the SMA material is given by

$$\boldsymbol{\sigma} = \mathbf{C}(\xi) : (\epsilon - \epsilon^{tr}), \tag{78}$$

where **C** is the standard isotropic stiffness tensor but its components depend upon the internal variable ξ in the form

$$E = E_A + \xi^m \left(E_M - E_A\right) \tag{79}$$

and

$$\nu = \nu_A + \xi^m \left(\nu_M - \nu_A\right), \tag{80}$$

where E_A, E_M, ν_A and ν_M are the Young's moduli and Poisson ratios of austenite and martensite phases, respectively, and m is a positive material parameter, which controls the slope of the stress-strain curve.

The stress-strain behavior of the material may be separated into several different branches. It is clear that if $\dot{F}_A < 0$, backward transformation occurs, while forward transformation occurs if $\dot{F}_M > 0$. Since only one internal variable is considered, $\dot{F}_A = \dot{F}_M$; hence, only one phase transformation is active at any given time. Consequently, the forward and backward transformations may be decoupled into two separate criteria simply by defining

$$h_M = \frac{\langle -F_{M_s} F_{M_f}\rangle}{|F_{M_s} F_{M_f}|}, \qquad h_A = \frac{\langle -F_{A_s} F_{A_f}\rangle}{|F_{A_s} F_{A_f}|}. \tag{81}$$

For a forward transformation,

$$\begin{array}{ll} \text{if, } h_M = 0 \text{ then} & \text{elastic state} \\ \text{if } h_M = 1, \text{ then} & \\ \quad \dot{F}_M < 0 & \text{elastic unloading} \\ \quad \dot{F}_M = 0 & \text{neutral loading} \\ \quad \dot{F}_M > 0 & \text{inelastic loading}. \end{array} \tag{82}$$

For a backward transformation,

$$\begin{array}{ll} \text{if } h_A = 0, \text{ then} & \text{elastic state} \\ \text{if } h_A = 1, \text{ then} & \\ \quad \dot{F}_A > 0 & \text{elastic unloading} \\ \quad \dot{F}_A = 0 & \text{neutral loading} \\ \quad \dot{F}_A < 0 & \text{inelastic loading}. \end{array} \tag{83}$$

From a computational point of view, the non-linear behavior of the model may be treated as an implicit time discrete strain driven problem. Accordingly, the time discrete equations of the model are integrated over a time interval $(t, t+\Delta t)$ using an implicit backward Euler scheme, which is unconditionally stable. In the following, the details of the computational procedure for the determination of the SMA response are presented.

During elastic deformation

$$\begin{aligned} \xi^{t+\Delta t} &= \xi^{t}, \\ \epsilon^{tr,t+\Delta t} &= \epsilon^{tr,t}, \end{aligned} \tag{84}$$

and the new stress state is determined using eq. (78). For both cases of inelastic loading, the discrete equations may be derived as

$$\begin{aligned} \xi^{t+\Delta t} &= \xi^{t} + \Delta\xi \\ \Delta\xi &= \frac{R_{\xi}^{t+\Delta t}}{f_f^{t+\Delta t}} \Delta F, \\ \Delta F &= \frac{3}{2} \frac{J_2^{t+\Delta t} - J_2^{t}}{\bar{\sigma}^{t+\Delta t}} - c(T^{t+\Delta t} - T^{t}), \\ J_2^{t+\Delta t} &= \frac{1}{2} \boldsymbol{s}^{t+\Delta t} : \boldsymbol{s}^{t+\Delta t}, \\ \boldsymbol{s}^{t+\Delta t} &= \frac{E^{t+\Delta t}}{1+\nu^{t+\Delta t}} \left(\boldsymbol{\epsilon}^{t+\Delta t} - \frac{1}{3} \boldsymbol{\epsilon}^{t+\Delta t} : \boldsymbol{I} - \boldsymbol{\epsilon}^{tr,t+\Delta t} \right), \\ \boldsymbol{\epsilon}^{tr,t+\Delta t} &= \boldsymbol{\epsilon}^{tr,t} + \Delta\xi \epsilon_L \boldsymbol{n}^{t+\Delta t}, \end{aligned} \tag{85}$$

where $\boldsymbol{I}$ is the identity matrix, $\boldsymbol{n}$ is given in Eqs. (76) and (71) for backward and forward transformations, respectively,

$$R_{\xi} = 1 - \xi, \qquad f_f = -F_{M_f}, \qquad c = C_M, \tag{86}$$

for the case of a forward transformation, and

$$R_{\xi} = \xi, \qquad f_f = F_{A_f}, \qquad c = C_A, \tag{87}$$

for a backward transformation. The increment of the martensite fraction $\Delta\xi$ is obtained by solving the nonlinear equation

$$\Delta\xi - \frac{3}{2} \frac{R_{\xi}^{t+\Delta t}}{f_f^{t+\Delta t}} \frac{J_2^{t+\Delta t} - J_2^{t}}{\bar{\sigma}^{t+\Delta t}} - c(T^{t+\Delta t} - T^{t}) = 0\,, \tag{88}$$

and the new stress state may be determined using Eqs. (85) and (78).

The response of the material during an isothermal strain-driven uniaxial loading-unloading test is shown in Figs. 1(a) and 1(b) of Ref. [41] for the cases when $M_s < T < A_s$ and $A_s < T < A_f$, respectively. In these figures, the Young's moduli were taken as $E_M = 500$ MPa and $E_A = 1500$ MPa, the Poisson's ratios $\nu_A = \nu_M = 0.3$, the transformation temperatures were $M_s = 70$°C, $M_f = 10$°C, $A_s = 90$°C, $A_f = 130$°C, the slopes $C_M = C_A = 1$ MPa/°C, $\epsilon_L = 0.1$ and $m = 1$. The obtained results agreed exactly with those of Panoskaltsis et al. [63].

4.2 Constitutive modeling of plain concrete with evolving damage

In the present investigation, the constitutive model of plain concrete suggested by Tao and Phillips [64], was chosen and implemented. This model presents a three-dimensional analysis of concrete, with a feature of simplicity and avoidance of convergence problems, often observed in strain softening plasticity. This model describes the strain softening of a plain concrete in a framework of a continuum damage mechanics, with a weighted damage parameter and a damage multiplier that eliminates potential convergence problems and reduce the effect of hydrostatic pressure on damage. It may be noted that only a brief derivation of the model is given here. For further details, the reader is directed to Tao and Phillips [64].

Under isothermal conditions, the Helmholtz free energy ψ may be written as

$$\psi = \frac{1}{2}\boldsymbol{\epsilon} : \mathbf{K} : \boldsymbol{\epsilon}, \tag{89}$$

where $\mathbf{K}$ is the fourth-order stiffness tensor of the material (with $\mathbf{K}^0$ being the stiffness tensor of the undamaged concrete), and $\boldsymbol{\epsilon}$ is the strain tensor. Since experimental observations suggest that the failure of concrete is different for hydrostatic and deviatoric loadings, the above potential may be decoupled as

$$\psi = \frac{1}{2}(1-D)\,\boldsymbol{\epsilon} : \mathbf{K}^0 : \boldsymbol{\epsilon} + \frac{(1-\beta)\,D}{2}\epsilon_m^2 \mathbf{I} : \mathbf{K}^0 : \mathbf{I}, \tag{90}$$

where the stiffness degradation induced by material damage $\mathbf{K} = (1-D)\mathbf{K}^0$ is considered, $\epsilon_m = \boldsymbol{\epsilon} : \mathbf{I}/3$ is the mean strain, and β denotes for a damage multiplier that reduces the influence of the hydrostatic loading as

$$\beta = 1 - \frac{1}{1 + cYe^{-dY}} \tag{91}$$

with c and d as material constants and Y as a thermodynamic force. Finally, D is a combined tension/compression damage parameter

$$D = \frac{\sum \sigma_p^+ D_t + \sum |\sigma_p^-| D_c}{\sum |\sigma_p|}, \tag{92}$$

where σ_p^+ and σ_p^- are the positive and negative parts of the principal stress tensor, respectively, $\sum |\sigma_p|$ is the sum of the absolute values of the principal stresses, and D_t and D_c are scalars that denotes for the damage in tension and compression, respectively. It should be noted that according to Eq. (92), the damage in tension does not induce damage in compression.

The thermodynamic force Y may be separated to Y_t for tension and Y_c for compression as

$$Y_t = \frac{\partial \psi}{\partial D_t} = \frac{\sum \sigma_p^+}{2\sum |\sigma_p|}\left(\boldsymbol{\epsilon} : \mathbf{K}^0 : \boldsymbol{\epsilon} - (1-\beta)\epsilon_m^2 \mathbf{I} : \mathbf{K}^0 : \mathbf{I}\right), \tag{93}$$

$$Y_c = \frac{\partial \psi}{\partial D_c} = \frac{\sum \sigma_p^-}{2\sum |\sigma_p|}\left(\boldsymbol{\epsilon} : \mathbf{K}^0 : \boldsymbol{\epsilon} - (1-\beta)\epsilon_m^2 \mathbf{I} : \mathbf{K}^0 : \mathbf{I}\right), \tag{94}$$

and

$$\sigma \quad = \quad \frac{\partial \psi}{\partial \epsilon} = (1 - D)\mathbf{K}^0 : \epsilon + \frac{(1-\beta)D\epsilon_m}{3}(\mathbf{I} : \mathbf{K}^0 : \mathbf{I})\mathbf{I} \tag{95}$$

Since no plastic strain is taken into account in this model, the material tangent and secant moduli may be determined as

$$\mathbf{K}_t = (1 - D)\mathbf{K}^0 + \frac{(1-\beta)D}{9}(\mathbf{I} : \mathbf{K}^0 : \mathbf{I})\mathbf{I} \tag{96}$$

Finally, limit functions for tension and compression are given by

$$f_t(Y_t, D_t) \quad = \quad Y_t - Z_t = 0 \tag{97}$$

$$f_c(Y_c, D_c) \quad = \quad Y_c - Z_c = 0 \tag{98}$$

where the softening parameters Z_t and Z_c are expressed as

$$Z_t \quad = \quad Y_t^0 + \frac{1}{a_t}\left(\frac{D_t}{1 \;\; D_t}\right)^{\frac{1}{b_t}} \tag{99}$$

$$Z_c \quad = \quad Y_c^0 + \frac{1}{a_c}\left(\frac{D_c}{1 - D_c}\right)^{\frac{1}{b_c}}, \tag{100}$$

in which parameters a_t and b_t are determined from a simple uniaxial tensile test, while a_c and b_c are determined from a uniaxial compression test. The initial values of damage energy release rates, Y_t^0 and Y_c^0, are material constants, which denotes for the onset of the damage in tension and compression, respectively.

The response of concrete in tension and compression is illustrated in Figs. 2a and 2b, respectively, of Ref. [41] where the Young's modulus was taken as $E = 31.8$ GPa, with $\nu = 0.18$, $a_t = 7000$ MPa^{-1}, $b_t = 1.1$, $a_c = 29$ MPa^{-1}, $b_c = 0.94$, $Y_t^0 = 1.9 \times 10^{-4}$ MPa, $Y_c^0 = 3 \times 10^{-4}$ MPa, $c = 2$ MPa^{-1} and $d = 0.7$ MPa^{-1}. These results are in excellent agreement with those of Tao and Phillips [64].

4.3 Applications

Concrete is currently the most important material in the building industry. However, it is very weak in tension, compared to its strength in compression. To overcome this problem, prestressed concrete is usually used. Prestressed concrete is plain concrete with reinforcement of steel, polymers or, in this case, shape memory alloys. The prestressing is usually introduced by applying tension to the reinforcement in the concrete. Consequently, initial compressive stresses are applied to the concrete matrix; the application of permanent compressive stress increases the apparent tensile strength of the concrete, since upon tensile loading, the compressive stresses must first be nullified. Consider a unidirectional composite that consists of a concrete matrix reinforced by SMA fibers which are oriented in the x_1–direction (see Fig. 12(a)); this is defined as the axial direction. The present analyses are carried out for the case of a uniaxial applied stress in which the composite is subjected to an external strain in the axial direction while keeping all stress components other than

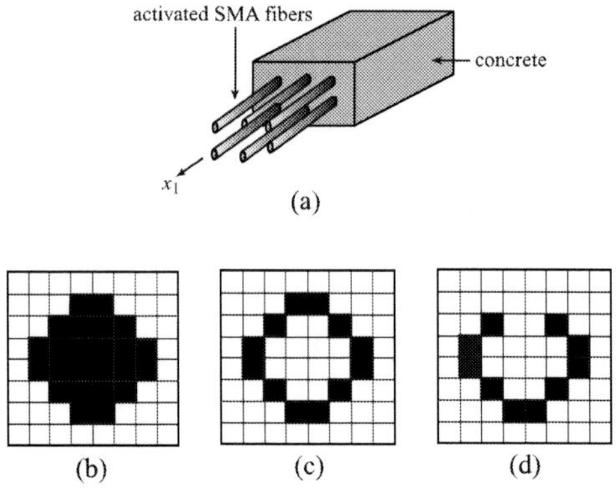

Figure 12. (a) Activated SMA fibers (oriented in the x_1-direction) embedded in a concrete matrix. The shapes of fibers are: (b) circle, (c) ring and (d) U-shaped.

$\bar{\sigma}_{11}$ equal to zero. To this end, several aspects of the thermo-micromechanical model are examined. First, the effect of the fiber shape is considered for two cases; in case (1), the coefficients of thermal expansion are set to zero, whereas for case (2), a mismatch between the coefficients of thermal expansion of the SMA fibers and the concrete matrix is assumed. Next, the relation between the response of the composite, the activation process (which is explained in the following), and the fiber volume fraction is investigated. Furthermore, the transverse response as a result of axial loading is studied in terms of the activation process and fiber volume fraction. Finally, the preferred orientation of the SMA fibers is examined. The material properties of the SMA fibers and concrete are given in Tables 6 and 7, respectively. These properties are taken from Panoskaltsis et al. [63] for the SMA fibers and Tao and Phillips [64] for the concrete,

In order to induce prestressing in the concrete, the shape memory effect is utilized. With this phenomenon, when the initial temperature of the SMA is lower than the austenite start temperature A_s (see Fig. 1(b)), residual strains are generated after a loading cycle. These strains may be recovered simply by heating to a temperature above the austenite final temperature A_f. Consequently, the activation process is modeled as follows. First, the SMA fibers are subjected to a loading cycle with a temperature below A_s. As a result, residual strains ϵ^0 (given in Table 6) are generated. These fibers are embedded in the concrete matrix. Upon subsequent heating of the composite, these strains are recovered, and compressive stresses are transmitted to the concrete matrix. By employing HFGMC, the algorithm for the prestressing process may be summarized as:

1. Fill all subcells with SMA.

2. Load and unload the SMA material to obtain the residual strains, ϵ^0, and the corresponding martensite fraction ξ.

3. Fill the subcells of the repeating unit cell according to the given composite configuration (see Figs. 12(b) through 12(d) for example).

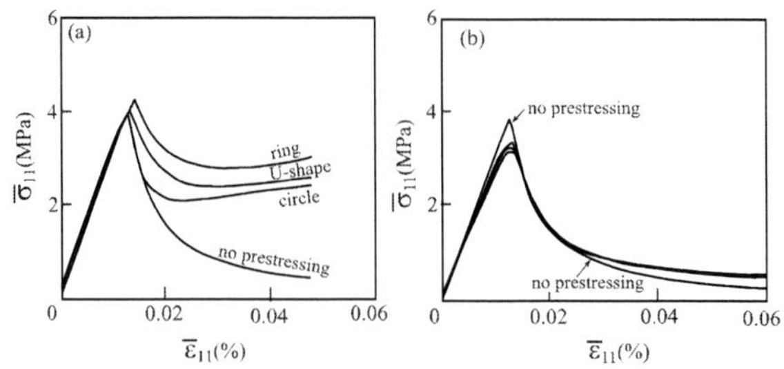

Figure 13. Effect of the shape of the SMA fibers on the behavior of the prestressed concrete: (a) Thermal expansion is not considered. (b) Thermal expansion is considered.

4. Assign zeroes for the initial strains and the internal damage variable D of the matrix subcells.

5. Set the initial global strain $\bar{\epsilon}$ to zero.

6. Increase the temperature of the body incrementally by a temperature change of ΔT. Use the relation $\bar{\epsilon}^{(\beta\gamma)} = \epsilon^0 + \mathbf{A}^{M(\beta\gamma)}\bar{\epsilon}$ (obtained from Eq. (16) for the present case of a partial TMC) which provides the strain in the subcells that are filled with SMA, where $\mathbf{A}^{M(\beta\gamma)}$ is the strain concentration tensor of the subcell which is incrementally provided by the micromechanical analysis, while ϵ^0 and the initial value of the internal variable ξ are those of step (2). It should be noted that for the micromechanical local-global relations, $\bar{\epsilon}^{(\beta\gamma)} - \epsilon^0$ is calculated.

Once the activation process is completed, compressive stresses are transmitted from the SMA fiber to the concrete matrix, while the effective strain vector $\bar{\epsilon}$ remains zero. Upon loading, the response of the prestressed concrete may be obtained.

The effect of the fiber shape on the behavior of the prestressed concrete is examined. To this end, 64 subcells in an 8×8 configuration are considered. Fibers with three different cross-sectional shapes: circular, ring, and U-shaped, as illustrated in Figs. 12(b) through 12(d), are examined. The volume fraction of the fiber is taken to be relatively small (in Moser et al. [65], for example, the fiber volume fraction was 1.2 %). Note that these figures show the shape of the fiber qualitatively; that is, the subcells in Figs. 12(b) through 12(d) do not have a uniform size. Specifically, subcells filled with SMA are significantly smaller than those of the concrete matrix.

Two cases are analyzed. In the first case, the coefficients of thermal expansion are taken to be zero for both the SMA and concrete constituents, that is, $\alpha_c = \alpha_s = 0$ where α_c and α_s are the coefficients of thermal expansion of concrete and SMA, respectively. In the second case, non-zero coefficients of thermal expansion (given in Tables 6 and 7), where $\alpha_c = 10 \times 10^{-6}$/K and $\alpha_s = 5 \times 10^{-6}$/K), are considered. With these two cases, the effect of the activation process may be examined in the absence and presence of thermal expansion. In all cases, the activation temperature change is taken as $\Delta T = 40$ K with fiber volume fraction $v_f = 1\%$. The results are illustrated in Fig. 13.

In Fig. 13(a), stress-strain curves are presented for different fiber shapes with zero thermal expansion. Hence, the temperature change ΔT contributes only to the activation of the fibers (with all thermal strains zero). From this graph, it may be observed that all fiber shapes exhibit better performance as compared to concrete which has not been prestressed containing circular fibers. In all cases, both the peak stress and the response in the damaged region show an increase in the stress level for a given strain. This increase is found to be approximately 400% to 530% for the different shapes with respect to a concrete with no prestressing. In particular, the ring shape, illustrated in Fig. 12(c), is the preferred reinforcement shape. However, it is interesting to observe that once the thermal expansion of the composite phases is incorporated, the difference between the response of the various fiber shapes is negligible, as shown in Fig. 13(b). Nevertheless, the prestressed composite still exhibits better performance in the damaged region, showing an increase of up to 75 % in the stress level. To understand this phenomenon, the reader may recall that the compressive stresses in the concrete result from the recovery of the residual strains in the SMA fibers. If the thermal expansion of the composite is considered as well, the additional thermal strains reduce the effect of the recovery of the SMA residual strains on the overall composite, since the volume fraction of the SMA fibers is relatively small. It may be shown that for higher values of v_f, the prestressed concrete exhibits an increase in the stress level when thermal expansion effects are included.

Table 6. Material properties of the SMA fibers (Panoskaltsis et al. [63]).

Property	Value
E_A	1500 MPa
E_M	500 MPa
ν_A	0.3
ν_M	0.3
m	1
α	5×10^{-6}/K
C_A	1 MPa/K
C_M	1 MPa/K
ϵ_L	0.1
A_s	363K
A_f	403K
M_s	343K
M_f	283K
ϵ^0_{11}	0.08168
ϵ^0_{22}	-0.04084
ϵ^0_{33}	-0.04084

As mentioned previously, the applied temperature has a significant effect on the behavior of the prestressed concrete. This effect is examined next for a range of activation temperature changes. From Table 6, it is observed that a temperature change of $\Delta T = 40$ K is sufficient for full recovery of the monolithic SMA studied here (since $A_f - A_s = 40$ K). However, since the fibers are embedded in the concrete matrix, there are obviously internal forces between the two phases which change the transformation temperatures (see

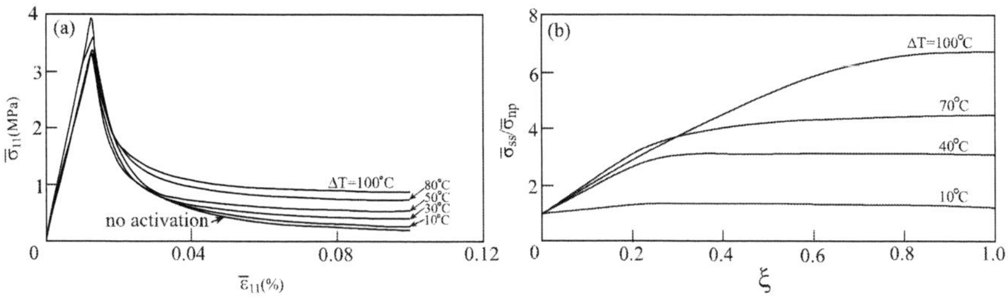

Figure 14. (a) Effect of activation of SMA fibers on the behavior of the prestressed concrete. (b) The ratio of the steady state axial stress $\bar{\sigma}_{ss}$ and the steady state axial stress in a SMA/concrete composite with no prestressing $\bar{\sigma}_{np}$ versus the martensite fraction ξ for different values of temperature changes.

Fig. 1(b)). Consequently, $\Delta T = 40$ K is not the optimal temperature change in the sense that full recovery is not necessarily achieved.

The behavior of the prestressed concrete at different activation temperature changes is shown in Fig. 14(a). In all cases, 9 square subcells in a 3×3 configuration, with fiber volume fraction of 1%, and non-zero coefficients of thermal expansion are considered. Since the fiber volume fraction is small, the SMA subcell is much smaller than the concrete subcells. The SMA fiber is located at $\beta = 2$ and $\gamma = 2$; that is, the subcell which contains the SMA fiber is in the middle of the repeating unit cell. From Fig. 14(a), it is clear that all activation temperatures produce an increase in the stress level in the damaged region. However, it may be observed that better results are achieved with an increase in the temperature change ΔT, from 33% for $\Delta T = 10$ K, up to 750% for $\Delta T = 100$ K.

Furthermore, the relation between the activation temperature change and the martensite fraction ξ is examined, as illustrated in Fig. 14(b). It may be noted that the martensite fraction ξ represents the amount of residual strains; $\xi = 1$ denotes the fully transformed martensite, which corresponds to the values of the residual strains $\epsilon^0_{11}, \epsilon^0_{22}, \epsilon^0_{33}$ given in Table 6. In this figure, σ_{ss} denotes the steady state axial stress in the damaged region, while $\bar{\sigma}_{np}$ denotes the steady state axial stress for SMA/concrete composite with no prestressing. The behavior exhibited by these curves is similar to that of Fig. 14(a). Namely, better results are achieved with an increase in the temperature change. However, it is observed that the same level of prestressing is reached for different values of martensite fraction ξ. For instance, in the case of $\Delta T = 10$ K, the same level of prestressing is achieved even for low values of martensite fraction ($\xi \cong 0.1$), whereas in the case of $\Delta T = 100$ K, steady state is reached only for $\xi \cong 0.9$.

Although it seems tempting to use the highest temperature change for the activation process, the reader may recall that the entire composite material is subjected to this temperature change, and not only the SMA fibers. Properties of concrete are altered significantly by exposure to high temperatures, mostly in terms of reduced stiffness and strength. Hence, the contribution of the prestressing process may be negligible, as compared to the damage degradation caused by the appearance of new micro-cracks in the heated concrete. A proper choice of the concrete and SMA fibers may allow use of a high temperature activation. In addition, the mismatch between the coefficients of thermal expansion of the concrete and the SMA fibers is taken as $\alpha_s/\alpha_c = 0.5$ in the analyses. It may be noted that the effect of the mismatch between the coefficients of thermal expansion of the SMA and concrete phases was examined within a range of $\alpha_s/\alpha_c = 0.5 - 2$. This effect turns out to be negligible. In the publication by Freed et al. [41], more discussions and results can be found.

5. SMA/Concrete composite with Plasticity-Damage Coupling in the Concrete Phase

In this section, shape memory alloy fibers are embedded within a concrete matrix. The constitutive behavior of the monolithic SMA fibers was already described in Subsection 4.1. The constitutive equations of the unreinforced concrete that exhibits plasticity effects coupled with evolving damage are presented. The macroscopic constitutive equations of the SMA/concrete are determined by employing the HFGMC model. Here too, the SMA fibers are activated in order to increase the tensile strength of the concrete.

5.1 Constitutive model of plain concrete with plasticity-damage coupling

In this subsection, a novel plasticity-damage constitutive model for plain concrete, which was developed by Wu et al. [66] for the isothermal case is briefly presented herein to incorporate thermal effects. This model is formulated in the framework of continuum damage mechanics to establish plasticity-damage constitutive equations with internal variables. A numerical procedure for the implementation of this model is given in Ref. [42].

The effective stress $\bar{\mathbf{\Sigma}}$ in the damaged material is defined as

$$\bar{\mathbf{\Sigma}} = \mathbf{C}_0 : (\boldsymbol{\epsilon} - \boldsymbol{\epsilon}^p - \boldsymbol{\epsilon}^T), \tag{101}$$

where $\mathbf{C}_0$ denotes the initial fourth order isotropic linear elastic stiffness tensor, $\boldsymbol{\epsilon}$ is the total strain tensor and $\boldsymbol{\epsilon}^p$ and $\boldsymbol{\epsilon}^T$ are the plastic and thermal strain tensors, respectively. The effective stress tensor is related to the Cauchy stress tensor $\mathbf{\Sigma}$ through the relation

$$\mathbf{\Sigma} = (\mathbf{I} - \mathbf{D})\bar{\mathbf{\Sigma}} \tag{102}$$

where $\mathbf{I}$ is the fourth order identity tensor and $\mathbf{D}$ is the fourth-order damage tensor expressed as

$$\mathbf{D} = d^+\mathbf{P}^+ + d^-\mathbf{P}^- \tag{103}$$

with the fourth order projection tensors $\mathbf{P}^+$ and $\mathbf{P}^-$ and d^+ and d^- are damage parameters, to be determined in the following. The projection tensors $\mathbf{P}^+$ and $\mathbf{P}^-$ are defined as

$$\begin{aligned} \mathbf{P}^+ &= \sum_{i=1}^{3} H(\bar{\Sigma}_i)\mathbf{p}^{(i)} \otimes \mathbf{p}^{(i)}, \\ \mathbf{P}^- &= \mathbf{I} - \mathbf{P}^+ \end{aligned} \tag{104}$$

where $H(\bar{\Sigma}_i)$ is the Heaviside unit function, $\bar{\Sigma}_i$ is the i−th principal effective stress and

$$\mathbf{p}^{(i)} = \mathbf{n}^{(i)} \otimes \mathbf{n}^{(i)}, \tag{105}$$

with $\mathbf{n}^{(i)}$ being the normalized eigenvector corresponding to $\bar{\Sigma}_i$.

The non-associative flow rule of the irreversible plastic strains is established as

$$\dot{\epsilon}^p = \dot{\lambda}\left(\frac{\bar{\mathbf{s}}}{\parallel \bar{\mathbf{s}} \parallel} + \alpha^p \mathbf{1}\right) \tag{106}$$

where a dot above a quantity denotes an increment, $\dot{\lambda}$ is a consistency parameter, $\parallel \bar{\mathbf{s}} \parallel = \sqrt{\bar{\mathbf{s}} : \bar{\mathbf{s}}}$ is the norm of the effective deviatoric stress $\bar{\mathbf{s}}$, α^p is a non negative material property and $\mathbf{1}$ is the second order identity tensor. In addition, an appropriate yield function for the concrete which depends upon the effective stress $\bar{\mathbf{\Sigma}}$ and the hardening $\boldsymbol{\kappa}$ is defined by

$$F(\bar{\mathbf{\Sigma}}, \boldsymbol{\kappa}) = \left(\alpha \bar{I}_1 + \sqrt{3\bar{J}_2} + \beta < \bar{\Sigma}_{i,\max} >\right) - (1-\alpha)c \tag{107}$$

with $< \cdot >$ denotes the McAuley bracket, $\bar{I}_1$ is the first invariant of the effective stress tensor, $\bar{J}_2$ is the second invariant of the effective deviatoric stress tensor, and $\bar{\Sigma}_{i,\max}$ represents the maximum principal effective stress. Here

$$\begin{aligned} \alpha &= \frac{\vartheta - 1}{2\vartheta - 1}, \\ \beta(\boldsymbol{\kappa}) &= \frac{(1-\alpha)\bar{f}^-(\boldsymbol{\kappa})}{\bar{f}^+(\boldsymbol{\kappa})} - (1+\alpha), \\ c(\boldsymbol{\kappa}) &= \bar{f}^-(\boldsymbol{\kappa}), \end{aligned} \tag{108}$$

where ϑ denotes the ratio between the yield strength of equibiaxial and uniaxial compressions, $\bar{f}^+(\boldsymbol{\kappa})$ and $\bar{f}^-(\boldsymbol{\kappa})$ are the current effective yield stresses for uniaxial tension and compression, respectively, which are given by

$$\begin{aligned} \bar{f}^+(\boldsymbol{\kappa}) &= \bar{f}_0^+ + E^{p+}\kappa^+ \\ \bar{f}^-(\boldsymbol{\kappa}) &= \bar{f}_0^- + E^{p-}\kappa^-. \end{aligned} \tag{109}$$

In Eq. (109), $\bar{f}_0^+$ and $\bar{f}_0^-$ denote the uniaxial effective tensile and compressive strengths, E^{p+} and E^{p-} are the effective plastic hardening modulus under uniaxial tensile and compressive loading, and κ^+ and κ^- are the equivalent plastic strains for uniaxial tension and compression, respectively. Finally, the evolution of the hardening vector $\boldsymbol{\kappa}$ is given by

$$\dot{\boldsymbol{\kappa}} = \begin{bmatrix} w\dot{\epsilon}^p_{\max} \\ \\ -(1-w)\dot{\epsilon}^p_{\min} \end{bmatrix} \tag{110}$$

with

$$w = \frac{\sum_{i=1}^{3} < \bar{\Sigma}_i >}{\sum_{i=1}^{3} |\bar{\Sigma}_i|}. \tag{111}$$

Once Eq. (107) is satisfied for given values of the effective stress and hardening, the damage function can be readily determined as follows. To this end, generic instant thresholds are given by Wu et al. [66] as

$$r^+ = \max\left\{r_0^+, \max_{history} Y^+\right\}, \quad r^- = \max\left\{r_0^-, \max_{history} Y^-\right\}, \tag{112}$$

where r_0^+ and r_0^- are the initial damage surfaces for tension and compression, respectively, and the current damage surfaces are given by

$$\begin{aligned} Y^+ &= \sqrt{E_0(\bar{\boldsymbol{\Sigma}}^+ : \mathbf{C}_0^{-1} : \bar{\boldsymbol{\Sigma}})}, \\ Y^- &= \alpha \bar{I}_1 + \sqrt{3\bar{J}_2} - \gamma < -\bar{\Sigma}_{i,\max} >, \\ \gamma &= \frac{3(1-K_c)}{2K_c - 1} \end{aligned} \tag{113}$$

with K_c denotes the ratio between the values of $\sqrt{J_2}$ for tensile and compressive tests and assumed to be constant. In Eq. (113), $\bar{\boldsymbol{\Sigma}}^+ = \mathbf{P}^+ : \bar{\boldsymbol{\Sigma}}$ is the positive component of the effective stress tensor and E_0 is the Young's modulus of the concrete which may be determined from uniaxial tensile tests.

The damage in the concrete is divided into two scalars, d^+ for tension and d^- for compression. These may be determined explicitly as

$$\begin{aligned} d^+ &= 1 - \frac{r_0^+}{r^+}\left\{(1-A^+) + A^+ \exp\left[B^+\left(1 - \frac{r^+}{r_0^+}\right)\right]\right\}, \\ d^- &= 1 - \left\{\frac{r_0^-}{r^-}(1-A^-) + A^- \exp\left[B^-\left(1 - \frac{r^-}{r_0^-}\right)\right]\right\}, \end{aligned} \tag{114}$$

where A^+ and B^+ are related to the percentage of reinforcement in the concrete. It should be noted that in this study, plain concrete is considered within the micromechanical framework. Therefore, $A^+ = 1$, which is suitable for plain concrete, is considered. Material parameters A^- and B^- may be determined from one dimensional experimental test. Once the damage functions d^+ and d^- are evaluated, the Cauchy stress tensor may be obtained from Eqs. (101)-(103). Thus the final form of the constitutive equation of the concrete can be written in the form

$$\boldsymbol{\Sigma} = \left[\left(1-d^+\right)\mathbf{P}^+ + \left(1-d^-\right)\mathbf{P}^-\right] : \mathbf{C}_0 : \left(\epsilon - \epsilon^p - \epsilon^{th}\right) \tag{115}$$

and the fourth order tensor

$$\mathbf{C} = \left[\left(1-d^+\right)\mathbf{P}^+ + \left(1-d^-\right)\mathbf{P}^-\right] : \mathbf{C}_0 \tag{116}$$

can be regarded as the current stiffness tensor of the material.

From a computational point of view, the non-linear behavior of the concrete model presented by Wu et al. [66] may be treated as an implicit time discrete strain driven problem. Accordingly, the time discrete equations of the model are integrated over a time interval $(t_n, t_{n+1} = t_n + \Delta t)$ using an implicit backward Euler scheme, which is unconditionally stable (Simo and Hughes [53]). In general, the solution may be divided into three stages. The first stage is an elastic trial state, which assumes an elastic increment. To this end, the trial state may be determined from

$$\bar{\mathbf{\Sigma}}^{tr} = \mathbf{C}_0 : (\boldsymbol{\epsilon} - \boldsymbol{\epsilon}^p_n - \boldsymbol{\epsilon}^T), \tag{117}$$

and the corresponding trial yield surface

$$F^{tr}(\bar{\mathbf{\Sigma}}^{tr}, \boldsymbol{\kappa}_n) = \alpha \bar{I}_1^{tr} + \sqrt{3\bar{J}_2^{tr}} + \beta_n < \bar{\Sigma}^{tr}_{i,\max} > -(1-\alpha)c_n. \tag{118}$$

It may be noted that the subscript $n+1$ (which denotes the value of the quantity at time t_{n+1}) is omitted in Eqs. (117) and (118) for simplicity. The subscript n denotes the value of the quantity at time t_n. If the trial state is admissible, that is, $F^{tr} \leq 0$, then the current step is elastic, and the elastic trial step does not need any correction. Otherwise, if $F^{tr} > 0$, this step is inelastic, and a new solution has to be evaluated according to a plastic corrector (which is the second stage). Here, the updated state is obtained using an iterative method to obtain an appropriate $\dot{\lambda}$ and a corresponding $\boldsymbol{\kappa}_{n+1}$. During the elastic trial state and the plastic corrector stages, the damage variables are held fixed. At the end of this stage, a new effective stress tensor is evaluated. The third stage is a damage-corrector, in which the effective stress is transformed to the Cauchy stress by Eq. (102). The plastic variables are kept fixed during the damage corrector stage. It is observed that the third stage is decoupled from the first two stages. The offered numerical procedure is summarized in Box 1 of Ref. [42].

The responses of plain concrete during an isothermal strain-driven uniaxial tension and compression tests are shown in Figs. 1a and b, respectively, of Ref. [42] where material properties of concrete are listed in Table 8. The parameters given in this table account for a particular aggregate size, constituent volume fraction and other mixture effects. As expected, residual plastic strains are observed upon unloading. These curves agree with those of Wu et al. [66].

5.2 Applications

The HFGMC micromechanical model in the absence of full TMC (where the temperature affects only the mechanical response), is implemented to predict the mechanical behavior of unidirectional shape memory alloy fibers which are embedded in a concrete matrix. The SMA fibers are oriented in the x_1-direction (see Fig. 12); which is defined as the axial direction. Note that the concrete is presently not prestressed. The material properties of the SMA fibers and concrete matrix are given in Tables 6 and 8, respectively. These properties are taken from Panoskaltsis et al. [63] for the SMA fibers and Wu et al. [66] for the concrete. The constitutive model of Panoskaltsis et al. [63] has been already summarized in subection 4.1.

Table 7. Material properties of the concrete matrix (Tao and Phillips [64]).

Property	Value
E	31.8 GPa
ν	0.18
α	10×10^{-6}/K
a_t	7000 MPa^{-1}
a_c	29 MPa^{-1}
b_t	1.1
b_c	0.94
Y_t^0	1.9×10^{-4} MPa
Y_c^0	3×10^{-4} MPa
c	2 MPa^{-1}
d	0.7 MPa^{-1}

Table 8. Material properties of the concrete matrix (Wu et al. [66]).

Property	Value
E_0	31 GPa
E^{p+}	20 GPa
E^{p-}	20 GPa
ν	0.2
$\bar{f}_0^+$	3.48 MPa
$\bar{f}_0^-$	15 MPa
ϑ	1.16
α^p	0.2
A^+	1
A^-	1
B^+	0.75
B^-	0.89
r_0^+	3.48 MPa
r_0^-	13.18 MPa
α_c	1×10^{-5} /K

The response of the composite material is examined for two different loading cases. The first analysis is carried out for the case of a uniaxial stress loading in which the composite is subjected to an external strain in the axial direction while keeping all stress components other than $\bar{\sigma}_{11}$ equal to zero. In the second analysis, the uniaxial stress is applied in the transverse direction, while keeping all stress components other than $\bar{\sigma}_{22}$ equal to zero. The resulting responses are shown in Figs. 15(a) and (b), respectively. For both cases, the macroscopic stress-strain behavior is compared to that of plain concrete. It may be noted that the Young's modulus of the shape memory alloy fibers is smaller than that of the concrete matrix. Consequently, different macroscopic stress-strain behavior is expected for these two loading situations in which fiber volume fraction of $v_f = 0.05$ is considered. The shape

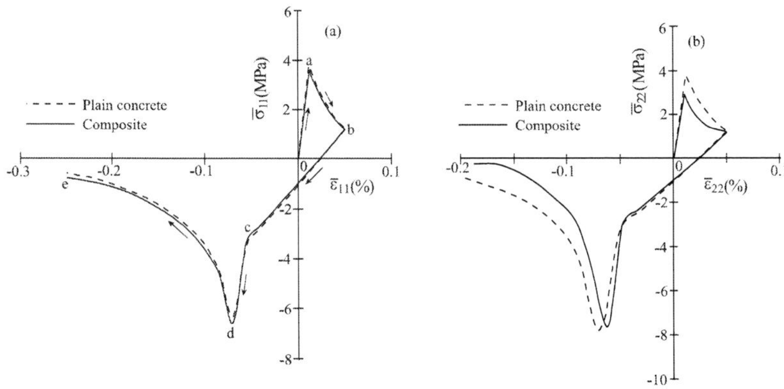

Figure 15. Uniaxial global stress-strain response of an SMA/concrete composite and plain concrete. (a) Loading in the axial (fiber) direction, (b) loading in the transverse direction.

memory alloy is in its martensite phase ($T = 363$ K) and the applied loadings are assumed to be isothermal.

Observing Figs. 15(a) and (b), five distinct points can be detected. The stress-strain behavior of the plain concrete, as well as the composite, starts with an elastic behavior, until a threshold at point a is reached. Upon an additional tensile loading, damage and plasticity are observed. At point b, the loading is reversed, producing a compression with constant plastic strains until point c. At this point, plastic strains are being generated while progressive damage and plastic deformations are observed from point d to e. Observing Fig 15(a), the curves of the plain concrete and the fiber reinforced material almost coincide through points a, b, c and d. At point d, some differences are observed. However, for a case of loading which is applied in the perpendicular to the fiber direction, Fig. 15(b), the curves of plain concrete and the composite material differ starting from point a. Recalling that the fiber volume fraction is rather low ($v_f = 0.05$), it appears that the effect of the SMA reinforcement is quite negligible in the case of axial loading whereas relatively significant effects can be observed in the transverse loading direction.

Next, initial yield surfaces of the SMA/concrete composite are established in terms of externally applied global stresses. It may be noted that when examining the evolution of inelastic strains, the transformation strains in the shape memory alloy fiber and the plastic strains in the concrete matrix should be considered. However, since the shape memory alloy fiber is initially at its martensite phase, and an isothermal loading is presently assumed, new transformation strains are not expected in the shape memory alloy fiber. Furthermore, for initial yield surfaces, the plastic strains in the concrete subcells are zero, namely, $\epsilon^{p(\beta\gamma)} = 0$. Hence under the present conditions, Eq. (16) reduces to $\epsilon^{(\beta\gamma)} = \boldsymbol{A}^{M(\beta\gamma)}\bar{\epsilon}$, which provides the local strain tensor in the concrete subcell in the form

$$\epsilon^{(con)} = \mathbf{A}^{M(con)}\bar{\epsilon} \tag{119}$$

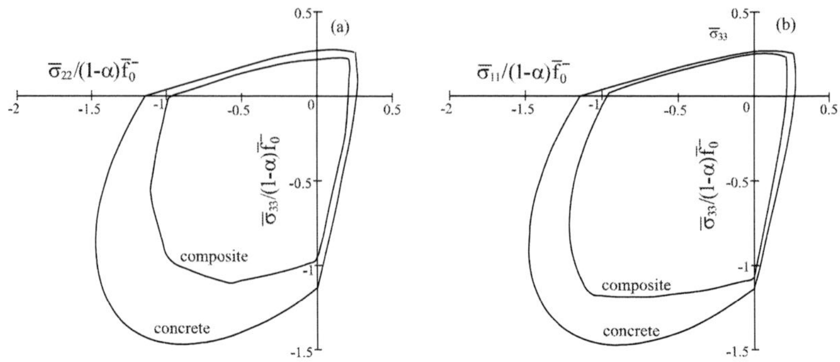

Figure 16. Initial yield surfaces of an SMA/concrete composite. (a) $\bar{\sigma}_{22}-\bar{\sigma}_{33}$, (b) $\bar{\sigma}_{11}-\bar{\sigma}_{33}$.

where a superscript (con) represents any concrete subcell. From Eq. (18),

$$\bar{\epsilon} = (\mathbf{C}^{*})^{-1}\bar{\boldsymbol{\sigma}}, \tag{120}$$

Substituting Eq. (120) into Eq. (119) results in

$$\epsilon^{(con)} = \mathbf{A}^{M(con)}(\mathbf{C}^{*})^{-1}\bar{\boldsymbol{\sigma}} \tag{121}$$

and the following relation between the local stress tensor in the concrete subcell and the global stress of the composite is established

$$\boldsymbol{\sigma}^{(con)} = \mathbf{C}^{(con)}\mathbf{A}^{M(con)}(\mathbf{C}^{*})^{-1}\bar{\boldsymbol{\sigma}}. \tag{122}$$

This relation is substituted into the yield condition which is given by Eq. (107) to obtain the initial yielding surface of the concrete matrix composite in terms of the externally applied global stresses. For a given combination of global stresses, this micromechanical analysis is performed on all concrete subcells of the composite and the initial yielding is determined by the first concrete subcell that fulfills Eq. (107).

In Figs. 16(a) and (b) the initial yield surfaces of the concrete matrix composite are illustrated in terms of the applied global stresses. Particularly, Fig. 16(a) demonstrates the effect of loading by the global in-plane stresses $\bar{\sigma}_{22}$ and $\bar{\sigma}_{33}$, while Fig. 16(b) illustrates the effect of loading by the global in-plane stress component $\bar{\sigma}_{22}$ and the axial component $\bar{\sigma}_{11}$. Here too, fiber volume fraction of 0.05 is considered. In both cases, due to the fact that the Young's modulus of the SMA fiber is lower than that of the concrete matrix, the initial yield surface of the composite material is smaller than that of plain concrete, especially in the region of compression. The yield surface of $\bar{\sigma}_{11}-\bar{\sigma}_{22}$ seems to be less smooth than that of $\bar{\sigma}_{22}-\bar{\sigma}_{33}$. which implies an appreciable anisotropic behavior of the composite for this type of loading.

The effect of the prestressing procedure on the initial yield surfaces of the concrete matrix composite is examined herein. To this end, the micromechanical procedure given in Eqs. (119) - (122) is generalized to include the thermal effects caused by the activation.

The micromechanically established macroscopic constitutive relation of the SMA/concrete composite is given by Eq. (18). Rewriting this equation in the form

$$\bar{\epsilon} = (\mathbf{C}^*)^{-1}\bar{\boldsymbol{\sigma}} + (\mathbf{C}^*)^{-1}\boldsymbol{\Gamma}^*\Delta T + \bar{\epsilon}^I \tag{123}$$

and substituting $\bar{\epsilon}$ into the local relation in the concrete subcells, Eq. (16), provides the strain tensor in these subcells

$$\epsilon^{(con)} = \mathbf{A}^{M(con)}\left[(\mathbf{C}^*)^{-1}\bar{\boldsymbol{\sigma}} + (\mathbf{C}^*)^{-1}\boldsymbol{\Gamma}^*\Delta T + \bar{\epsilon}^I\right] + \mathbf{V}^{T(con)}\Delta T. \tag{124}$$

It should be noted that for initial yielding in the concrete subcells, the inelastic effects there vanish, implying that $\mathbf{V}^{I(con)} = 0$. The corresponding local stress tensor in the concrete subcells is given by

$$\begin{aligned}\boldsymbol{\sigma}^{(con)} = \mathbf{C}^{(con)}\Big\{\mathbf{A}^{M(con)}&\left[(\mathbf{C}^*)^{-1}\bar{\boldsymbol{\sigma}} + (\mathbf{C}^*)^{-1}\boldsymbol{\Gamma}^*\Delta T + \bar{\epsilon}^I\right]\\ &+ \mathbf{V}^{T(con)}\Delta T\Big\} - \boldsymbol{\Gamma}^{(con)}\Delta T.\end{aligned} \tag{125}$$

Eq. (125) may be rewritten as

$$\boldsymbol{\sigma}^{(con)} = \mathbf{B}^{M(con)}\bar{\boldsymbol{\sigma}} + \mathbf{B}^{I(con)}\bar{\epsilon}^I + \mathbf{B}^{T(con)}\Delta T \tag{126}$$

where

$$\begin{aligned}\mathbf{B}^{M(con)} &= \mathbf{C}^{(con)}\mathbf{A}^{M(con)}(\mathbf{C}^*)^{-1},\\ \mathbf{B}^{I(con)} &= \mathbf{C}^{(con)}\mathbf{A}^{M(con)},\\ \mathbf{B}^{T(con)} &= \mathbf{C}^{(con)}\mathbf{A}^{M(con)}(\mathbf{C}^*)^{-1}\boldsymbol{\Gamma}^* + \mathbf{C}^{(con)}\mathbf{V}^{T(con)} - \boldsymbol{\Gamma}^{(con)}\end{aligned} \tag{127}$$

which may be referred to as the mechanical, inelastic and thermal stress concentration tensors. Note that $\bar{\epsilon}^I$ consists of the effect of the recovered transformation strains in the shape memory alloy fiber.

The initial yield surfaces of the SMA/concrete composite are determined by substituting the local stress tensor given by Eq. (126) into the yield surface of Eq. (107). As previously discussed, for a given combination of global stresses, the micromechanical analysis is performed on all concrete subcells of the composite and the initial yielding is determined by the first concrete subcell that fulfills Eq. (107). Three initial yield surfaces are illustrated in Fig. 17(a). In all cases, fiber volume fraction of $v_f = 0.01$ is assumed. In this figure, the yield surface of an SMA/concrete composite is compared to two other cases. In the first one, the temperature of the composite material is increased by $\Delta T = 40$ K when it is assumed that no recovery of transformation strains in the shape memory alloy fiber occurs. In this case, the initial yield surface of the composite is smaller than that of the SMA/concrete composite in the reference temperature in which $\Delta T = 0$. However, when considering recovery of transformation strains in the shape memory alloy fibers, the size of the initial yield surface of the composite in tension increases (see Fig. 17(b)), while the size of the initial yield surface in compression decreases. This exactly demonstrates the prestressing procedure, where the desired increase of the yielding point in tension is achieved. It is worth

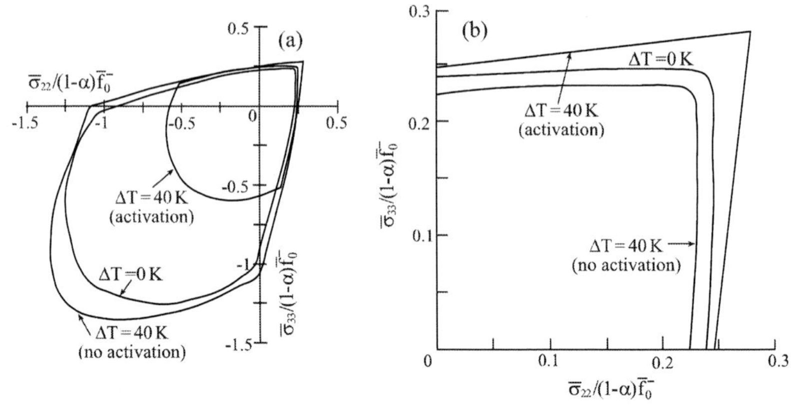

Figure 17. (a) Initial yield surfaces of an SMA/concrete composite and an activated SMA/concrete composite and the effect of the temperature change. (b) The corresponding initial yield surfaces in the tensile region.

mentioning that the prestressing procedure should be carried out only in the specific parts of the concrete specimen which are expected to be subjected to tensile loading. Consequently, the resulting strength decrease in compression does not affect the overall behavior of the specimen. More results can be found in Ref. [42].

6. Conclusion

This Chapter reviews a series of investigations for the micromodeling of unidirectional SMA composites. Both metallic and polymeric matrices with embedded continuous SMA fibers are investigated. The SMA fibers may exhibit various features such as pseudoelasticity, one and two-way shape memory effects, transformation induced plasticity and rate-dependence. In order to model these features, various multiaxial constitutive equations have been developed by various investigators. This Chapter employes these SMA models in order to micromechanically predict the response of the corresponding metal and polymer matrix composites with unidirectional SMA fibers. This is performed by employing the HFGMC micromechanical analysis which enables the modeling of both partial (in which the temperature affects only the mechanical response) and full TMC of the composite. In addtion to the coupled energy equations of standard metallic and polymeric materals, the full TMC analysis necessitates the establishment of the energy equation of the SMA material which provides the coupling terms between the mutual thermal and mechanical interaction effects.

In addition to the various multi-axial constitutive equations of SMA materials that have been employed in this Chapter as constituents in the HFGMC micromechanical analysis, other types of three-dimensional SMA equations can be also used in the HFGMC model. Such variety of three-dimensional SMA constitutive relations have been presented and reviwed in the recent book edited by Lagoudas [16]. Furthermore, the various approaches reviewed in this chapter can be extended for the accommodation in the HFGMC analysis of new types of more sophisticated SMA models (which might be necessary to represent new features that are exhibited by the SMA materials) to predict the behavior of the corresponding SMA composites.

References

[1] Perkins, J. Residual stresses and the origin of reversible (two-way) shape memory effect. Scripta Metall. 1974, vol. 8, 1469-1476.

[2] Stalmans, R.; Humbeeck, J. V.; Delaey, L. The two way memory effect in copper-based shape memory alloys - thermodynamics and mechanisms. Acta Metall. Mater. 1992, vol. 40, 2921-2931.

[3] Hebda, D. A.; White, S. R. Effect of training conditions and extended thermal cycling on Nitinol two-way shape memory behavior. Smart Mater. Struct. 1995, vol. 4, 298-304.

[4] Miller, D. A.; Lagoudas, D. C. Influence of cold work and heat treatment on shape memory effect and plastic strain development of NiTi. Mater. Sci. Eng. A. 2001, vol. 308, 161-175.

[5] Lahoz, R.; Puertolas, J. A. Training and two-way shape memory in NiTi alloys: influence on thermal parameters. J. Alloys Compounds 2004, vol. 381, 130-136.

[6] Patoor, E.; Lagoudas, D. C.; Entchev, P. B.; Brinson, L. C.; Gao, X. J. Shape memory alloys, Part I: General properties and modeling of single crystals. Mech. Mater. 2006, vol. 38, 391-429.

[7] Abel, E.; Luo, H. Issues for the optimal training of two-way shape memory alloy. In Progress in Smart Materials and Structures; Reece, P. L.; Ed.; Nova Science Publishers Inc., Hauppauge, NY, 2007; pp. 15-28.

[8] Liu, Y.; Liu, Y.; van Humbeeck, J. Two way shape memory effect developed by martensite deformation in NiTi. Acta Mater. 1999, vol. 47, 199-209.

[9] Janke, L.; Czaderski, C.; Motavalli, M.; Ruth, J. Applications of shape memory alloys in civil engineering structures - overview, limits and new ideas. Mater. Struct. 2005, vol. 38, 578-592.

[10] Song G.; Ma N.; Li H.-N. Applications of shape memory alloys in civil structures. Eng. Struct. 2006, vol. 28, 1266-1274.

[11] Kudva, J. Overview of the DARPA smart wing project. J. Intell. Mater. Syst. Struct. 2004, vol. 15, 261-267.

[12] Machado, L.; Savi, M. Medical applications of shape memory lloys. Braz. J. Med. Biol. Res. 2003, vol. 36, 683-691.

[13] Morgan, N. B. Medical shape memory alloy applications - the market and its products. Mater. Sci. Eng. A - Structural Materials Properties Microstructure and Processing, 2004, vol. 378, 16-23.

[14] Luo, Y.; Salvia, M. SMA actuators in medical applications. Ann. Chim.-Sci. Mater. 2004, vol. 29, 33-41.

[15] Razov, A. I. Application of titanium nickelide-based alloys in engineering. Phys. Met. Metallogr. 2004, vol. 97, S97-S126.

[16] Lagoudas, D. C.; Editor. Shape Memory Alloys. Modeling and Engineering Applications. Springer, New York, NY., 2008.

[17] Wei, Z. G.; Tang, C. Y.; Lee, W. B. Design and fabrication of intelligent composites based on shape memory alloys. J. Mater. Process. Tech. 1997, vol. 69, 68-74.

[18] Wei, Z. G.; Sandsrom, R.; Miyazaki, S. Shape memory materials and hybrid composites for smart systems. Part II shape-memory hybrid composites. J. Mater. Sci., 1998, vol. 33, 3763-3783.

[19] Lagoudas, D. C.; Moorthy, D.; Qidwai, M.A.; Reddy, J. N. Modeling of the thermomechanical response of active laminates with SMA strips using the layerwise finite element method. J. Intell. Mater. Syst. Struct. 1997, vol. 8, 476-488.

[20] Su, Z.; Mai, H.; Lu, M.; Ye, L. Thermo-mechanical behavior of shape memory alloy reinforced composite laminate (Ni-Ti/glass-fibre/epoxy). Compos. Struct. 1999, vol. 47, 705-710.

[21] Thompson, S. P.; Loughlan, J. The control of the post-buckling response in thin composite plates using smart technology. Thin Walled Struct. 2000, vol. 36, 231-263.

[22] Gilat, R.; Aboudi, J. Dynamic response of active composite plates: shape memory alloy fibers in polymeric/metallic matrices. Int. J. Solids Struct. 2004, vol. 41, 5717-5731.

[23] Gilat, R.; Aboudi, J. Thermal buckling of activated shape memory reinforced laminated plates. Smart Mater. Struct. 2006, vol. 15, 829-838.

[24] Lee, W. B.; Jie, M.; Tang, C .Y. Constitutive modeling of aluminum matrix fiber-reinforced smart composite. J. Mater. Process. Tech. 2001, vol. 116, 219-223.

[25] Wang, Y .F.; Wang, X. M.; Yue, Z. F. Micromechanical modelling of shape memory alloy composites. Materialwissenschaft Werkstofftechnik 2004, vol. 33, 158-162.

[26] Gall, K.; Mikulas, M.; Munshi, N. A.; Beavers, F.; Tupper, M. Carbon fiber reinforced shape memory polymer composites. J. Intell. Mater. Syst. Struct. 2000, vol. 11, 877-886.

[27] Taketa, I.; Amano, M.; Kobayahi, M.; Ogisu, T.; Okabe, Y.; Takeda, N. Modeling of thermo-mechanical behavior of Ti-Ni shape memory alloy foils embedded in carbon fiber reinforced plastic laminates. Adv. Compos. Mater. 2005, vol. 14, 25-42.

[28] Boyd, J. G.; Lagoudas, D. C. Thermomechanical response of shape memory composites. J. Intell. Mater. Syst. Struct. 1994, vol. 5, 333-346.

[29] Lagoudas, D. C.; Bo, Z.; Qidwai, M. A. A unified thermodynamic constitutive model for SMA and finite element analysis of active metal matrix composites. Mech. Compos. Mater. Struct. 1996, vol. 3, 153-179.

[30] Kawai, M.; Ogawa, H.; Baburaj, V.; Koga T. Micromechanical analysis for hysteretic behavior of unidirectional TiNi SMA Fiber Composite. J. Intell. Mater. Syst. Struct. 1996, vol. 10, 14-28.

[31] Carvelli, V.; Taliercio, A. Micromechanical model for the analysis of unidirectional elasto-plastic composites subjected to 3D stresses. Mech. Res. Comm. 1999, vol. 26, 547-553.

[32] Song, G. Q.; Sun, Q. P.; Cherkaoui, M. Role of microstructures in the thermomechanical behavior of SMA composites. J. Eng. Mater. Tech. 1999, vol. 121, 86-92.

[33] Kawai, M. Effects of matrix inelasticity on the overall hysteretic behavior of TiNi-SMA fiber composites. Int. J. Plasticity 2000, vol. 16, 263-282.

[34] Marfia, S. Micro-macro analysis of shape memory alloy composites. Int. J. Solids Struct. 2005, vol. 42, 3677-3699.

[35] Aboudi, J. Micromechanical analysis of fully coupled electro-magneto-thermo-elastic multiphase composites. Smart Mater. Struct. 2001, vol. 10, 867-877.

[36] Aboudi, J. The generalized method of cells and high-fidelity generalized method of cells micromechanical models - a review. Mech. Adv. Mater. Struct. 2004, vol. 11, 329-366.

[37] Aboudi, J.; Freed, Y. Two-way thermomechanically coupled micromechanical analysis of shape memory alloy composites. J. Mech. Mater. Struct. 2006, vol. 1, 937-955.

[38] Freed, Y.; Aboudi, J. Thermomechanically coupled micromechanical analysis of shape memory alloy composites undergoing transformation induced plasticity. J. In tell. Mater. Struct. 2008.

[39] Freed, Y.; Aboudi, J.; Gilat, R. Investigation of shape memory alloy honeycombs by means of a micromechanical analysis Model. Sim. Mater. Sci. Eng. 2008, vol. 16, 055002-22pp.

[40] Freed, Y.; Aboudi, J. Micromechanical perición of the two-way shape memory effect in shape memory alloy composites. Int. J. Solids Struct. (to appear).

[41] Freed, Y.; Aboudi, J.; Gilat, R. A thermomechanically micromechanical modeling of prestressed concrete reinforced with shape memory alloys fibers. Smart Mater. Struct. 2007, vol. 16, 717-727.

[42] Freed, Y.; Aboudi, J. Micromechanical investigation of plasticity-damage coupling of concrete reinforced by shape memory alloy fibers. Smart Mater. Struct. 2008, vol. 17, 015046-14pp.

[43] Auricchio, F. A robust integration-algorithm for a finite-strain shape-memory-alloy superelastic model. Int. J. Plasticity 2001, vol. 17, 971-990.

[44] Aboudi, J. Micromechanically based constitutive equations for shape-memory fiber composites undergoing large deformations. Smart Mater. Struct. 2004, vol. 13, 828-837.

[45] Aboudi, J. Themomechanically coupled micromechanical analysis of multiphase composites. J. Eng. Math. 2008, vol. 61, 111-132.

[46] Mitchell, A. R.; Griffiths, D. F. The Finite Difference Method in Partial Differential Equations; John Wiley, Chichester, 1980.

[47] Aboudi, J.; Pindera, M.-J.; Arnold, S. M. High-fidelity generalized method of cells for inelastic periodic multiphase materials. NASA TM-2002-211469, 2002.

[48] Aboudi, J.; Pindera, M.-J.; Arnold, S. M. Higher-order theory for periodic multiphase materials with inelastic phases. Int. J. Plasticity 2003, vol. 19, 805-847.

[49] Auricchio, F.; Petrini, L. Improvements and algorithmical considerations on a recent three-dimensional model describing stress-induced solid phase transformation. Int. J. Num. Meth. Eng. 2002, vol. 55, 1255-1284.

[50] Auricchio, F.; Petrini, L. A three-dimensional model describing stress-temperature induced solid phase transformations: thermomechanical coupling and hybrid composite application. Int. J. Num. Meth. Eng. 2004, vol. 61, 716-737.

[51] Auricchio, F.; Petrini, L. A three-dimensional model describing stress-temperature induced solid phase transformations: solution algorithm and boundary value problems. Int. J. Num. Meth. Eng. 2004, vol. 61, 807-836.

[52] Souza, A. C.; Mamiya, E. N.; Zouain, N. Three-dimensional model for solid undergoing stress-induced phase transformation. Europ. J. Mech. A/Solids 1998, vol. 17, 789-806.

[53] Simo, J. C.; Hughes, T. J. R. Computational Inelasticity, Springer-Verlag, New York, NY, 1998.

[54] Auricchio, F., Reali, A., Stefanelli, U. A three-dimensional model describing stress-induced solid phase transformation with permanent inelasticity, Int. J. Plasticity 2007, vol. 23, 207-226.

[55] Grummon, D. S.; Shaw, J. A.; Foltz, J. Fabrication of cellular shape memory alloy materials by reactive eutectic brazing using niobium. Mater. Sci. Eng. A - Struct. Mater., Properties, Microstructure Process. 2006, vol. 438, 1113-1118.

[56] Gibson, L. J.; Ashby, M. F. Cellular solids: structures and properties, second edition. Cambridge University Press, Cambrige, UK, 1997.

[57] Gibson, L. J,; Ashby, M. F.; Schajer, G. S.; Robertson, C. I. The mechanics of two dimensional cellular materials. Proc. Royal Soc. London 1982, vol. A382(1782), 25-42.

[58] Almgren, R. F. An isotropic three-dimensional structure with Poisson's ratio =-1. J. Elasticity 1985, vol. 15, 427-430.

[59] Masters, I. G.; Evans, K. E. Models for the elastic deformation of honeycombs. Compos. Struct. 1996, vol. 35, 403-422.

[60] Shaw, J. A.; Grummon, D. S.; Foltz, J. Superelastic NiTi honeycombs: fabrication and experiments. Smart Mater. Struct. 2007, vol. 16, S170-S178.

[61] Wang, A. J.; McDowell, D. L. Yield surfaces of various periodic metal honeycombs at intermediate relative density. Int. J. Plasticity 2005, vol. 21, 285-320.

[62] Helm, D.; Haupt, P. Shape memory behavior: modelling within continuum thermomechanics. Int. J. Solids Struct. 2003, vol. 40, 827-849.

[63] Panoskaltsis, V. P.; Bahuguna, S.; Soldatos, D. On the thermomechanical modeling of shape memory alloys. Int. J. Non-Linear Mech. 2004, vol. 39, 709-722.

[64] Tao, X. Y.; Phillips, D. V. A simplified isotropic damage model for concrete under bi-axial stress states. Cement Concrete Compos. 2005, vol. 27, 716-726.

[65] Moser, K.; Bergamini, A.; Christen, R.; Czaderski C. Feasibility of concrete prestressed by shape memory alloy short fibers. Mater. Struct. 2005, vol. 38, 593-600.

[66] Wu, J. Y.; Li, J.; Faria, R. An energy release rate-based plastic-damage model for concrete. Int. J. Solids Struc. 2006, vol. 43, 583-612.

In: Shape Memory Alloys
Editor: H. R. Chen, pp. 425-440

ISBN 978-1-60741-789-7

Chapter 15

The Numerical Analysis of Two - Shape Memory Model

N. Chemetov*
Centro de Matemática e Aplicações Fundamentais, Universidade de Lisboa
Av. Prof. Gama Pinto, 2 1649-003 Lisboa, Portugal

Abstract

In this chapter, we discuss the implicit time discretization for Frémond's model of two-shape memory phenomenon and the error estimates between the approximate solution and the exact solution of Frémond's model. We show optimal order error estimates on the whole time interval. Also, we formulate the continuous dependence result of the solution from the data of the problem.

Key Words: two-shape memory alloys, numerical discretization, a priori estimates.

AMS Subject Classification: 65 M 15, 35 K 60, 80 A 22

1. The Model of Two-Shape Memory Alloys

The shape memory alloy is a metallic material that is deformed at a low temperature without any external forces, and if we heat the alloy, a non deformed initial state will be recovered [8]. But, if we cool it again, the alloy cannot go back to the deformed form. Therefore, materials that remember just the non deformed configuration are called one-shape memory materials. There are shape memory alloys with a two-shape memory effect. They keep in memory two shapes: one at low temperature, the other at high temperature. Following the temperature evolution, it goes from one shape to the other without external action, i.e., the shape of the material changes reversiblely following the temperature. Hence, we say that such alloys can exhibit the two-shape memory effect. The macroscopic predictive model of this phenomenon was introduced by M. Frémond [8]; it can be written as the system of

*E-mail address: chemetov@ptmat.lmc.fc.ul.pt

differential equations in a bounded domain Ω of $\mathbb{R}^3$

$$\frac{\partial}{\partial t}[c_0\theta - L_0\chi_1 + L_e\chi_3 + (\alpha(\theta) - \theta\alpha'(\theta))\,\chi_2\, div(\mathbf{u})] - h\Delta\theta = F + \alpha(\theta)\chi_2\frac{\partial}{\partial t}div(\mathbf{u}), \quad (1.1)$$

$$div(\lambda div(\mathbf{u})J + 2\mu\varepsilon(\mathbf{u}) + \alpha(\theta)\chi_2 J - \nu\Delta(div\mathbf{u})J) = \mathbf{G}, \quad (1.2)$$

$$\frac{\partial}{\partial t}\boldsymbol{\chi} + \mathbf{W}(\theta, \mathbf{u}) + \partial I_P(\boldsymbol{\chi}) \ni 0, \quad (1.3)$$

for any $(\mathbf{x}, t) \in \Omega_T := \Omega \times (0, T)$. Here we denote by $\mathbf{W}(\theta, \mathbf{u}) := (\frac{L_0}{\theta_0}(\theta - \theta_0),\, \alpha(\theta)div(\mathbf{u}),\, -\frac{L_e}{\theta_e}(\theta - \theta_e))$ with θ being the absolute temperature of the alloy and $\mathbf{u} = (u_1, u_2, u_3) \in \mathbb{R}^3$ - the displacement vector, related with the strain tensor $\varepsilon(\mathbf{u}) = \frac{1}{2}(\frac{\partial \mathbf{u}}{\partial \mathbf{x}} + \frac{\partial \mathbf{u}}{\partial \mathbf{x}}^*)$. Let us denote by β_1, β_2 and β_3 the volumetric proportions of two martensitic variants and of the austenite with the internal constraints $\beta_1 + \beta_2 + \beta_3 = 1,\quad \beta_i \geqslant 0,\ i = 1, 2, 3,$ which are equivalent to $|\beta_2 - \beta_1| \leqslant \beta_1 + \beta_2 \leqslant 1.$ The vector $\boldsymbol{\chi}$ consists of three components $\chi_1 := \beta_1 + \beta_2, \chi_2 := \beta_2 - \beta_1$ and $\chi_3 := \delta$ being a so-called *education parameter* for the two-shape memory alloy. The set of admissible values for $\boldsymbol{\chi}$ is

$$P := \bigcup_{0\leqslant\delta\leqslant 1} C(\delta), \text{ where } C(\delta) = \{(\gamma_1, \gamma_2, \delta) : \quad |\gamma_2| \leqslant \gamma_1 \leqslant 1,$$

$$(\gamma_1 - \gamma_2) \leqslant (\gamma_1 + \gamma_2)\frac{1 - \delta(1 - \beta_{max})}{\delta(1 - \beta_{max})}\}, \quad (1.4)$$

where β_{max} $(0 < \beta_{max} < 1)$ is a known maximum of β_2 for the completely educated material $(\delta = 1)$. Let us remark that the pyramid P is $non\ convex$ set, but the generalized derivative ∂I_P can be well defined, that we will discuss in the following section.

We close the system (1.1)-(1.3) by the following conditions on a C^2-smooth boundary Γ of the domain Ω

$$h\frac{\partial\theta}{\partial\mathbf{n}} + \eta(\theta - f) = 0, \quad \text{on } \Gamma_T := \Gamma \times (0, T); \quad (1.5)$$

$$\frac{\partial(div(\mathbf{u}))}{\partial\mathbf{n}} = 0 \quad \text{on } \Gamma_T, \qquad \mathbf{u} = 0 \quad \text{on } \Gamma_T^0 := \Gamma^0 \times (0, T),$$

$$(\lambda div(\mathbf{u})J + 2\mu\varepsilon(\mathbf{u}) + \alpha(\theta)\chi_2 J - \nu\Delta(div(\mathbf{u}))J, \mathbf{n}) = \mathbf{g} \quad \text{on } \Gamma_T^1 := \Gamma^1 \times (0, T) \quad (1.6)$$

and the initial conditions

$$(\theta, \boldsymbol{\chi})(\mathbf{x}, 0) = (\theta^0, \boldsymbol{\chi}^0)(\mathbf{x}) \text{ for } \mathbf{x} \in \Omega. \quad (1.7)$$

Here $\Gamma := \Gamma^0 \cup \Gamma^1$ and $meas(\Gamma^0), meas(\Gamma^1) \neq 0$. The functions F and f describe distributed heat sources in Ω and surface heat sources on Γ; $\mathbf{G}$ and $\mathbf{g}$ are body forces in Ω and tractions on Γ^1, respectively; J is the unit matrix in $\mathbb{R}^3$; c_0, L_0, L_e, h, λ, μ, ν, θ_0, $\theta_e(\theta_0 > \theta_e)$ are given positive constants. For the physical meaning of these constants we refer to the book of M. Frémond [8]. The function $\alpha = \alpha(\xi)$ represents the thermal expansion of the medium.

The investigations of systems, which are close to (1.1)-(1.7), have been developed in many articles: for instance, see E. Bonetti [4], P. Colli, M. Frémond and A. Visintin [2], P. Colli and J. Sprekels [3], N. Shemetov [11]. The complete list of the literature on the Frémond model can be found in the excellent reviews of E. Bonetti [4], U. Stefanelli [13]. We also refer to recent developments of Frémond's model in [9]. In the present chapter, we continue the study of two-shape memory model, started in the article [7]system (1.1)-(1.7) have been proven. We note that the demonstration of the existence of the solution in [7] has been based on the time discretization procedure. Therefore, from the numerical point of view, it is interesting to know the order of convergence for the discretization problem. Let us point out that the first study on the discretization error for the three-dimensional Frémond model of one-shape memory alloy has been carried out by U. Stefanelli [12], but without a very important physical term $\alpha(\theta)\,\chi_2\,\frac{\partial}{\partial t}div(\mathbf{u})$ in the equation (1.1). This term presents one of the main difficulties in the investigation of our system (existence, uniqueness, numerical study, etc.). In the paper [13], optimal order error estimates have been obtained *just* for the *one-dimensional* version of the full system (1.1)-(1.7), retaining all the nonlinear terms in the energy balance equation. In articles [1]and [10], the numerical calculations for Frémond type models have been done.

The main aim of the present chapter is to study the time step discretization of this system in the *3-dimensional* case *without any simplification in the energy balance equation*. The chapter is presented as follows: In section 2, we give a precise variational formulation of our problem with the existence result. Section 3 contains the statement of the approximated problem. Sections 4, 5 and 7 are devoted to the deduction of *optimal order* error estimates between the exact solution and the approximate solution for (1.1)-(1.7). These estimates depend only on the data of the problem. In section 6, the continuous dependence result of the solution from the data of our problem is given.

2. Variational Formulation of the Problem

Let us remember results from non-convex analysis.

Definition. *Let $K \subset \mathbb{R}^3$ be a closed set. Define the metric projection of $z \in \mathbb{R}^3$ into K as $\pi_K(z) := \{\, x \in K : |x - z| = inf_{y\in K}|y - z|\}$. A vector v is a proximal normal to the set K at a point $x \in K$ if either there exists $r > 0$ such as $x \in \pi_K(x + rv/|v|)$ or it is equal to 0. The set of all proximal normals to K at the point x is signified by $N_K(x)$.*

Lemma 1. (G. Colombo and V.V. Goncharov [5]) *For any closed set K of $\mathbb{R}^3$ a vector $\mathbf{v} \in \mathbb{R}^3$ is a proximal normal to K if and only if there exists a positive σ such that $\mathbf{v}\cdot(\mathbf{y}-\mathbf{x}) \leqslant \sigma|\mathbf{y}-\mathbf{x}|^2$, $\forall \mathbf{y} \in K$. Moreover if there exists a constant $\phi_0 \geqslant 0$, such that $\sigma = \phi_0|\mathbf{v}|$ and for all $\mathbf{x}, \mathbf{y} \in K, \mathbf{v} \in \mathcal{N}_K(\mathbf{x})$*

$$\mathbf{v}\cdot(\mathbf{y}-\mathbf{x}) \leqslant \phi_0|\mathbf{v}||\mathbf{y}-\mathbf{x}|^2, \tag{2.1}$$

then $\partial I_K(\mathbf{x}) = \mathcal{N}_K(\mathbf{x})$.

Definition. *Any closed set K of $\mathbb{R}^3$ satisfying to the condition (2.1) is called as a φ_0-prox-regular set.*

The prox-regular condition (2.1) is equivalent to the following geometrical meaning: the external tangent ball with a small radius continuously depending on the tangent point $\mathbf{x}$, can be rolled around K with its closure touching K only at $\mathbf{x}$. By this reason, our set P, defined in (1.4), is a ϕ_0-prox-regular closed set and the generalized derivative ∂I_P of the indicator function I_P in (1.3) is well defined.

We assume that the data of our problem (1.1)-(1.7) fulfill the following conditions: P *is a ϕ_0 - prox-regular, closed and bounded set of $\mathbb{R}^3$, such that $\forall \chi \in P$: $\quad |\chi| \leqslant c_P$ for* *a positive constant c_P and*

$$\begin{array}{rcll} \mathcal{P} := & \{\gamma = & (\gamma_1,\gamma_2,\gamma_3) \in (L^\infty(\Omega_T))^3 : & \gamma(\mathbf{x},t) \in P \ a.e. \ in \, \Omega_T\}; \quad (2.2) \\ F & \in & L^2(0,T;L^\infty(\Omega)) \cap H^1(0,T;L^2(\Omega)), & \\ f & \in & H^1(0,T;L^2(\Gamma)) \cap L^\infty(\Gamma_T), & \\ \mathbf{G} & \in & H^1(0,T;(L^2(\Omega))^3), & \mathbf{g} \in H^1(0,T;(L^2(\Gamma^1))^3); \quad (2.3) \\ \theta^0 & \geqslant & 0 \quad a.e. \ in \, \Omega, & \\ F & \geqslant & 0 \quad a.e. \ in \, \Omega_T, & f \geqslant 0 \quad a.e. \ in \, \Gamma_T, \\ \theta^0 & \in & H^1(\Omega) \cap L^\infty(\Omega), & \chi^0 \in \mathcal{P} \quad (2.4) \end{array}$$

and the function $\alpha = \alpha(\xi) \in C^2(\mathbb{R})$ is positive function, such that

$$\alpha'(\xi) = 0, \ \ \forall \xi \leqslant 0 \quad \text{and} \quad \alpha(\xi) = 0 \text{ for } \xi > \theta_c, \tag{2.5}$$

where $c_\alpha = \|\alpha''\|_{L^\infty([0,\theta_c])}$ is sufficiently small, such that

$$0 < \mathbb{A} := c_0 - \ \mathbb{B}\theta_c c_\alpha c_P, \qquad (\theta_c^2 + \theta_c) c_\alpha c_P \leqslant \sqrt{\mathbb{A}}\sqrt{\lambda + 2\mu/3}, \tag{2.6}$$

where $\theta_c > \theta_0$ and the constant $\mathbb{B}$ is defined in Lemma 2.

Remark. The assumptions (2.5) imply the inequalities

$$|\alpha'(\xi)|, \ |\xi\alpha''(\xi)| \leqslant \theta_c c_\alpha; \ |\alpha(\xi)|, \ |\xi\alpha'(\xi)| \leqslant \theta_c^2 c_\alpha, \ \forall \xi \in \mathbb{R}. \tag{2.7}$$

In the sequel, the scalar product and the norm in $L^2(\Omega)$ are denoted by $(\cdot,\cdot)$ and $\|\cdot\|$, respectively. Let us introduce the Hilbert space

$$V := \{\mathbf{v} \in (H^1(\Omega))^3 : \mathbf{v} = 0 \quad \text{on } \Gamma^0, \ div(\mathbf{v}) \in H^1(\Omega)\}$$

with the norm $||\mathbf{v}||_V := ||\mathbf{v}||_{(H^1(\Omega))^3} + ||\nabla(div\,\mathbf{v})||$ and the scalar product, defined by the bilinear form

$$a(\mathbf{v},\mathbf{w}) := \int_\Omega \{\lambda div(\mathbf{v}) div(\mathbf{w}) + 2\mu \sum_{i,j=1}^{3} \varepsilon_{ij}(\mathbf{v})\varepsilon_{ij}(\mathbf{w}) + \nu \nabla div(\mathbf{v}) \cdot \nabla div(\mathbf{w})\} \, d\mathbf{x}$$

for $\mathbf{v}, \mathbf{w} \in V$.

Lemma 2. *Under assumptions (2.3), (2.5), for given functions $\theta, \chi_2 \in C^0(0,T; L^2(\Omega))$ with $|\chi_2| \leqslant c_P$ a.e. in Ω_T, there exists the unique solution $\mathbf{u} \in C^0(0,T;V)$ of the system (1.2), (1.6), satisfying*

$$||div(\mathbf{u})||_{C^0(\overline{\Omega})}(t) \leqslant \mathbb{B}, \ a.e. \ t \in [0,T]. \tag{2.8}$$

Moreover if $\frac{\partial}{\partial t}\theta, \frac{\partial}{\partial t}\chi_2 \in L^2(\Omega_T)$, then

$$||\frac{\partial}{\partial t} div(\mathbf{u})||_{L^2(0,T;C^0(\overline{\Omega}))} \leqslant C. \tag{2.9}$$

The constants $\mathbb{B}, C$ depend only on the data $\mathbf{G}$, $\mathbf{g}$, c_P and α.

The first estimate of Lemma 2 has been shown in [2] and the second one has been deduced in [6], p. 886. The following existence and uniqueness result has been proven in [7].

Theorem 1. *Under the assumptions (2.2)-(2.6) there exists the unique solution* (θ, u, χ) *of the problem (1.1)-(1.7), such that*

$$\begin{aligned} \theta &\in H^1(0,T;L^2(\Omega)) \cap L^\infty(0,T;H^1(\Omega)) \cap L^\infty(\Omega_T), \\ \mathbf{u} &\in (H^1(0,T;V))^3, \qquad \chi \in (H^1(0,T;L^2(\Omega)))^3 \cap \mathcal{P}, \end{aligned} \tag{2.10}$$

satisfying (2.8), (2.9) and

$$\theta \geqslant 0, \qquad |\frac{\partial}{\partial t}\chi| \leqslant 2Q \qquad a.e.\ in\ \Omega_T \tag{2.11}$$

with $Q := \sqrt{(\frac{L_0}{\theta_0}(d+\theta_0))^2 + (\mathbb{B}||\alpha||_{L^\infty([0,\theta_c])})^2 + (\frac{L_e}{\theta_e}(d+\theta_e))^2}$ *and* $d := ||\theta||_{L^\infty(\Omega_T)}$. *The equation (1.3) is equivalent to the inequality*

$$[\chi - \gamma]\cdot[\frac{\partial}{\partial t}\chi + \mathbf{W}(\theta,\mathbf{u})] \leqslant Q\phi_0|\chi-\gamma|^2, \qquad \forall\gamma\in\mathcal{P},\ \ a.e.\ in\ \Omega_T. \tag{2.12}$$

3. Approximated Problem

In this section, we propose an implicit time discretization of the problem (1.1)-(1.7) and give the existence result of the solution for this time discretization scheme. Let us fix a positive integer N and put the time step $\tau := T/N$ in our scheme. Taking into account (2.3), let us define for $i = 0, ..., N$

$$\begin{aligned} F^i &:= F(\cdot, i\tau) \in L^2(\Omega), \qquad && f^i := f(\cdot, i\tau) \in L^2(\Gamma), \\ \mathbf{G}^i &:= \mathbf{G}(\cdot, i\tau) \in (L^2(\Omega))^3, \qquad && \mathbf{g}^i := \mathbf{g}(\cdot, i\tau) \in (L^2(\Gamma^1))^3 \end{aligned} \tag{3.1}$$

Let us select approximated initial data $\Theta_{0,\tau} \in H^1(\Omega)$, $\mathbf{X}_{0,\tau} \in \mathcal{P}$, depending on τ and fulfilling the estimate

$$||\theta^0 - \Theta_{0,\tau}|| + ||\chi^0 - \mathbf{X}_{0.\tau}|| \leqslant C\tau. \tag{3.2}$$

Here and *in the following considerations*, C denotes positive constants depending on the data of our problem (1.1)-(1.7), but independent of the time step τ.

Let the functions $\mathbf{u}_0 \in V$ and $\mathbf{U}_{0,\tau} \in V$ be the solutions of the problems

$$\begin{aligned} a(\mathbf{u}^0, \mathbf{v}) + (\alpha(\theta^0)\chi_2^0, div(\mathbf{v})) &= (\mathbf{G}^0, \mathbf{v}) + \int_{\Gamma^1} \mathbf{g}^0\mathbf{v}\,d\mathbf{x}, \\ a(\mathbf{U}_{0,\tau}, \mathbf{v}) + (\alpha(\Theta_{0,\tau})\,(X_2)_{0,\tau}, div(\mathbf{v})) &= (\mathbf{G}^0, \mathbf{v}) + \int_{\Gamma^1} \mathbf{g}^0\mathbf{v}\,d\mathbf{x}, \quad \forall \mathbf{v} \subset V. \end{aligned}$$

Taking the difference of these equalities and choosing $\mathbf{v} := \mathbf{u}^0 - \mathbf{U}_{0,\tau}$, as a consequence of (3.2), we obtain

$$||\mathbf{u}^0 - \mathbf{U}_{0,\tau}||_V \leqslant C\tau. \tag{3.3}$$

Now we formulate the following *implicit* time discretization scheme.

Find $\Theta^1, ..., \Theta^N \in H^1(\Omega)$; $\mathbf{U}^1, ..., \mathbf{U}^N \in V$; $\mathbf{X}^1, ..., \mathbf{X}^N \in \mathcal{P}$, *which satisfy the following equations for* $i = 1, ..., N$

$$
\begin{aligned}
(\frac{\partial}{\partial \bar{t}}[c_0\Theta - L_0 X_1 + L_e X_3 + (\alpha(\Theta) &- \Theta\alpha'(\Theta))X_2 div(\mathbf{U})]^i, \varphi) + h(\nabla\Theta^i, \nabla\varphi) \\
+ \int_\Gamma \eta(\Theta^i - f^i)\varphi \, d\mathbf{x} &= (F^i + \alpha(\Theta^i)X_2^i \frac{\partial}{\partial \bar{t}} div(\mathbf{U}^i), \varphi), \quad \forall\varphi \in H^1(\Omega), \\
\Theta^0 &= \Theta_{0,\tau} \quad \text{a.e. in } \Omega; \qquad (3.4) \\
a(\mathbf{U}^i, \mathbf{v}) + (\alpha(\Theta^i)X_2^i, div(\mathbf{v})) &= (\mathbf{G}^i, \mathbf{v}) + \int_{\Gamma^1} \mathbf{g}^i \mathbf{v} \, d\mathbf{x}, \quad \forall \mathbf{v} \in V, \\
\mathbf{U}^0 &= \mathbf{U}_{0,\tau} \quad \text{a.e. in } \Omega; \qquad (3.5) \\
\frac{\partial}{\partial \bar{t}}\mathbf{X}^i + \mathbf{W}((\Theta^i)_d, \mathbf{U}^i) + \mathcal{N}_{\mathcal{P}}(\mathbf{X}^i) &\ni 0 \quad \text{a.e. in } \Omega, \\
\mathbf{X}^0 &= \mathbf{X}_{0,\tau} \quad \text{a.e. in } \Omega. \qquad (3.6)
\end{aligned}
$$

Here we use the notation $\frac{\partial}{\partial \bar{t}}\xi^i := \frac{1}{\tau}(\xi^i - \xi^{i-1})$ and $(\varphi)_d := max(0, min(\varphi, d))$.

In article [7], the existence result of the solution for a *semi-implicit* time discretization scheme has been shown, using Schauder's fixed point argument. We can apply the same argument and prove the following existence result for the problem (3.4) - (3.6).

Lemma 3. *Under the hypotheses (2.2) - (2.6) and for* τ *satisfying*

$$\tau(4\phi_0 Q + 1) < 1, \tag{3.7}$$

there exists the unique solution for the problem (3.4) - (3.6), such that for any $i = 1, ..., N$

$$
\begin{aligned}
|\mathbf{X}^i| &\leqslant c_{\mathcal{P}}, \qquad |\frac{\partial}{\partial \bar{t}}\mathbf{X}^i| \leqslant 2Q \quad \text{a.e. in } \Omega, \\
\|\mathbf{U}^i\|_V &\leqslant C, \qquad \|div(\mathbf{U}^i)\|_{C^0(\overline{\Omega})} \leqslant \mathbb{B}.
\end{aligned} \tag{3.8}
$$

Let us consider a sequence of functions $\psi^i := \psi^i(\mathbf{x})$, $i = 0, ..., N$ and define two functions

$$
\begin{aligned}
\psi_{\tau,lin}(\mathbf{x}, t) &:= \psi^{i-1}(\mathbf{x}) + \frac{t - (i-1)\tau}{\tau}(\psi^i(\mathbf{x}) - \psi^{i-1}(\mathbf{x})), \quad t \in [(i-1)\tau, i\tau] \qquad (3.9) \\
\psi_\tau(\mathbf{x}, t) &:= \psi^i(\mathbf{x}), \qquad t \in ((i-1)\tau, i\tau], \qquad (3.10)
\end{aligned}
$$

such that $\psi_\tau(\mathbf{x}, 0) = \psi^0(\mathbf{x})$ and for $i = 1, ..., N$. Then the system (3.4)-(3.6) can be rewritten as

$$
\begin{aligned}
(\frac{\partial}{\partial t}[c_0\Theta_{\tau,lin} - L_0(X_1)_{\tau,lin} + L_e(X_3)_{\tau,lin} &+ ((\alpha(\Theta) - \Theta\alpha'(\Theta))X_2 div(\mathbf{U}))_{\tau,lin}], \varphi) \\
+h(\nabla\Theta_\tau, \nabla\varphi) + \int_\Gamma \eta(\Theta_\tau &- f_\tau)\varphi\, d\mathbf{x} \\
= (F_\tau + \alpha(\Theta_\tau)(X_2)_\tau \frac{\partial}{\partial t} div(\mathbf{U}_{\tau,lin}), \varphi), &\quad \forall\varphi \in H^1(\Omega), \qquad \text{a.e. in } (0,T), \\
\Theta_{\tau,lin}(\mathbf{x}, 0) &= \Theta_{0,\tau}(\mathbf{x}) \quad \text{a.e. in } \Omega; \qquad (3.11) \\
a(\mathbf{U}_\tau, \mathbf{v}) + (\alpha(\Theta_\tau)(X_2)_\tau, div(\mathbf{v})) &= (\mathbf{G}_\tau, \mathbf{v}) + \int_{\Gamma^1} \mathbf{g}_\tau \mathbf{v}\, d\mathbf{x}, \quad \forall \mathbf{v} \in V, \\
\mathbf{U}_\tau(\mathbf{x}, 0) &= \mathbf{U}_{0,\tau}(\mathbf{x}) \quad \text{a.e. in } \Omega; \qquad (3.12) \\
\frac{\partial}{\partial t}\mathbf{X}_{\tau,lin} + \mathbf{W}((\Theta_\tau)_d, \mathbf{U}_\tau) + \mathcal{N}_P(\mathbf{X}_\tau) &\ni 0 \quad \text{a.e. in } \Omega_T, \\
\mathbf{X}_{\tau,lin}(\mathbf{x}, 0) &= \mathbf{X}_{0,\tau}(\mathbf{x}) \quad \text{a.e. in } \Omega. \qquad (3.13)
\end{aligned}
$$

In conformity with (2.3), (3.1), (3.9) and (3.10), we have

$$
\begin{aligned}
&F_{\tau,lin} \in L^2(0,T;L^\infty(\Omega)) \cap H^1(0,T;L^2(\Omega)), \qquad && f_{\tau,lin} \in H^1(0,T;L^2(\Gamma)) \cap L^\infty(\Gamma_T), \\
&\mathbf{G}_{\tau,lin} \in H^1(0,T;(L^2(\Omega))^3), && \mathbf{g}_{\tau,lin} \in H^1(0,T;(L^2(\Gamma^1))^3)
\end{aligned}
$$

and

$$
||F - F_\tau||_{L^2(\Omega_T)} + ||f - f_\tau||_{L^2(\Gamma_T)} + ||\mathbf{G} - \mathbf{G}_\tau||_{(L^2(\Omega_T))^3} + ||\mathbf{g} - \mathbf{g}_\tau||_{(L^2(\Gamma^1_T))^3} \leqslant C\tau. \quad (3.14)
$$

The following lemma is devoted to the formulation of a priori estimates for the solution of the problem (3.4) - (3.6). The result of this lemma can be obtained with the help of the estimates (3.8) and the method, suggested in *Lemma 4* of article [7].

Lemma 4. *The solution of the problem (3.4) - (3.6) fulfills*

$$
\begin{aligned}
|\mathbf{X}_\tau|, |\mathbf{X}_{\tau,lin}| \leqslant c_P, \ |\frac{\partial}{\partial t}\mathbf{X}_{\tau,lin}| &\leqslant 2Q \quad \text{a.e. in } \Omega_T; \qquad (3.15) \\
\|\mathbf{U}_\tau\|_V, \ \|\mathbf{U}_{\tau,lin}\|_V &\leqslant C, \\
\|div(\mathbf{U}_\tau)\|_{C^0(\overline{\Omega})}, \ \|div(\mathbf{U}_{\tau,lin})\|_{C^0(\overline{\Omega})} &\leqslant \mathbb{B}, \\
\int_0^t \{\|\frac{\partial}{\partial t}\Theta_{\tau,lin}\|^2 + \|\frac{\partial}{\partial t}\mathbf{U}_{\tau,lin}\|_V^2\}ds + \|\Theta_\tau\|^2_{H^1(\Omega)}(t) &\leqslant C \quad \text{a.e. in } [0,T]. \qquad (3.16)
\end{aligned}
$$

Moreover, *the equation of (3.13) is equivalent to the variational inequality*

$$
[\mathbf{X}_\tau - \gamma] \cdot [\frac{\partial}{\partial t}\mathbf{X}_{\tau,lin} + \mathbf{W}((\Theta_\tau)_d, \mathbf{U}_\tau)] \leqslant Q\phi_0|\mathbf{X}_\tau - \gamma|^2, \quad \forall\gamma \in \mathcal{P}, \ \text{ a.e. in } \Omega_T. \qquad (3.17)
$$

4. Auxiliary Error Estimates

In this section, we formulate auxiliary error estimates for the difference between the exact solutions of the problem (1.1)-(1.7) and the problem (3.4)-(3.6), which will be very useful to show our main result, given in Theorem 2. The proof of these auxiliary estimates are given in Appendix for the simplification of explications.

Coming from (3.9), (3.10), (3.16), it is easy to show

$$
||\Theta_\tau - \Theta_{\tau,lin}||_{L^2(\Omega_T)} + ||\mathbf{U}_\tau - \mathbf{U}_{\tau,lin}||_{L^2(0,T;V)} + ||\mathbf{X}_\tau - \mathbf{X}_{\tau,lin}||_{L^\infty(\Omega_T)} \leqslant C\tau. \qquad (4.1)
$$

Let $\widehat{\varphi}$ and $\overline{\varphi}$ represent the difference of two functions $\varphi_\tau,\ \varphi$ and $\varphi_{\tau,lin},\ \varphi$, i.e. $\widehat{\varphi} := \varphi_\tau - \varphi$, $\overline{\varphi} := \varphi_{\tau,lin} - \varphi$. We have

$$\widehat{\varphi} = \overline{\varphi} + (\varphi_\tau - \varphi_{\tau,lin}) \tag{4.2}$$

and in the sequel we often use the identities

$$\widehat{\varphi\psi} = \widehat{\varphi}\psi_\tau + \varphi\widehat{\psi} = \widehat{\varphi}\psi + \varphi_\tau\widehat{\psi}, \qquad \overline{\varphi\psi} = \overline{\varphi}\psi_{\tau,lin} + \varphi\overline{\psi} = \overline{\varphi}\psi + \varphi_{\tau,lin}\overline{\psi},$$

that allow us to have a necessary factor ψ or ψ_τ for $\widehat{\varphi}$ and ψ or $\psi_{\tau,lin}$ for $\overline{\varphi}$, respectively.

First, we give a priori estimates for the difference of γ, $\mathbf{u}$ and their approximations $\mathbf{X}_{\tau,lin}$, $\mathbf{U}_{\tau,lin}$, respectively.

Lemma 5. *For any* $t_0, t : 0 \leqslant t_0 < t \leqslant T$, *we have that*

$$\|\overline{\chi}\|^2(t) \leqslant \|\overline{\chi}\|^2(t_0) + C\int_{t_0}^{t} \{\|\overline{\theta}\|^2 + \|div(\overline{\mathbf{u}})\|^2\}\, ds + C\tau^2, \tag{4.3}$$

$$\frac{c_V}{8}\int_{t_0}^{t} \|\overline{\mathbf{u}}\|_V^2\, ds + \frac{5}{8}\int_{t_0}^{t} \|div(\overline{\mathbf{u}})\|_{**}^2\, ds \leqslant C\int_{t_0}^{t} \|\overline{\chi}\|^2\, ds + \theta_c c_\alpha c_P \int_{t_0}^{t} \|\overline{\theta}\|\,\|div(\overline{\mathbf{u}})\|\, ds + C\tau^2, \tag{4.4}$$

where $\|div(\overline{\mathbf{u}})\|_{**}^2(t) := \{(\lambda + 2\mu/3)\|div(\overline{\mathbf{u}})\|^2 + \nu\|\nabla(div(\overline{\mathbf{u}}))\|^2\}(t)$.

Second, we present a priori estimates of the difference of the temperature θ and the approximate solutions $\Theta_{\tau,lin},\ \Theta_\tau$. To formulate the following lemma, we define the function

$$N := \|\frac{\partial}{\partial t}\{\alpha(\theta)\,\chi_2\}\|,$$

such that $N(t) \in L^2(0,T)$ by (2.7), (2.10).

Lemma 6. *For any* $t_0, t : 0 \leqslant t_0 < t \leqslant T$ *we have that*

$$\frac{3\mathbb{A}}{4}\int_{t_0}^{t} \|\overline{\theta}\|^2\, ds + \frac{1}{2}\|\widehat{\theta}\|_*^2(t,t_0) \leqslant \Big(\frac{1}{8} + C\int_{t_0}^{t} N^2(s)\, ds\Big)\int_{t_0}^{t} \|div(\overline{\mathbf{u}})\|_{**}^2(s)ds$$
$$+C\int_{t_0}^{t}\Big\{\int_{t_0}^{r} \|\overline{\theta}\|^2 ds + N^2(r)\|\widehat{\theta}\|_*^2(r,t_0)\Big\}dr + C\int_{t_0}^{t} \|\overline{\chi}\|^2\, ds + \theta_c^2 c_\alpha c_P \int_{t_0}^{t} \|\overline{\theta}\|\,\|div(\overline{\mathbf{u}})\|\, ds$$
$$+C\{\|\overline{\theta}\|^2 + \|div(\overline{\mathbf{u}})\|^2 + \|\overline{\chi}\|^2\}(t_0) + C\tau^2, \tag{4.5}$$

where $\|\widehat{\theta}\|_*^2(t,t_0) := \frac{h}{2}\|\nabla(\int_{t_0}^{t} \widehat{\theta} ds)\|^2 + \frac{\eta}{2}\|\int_{t_0}^{t} \widehat{\theta} ds\|_{L^2(\Gamma)}^2$.

5. Optimal Error Estimates

Let us formulate and prove the main result of the article.

Theorem 2. *Under the assumptions (2.2)-(2.6) and (3.7) the difference of the solutions for the problems (1.1)-(1.7) and (3.4)-(3.6) is estimated as*

$$\|\theta - \Theta_{\tau,lin}\|^2_{L^2(\Omega_T)} + \|\mathbf{u} - \mathbf{U}_{\tau,lin}\|^2_{L^2(0,T;V)} \leqslant C\tau^2 \tag{5.1}$$

and

$$\|\nabla \int_0^t (\theta - \Theta_\tau)ds\|^2 + \| \int_0^t (\theta - \Theta_\tau)ds\|^2_{L^2(\Gamma)}, \ \|\boldsymbol{\chi} - \mathbf{X}_{\tau,lin}\|^2(t) \leqslant C\tau^2, \ \ \forall t \in [0,T], \tag{5.2}$$

where the constants C depend on the data of the problem (1.1)-(1.7), but independent of the discretization time step τ.

Proof. Concerning (2.6), we have

$$\begin{aligned}(\theta_c^2 + \theta_c)c_\alpha c_P \|div(\overline{\mathbf{u}})\| \|\overline{\theta}\| &\leqslant \frac{2\mathbb{A}}{3}\|\overline{\theta}\|^2 + \frac{3}{8}(\lambda + 2\mu/3)\|div(\overline{\mathbf{u}})\|^2 \\ &\leqslant \frac{2\mathbb{A}}{3}\|\overline{\theta}\|^2 + \frac{3}{8}\|div(\overline{\mathbf{u}})\|^2_{**},\end{aligned}$$

therefore summing (4.4) and (4.5) we infer

$$\begin{aligned}\frac{\mathbb{A}}{12}\int_{t_0}^t \|\overline{\theta}\|^2\, ds &+ \frac{1}{2}\|\widehat{\theta}\|^2_*(t,t_0) + (\frac{1}{8} - C\int_{t_0}^t N^2(s)\, ds)\int_{t_0}^t \|div(\overline{\mathbf{u}})\|^2_{**}(s)\, ds \\ &+ \frac{c_V}{8}\int_{t_0}^t \|\overline{\mathbf{u}}\|^2_V ds \leqslant C\int_{t_0}^t \{\|\overline{\boldsymbol{\chi}}\|^2(r) + \int_{t_0}^r \|\overline{\theta}\|^2 ds + N^2(r)\|\widehat{\theta}\|^2_*(r,t_0)\}dr \\ &+ C(\|\overline{\theta}\|^2 + \|div(\overline{\mathbf{u}})\|^2 + \|\overline{\boldsymbol{\chi}}\|^2)(t_0) + C\tau^2. \end{aligned} \tag{5.3}$$

In view of the absolute continuity of integral, there exists a small number $\overline{t} > 0$ such that

$$C\int_{t_0}^{t_0+2\overline{t}} N^2(s)\, ds < \frac{1}{16}$$

for any $t_0 : 0 \leqslant t_0 \leqslant T - 2\overline{t}$. We can select $\overline{t}$ such that $\frac{T}{\overline{t}}$ is integer. If we insert the estimate (4.3) into (5.3), we deduce that the function

$$y(t,t_0) := \int_{t_0}^t \{\|\overline{\theta}\|^2 + \|div(\overline{\mathbf{u}})\|^2_{**}\}(s)ds + \|\widehat{\theta}\|^2_*(t,t_0) \tag{5.4}$$

satisfies the inequality

$$y(t,t_0) + \int_{t_0}^t \|\overline{\mathbf{u}}\|^2_V ds \leqslant \int_{t_0}^t R(s)y(s,t_0)ds + C(\|\overline{\theta}\|^2 + \|div(\overline{\mathbf{u}})\|^2 + \|\overline{\boldsymbol{\chi}}\|^2)(t_0) + C\tau^2.$$

Here $R(t) := C\max\left(1,\ N^2(t)\right) \in L^1(0,T)$. Using Gronwall's inequality we ensure

$$y(t,t_0) \leqslant C(\|\overline{\theta}\|^2 + \|div(\overline{\mathbf{u}})\|^2 + \|\overline{\chi}\|^2)(t_0) + C\tau^2$$

for any $t \in [t_0, t_0 + 2\overline{t}]$ with $t_0 \in [0, T - 2\overline{t}]$. In view of (5.3), (5.4), we have

$$\int_{t_0}^{t} \{\|\overline{\theta}\|^2 + \|div(\overline{\mathbf{u}})\|_{**}^2 + \|\overline{\mathbf{u}}\|_V^2\}(s)ds \leqslant C\{(\|\overline{\theta}\|^2 + \|div(\overline{\mathbf{u}})\|^2 + \|\overline{\chi}\|^2)(t_0) + \tau^2\}, \tag{5.5}$$

$$\|\widehat{\theta}\|_*^2(t,t_0) \leqslant C(\|\overline{\theta}\|^2 + \|div(\overline{\mathbf{u}})\|^2 + \|\overline{\chi}\|^2)(t_0) + \tau^2, \tag{5.6}$$

for any $t_0 : 0 \leqslant t_0 \leqslant T - 2\overline{t}$ and $t \in [t_0, t_0 + 2\overline{t}]$. Let us show that these estimates, which are local in the time variable, imply the validation of the estimate (5.5) for $t_0 = 0,\quad t = T$ and the estimate (5.6) for $t_0 = 0$, $\forall t \in [0, T]$, which guarantees the estimates (5.1), (5.2).

1st *step.* If we put $t_0 = 0$ in (5.5), (5.6) and use (3.2), (3.3), we derive

$$\int_0^{2\overline{t}} \{\|\overline{\theta}\|^2 + \|div(\overline{\mathbf{u}})\|_{**}^2 + \|\overline{\mathbf{u}}\|_V^2\}(s)ds \quad\leqslant\quad C\tau^2, \tag{5.7}$$

$$\|\widehat{\theta}\|_*^2(t,0) \quad\leqslant\quad C\tau^2, \qquad \forall t \in [0, 2\overline{t}]. \tag{5.8}$$

The estimate (4.3) brings to

$$\|\overline{\chi}\|^2(t) \leqslant C(\overline{t})\tau^2, \qquad \forall t \in [0, 2\overline{t}], \tag{5.9}$$

where the constant $C(\overline{t})$ depends on the chosen value of $\overline{t}$, being independent of τ.

Unfortunately, for the next interval $\left[2\overline{t}, 4\overline{t}\right]$ we cannot repeat previous calculations, since we are not able to estimate the term $\left(\|\overline{\theta}\|^2 + \|div(\overline{\mathbf{u}})\|^2\right)(2\overline{t})$. To avoid **this principal difficulty,** let us use the following argument.

2nd *step.* We consider the inequality (5.5) on the interval $\left[\overline{t}, 3\overline{t}\right]$. If we take $t := 3\overline{t}$ and integrate this inequality with respect to the variable t_0 over $\left[\overline{t}, 2\overline{t}\right]$, using (5.7), (5.9), we deduce the inequalities

$$\begin{aligned}\overline{t}\int_{2\overline{t}}^{3\overline{t}} \{\|\overline{\theta}\|^2 &+ \|div(\overline{\mathbf{u}})\|_{**}^2 + \|\overline{\mathbf{u}}\|_V^2\}(s)ds \leqslant \int_{\overline{t}}^{2\overline{t}} \{\int_{t_0}^{3\overline{t}} [\|\overline{\theta}\|^2 + \|div(\overline{\mathbf{u}})\|_{**}^2 + \|\overline{\mathbf{u}}\|_V^2](s)ds\}dt_0 \\ &\leqslant \int_{\overline{t}}^{2\overline{t}} C\{[\|\overline{\theta}\|^2 + \|div(\overline{\mathbf{u}})\|^2 + \|\overline{\chi}\|^2](t_0) + \tau^2\}dt_0 \leqslant C(\overline{t})\tau^2,\end{aligned}$$

i.e.

$$\int_{2\overline{t}}^{3\overline{t}} \{\|\overline{\theta}\|^2 + \|div(\overline{\mathbf{u}})\|_{**}^2 + \|\overline{\mathbf{u}}\|_V^2\}(s)ds \leqslant C(\overline{t})\tau^2.$$

Combining the last estimate with (5.7), we obtain

$$\int_0^{3\overline{t}} \{\|\overline{\theta}\|^2 + \|div(\overline{\mathbf{u}})\|_{**}^2 + \|\overline{\mathbf{u}}\|_V^2\}(s)ds \leqslant C(\overline{t})\tau^2.$$

Hence, from (4.3), we have $\|\overline{\chi}\|^2(t) \leqslant C(\overline{t})\tau^2, \qquad \forall t \in [0, 3\overline{t}]$.

Moreover, taking into account (5.8) and using the definition of the function $\|\widehat{\theta}\|_*^2(t, t_0)$, we can write for any $t_0 \in [\overline{t}, 2\overline{t}]$

$$\|\widehat{\theta}\|_*^2(t,0) \leqslant 2[\|\widehat{\theta}\|_*^2(t,t_0) + \|\widehat{\theta}\|_*^2(t_0,0)] \leqslant 2\|\widehat{\theta}\|_*^2(t,t_0) + C\tau^2, \qquad \forall t \in [2\overline{t}, 3\overline{t}],$$

then integrating it with respect of t_0 over $[\overline{t}, 2\overline{t}]$ and with the help of (5.6), (5.7), (5.9), we get

$$\begin{aligned}\|\widehat{\theta}\|_*^2(t,0) &\leqslant \frac{1}{\overline{t}}\{2\int_{\overline{t}}^{2\overline{t}} \|\widehat{\theta}\|_*^2(t,t_0)\, dt_0 + \overline{t}C\tau^2\} \\ &\leqslant C(\overline{t})\int_{\overline{t}}^{2\overline{t}} \{[\|\overline{\theta}\|^2 + \|div(\overline{\mathbf{u}})\|^2 + \|\overline{\chi}\|^2](t_0) + \tau^2\}dt_0 \leqslant C(\overline{t})\tau^2\end{aligned}$$

for any $t \in [2\overline{t}, 3\overline{t}]$. By (5.8) the inequality $\|\widehat{\theta}\|_*^2(t,0) \leqslant C(\overline{t})\tau^2$ is valid for any $t \in [0, 3\overline{t}]$. Hence, we derive the estimates (5.5) and (5.6) for $t_0 = 0, \quad t = 3\overline{t}$ and for $t_0 = 0, \forall t \in [0, 3\overline{t}]$, respectively.

<u>*i step*</u>. Applying the previous argument for the intervals $[i\overline{t}, (i+2)\overline{t}]$ with $i = 2, ..., \frac{T}{\overline{t}} - 2$, we obtain the estimates (5.5) for $t_0 = 0, t = T$ and (5.6) for $t_0 = 0, \forall t \in [0, T]$. We just note that the constants $C(\overline{t})$ grow with the repetition of this iterative argument, continuing to be a finite number, including the last step, when i takes the value $\frac{T}{\overline{t}} - 2$. ⊡

6. Continuous Dependence on Data

In this section, we formulate the continuous dependence result of the solutions on data of the problem (1.1)-(1.7). We consider two families of the data

$$F_i,\ f_i,\ \mathbf{G}_i,\ \ \mathbf{g}_i,\ \theta_i^0,\ \chi_i^0,\ i = 1, 2, \tag{6.1}$$

satisfying the conditions (2.3)-(2.4). The following statement holds.

Theorem 3. *Under the assumptions (2.2)-(2.6), let $\theta_i, \mathbf{u}_i, \chi_i$ $(i = 1, 2)$ be the solutions of (1.1)-(1.7) corresponding to the data (6.1). The difference of these solutions for the problem (1.1)-(1.7) is estimated as*

$$\|\theta_1 - \theta_2\|_{L^2(\Omega_T)}^2 + \|\mathbf{u}_1 - \mathbf{u}_2\|_{L^2(0,T;V)}^2 \leqslant C\Delta_T$$

and

$$\|\nabla \int_0^t (\theta_1 - \theta_2)ds\|^2 + \|\int_0^t (\theta_1 - \theta_2)ds\|_{L^2(\Gamma)}^2, \ \|\chi_1 - \chi_2\|^2(t) \leqslant C\Delta_t$$

for any $t \in [0, T]$, where

$$\begin{aligned}\Delta_t := \quad & \|\theta_1^0 - \theta_2^0\|^2 + \|\ \chi_1^0 - \chi_2^0\|^2 + \|F_1 - F_2\|_{L^2(0,t;L^2(\Omega))}^2 \\ + \quad & \|f_1 - f_2\|_{L^2(0,t;L^2(\Gamma))}^2 + \|\mathbf{g}_1 - \mathbf{g}_2\|_{L^2(0,t;L^2(\Gamma^1))}^2 + \|\mathbf{G}_1 - \mathbf{G}_2\|_{L^2(0,t;L^2(\Omega))}^2.\end{aligned}$$

To prove this theorem, we can apply the same approach that has been used in Theorem 2, but instead of the approximation error $C\tau^2$, we have to work with the time dependent "error" Δ_t in all estimates.

7. Appendix

Proof of Lemma 5. If we select for $\gamma = \chi$ in the equation (3.17) and $\gamma = \mathbf{X}_{\tau,lin}$ in (2.12), respectively, then we have

$$\widehat{\chi} \cdot [\frac{\partial}{\partial t}\mathbf{X}_{\tau,lin} + \mathbf{W}((\Theta_\tau)_d, \mathbf{U}_\tau))] \leqslant Q\phi_0|\widehat{\chi}|^2,$$
$$-\overline{\chi} \cdot [\frac{\partial}{\partial t}\chi + \mathbf{W}(\theta, \mathbf{u})] \leqslant Q\phi_0|\overline{\chi}|^2$$

for a.e. $(\mathbf{x}, t) \in \Omega_T$. Taking into consideration (4.1), the sum of these inequalities gives

$$\begin{aligned}\overline{\chi} \cdot [\frac{\partial}{\partial t}\overline{\chi} + \widehat{\mathbf{W}}] + Z &\leqslant Q\phi_0(|\widehat{\chi}|^2 + |\overline{\chi}|^2) \\ &\leqslant 2\,Q\phi_0|\overline{\chi}|^2 + Q\phi_0|\mathbf{X}_\tau - \mathbf{X}_{\tau,lin}|^2 \leqslant 2\,Q\phi_0|\overline{\chi}|^2 + C\tau^2, \qquad (7.1)\end{aligned}$$

where $\widehat{\mathbf{W}} := \mathbf{W}((\Theta_\tau)_d, \mathbf{U}_\tau) - \mathbf{W}(\theta, \mathbf{u})$, $\qquad Z := [\mathbf{X}_\tau - \mathbf{X}_{\tau,lin}] \cdot [\frac{\partial}{\partial t}\mathbf{X}_{\tau,lin} + \mathbf{W}((\Theta_\tau)_d, \mathbf{U}_\tau)]$ for $t \in ((i-1)\tau, i\tau]$.

It is easy to check that the term $\widehat{\mathbf{W}}$ is estimated by $|\widehat{\mathbf{W}}| \leqslant C(|\widehat{\theta}| + |div(\widehat{\mathbf{u}})|$ a.e. in Ω_T and the second factor in the expression Z is bounded in $L^\infty(\Omega_T)$ by (2.7), (3.15), (3.16). Therefore, integrating the inequality (7.1) on the variables $(\mathbf{x}, s) \in \Omega \times (t_0, t)$, in view of (4.1), (4.2), we infer

$$\|\overline{\chi}\|^2(t) \leqslant \|\overline{\chi}\|^2(t_0) + C\int_{t_0}^{t} (\|\overline{\chi}\|^2 + \|\overline{\theta}\|^2 + \|div(\overline{\mathbf{u}})\|^2)\, ds + C\tau^2,$$

that implies (4.3) by Gronwall's lemma.

Now we derive the estimate (4.4). Let us take the difference of the integral form of (1.2), (1.6) and the equation (3.12) and take $\mathbf{v} := \widehat{\mathbf{u}}$, then we have

$$a(\widehat{\mathbf{u}}, \widehat{\mathbf{u}}) = \sum_{i=1}^{3} I_i \text{ a.e. in } (0, T),$$

where $I_1 := -\int_\Omega \widehat{\alpha(\theta)\,\chi_2}\, div(\widehat{\mathbf{u}})\, d\mathbf{x}$, $\qquad I_2 := \int_\Omega \widehat{\mathbf{G}} \cdot \widehat{\mathbf{u}}\, d\mathbf{x}$, $\qquad I_3 := \int_{\Gamma^1} \widehat{\mathbf{g}} \cdot \widehat{\mathbf{u}}\, d\,\mathbf{x}$. It is well known there exists a constant c_V, such that $c_V||\mathbf{v}||_V \leqslant a(\mathbf{v}, \mathbf{v}), \forall \mathbf{v} \in V$ and easy to check that $(\lambda + 2\mu/3)||div(\widehat{\mathbf{u}})||^2 + \nu||\nabla div(\widehat{\mathbf{u}})||^2 \leqslant a(\widehat{\mathbf{u}}, \widehat{\mathbf{u}})$. Therefore

$$\frac{c_V}{4}\|\widehat{\mathbf{u}}\|_V^2 + \frac{3}{4}\|div(\widehat{\mathbf{u}})\|_{**}^2(t) \leqslant a(\widehat{\mathbf{u}}, \widehat{\mathbf{u}}) \leqslant \sum_{i=1}^{3} |I_i|. \qquad (7.2)$$

By the estimations (2.7), $|\chi_2| \leqslant c_P$ a.e. in Ω_T and due to the inequality $ab \leqslant \epsilon\frac{a^2}{2} + \frac{b^2}{2\epsilon}$, the terms I_i are estimated as

$$\begin{aligned}|I_1| &\leqslant C\int_\Omega |\widehat{\chi}_2||div(\widehat{\mathbf{u}})|\, d\mathbf{x}| + \theta_c c_\alpha c_P|\int_\Omega |\widehat{\theta}||div(\widehat{\mathbf{u}})|\, d\mathbf{x}| \\ &\leqslant C\|\widehat{\chi}\|^2 + \frac{1}{8}\|div(\widehat{\mathbf{u}})\|_{**}^2 + \theta_c c_\alpha c_P\|\widehat{\theta}\|\|div(\widehat{\mathbf{u}})\|, \\ |I_2|, |I_3| &\leqslant \frac{c_V}{16}\|\widehat{\mathbf{u}}\|_V^2 + C\tau^2.\end{aligned}$$

Substituting these estimates in (7.2) and integrating (7.2) on the time interval $[t_0, t]$ with the help of (4.1) and (4.2), we deduce (4.4). ☐

Proof Lemma 6. To simplify notations, where it will be possible, we will omit the dependence of the functions from the parameter $\mathbf{x}$. Sometimes we will write $\xi(t)$, instead of $\xi(\mathbf{x}, t)$. Let us integrate the integral form of (1.1), (1.5) and (3.11) over the variable $t \in (t_0, r)$, where $0 \leqslant t_0 < r \leqslant T$. Taking the difference of the obtained identities, we have

$$\begin{aligned}
(\{c_0\bar{\theta} \quad &+ \quad \overline{(\alpha(\theta) - \theta\alpha'(\theta))\chi_2 div(\mathbf{u})}\}(r), \varphi) + \{h(\nabla(\int_{t_0}^{r} \widehat{\theta}\, ds), \nabla\varphi) + \eta \int_\Gamma (\int_{t_0}^{r} \widehat{\theta}\, ds)\varphi\, d\mathbf{x}\} \\
&= \quad ([(L\overline{\chi}_1 - L_e\overline{\chi}_3)(r) + \int_{t_0}^{r} \widehat{F}\, ds], \varphi) + \eta \int_\Gamma (\int_{t_0}^{r} \widehat{f}\, ds)\varphi\, d\mathbf{x} \\
&+ \quad (\int_{t_0}^{r} \{\alpha(\Theta_\tau)\, X_{2\tau} \frac{\partial}{\partial s}[div(\mathbf{U}_{\tau,lin})] - \alpha(\theta)\chi_2 \frac{\partial}{\partial s}\, [div(\mathbf{u})]\}(s)\, ds, \varphi) \\
&+ \quad (\{c_0\bar{\theta} - L\overline{\chi}_1 + L_e\overline{\chi}_3 + \overline{(\alpha(\theta) - \theta\alpha'(\theta))\chi_2 div(\mathbf{u})}\}(t_0), \varphi)
\end{aligned}$$

for arbitary function $\varphi \in L^\infty(0, T, H^1(\Omega))$. Therefore, integrating this equality on (t_0, t), we see that the function $\bar{\theta}$ fulfills the equality

$$\begin{aligned}
\int_{t_0}^{t} [(\{c_0\bar{\theta} \quad &+ \quad \overline{(\alpha(\theta) - \theta\alpha'(\theta))}\chi_2 div(\mathbf{U}_{\tau,lin})\}(r), \varphi(r)) \\
&+ \quad \{h(\nabla(\int_{t_0}^{r} \widehat{\theta}\, ds), \nabla\varphi(r)) + \eta \int_\Gamma (\int_{t_0}^{r} \widehat{\theta} ds)\varphi(r)\, d\mathbf{x}\}]dr = \sum_{i=1}^{6} J_i(\varphi), \qquad (7.3)
\end{aligned}$$

where

$$\begin{aligned}
J_1(\varphi) \quad &: \quad = \eta \int_{t_0}^{t} [\int_\Gamma (\int_{t_0}^{r} \widehat{f}\, ds)\varphi(r)\, d\mathbf{x}]\, dr, \\
J_2(\varphi) \quad &: \quad = \int_{t_0}^{t} ([\{L\overline{\chi}_1 - L_e\overline{\chi}_3 - (\alpha(\Theta_{\tau,lin}) - \Theta_{\tau,lin}\alpha'(\Theta_{\tau,lin}))div(\mathbf{U}_{\tau,lin})\overline{\chi}_2\}(r) \\
&\qquad + \int_{t_0}^{r} \widehat{F}\, ds], \varphi(r))\, dr, \\
J_3(\varphi) \quad &: \quad = \int_{t_0}^{t} (\{-(\alpha(\theta) - \theta\alpha'(\theta))\chi_2 div(\overline{\mathbf{u}})\}(r) + \int_{t_0}^{r} \alpha(\theta)\chi_2 \frac{\partial}{\partial s} div(\overline{\mathbf{u}})\, ds,\ \varphi(r))\, dr, \\
J_4(\varphi) \quad &: \quad = \int_{t_0}^{t} (\int_{t_0}^{r} [\{\alpha(\Theta_\tau)X_{2\tau} - \alpha(\theta)\chi_2\}(s) \frac{\partial}{\partial s} div(\mathbf{U}_{\tau,lin})\,]ds, \varphi(r))\, dr, \\
J_5(\varphi) \quad &: \quad = \int_{t_0}^{t} (\{H(\mathbf{z}_{\tau,lin}) - (H(\mathbf{z}))_{\tau,lin}\}(r), \varphi(r))\, dr, \\
J_6(\varphi) \quad &: \quad = \int_{t_0}^{t} (\{c_0\bar{\theta} - L\overline{\chi}_1 + L_e\overline{\chi}_3 + \overline{(\alpha(\theta) - \theta\alpha'(\theta))\chi_2 div\mathbf{u}}\}(t_0), \varphi(r))dr
\end{aligned}$$

in the term $J_6(\varphi)$, we denote by $H(\mathbf{z}) := (\alpha(\Theta) - \Theta\alpha'(\Theta))\,\chi_2\, div(\mathbf{U})$ and $\mathbf{z} := (\Theta,\, \chi_2,\, \mathbf{U})$.

Let us choose $\varphi := \widehat{\theta}$ in (7.3). The mean value theorem gives the existence of a function $\xi(\mathbf{x}, t)$, satisfying the identity $\xi\alpha''(\xi)\overline{\theta} = \overline{(\alpha(\theta) - \theta\alpha'(\theta))}$ for a.e. $(\mathbf{x}, t) \in \Omega$. Then, in view of (2.6), $\xi\alpha''(\xi), \chi_2, div(\mathbf{U}_{\tau,lin}) \in L^\infty(\Omega_T)$ and using (4.1), (4.2), we obtain

$$\begin{aligned}
\mathbb{A}\int_{t_0}^{t} \|\overline{\theta}\|^2\, ds + \|\widehat{\theta}\|_*^2(t, t_0) &\leqslant \int_{t_0}^{t}\int_{\Omega} (c_0 - \xi\alpha''(\xi)\chi_2 div(\mathbf{U}_{\tau,lin}))|\overline{\theta}|^2\, d\mathbf{x}ds + \|\widehat{\theta}\|_*^2(t, t_0) \\
&= \int_{t_0}^{t}\int_{\Omega} [c_0\overline{\theta} + \overline{(\alpha(\theta) - \theta\alpha'(\theta))}\chi_2 div(\mathbf{U}_{\tau,lin})]\widehat{\theta}\, d\mathbf{x}ds \\
&+ \|\widehat{\theta}\|_*^2(t, t_0) + C\tau^2 \leqslant \sum_{i=1}^{6} |J_i(\widehat{\theta})| + C\tau^2.
\end{aligned} \tag{7.4}$$

Now we estimate the terms $J_i(\widehat{\theta})$. Integrating $J_1(\widehat{\theta})$ by parts on the time variable t, we derive

$$\begin{aligned}
|J_1(\widehat{\theta})| &= |\eta \int_\Gamma [(\int_{t_0}^{t} \widehat{f}\, ds)(\int_{t_0}^{t} \widehat{\theta}\, ds) - \int_{t_0}^{t} \widehat{f}(r)(\int_{t_0}^{r} \widehat{\theta}\, ds)\, dr]\, d\mathbf{x}| \\
&\leqslant 2\eta \int_{t_0}^{t} [\int_\Gamma |\widehat{f}|(r)| \int_{t_0}^{t} \widehat{\theta}\, ds|\, d\mathbf{x}]dr \leqslant \frac{1}{4}\|\widehat{\theta}\|_*^2(t, t_0) + C\tau^2.
\end{aligned} \tag{7.5}$$

With the help of $(\alpha(\Theta_{\tau,lin}) - \Theta_{\tau,lin}\alpha'(\Theta_{\tau,lin}))$, $div(\mathbf{U}_{\tau,lin}) \in L^\infty(\Omega_T)$, (3.14), (4.1) and (4.2), we get

$$\begin{aligned}
|J_2(\widehat{\theta})| &\leqslant C \int_{t_0}^{t}\int_{\Omega} [|\overline{\chi}| + |\int_{t_0}^{t} \widehat{F}\, ds|]|\widehat{\theta}|\, d\mathbf{x}dr \\
&\leqslant \frac{\mathbb{A}}{20} \int_{t_0}^{t} \|\overline{\theta}\|^2 dr + C\int_{t_0}^{t} \|\overline{\chi}\|^2\, dr + C\tau^2.
\end{aligned} \tag{7.6}$$

The integration by parts on the time variable in the expression

$$\begin{aligned}
\{-(\alpha(\theta) &- \theta\alpha'(\theta))\chi_2 div(\overline{\mathbf{u}})\}(r) + \int_{t_0}^{r} \alpha(\theta)\chi_2 \frac{\partial}{\partial s} div(\overline{\mathbf{u}})\, ds \\
&= \theta\alpha'(\theta)\chi_2 div(\overline{\mathbf{u}})(r) - \alpha(\theta)\chi_2 div(\overline{\mathbf{u}})(t_0) - \int_{t_0}^{r} div(\overline{\mathbf{u}})\frac{\partial}{\partial s}(\alpha(\theta)\chi_2)\, ds
\end{aligned}$$

allows us to present $J_3(\widehat{\theta})$ as

$$\begin{aligned}
J_3(\widehat{\theta}) &= \int_{t_0}^{t} \Bigg(\widehat{\theta}(r),\ \theta\alpha'(\theta)\chi_2 div(\overline{\mathbf{u}})(r) - \alpha(\theta)\chi_2 div(\overline{\mathbf{u}})(t_0) \\
&- \int_{t_0}^{r} div(\overline{\mathbf{u}})\frac{\partial}{\partial s}[\alpha(\theta)\chi_2]\, ds\Bigg)\, dr = J_8(\widehat{\theta}) - J_9(\widehat{\theta}).
\end{aligned}$$

Thanks to (2.7), the term $J_8(\widehat{\theta})$ is estimated as

$$|J_8(\widehat{\theta})| \leqslant \theta_c^2 c_\alpha c_P \int_{t_0}^{t}\int_{\Omega} |div(\overline{\mathbf{u}})||\widehat{\theta}|\, d\mathbf{x}dr + C\int_{t_0}^{t}\int_{\Omega} |div(\overline{\mathbf{u}})(\mathbf{x},t_0)||\widehat{\theta}(\mathbf{x},r)|\, d\mathbf{x}dr \quad (7.7)$$

$$\leqslant \frac{\mathbb{A}}{20}\int_{t_0}^{t} \|\overline{\theta}\|^2 dr + \theta_c^2 c_\alpha c_P \int_{t_0}^{t} \|\overline{\theta}\|\|div(\overline{\mathbf{u}})\|\, dr + C\|div(\overline{\mathbf{u}})\|^2(t_0) + C\tau^2.$$

Since $\widehat{\theta}(r) = \frac{\partial}{\partial r}[\int_{t_0}^{r} \widehat{\theta}\, ds]$, the term $J_9(\widehat{\theta})$ can be rewritten as

$$J_9(\widehat{\theta}) = \int_{\Omega}(\int_{t_0}^{t} \widehat{\theta}\, ds \int_{t_0}^{t} \frac{\partial}{\partial s}[\alpha(\theta)\chi_2]\ div(\overline{\mathbf{u}})\, ds$$

$$- \int_{t_0}^{t}\int_{t_0}^{r} \widehat{\theta}\, ds\, \{\frac{\partial}{\partial r}[\alpha(\theta)\chi_2]\ div(\overline{\mathbf{u}})\}(r))\, dr d\mathbf{x} = J_{10}(\widehat{\theta}) - J_{11}(\widehat{\theta}).$$

Applying the Holder inequality and the embedding theorem $H^1(\Omega) \subset L_4(\Omega)$ we infer

$$|J_{10}(\widehat{\theta})| \leqslant \|\int_{t_0}^{t} \widehat{\theta}\, ds\|_{L^4(\Omega)} \int_{t_0}^{t} N(s)\|div(\overline{\mathbf{u}})\|_{L^4(\Omega)}\, ds$$

$$\leqslant \frac{1}{4}\|\widehat{\theta}\|_*^2(t,t_0) + C\int_{t_0}^{t} N^2(s)\, ds \int_{t_0}^{t} \|div(\overline{\mathbf{u}})\|_{**}^2(s)\, ds \quad (7.8)$$

and, analogously, we have

$$|J_{11}(\widehat{\theta})| \leqslant \int_{t_0}^{t} \|\int_{t_0}^{r} \widehat{\theta}\, ds\|_{L^4(\Omega)} N(r)\|div(\overline{\mathbf{u}})\|_{L^4(\Omega)}(r)\, dr$$

$$\leqslant C\int_{t_0}^{t} N^2(r)\|\widehat{\theta}\|_*^2(r,t_0)\, dr + \frac{1}{8}\int_{t_0}^{t} \|div(\overline{\mathbf{u}})\|_{**}^2(r)\, dr. \quad (7.9)$$

Let us establish the estimate for $J_4(\widehat{\theta})$. By the estimate (2.9), we have that $M(t) := \|\frac{\partial}{\partial t} div(\mathbf{U}_{\tau,lin})\|_{C^0(\overline{\Omega})}(t) \in L^2(0,T)$, such that $||M||_{L^2(0,T)} \leqslant C$. Therefore, we obtain

$$|J_4(\widehat{\theta})| \leqslant \int_{t_0}^{t} \|\widehat{\theta}\|\,(r)\,\{\int_{t_0}^{r} M(s)\|\alpha(\Theta_\tau)X_{2\,\tau} - \alpha(\theta)\chi_2\|(s)\, ds\}\, dr$$

$$\leqslant \frac{\mathbb{A}}{20}\int_{t_0}^{t} \|\overline{\theta}\|^2\, dr + C\int_{t_0}^{t}\int_{t_0}^{r} \|\overline{\theta}\|^2(s)\, ds dr + C\int_{t_0}^{t} \|\overline{\chi}\|^2\, ds + C\tau^2. \quad (7.10)$$

In the last inequality, we use (4.1), (4.2).

Let us consider $J_5(\widehat{\theta})$. We have

$$\begin{aligned}|H(\mathbf{z}_{\tau,lin})-(H(\mathbf{z}))_{\tau,lin}| &= |\gamma(t)\{H(\gamma(t)\mathbf{z}^{i-1}+(1-\gamma(t))\mathbf{z}^{i})-H(\mathbf{z}^{i-1})\}\\ &+ (1-\gamma(t))\{H(\gamma(t)\mathbf{z}^{i-1}+(1-\gamma(t))\mathbf{z}^{i})-H(\mathbf{z}^{i})\}|\\ &\leqslant C\gamma(t)(1-\gamma(t))|\mathbf{z}^{i}-\mathbf{z}^{i-1}|\leqslant C\tau|\mathbf{z}^{i}_{\overline{t}}|\end{aligned}$$

for the time interval $t\in[(i-1)\tau,i\tau]$ with $\gamma(t):=\frac{(t-(i-1)\tau)}{\tau}$, since the function $H(\mathbf{z})$ is a Lipshitz function on the parameter $\mathbf{z}$. By (3.15), (3.16) we conclude that

$$\int_0^T\int_\Omega |H(\mathbf{z}_{\tau,lin})-(H(\mathbf{z}))_{\tau,lin}|^2 d\mathbf{x}dt\leqslant C\tau^2.$$

Hence, we infer

$$|J_5(\widehat{\theta})|\leqslant\frac{\mathbb{A}}{20}\int_{t_0}^{t}\|\overline{\theta}\|^2\,dr+C\tau^2. \tag{7.11}$$

Finally the term $J_6(\widehat{\theta})$ is estimated by

$$|J_6(\widehat{\theta})|\leqslant\frac{\mathbb{A}}{20}\int_{t_0}^{t}\|\overline{\theta}\|^2dr+C(\|\overline{\theta}\|^2+\|div(\overline{\mathbf{u}})\|^2+\|\overline{\chi}\|^2)(t_0)+C\tau^2. \tag{7.12}$$

Collecting the deduced estimates (7.5)-(7.12) in (7.4) we get (4.5). ⊡

Acknowledgments

N.V. Chemetov is grateful for the support from FCT-Plurianual 2008 and the Project "F´ısica - Matemática", PTDC / MAT / 69635/ 2006, FCT, Portugal.

References

[1] Baêta-Neves A.P., Savi M.A., Pacheco P.M.C.L. Mechanics Research Communications. 2004, **31**, 677–688.

[2] Colli P., Frémond M., Visintin A. Quart. Appl. Math. 1990, **48**, 31-47.

[3] Colli P., Sprekels J. Math. Methods Appl. Sci. 1995, **18**, 371-385.

[4] Bonetti E. Rend. Sem. Mat. Univ. Pol. Torino. 2002, **6**, 115-128.

[5] Colombo G., Goncharov V.V. J. Convex Analysis. 2001, **8 (1)**, 197-221.

[6] Chemetov N.V. Z. Anal. Anwendungen. 1998, **17 (4)**, 877-892.

[7] Chemetov N.V. Math. Methods in the Applied Sciences. 2006, **29(2)**, 209-233.

[8] Frémond M. Non-Smooth thermo-mechanics; Springer-Verlag: Berlin Heidelberg, 2002.

[9] Paiva A., Savi M.A. Math. Problems in Engineering. 2006, Volume 2006, Article ID 56876, 1-30.

[10] Savi M.A., Sá M.A.N., Paiva A., Pacheco P.M.C.L. Chaos, Solitons and Fractals. 2008, **36**, 828–842.

[11] Shemetov N.V. Adv. Math. Sci. Appl. 1998, **8 (1)**, 157-172.

[12] Stefanelli U. Math. Comp. 2002, **71 (240)**, 1431-1453.

[13] Stefanelli U. Adv. Math. Sci. Appl. 2000, **10 (2)**, 917-936.

In: Shape Memory Alloys
Editor: H. R. Chen, pp. 441-485
ISBN 978-1-60741-789-7

Chapter 16

SHAPE MEMORY ALLOYS IN MICROPOSITIONING APPLICATIONS

E. Asua, J. Feuchtwanger, V. Etxebarria and A. García-Arribas
Departmento Electricidad y Electronica, UPV/EHU, Spain

Abstract

Shape memory alloys (SMA) are a special kind of smart materials whose dimensions can be modified as the result of a temperature-dependent structural phase transition. This property can be used to generate motion or force in electromechanical devices and micromachines. The use of shape memory alloys in actuators offers the opportunity to develop robust, simple and lightweight elements for application in a multitude of different industries. Despite all these advantages, the accuracy of SMA actuators is severely limited by their highly nonlinear stimulus-response characteristics. Traditionally, they have been used as "on-off" electromechanical actuators due to the non-linear and hysteretic nature of the martensite-austenite transformation. The ideal solution to this problem is to model the hysteresis mathematically to compensate for it in the control loop. However, useful models of phase transitions are difficult to obtain. In this chapter, a Nickel-Titanium alloy (Nitinol) wire is considered as the active element in micropositioning actuators. An electric current, applied through the wire, heats it to induce the phase transition and the consequent contraction. The purpose of this investigation is to finely control the wire contraction. The wire needs to be able to contract and relax to an intermediate position within its range of movement and the gradual contraction and relaxation needs to be controllable. An experimental micropositioning device, based on a Ni-Ti wire, is controlled using several control strategies. To account for the hysteresis effects, an inverse hysteresis model is obtained using two different compensating methods. When used in a controller, the inverse models nicely compensate the non-linear and hysteretic behavior of the wire, and a proportional-integral with antiwindup control works adequately, obtaining position accuracies around 3 micrometers. Once the real possibility of using these materials as micrometric actuators is analyzed and tested, a SMA based actuator prototype grip for a flexible robot is developed, so that it could be used in light applications. For this purpose, the bulky LVDT (linear variable differential transformer) used in the previous experiments to provide the position feedback to the controller must be replaced by lighter alternatives. The first solution consists of deducing the strain of the SMA wire from the readings of a set of strain gauges that are placed in the fingers of the grip. Although the precision is limited by the strain gauge accuracy, position errors about

30 micrometers are achieved. The second one is the implementation of a sensorless design, where the change in the resistivity of the SMA wire is used to determine the strain of the active element. Position errors about 60 micrometers are achieved, with the great advantage that the actuator is reduced to a single SMA element, specially important if the goal is to reduce the overall size and weight of the actuator, like is the trend in the miniaturization in mechatronics and robotics. The experiments presented show that SMA wires as active elements are serious alternatives to be used as precision actuators in micromachines.

1. Introduction

Shape Memory Alloys (SMAs) are attractive materials for use as the drive elements in actuators. This is because they can exhibit large strains when receiving an external stimulus, usually in the form of heat. At an specific temperature, these alloys undergo a phase transition, and as a result, at a microscopic level, the interatomic spacing in the alloy changes. At a macroscopic level, the phase transition is manifested as a change in the dimensions of the alloy. If the material is shaped in the form of a wire, its length will change as it is heated above the transformation temperature. Fortunately enough, the SMA material can conduct electricity, so the thermal transformation can be induced by making an electrical current, of adequate intensity, flow through it. This opens up the possibility for light-weight actuators, a matter of key importance if the goal is to reduce the overall weight, size and cost of the actuator.

1.1. Material Properties

SMAs are a subset of a class of materials known as "active" or "smart" materials. In particular, they respond to an external stimulus by changing their dimensions. SMAs strain (3 – 8%) when heated above a specific temperature, called the martensite transformation temperature. The change in dimensions in these materials is due to a diffusionless phase transformation from a lower symmetry, low-temperature phase to a higher symmetry, high-temperature one. In addition to the large strain, they also exhibit a "shape memory" effect. These materials can sustain large deformations in the low temperature phase, but unlike conventional metals, if they are heated into the high temperature phase, they recover their original shape.

When these materials transform from the high temperature phase into the low temperature phase, the transformation stresses are accommodated microscopically by twinning rather than by slip. In slip when the material is deformed inelastically, dislocations glide to the edges of the crystal causing a permanent deformation. When a material deforms by twinning, variants are formed. These are regions in the crystal where the unique axis of the crystal points in a specific direction. In adjacent variants the unique axis is rotated by an angle related to the difference in length between the longer and shorter axes in the crystal. The planes that separate adjacent variants are called twin-boundaries, and the crystal structure of the variant on one side of the twin-boundaries is the mirror image of the crystal structure of the adjacent variant.

The application of a shear stress transforms the single variant arrangement to a twinned structure with different variant orientations, then to a different single variant arrangement.

This mode of deformation is called pseudo-plasticity, because if the stress is removed, the deformation remains, but if a stress is applied in the opposite direction it can restore the material to its initial state. In twinning, even after large deformations, all atoms are connected to the same atoms they were connected to before the crystal was deformed. When the deformed material is heated into the higher symmetry phase the twins disappear and the original shape of the material is restored.

There is also a "two way shape memory", if a material is "trained", by cycling it repeatedly from one phase to the other while constrained, defects are created that cause twin-boundaries to form in the same place. By forcing the twin-boundaries reform in the same places, the same shape can be obtained every time the material is cooled into the low temperature phase, giving the material its "two way shape memory". The phase transformation in these alloys can also be achieved by applying a stress to them. If an alloy above its transformation temperature is subjected to a stress, it will transform into the lower temperature phase, where the deformation will take place by twin boundary motion. As soon as the stress is released, the alloy will re-transform into the high temperature phase and the deformation will be recovered. This behavior is known as the super-elastic effect. For their use as actuator materials, neither the super-elastic nor the pseudo-plastic effects are of interest, only the strain induced by the phase transition [1].

There are several alloys that show a shape memory effect, but only a few have found a use in practical applications. Cu–Al–Zn and Cu–Al–Ni are used when a narrow thermal hysteresis is required. But by far, the most widely used shape memory alloys are the Ni-Ti alloys, with compositions near 50% Ni atom. This is because they show better fatigue characteristics and corrosion resistance than the copper based ones [2]. A few of the main properties of these two types of SMAs are shown in Table 1.

Table 1. Summary of the main properties of SMAs (from Van Hubeeck [1]).

Property (units)	NiTi alloys	Cu-based alloys
Melting point (°C)	1300	1000
Density (g cm^3)	6.5	7.8
Young's modulus (austenite) (GPa)	80	80
Elongation at failure (%)	30–50	10–15
Transformation temperature (°C)	-100–100	-200–200
Hysteresis (°C)	5–50	5–20
Superelastic strain (%)	6–8	4
Work output (J/g)	4	1

1.2. Applications

In some robotic and mechanic systems, miniaturization has created the need of compact smart material-based actuators to power mechanical devices built at smaller scales. The use of SMA in actuators offers the opportunity to develop robust, simple and lightweight elements for application in a multitude of different industries. Those actuators, that have a high power to weight ratio, have been used to replace electric, hydraulic and pneumatic actuators and have found applications in many areas including micro-mechatronics and light robotics.

Among the early applications of SMA worth noting we find the use of SMAs in weldless pipe joints for aero-space and maritime applications. In the automotive industry we can find examples of their use in pressure control valves for the automatic transmissions. However, where they have found the larger number of applications is in the field of medicine, where they have been used as inter-cranial plates, stents and orthodontics, to cite a few examples. There have also been large efforts to incorporate them into composite materials to mitigate noise and vibrations, and for the active control of structures. Lastly, we should also mention the use of these alloys in some recently designed microactuators [3, 4, 5, 6].

However, in most of the applications mentioned before, SMAs have traditionally been only used as "on-off" electromechanical actuators, because of the intrinsic difficulty in accurately controlling the material's mechanical deformation. This is due to the microscopic rearrangements that occur during the martensite-austenite transformation which take place quite sharply and in a highly complex and nonlinear fashion.

1.3. Control

From the experimental point of view, the main difficulty in controlling these materials is caused by the hysteretic behavior of the phase transition. Despite their highly non linear character, there have been may attempts to control actuators based on these materials using several control techniques. Hashimoto, Takeda, Sagawa, Chiba and Sat applied a PD control scheme to the SMA wires used as actuators of a biped walking robot [7]. Ikuta, Tsukamoto y Hirose used active PID control on a segmented active endoscope made with SMA springs [8], and Troisfontaine, Bidaud y Dario applied PI control on SMA actuators with an additional thermal sensor [9].

A memoryless controller is in general not sufficient to achieve high precision positioning in a SMA actuator, due to the presence of memory in the system under control. One possible solution to this problem would be to develop an accurate model for the thermomechanical behavior of the material, based on already developed mathematical models of hysteresis [10], as it has been done for other materials [6]. From this point on, an inverse model could be generated and implemented in the controller, in order to compensate the system's hysteresis [11]. The principal advantage of this method is that the system compensator-actuator is quasilinear. However, good models of phase transitions, ready to be used in a control loop, are by no means easily obtainable.

1.4. Hysteresis compensation

Several models have been proposed for the behavior of hysteretic systems in general [12], and specifically for SMAs. One of the best known models is the Preisach model [13, 14]. Madill and Wang used a very simple control to verify the SMA system model, adapted it, and discussed the system stability [15]. Majima, modelled a SMA actuator based on the model proposed by Briston and controlled it using a simple PID [16]. Hasegawa modelled the hysteresis and proposed to compensate it by applying an inverse model, and showed experimentally the utility of this approach [17]. As discussed in the same paper, the static Preisach hysteresis model can not represent the dynamic behavior of the actuators so, in order to compensate the hysteresis of a dynamic plant, the hysteresis was divided into a dynamic and static part, and a controlled system which controls each part by using two reverse models separately was proposed.

The inverse algorithm needs to be as fast as possible, for a real time application. For this reason, even though the classic Preisach model is the most used, it proves to be impractical for real time control because of the large computation times it requires. In the bibliography, several works dealing with this problem can be found, implementing compensation methods such as a fuzzy method based on the Preisach model [18] or adaptive control algorithms [19, 20, 21]. For instance, Dickinson used an adaptive control to compensate directly the hysteresis of SMAs when they are used in vibration suppression applications. Weeb, Wilson, Lagoudas and Rediniotis took into account in their adaptive control algorithm the estimation of thermal exchange between the SMA actuators and the environment and calculated the compensating input using the established SMA model. Lee described and implemented experimentally the control of an SMA-based actuator, similar to the one that will be described in this work, using the LQR and H_2 techniques [22].

In a slightly different strategy, some work has been done where the hysteresis is treated as a phase lag [23], and by implementing a compensator called a "phaser". This operator is capable of pushing the phase forward, so that the non-linearities can be compensated. This approach has been applied to SMAs in [24].

The Bouc-Wen model has gained a lot of interest as a way of characterizing smooth hysteresis, manly because it is capable of reproducing, in an analytical form, a range of different shapes of hysteresis loops describing the behavior of a system. Numerical simulations have shown that, by combining it with an adaptive [25] or PID control [26], very satisfactory results can be obtained.

There are several works dealing with the fabrication and control of micropositioners based on SMAs for real applications. Among them we find the development of a gripper capable of handling objects 100-500 μm in size [27], the design of a microgripper for micromanipulation of microcomponents using SMA wires and flexible hinges [28], or the control of a structure by the use of counter-acting actuators [29].

1.5. Summary of the work presented

In our experiments, we use a Nickel-Titanium alloy (Nitinol) wire as the active element in a series of micropositioning actuator prototypes. An electric current applied through the wire is used to heat it, inducing the phase transition and the consequent contraction. When the material cools off, the wire stretches again, provided that a sufficient tensile stress is applied to the wire as a restoring force.

Our goal is to finely control the wire contraction, aiming to develop SMA-based actuators capable of being positioned with micrometric precision. There are two tasks that need to be accomplished: on the one hand, we must be able to command the wire to contract and relax to a fixed position. On the other, we must be able to control the velocity of the contraction and relaxation processes. Several control strategies are implemented on different experimental set-ups in order to analyze the actual possibilities of using SMA-s as precision actuators.

As it has already been mentioned, because of the nonlinearities of the Nitinol phase transformation, a simple control strategy is insufficient for micrometric control. An attempt has been made to mathematically model the hysteretic behavior of those materials. For that purpose, the hysteresis of the system is characterized and the influence of recovery force in its shape is analyzed. Then, the hysteresis is modelled using the Bouc-Wen model [30], obtaining a mathematical representation of system behavior, although for the moment, it is only valid for the major loop of the hysteresis.

Using a different perspective, two experimental approaches of hysteresis compensation have been implemented with the purpose of converting the nonlinear system in a quasilinear one. For this task, the hysteretic nonlinear input-output characteristics are first carefully described from experimental data. Then, the compensation is made, either introducing a phase shift in the system by means of a linear compensator, or using a neural network appropriately trained to learn the system's inverse hysteresis.

In the first approach, the system hysteresis is essentially considered as a phase lag between a periodic input signal and the corresponding output signal [23]. Therefore, we implement a compensating stage that shifts the necessary phase angle to approximately cancel the system nonlinearities.

The second strategy is based on producing a non-linear compensation by implementing empirically the behavior of the wire with a neural network (NN). NNs are usually employed within a control-systems context as modelling tools to approximate either the unknown plant's direct or inverse dynamics, exploiting the networks's well-known capability of nonlinear mapping to any desired accuracy [31, 32]. Thus, they are ideal tools to implement a nonlinear inverse hysteresis compensation [33].

Both compensation methods are closed-loop tested on an experimental SMA laboratory positioning device. In this set-up, a high resolution position sensor (namely, a linear variable differential transformer, LVDT) is used to measure the position and to provide the feedback signal. Experimental closed-loop results obtained with both compensation methods are presented and their respective performance merits are discussed.

Our results show that, despite their nonlinear hysteretic characteristics, SMA wires can be successfully controlled to be used in micropositioning applications with micrometer accuracy, by including hysteresis effects in a nonlinear control scheme. It is found that, appropriately trained NNs, inserted in a PI with antiwindup control loop, are able to obtain position errors around 2-3 μm.

Once the real possibility of using those materials as micrometric actuators was analyzed and experimentally tested, we designed and constructed a SMA based actuator prototype that could be used in light-weight applications. The accuracy of the actuation strongly depends on the sensor used to provide the feedback signal in the control loop. The LVDT position sensor used in previous control experiments results to be too costly and, most of all, too heavy for those applications.

Some previous attempts of obtaining a suitable feedback signal include the use of the temperature of the SMA wire [9], strain gages to measure the deformation caused by the wire strain in an associated structure, and the use of the wire resistance as a measure of its deformation [34]. Temperature control proved impractical due to the difficulty in measuring the temperature, so, among all the options analyzed, the other two solutions are tested in this work: the use of strain gauges to measure the structural deformation of the actuator; and the implementation of a sensorless design, where the change in the resistivity of the

SMA wire resulting from the phase transition is used to determine the strain of the active element.

The experimental results show that, with the first method, position errors around 30 μm are obtained, being the main drawback the limited accuracy of the strain gauges. The sensorless control, allows to obtain accuracies around 60 μm, with the great advantage that no additional position sensor is needed and the actuator can be reduced to a single, small, compact and light SMA element.

In summary, in this chapter we will show through experimental results obtained on three different prototype systems that SMAs are relevant contenders for use in micropositioning actuators. It will be proven that they can achieve resolutions of 2 to 3 μm when a sensitive position sensor (LVDT) is used to provide the feedback signal. Even when the change in resistance of the wire is used to infer the deformation of the wire, accuracies of the order of 60 μm can be achieved. A detailed description of our experiments and results can be found elsewhere [35, 36, 37, 38, 39]. The results that we show are complementary to those obtained by Song et al. in several of their publications [40, 34].

2. Shape Memory Alloys

SMAs are a particular kind of smart materials. Smart materials can be defined as materials capable of changing in a recoverable fashion their physical properties (i.e. stiffness, viscosity, shape, color, among others) in the presence of external stimuli. There are different kinds of smart materials depending on their nature, the material itself, or the external driving stimulus.

The shape memory effect found in SMAs is based on the diffusion-less transformation between two solid phases: a low temperature one (martensite), and another high temperature one (austenite). The martensitic transformation is reversible, in the sense that upon heating from low-temperature martensite phase, the onset of the parent phase will commence above A_s and will be complete when the temperature exceeds A_f. Upon cooling to below M_s from the high-temperature austenite phase, the transformation back to martensite will commence. That transformation will be complete when M_f is reached (see figures 1 and 2, [41, 3]). [1]

The fundamental difference between the martensite and austenite phases, and the basis for an explanation of the shape memory effect, is the crystallographic symmetry: the austenite phase exhibits greater symmetry that the martensite phase. The variation in its crystal structure induces shape changes in the material, specifically in its length if the material has wire form. This property can be used to generate motion or force in electromechanical devices by electrically heating the material.

Shape recovery and pseudoelasticity are probably the best-known macroscopic changes accompanying the phase transformation in SMAs. Nevertheless, a significant number of

[1] The terms martensite and austenite are used to describe the low temperature, low symmetry and the high temperature, high symmetry phases respectively. These terms are used because the phase transitions in SMAs are similar to those observed in steels, where the face centered cubic phase is named austenite after the British metallurgist Sir William Chandler Roberts-Austen, and the body centered tetragonal phase is called martensite after German metallurgist Adolf Martens. The transformation from austenite to martensite is called a martensitic transformation, it also occurs in steels, but unlike in SMAs, this transformation is not reversible.

other physical and chemical properties are also altered during the transformation [3].

- *Electrical Properties:* The electrical resistivity, thermoelectric power, electromotive force and other properties are modified during the transformation process.
- *Thermal Properties:* The thermal conductivity, heat capacity and latent heat of transformation are altered.
- *Mechanical Properties:* Yield strength, Young's modulus, damping and internal friction are some of the mechanical properties that are altered during the transformation.

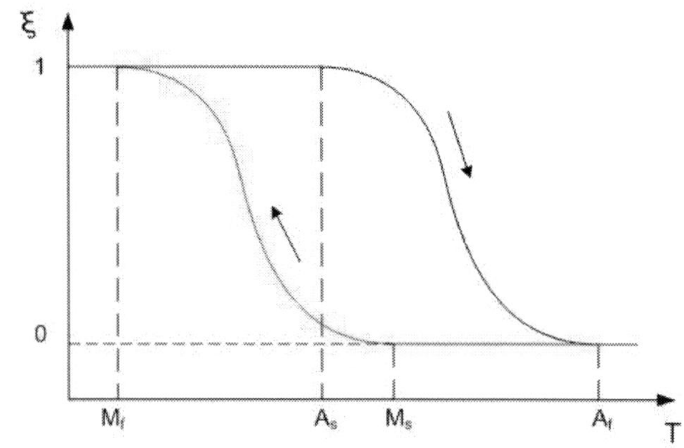

Figure 1. Strain-temperature curve of the phase transformations of SMA materials driven by thermal and mechanical stimulus.

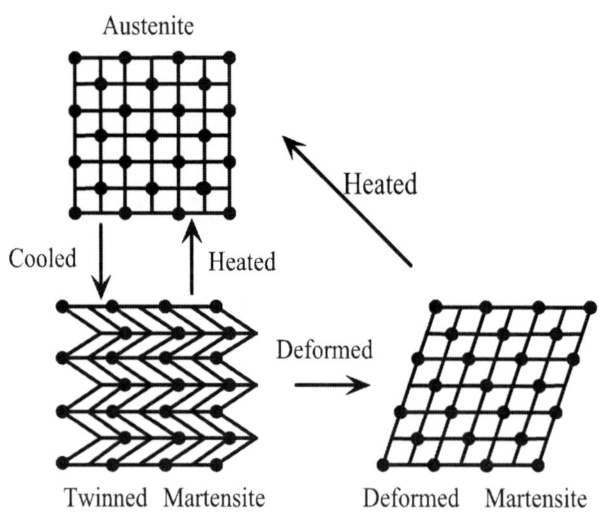

Figure 2. Scheme of the phase transformations of SMA materials driven by thermal and mechanical stimulus.

2.1. Nitinol wires

There are a lot of alloys that show shape memory, but only few of them have been commercially developed, as NiTi, NiTi-X (where X is a ternary element) and CuZnAl. Actually, 90% of applications are based in NiTi, NiTiCu and NiTiNb.

Nickel-titanium alloys (Nitinol) were first developed between 1962 and 1963, at the Naval Ordnance Laboratory (White Oak, Maryland USA) and has become one of the most useful variants of shape memory alloys, due to its mechanical properties and its capability for large strain recovery. During the 60's and 70's, it began to be produced on a commercial scale in different forms: sheets, wires, tubes...

NiTi alloys are thermally stable, they have an excellent corrosion resistance compared to the copper based SMAs, they have a high electric resistivity, they are easily alloyed and can be extruded, they have a low cyclic fatigue and they have a wide range of transformation temperatures. Moreover, the NiTi intermetallic has a narrow solubility range, making it easy to alloy them into ternary alloys [5].

As Nitinol has high electric conductivity, and wires have small transversal section, electrical dissipation because of Joule effect is used to heat the alloy. The applied current range can vary between 50 mA and 1 A.

Those wires produce large strains, typically on the order of 4% to 10%, but 5% should not be exceeded to avoid cyclic fatigue and obtain a repeatable behavior. The strain achieved depends on the load on the SMA in the low temperature phase, the restoring force.

Table 2. Nitinol wire's properties.

Property (units)	**Austenite**	**Martensite**
Density (g/cm^3)	6.45	
Resistivity ($\mu\Omega.cm$)	100 aprox.	70 aprox.
Thermal conductivity (W^oC/cm)	18	8.5
Breaking stress (MPa)	895	
Phase transformation temperature (^0C)	From -200 to 110	
Twining strain	8.5% máx.	

The disadvantage of Nitinol, and SMAs in general, is time response. The frequency response of the material has been characterized and measured by several authors (see for instance [42]). It is observed that typical SMA wires behave essentially like a low-pass filter with a cut-off frequency of around 1 Hz (which corresponds to a period of 1 s). As a result, SMA actuators are rarely used for applications requiring response times shorter than seconds or tenths of seconds. Therefore, the use of SMAs for suppression of high-frequency vibrations is limited by the slow response time of the material. In light of this, our experiments have been designed with position accuracy rather than speed in mind. Table 2 summarizes the main characteristics of Nitinol wires.

2.2. Applications based in shape memory alloys

During the last two decades, both industry and academia have evinced a growing interest in the application of Nitinol, and of SMAs in general, increasing the number of engineering systems being developed that utilize smart materials. This is basically because they are intrinsically susceptible of use as sensors and as actuators, which makes them suitable for integration in smart structures for these purposes. SMAs are particularly well suited to applications that require a large amount of deformation since SMAs can produce much larger strains than can conventional piezoelectric devices. The use of SMAs, though, is based on their large range of motion and silent operation.

SMAs are being increasingly considered for civil, aeronautic, aerospace and industrial applications which require significant passive damping or utilize the high work output densities exhibited by the materials. Some of these applications are not based on the shape memory effect, but on the superelastic or pseudoplastic effects. The advantage of SMAs for these applications is their ability to undergo large deformations without suffering from plastic deformation. Because the energy dissipated by the materials is proportional to the area of the hysteresis loop, superelastic operating regimes which maximize hysteresis are required when employing SMA as tendons to attenuate earthquake or wind-induced vibrations in buildings or as fibers to eliminate vibrations in articulated antennas or membrane mirrors [3].

In a number of early applications, the shape memory properties of SMA were exploited to create innovative and efficient fasteners, connectors, and clamps. For example, commercially available pipe couplers have been manufactured by employing a Nitinol sleeve whose inner diameter is smaller than that of the adjoining pipe. In the automotive industry, for example, we can find examples such as the pressure control valves for the automatic transmission or the safety clamps for the fog lights.

SMAs properties, along with biocompatibility, allow them to be used in medical applications to combat cardiovascular disease. Stents, which consist of cylindrical memory alloy mesh that expands when placed in an artery or vein, open the blood vessel and restore blood flow. Other medical applications include their use as tumor locators, blood filters, intracranial plates, orthodontics and many more.

Shape memory materials have also been used for active-passive vibration suppression systems and as positioning devices for systems such as biomimetic hydrofoils.

More recently, SMA have been considered as robotic actuators, fire detection, sprinkler and gas shutoff valves which sense and respond to elevated temperatures.

3. Experimental Set-Up

We have used a number of experimental set-ups in our lab to test the performance of SMA-based actuators. Here, we describe the experiments performed on three of them. The first one is a generic experimental platform that we have used to implement different control strategies for the linear motion control. It has been used to select the most suitable control schemes in terms of accuracy and control effort (the magnitude and frequency content of the control signals). The second one is a prototype of a SMA-based grip to be used in a light flexible robot. The challenges of controlling this grip lead us to design a simplified version, the third one in the list: a SMA-based actuator for angular deflection control experiments.

For all of our experiments, we have used a commercial SMA wire, sold under the name Flexinol. The actuation of the wires is triggered by Joule heating, that is, the phase transformation that causes the contraction of the wires is induced thermally by passing an electric current through them.

3.1. Linear motion control laboratory experimental set-up

The generic experimental set-up used in our investigation is depicted in figure 3. It is composed of a holding platform from which the Nitinol wire is suspended vertically. The lower part of the wire is attached to the core of a linear variable differential transformer (LVDT) which is free to move without friction. An additional weight hangs from the LVDT core, to provide the restoring force which makes the wire recover its original length.

A Nitinol wire 75 mm long (in the low temperature phase) with a diameter of 0.15 mm is used in this device.

The wire is heated by the current provided by a home-made current source whose output is controlled by a voltage signal. The actual current flowing through the wire is measured by the voltage drop in a 1 Ohm calibrated resistor. The displacement of the wire, that is, its contraction or elongation, is measured by the position of its lower end, using the LVDT position sensor. The excitation and signal conditioning circuit of the LVDT is based on the Analog Devices AD598 chip. It has been designed so the sensor provides an output of -10 V when the wire is not contracted and a value near 10 V when it is fully contracted (approximately 3 mm). A gauge situated on top of the holding platform allows to finely adjust the position of the wire in its relaxed state in order to get the initial -10 V reading.

For this set-up, all the input and output signals involved in the experiment are managed by a data acquisition card (National Instruments 6025E) inserted in a personal computer. They include the voltage signal that drives the current source, the voltage drop in the calibrated resistor, the output of the LVDT position sensor, and also the voltage drop in the wire which is used in the sensorless control experiments described in section 6. The 12 bit A/D converter of the acquisition card allows a position resolution of about $3/4$ of a micron when the full span (-10 V to 10 V) of the LVDT output is used. Most of the experiments performed with this experimental set-up are executed using Simulink^{TM} and Matlab^{TM}, although LabView^{TM} is used for some others.

3.2. SMA-based grip prototype

Figure 4 shows a SMA-based gripping actuator developed to be used in a light flexible robot. The grip "fingers" consist of two flexible stainless steel plates. The grip is closed when the Nitinol wire attached between them is contracted. Each "finger" is sensorized by a set of two strain gages, situated on both faces of the stainless steel plates. They are connected in a half-bridge configuration to a strain recorder (Vishay, model P3) that provides their deformation state. The deformation of the "fingers" is related to the strain of the SMA wire, using the classic Euler-Bernoulli beam theory [43].

The performance of this gripping actuator depends on the control of the flexion of the fingers. Figure 5 shows a typical result of the deformation of the grip fingers when a current passes through the SMA wire. The graph on top shows the magnitude of the current that flows through the wire. In this experiment, the current increases linearly until the SMA wire is fully contracted and then decreases, allowing its relaxation. The center and bottom graphs

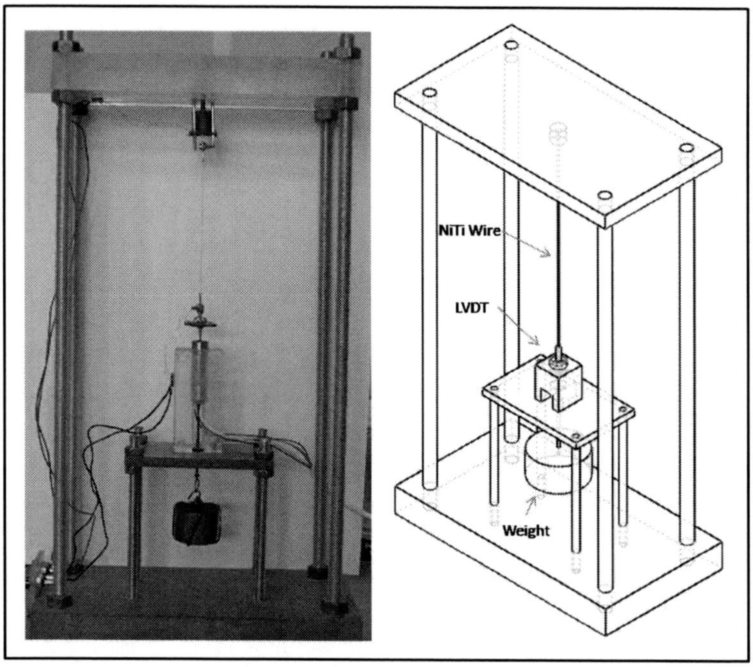

Figure 3. Linear positioning experimental platform and scheme.

show the measured deformation of the fingers. Both deformations are very similar, as can be expected. Therefore, to obtain a better understanding of the problem and to investigate the optimal control strategies for this kind of actuator, we have decided to simplify the problem by studying in detail the behavior of a single finger. An additional actuator prototype is designed and constructed, which is presented in the following section.

3.3. Angular motion control actuator

To simplify the problem addressed by the grip, we develop a new actuator prototype which can be considered as a single-fingered version of the former one. As the response of both "fingers" of the gripper are equivalent, as shown in figure 5, we expect that the results obtained in the control experiments performed on this second prototype are directly applicable to the full, two fingered, gripper. Evidently, the experiments are simpler because only one finger needs to be accounted for.

We call this device as "angular positioning actuator" since the effect of the contraction of the SMA wire is to displace angularly the flexible stainless steel plate (the finger).

The set-up for this actuator is shown in figure 6. The Nitinol wire used in this platform is 90 mm long and has a diameter of 0.15 mm. The prototype is composed by a flexible stainless steel plate which is rigidly fixed at one end, while the other one is pulled downwards by the SMA wire when contracted. The flexion of the plate is measured using four strain gauges (two on each side) in full-bridge configuration, connected to a strain indicator (Vishay. model P3). An inductive sensor (Carlo Gavazzi, model EI18051020) is situated near the end of the free end of the finger, to accurately measure the displacement produced by the contraction of the SMA wire.

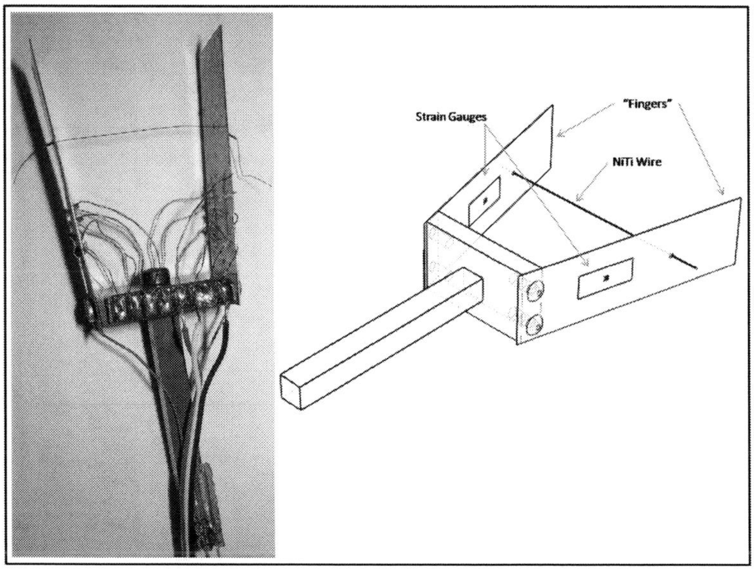

Figure 4. A SMA-based grip prototype for a light robot.

The instrumentation selected for this actuator includes a Source-Meter (Keithley, model 2602) to provide the drive current to the wire and to measure the output of the inductive sensor. In addition, it records the voltage drop across the wire to determine its resistance, although this data has not yet been used in the control of this actuator. The Source-Meter is connected via GPIB to a personal computer that also receives the data from the strain indicator via an USB connection. For this set-up, the control experiments are implemented in LabView™.

4. Hysteresis and Hysteresis Compensation Methods

An example of the SMA wire hysteresis measured in the linear motion control laboratory experimental set-up (figure 3) is shown in figure 7. This particular curve is obtained with a sinusoidal current that has a 40 s period, flowing through the wire. Its amplitude is limited to 0.4 A to avoid overheating and degradation of the wire. The maximum contraction experienced by the wire is 2.9 mm and sharp changes in length are observed at about 0.3 A in the ascending branch and 0.15 A in the descending one.

As described in the literature [3, 4], and mentioned in section 2, the strain that is possible to achieve depends on the load on the SMA in the low temperature phase, that we call the restoring force. In this section, we have experimentally verified it and analyzed how the hysteresis shape depends on the restoring force.

As it has already been mentioned in the introduction, the main difficulty in controlling these materials is the hysteretic behavior of the phase transition. For this reason, any procedure of hysteresis compensation is likely to improve the precision in control experiments. Our approach is to model empirically the behavior of the wire in order to be able to

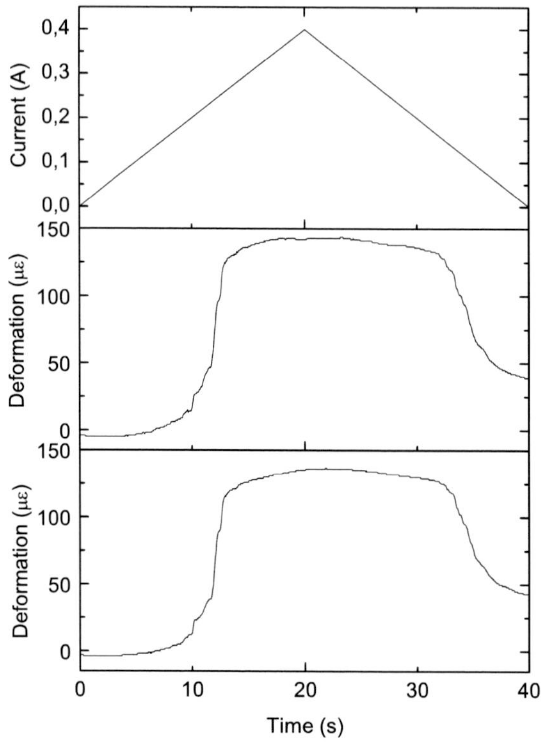

Figure 5. Experimentally measured finger deformation in the robot grip: *Top:* Current through the wire; *Center:* Deformation of one finger; *Bottom:* Deformation of the other finger.

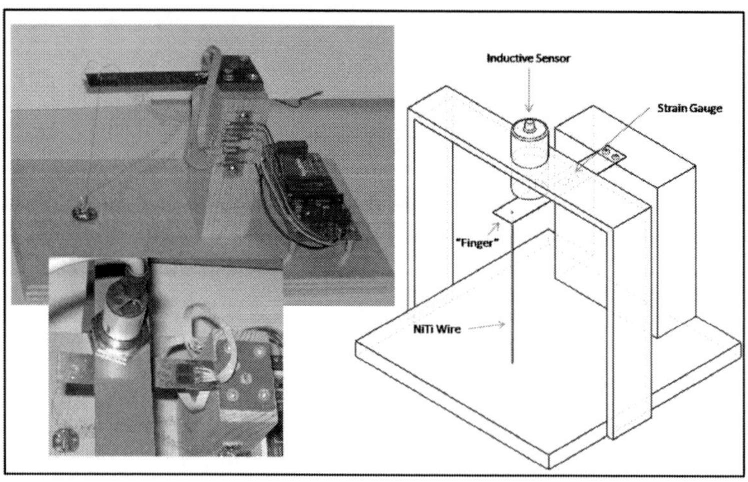

Figure 6. Angular positioning SMA based actuator prototype and scheme.

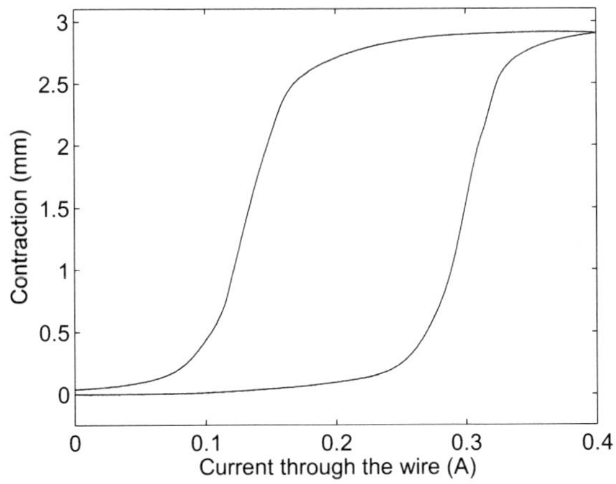

Figure 7. Experimentally measured hysteresis loop of the Ni-Ti wire.

compensate the hysteresis by different methods.

The detailed modelling of SMA materials (and other hysteretic systems) using first principles or basic laws of physics is an arduous task, and the obtained models are often too complex to be used in practical applications involving characterization of systems, identification or control. For this reason, alternative phenomenological models of these complex systems have been proposed. In this section, two experimental approaches are described to be able to implement a hysteretic nonlinear compensation control: a linear compensation based on introducing a phase shift in the system, and a nonlinear one based on neural networks (NNs) appropriately trained to learn the system's inverse hysteresis.

As an alternative approach and in order to study theoretically the system stability and analysis, we try to obtain a mathematical model of the hysteresis. Ikhouane and Rodellar, in several of their works [30], analyze, identify and control systems with hysteresis using the Bouc-Wen model from a theoretical point of view. We apply this theory for our particular experimental system.

4.1. Recovery force's influence in the shape of the hysteresis

When the wire is heated, in our case by an electrical current, a structural phase transition is induced that results in a contraction of the wire. When the material cools off, the load hanged provides the necessary restoring force so the wire stretches it back to its original length. The contraction achieved depends on the magnitude of the load on the wire.

The hysteretic stimulus-response curve of the system has been measured under hanging loads of $m1 = 19.03g$, $m2 = 60.12g$, $m3 = 88.81g$, $m4 = 160g$, $m5 = 220g$ and $m6 = 320g$. They provide increasing recovery forces to the SMA wire. As shown in figure 8, a heavier load narrows the hysteresis loop and displaces it to the right, meaning that higher temperatures (currents through the wire) are needed to induce both phase transitions, from austenite to martensite and from martensite to austenite. In the same way, the more force we apply, the greater is the contraction.

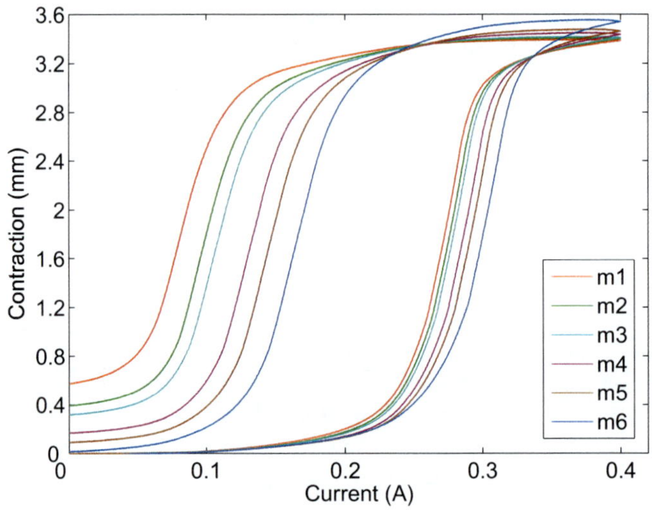

Figure 8. Hysteresis loops of the Nitinol wire at different recovery forces.

4.2. Phase shift based hysteresis compensation

The hysteresis effects observed in SMA materials can be seen as a phase lag between a periodic input and the corresponding output. Thus, the hysteresis compensation can be performed by introducing a compensating phase shift, for instance by means of the "phaser" operator L_{pa} proposed in [23]

$$L_{pa}(j\omega) = a + jb \quad \text{where} \quad |L_{pa}(j\omega)| = 1 \quad \text{and} \quad \angle L_{pa}(j\omega) = \phi. \tag{1}$$

Note that this operator can be viewed as the counterpart of a gain, which modifies the magnitude of the signal but not its phase.

If ϕ is the phase lag introduced by the hysteresis and ω is the frequency of the input signal, this ideal phase shift can be written as

$$L_{pa}(j\omega) = \cos(\phi) + \frac{sin(\phi)}{\omega}s \tag{2}$$

which resembles a Proportional Derivative (PD) controller $C(s) = k_p + k_d s$, with $k_p = \cos(\phi)$ and $k_d = sin(\phi)/\omega$.

This compensation approach is very simple, although it is not perfect due to several reasons. The obvious one is that a hysteresis is much more complex than a pure phase shift. Second, the phase lag introduced by the hysteretic behavior depends both on the amplitude and frequency of the signals involved, so the PD gains k_p and k_d are not constant. Finally, the PD controller of Eq. (2) is not causal and therefore it amplifies high frequency signals (and noise), although this can be easily solved by adding one additional pole about one decade apart from the zero. In this case, the transfer function of the compensator would read

$$FT_1 = \frac{K(1+\frac{k_d}{k_p}s)}{(1+\frac{k_d}{10k_p}s)}. \tag{3}$$

According with this formalism, we have determined experimentally that, when the wire is driven at 0.025 Hz (period of 40 s), the wire's hysteresis can approximately be seen as 30^0 phase lag between the current through the wire and its contraction. Using Eq. (3) for this particular frequency and phase angle, the transfer function of compensator FT_1 is obtained:

$$FT_1 = \frac{3.183s + 0.866}{0.3676s + 1} \tag{4}$$

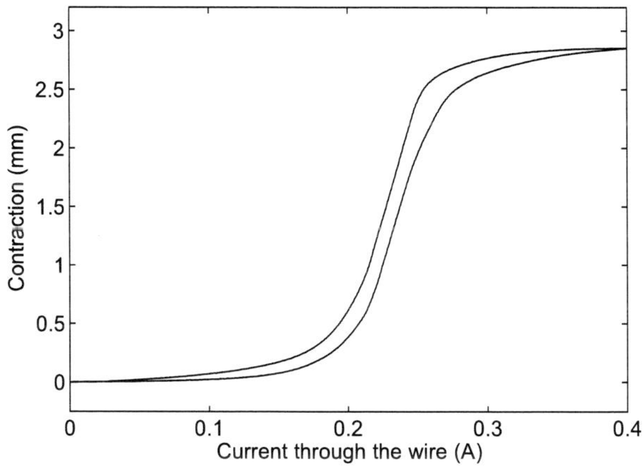

Figure 9. Hysteresis compensation using phase shift method in open loop.

As expected, when this compensator is used in an open-loop, it approximately cancels the system hysteresis (figure 9). However, we should note that the compensator FT_1 has relative order zero, so noise can still be a limiting factor. The performance, in principle, can be improved by inserting a second pole, so the high-frequency gain gets lower but without changing the phase shift of the desired frequency. The transfer function of the compensator (FT_2) is now

$$FT_2 = \frac{FT_1}{0.1s + 1} = \frac{3.183s + 0.866}{0.03676s^2 + 0.4676s + 1}, \tag{5}$$

and its Bode diagram, compared with FT_1 is displayed in Fig. 10.

4.3. Neural network based hysteresis compensation

Neural networks (NN) are well known for their capability of approximating nonlinear mappings to any desired accuracy [32]. Their properties make them invaluable tools for solving nonlinear control problems [31] and [33]. In our context, since an analytical description of hysteresis is not easily obtainable, an alternative approach to hysteresis compensation is to train a neural network to approximate an unknown inverse model of the hysteresis. In other words, a NN can learn the hysteresis characteristics by means of several sets of

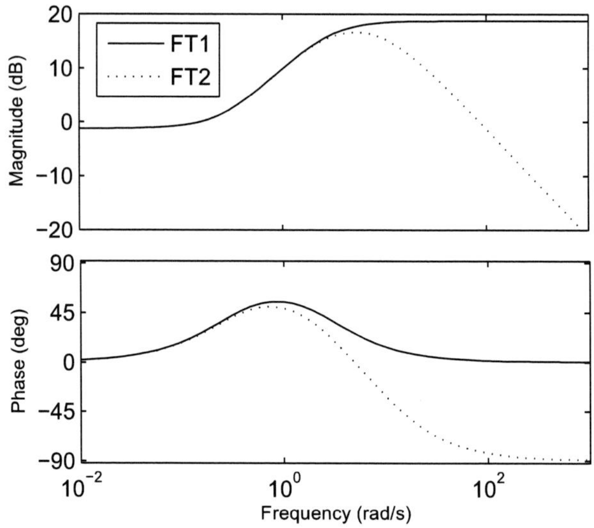

Figure 10. Bode diagrams: FT_1 and FT_2.

experimental input-output mappings performed under different conditions, measured using our experimental device.

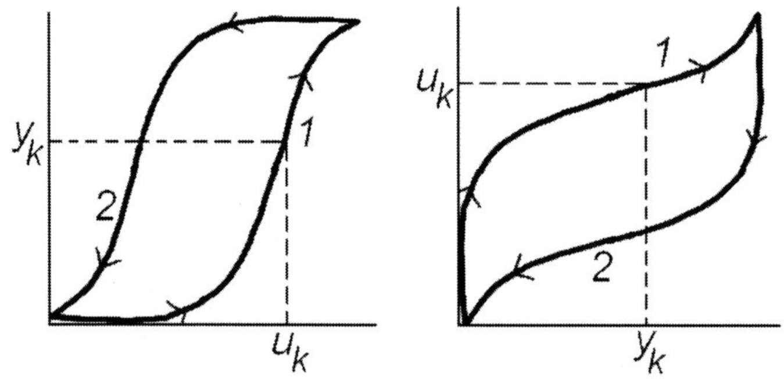

Figure 11. SMA hysteresis and its inverse.

The set of input-output data required for learning a nonlinear function depends on many factors. In the easiest case (a single-variable memoryless monovaluated scalar function), only single pairs of input-output samples (u_k, y_k) are needed for training. However, the hysteresis is a more complex case since, for a given input, the output depends not only on the input, but on the previous outputs as well (i.e. the function has memory). In this way, the set of data used for training the NN should contain information to determine whether the state of the wire lays in the ascending or descending branch of the loop. The comparison between y_k and y_{k-1} provides this information (see figure 11, if $y_k > y_{k-1}$, then select branch 1, and if $y_k < y_{k-1}$ then branch 2).

Moreover, the hysteric characteristics of SMAs, depend both on the amplitude and frequency of the involved signals. Thus, NN should receive also frequency information in the

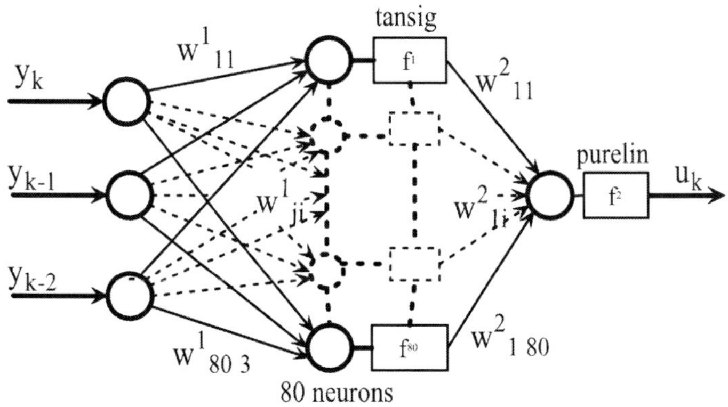

Figure 12. The Neural Network used to learn an inverse model of the hysteresis.

input-output training sets. This is done by including also previous signal samples (y_k, y_{k-1}, y_{k-2}) as inputs to the NN (i.e. augmenting its order). In this way the signal speed can be approximated by $(y_{k-1} - y_{k-2})/\Delta t$, where Δt is the sampling time.

Regarding the sampling time, it should be chosen in such a way that the NN receives as accurate information as possible from the sample signals. Note that this implies neither a too slow sampling, which loses information, nor a too high sample rate, which involves noisy and possibly contradictory readings. For instance, if the sampling time is too short, noise can produce measurements such that $y_k > y_{k-1}$ but $y_{k+1} < y_k$, so the NN doesn't receive a correct information on whether the wire is in the ascending or the descending branch of the hysteresis loop.

The NN proposed here learns off-line the dynamics of the system using the Levenberg-Marquardt algorithm (implemented in Matlab as function *trainlm*) which was designed to approach second-order training speed without having to compute the Hessian matrix like the quasi-Newton methods do [44].

When the performance function has the form of a sum of squares (as is typical in training feedforward networks), then the Hessian matrix can be approximated as

$$H = J^T(x_k)J(x_k), \tag{6}$$

where J is the Jacobian matrix that contains first derivatives of the network errors with respect to the weights (x_k).

The gradient g_k can be written in matrix form as

$$g_k = J^T e \tag{7}$$

being e a vector of network errors.

The Levenberg-Marquardt algorithm uses this approximation to the Hessian matrix in the following Newton-like update:

$$x_{k+1} = x_k - [J^T J + \mu I]^{-1} J^T e. \tag{8}$$

When the scalar μ is zero, this is just Newton's method, using the approximate Hessian matrix. When μ is large, this becomes gradient descent with a small step size.

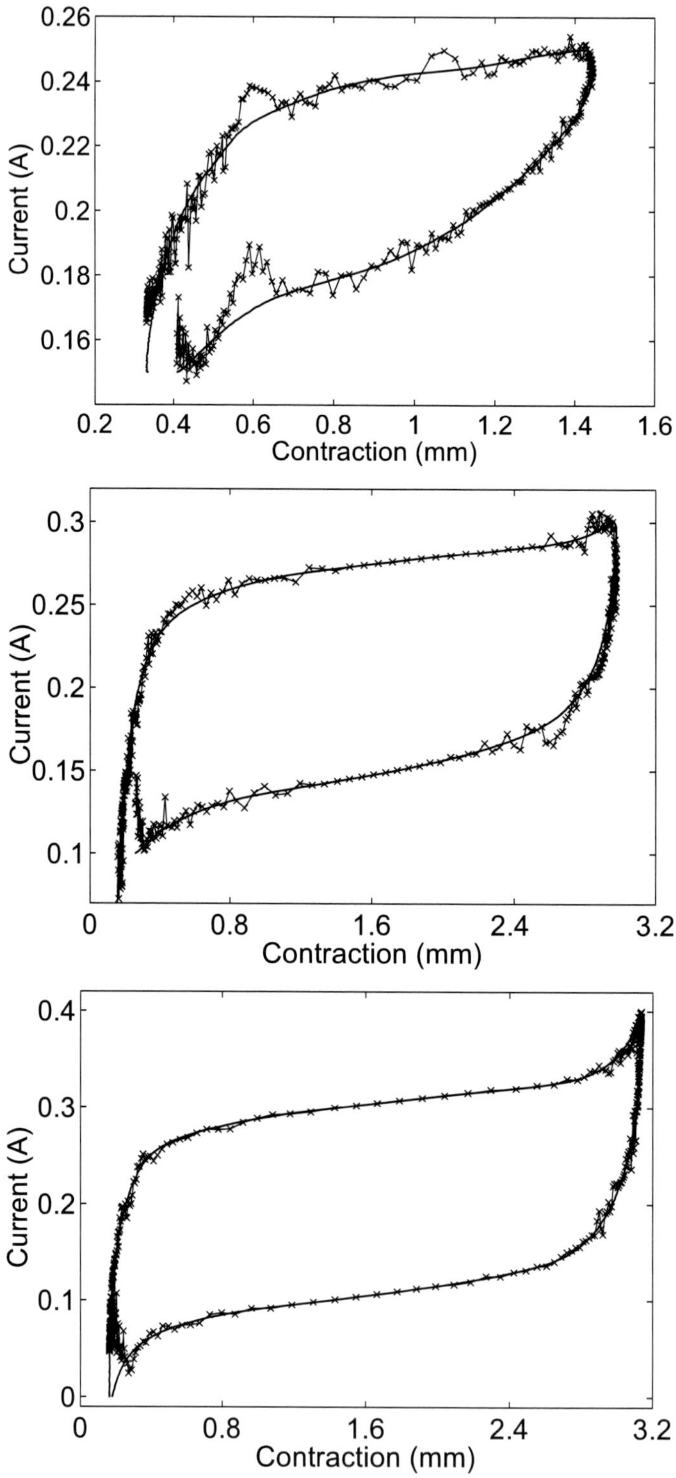

Figure 13. Comparation between experimental measured inverse hysteresis (solid line) and NN guessing (crosses) for minor and major loops. Inverse hysteresis under a 40 Hz input signal.

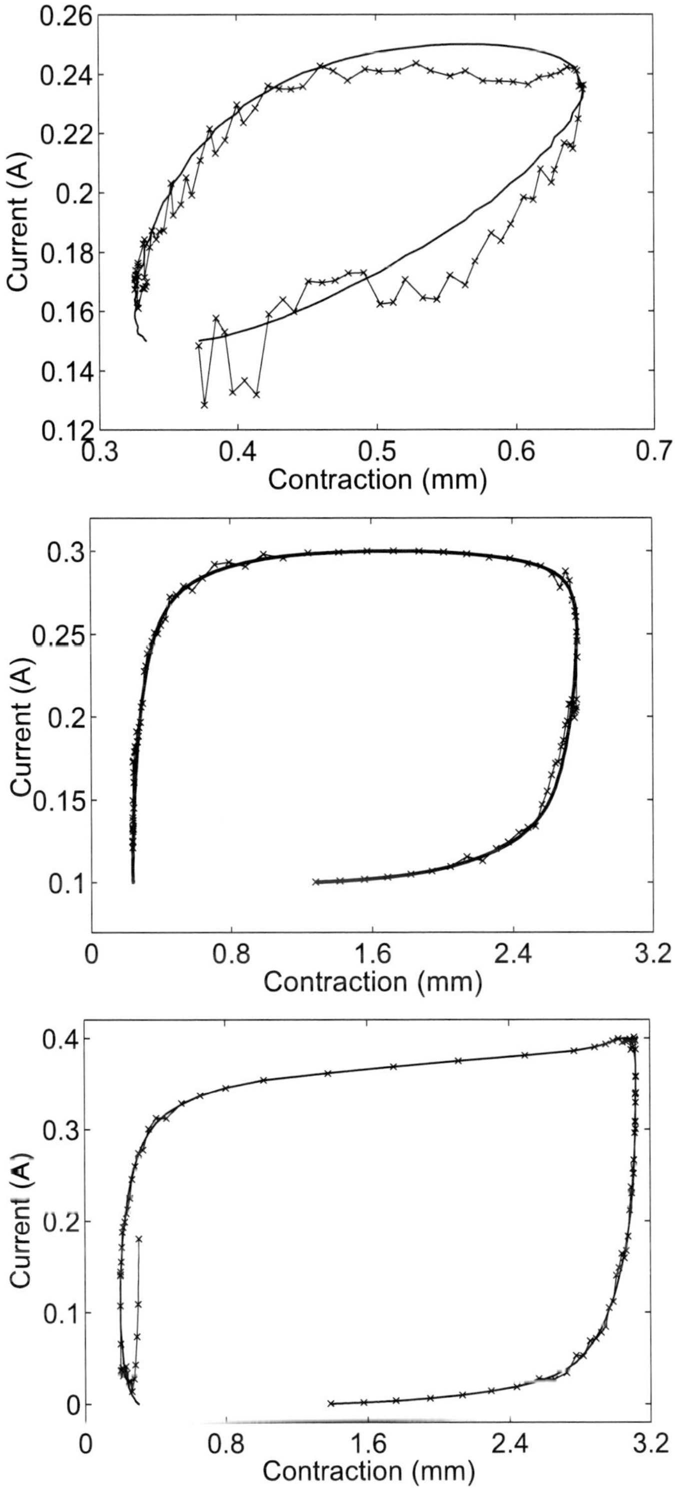

Figure 14. Comparation between experimental measured inverse hysteresis (solid line) and NN guessing (crosses) for minor and major loops. Inverse hysteresis under a 10 Hz input signal.

To train a neural network to approximate an unknown inverse model of the hysteresis, several sets of experimental input-output mappings of major and minor hysteresis loops are measured at different frequencies. As discussed before, the input layer of the NN receives three successive samples of the desired contraction signal and the output layer gives the estimated value of the necessary current that must be applied to the wire in order to obtain the desired contraction.

The NN shown in figure 12 is trained using the Levenberg-Marquardt algorithm and 3000 epoch have been found necessary to find a global minimum. The NN response and the experimental measured inverse hysteresis are compared and shown in figure 13 and 14, where the proposed NN model validation is illustrated by experiments. The NN predicts current values corresponding to a set of uniformly distributed useful contraction values for the wire. This NN prediction is then compared to the inverse contraction versus current hysteresis curves actually measured on the SMA wire. As it can be readily observed in the figures, the NN is able to learn correctly both the major and minor loops. These results appropriately validate the proposed NN model of inverse hysteresis.

4.4. Analytical Bouc-Wen model of hysteresis

The compensation methods described in the previous section have been established phenomenologically from experimental data. In this section, we intend to develop an analytical mathematical description of the system hysteresis based on the Bouc-Wen model. In principle, having an analytical description gives a way for proving the closed-loop system theoretical stability and, moreover, an inverse model of the hysteresis can be generated to be implemented in a control scheme.

According to the Bouc-Wen model, the hysteresis of a system can be described using the following first-order nonlinear differential equation:

$$\dot{z} + \gamma|\dot{x}||z|^{n-1}z + \beta\dot{x}|z|^{n} - A\dot{x} = 0 \tag{9}$$

that relates the input displacement (x) to the output restoring force (z) in a rate-independent hysteretic way. The parameters that appear in the differential equation (β, γ, A and n) have to be tuned to match the hysteresis loop of the system under study. That is, given a set of experimental input-output data, Bouc-Wen model parameters have to be adjusted so that the output of the model matches the experimental data.

As it is shown in figure 15, the hysteretic characteristic of SMAs depend on both the amplitude and frequency of the stimulus. To obtain a model that describes the system behavior for any stimulus, it is compulsory to adjust a set of different parameters for each loop. Evidently, it is a very tough job to determine all those parameter sets individually by trial and error. Also, it is complicated to establish an automated procedure to correctly assign the parameters that correspond to each minor loop. As a first approach, we have determine them for the major loop, in order to demonstrate the real possibilities of the Bouc-Wen model to describe the hysteresis of the Nitinol wire.

As it is done in most of the cases studied by Rodellar and Ikhouane, n and A are supposed to be 1. We analyze the evolution of the model with the variation of the other two parameters, β and γ. Figure 16 shows the hysteresis curves obtained for different γ, with β held constant and viceversa. The β parameter affects mainly to the width of the hysteresis

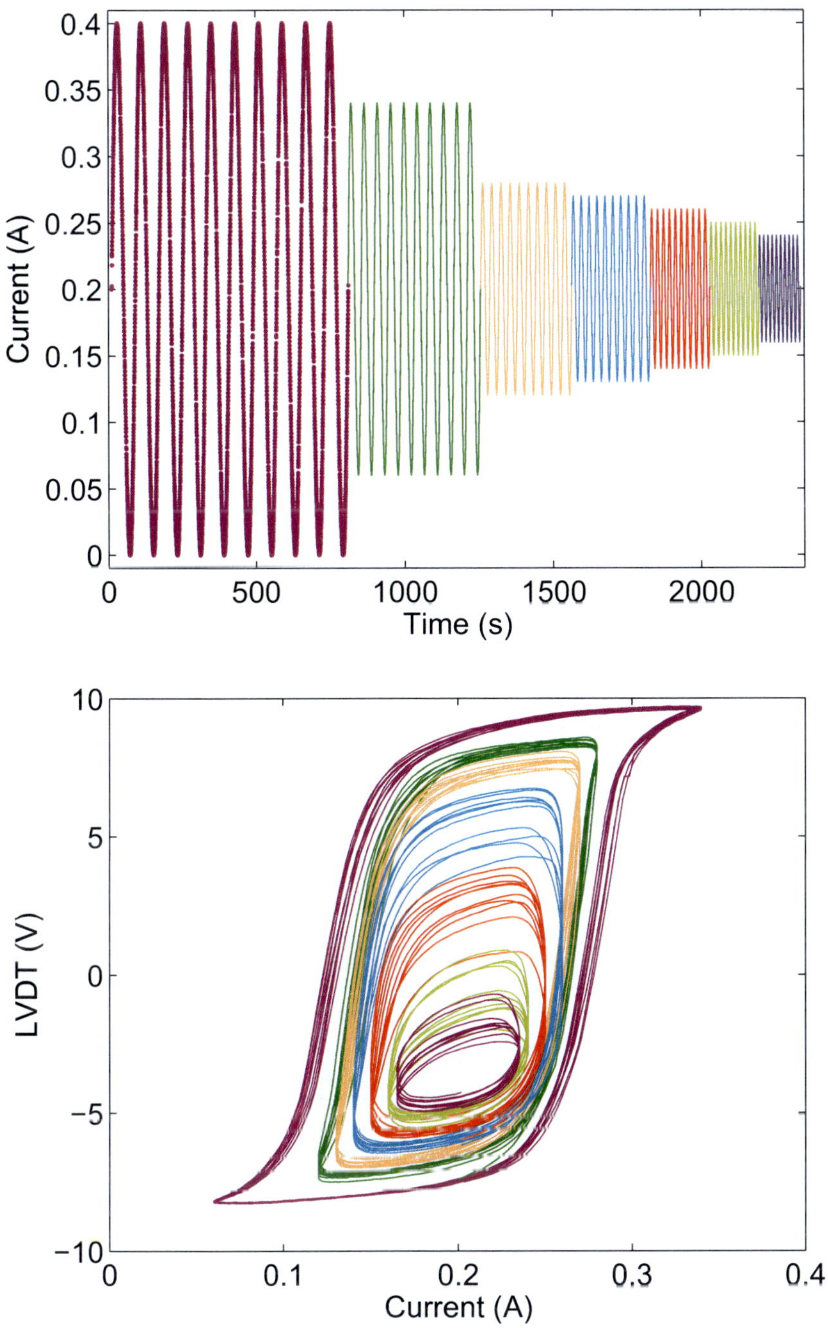

Figure 15. Nitinol wire hysteresis loop under different applied current signals.

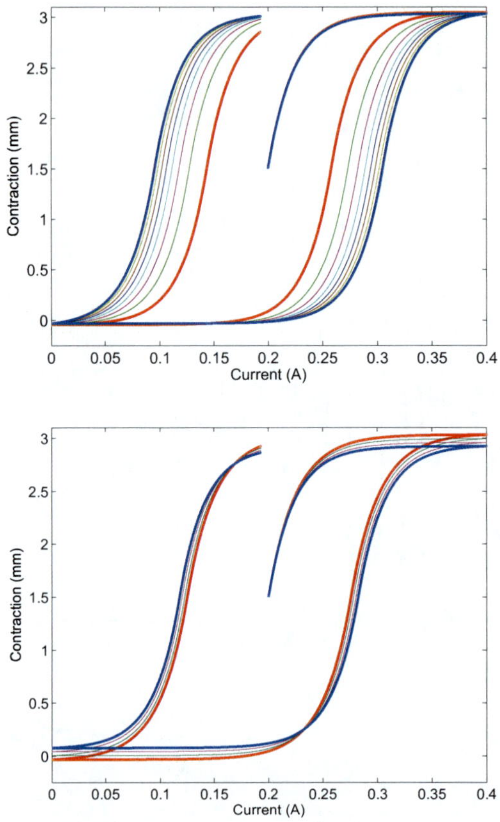

Figure 16. Analysis of the influence of *Top:* γ and *Bottom:* β parameters in the Bouc-Wen model.

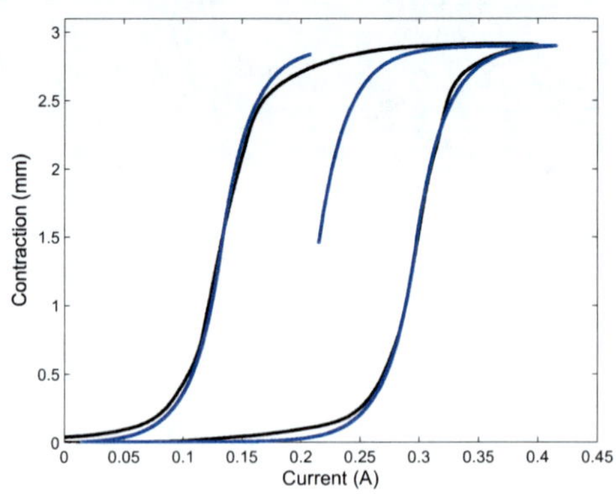

Figure 17. Comparation between experimental measured hysteresis (black) and Bouc-Wen model (blue) for major loop.

loop. Greater values of β implies a wider loop. γ also has influence on width, but it affects mainly to the amplitude of the hysteresis loop.

The best set of parameters (β=42 and γ=0.15) reproduces reasonably well the shape of the experimental hysteresis loop. They are compared in figure 17. It is important to clarify that the Bouc-Wen model as stated in Eq. 9, describes an hysteresis loop with an amplitude normalized to one and centered in the origin. To reproduce the experimental hysteresis of the SMA wire shown in figure 17, the output is shifted by a change of variable and scaled by a suitable gain ($K = 61$) inserted in the equation. Finally, the initial conditions of the integration process needs to be adjusted to force the first point of the calculation to be in its corresponding value and eliminate the discrepancy observed in figure 17.

5. Control Strategies

The control experiments presented in this section have been performed under two types of reference signals:

- **Square wave:** It represents a periodic contraction and relaxation of the wire between two intermediate contraction levels (0.5 and 2 mm). The length change is commanded to take place within 1 second.

- **Slope wave:** It represents a gradual contraction and relaxation of the wire (from 0 to 2.4 mm and back) at a constant speed.

In all the experiments a saturator is inserted to limit the current to a maximum of 0.4 A, so the wire is protected from overheating.

5.1. Linear positioning control experiments without hysteresis compensation

Several control strategies are implemented in the experimental platform shown in figure 3: Proportional, Proportional-Integral, Sliding-mode and PI with antiwindup control strategies, being this last variable-structure control method which displays the best results. The integral part is only switched on when the control signal is not saturated. It is found that by properly tuning the controller, it is possible to obtain position errors about 4 μm. However, the hysteresis of the system results in a control signal that has a very large amplitude and frequency, switching rapidly between the extreme values of the current to correct small errors (see figures 18 and 19). This clearly illustrates the difficulty in controlling a hysteretic system.

5.2. Linear positioning control experiments with hysteresis compensation

In order to provide greater reliability and longer life of the positioning device, a drastic decrease in the control effort, which can deteriorate the SMA wire, could be very beneficial. The great control signal activity observed in the previous experiments can be improved by including a method of compensating and reducing the system hysteresis in a nonlinear control scheme. Figure 20 shows a general diagram for this type of control.

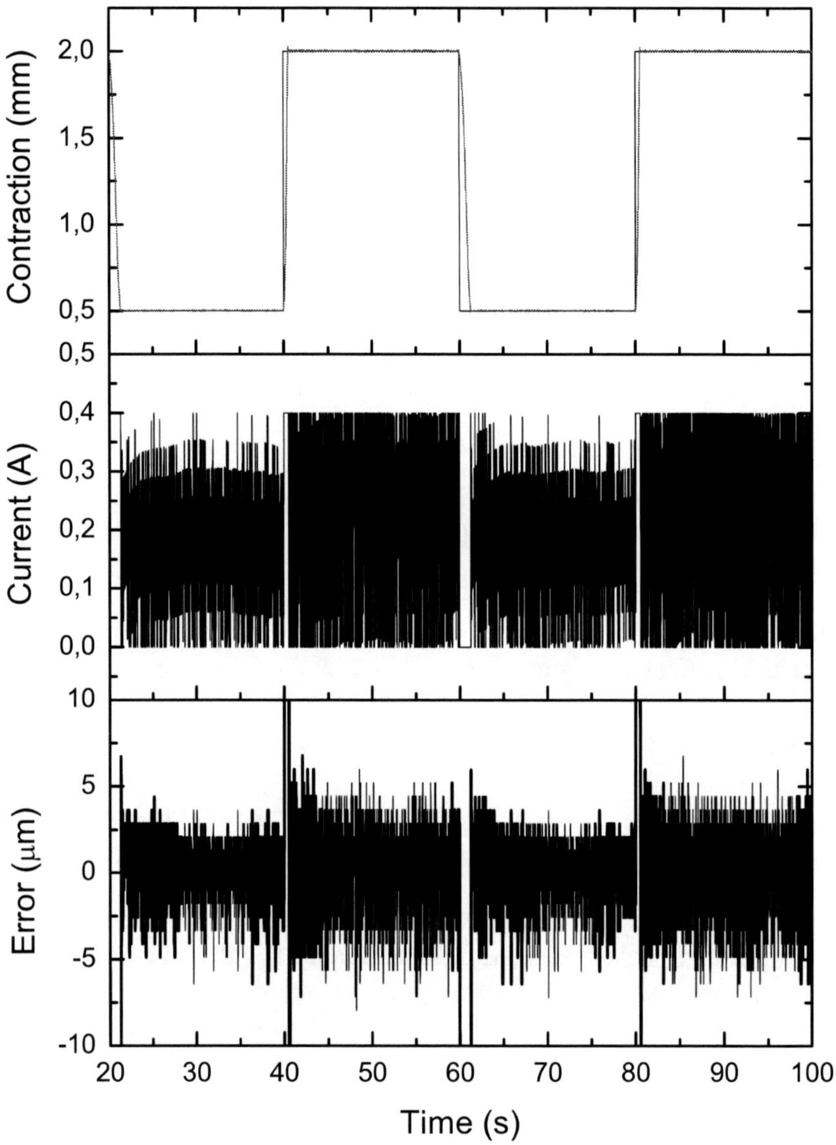

Figure 18. PI with antiwindup control strategy under square reference: *Top:* contraction, *Center:* control signal, *Bottom:* error.

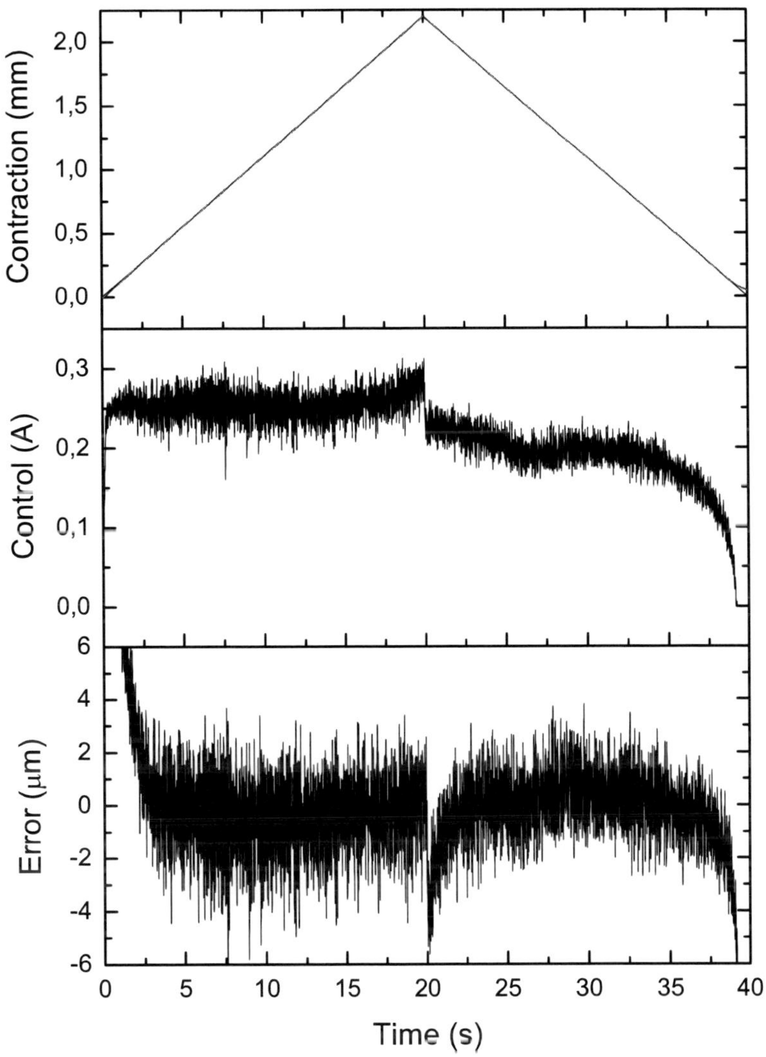

Figure 19. PI with antiwindup control strategy under slope reference: *Top:* contraction, *Center:* control signal, *Bottom:* error.

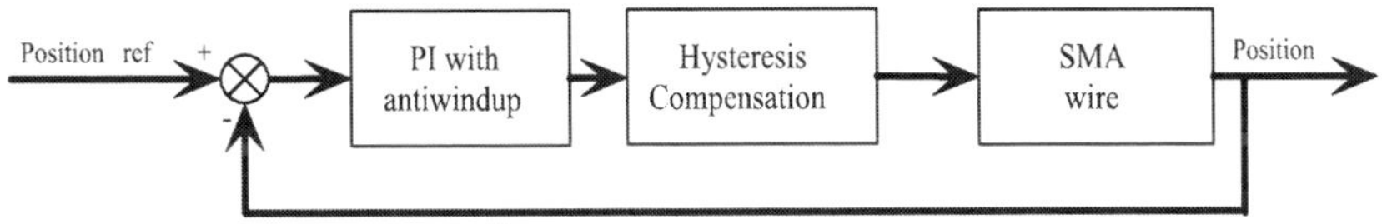

Figure 20. Simplified scheme of the block diagram implemented.

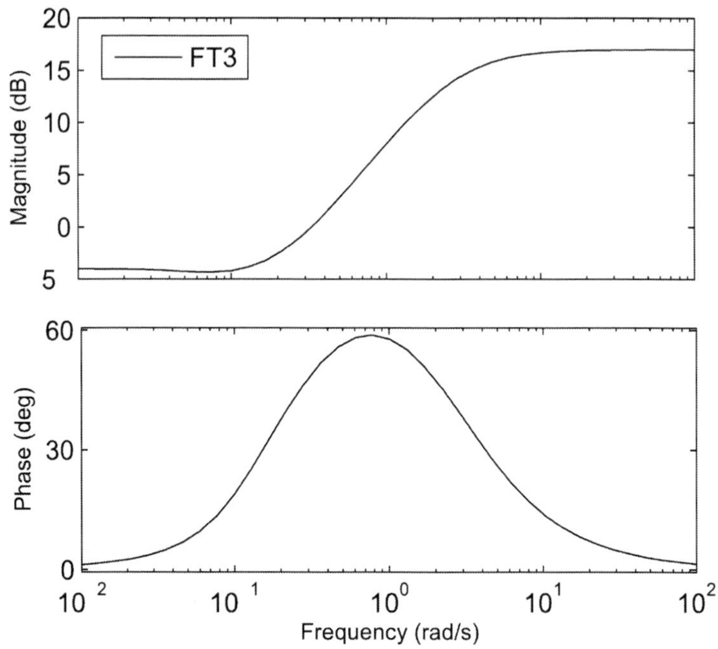

Figure 21. Required frequency response (FT_3).

5.2.1. Phase shift based hysteresis compensation

If we intend to refine the phase shift method developed in section 4, we must notice that the closed loop signal that feeds the hysteresis compensator is generated by the PI with anti-windup controller, as schematized in figure 20. It can be expected that this signal had many frequency components, so a useful compensator should be designed to shift the corresponding phase in the whole frequency range of this signal. In order to accomplish this goal, we have performed several measurements using input signals with different frequencies, determining the corresponding phase lag introduced by the hysteresis at each frequency. With all these phase lags, the required frequency response diagram for the compensator is sketched and a compensator that displays the demanded behavior is found using usual identification techniques. Its transfer function is approximated by Eq. (10), and its Bode diagram is represented in figure 21:

$$FT_3 = \frac{175.7s^3 + 83.43s^2 + 12.72s + 0.63}{24.9s^3 + 73.96s^2 + 18.18s + 1}. \tag{10}$$

Despite its increased complexity, when inserted in the PI with anti-windup loop, FT_3 is only slightly better than FT_1 and FT_2 (see section 4.2). Figure 22 shows the control results obtained for a square wave reference signal, closing the loop with the PI with antiwindup controller plus the hysteresis compensation using FT_3. It is shown that a significantly smaller control effort is needed to follow the reference signal with respect to the non compensated case (compare figures 18 and 22). An improvement in the obtained position error is also observed. Thus, the method of the phase-shift hysteresis compensation reveals its

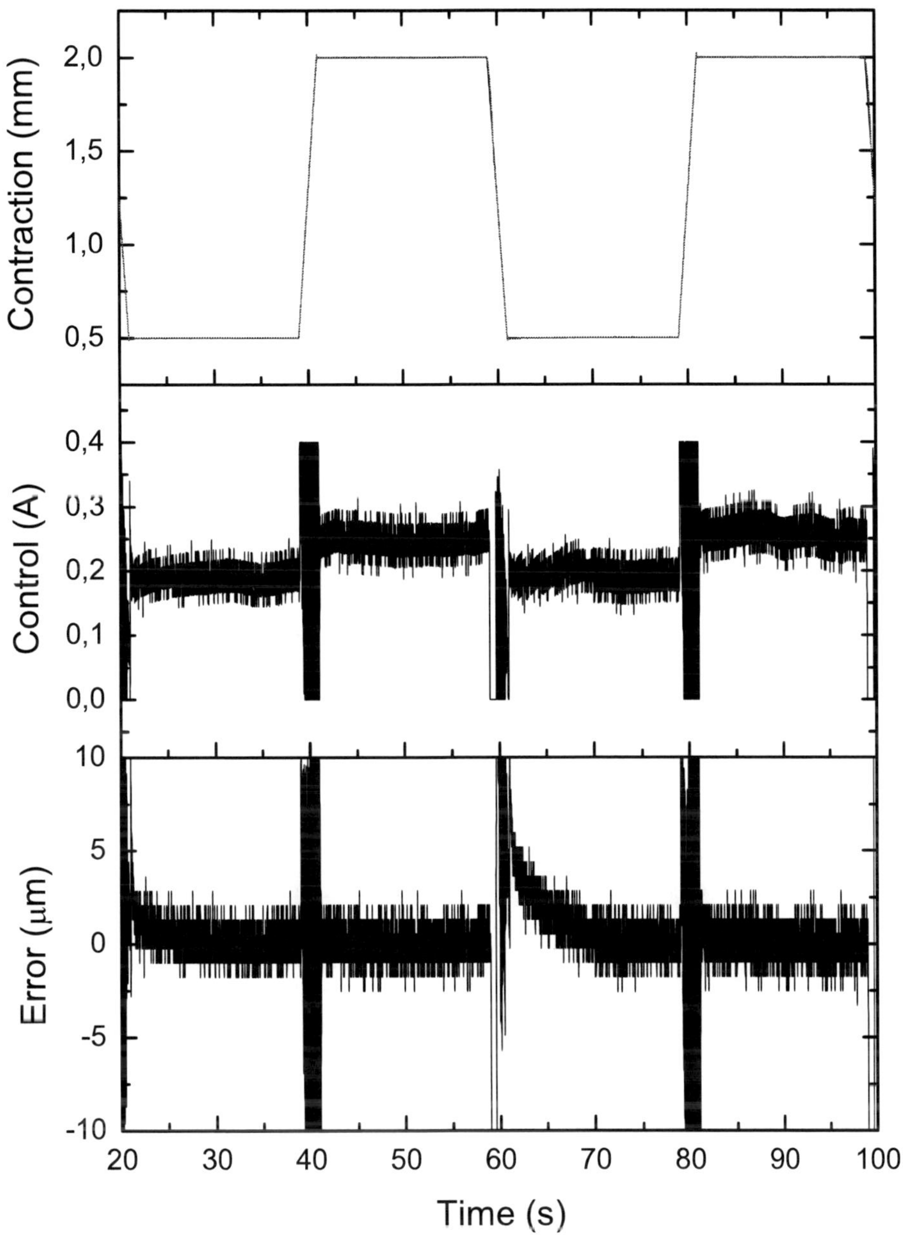

Figure 22. Phase lead compensate control [using FT_3, Eq. (10)] under square reference signal: *Top:* contraction, *Center:* control signal, *Bottom:* error.

usefulness when used inside the control loop.

5.2.2. Neural network based hysteresis compensation

The other compensation method proposed in section 4 is based on using a neural network as an hysteresis inverse model in order to reduce the burden on the control system.

In this section, we introduce in the control loop the NN trained and validated in the previous section, to carry out positioning experiments. The results of the experiments for square and slope reference signals are presented in figures 23 and 24. It is found that an appropriately trained NN introduced in a PI with antiwindup control loop is able to obtain position errors around 2-3 μm, very similar to the results obtained without using NN compensation, but as it can be deduced by observing figures 25 and 26, the control effort is far smaller when the NN is inserted in the loop. Again, it should be remarked that the reduction in the control effort is a very important issue for SMA actuators, which can deteriorate rather quickly if large control currents are applied.

5.3. Angular positioning control experiments

It has been experimentally demonstrated that SMAs are serious alternatives as micrometric actuators that can obtain accuracies of around 3 μm in linear position control if high-quality position sensors such as LVDTs are used to feedback position readings in the control loop. However, for angular motion actuator devices, if the idea is to use a small light actuator such as the robot gripper described in figure 4, the LVDT sensor used in the previous experiments must be replaced with a lighter and less bulky alternative. A possible solution can be based on inserting strain gauges to relate the deformation of the "fingers" to the strain of the SMA wire, using the classic Euler-Bernoulli beam theory [43]. This solution is adopted both for the gripper shown in figure 4 and the one-fingered device described in figure 6.

As it is explained and experimentally justified in section 3, in the angular control experiments we have focused on the single-fingered device, since control experiments on this prototype are directly applicable to the two-fingered gripper. Figure 27 shows the corresponding experimental results. The control error is practically the instrumental error due to the strain gauges accuracy, about 5 $\mu\varepsilon$, so inserting a NN to compensate the system hysteresis will not improve this error. The inductive proximity sensor added to the experimental platform allows us to report the deflexion at the end of the arm in units of displacement. The error in this position is approximately 30 μm. This is the limit imposed by the strain gauges.

6. Sensorless Control

Among all the options analyzed to replace the LVDT sensor, one of the most attractive is the implementation of a sensorless design, where the change in the resistivity of the SMA wire resulting from the phase transition is used to determine the strain of the active element. Therefore, if the relationship between the contraction and the resistance was known, a control loop like that represented in figure 28 would allow the control of the actuator's position without using a separate sensor. This is a very attractive proposal, since the actuator could

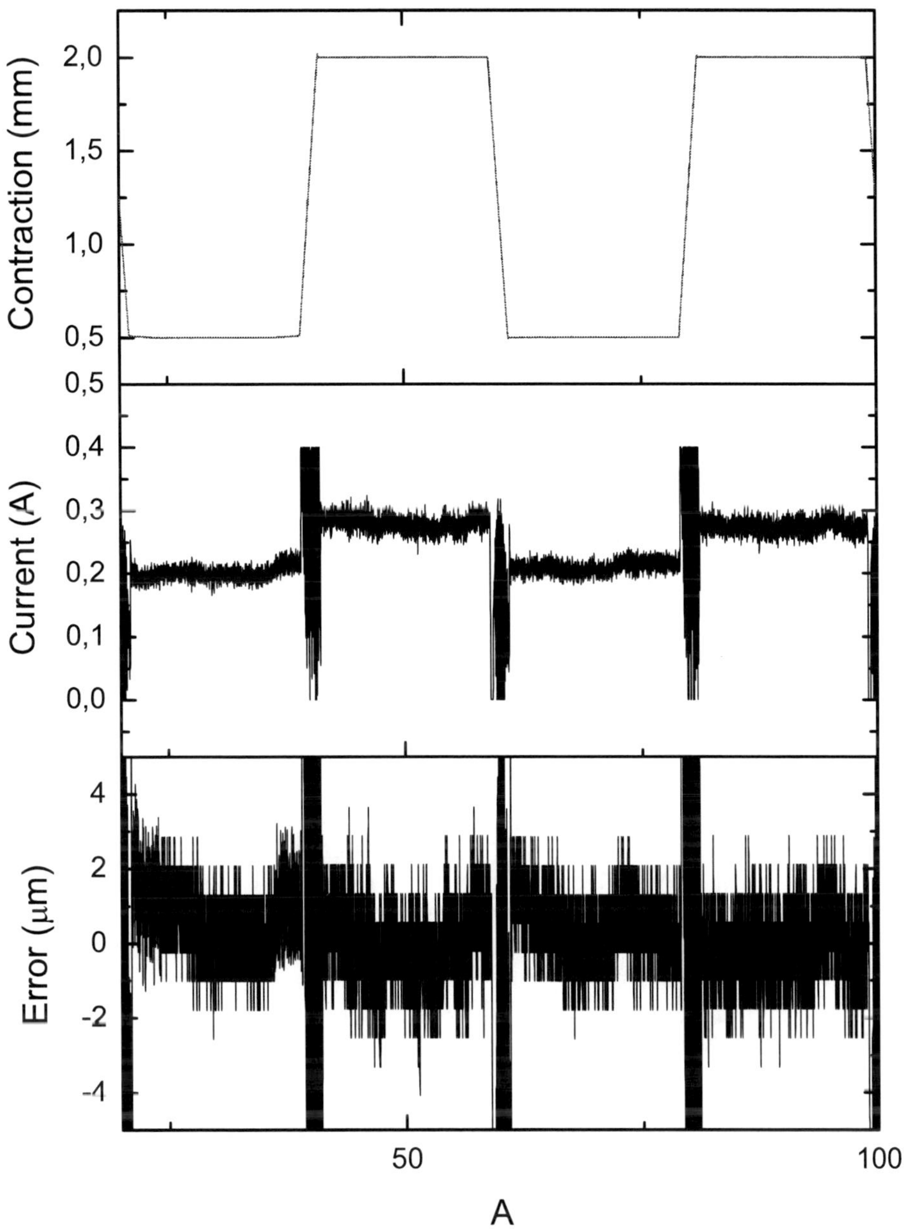

Figure 23. NN-based compensation combined with PI with antiwindup control strategy under square reference: *Top:* contraction, *Center:* control signal, *Bottom:* error.

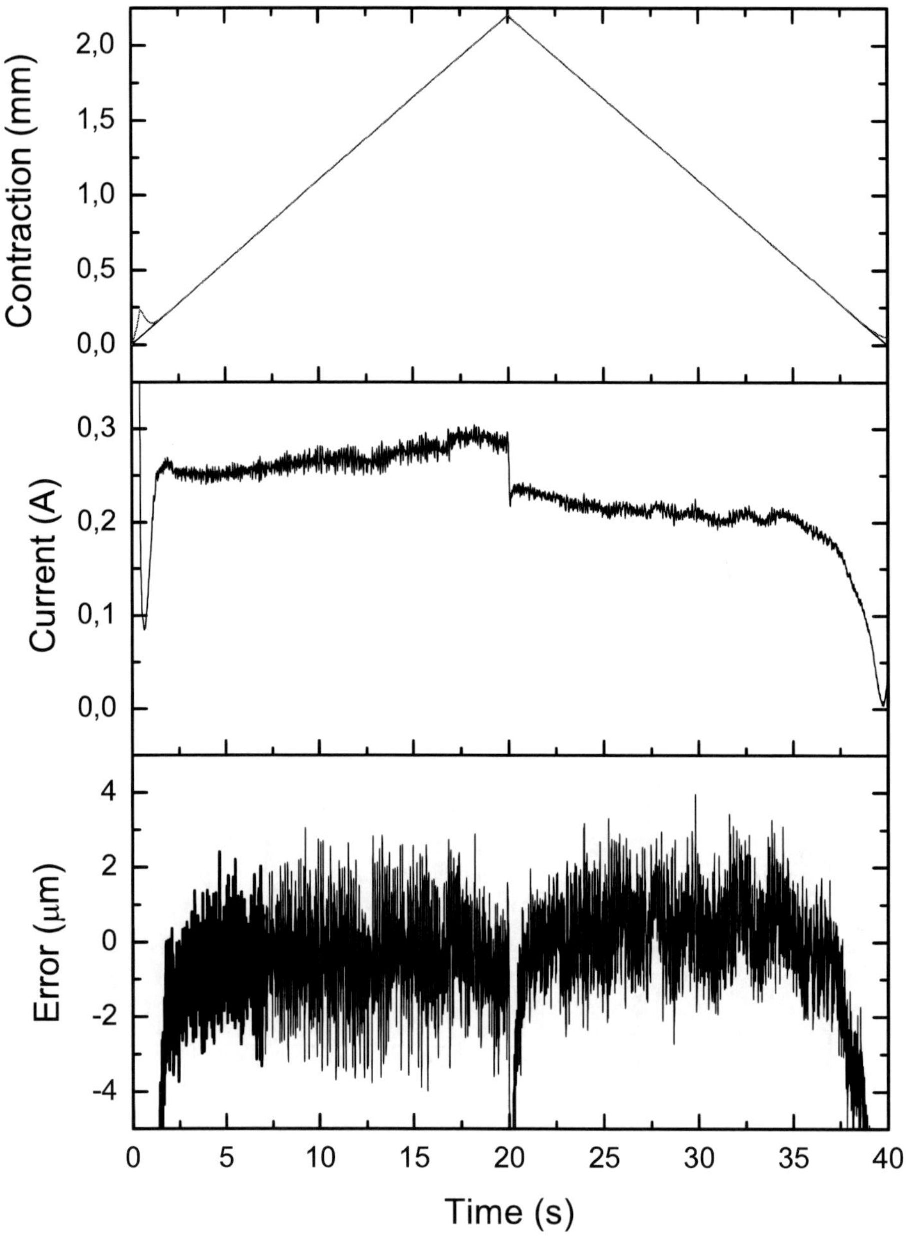

Figure 24. NN-based compensation combined with PI with antiwindup control strategy under slope reference: *Top:* contraction, *Center:* control signal, *Bottom:* error.

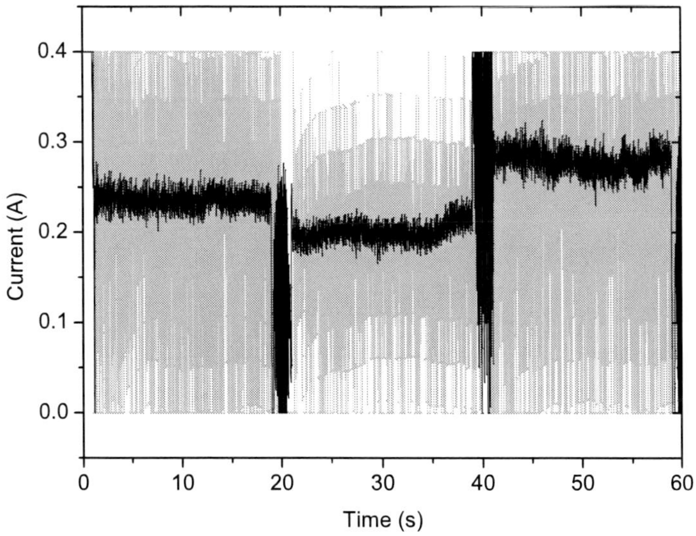

Figure 25. Difference between control signal with (black) and without (grey) NN-based compensation under square reference.

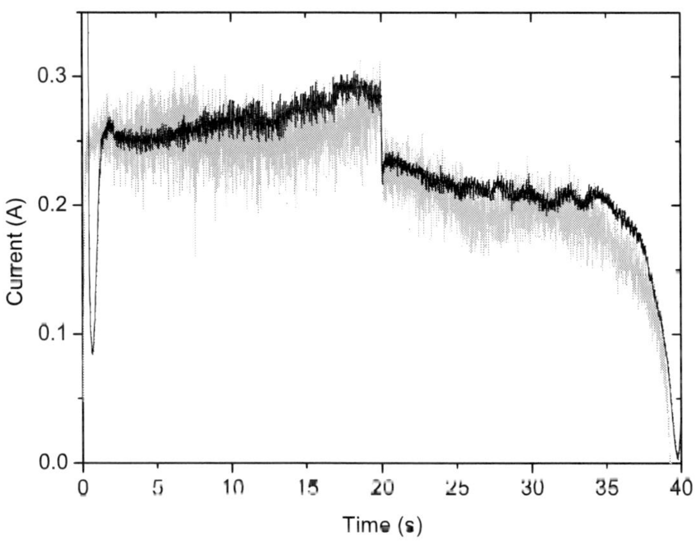

Figure 26. Difference between control signal with (black) and without (grey) NN-based compensation under slope reference.

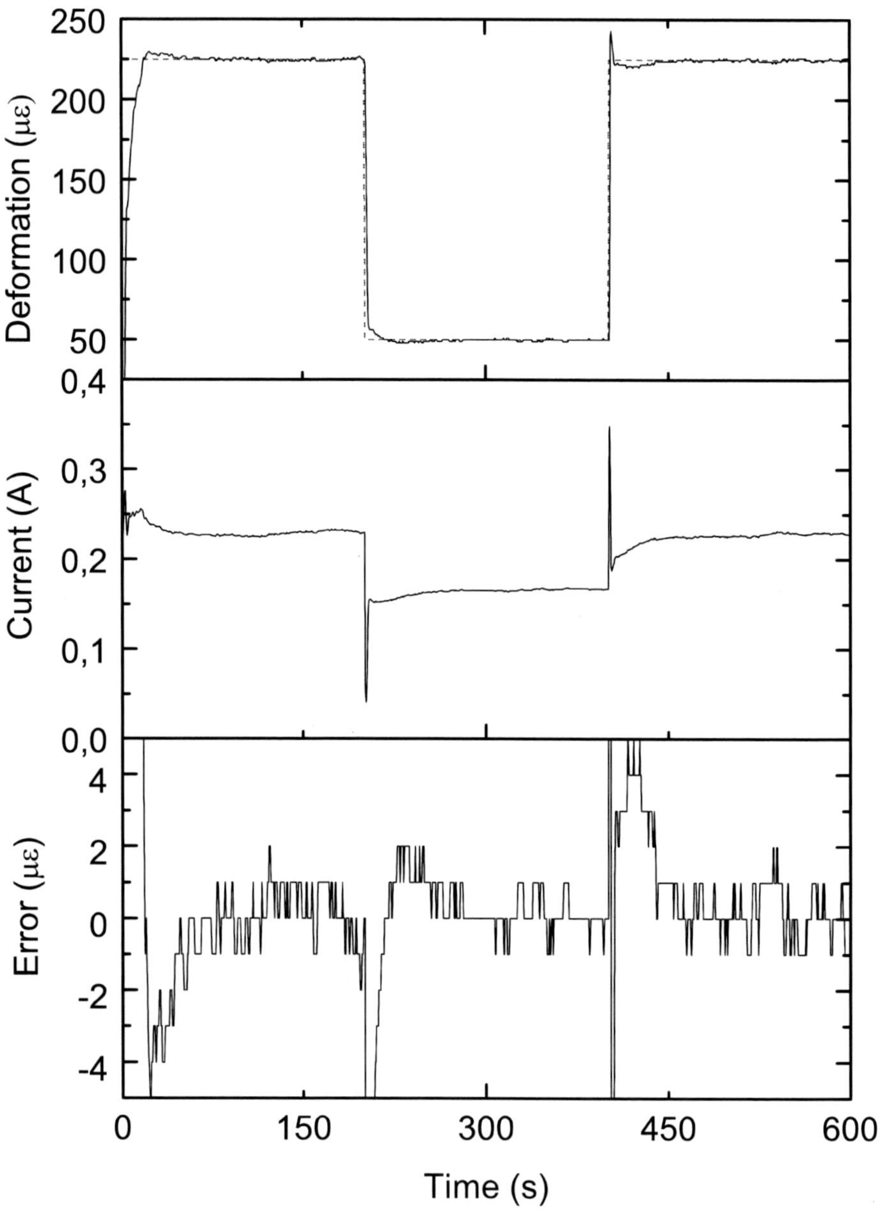

Figure 27. PID control results under square reference in the one-fingered actuator: *Top:* Deformation; *Center:* Control Signal; *Bottom:* Error.

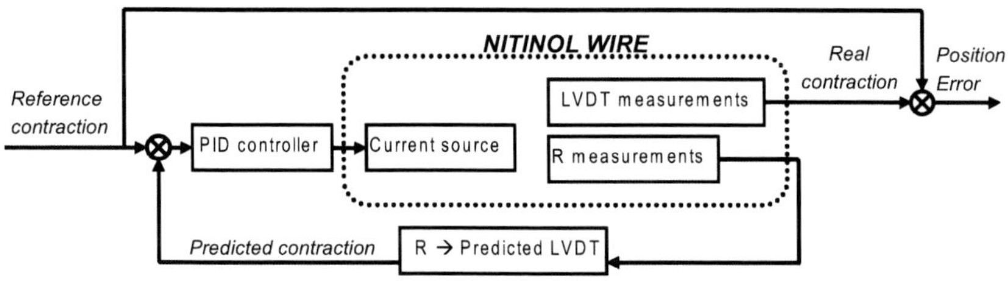

Figure 28. Simplified scheme of the sensorless block diagram implemented.

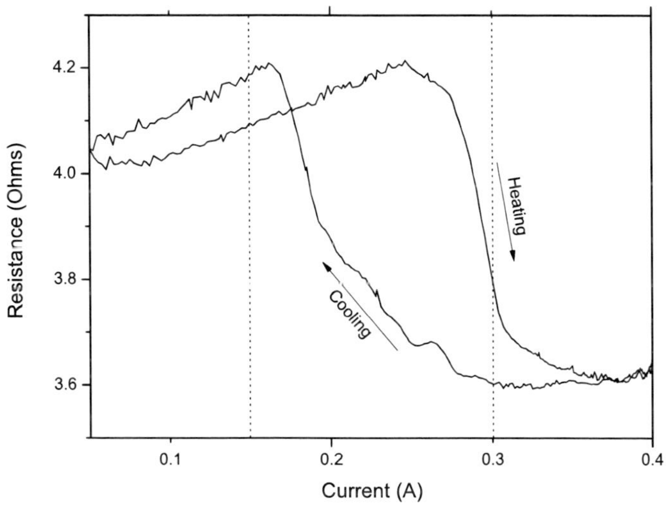

Figure 29. Experimentally measured resistance cycle of the Ni-Ti wire.

be reduced to a single SMA element (both sensing and actuating), which supposes a great advantage, specially if the main goal is to reduce the overall weight of the actuator, like is the trend in the miniaturization in mechatronics and robotics.

In this section, a relationship between the strain and resistance of a SMA Ni-Ti (Nitinol) wire is deduced, and once this characterization is done, it is inserted into a control loop. A common PID control strategy is implemented on the laboratory SMA actuator prototype shown in figure 3.

6.1. Resistance measurements and contraction vs. resistance relationship

A Nitinol wire's resistance changes with the structural phase transition. An example of this variation is shown in figure 29 (the wire is submitted to a heating-cooling cycle by applying a current profile as shown in figure 30). The control of the position using directly the resistance is complicated because of the nonlinear and hysteretic nature of the resistance behavior. Besides, in both the stretching and contraction branches, there are some resistance values that correspond to two different states of the wire. However, if we limit the current values applied to the wire (see dashed lines in figures 29 and 30), the comparison of the

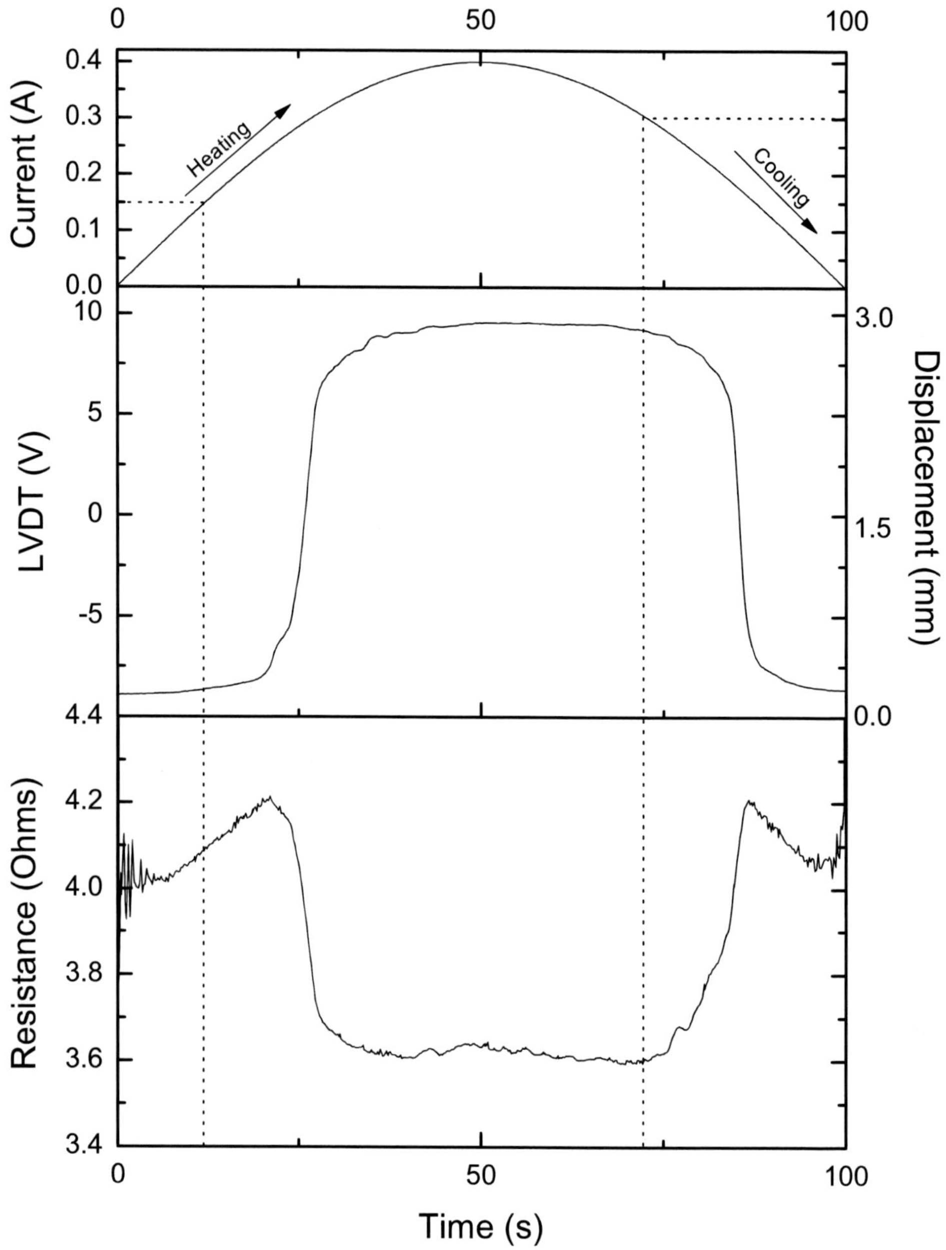

Figure 30. *Top:* Current through the wire; *Center:* LVDT signal variation of the Ni-Ti wire; *Bottom:* Resistance variation of the Ni-Ti wire.

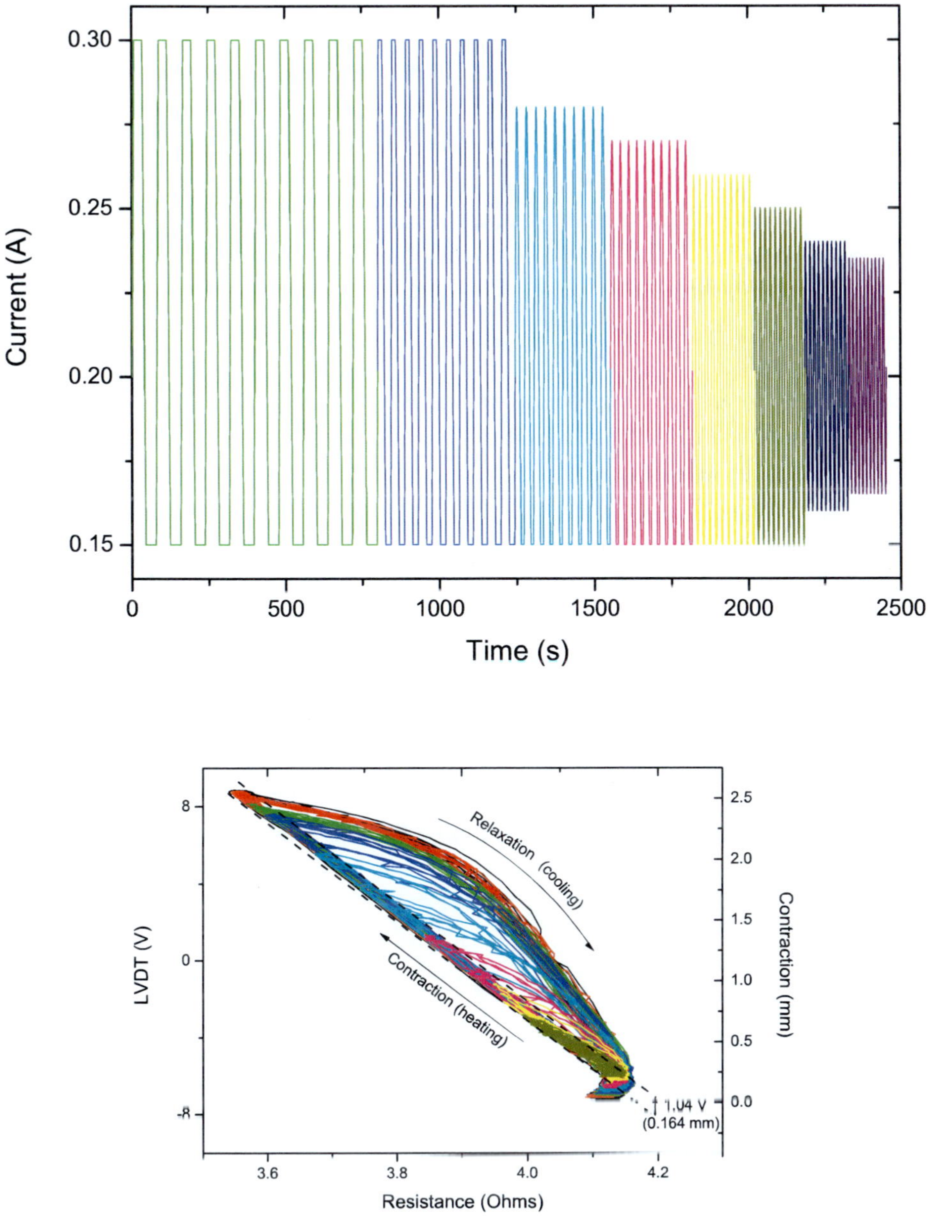

Figure 31. *Top:* Current applied through the Nitinol wire during characterization; *Bottom:* resulting contraction vs resistance curve.

LVDT reading and the resistance as a function of time (figure 30), suggest that the position could be controlled by first converting the resistance into contraction and then using this to feedback into the control loop.

To obtain the relationship between the resistance and the deformation, a series of open-loop tests is conducted, using a sinusoidal current to drive the SMA. Figure 31 shows the response of the wire to different applied current intensities and frequencies. As it can be seen, in order to obtain the wire's strain, one needs to know not only the current value of the resistance, but also the value of the resistance at previous times (i.e. the function has memory). Moreover, just as it is a case with the strain-current curve, its characteristic curves depend both on the amplitude and frequency of the input signals.

As said before, in order to simplify the problem, the current is limited to a minimum of 0.15 A and maximum of 0.30 A. In doing so, it is possible to ensure that the curve only consist of two branches, one ascending and one descending, except for a small range of resistance values (the curved tail between 4.10 and 4.15 Ω in the bottom graph of figure 31). The loops for the various amplitudes show the same overall shape, only varying in their maximum value, as can be seen in figure 31 (bottom). From these loops it should be possible to know the strain from the measured resistance. When the Nitinol wire is contracting, the contraction depends linearly with the wire's resistance, whereas in the relaxation, the relationship can be fitted as a second order polynomial. The same figure shows that on cooling, the fitting parameters for the equation depend strongly on the previous values of resistance and contraction, making this a non-trivial exercise. On contraction, this is less so than on the relaxation, making it easier to characterize the curve. However, while the slope of the contraction curves is for practical purposes the same, all the curves showed a slightly different y-intercept value. This leads to the existence of an intrinsic error in the resistance-position relation (see the two bounding lines in figure 31).

6.2. Control experiments

Using the results from the previous subsection as feedback into the controller, two control experiments are proposed, using different shapes of the reference signal:

- A steep contraction of the wire between two intermediate contraction levels (1 mm and 1.9 mm).
- A gradual contraction of the wire with constant speed (from 0.75 mm to 2 mm).

In all the experiments a saturator was inserted after the controller to limit the current to a maximum of 0.4 A, so, the wire would be protected from overheating.

Given the difficulty in characterizing the relaxation (cooling) curve, the experiments are limited to setting only positions that required the wire to contract (heating). Positions where the wire needed to extend are achieved by allowing the wire to fully relax, and then programming the new position, so this again would be achieved by contracting the wire.

To prevent abrupt oscillations in the control signal which could deteriorate the wire, the PID control gains are tuned in a somewhat conservative fashion, which leads to a relatively slow control system. In this way, along the control process, the minor loops experienced by

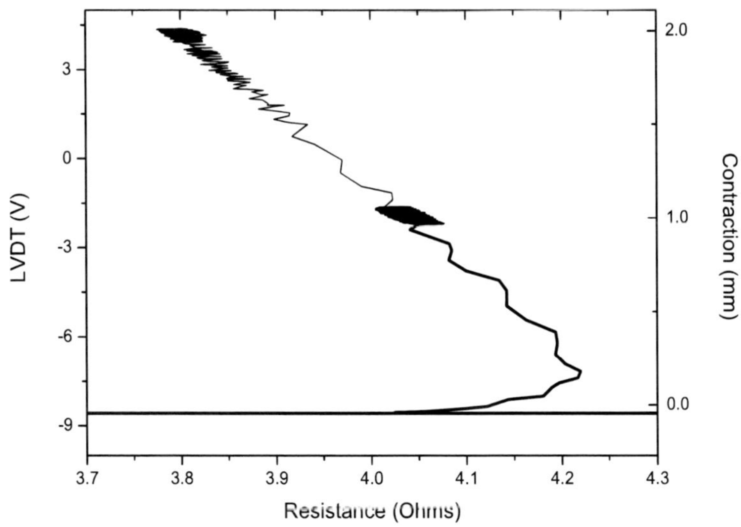

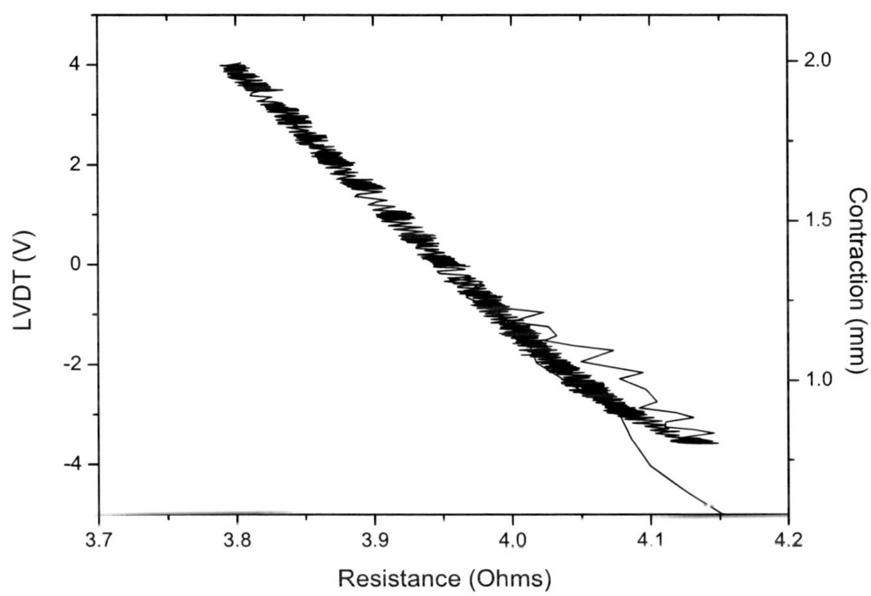

Figure 32. *Top:* Minor loops that appear during a steep contraction control experiments; *Bottom:* Minor loops that appear during a gradual contraction control experiments.

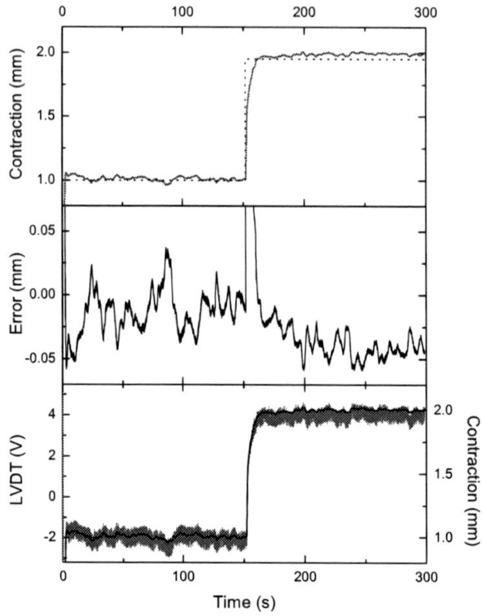

Figure 33. PID control under step reference: *Top:* reference and real contraction; *Center:* error; *Bottom:* measured real LVDT signal and predicted signal.

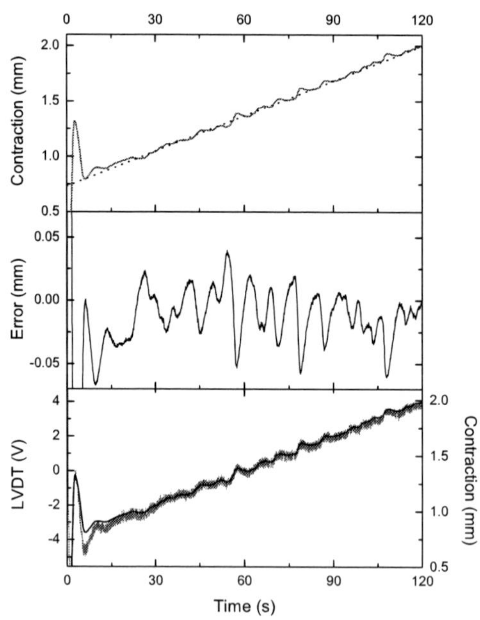

Figure 34. PID control under ramp reference: *Top:* reference and real contraction; *Center:* error; *Bottom:* measured real LVDT signal and predicted signal.

the system have small amplitude and tend to converge to the contraction ramp of the major loop, as showed in figure 32 for both types of experiments.

Figures 33 and 34 show the experimental results for both control cases mentioned previously. The curves at the top show the position reference and the actual position as measured by the LVDT. The difference between these values is displayed in the center graphs. In both cases, the position error is around 60 μm. It should be noted that the control performs very well, taking into account that the position feedback signal comes entirely from the conversion between resistance and position as shown in the graphs at the bottom of figures 33 and 34 where the black line is the measured position and the other is the position deduced from the resistance of the wire according with the characterization performed in figure 31.

7. Conclusions and Future Works

The difficulty in smoothly controlling the austenite-martensite proportion in SMAs (and thus the material deformation) implies these materials are mainly used in practice only as "on-off" actuators. In this work, it has been experimentally shown that, despite the nonlinear and hysteretic characteristics of the material phase transitions, SMA wires can be successfully controlled to be used in micropositioning applications.

Several linear motion control strategies (Proportional, Proportional - Integral, sliding-mode and Proportional-Integral with antiwindup) have been experimented in a laboratory SMA-based linear micropositioning platform and it has been found that the PI with antiwindup control strategy displays the best results. However, the hysteresis of the system translates into great efforts in the control signal to correct small errors.

This important question of too large control activity, which can deteriorate the SMA wire, has been addressed by including the hysteresis effects in a nonlinear control scheme. Two inverse models have been designed to compensate and reduce the SMA actuators hysteresis response. The first one considers the hysteresis as a phase lag between a periodic input and the corresponding output, which can be corrected using a linear compensator to shift back this phase angle. The second approach consists of using a Neural Network to learn the hysteresis from experimental input-output data. It has been experimentally tested that an appropriately trained NN inserted in a PI with antiwindup control loop result in smaller control efforts, which benefits durability of the positioning device and power consumption.

For light-weight applications of these type of actuators, the issue of replacing the relatively heavy LVDT position sensor with another alternative sensor is key. Among all options analyzed, using the temperature of the SMA wire, strain gauges or the resistance of the SMA wire as a measure of its deformation, those last two options are implemented.

Several angular motion control experiments have been carried out on a SMA-actuated light robot gripper and on a single-fingered experimental device. Using strain gauges to provide the feedback signal, position errors around 30 μm have been obtained, being the main drawback the limited accuracy of the strain gauges.

The experimental results using sensorless strategy, that is, using the resistance of the SMA wire as a measure of its deformation, show that it is possible to obtain accuracies around 60 μm, with the great advantage that no separate position sensor is needed and the actuator could be reduced to a single, small, compact and light SMA element.

As a general conclusion of the research, it can be stated that the experiments presented have demonstrated the possibility of achieving micrometric control using SMA wires as active elements. This makes them a serious alternative to other driving mechanisms in micromachines.

7.1. Future works

Further research should be carried out improving the fit of the wire's resistance and deformation using more powerful nonlinear optimization methods, such as neural networks for example. In sensorless control, due to the difficulty in characterizing the relaxation curve, the experiments are limited to setting only positions that requires the wire to contract. Future works would be focused in characterizing both the contraction and relaxation curves using neural networks.

Moreover, SMA actuators are not only promising in positioning applications, since they can also exert significant forces compared to its weight and size. An interesting future work would be the implementation of force control on SMA wires, which would significantly widen its scope and interest for micro-mechatronic applications.

The goal of obtaining higher drive frequencies while maintaining high work densities has motivated the development of a ferromagnetic shape memory alloy (FSMA) based actuator, which relies on magnetic field-induced phase transitions to provide actuator inputs.

Acknowledgments

The authors are grateful to CYCYT for partial support of this work though projects DPI2005-03121 and DPI2008-03749. They also wish to thank The Basque Government for grant BFI05.283 and support under ACTIMAT and GIC07/49-208 projects. The EU Marie Curie IIF actions under FP6.

References

[1] J.V. Humbeeck. Shape memory alloys: A material and a technology. *Advanced Engineering Materials*, 3(11):837–850, 2001.

[2] Schwartz. *Mel Encyclopedia of Smart Materials*. John Wiley, 2002.

[3] J.L. Pons. *Emerging Actuator Technologies: A Micromechatronic Approach*. John Wiley, 2005.

[4] D.J. Leo. *Engineering Analysis of Smart Material Systems*. John Wiley, 2007.

[5] A. Leppaniemi. Shape memory alloys-applications and commercial aspects. In *Shape memory alloys*, 1998.

[6] R.C. Smith. *Smart Material Systems. Model Development*. SIAM press, 2005.

[7] M. Hashimoto, M. Takeda, H. Sagawa, J. Chiba, and K. Sato. Shape memory alloy and robotic actuator. *Journal of Robotic System*, 2(1):3–25, 1985.

[8] K. Ikuta, M. Tsukamoto, and S. Hirose. Shape memory servo actuator system with electric resistance feedback and its application to active endoscope. In *Proceedings of the 1988 IEEE International Conference on Robotics and Automation* , pages 427–430, Philadelphia, PA, USA, 1998.

[9] N. Troisfontaine, P. Bidaud, and P. Dario. Control experiments on two sma based micro-actuators. In *Experimental Robotics V*, pages 490–499, London, 1998.

[10] I. Mayergoyz. *Mathematical models of hysteresis and their applications* . Elsevier, 2003.

[11] G. Tao and P.V. Kokotović. *Adaptative Control of Systems with Actuator and Sensor Nonlinearities*. John Wiley, 1996.

[12] J.A. Burns and R.D. Spies. Modelling for control of shape memory alloys. In *Proceedings of the 30th Conference on Decision and Control*, pages 2334–2339, Brighton, England, 1991.

[13] D. Hughes and T.J. Wen. Preisach modeling and compensation for smart material hysteresis. *SPIE, Active Materials and Smart Structures*, 2427:50–64, 1995.

[14] D. Hughes and T.J. Wen. Preisach modeling of a piezoceramic and shape memory alloy hysteresis. *IEEE Conference on Control Applications*, pages 1086–1091, 1995.

[15] D.R. Madill and D. Wang. Modeling and l2-stability of a shape memory alloy position control system. *IEEE Trans. on Control Systems Technology*, 6(4):473–481, 1998.

[16] S. Majima, K.Kodama, and T. Hasegawa. Modeling of shape memory alloy actuator and tracking control system with the model. *IEEE Trans. on Control Systems Technology*, 9(1):54–59, 2001.

[17] T. Hasegawa and S. Majima. A control sustem to compensate the hysteresis by preisach model on sma actuator. In *1998 International Symposium on Micromechatronics and Human Science (MHS'98)*, pages 171–176, 1998.

[18] K.A. Kyoung and B.K. Nguyen. Internal model control for shape memory alloy actuators using fuzzy based preisach model. *Sensors and Actuators*, 136:730–741, 2007.

[19] C.A. Dickinson and J.T. Wen. Feedback control using shape memory alloy actuators. *Journal of Intelligent Material Systems and Structures* , 9(4):242–250, 1998.

[20] C.A. Dickinson. Feedback compensation of shape memory alloy hysteresis. In *Doctoral tesis*, 1997.

[21] G. Weeb, L. Wilson, D. Lagoudas, and O. Rediniotis. Adaptive control of shape memory alloy actuators for underwater biomimetic applications. *AIAA Journal*, 38(2):325–334, 2000.

[22] C. Lee and C. Mavroidis. Analytical dynamic model and experimental robust and optimal control of shape-memory-alloy bundle actuators. In *Proceedings of the 2002 Sysmposium on Advances in Robot Dynamics and Control, ASME International Mechanical Engineering Congress and Exposition*, New Orleans, LA, USA, 2002.

[23] J.M. Cruz-Hernández and V. Hayward. Phase control approach to hysteresis reduction. *IEEE Trans. on Control Systems Technology*, 9(1):17–26, 2001.

[24] J.M. Cruz-Hernández and V. Hayward. An approach to reduction of hysteresis in smart material. In *Proceedings of the 1988 IEEE International Conference on Robotics and Automation*, Leuven, Belgium, 1998.

[25] F. Ikhouane, V. Ma nosa, and J. Rodellar. Adaptative control of a hysteretic structural system. *Automatica*, 41:225–231, 2006.

[26] F. Ikhouane and J. Rodellar. A linear controller for hysteretic systems. *IEEE Transactions on automatic control*, 51(2):340–344, 2006.

[27] Z.W. Zhong and C.K. Yeong. Development of a gripper using sma wire. *Sensors and Actuators*, 126:375–381, 2006.

[28] J.H. Kyung, B.G. Ko, Y.H. Ha, and G.J. Chung. Design of a microgripper for micromanipulation of microcomponents using sma wires and flexible hinges. *Sensors and Actuators*, pages 144–150, 2008.

[29] D. Grant and V. Hayward. Variable structure control of shape memory alloy actuators. *IEEE Systems and Control Magazine*, 17(3):80–88, 1997.

[30] F. Ikhouane and J. Rodellar. *Systems with hysteresis. Analysis, Identification and Control using the Bouc-Wen model*. John Wiley, 2007.

[31] K.S. Narendra and K. Parthasarathy. Identification and control of dynamical systems using neural networks. *IEEE Trans. Neural Networks*, 1:4–27, 1990.

[32] K. Funahashi. On the approximate realization of continuous mappings by neural networks. *Neural Networks*, 2:183–192, 1989.

[33] P.J. Antsakis. Neural networks in control systems. *IEEE Trans. Neural Networks*, 10:3–5, 1990.

[34] N. Ma, G. Song, and H.J. Lee. Position control of shape memory alloy actuators with internal electrical resistance feedback using neural networks. *Smart Materials and Structures*, 13(4):777–783, 2004.

[35] E. Asua, V. Etxebarria, and A. García-Arribas. Micropositioning control using shape memory alloys. In *Proceedings of IEEE Conference on Control Applications CCA 2006*, pages 3229–3234, Munich, Germany, 2006.

[36] E. Asua, V. Etxebarria, and A. García-Arribas. Diseño y control de actuadores basados en aleaciones con memoria de forma (sma). In *XXVII Jornadas de Automtica*, Huelva, Spain, 2007.

[37] E. Asua, V. Etxebarria, and A. García-Arribas. Neural network-based micropositioning control of smart shape memory alloy actuators. *Engineering Applications of Artificial Intelligence*, 21(5):796–804, 2008.

[38] E. Asua, A. García-Arribas, and V. Etxebarria. Micropositioning using shape memory alloy actuators. *European Physical Journal*, 158(1):231–236, 2007.

[39] N. Vidal, E. Asua, J. Feuchwanger, A. García-Arribas, J. Gutiérrez, and J.M. Barandiarán. Fem simulation of the nitinol wire. *European Physical Journal*, 158(1):39–44, 2007.

[40] G. Song. Robust position regulation of shape memory alloy wire actuactor. *Proc. Inst. Mech, Eng*, 216:301–308, 2002.

[41] J. Ryhnen. Biocompativility evaluation of nickel-titanium shape memory metal alloy. In *Doctoral thesis*, 1999.

[42] Y.H. Teh and R. Featherstone. Frequency response analysis of shape memory alloy actuators. In *Proceedings of the SPIE*, volume 6423, page 64232J, 2007.

[43] R.P. Feynman, R.B. Leighton, and M. Sands. *Lectures on physics*. Addison-Wesley, 1977.

[44] M. Beale H. Demuth. *Neural Network Toolbox*. Math Works Inc., 1998.

In: Shape Memory Alloys
Editor: H. R. Chen, pp. 487-536
ISBN 978-1-60741-789-7

Chapter 17

Analysis of a Thermomechanical Model of Shape Memory Alloys

***Aida Timofte** and Vlad Timofte*†**
Institute of Mathematics of the Romanian Academy
P.O. Box 1-764, RO-014700 Bucharest, Romania

Abstract

This chapter is devoted to the mathematical study of a thermomechanical model describing the macroscopic behavior of shape memory alloys (SMA). Various experiments (slow uniaxial traction-compression tests on cylindrical SMA samples) emphasized the temperature changes during hysteresis loops and relaxation and creep phenomena. They also established the absence of permanent deformations at the end of a loading-unloading cycle, the so-called pseudoelastic behavior of SMA. The macroscopic phenomena observed during the experiments can be predicted by using a classical (Gibbs) thermomechanical model founded on a free energy, which is convex with respect to the strain and to the martensitic volume fraction, and concave with respect to the temperature. The model takes into account the non-isothermal character of the phase transition, as well as the existence of the intrinsic dissipation. The first law of thermodynamics, the balance of momentum in its quasistatic form, together with the evolution equation for the internal variables, form a partial differential equation system, whose solutions fulfill the second principle of thermodynamics (the entropy inequality). In the circular cylindrical case with non-negligible radius, this system reduces to an ordinary differential system. In this case, we prove the uniqueness of solutions in a large class of spaces (abstract differential structures). Existence and regularity of solutions are then established in various functions spaces endowed with natural derivatives. At the end of the chapter, for a particular traction compression test we present the exact solution, together with the associated hysteresis loop.

1. Introduction

The term Shape Memory Alloys (SMA) is applied to a group of metallic materials having the ability to return to some previously defined shape or size when subjected to the appro-

*E-mail address: aida.timofte@imar.ro
†E-mail address: vlad.timofte@imar.ro

priate thermal procedure. Arne Olander first observed this unusual property in 1938, but until the 1960's no serious research advances were made in the field of shape memory alloys. Generally, these materials can be plastically deformed at a relatively low temperature, but will return to their original shape upon exposure to some higher temperature. This is called the "shape memory effect".

Although a relatively wide variety of alloys are known to exhibit the shape memory effect, only those recovering substantial amounts of strain or generating significant force upon changing shape are of commercial interest. The most commonly known are the nickel-titanium, copper-zinc-aluminum, copper-aluminum-nickel, and iron-manganese-silicon alloys. The generic name for the family of nickel-titanium alloys is Nitinol, which stands for Nickel Titanium Naval Ordnance Laboratory. In 1961, William J. Buehler, a researcher at the Naval Ordnance Laboratory in White Oak, Maryland, discovered accidentally this shape memory alloy. At a laboratory management meeting, a strip of Nitinol was bent out of shape many times. One of the people present there, David Muzzey, heated it with his pipe lighter, and surprisingly, the strip stretched back to its original form.

1.1. Applications

SMA have been used for military, medical, safety, and robotics applications. Many of their current applications are in the field of medicine. Micro tools were used in minimally invasive surgeries like miniature forceps or manipulators that can close/open, twist, push/pull, or perform similar actions by applying a small amount of heat to the tool. Bone plates/stents need to fuse together or expand. Stents are extremely popular for heart surgeries, when arteries need to be expanded. A stent of a small diameter is easily placed within a partially clogged artery; then heat causes it to expand and push on deposits, thus opening the artery and increasing blood flow. Flexible, active endoscopes can be used for stomach and colon assessment. Wires that expand or contract may replace traditional wires used by orthodontists and dentists when straightening teeth. SMA wires can expand or contract at a gradual constant rate, thus providing the desired results with very little discomfort, rather than dramatic expansion/contraction performed by dentists or orthodontists at irregular intervals. A remarkable property of SMA is their transparency to magnetic resonance tomography, which is not the case of conventional implant materials such as stainless steel. This feature makes it much easier to monitor the devices inside the human body after implantation.

The military has been using Nitinol couplers in F-14 fighter planes since the late 1960's. These couplers join hydraulic lines tightly and easily.

Nitinol is being used in robotics actuators and micromanipulators to simulate human muscle motion. The main advantage of Nitinol is the smooth, controlled force it exerts upon activation. Six-legged robot (similar to a spider) are designed with shape-memory alloys leg components. As an increase in heat occurs, the legs contract; as the material cools, the legs expand. Thus, the leg components function like a muscle to yield a walking motion. Other examples of SMA used for joint control include a winged robot whose wings expand and contract to produce a flight motion, a fish robot that swims, and a snake robot that slithers.

Other miscellaneous applications of SMA include use in household appliances, in clothing, and in structures. There are examples of SMA used in safety devices which will save lives in the future. Anti-scalding devices and fire-sprinklers utilizing SMA are already on the market. The anti-scalding valves can be used in water faucets and shower heads. After a certain temperature, the device automatically shuts off the water flow. The main advantage of Nitinol-based fire sprinklers is the decrease in response time. A deep fryer utilizes the thermal sensitivity by lowering the basket into the oil at the correct temperature. Glasses can be made from superelastic metal alloy. Therefore, they can be bended quite drastically without permanent damage.

1.2. Advantages and Disadvantages of Shape Memory Alloys

Some of the main advantages of shape memory alloys include: bio-compatibility, diversity of fields of application, good mechanical properties including corrosion resistance.

There are still some difficulties to overcome before that shape memory alloys can live up to their full potential. These alloys are still relatively expensive to manufacture and machine, compared to other materials such as steel and aluminum. Most SMA have poor fatigue properties; this means that under the same loading conditions (i.e. twisting, bending, compressing) a steel component may survive for more than one hundred times more cycles than an SMA element.

The many uses and applications of shape memory alloys ensure a bright future for these metals. Research is currently carried out at many robotics departments and materials science departments. With the innovative ideas for applications of SMA and the number of products on the market using SMA continually growing, advances in the field of shape memory alloys for use in many different fields of study seem very promising.

1.3. Modeling of Shape Memory Alloys

A great deal of effort has been devoted in the last four-decade history of SMA within both mechanics and metallurgy communities, to the understanding and the modeling of the distinctive mechanical behavior of these materials. The source of this behavior is a crystalline phase transformation between a high symmetry, highly ordered parent phase (austenite), and a low symmetry, less ordered product phase (martensite). Martensitic structure is obtained from austenite by applying a mechanical load or by decreasing the temperature. Upon heating or stress reduction, the austenitic structure is recovered. This is the cause of the two most significant phenomena that characterize the mechanical behavior of SMA: the shape memory effect and the pseudoelastic behavior. Pseudoelasticity refers to the material's ability to be strained significantly, and then to return via a hysteresis loop to its unstrained configuration upon unloading.

There are several viewpoints in the constitutive modeling of these materials and a rich literature on this subject. A natural distinction can be made between macro-, meso-, and micro-models for shape memory alloys. Macro-models are based on the principles of thermodynamics. The variables that are sought after are the temperature and the volume fractions of the austenite and martensite. These models are mainly used to compute macroscopic effects of the phase transition, that is, shape changes of the specimen. Models to study measure-valued solutions solve the problem on a intermediate meso-scale. The problems that arise from using a nonconvex energy functional are treated by giving solutions as a probability (Young) measure describing infinitely fine microstructure.

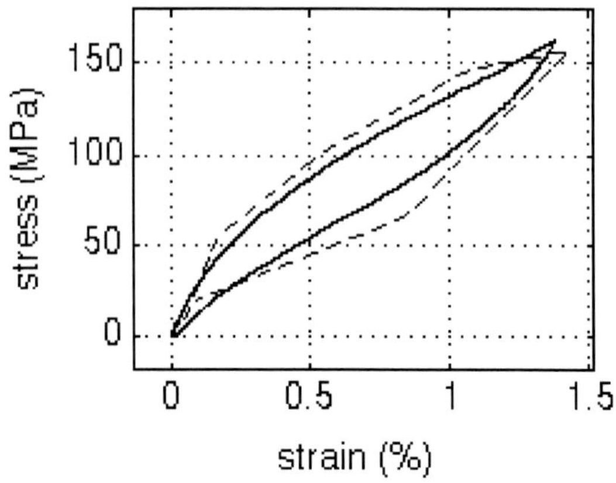

Figure 1. Experimental (dotted curve) and theoretical (continuous curve) hysteresis loops.

Many micromechanical aspects of SMA's behavior have been elucidated, including for example the capture of crystallographic aspects, dislocation microstructures, and twin boundaries. Starting with the paper by Ericksen [6], several important studies on phase transformations have been carried out; they make appeal to the minimization of the stored energy of the mixed phases and use subtle mathematical techniques (see for instance [4, 7]) in order to determine the amount of phases that coexist and the way they are juxtaposed. In fact this approach does not address the process of phase transformation rather it describes the coexistence of the phases after the transformation has taken place.

This chapter is devoted to the mathematical study of a thermomechanical model describing the macroscopic behavior of shape memory alloys. This model was first published in [3] (then slightly corrected in [12]) and was inspired by Pham's model from [8] and by A. Chrysochoos [5]. Qualitative properties of the solutions were established in [13]. A simplified version (without dissipation) of the model was used in [1, 2] in order to describe rheological properties of SMA during direct and inverse phase transformations. Figure 1 (taken from [3, Fig.7.2, p.354]) shows the experimental and theoretical (according to the model analyzed in this chapter) hysteresis loops for a slow uniaxial traction-compression test on a cylindrical CuZnAl sample. Various such experiments emphasized the temperature changes during hysteresis loops and relaxation and creep phenomena. They also established the absence of permanent deformations at the end of a loading-unloading cycle, the so-called pseudoelastic behavior of SMA. As shown in [3], the macroscopic phenomena observed during the experiments can be predicted by using a classical (Gibbs) thermomechanical model founded on a free energy, which is convex with respect to the strain and to the martensitic volume fraction, and concave with respect to the temperature. The proposed three dimensional thermomechanical model takes into account the non-isothermal character of the phase transition, as well as the existence of the intrinsic dissipation. The first law of thermodynamics, the balance of momentum in its quasistatic form, together with the evolution equation for the internal variables, form a partial differential equation system, whose solutions fulfill the second principle of thermodynamics (the entropy inequality). In the circular cylindrical case with non-negligible radius, this system reduces to an ordinary differential system. The remaining material from this chapter is organized as follows:

Section 2. contains the description of the model from [3] and the differential system obtained in the circular cylindrical case. The main difficulties of the problem are outlined in Section 2.3..

In Section 3., by defining some new notions (increment/decrement points of a function and abstract differential spaces and structures) we get an equivalent derivative-free version of the differential system. This will allow a unitary study of our mathematical problem in various functions spaces endowed with natural derivatives.

In Section 4. we define the SMA linear operators and we use them in order to obtain an optimum principle and the uniqueness of the solutions in every abstract differential structure.

In Sections 5. and 6. we get the existence of the solutions first in the space of all continuous functions, and then in various abstract differential structures.

The results from Sections 3.–6. were first published in [9, 10, 11].

2. The Thermomechanical Model of SMA

In this section we present the thermomechanical model of shape memory alloys introduced in [3] (including the correction from [12]). For reference purposes and with the kind permission of the authors, our exposition follows very closely the original publication.

2.1. Intrinsic Dissipation and Complementary Laws

The classical thermomechanical model we use here in order to describe the behavior of the SMA contains three independent state variables. Along with the strain tensor ϵ and the absolute temperature T, we consider an internal variable β, which is an adimensional scalar field representing the volume fraction of the martensitic phase. Obviously, we must satisfy the restriction

$$0 \leq \beta \leq 1. \tag{$\mathcal{R}$}$$

We assume that the system has a thermodynamic potential, and that this potential is its specific free energy,

$$\psi = e - Ts. \tag{1}$$

Consequently, we may express the constitutive thermomechanical relations by using the C^2-class function $\psi(\epsilon, T, \beta)$:

$$\sigma = \rho \frac{\partial \psi}{\partial \epsilon}, \qquad s = -\frac{\partial \psi}{\partial T}, \qquad B(\epsilon, T, \beta) = -\frac{\partial \psi}{\partial \beta}.$$

Here ρ denotes the mass density, e the specific internal energy, $\boldsymbol{\sigma}$ the stress tensor, s the specific entropy, and B the thermodynamical force associated to β.

The set of constitutive equations is completed by assuming for the heat flux vector $\mathbf{q}$ a classical Fourier law:

$$\mathbf{q} = -K\nabla T, \tag{2}$$

where $K > 0$ is a material constant.

The evolution of the SMA is governed by the balance of momentum equation, considered here in its quasistatic form,

$$\operatorname{div} \boldsymbol{\sigma} = 0, \tag{3}$$

and by the first law of thermodynamics,

$$\rho \dot{e} = \boldsymbol{\sigma} \cdot \dot{\boldsymbol{\epsilon}} - \operatorname{div} \mathbf{q}. \tag{4}$$

A superposed dot denotes the time derivative.

We neglect external body forces and external heat sources.

In order to describe the evolution of the SMA, the previous system must be completed with an equation for the internal variable β. This equation must be chosen such that any thermomechanical process satisfying the accordingly completed system, identically satisfies the restriction $(\mathcal{R})$ together with the second law of thermodynamics,

$$\rho \dot{s} \geq -\operatorname{div} \frac{\mathbf{q}}{T}. \tag{5}$$

By using (1) and (4) in (5), we get

$$\boldsymbol{\sigma} \cdot \dot{\boldsymbol{\epsilon}} - \rho s \dot{T} - \rho \dot{\psi} - \frac{1}{T} \mathbf{q} \cdot \nabla T \geq 0. \tag{6}$$

Since

$$\delta_t := -\frac{1}{T} \mathbf{q} \cdot \nabla T = \frac{K}{T} (\nabla T)^2 \geq 0,$$

the inequality (6), and hence the second principle of thermodynamics, is satisfied whenever

$$\delta_i := \boldsymbol{\sigma} \cdot \dot{\boldsymbol{\epsilon}} - \rho s \dot{T} - \rho \dot{\psi} = -\rho \frac{\partial \psi}{\partial \beta} \dot{\beta} = \rho B \dot{\beta} \geq 0,$$

where δ_i is the intrinsic dissipation.

Dissipationless models have been extensively studied (see [1, 2]) and their predictions are in good accord with the experimental readings for a number of cases. However, such a model predicts that the size of the hysteresis loop goes to zero as fast as the strain (or stress) rate of the traction-compression experiment, while significant hysteresis loops have been observed experimentally for near-zero rates.

We shall consider evolution equations leading to a nonzero internal dissipation. Let $\Phi : \mathbb{R} \to \mathbb{R}$ be a convex, continuous, non-negative, and positively homogeneous of degree $r > 0$ function, such that $\Phi(0) = 0$. We consider evolution equations for β based on the dissipation potential Φ:

$$r \rho B \in \partial \Phi(\dot{\beta}) + \partial \mathrm{I}(\beta), \tag{7}$$

where $\mathrm{I} : \mathbb{R} \to \{0, \infty\}$ denotes the indicator function of the segment $[0, 1]$,

$$\mathrm{I}(\beta) = \begin{cases} \infty, & \text{if } \beta \notin [0,1], \\ 0, & \text{if } \beta \in [0,1], \end{cases}$$

and $\partial \mathrm{I}$ its subdifferential,

$$\partial \mathrm{I}(\beta) = \{x \in \mathbb{R} \mid \mathrm{I}(\alpha) \geq \mathrm{I}(\beta) + x(\alpha - \beta) \text{ for every } \alpha \in \mathbb{R}\}.$$

The energy balance law (4) may be written as

$$\rho C\dot{T} + \operatorname{div}\mathbf{q} = \delta_i + \rho T\frac{\partial^2\psi}{\partial\epsilon\,\partial T}\cdot\dot{\epsilon} + \rho T\frac{\partial^2\psi}{\partial\beta\,\partial T}\cdot\dot{\beta},$$

where $C = -T\frac{\partial^2\psi}{\partial T^2}$ is the specific heat of the system. Accordingly, any thermodynamic process which fulfills (7) and (4) will also satisfy

$$\rho C\dot{T} + \operatorname{div}\mathbf{q} = \Phi(\dot{\beta}) + \rho T\frac{\partial^2\psi}{\partial\epsilon\,\partial T}\cdot\dot{\epsilon} + \rho T\frac{\partial^2\psi}{\partial\beta\,\partial T}\cdot\dot{\beta}, \tag{8}$$

and any such process which fulfills (7) and (8) will also satisfy (7) and (4). We may therefore replace (4) by (8) in the system of evolution equations.

2.2. The Choice of the Tthermodynamical and Dissipation Potentials

Let T_0 be the absolute temperature (supposed constant) of the room in which the SMA is placed; the difference between the absolute temperature T of the sample and T_0 is denoted by θ, that is,

$$T = T_0 + \theta.$$

Our model is intended to describe the behavior of the SMA during quasistatic traction-compression tests. In this case, experimental data show that the variation of temperature θ is small compared to T_0, and that the norm of the strain is small with respect to 1. It has also been noticed that the classical thermoelastic coupling is characterized by a heat dilatation coefficient α of order $10^{-6}/^{\circ}$C, that is, negligible with respect to the other terms from the heat propagation law.

Accordingly, we may attempt to describe the behavior of the SMA by using a general quadratic specific free energy with zero thermal dilatation coefficient,

$$\begin{aligned}\rho\psi(\epsilon,\theta,\beta) = \frac{\lambda}{2}(\operatorname{tr}\epsilon)^2 + \mu\epsilon\cdot\epsilon - \rho C\frac{\theta^2}{2T_0} - 2\mu\beta\boldsymbol{R}\cdot\epsilon + \mu\beta^2\boldsymbol{R}\cdot\boldsymbol{R}\\ +\rho\frac{L}{T_0}(T_0+\theta-T_{\mathrm{a}})\beta + \rho\frac{L}{2T_0}\beta^2(T_{\mathrm{a}}-T_{\mathrm{m}}),\end{aligned}$$

where λ and μ are Lamé's coefficients, $\boldsymbol{R}$ is the structural tensor of the system, L represents the latent heat of the body, and T_{a} and $T_{\mathrm{m}} < T_{\mathrm{a}}$ are characteristic temperatures for the material. All the above scalars are strictly positive[1], and the structural tensor is constant, symmetric, and traceless,

$$\boldsymbol{R}^{\mathrm{T}} = \boldsymbol{R}, \qquad \operatorname{tr}\boldsymbol{R} = 0. \tag{9}$$

For the dissipation potential Φ we let the simplest form ($r = 1$),

$$\Phi(\dot{\beta}) = K_{\mathrm{p}}|\dot{\beta}|, \tag{10}$$

[1]By *positive* and *strictly positive*, we mean "≥ 0", respectively "> 0".

where $K_{\mathrm{p}} > 0$ is a constant. The constitutive equations are now:

$$\begin{cases} \boldsymbol{\sigma} = \lambda \operatorname{tr} \boldsymbol{\epsilon}\, \mathbf{1} + 2\mu\boldsymbol{\epsilon} - 2\mu\beta\boldsymbol{R}, \\ \rho s = \rho C \dfrac{\theta}{T_0} - \rho \dfrac{L}{T_0}\beta, \\ \rho B = 2\mu\boldsymbol{R}\cdot\boldsymbol{\epsilon} - 2\mu\beta\boldsymbol{R}\cdot\boldsymbol{R} - \rho\dfrac{L}{T_0}(T_0 + \theta - T_{\mathrm{a}} + \beta\Delta T), \end{cases} \tag{11}$$

where

$$\Delta T := T_{\mathrm{a}} - T_{\mathrm{m}} > 0. \tag{12}$$

The energy balance law (8) becomes

$$\rho C\dot{T} + \operatorname{div}\mathbf{q} = K_{\mathrm{p}}|\dot{\beta}| + \rho T\frac{L}{T_0}\dot{\beta}. \tag{13}$$

In order to determine the form of the evolution equation for β, by using (9) and the first equality from (11), we express B as a function of $(\boldsymbol{\sigma}, T, \beta)$ as follows:

$$\rho B = \boldsymbol{\sigma}\cdot\boldsymbol{R} - \rho\frac{L}{T_0}(T_0 + \theta - T_{\mathrm{a}} + \beta\Delta T). \tag{14}$$

Let us put[2]

$$\sigma_R := \boldsymbol{\sigma}\cdot\boldsymbol{R}, \qquad \sigma_R^{\pm} := \rho\frac{L}{T_0}(T_0 + \theta - T_{\mathrm{a}} + \beta\Delta T) \pm K_{\mathrm{p}}.$$

By using (7), (10), and (14), we obtain the evolution system [3]

$$(\mathcal{E})^3 : \begin{cases} \text{If } \beta = 0, \text{ then } \sigma_R \le \sigma_R^+ \text{ and} \\ \qquad \dot{\beta} < 0 \Rightarrow \sigma_R \le \sigma_R^- \\ \text{If } 0 < \beta < 1, \text{ then } \sigma_R^- \le \sigma_R \le \sigma_R^+ \text{ and} \\ \qquad \begin{cases} \dot{\beta} < 0 \Rightarrow \sigma_R = \sigma_R^- \\ \dot{\beta} > 0 \Rightarrow \sigma_R = \sigma_R^+ \end{cases} \\ \text{If } \beta = 1, \text{ then } \sigma_R \ge \sigma_R^- \text{ and} \\ \qquad \dot{\beta} > 0 \Rightarrow \sigma_R \ge \sigma_R^+ \end{cases}$$

2.3. The Circular Cylindrical Case. Associated Differential System

All mechanical processes considered here are traction-compression tests on a cylindrical specimen of non-negligible radius $r > 0$. Accordingly, we consider a frame with the origin O at the fixed end of the specimen and the axis Ox_3 parallel to the generator of the sample. The evolution equations (3), (7), and (8), should be satisfied in the sense described in Section 2., at every point of the cylinder $\Omega = S \times [0, l]$ with circular section S and lateral surface L given by

$$S := \{(x_1, x_2) \in \mathbb{R}^2 \mid x_1^2 + x_2^2 \le r^2\}, \qquad L = \partial S \times [0, l].$$

[2]In expressions containing several times the symbol "$\pm$", the signs "$+$" and "$-$" correspond.

[3]In [3, p.320] the evolution system was written with equivalences instead of the implications "$\Rightarrow$"; this error was corrected in [12].

We may therefore assume that only the diagonal components of the structural tensor $\boldsymbol{R}$ are nonzero, and that

$$R_{11} = R_{22} = -\frac{g}{2}, \qquad R_{33} = g.$$

Here the material constant g represents the maximal axial strain in the Ox_3 direction due to the complete martensitic phase transformation. This assumption is justified by the fact that several traction-compression cycles (the so called training period) are performed before the start of the experiment, and it is experimentally proved that the initial isotropy of the sample with respect to the phase transformation is now replaced by this particular type of orthotropic behavior.

A more elaborate model should consider the structural tensor $\boldsymbol{R}$ as an internal variable, and deduce its evolution during the training cycles, from its initial spherical value at the beginning of the period, to the above described form at its end.

The lateral surface of the sample is free of tensions and, since the temperature difference θ is not very important, we may assume that the exchange of heat through the lateral surface is governed by a classical Newton law. Consequently, the P.D.E. system formed by (3), (7), and (8), should be completed by the conditions

$$\boldsymbol{\sigma}[\mathbf{n}] = 0 \quad \text{on } L, \tag{15}$$

$$\mathbf{q} \cdot \mathbf{n} = h\theta \quad \text{on } L, \tag{16}$$

where $\mathbf{n}$ is the outward normal at L and $h > 0$ is a material constant. Since in (7) no derivatives with respect to the spatial variables are involved, we should not impose any limit conditions for β and B. At the beginning of the experiment the SMA is in a stress-free configuration in the austenitic state, and at thermal equilibrium with the environment. Consequently,

$$T_{\mathrm{m}} < T_{\mathrm{a}} < T_0,$$

and the initial conditions of our system are

$$\left\{ \begin{array}{l} \boldsymbol{\epsilon}(x,0) = 0 \\ \theta(x,0) = \beta(x,0) = 0 \end{array} \right| \quad \text{for every } x \in \Omega. \tag{17}$$

Let us consider the cylindrical (Bessel) equation with bilocal conditions

$$\left\{ \begin{array}{l} x^2 f''(x) + x f'(x) + x^2 f(x) = 0, \\ f(\gamma r) = 1, \quad f'(0) = 0, \end{array} \right. \tag{18}$$

where $\gamma > 0$ is a constant. Let $J_{0,\gamma}$ denote the unique Bessel function satisfying (18). As a consequence of standard properties of Bessel functions, the map $(0,\infty) \ni \gamma \longmapsto \gamma J'_{0,\gamma}(\gamma r) \in (-\infty, 0)$ is a bijection. Let γ^* denote the unique solution of the equation $\gamma J'_{0,\gamma}(\gamma r) = -\frac{h}{K}$. It is obvious that the map $\phi : S \to \mathbb{R}$, defined by

$$\phi(x_1, x_2) := J_{0,\gamma^*}\left(\gamma^* \sqrt{x_1^2 + x_2^2}\right),$$

satisfies the conditions

$$\Delta\phi = -(\gamma^*)^2 \phi \quad \text{on } S, \tag{19}$$

$$\left\{ \begin{array}{l} \phi = 1 \\ \nabla\phi[\mathbf{n}] = -\frac{h}{K} \end{array} \right| \quad \text{on } \partial S,$$

where $\mathbf{n}$ is the outward $\mathbb{R}^2$ normal to ∂S.

The form of the differential operators of the system (3), (7), (8), of the limit conditions (15), (16), and of the initial conditions (17), as well as the particular shape of the body, suggest that we search for solutions by using the classical method of separation of variables. Let

$$\theta(x,t) = \phi(x_1,x_2)\bar{\theta}(t), \qquad \beta(x,t) = \phi(x_1,x_2)\bar{\beta}(t),$$

and

$$[\boldsymbol{\epsilon}] = \begin{bmatrix} \epsilon_1 & 0 & 0 \\ 0 & \epsilon_1 & 0 \\ 0 & 0 & \epsilon \end{bmatrix},$$

where $\epsilon(x,t) = \phi(x_1,x_2)\bar{\epsilon}(t)$ and

$$\epsilon_1 = -\frac{\lambda}{2(\lambda+\mu)}\epsilon + \frac{\mu g}{2(\lambda+\mu)}\beta.$$

By (11) we deduce that the only nonzero component of $\boldsymbol{\sigma}$ is σ_{33}, and so the limit condition (15) is satisfied, and that

$$\sigma_{33}(x,t) = \phi(x_1,x_2)E\big(\bar{\epsilon}(t) - g\bar{\beta}(t)\big) = \phi(x_1,x_2)\bar{\sigma}(t), \tag{20}$$

where E is the Young modulus. Hence $\frac{\partial\sigma_{33}}{\partial x_3} = 0$, and so the condition (3) is satisfied. An important consequence of (20) is:

$$\bar{\sigma} = E\big(\bar{\epsilon} - g\bar{\beta}\big). \tag{$\mathcal{C}$}$$

By (2) we get successively

$$\mathbf{q}(x,t) = -K\bar{\theta}(t)\nabla\phi(x_1,x_2), \qquad \mathbf{q}\cdot\mathbf{n} = -K(\nabla\phi[\mathbf{n}])\bar{\theta}.$$

By (19) it follows that the limit condition (16) is satisfied. By (2) and (19) we get

$$\operatorname{div}\mathbf{q} = -K\operatorname{div}(\nabla\theta) = -K\bar{\theta}\Delta\phi = K(\gamma^*)^2\phi\theta,$$

and, as $J_{0,\gamma}$ is positive, $|\dot{\beta}| = \phi|\dot{\bar{\beta}}|$. The heat propagation equation (13) is satisfied if and only if

$$\dot{\bar{\theta}} + \frac{\bar{\theta}}{\tau} = \Gamma|\dot{\bar{\beta}}| + \frac{L}{C}\dot{\bar{\beta}}, \tag{$\mathcal{H}$}$$

where

$$\tau := \frac{\rho C}{K(\gamma^*)^2}, \qquad \Gamma := \frac{K_{\mathrm{p}}}{\rho C}. \tag{21}$$

Experimental data show that Γ is small and $\Gamma < \frac{L}{C}$. With the notations

$$p := \frac{\rho L}{gT_0}, \qquad q := \frac{K_{\mathrm{p}}}{g}, \tag{22}$$

and

$$\sigma^{\pm} := p(T_0 - T_{\mathrm{a}} + \theta + \beta\Delta T) \pm q, \tag{23}$$

the evolution system $(\mathcal{E})^3$ takes the simpler form

$$(\mathcal{E}):\begin{cases}\text{If } \bar{\beta}=0, \text{ then } \bar{\sigma}\le\sigma^{+} \text{ and}\\ \qquad \dot{\bar{\beta}}<0 \Rightarrow \bar{\sigma}\le\sigma^{-}\\ \text{If } 0<\bar{\beta}<1, \text{ then } \sigma^{-}\le\bar{\sigma}\le\sigma^{+} \text{ and}\\ \qquad \begin{cases}\dot{\bar{\beta}}<0 \Rightarrow \bar{\sigma}=\sigma^{-}\\ \dot{\bar{\beta}}>0 \Rightarrow \bar{\sigma}=\sigma^{+}\end{cases}\\ \text{If } \bar{\beta}=1, \text{ then } \bar{\sigma}\ge\sigma^{-} \text{ and}\\ \qquad \dot{\bar{\beta}}>0 \Rightarrow \bar{\sigma}\ge\sigma^{+}\end{cases}$$

The initial conditions (17) imply

$$\bar{\sigma}(0)=\bar{\epsilon}(0)=\bar{\beta}(0)=\bar{\theta}(0)=0. \tag{$\mathcal{I}$}$$

For simplicity of notation, from now on we shall write $\sigma, \theta, \epsilon, \beta$ instead of $\bar{\sigma}, \bar{\theta}, \bar{\epsilon}, \bar{\beta}$.

Now the heat propagation equation $(\mathcal{H})$, the constitutive equation $(\mathcal{C})$, the restriction $(\mathcal{R})$, the evolution system $(\mathcal{E})$, and the initial conditions $(\mathcal{I})$, lead us to the following system:

$$(\mathcal{T}):\begin{cases}(\mathcal{H}):\ \dot{\theta}+\dfrac{\theta}{\tau}=\Gamma|\dot{\beta}|+\dfrac{L}{C}\dot{\beta},\\ (\mathcal{C}):\ \sigma=E(\varepsilon-g\beta),\\ (\mathcal{R}):\ 0\le\beta\le 1,\\ (\mathcal{E}):\begin{cases}\text{If } \beta=0, \text{ then } \sigma\le\sigma^{+} \text{ and}\\ \qquad \dot{\beta}<0 \Rightarrow \sigma\le\sigma^{-}\\ \text{If } 0<\beta<1, \text{ then } \sigma^{-}\le\sigma\le\sigma^{+} \text{ and}\\ \qquad \begin{cases}\dot{\beta}<0 \Rightarrow \sigma=\sigma^{-}\\ \dot{\beta}>0 \Rightarrow \sigma=\sigma^{+}\end{cases}\\ \text{If } \beta=1, \text{ then } \sigma\ge\sigma^{-} \text{ and}\\ \qquad \dot{\beta}>0 \Rightarrow \sigma\ge\sigma^{+}\end{cases}\\ (\mathcal{I}):\ \sigma(0)=\varepsilon(0)=\beta(0)=\theta(0)=0.\end{cases}$$

All constants $\tau, \Gamma, L, C, E, g, p, q, T_0, T_{\mathrm{a}}, \Delta T$ are strictly positive, with $T_0 > T_{\mathrm{a}}$ and $\Gamma < \frac{L}{C}$.

Let us consider a non-degenerate interval $J \subset \mathbb{R}$ with $\min J = 0$, and the functions spaces

$$\mathrm{C}(J) := \{u : J \to \mathbb{R} \mid u \text{ is continuous}\},$$

$$\mathrm{D}_{\mathrm{side}}(J) := \left\{ u \in \mathrm{C}(J) \,\middle|\, \begin{array}{l} u \text{ is semi-differentiable (has finite} \\ \text{one-sided derivatives at every point)} \end{array} \right\}.$$

J is the time interval of the process' evolution. The known data is an arbitrarily given function

$$\sigma \in \mathrm{C}(J), \text{ with } \sigma(0)=0.$$

The system $(\mathcal{T})$ is initially considered for unknown functions

$$\beta, \theta \in \mathrm{D}_{\mathrm{side}}(J), \qquad \varepsilon \in \mathrm{C}(J),$$

since β and θ need to satisfy the heat propagation equation $(\mathcal{H})$ and the evolution system $(\mathcal{E})$. The time derivatives from $(\mathcal{H})$ and $(\mathcal{E})$ are considered in the sense of both one-sided derivatives. Some comments are necessary for a better understanding of the mathematical problem raised by the system $(\mathcal{T})$:

1. We have three unknown functions $(\beta, \theta, \varepsilon)$ and two equations (heat, constitutive). It is not clear at all if the evolution system $(\mathcal{E})$ is equivalent to another equation.

2. If the given function σ is such that[4] $\dot{\beta}_{\mathrm{b}}(t_0) > 0$ and $\beta(t_0) = 1$, then β cannot be differentiable at t_0, since $\beta \leq 1 = \beta(t_0)$ forces $\dot{\beta}_{\mathrm{f}}(t_0) \leq 0$. This may occur even if σ is analytic on J. Thus, β may be less regular than σ. This is the reason for insisting on one-sided differentiability and introducing the functions space $\mathrm{D}_{\mathrm{side}}(J)$.

3. If β is strictly increasing[5] on some open subinterval $J_0 \subset J$, then $\{t \in J_0 \mid \dot{\beta}_{\mathrm{f}}(t) > 0\}$ is a dense subset of J_0, and so $\sigma = \sigma^+ = p(T_0 - T_{\mathrm{a}} + \theta + \beta \Delta T) + q$ on J_0, by $(\mathcal{E})$. In this case, we should have $\sigma|_{J_0} \in \mathrm{D}_{\mathrm{side}}(J_0)$, since $\beta, \theta \in \mathrm{D}_{\mathrm{side}}(J)$. This poses a compatibility problem for our system if the given σ does not have such derivatives (for instance, if $\sigma \in \mathrm{C}(J)$ is nowhere differentiable).

4. There exist increasing functions $u \in \mathrm{C}(J)$, such that

$$\int_0^t \dot{u}(s)\mathrm{d}s = 0 < u(t) - u(0) \quad \text{for every } t > 0.$$

Since the usual derivative sometimes fails to characterize (up to an additive constant) continuous and almost everywhere differentiable functions, the system $(\mathcal{T})$ written in terms of usual derivatives may not guarantee the uniqueness of the solutions.

5. Since for arbitrarily given $\sigma \in \mathrm{C}(J)$ a pronounced non-differentiability of solutions may occur, it would be natural to study $(\mathcal{T})$ in the larger space $\mathrm{C}(J)$ endowed with the derivative in the sense of distributions:

$$\mathrm{C}(J) \ni u \longmapsto u' \in \mathcal{D}'(\overset{\circ}{J}).$$

This is related to serious difficulties: what is the meaning of the absolute value $|\dot{\beta}|$ from $(\mathcal{H})$ and of $\dot{\beta}(t)$ from $(\mathcal{E})$, if $\dot{\beta}$ is a distribution, but not a function?

3. The Mathematical Problem in the Circular CylindricalCase

In this section we give a precise meaning of the mathematical problem posed by the system $(\mathcal{T})$ in a large class of functions spaces endowed with natural differential structures (see Definitions 3..11 and 3..14). By obtaining a derivative-free equivalent problem, we will be able to perform a unitary study of it in arbitrary abstract differential structures.

---

[4] We denote by $\dot{u}_{\mathrm{f}}(t)$ and $\dot{u}_{\mathrm{b}}(t)$ the one-sided (forward and backward) derivatives of the function u at t.

[5] We use the terms *increasing* and *strictly increasing* with the obvious meaning.

3.1. The Derivative-free Form of the Evolution System

In order to remove the derivatives of β from the evolution system $(\mathcal{E})^3$, we next introduce and study a new notion.

Definition 3..1 (increment point, decrement point) *Consider a non-degenerate interval $J_0 \subset \mathbb{R}$ and $u \in \mathrm{C}(J_0)$. A point $t \in J_0$ is said to be an* increment point *(respectively a* decrement point*) of u, if and only if for every neighborhood V of t, there exist $t_1, t_2 \in V \cap J_0$, such that*

$$t_1 < t_2, \qquad u(t_1) < u(t_2)$$

(respectively, $t_1 < t_2$ and $u(t_1) > u(t_2)$). Let $M^+(u)$ (respectively $M^-(u)$) denote the subset of J_0 consisting of all such points.

Note that $u \in M^+(u)$ does not mean that u is increasing on some neighborhood of t. Indeed, the function $u \in \mathrm{C}(\mathbb{R})$ defined by

$$u(t) = \begin{cases} t \sin \frac{1}{t}, & \text{for } t \neq 0, \\ 0 \quad , & \text{for } t - 0, \end{cases}$$

is neither decreasing, nor increasing on any neighborhood of 0, but $0 \in M^+(u) \cap M^-(u)$. Another example: if $v \in \mathrm{C}(J_0)$ is nowhere differentiable, then $M^+(v) = M^-(v) = J_0$.

Notation 3..2 *Let us consider a non-degenerate interval $J_0 \subset \mathbb{R}$. For every $u \in \mathrm{C}(J_0)$, set*

$$\begin{aligned} N_{\mathrm{f}}(u) &:= \{t \in J_0 \mid u \textit{ does not have a finite forward derivative at } t\}, \\ N_{\mathrm{b}}(u) &:= \{t \in J_0 \mid u \textit{ does not have a finite backward derivative at } t\}, \\ N(u) &:= \{t \in J_0 \mid u \textit{ is not differentiable at } t\}. \end{aligned}$$

Let us consider the functions spaces

$$\begin{aligned} \mathrm{D}_{\mathrm{f}}^{\aleph_0}(J_0) &:= \{u \in \mathrm{C}(J_0) \mid N_{\mathrm{f}}(u) \textit{ is at most countable}\}, \\ \mathrm{D}_{\mathrm{b}}^{\aleph_0}(J_0) &:= \{u \in \mathrm{C}(J_0) \mid N_{\mathrm{b}}(u) \textit{ is at most countable}\}, \\ \mathrm{D}^{\aleph_0}(J_0) &:= \{u \in \mathrm{C}(J_0) \mid N(u) \textit{ is at most countable}\}, \\ \mathrm{AC}_{\mathrm{loc}}(J_0) &:= \{u \in \mathrm{C}(J_0) \mid u \textit{ is locally absolutely continuous}\}. \end{aligned}$$

We next point out some useful properties of the sets $M^+(u)$ and $M^-(u)$. The fourth property below will be the key ingredient in obtaining the derivative-free form of the evolution system.

Proposition 3..3 *Let a non-degenerate interval $J_0 \subset \mathbb{R}$ and $u \in \mathrm{C}(J_0)$.*

(i) *For every $r \in J_0$, we have the equivalences*

$$\begin{aligned} r \notin M^-(u) &\iff u \textit{ is increasing on some neighborhood of } r, \\ r \notin M^+(u) &\iff u \textit{ is decreasing on some neighborhood of } r. \end{aligned}$$

(ii) *If J_1 is a subinterval of J_0, then*

$$u \text{ is increasing on } J_1 \iff M^-(u) \cap \overset{o}{J}_1 = \emptyset,$$
$$u \text{ is decreasing on } J_1 \iff M^+(u) \cap \overset{o}{J}_1 = \emptyset.$$

(iii) *$M^+(u)$ and $M^-(u)$ are perfect sets in J_0, and $M^-(u) = M^+(-u)$.*

(iv) *We have the equivalence*[6]

$$u \text{ is locally monotone} \iff M^+(u) \cap M^-(u) = \emptyset.$$

(v) *If $J_1, J_2 \subset J_0$ are closed intervals such that $J_0 = J_1 \cup J_2$, then*

$$M^+(u) = M^+(u|_{J_1}) \cup M^+(u|_{J_2}), \qquad M^-(u) = M^-(u|_{J_1}) \cup M^-(u|_{J_2}).$$

(vi) *We have the inclusion*

$$M^+(u) \supset \overline{\{t \in J_0 \setminus N_{\mathrm{f}}(u) \mid \dot{u}_{\mathrm{f}}(t) > 0\}} \cap J_0,$$

with equality if $u \in \mathrm{D}_{\mathrm{f}}^{\aleph_0}(J_0) \cup \mathrm{AC}_{\mathrm{loc}}(J_0)$. The inclusion also holds with $N_{\mathrm{b}}(u)$, $\dot{u}_{\mathrm{b}}(t)$ instead of $N_{\mathrm{f}}(u)$, $\dot{u}_{\mathrm{f}}(t)$, with equality if $u \in \mathrm{D}_{\mathrm{b}}^{\aleph_0}(J_0) \cup \mathrm{AC}_{\mathrm{loc}}(J_0)$. Similar statements hold for the set $M^-(u)$.

Proof. (i) and (ii) are immediate.
(iii). We need to show that $(M^+(u))' \cap J_0 = M^+(u)$, where A' denotes the set of all accumulation points of an arbitrary subset $A \subset \mathbb{R}$.
"$\subset$". Suppose that $t \notin M^+(u)$ for some $t \in (M^+(u))' \cap J_0$. According to (i), there exists $\delta > 0$, such that u is decreasing on the interval $J_0 \cap (t-\delta, t+\delta)$. By (ii) we get $M^+(u) \cap (t-\delta, t+\delta) = \emptyset$, which contradicts $t \in (M^+(u))'$.
"$\supset$". Suppose that $t \notin (M^+(u))' \cap J_0$ for some $t \in M^+(u) \subset J_0$. Hence $t \notin (M^+(u))'$, and so $M^+(u) \cap (t-\delta, t+\delta) \setminus \{t\} = \emptyset$ for some $\delta > 0$. Thus, $M^+(u) \cap (t-\delta, t) = M^+(u) \cap (t, t+\delta) = \emptyset$. According to (ii), u is decreasing on both intervals $J_0 \cap (t-\delta, t]$ and $J_0 \cap [t, t+\delta)$, and hence on their union $J_0 \cap (t-\delta, t+\delta)$. By (i) we get $t \notin M^+(u)$, a contradiction.
By the above inclusions we conclude that $(M^+(u))' \cap J_0 = M^+(u)$, and hence that $M^+(u)$ is a perfect set in J_0. It is clear that $M^-(u) = M^+(-u)$. Therefore, $M^-(u)$ is a perfect set in J_0.
(iv). The equivalence is immediate, by (i).
(v). Let us show the equality for $M^+(u)$ (then the other will follow by the equality $M^-(u) = M^+(-u)$).
"$\supset$". The inclusion is obvious, by Definition 3..1.
"$\subset$". Let us fix $t \in M^+(u)$. According to (iii), there is a sequence $(t_n)_{n\in\mathbb{N}} \subset M^+(u) \setminus \{t\}$, such that $\lim_{n\to\infty} t_n = t$. Since $t \in J = J_1 \cup J_2$, by taking a subsequence if necessary, we can assume that $(t_n)_{n\in\mathbb{N}} \subset M^+(u) \cap \overset{o}{J}_1 \subset M^+(u|_{J_1})$. By (iii) we get

[6] Locally monotone: for every $t \in J_0$, the function u is monotone on some neighborhood of t.

$t = \lim_{n\to\infty} t_n \in M^+(u|_{J_1})$.

(vi). By the definition of the forward derivative, we see that $\{t \in J_0 \setminus N_{\mathrm{f}}(u) \,|\, \dot{u}_{\mathrm{f}}(t) > 0\} \subset M^+(u)$. Since by (iii) the set $M^+(u)$ is closed in J_0, we have $\overline{\{t \in J_0 \setminus N_{\mathrm{f}}(u) \,|\, \dot{u}_{\mathrm{f}}(t) > 0\}} \cap J_0 \subset M^+(u)$, which is the claimed inclusion. For arbitrarily fixed $u \in \mathrm{D}_{\mathrm{f}}^{\aleph_0}(J_0) \cup \mathrm{AC}_{\mathrm{loc}}(J_0)$, suppose that

$$M_+(u) \not\subset \overline{\{t \in J_0 \setminus N_{\mathrm{f}}(u) \,|\, \dot{u}_{\mathrm{f}}(t) > 0\}}.$$

Hence there exist $t \in M^+(u)$ and $\delta > 0$, such that

$$\dot{u}_{\mathrm{f}}(s) \le 0 \quad \text{for every } s \in J_0 \cap (t-\delta, t+\delta) \setminus N_{\mathrm{f}}(u).$$

We deduce that u is decreasing on the interval $J_0 \cap (t-\delta, t+\delta)$. Indeed, this follows by Denjoy-Bourbaki's theorem if $u \in \mathrm{D}_{\mathrm{f}}^{\aleph_0}(J_0)$, or by the Leibniz-Newton formula if $u \in \mathrm{AC}_{\mathrm{loc}}(J_0)$. By (i) we get $t \notin M^+(u)$, a contradiction. □

The evolution systems from Sections 2.2. and 2.3. were initially considered for unknown functions $\beta, \theta \in \mathrm{D}_{\mathrm{side}}(J)$. Now Proposition 3..3(vi) and the inclusion $\mathrm{D}_{\mathrm{side}}(J) \subset \mathrm{D}_{\mathrm{f}}^{\aleph_0}(J)$ make it obvious that the system $(\mathcal{E})^3$ has the equivalent derivative-free form

$$(\mathcal{E})^3_{\mathrm{D_{side}}(J)} : \begin{cases} \text{If } \beta(t) = 0, \text{ then } \sigma_R(t) \le \sigma_R^+(t) \text{ and} \\ \qquad t \in M^-(\beta) \Rightarrow \sigma_R(t) \le \sigma_R^-(t) \\ \text{If } 0 < \beta(t) < 1, \text{ then } \sigma_R^-(t) \le \sigma_R(t) \le \sigma_R^+(t) \text{ and} \\ \qquad \begin{cases} t \in M^-(\beta) \Rightarrow \sigma_R(t) = \sigma_R^-(t) \\ t \in M^+(\beta) \Rightarrow \sigma_R(t) = \sigma_R^+(t) \end{cases} \\ \text{If } \beta(t) = 1, \text{ then } \sigma_R(t) \ge \sigma_R^-(t) \text{ and} \\ \qquad t \in M^+(\beta) \Rightarrow \sigma_R(t) \ge \sigma_R^+(t) \end{cases}$$

Since $(\mathcal{E})^3_{\mathrm{D_{side}}(J)}$ does not involve the derivatives of β, we can write the evolution system for functions β, θ in an arbitrary vector subspace $\mathrm{X}(J) \subset \mathrm{C}(J)$, thus obtaining a system which will be denoted by $(\mathcal{E})^3_{\mathrm{X}(J)}$. The corollary of Theorem 3..4 below will provide a much simpler form of the evolution system $(\mathcal{E})^3_{\mathrm{X}(J)}$.

Theorem 3..4 *Let $\beta, \theta \in \mathrm{X}(J)$ satisfy the evolution system $(\mathcal{E})^3_{\mathrm{X}(J)}$. Then*

$$\sigma_R(t) = \sigma_R^+(t) \quad \textit{for every } t \in M^+(\beta),$$
$$\sigma_R(t) = \sigma_R^-(t) \quad \textit{for every } t \in M^-(\beta).$$

Furthermore, if all β, θ, σ are differentiable at some $t \in M^+(\beta) \cup M^-(\beta)$, then

$$\dot{\sigma}_R(t) = \dot{\sigma}_R^+(t) = \dot{\sigma}_R^-(t).$$

Proof. Suppose that $\sigma_R(t) \ne \sigma_R^+(t)$ for some $t \in M^+(\beta)$. As $\sigma_R, \sigma_R^+ \in \mathrm{C}(J)$, there exists $\delta > 0$, such that $\sigma_R(s) \ne \sigma_R^+(s)$ for every $s \in J \cap (t-\delta, t+\delta)$. Since $t \in M^+(\beta)$, according to Definition 3..1 we have $t_1 < t_2$ and $\beta(t_1) < \beta(t_2)$ for some $t_1, t_2 \in J \cap (t-\delta, t+\delta)$. Set

$$s_1 := \sup\{s \in [t_1, t_2] \,|\, \beta(s) \le \beta(t_1)\}, \qquad s_2 := \inf\{s \in [s_1, t_2] \,|\, \beta(s) \ge \beta(t_2)\}.$$

By the construction of s_1 and s_2, we see that $t_1 \le s_1 < s_2 \le t_2$ and

$$\begin{aligned}&\beta(s_1) = \beta(t_1) < \beta(t_2) = \beta(s_2),\\&\beta(s_1) < \beta(s) < \beta(s_2) \quad \text{for every } s \in (s_1, s_2).\end{aligned}$$

Since $\beta(s_1) < \beta(s_2)$, by Proposition 3..3(ii) we can find $s \in M^+(\beta) \cap (s_1, s_2) \neq \emptyset$. It follows that $s \in (s_1, s_2) \subset (t_1, t_2) \subset J \cap (t-\delta, t+\delta)$, which yields $\sigma_R(s) \neq \sigma_R^+(s)$. As $\beta(s) \in (\beta(s_1), \beta(s_2)) \subset (0,1)$ and $s \in M^+(\beta)$, by $(\mathcal{E})^3_{\mathrm{X}(J)}$ we get $\sigma_R(s) = \sigma_R^+(s)$, a contradiction. We conclude that $\sigma_R(t) = \sigma_R^+(t)$ for every $t \in M^+(\beta)$. For $t \in M^-(\beta)$, the proof of the equality $\sigma_R(t) = \sigma_R^-(t)$ is similar.
Now assume that all β, θ, σ are differentiable at some $t \in M^+(\beta)$. Hence $\sigma_R, \sigma_R^+, \sigma_R^-$ are also differentiable at t. By Proposition 3..3(iii), there is a sequence $(t_n)_{n\in\mathbb{N}} \subset M^+(\beta)\backslash\{t\}$, with $\lim_{n\to\infty} t_n = t$. We have $\sigma_R(t_n) = \sigma_R^+(t_n)$ for every $n \in \mathbb{N}$, and so

$$\dot\sigma_R(t) = \lim_{n\to\infty} \frac{\sigma_R(t_n) - \sigma_R(t)}{t_n - t} = \lim_{n\to\infty} \frac{\sigma_R^+(t_n) - \sigma_R^+(t)}{t_n - t} = \dot\sigma_R^+(t) = \dot\sigma_R^-(t).$$

The proof in the case $t \in M^-(\beta)$ is very similar. □

Corollary 3..5 (the derivative-free form of the evolution system) *Let an arbitrary vector subspace* $\mathrm{X}(J) \subset \mathrm{C}(J)$.

(a) *The system* $(\mathcal{E})^3_{\mathrm{X}(J)}$ *has the equivalent form*

$$(\mathcal{E})^3_{\mathrm{X}(J)} : \begin{cases} \beta(t) > 0 \Rightarrow \sigma_R(t) \ge \sigma_R^-(t) \\ \beta(t) < 1 \Rightarrow \sigma_R(t) \le \sigma_R^+(t) \\ t \in M^+(\beta) \Rightarrow \sigma_R(t) = \sigma_R^+(t) \\ t \in M^-(\beta) \Rightarrow \sigma_R(t) = \sigma_R^-(t) \end{cases}$$

(b) *In the circular cylindrical case the evolution system* $(\mathcal{E})^3_{\mathrm{X}(J)}$ *becomes*

$$(\mathcal{E})_{\mathrm{X}(J)} : \begin{cases} \beta(t) > 0 \Rightarrow \sigma(t) \ge \sigma^-(t) \\ \beta(t) < 1 \Rightarrow \sigma(t) \le \sigma^+(t) \\ t \in M^+(\beta) \Rightarrow \sigma(t) = \sigma^+(t) \\ t \in M^-(\beta) \Rightarrow \sigma(t) = \sigma^-(t) \end{cases}$$

Proof. The conclusion is immediate, by Theorem 3..4. □

Remark 3..6 *The last two implications from the evolution system* $(\mathcal{E})_{\mathrm{X}(J)}$ *may be written as*

$$M^+(\beta) \subset (\sigma^+ - \sigma)^{-1}(\{0\}), \qquad M^-(\beta) \subset (\sigma^- - \sigma)^{-1}(\{0\}).$$

Corollary 3..7 (phase transformation inertia) *If* $\beta, \theta \in \mathrm{X}(J)$ *satisfy the evolution system* $(\mathcal{E})^3_{\mathrm{X}(J)}$*, then the function* β *is locally monotone.*

Proof. As $\sigma_R^+(t) \neq \sigma_R^-(t)$ for every $t \in J$, by Corollary 3..5(a) we deduce that $M^+(\beta) \cap M^-(\beta) = \emptyset$. This is equivalent to the local monotony of β, by Proposition 3..3(iv). □

Corollary 3..7 will prove extremely useful in the study of the heat propagation equation $(\mathcal{H})$. Note that the corollary still holds for β, θ satisfying the "cylindrical" evolution system $(\mathcal{E})_{\mathrm{X}(J)}$. More precisely, if $\beta, \theta \in \mathrm{X}(J)$ satisfy the evolution system $(\mathcal{E})_{\mathrm{X}(J)}$, then for every $r \in J$ we have the implications

$$\sigma(r) \neq \sigma^-(r) \Longrightarrow \beta \text{ is increasing on a neighborhood of } r,$$
$$\sigma(r) \neq \sigma^+(r) \Longrightarrow \beta \text{ is decreasing on a neighborhood of } r.$$

Remark 3..8 *For $\beta, \theta \in \mathrm{X}(J)$ satisfying the evolution system $(\mathcal{E})_{\mathrm{X}(J)}$, the monotony of β around $r \in J$ is determined by the values $\beta(r)$ and $\theta(r)$.*

Indeed $\sigma \in \mathrm{C}(J)$ is given (the initial data) and the values $\sigma^{\pm}(r)$ can be found by (23).

Corollary 3..9 *Assume that a subinterval $J_0 \subset J$ has one endpoint in $M^+(\beta)$ and the other in $M^-(\beta)$. Then β is constant on some nondegenerate subinterval of J_0.*

Proof. The proof by Corollary 3..7 is immediate. □

3.2. Abstract Differential Structures and the Heat Propagation Equation

For $\beta, \theta \in \mathrm{D_{side}}(J)$, under the assumptions of the evolution system, the function β is locally monotone, and so the heat propagation equation $(\mathcal{H})$ from the system $(\mathcal{T})$ has the equivalent form

$$(\mathcal{H})_{\mathrm{D_{side}}(J)}: \begin{cases} \text{For every subinterval } J_0 \subset J \text{ of monotony for } \beta, \text{ we have} \\ \quad \dot{\theta} + \dfrac{\theta}{\tau} = \Lambda \dot{\beta} \quad \text{on } \overset{o}{J_0}, \\ \text{where } \Lambda := \Gamma_{\pm} + \frac{L}{C} \text{ and } \Gamma_{\pm} = \begin{cases} \Gamma, & \text{if } \beta \text{ is increasing on } J_0, \\ -\Gamma, & \text{if } \beta \text{ is decreasing on } J_0. \end{cases} \\ \text{If } \beta \text{ is constant on } J_0, \text{ we make an arbitrary choice for } \Gamma_{\pm} \in \{\Gamma, -\Gamma\}. \end{cases}$$

From now on, we will keep the above convention for $\Gamma_{\pm}$ and Λ. Let us recall that for every solution of the system $(\mathcal{T})_{\mathrm{X}(J)}$, the monotony of β around any point $r \in J$ is uniquely determined by the values $\beta(r)$ and $\theta(r)$ (see Remark 3..8).

The condition $(\mathcal{H})_{\mathrm{D_{side}}(J)}$ may be rephrased by replacing the space $\mathrm{D_{side}}(J)$ and the one-sided derivatives by any of the spaces and derivatives from the following example.

Example 3..10 (functions spaces with derivations)

(i) $\mathrm{C}(J)$*, with the derivative in the sense of distributions in $\mathcal{D}'(\overset{o}{J})$. We have the natural inclusions $\mathrm{C}(J) \subset \mathrm{C}(\overset{o}{J}) \subset \mathcal{D}'(\overset{o}{J})$. Let us note that a function $u \in \mathrm{C}(J)$ is increasing, if and only if its derivative $u' \in \mathcal{D}'(\overset{o}{J})$ is a positive distribution (see* Theorem 3..12 *below). Hence for any locally monotone function $\beta \in \mathrm{C}(J)$, on monotony intervals $J_0 \subset J$ the absolute value $|\dot{\beta}|$ makes sense in $\mathcal{D}'(\overset{o}{J_0})$.*

(ii) $\mathrm{BV}_{\mathrm{loc}}(J) := \{u \in \mathrm{C}(J) \mid u$ *has locally bounded variation*$\}$, *with the derivative in the sense of distributions.*

(iii) $\mathrm{AC}_{\mathrm{loc}}(J)$, *with the derivative almost everywhere (a.e.).*

(iv) $\mathrm{Lip}_{\mathrm{loc}}(J) := \{u \in \mathrm{C}(J) \mid u$ *is locally Lipschitz*$\}$, *with the derivative almost everywhere.*

(v) *For* $p \in [1, \infty]$, *the Sobolev space* $\mathrm{W}^{1,p}_{\mathrm{loc}}(J) = \big\{u \in \mathrm{AC}_{\mathrm{loc}}(\overline{J}) \big| \, \widehat{u} \in \mathrm{L}^p_{\mathrm{loc}}(\overline{J})\big\}$, *with the derivative almost everywhere.*

(vi) *For fixed at most countable subsets* $A, B \subset \overline{\mathbb{R}}$, *let us consider the spaces:*

- **(a)** $\mathrm{D}^A_{\mathrm{f}}(J) := \{u \in \mathrm{C}(J) \mid u$ *is differentiable to the right on* $J \setminus A\}$ *(respectively, the space* $\mathrm{D}^B_{\mathrm{b}}(J)$*), with the forward (respectively backward) derivative.*
- **(b)** $\mathrm{D}^{A,B}_{\mathrm{side}}(J) := \mathrm{D}^A_{\mathrm{f}}(J) \cap \mathrm{D}^B_{\mathrm{b}}(J)$, *with both one-sided derivatives.*

In particular, for $\mathrm{D}_{\mathrm{f}}(J) := \mathrm{D}^{\{\sup J\}}_{\mathrm{f}}(J)$ *and* $\mathrm{D}_{\mathrm{b}}(J) := \mathrm{D}^{\{\inf J\}}_{\mathrm{b}}(J)$, *we have*

$$\mathrm{D}_{\mathrm{side}}(J) = \mathrm{D}_{\mathrm{f}}(J) \cap \mathrm{D}_{\mathrm{b}}(J).$$

(vii) $\mathrm{D}^{\aleph_0}_{\mathrm{f}}(J)$ *with the forward derivative (see* Notation 3..2*),* $\mathrm{D}^{\aleph_0}_{\mathrm{b}}(J)$ *with the backward derivative, and* $\mathrm{D}^{\aleph_0}_{\mathrm{side}}(J) := \mathrm{D}^{\aleph_0}_{\mathrm{f}}(J) \cap \mathrm{D}^{\aleph_0}_{\mathrm{b}}(J)$ *with both one-sided derivatives.*

(viii) $\mathrm{D}^{\aleph_0}(J)$, *with the usual derivative.*

(ix) *Let us consider the following three spaces of one-sided analytic functions:*

- **(a)** $\mathrm{A}_{\mathrm{f}}(J) := \{u \in \mathrm{C}(J) \mid u$ *is forward-analytic*$\}$, *with the forward derivative. A function* $u \in \mathrm{C}(J)$ *is called forward-analytic at* $t \in J \setminus \{\sup J\}$, *if and only if* u *is analytic on some* $[t, s) \subset J$ *(*$s > t$*). We call* u *a* forward-analytic function, *if and only if* u *is forward-analytic at every point* $t \in J \setminus \{\sup J\}$.
- **(b)** $\mathrm{A}_{\mathrm{b}}(J) := \{u \in \mathrm{C}(J) \mid u$ *is backward-analytic*$\}$, *with the backward derivative (the definitions are similar to those for* $\mathrm{A}_{\mathrm{f}}(J)$*).*
- **(c)** $\mathrm{A}_{\mathrm{side}}(J) := \mathrm{A}_{\mathrm{f}}(J) \cap \mathrm{A}_{\mathrm{b}}(J)$, *with both one-sided derivatives.*

Any of the above functions spaces may be replaced by a vector subspace, while keeping the sense of the derivation. In order to deal simultaneously with various possible such spaces, we next introduce the notions of *abstract differential space* and *abstract differential structure*. Let

$$\mathrm{A}(\mathbb{R}) := \{u \in \mathrm{C}(\mathbb{R}) \mid u \text{ is analytic}\}, \qquad \mathrm{A}(\mathbb{R})|_J = \big\{a|_J : J \to \mathbb{R} \,\big|\, a \in \mathrm{A}(\mathbb{R})\big\}.$$

Definition 3..11 (abstract differential spaces) *Let us consider*

(a) $\mathrm{X}(J)$, *an* $\mathrm{A}(\mathbb{R})|_J$*-submodule*[7] *of* $\mathrm{C}(J)$, *containing the antiderivatives of all functions from* $\mathrm{X}(J)$,

[7] Submodule: $\mathrm{X}(J) \subset \mathrm{C}(J)$ is a vector subspace, such that $a \cdot u \in X(J)$ for all $a \in \mathrm{A}(\mathbb{R})|_J$ and $u \in \mathrm{X}(J)$.

(b) *Y, an $\mathrm{A}(\mathbb{R})|_J$-module,*

(c) *$\zeta : \mathrm{X}(J) \to Y$, an injective morphism (embedding) of $\mathrm{A}(\mathbb{R})|_J$-modules.*

We call an Y-valued derivation *on $\mathrm{X}(J)$ any function $\partial : \mathrm{X}(J) \to Y$ with the following properties:*

(i) *∂ is $\mathbb{R}$-linear and $\ker \partial \subset \mathbb{R}$,*

(ii) *$\partial(au) = \dot{a} \cdot \zeta(u) + a \cdot \partial u$ for all $a \in \mathrm{A}(\mathbb{R})|_J$ and $u \in \mathrm{X}(J)$,*

(iii) *$\partial u = \zeta(\dot{u})$ for every $u \in \mathrm{C}^1(J)$, such that $\dot{u} \in \mathrm{X}(J)$.*

Under the above assumptions, the triplet $(\mathrm{X}(J), \zeta, \partial)$ is said to be an abstract differential space.

Let us note that every abstract differential space $(\mathrm{X}(J), \zeta, \partial)$ contains the constant functions, which are the antiderivatives of $0 \in \mathrm{X}(J)$. Since $\mathrm{X}(J)$ is an $\mathrm{A}(\mathbb{R})|_J$-submodule of $\mathrm{C}(J)$, we have

$$\mathrm{A}(\mathbb{R})|_J \subset \mathrm{X}(J).$$

The condition $\ker \partial \subset \mathbb{R}$ from the above definition means that $\partial u \in Y$ characterizes the function $u \in X(J)$ up to an additive constant. The next theorem related to Example 3..10(i) shows in particular that for the derivation $(\cdot)' : \mathrm{C}(J) \to \mathcal{D}'(\overset{o}{J})$ in the sense of distributions, we have indeed $\ker(\cdot)' \subset \mathbb{R}$.

Theorem 3..12 (monotone continuous functions) *Let a function $u \in \mathrm{C}(J)$. Then u is increasing, if and only if $u' \in \mathcal{D}'(\overset{o}{J})$ is a positive distribution (takes positive values on all positive test functions).*

Proof. "$\Rightarrow$". Let $\varphi \in \mathrm{C}_c^\infty(\overset{o}{J})$, with $\varphi \geq 0$. Let us choose $a, b \in \overset{o}{J}$, such that $\mathrm{supp}\varphi \subset [a, b] \subset \overset{o}{J}$. By using the definition of the derivative in the sense of distributions and the Riemann-Stieltjes integral, it follows that

$$\langle u', \varphi\rangle = -\langle u, \varphi'\rangle = -\int_{[a,b]} u\varphi' = -\int_a^b u\mathrm{d}\varphi = -u\varphi\Big|_a^b + \int_a^b \varphi\mathrm{d}u = \int_a^b \varphi\mathrm{d}u \geq 0$$

(the last Riemann-Stieltjes integral is positive, since $\varphi \geq 0$ and u is increasing).
"$\Leftarrow$". Let us fix $a, b \in \overset{o}{J}$, such that $a < b$. Let $\varepsilon > 0$. As $u \in \mathrm{C}(J)$, there exists $\delta > 0$, such that

$$\begin{cases} \delta < \frac{b-a}{2}, & [a-\delta, b+\delta] \subset \overset{o}{J}, \\ |u(t) - u(a)| < \frac{\varepsilon}{2} & \text{for every } t \in [a-\delta, a+\delta], \\ |u(t) - u(b)| < \frac{\varepsilon}{2} & \text{for every } t \in [b-\delta, b+\delta]. \end{cases}$$

Let us choose a function $\psi \in \mathrm{C}_c^\infty(\mathbb{R})$, such that

$$\psi \geq 0, \qquad \mathrm{supp}\psi \subset [-\delta, \delta], \qquad \int_{\mathbb{R}} \psi = 1.$$

Now let us define the function

$$\varphi : \overset{o}{J} \to \mathbb{R}, \qquad \varphi(t) = \int_{a-\delta}^{t} [\psi(s-a) - \psi(s-b)]\mathrm{d}s.$$

Obviously, $\operatorname{supp}\psi \subset [a-\delta, b+\delta] \subset \overset{o}{J}$ and $0 \le \varphi \in \mathrm{C}_{\mathrm{c}}^{\infty}(\overset{o}{J})$. As $u' \ge 0$ in $\mathcal{D}'(\overset{o}{J})$, we have

$$0 \ge -\langle u', \varphi\rangle = \langle u, \varphi'\rangle = \int_{\overset{o}{J}} u\varphi' = \int_{\overset{o}{J}} u(t)[\psi(t-a) - \psi(t-b)]\mathrm{d}t$$
$$= \int_{a-\delta}^{a+\delta} u(t)\psi(t-a)\mathrm{d}t - \int_{b-\delta}^{b+\delta} u(t)\psi(t-b)\mathrm{d}t$$
$$\ge \int_{a-\delta}^{a+\delta} \left(u(a) - \frac{\varepsilon}{2}\right)\psi(t-a)\mathrm{d}t - \int_{b-\delta}^{b+\delta} \left(u(b) + \frac{\varepsilon}{2}\right)\psi(t-b)\mathrm{d}t = u(a) - u(b) - \varepsilon.$$

We thus get $u(a) - u(b) \le \varepsilon$ for every $\varepsilon > 0$, which yields $u(a) \le u(b)$. As $a, b \in \overset{o}{J}$ were arbitrarily fixed, it follows that u is increasing on $\overset{o}{J}$, and hence on J, since u is continuous.
□

Proposition 3..13 (concrete abstract differential spaces) *All spaces with derivations from* Example 3..10 *have natural structures of abstract differential spaces.*

Proof. In each case we only give the construction of the functions $\zeta, \partial : \mathrm{X}(J) \to Y$. Verifying the properties from Definition 3..11 will be left to the reader. Let $\mathrm{F}(T)$ denote the space of all real functions on a set $T \neq \emptyset$. For simplicity, we will use the same symbol to denote a continuous function on the interval J and its restriction to $\overset{o}{J}$.
(i,ii). Consider $\zeta, \partial : \mathrm{C}(J) \to \mathcal{D}'(\overset{o}{J})$, where ζ is the natural inclusion morphism and $\partial u = u'$ is the derivative of u in the sense of distributions. The construction for (ii) is similar.
(iii,iv). Set $Y := \{\widehat{u} \mid u \in \mathrm{F}(J)\}$, where $\widehat{u}$ denotes the equivalence class of the function u with respect to the equality a.e.. Consider $\zeta, \partial : \mathrm{AC}_{\mathrm{loc}}(J) \to Y$, defined by

$$\zeta(u) = \widehat{u}, \qquad \partial u = \widehat{\dot{u}}.$$

Note that ∂ is well-defined, since the derivative $\dot{u}$ exists a.e. for every $u \in \mathrm{AC}_{\mathrm{loc}}(J)$. The construction for (iv) is similar.
(v). Consider $\zeta, \partial : \mathrm{W}_{\mathrm{loc}}^{1,p}(J) \to \mathrm{L}_{\mathrm{loc}}^{p}(\overline{J})$, defined by

$$\zeta(u) = \widehat{u}, \qquad \partial\widehat{u} = \widehat{\dot{u}}.$$

(vi). Consider $\zeta, \partial : \mathrm{D}_{\mathrm{f}}^{A}(J) \to \mathrm{F}(J \setminus A)$, defined by

$$\zeta(u) = u|_{J\setminus A}, \qquad \partial u = \dot{}_{\mathrm{f}}u$$

The construction for $\mathrm{D}_{\mathrm{b}}^{B}(J)$ is similar. For $\mathrm{D}_{\mathrm{side}}^{A,B}(J)$ (hence in particular for $\mathrm{D}_{\mathrm{side}}(J)$), consider the functions $\zeta, \partial : \mathrm{D}_{\mathrm{side}}^{A,B}(J) \to \mathrm{F}(J \setminus A) \times \mathrm{F}(J \setminus B)$, defined by

$$\zeta(u) = (u|_{J\setminus A}, u|_{J\setminus B}), \qquad \partial u = (\dot{u}_{\mathrm{f}}, \dot{u}_{\mathrm{b}}).$$

(vii,viii). Set $Y := \{\widehat{u} \mid u \in \mathrm{F}(J)\}$, where $\widehat{u}$ denotes the class of the function u, with respect to the equivalence relation defined on $\mathrm{F}(J)$ by: $v \sim w \Leftrightarrow \{t \in J \mid v(t) \neq w(t)\}$ is an at most countable set. Consider $\zeta, \partial : \mathrm{D}_{\mathrm{f}}^{\aleph_0}(J) \to Y$, defined by

$$\zeta(u) = \widehat{u}, \qquad \partial u = \widehat{\dot{u}_{\mathrm{f}}}.$$

The construction for $\mathrm{D}_{\mathrm{b}}^{\aleph_0}(J)$ is similar. For $\mathrm{D}_{\mathrm{side}}^{\aleph_0}(J)$, define $\zeta, \partial : \mathrm{D}_{\mathrm{side}}^{\aleph_0}(J) \to Y \times Y$ by

$$\zeta(u) = (\widehat{u}, \widehat{u}), \qquad \partial u = \left(\widehat{\dot{u}_{\mathrm{f}}}, \widehat{\dot{u}_{\mathrm{b}}}\right).$$

For $\mathrm{D}^{\aleph_0}(J)$, define $\zeta, \partial : \mathrm{D}^{\aleph_0}(J) \to Y$ by

$$\zeta(u) = \widehat{u}, \qquad \partial u = \widehat{\dot{u}}.$$

(ix). Consider $\zeta, \partial : \mathrm{A}_{\mathrm{f}}(J) \to \mathrm{F}(J \setminus \{\sup J\})$, defined by

$$\zeta(u) = u|_{J\setminus\{\sup J\}}, \qquad \partial u = \dot{}_{\mathrm{f}} u$$

The constructions for $\mathrm{A}_{\mathrm{b}}(J) \subset \mathrm{D}_{\mathrm{b}}(J)$ and $\mathrm{A}_{\mathrm{side}}(J) \subset \mathrm{D}_{\mathrm{side}}(J)$ are similar. □

There are many other abstract differential spaces that can be imagined. For instance, if we want to emphasize the differentiability of the functions at some fixed $t_0 \in J$, we may consider

$$\begin{aligned}
&\mathrm{X}(J) = \{u \in \mathrm{C}(J) \mid u \text{ is differentiable at } t_0\},\\
&\zeta, \partial : \mathrm{X}(J) \to \mathcal{D}'(\overset{o}{J}) \times \mathbb{R},\\
&\zeta(u) = (u, u(t_0)), \quad \partial u = (u', \dot{u}(t_0)).
\end{aligned}$$

Definition 3..14 (abstract differential structures) *Let us consider* $\mathrm{X}(J) \subset \mathrm{C}(J)$ *as in* Definition 3..11(a). *For every subinterval* $J_0 \subset J$, *set*

$$\mathrm{X}(J_0) = \mathrm{X}(J)|_{J_0} := \left\{u|_{J_0} \,\middle|\, u \in \mathrm{X}(J)\right\}.$$

Assume that on each $\mathrm{X}(J_0)$ *an abstract derivation* ∂_{J_0} *is specified*[8]. *For simplicity, we will write* ∂ *and* ζ *instead of* ∂_{J_0} *and* ζ_{J_0}. *From now on,* $\mathrm{X}(J)$ *denotes such a structure (with a given abstract derivation on each* $\mathrm{X}(J_0)$*), which will be called an* abstract differential structure.

We can now consider the heat propagation equation in an arbitrary abstract differential structure $\mathrm{X}(J)$, by replacing $(\mathcal{H})_{\mathrm{D}_{\mathrm{side}}(J)}$ by

$$(\mathcal{H})_{\mathrm{X}(J)} : \begin{cases} \text{For every subinterval } J_0 \subset J \text{ of monotony for } \beta, \text{ we have} \\ \partial\theta + \dfrac{\zeta(\theta)}{\tau} = \Lambda\partial\beta \quad \text{in } (\mathrm{X}(J_0), \zeta, \partial). \end{cases}$$

[8]The $\mathrm{A}(\mathbb{R})|_{J_0}$-submodule $\mathrm{X}(J_0)$ of $\mathrm{C}(J_0)$ also satisfies the requirements of Definition 3..11(a), and so an abstract derivation always exists at least in the sense of distributions (see Example 3..10(i)).

3.3. The Mathematical Problem in Abstract Differential Structures

The heat propagation equation $(\mathcal{H})_{\mathrm{X}(J)}$, the restriction $(\mathcal{R})$ together with the evolution system $(\mathcal{E})_{\mathrm{X}(J)}$, the constitutive equation $(\mathcal{C})$, and the initial conditions, lead us to a system which will be denoted by $(\mathcal{T})_{\mathrm{X}(J)}$. Note that $(\mathcal{T})_{\mathrm{X}(J)}$ is a more general setting of the system $(\mathcal{T})$ from Section 2.3., since the latter is equivalent to $(\mathcal{T})_{\mathrm{D_{side}}(J)}$.

Our mathematical problem is the following: for a fixed abstract differential structure $\mathrm{X}(J)$ and an arbitrarily given $\sigma \in \mathrm{C}(J)$ with $\sigma(0) = 0$, we wish to investigate the solutions $(\beta, \theta, \varepsilon)$ of the system $(\mathcal{T})_{\mathrm{X}(J)}$. We require that

$$\beta, \theta \in \mathrm{X}(J), \qquad \varepsilon \in \mathrm{C}(J).$$

It is easily seen that we can ignore the constitutive equation $(\mathcal{C})$ and the initial condition $\varepsilon(0) = 0$. Indeed, for $(\beta, \theta) \in \mathrm{X}(J)^2$ satisfying all the other conditions from the system, we get a solution $(\beta, \theta, \varepsilon)$ of $(\mathcal{T})_{\mathrm{X}(J)}$ with $\varepsilon = \frac{\sigma}{E} + g\beta \in \mathrm{C}(J)$. Therefore, our mathematical problem has the following reduced form:

$$(\mathcal{T})_{\mathrm{X}(J)} : \begin{cases} (\mathcal{H})_{\mathrm{X}(J)} : \begin{cases} \text{For every subinterval } J_0 \subset J \text{ of monotony for } \beta, \text{ we have} \\ \partial\theta + \dfrac{\zeta(\theta)}{\tau} = \Lambda \partial\beta \quad \text{in } (\mathrm{X}(J_0), \zeta, \partial), \end{cases} \\ (\mathcal{R}) : \ 0 \le \beta \le 1, \\ (\mathcal{E})_{\mathrm{X}(J)} : \begin{cases} \beta(t) > 0 \Rightarrow \sigma(t) \ge \sigma^-(t) \\ \beta(t) < 1 \Rightarrow \sigma(t) \le \sigma^+(t) \\ t \in M^+(\beta) \Rightarrow \sigma(t) = \sigma^+(t) \\ t \in M^-(\beta) \Rightarrow \sigma(t) = \sigma^-(t) \end{cases} \\ (\mathcal{I}) : \ \beta(0) = \theta(0) = 0. \end{cases}$$

Let us recall that $\Lambda = \Gamma_\pm + \frac{L}{C}$ is subject to the convention from $(\mathcal{H})_{\mathrm{D_{side}}(J)}$.

By a solution of the above system we mean any pair of functions $(\beta, \theta) \in \mathrm{X}(J)^2$ satisfying $(\mathcal{T})_{\mathrm{X}(J)}$. Let us note that *our problem depends on the abstract differential structure, that is, on its underlying set* $\mathrm{X}(J)$ *and on the derivations defining the differential structure* (see Definition 3..14).

3.4. The Derivative-free Form of the Heat Propagation Equation

Our next result provides an integral form of the heat propagation equation $(\mathcal{H})_{\mathrm{X}(J)}$. Roughly speaking, the system $(\mathcal{T})_{\mathrm{X}(J)}$ is reduced to the problem of finding the function $\beta \in \mathrm{X}(J)$ (which must be locally monotone, according to Corollary 3..7); then $\theta \in \mathrm{X}(J)$ is uniquely determined by the integral formula from the theorem below[9].

Theorem 3..15 (the derivative-free form of the heat propagation equation)
For $\beta, \theta \in \mathrm{X}(J)$, the following two conditions are equivalent.

(i) (β, θ) *satisfies the heat propagation equation* $(\mathcal{H})_{\mathrm{X}(J)}$.

[9]One can find θ by applying the formula piecewise, on successive intervals of monotony for β.

(ii) *For every subinterval $J_0 \subset J$ of monotony for β and for fixed $r \in J_0$, we have*

$$\theta(t) = \theta(r)\mathrm{e}^{(r-t)/\tau} + \Lambda\left[(\beta(t)-\beta(r)) - \frac{\mathrm{e}^{-t/\tau}}{\tau}\int_r^t \mathrm{e}^{s/\tau}(\beta(s)-\beta(r))\mathrm{d}s\right] \quad \textit{for every } t \in J_0.$$

Proof. Let us fix an interval J_0 as in (ii) and $r \in J_0$. By the properties of the abstract differential space $(\mathrm{X}(J_0), \zeta, \partial)$, we get the equivalences

$$\begin{aligned}
\partial\theta + \frac{\zeta(\theta)}{\tau} = \Lambda\partial\beta &\iff \mathrm{e}^{t/\tau}\partial\theta + (\dot{\overline{\mathrm{e}^{t/\tau}}})\zeta(\theta) = \Lambda\mathrm{e}^{t/\tau}\partial\beta = \Lambda\mathrm{e}^{t/\tau}\partial(\beta - \beta(r)) \\
&\iff \partial(\mathrm{e}^{t/\tau}\theta) = \Lambda\left[\partial\left(\mathrm{e}^{t/\tau}(\beta - \beta(r))\right) - \frac{1}{\tau}\zeta\left(\mathrm{e}^{t/\tau}(\beta - \beta(r))\right)\right] \\
&\qquad = \Lambda\left[\partial\left(\mathrm{e}^{t/\tau}(\beta - \beta(r))\right) - \frac{1}{\tau}\partial\left(\int_r^t \mathrm{e}^{s/\tau}(\beta(s) - \beta(r))\mathrm{d}s\right)\right] \\
&\qquad = \Lambda \cdot \partial\left(\mathrm{e}^{t/\tau}(\beta - \beta(r)) - \frac{1}{\tau}\int_r^t \mathrm{e}^{s/\tau}(\beta(s) - \beta(r))\mathrm{d}s\right) \\
&\iff \partial\left(\mathrm{e}^{t/\tau}\theta - \Lambda\left[\mathrm{e}^{t/\tau}(\beta - \beta(r)) - \frac{1}{\tau}\int_r^t \mathrm{e}^{s/\tau}(\beta(s) - \beta(r))\mathrm{d}s\right]\right) = 0 \\
&\iff \mathrm{e}^{t/\tau}\theta - \Lambda\left[\mathrm{e}^{t/\tau}(\beta - \beta(r)) - \frac{1}{\tau}\int_r^t \mathrm{e}^{s/\tau}(\beta(s) - \beta(r))\mathrm{d}s\right] \equiv \mathrm{e}^{r/\tau}\theta(r).
\end{aligned}$$

The last above equivalence follows the property $\ker\partial \subset \mathbb{R}$ of the abstract derivation ∂. □

According to Theorem 3..15, the heat propagation equation (equivalent to its integral form) does not depend on the derivations defining the abstract differential structure, but only on its underlying set $\mathrm{X}(J)$. Consequently, *the whole system $(\mathcal{T})_{\mathrm{X}(J)}$ only depends on the set* $\mathrm{X}(J)$. This fact and the next corollary will allow a unitary study of the system $(\mathcal{T})_{\mathrm{X}(J)}$.

Corollary 3..16 *For $\beta, \theta \in \mathrm{C}(J)$, we have the equivalence*

$$(\beta,\theta) \textit{ is a solution of } (\mathcal{T})_{\mathrm{X}(J)} \iff \begin{cases} (\beta,\theta) \textit{ is a solution of } (\mathcal{T})_{\mathrm{C}(J)}, \\ \beta,\theta \in \mathrm{X}(J). \end{cases}$$

Proof. The equivalence is immediate, by the derivative-free forms of the evolution system $(\mathcal{E})_{\mathrm{X}(J)}$ (Corollary 3..5(b)) and of the heat propagation equation $(\mathcal{H})_{\mathrm{X}(J)}$ (Theorem 3..15).
□

In the heat propagation equation $(\mathcal{H})_{\mathrm{X}(J)}$ we may replace the abstract derivations by those in the sense of distributions, which always exist. If a solution of the system $(\mathcal{T})_{\mathrm{X}(J)}$ is differentiable, then the heat propagation equation is automatically satisfied in the usual sense:

Corollary 3..17 (classical solutions) *Let (β,θ) be a solution of the system $(\mathcal{T})_{\mathrm{X}(J)}$. If one of the functions β and θ has a finite forward derivative at some $t \in J$, then so does the other and*

$$\dot\theta_{\mathrm{f}}(t) + \frac{\theta(t)}{\tau} = \Gamma|\dot\beta_{\mathrm{f}}(t)| + \frac{L}{C}\dot\beta_{\mathrm{f}}(t).$$

The statement also holds for the backward derivative.

Proof. Let $t_0 \in J \setminus \{\sup J\}$. According to Corollary 3..7, there exists $\delta > 0$, such that β is monotone on $J_0 := J \cap (t_0 - \delta, t_0 + \delta)$. By the equality from Theorem 3..15(ii), we see that one of the functions β, θ has a finite forward derivatives at t_0, if and only if so does the other. If both β and θ have such derivatives, then multiplying the formula from Theorem 3..15(ii) by $\mathrm{e}^{t/\tau}$ and considering the forward derivative at t_0 leads to the claimed equality. A similar proof works for the statement on the backward derivatives at some $t_0 \in J \setminus \{0\}$. □

For technical reasons (for the proofs of Theorem 4..11 and Corollary 4..12) we will need another equivalent form of the heat propagation equation. This requires us to define some new auxiliary constants and functions.

Notation 3..18 (auxiliary constants and functions)

(a) *For fixed* $\Gamma_\pm \in \{\Gamma, -\Gamma\}$, *set*

$$\Lambda := \Gamma_\pm + \frac{L}{C}, \qquad \alpha := \frac{\Delta T}{\tau(\Lambda + \Delta T)}, \qquad \gamma := \frac{1}{\tau} - \alpha, \qquad c := \frac{p\Delta T}{\tau\alpha}.$$

We have $\Lambda + \Delta T = \frac{\Delta T}{\tau\alpha} = \frac{c}{p}$ *and* $\Lambda = \frac{\gamma\Delta T}{\alpha}$. *Since* $\Gamma < \frac{L}{C}$, *we have* $\Lambda, \alpha, \gamma, c > 0$.

(b) *For* $r \in J$ *and* $\beta \in \mathrm{X}(J)$, *we may define the auxiliary function*

$$\chi_r := c(\beta - \beta(r)) \in \mathrm{X}(J).$$

(c) *By setting* $\eta_\pm := (T_0 - T_{\mathrm{a}}) \pm \frac{q}{p}$, *the notation* (23) *becomes*

$$\sigma^\pm = p(\theta + \beta\Delta T + \eta_\pm).$$

(d) *Let* β *and* θ *satisfy* $(\mathcal{H})_{\mathrm{X}(J)}$. *For every* $r \in J$, *let us define the functions*

$$\sigma_r^\pm \in \mathrm{A}(\mathbb{R})|_J, \quad \sigma_r^\pm(t) = p\left(\theta(r)\mathrm{e}^{(r-t)/\tau} + \beta(r)\Delta T + \eta_\pm\right).$$

Note that if β *is constant on a subinterval* $J_0 \subset J$ *with* $r \in J_0$, *then on* J_0 *we have*

$$\theta(t) = \theta(r)\mathrm{e}^{(r-t)/\tau}, \qquad \sigma^\pm = \sigma_r^\pm.$$

All above auxiliary constants and functions have two versions, corresponding to the possible values of $\Gamma_\pm \in \{\Gamma, -\Gamma\}$. The results from Section 4.1. will be valid for both versions. For all results related to intervals of monotony for β (such as Corollary 3..19 below), the constant $\Gamma_\pm$ is subject to the convention from $(\mathcal{H})_{\mathrm{D_{side}}(J)}$.

Corollary 3..19 *For* $\beta, \theta \in \mathrm{X}(J)$, *the following two conditions are equivalent.*

(i) (β, θ) *satisfies the heat propagation equation* $(\mathcal{H})_{\mathrm{X}(J)}$.

(ii) *For every subinterval* $J_0 \subset J$ *of monotony for* β *and for fixed* $r \in J_0$, *we have*[10]

$$(\sigma^+ - \sigma_r^+)(t) = (\sigma^- - \sigma_r^-)(t) = \chi_r(t) - \gamma\mathrm{e}^{-t/\tau}\int_r^t \mathrm{e}^{s/\tau}\chi_r(s)\mathrm{d}s \quad \textit{for every } t \in J_0.$$

[10] By using the linear operator V_r from Definition 4..1, we may write the equality as $\sigma^+ - \sigma_r^+ = V_r\chi_r$.

Proof. According to Notation 3..18(c,d), we have

$$\frac{(\sigma^{\pm}-\sigma_r^{\pm})(t)}{p}=\theta(t)-\theta(r)\mathrm{e}^{(r-t)/\tau}+\Delta T(\beta(t)-\beta(r))\quad \text{for every } t\in J.$$

Let us fix $J_0 \subset J$ and $r \in J_0$ as in (ii). For every $t \in J_0$, the equality from Theorem 3..15(ii) is equivalent to

$$\begin{aligned}\frac{(\sigma^{\pm}-\sigma_r^{\pm})(t)}{p}&=(\Lambda+\Delta T)(\beta(t)-\beta(r))-\frac{\Lambda\mathrm{e}^{-t/\tau}}{\tau}\int_r^t \mathrm{e}^{s/\tau}(\beta(s)-\beta(r))\mathrm{d}s\\&=\frac{1}{p}\Big(\chi_r(t)-\gamma\mathrm{e}^{-t/\tau}\int_r^t \mathrm{e}^{s/\tau}\chi_r(s)\mathrm{d}s\Big).\end{aligned}$$

Thus, the conclusion follows by Theorem 3..15. □

4. The Uniqueness Theorem in Abstract Differential Structures

In this section we prove the uniqueness of the solutions of the system $(\mathcal{T})_{\mathrm{X}(J)}$, for every abstract differential structure $\mathrm{X}(J)$. The uniqueness will be obtained via an optimum result (Theorem 4..11 below) stating that whenever $\beta|_{[r,s]}$ is monotone, the difference $\sigma^{\pm}-\sigma_r^{\pm}$ is uniquely determined by $\sigma-\sigma_r^{\pm}$, that is, when knowing the particular values $\beta(r)$ and $\theta(r)$. Then $\beta|_{[r,s]}$ and $\theta|_{[r,s]}$ may be found by "inverting" the formula from Corollary 3..19(ii), and respectively by Theorem 3..15.

4.1. The SMA Linear Operators

We next introduce two linear operators playing an important role in our unitary study of the system $(\mathcal{T})_{\mathrm{X}(J)}$. Subsequently, J_0 denotes a non-degenerate interval and $r \in J_0$ a fixed element.

The definition below of the linear operator V_r (and of its inverse U_r) is inspired by the integral formula from Corollary 3..19(ii).

Definition 4..1 (the SMA linear operators on J_0) *Let us consider the linear operators $U_r, V_r : \mathrm{C}(J_0) \to \mathrm{C}(J_0)$, defined by*

$$U_r v(t)=v(t)+\gamma\mathrm{e}^{-\alpha t}\int_r^t \mathrm{e}^{\alpha s}v(s)\mathrm{d}s,$$

$$V_r u(t)=u(t)-\gamma\mathrm{e}^{-t/\tau}\int_r^t \mathrm{e}^{s/\tau}u(s)\mathrm{d}s,$$

for all $u, v \in \mathrm{C}(J_0)$ and $t \in J_0$.

Let us note that the expressions of the operators U_r and V_r depend on the value of $\Gamma_{\pm} \in \{\Gamma, -\Gamma\}$, since so do the constants α and γ introduced by Notation 3..18(a).

The results from the remainder of this section are purely technical, are only designed for the needs of some forthcoming proofs, and may be skipped at a first lecture.

Remark 4..2 *We have* $U_r w(r) = V_r w(r) = w(r)$ *for every* $w \in \mathrm{C}(J_0)$.

Proposition 4..3 U_r *and* V_r *are invertible, with* $(U_r)^{-1} = V_r$. *Moreover,* U_r *and* V_r *are continuous with respect to the compact-open topology* [11] *on* $\mathrm{C}(J_0)$.

Proof. It is easily seen that U_r and V_r are linear and continuous with respect to the compact-open topology on $\mathrm{C}(J_0)$. For arbitrary $v \in \mathrm{C}(J_0)$ and $t \in J_0$, we have

$$\begin{aligned}
V_r U_r v(t) &= U_r v(t) - \gamma \mathrm{e}^{-t/\tau} \int_r^t \mathrm{e}^{s/\tau} U_r v(s) \mathrm{d}s \\
&= U_r v(t) - \gamma \mathrm{e}^{-t/\tau} \left[\int_r^t \mathrm{e}^{s/\tau} v(s) \mathrm{d}s + \gamma \int_r^t \mathrm{e}^{s/\tau} \mathrm{e}^{-\alpha s} \left(\int_r^s \mathrm{e}^{\alpha\xi} v(\xi) \mathrm{d}\xi \right) \mathrm{d}s \right] \\
&= U_r v(t) - \gamma \mathrm{e}^{-t/\tau} \int_r^t \mathrm{e}^{s/\tau} v(s) \mathrm{d}s - \gamma \mathrm{e}^{-t/\tau} \int_r^t \dot{(\mathrm{e}^{\gamma s})} \left(\int_r^s \mathrm{e}^{\alpha\xi} v(\xi) \mathrm{d}\xi \right) \mathrm{d}s \\
&= U_r v(t) - \gamma \mathrm{e}^{-t/\tau} \int_r^t \mathrm{e}^{s/\tau} v(s) \mathrm{d}s \\
&\quad - \gamma \mathrm{e}^{-t/\tau} \left[\left(\mathrm{e}^{\gamma s} \int_r^s \mathrm{e}^{\alpha\xi} v(\xi) \mathrm{d}\xi \right) \Big|_r^t - \int_r^t \mathrm{e}^{\gamma s} \mathrm{e}^{\alpha s} v(s) \mathrm{d}s \right] \\
&= U_r v(t) - \gamma \mathrm{e}^{-\alpha t} \int_r^t \mathrm{e}^{\alpha s} v(s) \mathrm{d}s = v(t).
\end{aligned}$$

Hence $V_r U_r v = v$ for every $v \in \mathrm{C}(J_0)$. A similar argument shows that $U_r V_r u = u$ for every $u \in \mathrm{C}(J_0)$. We conclude that U_r and V_r are invertible, with $(U_r)^{-1} = V_r$. □

Remark 4..4 (invariant subspaces) *Every abstract differential space* $\mathrm{X}(J_0) \subset \mathrm{C}(J_0)$ *is an invariant vector subspace for the operators* U_r *and* V_r.

Indeed, this follows by Definition 3..11(a).

Proposition 4..5 *For arbitrary* $u, v \in \mathrm{C}(J_0)$, *we have the equivalences*

$$v = V_r u \iff u = U_r v \iff \begin{cases} u' + \alpha u = v' + \frac{v}{\tau} \text{ in } \mathcal{D}'(\overset{o}{J}_0), \\ u(r) = v(r). \end{cases}$$

Proof. The first equivalence is clear, by Proposition 4..3. In order to prove the second equivalence, let us fix $u, v \in \mathrm{C}(J_0)$. Let $u_1, v_1 \in \mathrm{C}(J_0)$, defined by $u_1(t) = \mathrm{e}^{\alpha t} u(t)$ and $v_1(t) = \mathrm{e}^{\alpha t} U_r v(t)$. Thus,

$$u = U_r v \iff u_1 = v_1 \iff \begin{cases} u_1' = v_1' \text{ in } \mathcal{D}'(\overset{o}{J}_0), \\ u_1(r) = v_1(r) \ (\Leftrightarrow u(r) = v(r)). \end{cases}$$

An easy computation in $\mathcal{D}'(\overset{o}{J}_0)$ shows that

$$u_1' = v_1' \iff u' + \alpha u = v' + \frac{v}{\tau},$$

[11] The compact-open topology on $\mathrm{C}(J_0)$ is that of uniform convergence on compact subsets of J_0.

and the second claimed equivalence follows. □

In the remainder of this section, we assume that

$$r = \inf J_0.$$

Lemma 4..6 *For $v_1, v_2 \in \mathrm{D_f}(J_0)$, let*[12] $v_\wedge = v_1 \wedge v_2$ *and* $v_\vee = v_1 \vee v_2$. *Set*

$$u_1 = U_r v_1, \qquad u_2 = U_r v_2, \qquad u_\wedge = U_r v_\wedge, \qquad u_\vee = U_r v_\vee.$$

Then for every $t \in J_0 \setminus \{\sup J_0\}$, we have

$$(\dot{\overline{u_\wedge}})_\mathrm{f}(t) \geq (\dot{u}_1)_\mathrm{f}(t) \wedge (\dot{u}_2)_\mathrm{f}(t), \qquad (\dot{\overline{u_\vee}})_\mathrm{f}(t) \leq (\dot{u}_1)_\mathrm{f}(t) \vee (\dot{u}_2)_\mathrm{f}(t).$$

Proof. Let us first observe that $v_\wedge, v_\vee \in \mathrm{D_f}(J_0)$. By Proposition 3..13 and Remark 4..4, we see that $u_\wedge, u_\vee \in \mathrm{D_f}(J_0)$. For every $j \in \{1, 2\}$ we have $u_j = U_r v_j$, and so

$$(\dot{u}_j)_\mathrm{f}(t) = (\dot{v}_j)_\mathrm{f}(t) + \gamma v_j(t) - \alpha\gamma \mathrm{e}^{-\alpha t} \int_r^t \mathrm{e}^{\alpha s} v_j(s) \mathrm{d}s \quad \text{for every } t \in J_0 \setminus \{\sup J_0\}.$$

The same equality holds with $u_\wedge$ and $v_\wedge$ instead of u_j and v_j. Let us fix $t \in J_0 \setminus \{\sup J_0\}$. As $v_\wedge = v_1 \wedge v_2$, there exists $i \in \{1, 2\}$, such that $v_\wedge(t) = v_i(t)$ and $(\dot{\overline{v_\wedge}})_\mathrm{f}(t) = (\dot{v}_i)_\mathrm{f}(t)$. Hence

$$(\dot{\overline{u_\wedge}})_\mathrm{f}(t) = (\dot{u}_i)_\mathrm{f}(t) + \alpha\gamma \mathrm{e}^{-\alpha t} \int_r^t \mathrm{e}^{\alpha s}[v_i(s) - v_\wedge(s)] \mathrm{d}s \geq (\dot{u}_i)_\mathrm{f}(t) \geq (\dot{u}_1)_\mathrm{f}(t) \wedge (\dot{u}_2)_\mathrm{f}(t).$$

The second inequality of the lemma follows by applying the first to the functions $-v_1$ and $-v_2$. □

Proposition 4..7 *Let $v_1, v_2 \in \mathrm{C}(J_0)$. If both $U_r v_1$ and $U_r v_2$ are increasing (respectively decreasing) functions, then so is $U_r(v_1 \wedge v_2)$ (respectively $U_r(v_1 \vee v_2)$).*

Proof. Assume that both $u_1 := U_r v_1$ and $u_2 := U_r v_2$ are increasing. For each $j \in \{1, 2\}$, suitable approximations of the plot of the function u_j by polygonal chains provides a sequence of increasing functions $(u_j^n)_{n \in \mathbb{N}} \subset \mathrm{A_{side}}(J_0) \subset \mathrm{D_f}(J_0)$, such that $u_j^n \xrightarrow{\text{u.c.}} u_j$. Here "$\xrightarrow{\text{u.c.}}$" means the uniform convergence on compact subsets of J_0, that is, the convergence in the compact-open topology on $\mathrm{C}(J_0)$. By Proposition 4..3, we get successively the convergences

$$\begin{aligned} v_j^n := V_r u_j^n &\xrightarrow{\text{u.c.}} V_r u_j = v_j \quad (j \in \{1, 2\}), \\ v_1^n \wedge v_2^n &\xrightarrow{\text{u.c.}} v_1 \wedge v_2, \\ U_r(v_1^n \wedge v_2^n) &\xrightarrow{\text{u.c.}} U_r(v_1 \wedge v_2). \end{aligned}$$

By Remark 4..4 and Lemma 4..6, for $n \in \mathbb{N}$ we deduce that $U_r(v_1^n \wedge v_2^n) \in \mathrm{A_{side}}(J_0)$ and

$$(\dot{\overline{U_r(v_1^n \wedge v_2^n)}})_\mathrm{f}(t) \geq (\dot{u}_1^n)_\mathrm{f}(t) \wedge (\dot{u}_2^n)_\mathrm{f}(t) \geq 0 \quad \text{for every } t \in J_0 \setminus \{\sup J_0\}.$$

Since all $U_r(v_1^n \wedge v_2^n)$ are increasing, so is the limit $U_r(v_1 \wedge v_2)$. □

---

[12] Pointwise minimum/maximum: $(v_1 \wedge v_2)(t) := \min\{v_1(t), v_2(t)\}$ and $(v_1 \vee v_2)(t) := \max\{v_1(t), v_2(t)\}$ for $t \in J_0$.

4.2. The Optimum Principle and the Uniqueness Theorem

In this section we consider a fixed solution (β, θ) of $(\mathcal{T})_{\mathrm{X}(J)}$ and a non-degenerate subinterval $J_0 \subset J$. Unless otherwise stated, we assume that β is monotone on J_0, and that

$$r := \inf J_0 \in J_0.$$

We shall use the same symbol to denote a function from $\mathrm{C}(J)$ and its restriction to J_0.

Let us recall that the operators U_r and V_r defined on $\mathrm{C}(J_0)$ depend on the value of the constant $\Gamma_\pm \in \{\Gamma, -\Gamma\}$, which is assumed to be as in $(\mathcal{H})_{\mathrm{D_{side}}(J)}$.

Remark 4..8 *We have* $\sigma^+ - \sigma_r^+ = \sigma^- - \sigma_r^- = V_r\chi_r$ *on* J_0 *(see* Corollary 3..19*). By* Notation 3..18(c) *and* Proposition 4..3*, it follows that*

$$\beta = \beta(r) + \frac{U_r(\sigma^+ - \sigma_r^+)}{c} \quad on\ J_0. \tag{24}$$

Thus, $\beta \in \mathrm{C}(J_0)$ *is uniquely determined by the value* $\beta(r)$ *and the function* $\sigma^+ - \sigma_r^+ \in \mathrm{C}(J_0)$.

Theorem 4..11 below will show that the function $\sigma^+ - \sigma_r^+ \in \mathrm{C}(J_0)$ is uniquely determined by $\sigma - \sigma_r^+ \in \mathrm{C}(J_0)$, and hence by the values $\beta(r)$ and $\theta(r)$. According to Corollary 3..8, the same two values determine the monotony of β around r, and hence the value of the constant $\Gamma_\pm$. Roughly speaking, one can say that "*a partial solution* $\beta \in \mathrm{C}([0, r])$ *has a unique continuation* $\beta \in \mathrm{C}(J_0)$ *given by* (24)". This is the main idea leading to the existence and uniqueness of the solutions of $(\mathcal{T})_{\mathrm{X}(J)}$.

Notation 4..9 *For a given function* $\omega \in \mathrm{C}(J_0)$*, set*

$$A_r^+(\omega) := \left\{ v \in \mathrm{C}(J_0) \;\middle|\; \begin{array}{l} v(r) \geq 0 \textit{ and } v \geq \omega, \\ U_r v \textit{ is increasing} \end{array} \right\},$$

$$A_r^-(\omega) := \left\{ v \in \mathrm{C}(J_0) \;\middle|\; \begin{array}{l} v(r) \leq 0 \textit{ and } v \leq \omega, \\ U_r v \textit{ is decreasing} \end{array} \right\}.$$

Remark 4..10 *We have*

$$\begin{aligned} v_1 \wedge v_2 \in A_r^+(\omega) \quad & \textit{for all } v_1, v_2 \in A_r^+(\omega), \\ v_1 \vee v_2 \in A_r^-(\omega) \quad & \textit{for all } v_1, v_2 \in A_r^-(\omega). \end{aligned}$$

Indeed, these follow at once by Proposition 4..7.

Theorem 4..11 (the optimum principle)

(i) *If on* J_0 *we have* $\sigma \leq \sigma^+$ *and* β *is increasing, then* [13] *[*$\sigma^+ - \sigma_r^+ = \min A_r^+(\sigma - \sigma_r^+)$ *(where* $\Gamma_\pm = \Gamma$*).*(24)

(ii) If on J_0 *we have* $\sigma \geq \sigma^-$ *and* β *is decreasing, then [*$\sigma^- - \sigma_r^- = \max A_r^-(\sigma - \sigma_r^-)$ *(where* $\Gamma_\pm = -\Gamma$*).*

[13] Both hypotheses of (i) (respectively (ii)) hold if β is strictly increasing (respectively, strictly decreasing).

Proof. By Corollary 3..16, we see that (β, θ) is a solution of the system $(\mathcal{T})_{C(J)}$.
(i). To shorten notation, set $\omega^+ := \sigma^+ - \sigma_r^+ \in C(J_0)$. According to Corollary 3..19, we have $\omega^+ = V_r\chi_r$. By Proposition 4..3 we deduce that $U_r\omega^+ = \chi_r$ is increasing on J_0. Since $\omega^+ \geq \sigma - \sigma_r^+$ and $\omega^+(r) = 0$, it follows that

$$\omega^+ \in A_r^+(\sigma - \sigma_r^+).$$

It remains to show that ω^+ is a lower bound for $A_r^+(\sigma - \sigma_r^+)$ with respect to the pointwise ordering on $C(J_0)$. In order to prove this, let us fix $v \in A_r^+(\sigma - \sigma_r^+)$. We need to show that $\omega^+ \leq v$. According to Remark 4..10, we have

$$w := \omega^+ \wedge v \in A_r^+(\sigma - \sigma_r^+).$$

Clearly, $w(r) = 0$ and $u := U_r w$ is increasing. It is easily seen that U_r is a positive linear operator[14]. This and $\sigma - \sigma_r^+ \leq w \leq \omega^+$ yield

$$u = U_r w \leq U_r\omega^+ = \chi_r.$$

We next show that $u - \chi_r$ is increasing. This will follow if we prove that

$$(u - \chi_r)' \geq 0 \quad \text{in } \mathcal{D}'(\overset{o}{J_0}).$$

Let a test function $\varphi \in C_c^\infty(\overset{o}{J_0})$, with $\varphi \geq 0$. As $u - \chi_r = U_r(w - \omega^+)$, by Proposition 4..5 we get

$$(u - \chi_r)' + \alpha(u - \chi_r) = (w - \omega^+)' + \frac{w - \omega^+}{\tau} \quad \text{in } \mathcal{D}'(\overset{o}{J_0}).$$

It follows that

$$\begin{aligned}\langle (u - \chi_r)', \varphi\rangle &= \alpha \int_{J_0} (\chi_r - u)\varphi - \int_{J_0} (w - \omega^+)\left(\dot{\varphi} - \frac{\varphi}{\tau}\right)\\ &= \alpha \int_{J_0} (\chi_r - u)\varphi - \int_{J_0 \setminus M^+(\beta)} (w - \omega^+)\left(\dot{\varphi} - \frac{\varphi}{\tau}\right),\end{aligned}$$

because for every $t \in J_0 \cap M^+(\beta)$ we have $\sigma(t) = \sigma^+(t)$, and hence $w(t) = \omega^+(t)$. According to Proposition 3..3(iii), $\overset{o}{J_0} \setminus M^+(\beta)$ is an open subset of $\mathbb{R}$. By Lindelöff's theorem, we have

$$\overset{o}{J_0} \setminus M^+(\beta) = \bigcup_{i \in I} (a_i, b_i),$$

for an at most countable set $\{(a_i, b_i) \mid i \in I\}$ of mutually disjoint open intervals with end-points in the set $(J_0 \cap M^+(\beta)) \cup (\overline{J}_0 \setminus \overset{o}{J_0})$. Let (a, b) denote one of these intervals. We

[14] That is, $U_r v \geq 0$ for every $v \in C(J_0)$, with $v \geq 0$.

have

$$\begin{aligned}\int_a^b (w-\omega^+)\left(\dot\varphi-\frac{1}{\tau}\varphi\right) &= \int_a^b (w-\omega^+)\mathrm{d}\varphi - \frac{1}{\tau}\int_a^b (w-\omega^+)\varphi \\ &= (w-\omega^+)\varphi\Big|_a^b - \int_a^b \varphi\mathrm{d}(w-\omega^+) - \frac{1}{\tau}\int_a^b (w-\omega^+)\varphi \\ &= -\int_a^b \varphi\mathrm{d}w - \frac{1}{\tau}\int_a^b w\varphi + \int_a^b\left(\dot\omega^+ + \frac{\omega^+}{\tau}\right)\varphi \\ &= -\int_a^b \mathrm{e}^{-t/\tau}\varphi(t)\mathrm{d}\big(\mathrm{e}^{t/\tau}w(t)\big) + \alpha\int_a^b \chi_r\varphi,\end{aligned}$$

because β is constant on (a,b), which yields first $\sigma^+ = \sigma_a^+ \in \mathrm{A}(\mathbb{R})|_{(a,b)}$, and then the equality $\dot\omega^+ + \frac{\omega^+}{\tau} = \dot\chi_r + \alpha\chi_r = \alpha\chi_r$ (where $\dot\omega^+$ denotes the derivative of ω^+), by Proposition 4..5. We get successively

$$\mathrm{e}^{t/\tau}w(t) = \mathrm{e}^{t/\tau}u(t) - \gamma\int_r^t \mathrm{e}^{s/\tau}u(s)\mathrm{d}s \quad \text{for every } t\in J_0,$$

$$\int_a^b \mathrm{e}^{-t/\tau}\varphi(t)\mathrm{d}\big(\mathrm{e}^{t/\tau}w(t)\big) = \int_a^b \varphi\mathrm{d}u + \alpha\int_a^b u\varphi,$$

$$\int_a^b (w-\omega^+)\left(\dot\varphi-\frac{1}{\tau}\varphi\right) = -\int_a^b \varphi\mathrm{d}u + \alpha\int_a^b(\chi_r-u)\varphi.$$

It follows that

$$\int_{J_0\setminus M^+(\beta)} (w-\omega^+)\left(\dot\varphi-\frac{1}{\tau}\varphi\right) = -\int_{J_0\setminus M^+(\beta)}\varphi\mathrm{d}u + \alpha\int_{J_0\setminus M^+(\beta)}(\chi_r-u)\varphi.$$

We conclude that

$$\langle (u-\chi_r)',\varphi\rangle = \int_{J_0\setminus M^+(\beta)}\varphi\mathrm{d}u + \alpha\int_{J_0\cap M^+(\beta)}(\chi_r-u)\varphi \geq 0.$$

We thus have proved that $(u-\chi_r)'\geq 0$ in $\mathcal{D}'(\overset{o}{J_0})$, and hence that $u-\chi_r$ is an increasing function. As $(u-\chi_r)(r) = u(r) = w(r) = 0$, we have $u-\chi_r\geq 0$. This and $u\leq\chi_r$ yield $u=\chi_r$. It follows that $U_r\omega^+ = \chi_r = u = U_r w$. By Proposition 4..3, we get $\omega^+ = w = \omega^+\wedge v$, which forces $v\geq\omega^+$. Hence ω^+ is a lower bound for the set $A_r^+(\sigma-\sigma_r^+)$. We conclude that

$$\omega^+ = \min A_r^+(\sigma-\sigma_r^+).$$

(ii). Taking $\omega^- := \sigma^- - \sigma_r^- \in \mathrm{C}(J_0)$ and $w := \omega^-\vee v$ for fixed $v\in A_r^-(\sigma-\sigma_r^-)$, leads by an argument similar to that from the first case to the claimed equality. □

Corollary 4..12

(i) *If on J_0 we have $\sigma\leq\sigma^+$ and β is increasing, then*

$$\beta = \beta(r) + \frac{1}{c}\cdot\min\left\{u\in\mathrm{C}(J_0)\;\middle|\; \begin{array}{l} u(r)\geq 0 \textit{ and } V_r u\geq \sigma-\sigma_r^+,\\ u \textit{ is increasing}\end{array}\right\} \quad (\textit{where } \Gamma_\pm = \Gamma).$$

(ii) *If on J_0 we have $\sigma \geq \sigma^-$ and β is decreasing, then*

$$\beta = \beta(r) + \frac{1}{c} \cdot \max\left\{ u \in \mathrm{C}(J_0) \;\middle|\; \begin{array}{l} u(r) \leq 0 \textit{ and } V_r u \leq \sigma - \sigma_r^-, \\ u \textit{ is decreasing} \end{array} \right\} \quad (\textit{where } \Gamma_\pm = -\Gamma).$$

Proof. (i). Let us consider the sets

$$A := A_r^+(\sigma - \sigma_r^+), \qquad B := \left\{ u \in \mathrm{C}(J_0) \;\middle|\; \begin{array}{l} u(r) \geq 0 \text{ and } V_r u \geq \sigma - \sigma_r^+, \\ u \text{ is increasing} \end{array} \right\}.$$

By Proposition 4..3, we see that $B = U_r(A)$. According to Corollary 3..19 and Theorem 4..11(i), we have the equalities

$$V_r \chi_r = \sigma^+ - \sigma_r^+ = \min A.$$

Since $V_r^{-1} = U_r$ is a positive linear operator, we get successively

$$c(\beta - \beta(r)) = \chi_r = U_r(\min A) = \min(U_r(A)) = \min B, \qquad \beta = \beta(r) + \frac{\min B}{c}.$$

(ii). The proof runs as in the first case. □

Theorem 4..13 (uniqueness of the solutions) *For every abstract differential structure* $\mathrm{X}(J)$*, the system* $(\mathcal{T})_{\mathrm{X}(J)}$ *has at most one solution.*

Proof. Let (β_1, θ_1) and (β_2, θ_2) be solutions of $(\mathcal{T})_{\mathrm{X}(J)}$. Set

$$r := \sup\{s \in J \mid \beta_1 = \beta_2 \text{ on } [0, s]\}.$$

Suppose that $r < \sup J$. Thus, $J_\delta := [r, r + \delta] \subset J$ for sufficiently small $\delta > 0$. By the construction of r, we see that $\beta_1 = \beta_2$ on $[0, r]$. By Corollary 3..7 and Theorem 3..15 it follows that $\theta_1 = \theta_2$ on $[0, r]$ (this compact contains finitely many maximal intervals of monotony for $\beta_1 = \beta_2$). Therefore[15], $(\sigma^+)_1 = (\sigma^+)_2$ on $[0, r]$ and $(\sigma_r^+)_1 = (\sigma_r^+)_2$ on J. We need to analyze three cases.

Case 1. If $(\sigma^-)_1(r) \neq \sigma(r) \neq (\sigma^+)_1(r)$, then by the continuity of σ and $(\sigma^\pm)_j$, there exists $\delta > 0$, such that on $J_\delta \subset J$ we have

$$(\sigma^-)_j \neq \sigma \neq (\sigma^+)_j \quad (j \in \{1, 2\}).$$

By Corollary 3..5(b), it follows that $J_\delta \cap M^+(\beta_j) = \emptyset = J_\delta \cap M^-(\beta_j)$. Hence both β_1 and β_2 are constant on J_δ, by Proposition 3..3(ii). We thus get $\beta_1 = \beta_2$ on J_δ, a contradiction.

Case 2. If $\sigma(r) = (\sigma^+)_1(r) = (\sigma^+)_2(r)$, then by the continuity of σ and $(\sigma^-)_j$, there exists $\delta > 0$, such that on $J_\delta \subset J$ we have

$$\sigma > (\sigma^-)_j \quad (j \in \{1, 2\}).$$

[15] For the solution (β_j, θ_j) of $(\mathcal{T})_{\mathrm{X}(J)}$, we denote by $(\sigma^\pm)_j$ and $(\sigma_r^\pm)_j$ the corresponding auxiliary functions from Notation 3..18(c,d).

By Corollary 3..5(b), it follows that $J_\delta \cap M^-(\beta_j) = \emptyset$. Hence both β_1 and β_2 are increasing on J_δ, by Proposition 3..3(ii). We next consider two subcases.
Subcase 2a. If $\beta_1(r) = \beta_2(r) = 1$, it follows that $\beta_1 = \beta_2 \equiv 1$ on J_δ, a contradiction.
Subcase 2b. If $\beta_1(r) = \beta_2(r) < 1$, then by shrinking $\delta > 0$ if necessary, we can assume that $\beta_1, \beta_2 < 1$ on J_δ. By Corollary 3..5(b), it follows that $\sigma \le (\sigma^+)_j$. By Corollary 4..12(i) we get $(\chi_r)_1 = (\chi_r)_2$ on J_δ, which forces $\beta_1 = \beta_2$ on J_δ, a contradiction.
Case 3. If $\sigma(r) = (\sigma^-)_1(r) = (\sigma^-)_2(r)$, then an argument similar to that from the previous case leads to a contradiction.
Since all three cases led to contradictions, we must have $r = \sup J$, which yields $\beta_1 = \beta_2$. By Theorem 3..15, it follows that $\theta_1 = \theta_2$. We thus conclude that $(\beta_1, \theta_1) = (\beta_2, \theta_2)$. □

5. The Existence and Uniqueness Theorem in $\mathrm{C}(J)$

In this section, $J_0 \subset \mathbb{R}$ will denote a non-degenerate interval, such that

$$r := \inf J_0 \in J_0.$$

The constants $\alpha, \gamma, \tau > 0$, are as in Notation 3..18(a). Let us recall that $\alpha + \gamma = \frac{1}{\tau}$.

The following three technical results will be needed in the proof of the existence of the minimum from Theorem 4..11; this will then lead to the existence theorem for the system $(\mathcal{T})_{\mathrm{C}(J)}$.

Proposition 5..1 *Let $v \in \mathrm{C}(J_0)$, such that $v(r) \ge 0$ and $u := U_r v$ is increasing. For every $a \in J_0$, let us consider the function $\underline{v}_a \in \mathrm{A}(\mathbb{R})|_{J_0}$, defined by*

$$\underline{v}_a(t) = \tau\alpha u(a) + [v(a) - \tau\alpha u(a)]\mathrm{e}^{(a-t)/\tau}.$$

Then

(i) *The function $\underline{v}_a$ is decreasing and convex, and*

$$v(a) - \tau\alpha u(a) \ge 0.$$

The function $J_0 \times J_0 \ni (a, t) \longmapsto \underline{v}_a \in \mathbb{R}$ is continuous.

(ii) *For every fixed $t \in J_0$, the function $J_0 \ni a \longmapsto \underline{v}_a(t) \in \mathbb{R}$ is increasing.*

(iii) *We have*

$$v \le \underline{v}_a \text{ on } [r, a], \qquad v \ge \underline{v}_a \text{ on } [a, \infty) \cap J_0, \qquad U_r(v \wedge \underline{v}_a) = u \wedge u(a).$$

(iv) *For every subinterval $J_1 \subset J_0$ and every fixed $a \in J_1$, we have the equivalence*

$$u \text{ is constant on } J_1 \iff v = \underline{v}_a \text{ on } J_1.$$

Proof. (i). As $u = U_r v$, by Proposition 4..3 we have $v = V_r u$. It follows that

$$\begin{aligned}
v(a) - \tau\alpha u(a) = V_r u(a) - \tau\alpha u(a) &= (1-\tau\alpha)u(a) - \gamma \mathrm{e}^{-a/\tau}\int_r^a \mathrm{e}^{s/\tau}u(s)\mathrm{d}s \\
&= \tau\gamma \mathrm{e}^{-a/\tau}\left(\mathrm{e}^{a/\tau}u(a) - \int_r^a u(s)\mathrm{d}(\mathrm{e}^{s/\tau})\right) \\
&= \tau\gamma \mathrm{e}^{-a/\tau}\left(\mathrm{e}^{a/\tau}u(a) - \mathrm{e}^{s/\tau}u(s)\Big|_r^a + \int_r^a \mathrm{e}^{s/\tau}\mathrm{d}u(s)\right) \\
&= \tau\gamma \mathrm{e}^{-a/\tau}\left(\mathrm{e}^{r/\tau}u(r) + \int_r^a \mathrm{e}^{s/\tau}\mathrm{d}u(s)\right) \geq 0,
\end{aligned}$$

since $u(r) = v(r) \geq 0$ and u is increasing (hence the last Riemann-Stieltjes integral is positive). Hence $\underline{v}_a$ is decreasing and convex, since so is the function $\mathbb{R} \ni t \mapsto \mathrm{e}^{(a-t)/\tau} \in \mathbb{R}$. The last statement from (i) is obvious, since $u, v \in \mathrm{C}(J_0)$.
(ii). By the definition of the function $\underline{v}_a$ and our previous computation from (i), we see that

$$\underline{v}_a(t) = \tau\alpha u(a) + \tau\gamma \mathrm{c}^{-t/\tau}\left(\mathrm{c}^{r/\tau}u(r) + \int_r^a \mathrm{e}^{s/\tau}\mathrm{d}u(s)\right) \quad \text{for all } t, a \in J_0,$$

and the conclusion follows.
(iv). "⇒". Assume $u \equiv u(a)$ on J_1. As $v = V_r u$, for every $t \in J_1$ we have

$$\begin{aligned}
v(t) = u(t) - \gamma \mathrm{e}^{-t/\tau}\int_r^t \mathrm{e}^{s/\tau}u(s)\mathrm{d}s &= u(a) - \gamma \mathrm{e}^{-t/\tau}\left(\int_r^a \mathrm{e}^{s/\tau}u(s)\mathrm{d}s + u(a)\int_a^t \mathrm{e}^{s/\tau}\mathrm{d}s\right) \\
&= u(a)\left[1 - \tau\gamma\left(1 - \mathrm{e}^{(a-t)/\tau}\right)\right] - \mathrm{e}^{(a-t)/\tau}\cdot\gamma \mathrm{e}^{-a/\tau}\int_r^a \mathrm{e}^{s/\tau}u(s)\mathrm{d}s \\
&= \tau u(a)\left(\alpha + \gamma \mathrm{e}^{(a-t)/\tau}\right) + \mathrm{e}^{(a-t)/\tau}(v(a) - u(a)) \\
&= \tau\alpha u(a) + [v(a) - \tau\alpha u(a)]\mathrm{e}^{(a-t)/\tau} = \underline{v}_a(t).
\end{aligned}$$

"⇐". Assume $v = \underline{v}_a$ on J_1. Hence both v and $u = U_r v$ are differentiable on $\overset{o}{J}_1$, and

$$\dot{u} + \alpha u = \dot{\ } + \frac{v}{\tau} = \dot{\underline{v}}_a + \frac{\underline{v}_a}{\tau} \equiv \alpha u(a) \quad \text{on } \overset{o}{J}_1,$$

by Proposition 4..5. By solving the above differential equation we get $u \equiv u(a)$ on J_1.
(iii). By (ii) and the definition of the functions $\underline{v}_a$, we get

$$(t-a)(v(t) - \underline{v}_a(t)) = (t-a)(\underline{v}_t(t) - \underline{v}_a(t)) \geq 0 \quad \text{for every } t \in J_0.$$

Hence, $v \leq \underline{v}_a$ on $[r, a]$ and $v > v_a$ on $J_1 := [a, \infty) \cap J_0$ Let us consider the functions $u_a, w \in \mathrm{C}(J_0)$, defined by

$$u_a := u \wedge u(a), \qquad w := V_r u_a.$$

Hence $U_r w = u_a$, by Proposition 4..3. As u is increasing, we have $u_a|_{[r,a]} = u|_{[r,a]}$. This and the definition of V_r lead to

$$w|_{[r,a]} = (V_r u_a)|_{[r,a]} = (V_r u)|_{[r,a]} = v|_{[r,a]}.$$

Let us observe that $u_a(a) = u(a)$ and $w(a) = v(a)$ yield $\underline{w}_a = \underline{v}_a$. Since $u \equiv u(a)$ on J_1, using (iv) for u_a and w (instead of u and v) leads to

$$w|_{J_1} = \underline{w}_a|_{J_1} = \underline{v}_a|_{J_1}.$$

It follows that $w = v \wedge \underline{v}_a$, which implies $U_r(v \wedge \underline{v}_a) = U_r w = u_a$. □

Proposition 5..2 *Let $u, v \in \mathrm{C}(J_0)$, such that $u = U_r v$.*

(i) *If $u(r) \geq 0$ and u is increasing, then so is the function*

$$J_0 \ni t \longmapsto \mathrm{e}^{\gamma t} v(t) \in \mathbb{R}.$$

(ii) *If $v(r) \geq 0$ and the restrictions $u|_{[r,a]}$ and $v|_{[a,\infty)\cap J_0}$ are increasing (for some fixed $a \in J_0$), then u is increasing.*

Proof. (i). We have

$$u = U_r v \iff v = V_r u \iff v(t) = u(t) - \gamma \mathrm{e}^{-t/\tau} \int_r^t \mathrm{e}^{s/\tau} u(s) \mathrm{d}s \quad \text{for every } t \in J_0.$$

Let us consider the functions $w, \varphi \in \mathrm{C}(J_0)$, defined by

$$w(t) = \mathrm{e}^{\gamma t} u(t) - \gamma \int_r^t \mathrm{e}^{\gamma s} u(s) \mathrm{d}s, \qquad \varphi(t) = \mathrm{e}^{\gamma t} v(t) - w(t).$$

We claim that w and φ are increasing. Indeed, w is increasing, since so is u and

$$\begin{aligned} w(t) = \mathrm{e}^{\gamma t} u(t) - \int_r^t u(s) \mathrm{d}(\mathrm{e}^{\gamma s}) &= \mathrm{e}^{\gamma t} u(t) - \mathrm{e}^{\gamma s} u(s)\Big|_r^t + \int_r^t \mathrm{e}^{\gamma s} \mathrm{d}u(s) \\ &= \mathrm{e}^{\gamma r} u(r) + \int_r^t \mathrm{e}^{\gamma s} \mathrm{d}u(s) \quad \text{for every } t \in J_0. \end{aligned}$$

By the definitions of φ and w, we see that

$$\begin{aligned} \varphi(t) &= \mathrm{e}^{\gamma t}(v(t) - u(t)) + \gamma \int_r^t \mathrm{e}^{\gamma s} u(s) \mathrm{d}s \\ &= \gamma \int_r^t \mathrm{e}^{\gamma s} u(s) \mathrm{d}s - \gamma \mathrm{e}^{-\alpha t} \int_r^t \mathrm{e}^{s/\tau} u(s) \mathrm{d}s \quad \text{for every } t \in J_0. \end{aligned}$$

Hence φ is differentiable and

$$\varphi'(t) = \gamma \alpha \mathrm{e}^{-\alpha t} \int_r^t \mathrm{e}^{s/\tau} u(s) \mathrm{d}s \geq 0 \quad \text{for every } t \in J_0.$$

Therefore, φ is increasing. Our claim is proved. We thus conclude that $w + \varphi$ is increasing.
(ii). As $u = U_r v$, by the definition of U_r we see that $u - v$ is differentiable and

$$\frac{(u-v)'(t)}{\gamma} = v(t) - \alpha \mathrm{e}^{-\alpha t} \int_r^t \mathrm{e}^{\alpha s} v(s) \mathrm{d}s \quad \text{for every } t \in J_0.$$

For $t \in J_a := [a, \infty) \cap J_0$, we have

$$\begin{aligned}
\frac{e^{\alpha t}}{\gamma}(u-v)'(t) &= e^{\alpha t}v(t) - \alpha\left(\int_r^a e^{\alpha s}v(s)ds + \int_a^t e^{\alpha s}v(s)ds\right) \\
&\geq e^{\alpha t}v(t) - \alpha\left(\int_r^a e^{\alpha s}v(s)ds + v(t)\int_a^t e^{\alpha s}ds\right) \\
&= e^{\alpha a}v(t) - \frac{\alpha e^{\alpha a}}{\gamma}(u(a) - v(a)) \\
&\geq e^{\alpha a}\left[v(a) + \frac{\alpha}{\gamma}(v(a) - u(a))\right] = \frac{e^{\alpha a}}{\tau\gamma}(v(a) - \tau\alpha u(a)) \geq 0.
\end{aligned}$$

Hence $u - v$ is increasing on J_a. It follows that $u = (u - v) + v$ is increasing on J_a, and hence on the union $[r, a] \cup J_a = J_0$. □

Corollary 5..3 *Let $\omega \in C(J_0)$ and $a \in J_0$. For $v \in A_r^+(\omega|_{[r,a]})$ (see* Notation 4..9*), let us consider the function $\widetilde{v}_a \in C(J_0)$, defined by*

$$\widetilde{v}_a(t) = \begin{cases} v(t) & , \quad \textit{for } t \in [r, a], \\ v(a) + \max_{s\in[a,t]}(\omega(s) - \omega(a)), & \textit{for } t \in [a, \infty) \cap J_0. \end{cases} \tag{25}$$

Then $\widetilde{v}_a \in A_r^+(\omega)$. We have the equality

$$A_r^+(\omega|_{[r,a]}) = A_r^+(\omega)|_{[r,a]}.$$

Proof. Set $u := U_r\widetilde{v}_a \in C(J_0)$. According to the definition of $A_r^+(\omega|_{[r,a]})$, the function $u|_{[r,a]} = U_r v$ is increasing. By the construction of $\widetilde{v}_a$, we have $\widetilde{v}_a(r) = v(r) \geq 0$ and $\widetilde{v}_a \geq \omega$, and $\widetilde{v}_a|_{[a,\infty)\cap J_0}$ is increasing. Hence u is increasing, by Proposition 5..2(ii). We conclude that $\widetilde{v}_a \in A_r^+(\omega)$. Let us finally show the equality of the sets.
"$\supset$". The inclusion is immediate, by the definition of the sets $A_r^+(\cdot)$.
"$\subset$". Let us fix $v \in A_r^+(\omega|_{[r,a]})$. For the function $\widetilde{v}_a \in C(J_0)$ defined by (25), we have $\widetilde{v}_a \in A_r^+(\omega)$. Hence $v = \widetilde{v}_a|_{[r,a]} \in A_r^+(\omega)|_{[r,a]}$. □

Lemma 5..4 (existence of the optimal function) *Let $\omega \in C(J_0)$. Then for the function*

$$\omega^+ : J_0 \to \mathbb{R}, \quad \omega^+(t) = \inf_{v\in A_r^+(\omega)} v(t),$$

(the pointwise infimum of $A_r^+(\omega)$), we have

$$\omega^+ \in A_r^+(\omega), \qquad \omega^+(r) = \omega(r) \vee 0, \qquad M^+(U_r\omega^+) \subset (\omega^+ - \omega)^{-1}(\{0\}).$$

We call ω^+ the optimal function associated to ω.

Proof. We have divided the proof into a sequence of four steps.
Step 1. We first show that $\omega^+ \in C(J_0)$. According to Proposition 5..2(i), for every $v \in A_r^+(\omega)$ the function $J_0 \ni t \mapsto e^{\gamma t}v(t) \in \mathbb{R}$ is increasing. Thus, their pointwise infimum $J_0 \ni t \mapsto e^{\gamma t}\omega^+(t) \in \mathbb{R}$ is increasing and upper semi-continuous, and hence continuous to

the right. Therefore, ω^+ is continuous to the right and has one-sided limits everywhere on J_0. In order to prove that $\omega^+ \in \mathrm{C}(J_0)$, let us show that

$$\omega^+(b) - \omega^+(a) \leq \max_{s\in[a,b]} (\omega(s) - \omega(a)) \quad \text{for all } a, b \in J_0, \text{ with } a < b. \tag{26}$$

For a, b as above, let us fix $v \in A_r^+(\omega)$. According to Corollary 5..3, for the function $\widetilde{v}_a \in \mathrm{C}(J_0)$ defined by (25), we have $\widetilde{v}_a \in A_r^+(\omega)$. We thus get $\widetilde{v}_a \geq \omega^+$, which yields

$$v(a) + \max_{s\in[a,b]} (\omega(s) - \omega(a)) = \widetilde{v}_a(b) \geq \omega^+(b).$$

Taking here the infimum for $v \in A_r^+(\omega)$ yields the inequality (26). For every $b \in J_0$ such that $b > r$, using (26) leads to

$$0 \leq \omega^+(b) - \lim_{t\nearrow b} \omega^+(t) \leq \lim_{t\nearrow b} \left(\max_{s\in[t,b]} (\omega(s) - \omega(t)) \right) = 0.$$

It follows that ω^+ is continuous to the right, and hence that $\omega^+ \in \mathrm{C}(J_0)$.
Step 2. We next show that $\omega^+(r) = \omega(r) \vee 0$. It is clear that $\omega^+(r) \geq \omega(r) \vee 0$, by the definitions of $A_r^+(\omega)$ and ω^+. Let us consider the function $v \in A_r^+(\omega|_{\{r\}})$, defined by $v(r) = \omega(r) \vee 0$. According to Corollary 5..3, for the function $\widetilde{v}_r \in \mathrm{C}(J_0)$ defined by (25), we have $\widetilde{v}_r \in A_r^+(\omega)$. We thus get $\widetilde{v}_r \geq \omega^+$, which yields $\omega^+(r) \leq \widetilde{v}_r(r) = \omega(r) \vee 0$. We thus conclude that $\omega^+(r) = \omega(r) \vee 0$.
Step 3. We show that $U_r\omega^+$ *is increasing*. Let us fix $a \in J_0$ and $\delta > 0$. By the definition of ω^+ (pointwise infimum of $A_r^+(\omega)$), for every $t \in [r, a]$, we have $v_t(t) < \omega^+(t) + \delta$ for some $v_t \in A_r^+(\omega)$. By continuity, for every $t \in [r, a]$, the inequality $v_t < \omega^+ + \delta$ holds on some open neighborhood $W_t \in \mathcal{V}(t)$. Since $[r, a]$ is a compact set, we have $[r, a] \subset \bigcup_{t\in F} W_t$ for some finite subset $F \subset [r, a]$. According to Remark 4..10, we have

$$v := \bigwedge_{t\in F} v_t \in A_r^+(\omega).$$

On $[r, a]$, we have the inequalities $\omega^+ \leq v < \omega^+ + \delta$. Since $\delta > 0$ was arbitrary, it follows that there is a sequence $(v_n)_{n\in\mathbb{N}} \subset A_r^+(\omega)$, such that $v_n \xrightarrow{\text{u.}} \omega^+$ (converges uniformly) on $[r, a]$. Hence $U_r v_n \xrightarrow{\text{u.}} U_r\omega^+$ on $[r, a]$. It follows that $U_r\omega^+$ is increasing on $[r, a]$, since so are all $U_r v_n$. We conclude that $U_r\omega^+$ is increasing on J_0. By the conclusions of the above three steps, we deduce that

$$\omega^+ \in A_r^+(\omega).$$

Step 4. For $u := U_r\omega^+$, *we show that* $M^+(u) \subset (\omega^+ - \omega)^{-1}(\{0\})$. Suppose that $\omega^+(t_0) > \omega(t_0)$ for some $t_0 \in M^+(u) \cap \overset{o}{J_0}$. By Proposition 5..1(i), we see that[16]

$$\lim_{a,t\to t_0} \underline{\omega}_a^+(t) = \underline{\omega}_{t_0}^+(t_0) = \omega^+(t_0) > \omega(t_0).$$

Hence there exist $a, b \in J_0$, such that $t_0 \in (a, b)$ and $\underline{\omega}_a^+(b) \geq \omega(t)$ for every $t \in [a, b]$. Let us consider the function $v := \omega^+ \wedge \underline{\omega}_a^+ \in \mathrm{C}(J_0)$. Thus, $v \leq \omega^+$. By Proposition 5..1(i,iii)

[16]By $\underline{\omega}_a^+$ we denote the function $\underline{v}_a$ associated to $v = \omega^+$ (see Proposition 5..1).

we deduce that $v(r) = \omega^+(r) \geq 0$ and $U_r v = u \wedge u(a)$ is increasing, and that $v \geq \omega$ on $[r, b]$, since $v = \omega^+ \geq \omega$ on $[r, a]$ and

$$v(t) = \underline{\omega}_a^+(t) \geq \underline{\omega}_a^+(b) \geq \omega(t) \quad \text{for every } t \in [a, b].$$

According to Corollary 5..3, we have $v|_{[r,b]} \in A_r^+ \left(\omega|_{[r,b]}\right) = A_r^+(\omega)|_{[r,b]}$. Consequently, on $[r, b]$ we have the inequalities $v \geq \omega^+ \geq v$, and hence the equality $\omega^+ = v$. It follows that

$$u|_{[r,b]} = (U_r v)|_{[r,b]} = (u \wedge u(a))|_{[r,b]}.$$

Since u is increasing, we get $u|_{[a,b]} \equiv u(a)$. This forces $t_0 \notin M^+(u)$, a contradiction. We thus have proved the inclusion

$$M^+(u) \cap \overset{o}{J_0} \subset (\omega^+ - \omega)^{-1}(\{0\}).$$

As $M^+(u)$ is a perfect subset of J_0 (by Proposition 3..3(iii)) and $(\omega^+ - \omega)^{-1}(\{0\})$ is closed in J_0, by the above inclusion it follows that $M^+(u) \subset (\omega^+ - \omega)^{-1}(\{0\})$. □

Theorem 5..5 (existence and uniqueness of continuous solutions) *For every given $\sigma \in$ $C(J)$ with $\sigma(0) = 0$, the system $(\mathcal{T})_{C(J)}$ has a unique solution.*

Proof. Let us consider the number

$$r := \sup\{t \in J \,|\, (\mathcal{T})_{C([0,t])} \text{ has at least one solution}\}.$$

Step 1. We first show that $r > 0$. Since $\sigma \in C(J)$ and $\sigma(0) = 0 < p\eta_+$, there exists $t_0 \in J \setminus \{0\}$, such that $\sigma < p\eta_+$ on $[0, t_0]$. It it easy to check that $(0, 0)$ $(\beta = \theta \equiv 0)$ is a solution of the system $(\mathcal{T})_{C([0,t_0])}$. Hence $r \geq t_0 > 0$. By Theorem 4..13, we see that for all $0 < t_1 < t_2 < r$, the unique solution of the system $(\mathcal{T})_{C([0,t_2])}$ extends that of $(\mathcal{T})_{C([0,t_1])}$. This fact and the definition of r yield the existence of a unique solution $(\beta, \theta) \in C([0, r))^2$ of $(\mathcal{T})_{C([0,r))}$. We can assume that $r \in J$, since otherwise $J = [0, r)$ and (β, θ) is a solution of $(\mathcal{T})_{C(J)}$.
Step 2. We next prove that β is monotone on $[r_0, r)$, for some $r_0 \in [0, r)$. On the contrary, suppose this statement false. Since $\sigma \in C(J)$, there exists $\xi \in (0, r)$, such that

$$|e^{t/\tau}\sigma(t) - e^{r/\tau}\sigma(r)| \leq q \quad \text{for every } t \in [\xi, r].$$

According to Corollary 3..7, we have $M^+(\beta) \cap M^-(\beta) = \emptyset$. As β is not increasing on $[\xi, r)$, there exists $t_0 \in M^-(\beta) \cap (\xi, r)$. As β is not decreasing on $[t_0, r)$, there exists $s := \inf(M^+(\beta) \cap (t_0, r))$. Hence

$$s \in M^+(\beta), \qquad \xi < t_0 < s < r, \qquad M^+(\beta) \cap (t_0, s) = \emptyset.$$

Since $t_0 \in M^-(\beta)$, for $t := \sup(M^-(\beta) \cap (\xi, s))$ we have

$$t \in M^-(\beta), \qquad t_0 \leq t < s, \qquad M^-(\beta) \cap (t, s) = \emptyset.$$

According to Corollary 3..5(b), we have $\sigma(t) = \sigma^-(t)$ and $\sigma(s) = \sigma^+(s)$. Since

$$M^+(\beta) \cap (t, s) = M^-(\beta) \cap (t, s) = \emptyset,$$

the function β is constant on $[t,s]$, by Proposition 3..3(ii). By the formula from Theorem 3..15(ii), we see that $\mathrm{e}^{t/\tau}\theta(t) = \mathrm{e}^{s/\tau}\theta(s)$. We thus get

$$\begin{aligned} 2q &\geq \mathrm{e}^{s/\tau}\sigma(s) - \mathrm{e}^{t/\tau}\sigma(t) = \mathrm{e}^{s/\tau}\sigma^{+}(s) - \mathrm{e}^{t/\tau}\sigma^{-}(t) \\ &= p\big(\beta(t)\Delta T + T_0 - T_{\mathrm{a}}\big)\big(\mathrm{e}^{s/\tau} - \mathrm{e}^{t/\tau}\big) + q\big(\mathrm{e}^{s/\tau} + \mathrm{e}^{t/\tau}\big) > q\big(\mathrm{e}^{s/\tau} + \mathrm{e}^{t/\tau}\big) > 2q, \end{aligned}$$

a contradiction. We thus conclude that β is monotone on some $[r_0, r)$. Therefore, the limits

$$\lim_{t\nearrow r} \beta(t) =: \beta(r) \in [0,1], \qquad \lim_{t\nearrow r} \theta(t) =: \theta(r) \in \mathbb{R},$$

exist (for the latter we use the formula from Theorem 3..15(ii)). We thus get the continuous extensions $\beta, \theta \in \mathrm{C}([0,r])$. It is easy to check that (β,θ) is a solution of $(\mathcal{T})_{\mathrm{C}([0,r])}$.

Step 3. We finally show that $r = \sup J$. On the contrary, suppose that $r < \sup J$. We next need to consider three cases.

Case 1. If $\sigma(r) \notin \{\sigma^-(r), \sigma^+(r)\}$, then $\sigma(r) \neq \sigma_r^{\pm}(r)$. Since $\sigma, \sigma_r^{\pm} \in \mathrm{C}(J)$, there exists $t_1 \in J$, such that $t_1 > r$ and $\sigma(t) \neq \sigma_r^{\pm}(t)$ for every $t \in [r, t_1]$. Let us consider $\widetilde{\beta}, \widetilde{\theta} \in \mathrm{C}([0,t_1])$, defined by

$$\widetilde{\beta}(t) = \begin{cases} \beta(t), & \text{for } t \in [0,r], \\ \beta(r), & \text{for } t \in [r,t_1], \end{cases} \qquad \widetilde{\theta}(t) = \begin{cases} \theta(t) & , \quad \text{for } t \in [0,r], \\ \theta(r)\mathrm{e}^{(r-t)/\tau}, & \text{for } t \in [r,t_1]. \end{cases} \tag{27}$$

It is easy to check that $(\widetilde{\beta}, \widetilde{\theta})$ is a solution of $(\mathcal{T})_{\mathrm{C}([0,t_1])}$, which contradicts the definition of the number r.

Case 2. If $\sigma(r) = \sigma^+(r)$, then $r \notin M^-(\beta)$, by Corollary 3..5(b). According to Proposition 3..3(i), the function β is increasing on $[r_1, r]$, for some $r_1 \in [r_0, r)$. We need to consider two subcases.

Subcase 2a. If $\beta(r) = 1$, let us choose $t_1 \in J$, such that $t_1 > r$ and $\sigma \geq \sigma_r^-$ on $[r, t_1]$. For the functions $\widetilde{\beta}, \widetilde{\theta} \in \mathrm{C}([0,t_1])$ defined as in (27), it is easy to check that $(\widetilde{\beta}, \widetilde{\theta})$ is a solution of $(\mathcal{T})_{\mathrm{C}([0,t_1])}$. This contradicts the definition of r.

Subcase 2b. If $\beta(r) < 1$, let $J_r := [r,\infty) \cap J$. For $\omega := \sigma - \sigma_r^+ \in \mathrm{C}(J_r)$ we have $\omega(r) = 0$, since $\sigma(r) = \sigma^+(r) = \sigma_r^+(r)$. According to Lemma 5..4, the function $\omega^+ = \min A_r^+(\omega)$ exists and

$$\omega^+(r) = \omega(r) \vee 0 = 0, \qquad M^+(U_r\omega^+) \subset (\omega^+ - \omega)^{-1}(\{0\}). \tag{28}$$

Let us consider the functions $\beta^*, \theta^* \in \mathrm{C}(J_r)$, defined by

$$\beta^* = \beta(r) + \frac{U_r\omega^+}{c}, \qquad \theta^* = \frac{\omega^+ + \sigma_r^+}{p} - \beta^*\Delta T - \eta_+.$$

As $U_r\omega^+(r) = \omega^+(r) = 0$, we have

$$\beta^*(r) = \beta(r) < 1, \qquad \theta^*(r) = \theta(r), \qquad (\omega^+ - \omega)(r) = 0 < 2q.$$

Hence there exists $t_1 \in J_r \setminus \{r\}$, such that $\beta^*(t_1) \leq 1$ and $\omega^+ - \omega \leq 2q$ on $[r, t_1]$. By shrinking J if necessary, we can assume that

$$J = [0, t_1], \qquad J_r = [r, t_1].$$

Thus, $J = [0, r] \cup J_r$. Let us consider the functions $\widetilde{\beta}, \widetilde{\theta} \in \mathrm{C}(J)$, defined by

$$\widetilde{\beta}(t) = \begin{cases} \beta(t), & \text{for } t \in [0, r], \\ \beta^*(t), & \text{for } t \in J_r, \end{cases} \qquad \widetilde{\theta}(t) = \begin{cases} \theta(t), & \text{for } t \in [0, r], \\ \theta^*(t), & \text{for } t \in J_r. \end{cases}$$

We claim that $(\widetilde{\beta}, \widetilde{\theta})$ is a solution of the system $(\mathcal{T})_{\mathrm{C}(J)}$. The function β^* is increasing, since so is $U_r\omega^+$. Hence $\beta(r) \le \beta^* \le \beta^*(t_1) \le 1$, and so $0 \le \widetilde{\beta} \le 1$. We also have

$$\begin{aligned} \sigma^+|_{J_r} &= p(\theta^* + \beta^*\Delta T + \eta_+) = \omega^+ + \sigma_r^+, \\ \sigma^-|_{J_r} &= \omega^+ + \sigma_r^+ - 2q \le \sigma|_{J_r} \le \omega^+ + \sigma_r^+ = \sigma^+|_{J_r}, \end{aligned}$$

$$\chi_r|_{J_r} = c(\beta^* - \beta(r)) = U_r\omega^+ = U_r\big((\sigma^+ - \sigma_r^+)|_{J_r}\big). \tag{29}$$

As β^* is increasing, we have $M^-(\beta^*) = \emptyset$. By (28) and the definition of β^*, we see that

$$M^+(\beta^*) = M^+(U_r\omega^+) \subset (\omega^+ - \omega)^{-1}(\{0\}) = (\sigma^+ - \sigma)^{-1}(\{0\}) \cap J_r.$$

Since (β, θ) is a solution of $(\mathcal{T})_{\mathrm{C}([0,r])}$, by Proposition 3..3(v) and Remark 3..6, it follows that

$$\begin{aligned} M^+(\widetilde{\beta}) &= M^+(\beta) \cup M^+(\beta^*) \subset (\sigma^+ - \sigma)^{-1}(\{0\}), \\ M^-(\widetilde{\beta}) &= M^-(\beta) \cup M^-(\beta^*) \subset (\sigma^- - \sigma)^{-1}(\{0\}), \end{aligned}$$

Hence $(\widetilde{\beta}, \widetilde{\theta})$ satisfies the evolution system $(\mathcal{E})_{\mathrm{C}(J)}$. By (29) and Proposition 4..3, we see that

$$(\sigma^+ - \sigma_r^+)|_{J_r} = V_r\big(\chi_r|_{J_r}\big).$$

By Corollary 3..19, it follows that $(\widetilde{\beta}, \widetilde{\theta})$ satisfies the heat propagation equation $(\mathcal{H})_{\mathrm{C}(J)}$. We conclude that $(\widetilde{\beta}, \widetilde{\theta})$ is a solution of $(\mathcal{T})_{\mathrm{C}([0,t_1])}$, which contradicts the definition of r.
Case 3. If $\sigma(r) = \sigma^-(r)$, a construction similar to that from *Case 2* leads to a contradiction. Since all above three cases led to contradictions, we conclude that $\sup J = r \in J$. Hence $J = [0, r]$, and so (β, θ) is the unique solution of the system $(\mathcal{T})_{\mathrm{C}(J)}$. The proof is complete. □

6. Regularity of Solutions

In this section we consider the system $(\mathcal{T})_{\mathrm{X}(J)}$ for several important classes of abstract differential structures $\mathrm{X}(J)$ (see Setting 6..12 below). We shall prove that this system has a unique solution whenever $\sigma \in \mathrm{X}(J)$. Roughly speaking, this means that the solution of the problem has regularity properties similar to those of the initial data. We first need to prove some regularity results for the optimal function from Lemma 5..4.

6.1. Regularity Lemmas for the Optimal Function

Let us consider a non-degenerate interval $J_0 \subset \mathbb{R}$, with

$$r := \inf J_0 \in J_0,$$

a given function $\omega \in \mathrm{C}(J_0)$, the associated set $A_r^+(\omega)$ from Notation 4..9, and the functions

$$\omega^+ := \min A_r^+(\omega), \qquad u := U_r\omega^+.$$

In Lemma 5..4 we obtained the existence and some useful properties of the optimal function ω^+. Let us recall that

$$\omega^+ \in A_r^+(\omega), \qquad \omega^+(r) = \omega(r) \vee 0, \qquad M^+(u) \subset (\omega^+ - \omega)^{-1}(\{0\}). \tag{30}$$

The function u is increasing, since $\omega^+ \in A_r^+(\omega)$.

Remark 6..1 *For every $a \in J_0$, the optimal function of the restriction $\omega|_{[r,a]}$ is the restriction to $[r,a]$ of the optimal function ω^+, that is,*

$$\left(\omega|_{[r,a]}\right)^+ = \omega^+|_{[r,a]}.$$

Indeed, this is easily seen by the equality $A_r^+\left(\omega|_{[r,a]}\right) = A_r^+(\omega)|_{[r,a]}$ from Corollary 5..3.

The following sequence of lemmas show that the optimal function ω^+ has regularity properties similar to those of ω.

Lemma 6..2 (bounded variation) *We have $\omega^+ \in \mathrm{BV}_{\mathrm{loc}}(J_0)$.*

Proof. We have $U_r\omega^+ = u \in \mathrm{BV}_{\mathrm{loc}}(J_0)$, since u is increasing. As $\mathrm{BV}_{\mathrm{loc}}(J_0)$ is an abstract differential space, by Proposition 4..3 and Remark 4..4 it follows that $\omega^+ = V_r u \in \mathrm{BV}_{\mathrm{loc}}(J_0)$. □

Proposition 6..3 *For all $a, b \in J_0$ with $a \leq b$, we have the inequalities*

$$-\gamma\omega^+(a)(b-a) \leq \omega^+(b) - \omega^+(a) \leq \max_{t\in[a,b]}(\omega(t) - \omega(a)). \tag{31}$$

Proof. The second inequality from (31) was already stated as (26) and was proved in Section 5.. For $a, b \in J_0$ with $a \leq b$, by Proposition 5..2(i) we get $\mathrm{e}^{\gamma a}\omega^+(a) \leq \mathrm{e}^{\gamma b}\omega^+(b)$, which yields

$$\omega^+(a) - \omega^+(b) \leq \omega^+(a)\left(1 - \mathrm{e}^{\gamma(a-b)}\right) \leq \gamma\omega^+(a)(b-a).$$

□

Lemma 6..4 (Lipschitz condition, absolute continuity)

(a) *If $\omega \in \mathrm{Lip}_{\mathrm{loc}}(J_0)$, then $\omega^+ \in \mathrm{Lip}_{\mathrm{loc}}(J_0)$.*

(b) *If $\omega \in \mathrm{AC}_{\mathrm{loc}}(J_0)$, then $\omega^+ \in \mathrm{AC}_{\mathrm{loc}}(J_0)$.*

Proof. For fixed $c \in J_0$ with $c > r$, set $I := [r, c]$. Let X denote one of the vector spaces $\mathrm{Lip}(I)$ and $\mathrm{AC}(I)$. We need to prove that $\omega^+ \in X$ whenever $\omega \in X$. Assume $\omega \in X$. By (31), we see that

$$\omega^+|_I \leq M := \omega^+(r) + \max_{t\in[r,c]}(\omega(t) - \omega(r)) \quad \text{on } I.$$

This leads by (31) to

$$|\omega^+(b) - \omega^+(a)| \leq \gamma M(b-a) + \max_{t\in[a,b]} |\omega(t) - \omega(a)| \quad \text{for all } a, b \in I,\ a \leq b. \tag{32}$$

We next consider two cases.
Case 1. If $X = \mathrm{Lip}(I)$ and L is a Lipschitz constant for ω, then by (32) we get

$$|\omega^+(b) - \omega^+(a)| \leq (\gamma M + L)|b - a| \quad \text{for all } a, b \in I.$$

Hence $\gamma L + M$ is a Lipschitz constant for ω^+, and so $\omega^+ \in \mathrm{Lip}(I) = X$.
Case 2. If $X = \mathrm{AC}(I)$, let us fix $\epsilon > 0$. Since $\omega \in \mathrm{AC}(I)$, there exists $\delta > 0$, such that for every finite set $\mathcal{I}_S = \{(a_s, b_s) \mid s \in S\}$ of pairwise disjoint subintervals of I, we have

$$\sum_{s\in S}(b_s - a_s) < \delta \Longrightarrow \sum_{s\in S} |\omega(b_s) - \omega(a_s)| < \frac{\epsilon}{2}. \tag{33}$$

Obviously, we can choose δ with the above property, such that $2\gamma M\delta < \epsilon$. For any $\mathcal{I}_S$ as above, by (32) and (33) we get the implication

$$\sum_{s\in S}(b_s - a_s) < \delta \Longrightarrow \sum_{s\in S} |\omega^+(b_s) - \omega^+(a_s)| < \gamma M\delta + \frac{\epsilon}{2} < \epsilon.$$

We conclude that $\omega^+ \in \mathrm{AC}(I) = X$. □

Lemma 6..5 (differentiability to the right) *If ω is differentiable to the right at some point $a \in J_0$, then so is the optimal function ω^+ and*

$$\dot{\omega}_{\mathrm{f}}^+(a) = \begin{cases} \dot{\underline{\omega}}_a^+(a), & \text{if } u \equiv u(a) \text{ on some } [a,c] \subset J_0\ (c > a), \\ \dot{\omega}_{\mathrm{f}}(a), & \text{if } u(t) > u(a) \text{ for every } t > a. \end{cases}$$

Proof. Assume $\dot{\omega}_{\mathrm{f}}(a) \in \mathbb{R}$ exists at $a \in J_0 \setminus \{\sup J_0\}$. We need to consider two cases.
Case 1. If $u \equiv u(a)$ on some $[a, c] \subset J_0$ with $c > a$, then by Proposition 5..1(iv) it follows that $\omega^+ = \underline{\omega}_a^+$ on $[a, c]$. Hence there exists $\dot{\omega}_{\mathrm{f}}^+(a) = \dot{\underline{\omega}}_a^+(a) = \alpha u(a) - \frac{\omega^+(a)}{\tau}$.
Case 2. If $u(t) > u(a)$ for every $t > a$, then for $I := [a, \infty) \cap J_0$ we have $a \in M^+(u|_I)$. Since by Proposition 3..3(iii) the set $M^+(u|_I)$ is perfect in I, we have $\lim_{n\to\infty} t_n = a$ for some sequence

$$(t_n)_{n\in\mathbb{N}} \subset M^+(u|_I) \setminus \{a\} \subset M^+(u) \cap (a, \infty).$$

According to (30) and Proposition 5..1(iii), we have $\omega(t_n) = \omega^+(t_n) \geq \underline{\omega}_a^+(t_n)$ for every $n \in \mathbb{N}$. This and $\omega(a) = \omega^+(a) = \underline{\omega}_a^+(a)$ lead to

$$\dot{\omega}_{\mathrm{f}}(a) = \lim_{n\to\infty} \frac{\omega(t_n) - \omega(a)}{t_n - a} \geq \lim_{n\to\infty} \frac{\underline{\omega}_a^+(t_n) - \underline{\omega}_a^+(a)}{t_n - a} = \dot{\underline{\omega}}_a^+(a).$$

Now let us fix $\xi > \dot\omega_{\rm f}(a) \geq \dot{\underline{\omega}}_a^+(a)$. Choose $c \in J_0$, such that $c > a$ and

$$\left\{\begin{array}{l} \omega(t) < \omega(a) + \xi(t-a) \\ \dot{\underline{\omega}}_a^+(t) < \xi \end{array}\right| , \quad \text{for every } t \in (a, c].$$

Let us consider the function $v \in \mathrm{C}([r, c])$, defined by

$$v(t) = \left\{\begin{array}{ll} \omega^+(t) & , \quad \text{for } t \in [r, a], \\ \omega(a) + \xi(t-a), & \quad \text{for } t \in [a, c]. \end{array}\right.$$

Obviously, $v(r) = \omega^+(r) \geq 0$ and $v \geq \omega|_{[r,c]}$. We claim that

$$U_r v \text{ is increasing on } [r, c]. \tag{34}$$

In order to prove this, let us consider $w = v - \omega^+ \wedge \underline{\omega}_a^+ \in \mathrm{C}([r, c])$. By Proposition 5..1(iii), we see that $w \equiv 0$ on $[r, a]$ and $w = v - \underline{\omega}_a^+$ on $[a, c]$. Since $\dot w \equiv 0$ on $[r, a)$ and $\dot w = \xi - \dot{\underline{\omega}}_a^+ > 0$ on $(a, c]$, it follows that w is increasing and $w \geq 0$. By Proposition 5..2(ii) we deduce that $U_r w$ is increasing. Since by Proposition 5..1(iii) we have

$$U_r v = U_r w + U_r(\omega^+ \wedge \underline{\omega}_a^+) = U_r w + u \wedge u(a) \quad \text{on } [r, c],$$

the function $U_r v$ is increasing. Our claim (34) is proved. We thus get $v \in A_r^+\left(\omega|_{[r,c]}\right)$. By Remark 6..1, it follows that $\omega \leq \omega^+ \leq v$ on $[r, c]$. This and $\omega(a) = \omega^+(a) = v(a)$ yield

$$\dot\omega_{\rm f}(a) \leq \liminf_{t \searrow a} \frac{\omega^+(t) - \omega^+(a)}{t-a} \leq \limsup_{t \searrow a} \frac{\omega^+(t) - \omega^+(a)}{t-a} \leq \dot v_{\rm f}(a) = \xi.$$

Since $\xi > \dot\omega_{\rm f}(a)$ was arbitrary, we conclude that ω^+ is differentiable to the right at a, with the forward derivative $\dot\omega_{\rm f}^+(a) = \dot\omega_{\rm f}(a)$. □

Lemma 6..6 (differentiability to the left) *If ω is differentiable to the left at some point $a \in J_0$, then so is the optimal function ω^+ and*

$$\dot\omega_{\rm b}^+(a) = \left\{\begin{array}{ll} \dot{\underline{\omega}}_a^+(a), & \textit{if } u \equiv u(a) \textit{ on some } [c, a] \subset J_0 \ (c < a), \\ \dot\omega_{\rm b}(a)\ , & \textit{if } u(t) < u(a) \textit{ for every } t \in [r, a). \end{array}\right.$$

Proof. Assume $\dot\omega_{\rm b}(a)$ exists at $a \in J_0 \setminus \{r\}$. We need to consider two cases.
Case 1. If $u \equiv u(a)$ on some $[c, a] \subset J_0$ with $c < a$, then by Proposition 5..1(iv) it follows that $\omega^+ = \underline{\omega}_a^+$ on $[c, a]$. Hence there exists $\dot\omega_{\rm b}^+(a) = \dot{\underline{\omega}}_a^+(a) = \alpha u(a) - \frac{\omega^+(a)}{\tau}$.
Case 2. If $u(t) < u(a)$ for every $t \in [r, a)$, then $a \in M^+(u|_{[r,a]}) \subset M^+(u)$. According to (30), we have $\omega^+(a) = \omega(a)$. Let us fix $\xi < \dot\omega_{\rm b}(a)$. Let us consider $\varphi, \chi \in \mathrm{C}([r, a))$, defined by

$$\varphi(t) = \frac{\omega^+(a) - \omega^+(t)}{a-t}, \qquad \chi(t) = \frac{\omega(a) - \omega(t)}{a-t}.$$

As $\omega^+ \geq \omega$, we have $\varphi \leq \chi$. Since by Proposition 3..3(iii) the $M^+(u|_{[r,a]})$ is perfect in $[r, a]$, there exists $c \in M^+(u) \cap [r, a)$, such that $\chi(t) > \xi$ for every $t \in [c, a)$. We claim that

$$\varphi(t) > \xi \quad \text{for every } t \in [c, a). \tag{35}$$

In order to prove this, let us fix $t_0 \in [c, a)$. We need to consider two subcases.
Subcase 2a. If $t_0 \in M^+(u)$, then $\omega^+(t_0) = \omega(t_0)$ yields $\varphi(t_0) = \chi(t_0) > \xi$.
Subcase 2b. If $t_0 \notin M^+(u)$, set $d := \sup(M^+(u) \cap [c, t_0])$. By Proposition 3..3(iii) and (30), we get

$$d \in M^+(u), \qquad \omega^+(d) = \omega(d), \qquad c \le d < t_0 < a.$$

Since u is increasing and $M^+(u) \cap (d, t_0) = \emptyset$, by Proposition 3..3(ii) we deduce that $u \equiv u(d)$ on $[d, t_0]$. According to Proposition 5..1(iv), we have $\omega^+ = \underline{\omega}_d^+$ on $[d, t_0]$. Therefore, $\varphi \in \mathrm{A}(\mathbb{R})|_{(d,t_0)}$. By Proposition 5..1(ii), we see that $\omega^+(a) = \underline{\omega}_a^+(a) \ge \underline{\omega}_d^+(a)$. Consequently, on (d, t_0) we have

$$\dot{\varphi}(t) = \frac{\omega^+(a) - \omega^+(t) - \dot{\omega}^+(t)(a-t)}{(a-t)^2} \ge \frac{\underline{\omega}_d^+(a) - \underline{\omega}_d^+(t) - \dot{\underline{\omega}}_d^+(t)(a-t)}{(a-t)^2} \ge 0,$$

because $\underline{\omega}_d^+$ is a convex map, by Proposition 5..1(i). We thus get $\varphi(t_0) \ge \varphi(d) = \chi(d) > \xi$. From the above subcases, we conclude that $\varphi > \xi$ on $[c, a)$. Our claim (35) is proved. As $\chi \ge \varphi$, it follows that

$$\xi \le \liminf_{t \nearrow a} \varphi(t) \le \limsup_{t \nearrow a} \varphi(t) \le \limsup_{t \nearrow a} \chi(t) = \dot{\omega}_{\mathrm{b}}(a).$$

Since $\xi < \dot{\omega}_{\mathrm{b}}(a)$ was arbitrary, we conclude that ω^+ is differentiable to the left at a, with the backward derivative $\dot{\omega}_{\mathrm{b}}^+(a) = \dot{\omega}_{\mathrm{b}}(a)$. □

Corollary 6..7 *The difference set $N(\omega^+) \setminus N(\omega)$ is at most countable.*

Proof. Let us fix $a \in N(\omega^+) \setminus N(\omega)$. Thus, ω is differentiable at a, but ω^+ is not. By Lemmas 6..5 and 6..6, we have $a \in \overset{o}{J_0}$ and $\dot{\omega}_{\mathrm{f}}^+(a) \ne \dot{\omega}_{\mathrm{b}}^+(a)$, and one of the following two conditions holds:

$$\left\{ \begin{array}{l} u|_{[a,c]} \equiv u(a) \text{ for some } c > a, \\ u(t) < u(a) \text{ for every } t \in [r, a). \end{array} \right| , \qquad \left\{ \begin{array}{l} u|_{[c,a]} \equiv u(a) \text{ for some } c < a, \\ u(t) > u(a) \text{ for every } t > a. \end{array} \right| .$$

Hence $a \in I_a$ for some maximal non-degenerate subinterval $I_a \subset J_0$, on which u is constant (I_a is maximal with this property). Since by Lindelöff's theorem the set $\{I_a \,|\, a \in A\}$ of all such intervals is at most countable, so is $N(\omega^+) \setminus N(\omega)$. □

Lemma 6..8 *If ω belongs to one of the spaces (see* Example 3..10*),*

$$\mathrm{D}_{\mathrm{f}}^A(J_0), \mathrm{D}_{\mathrm{b}}^B(J_0), \mathrm{D}_{\mathrm{side}}^{A,B}(J_0), \mathrm{D}_{\mathrm{f}}^{\aleph_0}(J_0), \mathrm{D}_{\mathrm{b}}^{\aleph_0}(J_0), \mathrm{D}_{\mathrm{side}}^{\aleph_0}(J_0), \mathrm{D}^{\aleph_0}(J_0),$$

then the optimal function ω^+ belongs to the same space.

Proof. For all spaces defined in terms of one-sided differentiability, the conclusion follows at once by Lemmas 6..5 and 6..6. We only need to prove the lemma for $\mathrm{D}^{\aleph_0}(J_0)$. Assume $\omega \in \mathrm{D}^{\aleph_0}(J_0)$. Thus, $N(\omega)$ is at most countable. According to Corollary 6..7, the set $N(\omega^+) \setminus N(\omega)$ is at most countable, and consequently, so is

$$N(\omega^+) \subset N(\omega) \cup [N(\omega^+) \setminus N(\omega)].$$

Hence $\omega^+ \in \mathrm{D}^{\aleph_0}(J_0)$. □

Lemma 6..9 (Sobolev spaces) *If* $\omega \in \mathrm{W}^{1,p}_{\mathrm{loc}}(J_0)$ *for some* $p \in [1, \infty]$, *then* $\omega^+ \in \mathrm{W}^{1,p}_{\mathrm{loc}}(J_0)$.

Proof. As $\omega \in \mathrm{W}^{1,p}_{\mathrm{loc}}(J_0) \subset \mathrm{AC}_{\mathrm{loc}}(\overline{J_0})$, we have $\omega^+ \in \mathrm{AC}_{\mathrm{loc}}(\overline{J_0})$, by Lemma 6..4(b). Hence ω^+ is differentiable almost everywhere. According to Lemmas 6..5 and 6..6, we have

$$\dot{\omega}^+(t) \in \left\{\dot{\omega}(t), \alpha u(t) - \frac{\omega^+(t)}{\tau}\right\} \quad \text{for every } t \in J_0 \setminus [N(\omega^+) \cup N(\omega)].$$

For every compact subinterval $K \subset \overline{J_0}$ we have $\dot{\omega} \in \mathrm{L}^p(K)$ and $\alpha u - \frac{\omega^+}{\tau} \in \mathrm{C}(K)$, and so

$$\|\dot{\omega}^+\|_{\mathrm{L}^p(K)} \leq \|\dot{\omega}\|_{\mathrm{L}^p(K)} + \left\|\alpha u - \frac{\omega^+}{\tau}\right\|_{\mathrm{L}^p(K)} < \infty.$$

Therefore, $\dot{\omega}^+ \in \mathrm{L}^p_{\mathrm{loc}}(\overline{J_0})$. We thus conclude that $\omega^+ \in \mathrm{W}^{1,p}_{\mathrm{loc}}(J_0)$. □

Lemma 6..10 (analyticity to the right) *If* ω *is forward-analytic at some point* $a \in J_0$, *then so is the optimal function* ω^+.

Proof. Assume ω is forward-analytic at $a \in J_0 \setminus \{\sup J_0\}$. We need to consider two cases.
Case 1. If $u \equiv u(a)$ on some $[a, c] \subset J_0$ with $c > a$, then by Proposition 5..1(iv) it follows that $\omega^+ = \underline{\omega}^+_a$ on $[a, c]$. Hence ω^+ is forward-analytic at a.
Case 2. If $u(t) > u(a)$ for every $t > a$, then $a \in M^+(u|_{[a,\infty)\cap J_0}) \subset M^+(u)$. According to (30), we have $\omega^+(a) = \omega(a)$. As ω is forward-analytic at a, the function $\omega - \underline{\omega}^+_a$ is analytic and monotone on some $[a, c] \subset J_0$, with $c > a$. We claim that

$$(\omega - \underline{\omega}^+_a)|_{[a,c]} \text{ is increasing.} \tag{36}$$

On the contrary, suppose this map is decreasing. As $(\omega - \underline{\omega}^+_a)(a) = 0$, we have $\omega \leq \underline{\omega}^+_a$ on $[a, c]$. Let us consider the function

$$v := (\omega^+ \wedge \underline{\omega}^+_a)|_{[r,c]} \in \mathrm{C}([r, c]).$$

By Remark 4..10 and Proposition 5..1(iii), it follows that $v \in A^+_r(\omega|_{[r,c]})$. Hence $v \geq \omega^+|_{[r,c]}$, by Remark 6..1. On $[a, c]$ we have $\underline{\omega}^+_a \leq \omega^+ \leq v = \underline{\omega}^+_a$, and hence $\omega^+ = \underline{\omega}^+_a$. By Proposition 5..1(iv) this yields $u \equiv u(a)$, a contradiction. We conclude that $(\omega - \underline{\omega}^+_a)|_{[a,c]}$ is increasing. Our claim (36) is proved. Now let us consider the function $w \in \mathrm{C}([r, c])$, defined by

$$w(t) = \begin{cases} \omega^+(t), & \text{for } t \in [r, a], \\ \omega(t)\;, & \text{for } t \in [a, c]. \end{cases}$$

Clearly, $w(r) = \omega^+(r) \geq 0$ and $w \geq \omega|_{[r,c]}$. For $v \in \mathrm{C}([r, c])$ defined as above, we have

$$(w - v)|_{[r,a]} \equiv 0, \qquad (w - v)|_{[a,c]} = (\omega - \underline{\omega}^+_a)|_{[a,c]}.$$

Therefore, $w - v$ is increasing and positive. By Proposition 5..2(ii), it follows that $U_r(w - v)$ is increasing. Hence $U_r w = U_r(w - v) + U_r v$ is increasing. We thus get $w \in A^+_r(\omega|_{[r,c]}) = A^+_r(\omega)|_{[r,c]}$, which yields $w \geq \omega^+|_{[r,c]}$. On $[a, c]$ we have $\omega \leq \omega^+ \leq w = \omega$, and hence $\omega^+ = \omega$. We conclude that ω^+ is forward-analytic at a. □

Lemma 6..11 (analyticity to the left) *If ω is backward-analytic at some point $a \in J_0$, then so is the optimal function ω^+.*

Proof. Assume ω is backward-analytic at $a \in J_0 \setminus \{r\}$. We need to consider two cases.
Case 1. If $u \equiv u(a)$ on some $[c, a] \subset J_0$ with $c < a$, then by Proposition 5..1(iv) it follows that $\omega^+ = \underline{\omega}_a^+$ on $[c, a]$. Hence ω^+ is backward-analytic at a.
Case 2. If $u(t) < u(a)$, for every $t \in [r, a)$, then $a \in M^+(u|_{[r,a]}) \subset M^+(u)$. According to (30), we have $\omega^+(a) = \omega(a)$. On $[r, a]$ we have $\omega \le \omega^+ \le \underline{\omega}_a^+$, by Proposition 5..1(iii). Let us consider the function $\varphi \in \mathrm{C}([r, a])$, defined by

$$\varphi(t) = \mathrm{e}^{t/\tau}\left(\underline{\omega}_a^+(t) - \omega(t)\right).$$

As $\varphi \ge 0 = \varphi(a)$ and φ is backward-analytic at a, the function φ is analytic and decreasing on some $[c, a] \subset J_0$, with $c < a$. As $M^+(u|_{[r,a]})$ is a perfect set in $[r, a]$, there exists $d \in M^+(u) \cap [c, a) \neq \emptyset$. According to (30), we have $\omega^+(d) = \omega(d)$. Now let us consider $v_0 \in \mathrm{C}([r, a])$, defined by

$$v_0(t) = \begin{cases} \omega^+(t), & \text{for } t \in [r, d], \\ \omega(t)\ , & \text{for } t \in [d, a]. \end{cases}$$

Clearly, $v_0(r) = \omega^+(r) \ge 0$ and $v_0 \ge \omega|_{[r,a]}$. We claim that

$$u_0 := U_r v_0 \text{ is increasing.}$$

The restriction $u_0|_{[r,d]} = (U_r\omega^+)|_{[r,d]} = u|_{[r,d]}$ is increasing. On (d, a), we have the equivalences

$$\varphi \text{ is decreasing} \iff \dot{\varphi} \le 0 \iff \dot{\omega} + \frac{\omega}{\tau} \ge \dot{\underline{\omega}}_a^+ + \frac{\underline{\omega}_a^+}{\tau} \equiv \alpha u(a),$$

the last equality being easily checked. Since $u_0 = U_r v_0$, by Proposition 4..5 we see that

$$\dot{u}_0 + \alpha u_0 = \dot{v}_0 + \frac{v_0}{\tau} = \dot{\omega} + \frac{\omega}{\tau} \ge \alpha u(a) > \alpha u \quad \text{on } (d, a).$$

Consequently, we get the inequality

$$\dot{u}_0 > \alpha(u - u_0) = \alpha U_r(\omega^+ - v_0) \ge 0 \quad \text{on } (d, a),$$

since $\omega^+|_{[r,a]} \ge v_0$ and U_r is a positive operator on $\mathrm{C}([r, a])$. Hence the function $u_0 \in \mathrm{C}([r, a])$ is increasing, since so are its restrictions to both $[r, d]$ and $[d, a]$ It follows that $v_0 \in A_r^{!}(\omega|_{[r,a]})$, which yields $v_0 \ge \omega^+|_{[r,a]}$, by Remark 6..1. On $[d, a]$ we have $\omega \le \omega^+ \le v_0 = \omega$, and so $\omega^+ = \omega$. We conclude that ω^+ is backward-analytic at a. □

6.2. Existence and Regularity of Solutions

Let us consider an abstract differential structure $\mathrm{X}(J)$ and a function $\sigma \in \mathrm{X}(J)$, with $\sigma(0) = 0$. By Theorem 5..5, there is a unique solution (β_0, θ_0) of the system $(\mathcal{T})_{\mathrm{C}(J)}$ (the problem considered in the larger space $\mathrm{C}(J) \supset \mathrm{X}(J)$). According to Corollary 3..16, this is *the only possible solution* of $(\mathcal{T})_{\mathrm{X}(J)}$. Therefore, we have the equivalences

$$(\mathcal{T})_{\mathrm{X}(J)} \text{ has solutions} \iff (\beta_0, \theta_0) \text{ is a solution of } (\mathcal{T})_{\mathrm{X}(J)} \iff \beta_0, \theta_0 \in \mathrm{X}(J).$$

Thus, for arbitrary abstract differential structure $\mathrm{X}(J)$ and initial data $\sigma \in \mathrm{X}(J)$, the existence (and uniqueness) of solutions of $(\mathcal{T})_{\mathrm{X}(J)}$ is equivalent to the regularity of the solution (β_0, θ_0) of $(\mathcal{T})_{\mathrm{C}(J)}$.

Setting 6..12 *Let* $\mathrm{X}(J)$ *denote any of the abstract differential structures*

$$\begin{gathered}\mathrm{BV}_{\mathrm{loc}}(J),\ \mathrm{AC}_{\mathrm{loc}}(J),\ \mathrm{Lip}_{\mathrm{loc}}(J),\ \mathrm{W}^{1,p}_{\mathrm{loc}}(J),\\ \mathrm{D}^{A}_{\mathrm{f}}(J),\ \mathrm{D}^{A}_{\mathrm{b}}(J),\ \mathrm{D}^{A,B}_{\mathrm{side}}(J),\\ \mathrm{D}^{\aleph_0}_{\mathrm{f}}(J),\ \mathrm{D}^{\aleph_0}_{\mathrm{b}}(J),\ \mathrm{D}^{\aleph_0}_{\mathrm{side}}(J),\ \mathrm{D}^{\aleph_0}(J),\\ \mathrm{A}_{\mathrm{f}}(J),\ \mathrm{A}_{\mathrm{b}}(J),\ \mathrm{A}_{\mathrm{side}}(J)\end{gathered}$$

(see Example 3..10 *and* Proposition 3..13*).*

By Lemmas 6..2–6..11, we see that $\mathrm{X}(J)$ has the following property:

$$\omega^+ \in \mathrm{X}(J_0) \quad \text{for all } \begin{cases} J_0 \subset J \text{ subinterval,} \\ \omega \in \mathrm{X}(J_0). \end{cases} \tag{37}$$

The next proposition points out some other common properties of all spaces from the above list.

Proposition 6..13 *Let* $u \in \mathrm{C}(J)$.

(i) *If* $u \in \mathrm{X}(J)$, *then* $u \in \mathrm{X}(J_0)$ *for every subinterval* $J_0 \subset J$.

(ii) *If* $u \in \mathrm{X}(J \cap (-\infty, r])$ *and* $u \in \mathrm{X}(J \cap [r, \infty))$ *for some* $r \in J$, *then* $u \in \mathrm{X}(J)$.

(iii) $u \in \mathrm{X}(J)$, *if and only if for every* $t_0 \in J$, *we have* $u \in \mathrm{X}(W)$ *for some* $W = J \cap [a, b]$, *with* $a, b \in \mathbb{R}$ *and* $a < t_0 < b$.

Proof. The proof is straightforward for all spaces from Setting 6..12. □

Theorem 6..14 (existence and uniqueness of regular solutions) *For every given* $\sigma \in \mathrm{X}(J)$ *with* $\sigma(0) = 0$, *the system* $(\mathcal{T})_{\mathrm{X}(J)}$ *has a unique solution.*

Proof. By Theorem 5..5, there is a unique solution $(\beta, \theta) \in \mathrm{C}(J)^2$ of the system $(\mathcal{T})_{\mathrm{C}(J)}$. According to Corollary 3..16, it suffices to show that $\beta, \theta \in \mathrm{X}(J)$. Let us fix $t_0 \in J$. We claim that $\beta \in \mathrm{X}(W)$ for some W as in Proposition 6..13(iii). We need to analyze three cases.
Case 1. If $t_0 \notin M^+(\beta) \cup M^-(\beta)$, then by Proposition 3..3(i) it follows that β is constant on some neighborhood $W = J \cap [t_0 - \delta, t_0 + \delta]$, with $\delta > 0$. Hence $\beta \in \mathrm{X}(W)$.
Case 2. If $t_0 \in M^+(\beta)$, then by Corollary 3..5(b) we get $\sigma(t_0) = \sigma^+(t_0)$. This yields $t_0 > 0$, since $\sigma(0) = 0 < \sigma^+(0)$. As $\sigma(t_0) \neq \sigma^-(t_0)$, we have $t_0 \notin M^-(\beta)$. By

Proposition 3..3(i), it follows that β is increasing on some $W = J \cap [r,s]$, with $0 \le r < t_0 < s$. We next consider two subcases.
Subcase 2a. If $\beta(t_0) < 1$, then by the continuity of β we can assume that $\beta < 1$ on W. Thus, we have $\sigma \le \sigma^+$ on W and $\beta|_W$ is increasing. As $\sigma \in \mathrm{X}(J)$, we have $\omega := (\sigma - \sigma_r^+)|_W \in \mathrm{X}(W)$, by Proposition 6..13(i). Since $\min W = r$, by Theorem 4..11(i) it follows that

$$(\sigma^+ - \sigma_r^+)|_W = \min A_r^+(\omega) = \omega^+.$$

According to (37), we have $\omega^+ \in \mathrm{X}(W)$. By Corollary 3..19 and Proposition 4..3, we get on W the equalities

$$\omega^+ = cV_r\big(\beta|_W - \beta(r)\big), \qquad \beta|_W = \beta(r) + \frac{U_r\omega^+}{c}.$$

That $\beta \in \mathrm{X}(W)$ now follows by Remark 4..4.
Subcase 2b. If $\beta(t_0) = 1$, then $\beta \equiv 1$ on $W \cap [t_0, s]$, since $\beta|_W$ is increasing. Hence $\beta \in \mathrm{X}(W \cap [t_0,s])$. As $t_0 \in M^+(\beta)$, we have $\beta < 1$ on $[r, t_0) \subset W$. By Corollary 3..5(b), it follows that $\sigma < \sigma^+$ on $[r, t_0]$. As in the previous subcase, it follows that $\beta \subset \mathrm{X}([r, t_0])$. This and $\beta \in \mathrm{X}(W \cap [t_0, s])$ lead by Proposition 6..13(ii) to $\beta \in \mathrm{X}(W)$.
Case 3. If $t_0 \in M^-(\beta)$, an analysis similar to that in the previous case shows that $\beta \in \mathrm{X}(W)$ for some $W = J \cap [r,s]$, with $0 \le r < t_0 < s$.
From the above three cases, we conclude that for some neighborhood $W = [r, b]$ of t_0 in J, we have

$$\beta \in \mathrm{X}(W), \qquad \beta|_W \text{ is monotone.}$$

Since $\sigma^+ = p(\theta + \beta\Delta T + \eta_+)$, by Corollary 3..19 we deduce that

$$\theta = \frac{V_r\chi_r + \sigma_r^+}{p} - \beta\Delta T - \eta_+ \quad \text{on } W.$$

Therefore, $\theta \in \mathrm{X}(W)$. Since t_0 was arbitrary, by Proposition 6..13(iii) we get $\beta, \theta \in \mathrm{X}(J)$. Hence (β, θ) is the unique solution of the system $(\mathcal{T})_{\mathrm{X}(J)}$. □

The above existence and uniqueness theorem shows that all abstract differential structures $\mathrm{X}(J)$ from Setting 6..12 are natural spaces for the system $(\mathcal{T})_{\mathrm{X}(J)}$. It is worth pointing out that, in all of these cases, the unique solution (β, θ) of the system $(\mathcal{T})_{\mathrm{X}(J)}$ satisfy the evolution system $(\mathcal{E})_{\mathrm{X}(J)}$ and the heat propagation equation $(\mathcal{H})_{\mathrm{X}(J)}$ with respect to the derivations defining the natural differential structure of $\mathrm{X}(J)$.

Remark 6..15 Theorem 6..14 *also holds for every abstract differential structure satisfying* (37) *and the three properties from* Proposition 6..13.

7. A Numerical Traction-compression Test

In this section we consider a particular traction-compression test with constant stress rate (soft test) on a cylindrical CuZnAl sample. The material parameters (determined in [1, 2] and [3, Section 7]) are:

$T_{\rm a} = 305$ K	$T_{\rm m} = 267$ K	$C = 390$ J/(Kg K)
$L = 5900$ J/Kg	$\rho = 7800$ Kg/m^3	$\tau = 5$ s
$E = 27 \cdot 10^9$ Pa	$q = 8 \cdot 10^6$ Pa	$g = 0.026$

According to (12), (21), and (22), we have

$$\Delta T = T_{\rm a} - T_{\rm m} = 38 \text{ K}, \qquad \Gamma = \frac{qg}{\rho C} \simeq 0.0684 \text{ K}.$$

Our theoretical experiment considers for the room in which the CuZnAl sample is placed the absolute temperature $T_0 = 313$ K. By (22) we get

$$p = \frac{\rho L}{g T_0} \simeq 5.655 \cdot 10^6 \text{ J/(m}^3\text{K)}.$$

For $R = 30$ s and $a = \pm 10^7$ Pa/s, we analyze the soft test associated to the stress function

$$\sigma : [0, 2R] \to \mathbb{R}, \qquad \sigma(t) = \begin{cases} at &, \ \text{for } t \in [0, R], \\ a(2R - t), & \text{for } t \in [R, 2R]. \end{cases}$$

Since $\sigma \in \mathrm{A_{side}}([0, 2R])$, according to Theorem 6..14 the behavior of the sample is described by the unique solution of the system $(\mathcal{T})_{\mathrm{A_{side}}([0,2R])}$, expressed by the functions $\beta, \theta \in \mathrm{A_{side}}([0, 2R])$ defined piecewise as follows:

(L1) The first loading interval $[0, r_1]$, with elastic behavior in pure austenitic phase ($\beta \equiv 0$ and $\sigma = E\varepsilon$)

$$\left\{ \begin{array}{l} \beta(t) = 0 \\ \theta(t) = 0 \end{array} \right| \quad \text{for every } t \in [0, r_1],$$

where $r_1 = \frac{p\eta_+}{a} \simeq 5.324$ $\big($since $\eta_+ := (T_0 - T_{\rm a}) + \frac{q}{p}$, by Notation 3..18(c)$\big)$.

(L2) The second loading interval $[r_1, r_2]$, on which we have a complete austenite-martensite phase transition (β strictly increases from 0 to 1)

$$\left\{ \begin{array}{l} \beta(t) = \dfrac{a}{p\Delta T}\left[(t - r_1) - \dfrac{\tau\Lambda}{\Delta T}\left(1 - \mathrm{e}^{\alpha(r_1 - t)}\right)\right] \\ \theta(t) = \dfrac{a\tau\Lambda}{p\Delta T}\left(1 - \mathrm{e}^{\alpha(r_1 - t)}\right) \end{array} \right| \quad \text{for every } t \in [r_1, r_2],$$

where $\Lambda = \Gamma + \frac{L}{C}$ and $r_2 \simeq 28.742$ is such that $\beta(r_2) = 1$.

(L3-U1) The third loading interval $[r_2, R]$ followed by the first unloading interval $[R, r_3]$, both in pure martensitic phase ($\beta \equiv 1$)

$$\left\{ \begin{array}{l} \beta(t) = 1 \\ \theta(t) = \theta(r_2)\mathrm{e}^{(r_2 - t)/\tau} \end{array} \right| \quad \text{for every } t \in [r_2, r_3],$$

where $r_3 \simeq 34.131$ is such that $\sigma(r_3) = \sigma^-(r_3)$.

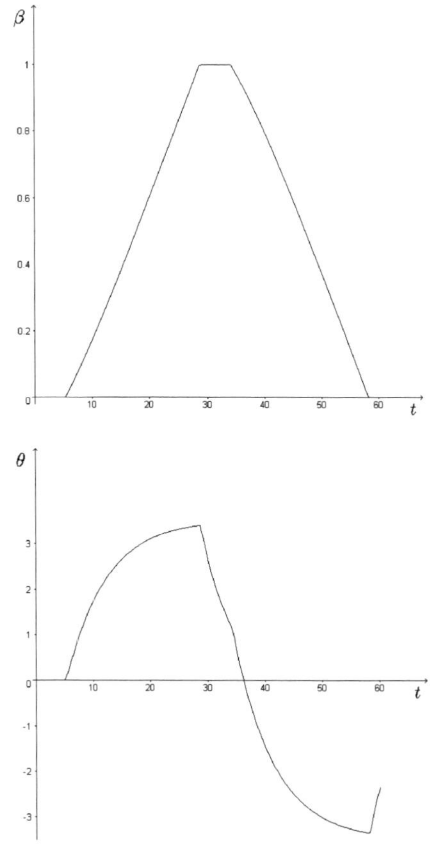

Figure 2. Plots of the components $\beta, \theta \in \mathrm{A}_{\mathrm{side}}([0, 2R])$ of the solution.

(U2) The second unloading interval $[r_3, r_4]$, on which we have a complete martensite-austenite phase transition (β strictly decreases from 1 to 0)

$$\left\{\begin{array}{l} \beta(t) = 1 - \dfrac{a}{p\Delta T}\left[(t - r_3) - \left(\dfrac{p\theta(r_3)}{a} + \dfrac{\tau\Lambda}{\Delta T}\right)\left(1 - \mathrm{e}^{\alpha(r_3 - t)}\right)\right] \\ \theta(t) = \theta(r_3)\mathrm{e}^{\alpha(r_3 - t)} - \dfrac{a\tau\Lambda}{p\Delta T}\left(1 - \mathrm{e}^{\alpha(r_3 - t)}\right) \end{array}\right. \quad \Bigg| \quad \begin{array}{l}\text{for every} \\ t \in [r_3, r_4],\end{array}$$

where $\Lambda = -\Gamma + \frac{L}{C}$ and $r_4 \simeq 58.173$ is such that $\beta(r_4) = 0$.

(U3) The third unloading interval $[r_4, 2R]$, with elastic behavior in pure austenitic phase ($\beta = 0$ and $\sigma = E\varepsilon$)

$$\left\{\begin{array}{l} \beta(t) = 0 \\ \theta(t) = \theta(r_4)\mathrm{e}^{(r_4 - t)/\tau} \end{array}\right. \quad \Bigg| \quad \text{for every } t \in [r_4, 2R].$$

Figures 2 and 3 illustrate the behavior of the solution (β, θ), as well as the hysteresis loop

$$[0, 2R] \ni t \longmapsto (\varepsilon(t), \sigma(t)) \in \mathbb{R}^2.$$

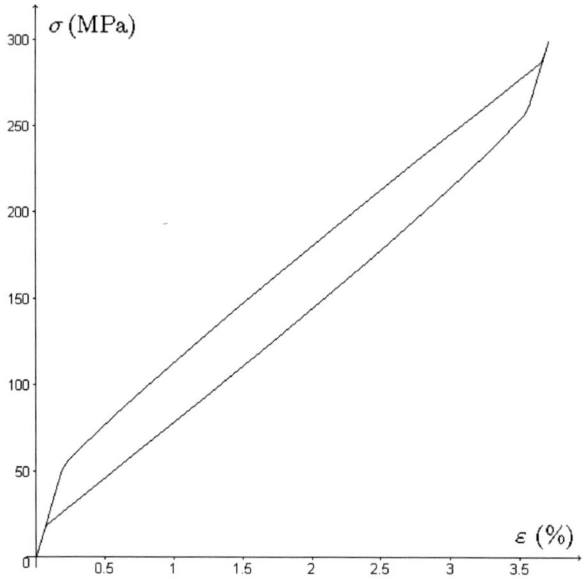

Figure 3. The hysteresis loop.

References

[1] Balandraud, X.; Ernst, E.; Soós, E. *Z. Angew. Math. Phys. (ZAMP)* 2000, *51*, 171–203.

[2] Balandraud, X.; Ernst, E.; Soós, E. *Z. Angew. Math. Phys. (ZAMP)* 2000, *51*, 419–448.

[3] Balandraud, X.; Ernst, E.; Soós, E. *Z. Angew. Math. Phys. (ZAMP)* 2005, *56*, 304–356.

[4] Ball, J. M.; James R. D. *Arch. Rat. Mech. Anal.* 1987, *100*, 13-52.

[5] Chrysochoos, A. *Rev. Roum. Sci. Techn. - Mec. Appl.* 1998, *43*, 649–666.

[6] Ericksen, J. L. *J. Elasticity* 1975, *5*, 191-201.

[7] James, R. D. *Arch. Rat. Mech. Anal.* 1981, *77*, 143-176.

[8] Pham, H. *Analyse thermomécanique du comportement d'un alliage mémoire de forme de type CuZnAl*; PhD thesis; Université Montpellier II, France, 1994; 1–188.

[9] Timofte, A.; Timofte, V. *Math. Mech. Solids* 2001, *6*, 447–466.

[10] Timofte, A.; Timofte, V. *Math. Mech. Solids* 2001, *6*, 541–545.

[11] Timofte, A.; Timofte, V.; Ratiu, T. *Math. Mech. Solids* 2006, *11*, 563–574.

[12] Timofte, A.; Timofte, V. *Z. Angew. Math. Phys. (ZAMP)* 2008, *59*, 181–185.

[13] Timofte, A.; Timofte, V. *Rev. Roumaine Math. Pures Appl.* 2002, *47*, 121–133.

INDEX

A

A_β, 504, 505, 507, 510, 516, 518, 529
accuracy, xiv, 38, 111, 189, 192, 218, 441, 446, 447, 449, 450, 457, 470, 481
acid, 85, 149
activation, 40, 43, 184, 399, 406, 407, 408, 409, 410, 416, 488
activation energy, 184
actuation, xiv, 224, 361, 373, 446, 450
actuators, ix, x, xiv, 113, 139, 141, 232, 318, 320, 366, 373, 420, 441, 442, 443, 444, 445, 446, 447, 449, 450, 470, 481, 482, 483, 484, 485, 488
adaptive control, 445
aerospace, ix, 108, 373, 450
AFM, 329, 330
algorithm, ix, 4, 35, 37, 41, 250, 261, 382, 406, 422, 445, 459, 462
alternative, ix, 41, 96, 455, 457, 470, 481, 482
alternatives, x, xv, 113, 441, 442, 470
alters, 306, 366
aluminum, 150, 374, 382, 383, 384, 385, 388, 389, 390, 420, 488, 489
ammonium, 85
amplitude, 193, 205, 383, 453, 456, 458, 462, 465, 478, 481
AMS, 425
anisotropy, 16, 106, 198, 203, 290, 305
annealing, 320, 328
annihilation, 203
anther, 302
antiferromagnet, 289, 292, 305, 306, 307, 308, 312
application, 379, 405, 422, 441, 442, 443, 445, 449, 483, 489
argon, 78, 80, 83, 149, 187, 189, 205, 206, 226, 249
argument, 380, 430, 434, 435, 512, 516, 518
arrest, 181, 183, 186, 210, 226
artery, 450, 488
atoms, 86, 96, 114, 120, 193, 233, 234, 235, 268, 269, 294, 354, 443

B

beams, 108, 308
bending, 76, 300, 302, 303, 393, 489
benefits, 481
binding energies, 94
biocompatibility, 76, 77, 86, 102, 103, 105, 107, 110, 450
biological activity, x, 69
biological behavior, 102
biomaterials, ix, 69, 71, 72, 76, 77, 102, 104
biomedical applications, 107, 108
biomimetic, 450, 483
blocks, 117, 121, 127, 129, 139, 140
blood, 70, 73, 450, 488
blood flow, 450, 488
blood vessels, 70, 73
body fluid, 70, 73, 91, 98, 111
bonding, 233, 239, 340, 353, 354, 356
bonds, 340, 353, 354
bone growth, 71
bone marrow, 103, 109
bone mass, 70
bone resorption, 77
boundary conditions, 377, 378, 384
boundary value problem, 384, 422
by-products, 86

C

cadmium, 140
calcium, ix, 69, 93, 109
calculus, 171
calorimetry, 217
canals, 73
capsule, 106, 107
carbides, 96, 268
carbon, 108, 268, 269, 328, 374, 420

carbon film, 108
cardiovascular disease, 450
cartilaginous, 103
casting, 271, 284
CAT, 401
categorization, 240
cell, 2, 4, 28, 37, 41, 43, 74, 76, 98, 100, 101, 103, 105, 149, 184, 233, 235, 248, 259, 267, 375, 376, 377, 378, 380, 384, 406, 409
cell culture, 100, 103
ceramic, 84, 198
channels, 80, 81
chemical etching, 328
chemical properties, 448
chemical reactions, 81
chemical vapor deposition, 96
circulation, 321, 328
civil engineering, 373, 419
civil structures, 419
classes, 248, 525
classical, 376, 387, 394, 490, 491, 493, 495, 496, 509
classification, 263
cleavage, 137
closure, 427
clustering, 258
clusters, 258
coatings, 96, 107, 109
cobalt, 98, 110
collaboration, 140
collagen, 70, 72
combustion, 80, 105, 106
communication, xiv, 361
communities, 489
compatibility, 29, 30, 86, 363, 364, 365, 374, 489, 498
compensation, 444, 445, 446, 453, 455, 456, 457, 462, 465, 468, 470, 471, 472, 473, 483
complexity, 15, 30, 319, 468
components, xii, 5, 10, 12, 29, 30, 35, 36, 45, 86, 89, 107, 172, 173, 198, 247, 263, 394, 402, 405, 414, 426, 468, 488, 495, 535
composites, xiv, 76, 98, 198, 371, 374, 375, 380, 385, 388, 396, 418, 419, 420, 421, 422
composition, xiii, 20, 78, 98, 114, 115, 116, 122, 140, 146, 147, 151, 155, 156, 183, 199, 203, 244, 249, 251, 267, 294, 297, 298, 314, 319, 320, 328, 339, 340, 364
compounds, 86, 94, 182, 292, 328
compressibility, 288
compressive strength, 411
computation, 16, 17, 23, 24, 35, 36, 38, 445, 512, 519
concentration, xiii, 85, 86, 94, 98, 100, 199, 294, 295, 303, 339, 340, 350, 351, 352, 353, 354, 355, 356, 362, 368, 379, 393, 407, 417
concrete, xiv, 371, 375, 399, 400, 404, 405, 406, 407, 408, 409, 410, 411, 412, 413, 414, 415, 416, 417, 418, 421, 422, 423, 506
conditioning, 451
conduction, 190, 206, 208, 218, 221, 223, 224, 225, 227, 230, 383
conductivity, xi, 181, 182, 185, 186, 192, 205, 206, 208, 218, 219, 221, 222, 223, 224, 226, 227, 377, 448, 449
configuration, 30, 41, 43, 108, 173, 236, 355, 382, 407, 409, 425, 451, 452, 489, 495
confinement, 249
conformity, 431
conjecture, 22
connectivity, 106
consolidation, 81, 82, 106
constant load, 320
constant rate, 488
constraints, 426
construction, 72, 502, 506, 507, 517, 521, 525
consumption, xiv, 361
continuity, 377, 378, 433, 517, 522, 526, 533
convergence, 404, 427, 512, 513
convex, 393, 426, 427, 487, 490, 492, 518, 519, 529
cooling process, 120
copper, xii, 115, 116, 146, 169, 170, 267, 280, 314, 419, 443, 449, 488
correlation, 238, 248
correlations, xiii, 339, 356
corrosion, 76, 86, 98, 99, 107, 108, 109, 110, 111, 115, 313, 374, 443, 449, 489
coupling, xiv, 287, 296, 297, 299, 300, 301, 303, 304, 305, 306, 311, 314, 315, 371, 374, 375, 382, 383, 384, 385, 386, 389, 390, 393, 396, 410, 418, 422, 493
crack, 71, 136, 137, 140
creep, xv, 83, 84, 487, 490
critical value, 2, 150, 152, 153, 340, 353
crystal growth, 81, 106
crystal structure, 146, 156, 194, 341
crystalline, 114, 115, 120, 184, 193, 232, 243, 244, 248, 251, 308, 489
crystallites, 193, 276, 277
crystals, 73, 86, 147, 149, 178, 184, 243, 245, 354, 393
cubic system, 290
culture, 100
CVD, 70, 96
cytocompatibility, 100
cytotoxicity, 110

D

damping, 25, 289, 312, 313, 315, 374, 448, 450
DARPA, 419
database, 321
decomposition, 114, 268, 269, 394
deduction, 427
defect formation, 149, 361, 365, 366
defects, xi, 116, 125, 130, 145, 146, 149, 158, 160, 163, 171, 173, 176, 177, 178, 194, 204, 205, 319, 328, 363, 364, 365, 366, 373, 443
definition, 240, 379, 435, 501, 505, 511, 519, 520, 521, 522, 523, 524, 525
deformability, 115
degenerate, 497, 499, 511, 514, 518, 525, 529
degradation, 270, 319, 386, 389, 404, 410, 453
density, xii, xiii, 73, 74, 76, 110, 121, 160, 169, 181, 186, 192, 193, 203, 204, 218, 219, 221, 223, 233, 261, 283, 318, 319, 329, 336, 377, 380, 386, 394, 423, 491
deposition, 70, 96, 98, 109
derivatives, xv, 377, 381, 387, 459, 487, 491, 495, 497, 498, 499, 501, 503, 504, 510
detection, 235, 450
detwinning, 3, 203, 204, 226
deviation, 5, 30, 41, 43, 116, 276, 352, 354, 364, 365
differential equations, 426
differential scanning, x, xii, 20, 113, 116, 155, 267, 271
differential scanning calorimetry, x, xii, 20, 113, 155, 267, 271
differentiation, 100, 109, 377
diffraction, x, 113, 210, 267, 276, 289, 296, 299, 320, 321, 324
diffusion, x, 86, 145, 178, 208, 230, 267, 268, 269, 270, 300, 447
diffusion process, 178
diffusivities, 190
diffusivity, 86, 181, 186, 190, 191, 192, 193, 218, 219, 220, 221, 223, 227, 268
direct observation, 117
discretization, xiv, 425, 427, 429, 430
disorder, 248, 267, 271, 279, 281, 283, 305
displacement, 18, 20, 138, 139, 149, 190, 193, 248, 249, 250, 260, 261, 262, 271, 289, 290, 291, 376, 378, 426, 451, 452, 462, 470
distortions, 363, 365, 366
distribution, 79, 83, 84, 92, 98, 109, 168, 258, 268, 367, 498, 503, 505
diversity, 489
division, 236
domain structure, 305, 306
DSM, 117
ductility, 77, 115
durability, 481
duration, 86, 217
dynamical systems, 484
dysprosium, 292, 305, 313

E

eigenvector, 411
elastic deformation, x, 114, 115, 139, 147, 163, 251, 403, 423
elasticity, 2, 76, 143, 250, 296
electric conductivity, 449
electric current, xiv, 441, 445
electric field, 315
electrical resistance, 147, 149, 150, 158, 484
electricity, 442
electrochemical impedance, 98
electrodeposition, 109
electrodes, 81, 98, 100
electrolyte, 98
electron, x, xiii, 20, 40, 70, 113, 161, 162, 199, 268, 271, 319, 339, 340, 341, 350, 351, 352, 353, 354, 355, 356, 362
electron microscopy, x, 70, 113, 161, 162, 268, 319
electronic structure, 339, 340, 353, 355
electrons, xiii, 89, 96, 117, 339, 340, 341, 350, 352, 353, 354, 355, 356, 362, 368
electroplating, 96
elongation, 20, 21, 141, 149, 189, 192, 194, 195, 196, 197, 198, 199, 203, 213, 214, 215, 218, 221, 251, 255, 451
embedded atom method, xi, 231
endoscope, 444, 483
endothermic, xii, 116, 120, 126, 139, 183, 210, 267, 271, 272, 282, 283
energy density, 320, 386, 394
entropy, xv, 146, 147, 155, 163, 171, 177, 380, 381, 386, 387, 394, 487, 490, 491
environment, 3, 70, 96, 98, 105, 108, 147, 445, 495
epoxy, 382, 383, 385, 388, 389, 390, 396, 397, 398, 399, 400, 420
equality, 437, 494, 500, 502, 506, 510, 511, 513, 516, 521, 523, 526, 531
equilibrium, xii, 29, 86, 87, 140, 147, 155, 164, 165, 166, 182, 193, 194, 209, 224, 239, 267, 269, 270, 271, 283, 290, 377, 378, 495
etching, 268
eukaryotic cell, 110
evaporation, 96
exothermic peaks, 116
experimental condition, 184, 209, 226
exposure, 409, 488
extrusion, 178

F

fabrication, x, 69, 71, 77, 102, 103, 115, 140, 328, 374, 420, 423, 445
fatigue, 142, 248, 364, 374, 443, 449, 489
FeMn, 341
femur, 102
ferrite, 268, 269
ferromagnetic, 482
ferromagnetism, xiii, 339, 354, 356
Feynman, 485
fiber, 374, 382, 383, 385, 396, 398, 399, 400, 406, 407, 408, 409, 414, 415, 416, 417, 420, 421, 422
fibers, xiv, 70, 72, 73, 371, 374, 375, 382, 383, 384, 385, 386, 388, 389, 390, 396, 398, 399, 400, 405, 406, 407, 408, 409, 410, 413, 414, 417, 418, 420, 421, 422, 423, 450
fibroblasts, 98, 110
fibrosis, 101, 103
fibrous tissue, 103
fidelity, 374, 421, 422
filament, 89
films, 100, 111, 366, 369
filters, 450
financial support, 46, 227, 263
finite element method, 374
fire, 450, 489
fixation, 77, 102
flow, 376, 387, 395, 401, 411, 442, 450, 488, 489
fluctuations, 205, 269
foams, 100, 107
focusing, 102, 181, 216, 217, 226, 232
foils, 271, 420
forceps, 488
Fourier, 377, 382, 491
fractures, 71, 75, 103, 193
free energy, xv, 2, 163, 165, 169, 170, 171, 177, 178, 182, 290, 304, 380, 381, 387, 394, 404, 487, 490, 491, 493
friction, 296, 314, 315, 451
fusion, 102, 103, 111

G

gallium, 314
gas, 450
gases, 89, 96, 205
gauge, 441, 451
gel, 117
generation, 106, 146, 156, 163, 166, 170, 171, 176, 178, 203, 256, 259, 262, 263, 373
germanium, 314
Gibbs, 487, 490
Gibbs energy, 2, 114, 163, 164
glass, 420
gold, 140
grain boundaries, 115, 147, 261, 275
grains, xii, 181, 187, 188, 189, 193, 194, 195, 199, 200, 201, 202, 205, 221, 247, 248, 249, 250, 251, 252, 253, 254, 255, 258, 261, 263, 274, 275, 276, 277, 283, 307, 328, 336, 373
grants, 263
graph, 255, 389, 408, 451, 478
graphite, 81, 98, 192
groups, xiii, 9, 36, 96, 99, 101, 102, 116, 160, 241, 244, 250, 256, 258, 294, 295, 320, 339, 341, 352, 356
growth, ix, 20, 22, 40, 41, 69, 81, 86, 98, 102, 103, 165, 177, 252, 255, 261, 267, 268, 269, 270, 319, 324
growth rate, 86, 98, 268
guessing, 460, 461

H

hardening, 385, 386, 389, 390, 411, 412
hardness, 268
heat capacity, 171, 186, 206, 380, 383, 386, 448
heat loss, 190
heat transfer, xi, 182, 189, 206, 384
heating rate, 116, 206, 208, 209, 223, 271, 272, 273
height, 174, 184
helium, 189, 190, 206, 208, 226
Hessian matrix, 459
heterogeneity, 25
high resolution, 446
high temperature, 398, 409, 410, 425, 442, 443
Hilbert space, 428
holmium, 292, 305, 313
HRTEM, 308, 309, 315
hybrid, 76, 420, 422
hydrogen, 91, 109, 184
hydrogen peroxide, 91, 109
hydrostatic pressure, 387, 404
hydroxyapatite, ix, 69, 70, 72, 92, 104, 107, 109, 110
hypersensitivity, 85
hypothesis, 184, 205
hysteresis loop, xv, 14, 122, 372, 373, 445, 450, 455, 459, 462, 463, 465, 487, 489, 490, 492, 536

I

illumination, 250, 252
image analysis, xi, 247, 249, 250, 252, 256, 260
images, xi, 79, 117, 161, 247, 249, 250, 251, 260, 261, 263, 275, 276, 308

immersion, 70, 88, 89, 90, 91, 92, 93, 98, 100, 108, 109, 111, 321
immune response, 107
implementation, 410, 442, 446, 470, 482
in transition, 356
in vitro, 98, 100, 102, 103, 105, 109, 110
in vivo, 98, 101, 102, 103
incidence, 133, 135, 140
inclusion, 384, 500, 501, 506, 521, 523
incompatibility, 363
incubation period, 269
incubation time, 205
indication, 120, 130
indices, 235, 242, 375
induction, 249, 328
industrial, 373, 450
industry, 319, 405, 444, 449, 450
inelastic, 2, 5, 7, 12, 14, 15, 16, 25, 26, 28, 33, 35, 36, 37, 374, 376, 377, 378, 379, 380, 382, 385, 386, 387, 389, 393, 394, 395, 396, 400, 401, 402, 403, 413, 415, 417, 422
inequality, xv, 26, 429, 431, 433, 434, 435, 436, 439, 487, 490, 492, 513, 522, 526, 531
inertia, 502
infection, 97, 110
inhibition, 98, 110, 146, 158, 161, 163, 165, 166, 167, 170, 171, 172, 174, 176, 177, 178
initial state, x, 114, 117, 118, 119, 120, 122, 123, 125, 126, 127, 130, 133, 134, 135, 136, 137, 139, 425, 443
initiation, 20, 137, 224, 391
insight, 281
instability, 1, 4, 20, 22, 23, 24, 30, 35, 36, 39, 40, 41
Instron, 117, 136, 149, 190
instruments, 77
integrated circuits, 90
integration, 84, 174, 175, 176, 400, 422, 438, 450, 465
interaction, xi, xii, 3, 30, 145, 146, 156, 166, 167, 169, 170, 171, 172, 174, 175, 176, 177, 178, 247, 252, 260, 261, 290, 374
interactions, 3, 171
interface, xi, 103, 114, 134, 146, 169, 170, 182, 216, 223, 224, 231, 236, 239, 245, 246, 271, 378
interface energy, 182
intermetallic compounds, 80, 353
intermetallics, 80
interphase, 21, 22, 30, 36
interval, x, xiv, 113, 114, 115, 116, 117, 120, 122, 125, 126, 133, 137, 138, 139, 174, 190, 361, 382, 403, 412, 425, 434, 436, 440, 497, 499, 500, 501, 506, 509, 511, 518, 525, 534, 535
intracranial, 450
intrinsic, 373, 444, 478, 487, 490, 492
intrusions, 204
ion implantation, 70, 89, 90, 91, 107, 108, 109
ions, 85, 86, 90, 91, 96, 107
iron, 294, 488
irradiation, 181, 190, 192, 219, 220, 221, 224
isothermal, 403, 404, 410, 413, 415, 487, 490
isotherms, 292
isotropic, 377, 382, 383, 399, 402, 410, 423
isotropy, 495
iteration, 37

J

Jacobian matrix, 459
joints, 444
Joule heating, 450

K

kinetics, 206

L

lamella, 23, 29, 30, 38, 39, 73
lamination, 374
lattice parameters, 155, 174, 233, 299, 301, 303, 304, 321, 363, 364
lattices, xi, 114, 231, 364
law, 374, 377, 379, 381, 382, 387, 402, 487, 490, 491, 492, 493, 494, 495
laws, 2, 6, 455
leaching, x, 69, 90, 91, 98, 110
learning, 458
Leibniz, 501
limitations, 374
linear, 378, 402, 410, 412, 441, 444, 446, 450, 451, 453, 455, 470, 481, 484, 491, 505, 510, 511, 512, 515, 517
linear dependence, 194, 203, 208
linear model, 250
lipids, 72
loading, 372, 373, 378, 383, 384, 385, 388, 389, 390, 391, 393, 396, 397, 398, 399, 400, 401, 402, 403, 404, 405, 406, 407, 411, 414, 415, 416, 418, 487, 489, 490, 534
localization, 40, 41, 378
low temperatures, 150, 185, 398

M

magnetic, 482, 488
magnetic effect, 340
magnetic field, 182, 287, 289, 290, 292, 294, 305, 306, 310, 311, 312, 318, 482

magnetic moment, 287, 293, 294, 354
magnetic properties, 294
magnetic resonance, 488
magnetic structure, 287, 293, 294
magnetism, 287, 313
magnetization, 196, 287, 289, 305, 307
magnetostriction, 292, 305, 311, 313, 314
management, 488
manganese, 314, 315, 488
manufacturing, 194, 196, 197
mapping, 382, 446
maritime, 444
market, 71, 72, 103, 104, 420, 489
market segment, 72
markets, 71
marrow, 73
martensitic transformations, x, xi, xii, xiii, 113, 114, 125, 138, 142, 145, 146, 149, 150, 198, 199, 205, 264, 294, 317, 318, 328, 368
mass spectrometry, 90
measurement, 15, 149, 156, 183, 184, 185, 189, 191, 192, 193, 194, 205, 206, 209, 210, 218, 219, 221, 226, 249, 259, 308, 310, 312, 329
measures, 184, 199
mechanical behavior, 374, 413, 420, 444, 489
mechanical properties, ix, 18, 69, 71, 76, 77, 78, 84, 115, 232, 319, 357, 373, 448, 449, 489
medicine, 444, 488
melt, 82, 365, 367, 368, 369
melting, 77, 149, 187, 223, 249
metabolism, 110
metal oxides, 96
metallurgy, 70, 268, 489
metals, 76, 81, 109, 140, 193, 205, 248, 339, 354, 369, 371, 374, 442, 489
methanol, 149
metric, 427
microdefects, 25
microelectronics, 109
micrometer, 90, 446
microscope, 20, 40, 116, 117, 149, 150, 249, 271, 289, 307, 308
microscopy, 70
military, 319, 488
miniaturization, 442, 443, 475
mining, 468
modeling, xiv, 2, 42, 142, 371, 374, 375, 400, 404, 418, 419, 420, 421, 423, 483, 489
models, ix, xiv, 1, 2, 3, 18, 25, 35, 42, 43, 232, 234, 235, 239, 243, 244, 245, 259, 374, 376, 386, 418, 419, 421, 427, 441, 444, 445, 455, 481, 483, 489, 492
modules, 141, 505
modulus, ix, 7, 8, 9, 15, 22, 25, 27, 31, 35, 36, 38, 69, 70, 75, 76, 77, 137, 138, 173, 251, 256, 259, 263, 296, 297, 298, 314, 353, 355, 380, 383, 386, 389, 391, 405, 411, 412, 414, 416, 443, 448, 496
moisture, 318
molar volume, 163
molecular dynamics, xi, 231, 232, 245
MOM, 271
momentum, xv, 487, 490, 492
monotone, 500, 502, 503, 505, 508, 510, 511, 514, 523, 524, 530, 533
morphology, 19, 20, 21, 22, 71, 86, 98, 99, 111, 117, 118, 120, 121, 122, 149, 150, 158, 194, 267, 268, 270, 275, 283, 289
mosaic, 251, 252, 254, 255, 260, 261
motion, xi, xiv, 184, 231, 245, 290, 291, 292, 305, 306, 307, 311, 312, 313, 441, 443, 447, 450, 451, 452, 453, 470, 481, 488
motion control, 450, 451, 452, 453, 481
movement, xv, 86, 96, 137, 223, 239, 287, 289, 290, 313, 363, 441
MTS, 249
multiphase materials, 422
multiplier, 395, 404
muscle, 488

N

NaCl, 85, 99, 107, 117
nano-crystals, 70
natural, 487, 489, 491, 498, 503, 506, 533
necrosis, 98
neglect, 221, 386, 492
network, 73, 176, 328, 329, 446, 457, 459, 462, 470, 485
neural network, 446, 457, 462, 470, 482, 484
neural networks, 482, 484
neurons, 459
Newton, 459, 495, 501
nickel (Ni), ix, 69, 78, 85, 86, 88, 89, 90, 91, 96, 98, 100, 102, 103, 105, 107, 108, 110, 111, 313, 420, 441, 443, 455, 475, 476, 485, 488
niobium, 422
NiTi shape memory, xii, xiii, 4, 77, 86, 105, 106, 107, 108, 109, 110, 111, 185, 317
nitrides, 96
nitrogen, 116, 184, 328
noise, 444, 456, 457, 459
nonlinearities, 445, 446
nucleation, xi, 2, 3, 22, 137, 140, 147, 150, 151, 231, 232, 236, 242, 243, 245, 268, 269, 270, 276, 300, 328, 329
nucleus, 165, 171

O

observable behavior, 372
observations, 122, 150, 166, 168, 170, 173, 178, 268, 271, 284, 361, 404
oil, 321, 489
one dimension, 224, 412
operator, 445, 456, 510, 511, 515, 517, 531
optical microscopy, 120, 137, 160
optimization, 482
optimization method, 482
orthodontists, 488
oscillations, 478
osteoporosis, 71, 103, 104
oxidation, 86, 87, 88, 91, 93, 97, 100, 103, 108, 127, 205, 313, 321, 328
oxides, 86
oxygen, 86, 89, 328

P

palladium, 314
parameter, 7, 10, 26, 182, 184, 186, 205, 206, 221, 224, 227, 236, 239, 255, 256, 259, 263, 289, 301, 356, 363, 377, 381, 386, 387, 391, 402, 404, 411, 426, 437, 440, 462
particles, 92, 120, 198, 375
partition, 377
passivation, x, 69
passive, 450
pathways, xi, 231, 236, 238, 239, 245
periodic, 375, 376, 377, 378, 384, 385, 393, 422, 423, 446, 456, 465, 481
periodicity, 276, 309, 377
peripheral blood, 100, 101
peroxide, 92
pH, 318
phase diagram, 77, 78, 86, 87, 264, 295, 296
phase transitions, xiv, 2, 155, 199, 271, 273, 279, 282, 441, 444, 447, 455, 481, 482
phonons, 185, 221
phosphorus, 93
photoelectron spectroscopy, 70
photographs, 152, 251
physical and mechanical properties, 115
physical mechanisms, 178
physical properties, 205, 318, 447
physics, 455, 485
piezoelectric, 450
plasma, 70, 81, 82, 89, 90, 91, 106, 107, 108, 109
plastic, 373, 374, 381, 383, 405, 410, 411, 413, 415, 419, 420, 421, 423, 443, 450
plastic deformation, x, 15, 113, 115, 135, 137, 138, 140, 141, 160, 178, 243, 251, 301, 319, 329, 415, 450
plastic strain, 373, 405, 411, 413, 415, 419
plasticity, xiv, 2, 3, 4, 16, 43, 371, 375, 376, 386, 387, 388, 389, 390, 391, 400, 404, 410, 415, 418, 421, 422, 443
Poisson, 383, 389, 390, 402, 403, 423
Poisson ratio, 402
Poland, 142, 180
polarization, 98
polymer, xiv, 103, 371, 380, 385, 418, 420
polymer composites, 420
polymeric matrices, 375, 388, 418
polymers, 71, 81, 248, 405
polynomial, 478
porosity, 78, 79, 80, 81, 82, 83, 84, 85, 98, 101, 111
porous materials, 84
Portugal, 141, 425, 440
power, xiv, 156, 361, 443, 448, 481
Prandtl, 376
precipitation, xii, 267, 268, 271, 274, 276, 279, 281, 283, 362, 367, 369
prediction, 386, 462
predictive model, 425
pressure, 106, 183, 186, 194, 206, 218, 232, 387, 404, 444, 450
probability, 147, 489
probe, 99
production, 382
program, 227, 263
programming, 478
proliferation, 88, 100
propagation, xi, 3, 18, 75, 137, 182, 208, 223, 224, 231, 232, 236, 493, 496, 497, 498, 503, 507, 508, 509, 510, 525, 533
proportionality, 259, 263
prostheses, 141
proteins, 72
prototype, xv, 441, 446, 447, 450, 451, 452, 453, 454, 470, 475
pseudo, 371, 418, 443
pseudoelastic cycling, 147, 149, 156, 158, 161, 163
pulse, 185, 221

Q

quartz, 149
quasilinear, 444, 446

R

radiation, 206, 221, 232, 359
radio, 89

radius, xv, 108, 255, 263, 361, 362, 381, 427, 487, 491, 494
raw materials, 78, 96
reaction mechanism, 106
reading, 261, 451, 478
real time, 445
reasoning, 161, 170
recall, 397, 408, 409, 503, 508, 514, 518, 526
recalling, 397
reconstruction, 194
recovery, 5, 7, 10, 77, 78, 203, 254, 301, 303, 319, 320, 372, 373, 384, 408, 409, 417, 446, 447, 449, 455, 456
recrystallization, 319
refining, 275, 276
reflection, 149, 321
regular, 427, 498, 532
regulation, 485
reinforcement, 405, 412, 415
relationship, ix, 2, 8, 26, 28, 34, 41, 146, 147, 151, 154, 170, 245, 294, 353, 373, 470, 475, 478
relationships, 26, 27, 30
relaxation, xv, 134, 182, 235, 244, 245, 441, 445, 451, 465, 478, 482, 487, 490
relaxation process, 134, 445
relaxation processes, 445
relevance, x, 114, 140
reliability, 465
REM, 268
repair, 102, 103
reparation, 106
replication, 107
resistance, x, 76, 86, 99, 107, 111, 114, 115, 140, 147, 148, 149, 158, 159, 310, 340, 353, 374, 443, 446, 447, 449, 453, 470, 475, 477, 478, 481, 482, 483, 484, 489
resistivity, 442, 446, 448, 449, 470
resolution, 94, 95, 249, 262, 263, 268, 446, 451
resources, 263
response time, 449, 489
rheological properties, 490
rings, 73
risk, 104
robotics, xiv, 361, 442, 443, 444, 450, 475, 482, 488, 489
rods, 73, 74, 98, 276, 277, 283, 284
rolling, 184
room temperature, x, xi, 20, 36, 39, 82, 85, 116, 117, 134, 139, 145, 146, 149, 150, 178, 181, 182, 186, 189, 190, 192, 196, 197, 198, 206, 210, 214, 218, 221, 223, 271, 301, 302, 306, 310, 311, 321, 322
roughness, 98, 252
rutile, 86, 88

S

safety, 319, 450, 488
salt, 70
saturation, 292, 386, 389
scalar, 400, 428, 458, 459, 491
scaling, 234, 235, 251, 255
scanning electron microscopy, 117
scatter, 321
schema, 191, 217, 224
science department, 489
scoliosis, 104
search, 235, 368, 496
SEM micrographs, 84, 85
semiconductor, 108
sensing, 475
sensitivity, xii, xiii, 114, 199, 317, 319, 324, 354, 489
sensors, 318, 450, 470
separation, 291, 334, 496
series, 418, 445, 478
shape memory alloys, 372, 374, 393, 396, 397, 400, 405, 419, 420, 421, 423, 425, 441, 443, 449, 483, 487, 489, 490, 491
shape memory properties, 450
shape-memory, 141, 306, 315, 368, 369
sharing, 120
shear, xi, 7, 8, 9, 10, 15, 16, 17, 21, 22, 24, 25, 35, 36, 38, 39, 40, 41, 43, 114, 120, 137, 146, 155, 173, 194, 204, 231, 232, 235, 239, 242, 243, 244, 245, 250, 258, 268, 296, 300, 304, 340, 353, 365, 366, 380, 386, 390, 394, 442
shear deformation, xi, 8, 231, 245
signals, 93, 450, 451, 456, 458, 459, 463, 465, 468, 470, 478
signs, 101, 167, 171, 301, 303, 306, 494
silicon, 328, 488
simulation, xi, 23, 24, 40, 41, 89, 231, 232, 233, 234, 235, 237, 239, 243, 244, 245, 485
Singapore, 180, 339, 361
single crystals, xi, 115, 145, 146, 147, 149, 150, 157, 160, 169, 177, 178, 182, 203, 219, 224, 292, 313, 419
sintering, 70, 78, 81, 82, 85, 99, 105, 106, 107
smart com, 420
smart materials, xiv, 318, 371, 374, 441, 447, 450
SME, 372, 373
smooth muscle cells, 98, 110
Sobolev space, 504, 530
sodium, 92, 98
sodium hydroxide, 92
software, 89, 261, 281, 321
solid phase, 422, 447

solid state, 248
solidification, 205
solubility, 449
spatial, 377, 495
species, 89, 98, 218, 233
specific heat, 163, 181, 184, 185, 186, 192, 193, 217, 219, 221, 222, 288, 377, 394, 493
spectroscopy, 70, 98
spectrum, xiv, 88, 361
speed, 117, 190, 234, 321, 329, 449, 459, 465, 478
spin, 290, 293, 294, 305, 312
springs, 444
square wave, 468
stability, ix, 99, 109, 111, 130, 137, 138, 139, 165, 177, 199, 210, 248, 264, 300, 313, 319, 355, 399, 444, 455, 462, 483
stabilization, 130, 139, 160, 329
stages, 2, 86, 281, 367, 413
stainless steel, 451, 452, 488
standard deviation, 259
standards, 71
stars, 353
steel, 268, 269, 374, 405, 451, 452, 488, 489
steel plate, 451, 452
stent, 107, 110, 488
Stieltjes, 505, 519
stiffness, 374, 376, 379, 402, 404, 409, 410, 412, 447
stimulus, xiv, 441, 442, 447, 448, 455, 462
strains, 372, 373, 374, 378, 385, 396, 397, 399, 406, 407, 408, 409, 411, 413, 415, 417, 442, 449, 450
strategies, xv, 441, 445, 450, 452, 465, 481
strength, 22, 73, 76, 77, 79, 82, 83, 104, 105, 106, 133, 174, 340, 354, 355, 371, 399, 405, 409, 410, 411, 418, 448
stress fields, 171, 334, 394
stress level, 408, 409
stress-strain curves, 16, 75, 133, 134, 146, 157, 177
stretching, 393, 475
structural changes, xi, 142, 231, 235
structural characteristics, 120
structural defects, 125, 139
structural modifications, 126, 127
structural relaxation, 235, 243, 244
substitutes, 71
substitution, 106, 183
substrates, 109
superconductivity, 306
superlattice, 276
suppression, 445, 449, 450
surface area, 86, 96, 98, 99, 182
surface layer, 86, 88, 89, 93, 94, 98, 99, 102, 103
surface layer structure, 86, 98
surface modification, x, 69, 86, 96, 99, 100, 102, 103, 111
surface structure, 121
surface treatment, 98, 100, 108
surgeries, 488
surgical, 373
symmetry, 114, 146, 164, 167, 168, 182, 210, 293, 442, 443, 447, 489
synthesis, 70, 79, 80, 105, 106

T

tangles, 204
tau, 371
TEM, 150, 160, 267, 271, 274, 276, 278, 279, 283, 284, 307, 308, 369
temperature annealing, 367
temperature dependence, 181, 184, 185, 186, 192, 194, 197, 198, 199, 210, 214, 217, 219, 221, 227
temperature gradient, 377
tendons, 450
tensile, 371, 373, 399, 405, 410, 411, 412, 415, 418, 445
tensile strength, xiv, 77, 81, 371, 399, 405, 410
tensile stress, 445
tension, xi, xii, 1, 2, 4, 14, 15, 16, 17, 18, 19, 20, 21, 22, 23, 24, 25, 30, 35, 36, 37, 38, 39, 40, 41, 42, 43, 198, 203, 204, 247, 249, 250, 389, 390, 404, 405, 412, 417
tetragonal lattice, 304, 309
thermal analysis, xii, 137, 267, 271, 272
thermal conduction, 383
thermal energy, 190
thermal equilibrium, 495
thermal expansion, 8, 171, 181, 184, 186, 189, 193, 194, 206, 216, 226, 273, 283, 380, 383, 386, 394, 399, 406, 407, 408, 409, 410
thermal load, 373
thermal properties, xi, 181, 183, 185, 186, 209
thermal stability, 182
thermal treatment, 127
thermodynamic, 404, 421, 491, 493
thermodynamic equilibrium, 138
thermodynamic parameters, 218
thermodynamics, xv, 3, 290, 304, 374, 381, 387, 419, 487, 490, 492
thermograms, 271, 272, 281, 324, 325, 327, 329, 332, 333, 334, 335, 336
thin films, 232, 364, 366
threshold, 256, 290, 291, 305, 310, 311, 415
tibia, 101, 102
tics, 441
tissue, ix, 69, 72, 76, 77, 84, 102, 103, 104, 105, 110
titania, 91, 109

titanium, ix, 69, 78, 85, 86, 91, 98, 102, 103, 105, 107, 108, 109, 111, 141, 420,441. 445, 449, 485, 488
TMC, 374, 375, 377, 379, 380, 383, 386, 396, 407, 413, 418
toluene, 192
topology, 390, 391, 393, 512, 513
total energy, 232
toxic effect, 86
toxicity, 77, 85, 103
tracking, 259, 262, 483
traction, 445, 487, 490, 492, 493, 494, 495, 533
training, 320, 373, 419, 458, 459, 495
training speed, 459
trans, 372, 381, 386, 389, 393, 395, 402, 443, 448, 481
transducer, 271
transfer, 384, 456, 457, 468
transformation processes, 35, 126
transformation product, 194
transformations, xi, xii, xiii, 114, 137, 140, 145, 146, 150, 154, 183, 184, 194, 203, 205, 268, 271, 279, 294, 317, 319, 322, 369, 373, 402, 403, 422, 490
transition metal, 293, 314, 340, 341, 354, 355
transition temperature, 219, 221, 303
transitions, 125, 147, 150, 178, 199, 272, 283
translation, 141, 164, 167, 168, 250
transmission, 170, 173, 267, 271, 450
transmission electron microscopy, 170, 173, 267
transparency, 488
transport, 77, 186, 205, 206, 226
transportation, 96
trial, 104, 413, 462
triggers, 391
tumor, 450
twinning, 146, 170, 171, 174, 307, 308, 309, 315, 365, 366, 442, 443
twins, 194, 203, 294, 296, 302, 305, 306, 307, 308, 311, 363, 364, 365, 366, 443

U

uniaxial tension, xi, 1, 4, 18, 247, 248, 249, 251, 258, 411, 413

V

vacuum, 78, 83, 96, 185, 271, 321
valence, xiii, 339, 340, 341, 350, 351, 352, 353, 354, 355, 356, 362, 368
validation, 434, 462
validity, 4, 41, 43
vapor, 70, 96
variables, xv, 2, 16, 386, 410, 413, 436, 487, 489, 490, 491, 495, 496
variance, 160, 177
variation, 2, 4, 12, 19, 23, 30, 33, 114, 115, 122, 125, 137, 271, 281, 301, 303, 318, 320, 324, 327, 341, 350, 353, 366, 447, 462, 475, 476, 493, 504, 526
vector, 160, 164, 166, 168, 170, 173, 174, 176, 178, 204, 248, 250, 260, 261, 289, 312, 376, 377, 378, 407, 411, 426, 427, 459, 491, 501, 504, 512, 527
velocity, 223, 224, 226, 234, 235, 445
vibration, 193, 197, 445, 450
viscoelastic properties, 104
viscosity, 447

W

weight ratio, 443
white blood cells, 107
wind, 450
wires, xiv, 77, 107, 328, 371, 442, 444, 445, 446, 449, 450, 451, 481, 482, 484, 488

X

xenon, 190
XPS, 70, 86, 88, 89, 91, 94, 95, 109, 110
X-ray diffraction (XRD), xii, xiii, 70, 88, 93, 94, 116, 117, 118, 119, 120, 123, 124, 125, 127, 128, 129, 130, 131, 133, 134, 135, 271, 317, 320, 321, 323, 324, 325, 326, 328, 336

Y

yield, x, 6, 9, 15, 22, 104, 114, 117, 130, 133, 182, 262, 263, 320, 383, 387, 391, 393, 395, 411, 413, 415, 416, 417, 418, 488, 515, 516, 520, 523, 528

Z

zinc (Zn), 280, 282, 443, 488